Tutorials on Mathematics to MATLAB

Mohammad Nuruzzaman
Electrical Engineering Department
King Fahd University
Dhahran, Saudi Arabia

1st Books - rev. 03/20/03

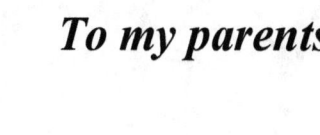

Preface

Main purpose of this book is to explore the built-in and toolbox functions that are available in MATLAB 6. Most of the commands presented in this book are also implementable in MATLAB 5. Expanded form of MATLAB is matrix laboratory. In order to be familiar with MATLAB functions, for the sake of brevity, mathematical notations and concept behind each function are presented. Tremendous features are being included in this software ever since it was evolved. Token examples are furnished for different class of problems. The book begins with the most elementary examples and develops the underlying concepts required for MATLAB solution. Sufficient mathematical details enable the average reader to follow without too much difficulty. A substantial number of problems have been covered, which can be implemented in the MATLAB Command Window just by typing few lines even if someone starts from the scratch. Each chapter is subdivided into articles instead of sections. Article headings describe the problems to be solved. Most problems are chosen from pure and applied mathematics. Almost all disciplines of science and engineering require quantitative research. Quantitative research is always confronted with huge amounts of numerical data that are collected from explorations, experiments, and surveys. Occasionally, data is even created virtually as it is done in random analysis. Data from various disciplines can be organized as matrices, manipulated in MATLAB workspace, stored as *.MAT files, and exported or imported for a particular class of problem.

The whole book of explaining MATLAB functions has been touched on in terms of the simplest examples. We believe that this is an effective introduction to the general ideas involved. They can immediately be extended to more complicated problems. Usage of any function in MATLAB needs the knowledge of input and output arguments of the function. The hidden arithmetic behind a function, of coarse, in no way new, but very often its clarification is encompassed with the traps of mathematical acuity. That digestion may be difficult for readers who have limited background or training in mathematics. We presuppose that the reader has some background in various disciplines of mathematics and a sound familiarity with the notations of those disciplines. At least the reader is supposed to be done with the twelfth grade science.

Toolboxes are a family of functions built on the wings of MATLAB platform. They can be applied to a particular area of interest. To mention some examples, there are *Symbolic Math Toolbox, Statistical Toolbox, Signal Processing Toolbox, Image Processing Toolbox, Optimization Toolbox, System Identification Toolbox*….. and many more. A broad overview of all Toolboxes is not possible in this short context. Elaborate discussion of each toolbox needs writing another book.

MATLAB's familiarity has continued to increase since Cleve Moler wrote this software. Recent MATLAB compilers are providing easy to use workspace, debugging convenience, fully featured graphical environment, and extensive help facilities. Throughout the text, MATLAB's interactive programming approach has been emphasized. One word or line MATLAB statement without any traditional programming, compiling, and debugging can execute dozens of program statements written in FORTRAN, PASCAL, or C. Broadly speaking, access features of MATLAB can be divided into two titles — MATLAB Command Window and the MATLAB's Graphics Window. Unlike traditional high-level structured programming software such as FORTRAN, PASCAL, or C, variables of MATLAB are matrices. Once the data of any particular discipline is organized in a matrix, it can be manipulated very easily and efficiently in MATLAB. One disadvantage of MATLAB is that the package is written in C that requires longer execution time to execute some programs than it does in C due to the conversion of codes. An M–file which contains complicated algorithms, calls hundreds of functions, manipulates matrices of higher dimensions, and deals with multi-loops suffers the shortcomings of execution time. But now a days cheaper processors with higher speed such as 1 GHz are being manufactured. To this context, real time realization of MATLAB can be accomplished.

6

This book is an outgrowth of my Master's Thesis and teaching Electrical Circuits and Electromagnetics in King Fahd University of Saudi Arabia. By the by, my research topic's title was "*Null Steering In Planar Antenna Arrays Using Genetic Algorithms*". While I was writing the MATLAB codes of the algorithms, I came across numerous built-in functions of MATLAB. I had to study the mathematics of some functions before their use in the algorithms. That inspired me writing this book. We hope explanatory nature of the book will make the reader a good user of MATLAB functions. Throughout the entire book, no discussions (other than introduction) of MATLAB's graphical capability have been mentioned. MATLAB is rich in easy visualization of scientific and technical problems in terms of various one/two/three dimensional plots. Apart from this, Graphical User Interface (GUI) can be implemented to simulate a technical problem. I hope our next effort will be devoted solely to MATLAB's graphical versatility.

Frankly speaking, computationally subjectiveness is the approach of the book. The book is intended for undergraduate as well as graduate students of science and engineering. Researchers and scientists who want to conduct their work in MATLAB can be beneficial from the book. Through out the text, we invariably called a small modular program for a specific purpose as procedure, subroutine, or function.

My words of acknowledgement are due to the King Fahd University of Petroleum and Minerals (KFUPM). I am especially appreciative of the printing and library facilities that I received from King Fahd University. All illustrative drill type problems and problems of varying complicity have been given MATLAB codes and implemented by a Pentium personal computer on Microsoft Windows operated system.

Mohammad Nuruzzaman

CONTENTS

Chapter 1
Introduction to MATLAB

Chapter 2
Matrix Fundamentals

Chapter 3
Solutions to Algebraic, Trigonometric, and Geometric Problems

Chapter 4
Matrix Algebra

Chapter 5

Problems of Differential Calculus

10

Chapter 6
Problems of Integral Calculus

Chapter 7
Problems of Complex Variables and Differential Equations

Chapter 8
Problems of Fourier, Laplace, and Z Transforms

12

Chapter 9
Problems of Statistics

Chapter 10
Miscellaneous Functions

Chapter 11

M–file Programming and Some Utilities

References

14

Chapter 1

Introduction to MATLAB

We welcome you to MATLAB. This is the quickest and easiest way to compute and visualize scientific and technical problems and solutions. As worldly standard for simulation and analysis, engineers, scientists, and researchers are becoming more and more affiliated with MATLAB. The general questionnaires before one gets started with MATLAB are the contents of this chapter. We explain the MATLAB Command Window so that you can be acquainted with its features and build yourself a good and efficient user of MATLAB. Our highlights will cover the command prompt, workspace, menu bar, help facilities, utility commands...etc that are available in the command window of MATLAB.

1.1 What is MATLAB?

MATLAB is mainly a scientific and technical computing software. Elaboration of MATLAB is matrix laboratory. Command prompt of MATLAB (>>) provides an interactive system. In the workspace of MATLAB, data element is dealt as a matrix without dimensioning. Cleve Moler wrote the first MATLAB in FORTRAN. The recent MATLAB is written in C by the MathWorks. MATLAB's easy to use platform enables us to compute matrix manipulations, perform numerical analysis, and visualize different variety of one/two/three dimensional graphics in a matter of second or seconds without conventional programming in FORTRAN, PASCAL, or C.

1.2 MATLAB command window

If you do not have MATLAB in your system, contact MathWorks for the installation CD. Have MATLAB installed in your personal computer. A new user of MATLAB should go through this chapter completely. After installation of MATLAB, go to the command window. We assume that the reader is familiar with Microsoft Windows operating system and handling a mouse. We also presume that you know how to select some text or object using a mouse. Let us get familiarized with MATLAB command window. Figure 1.1 shows a typical MATLAB command window. Following features are there in the command window.

Command prompt of MATLAB: Like DOS (Disc Operating System) prompt 'C:\>' and Unix prompt '$', MATLAB also has its command prompt, which is '>>' (see the command prompt in figure 1.1). Command prompt means that you tell MATLAB to do something or you enter executable commands from here. From the prompt, M–file as well as MATLAB graphics window can be run. As an interactive system,

2

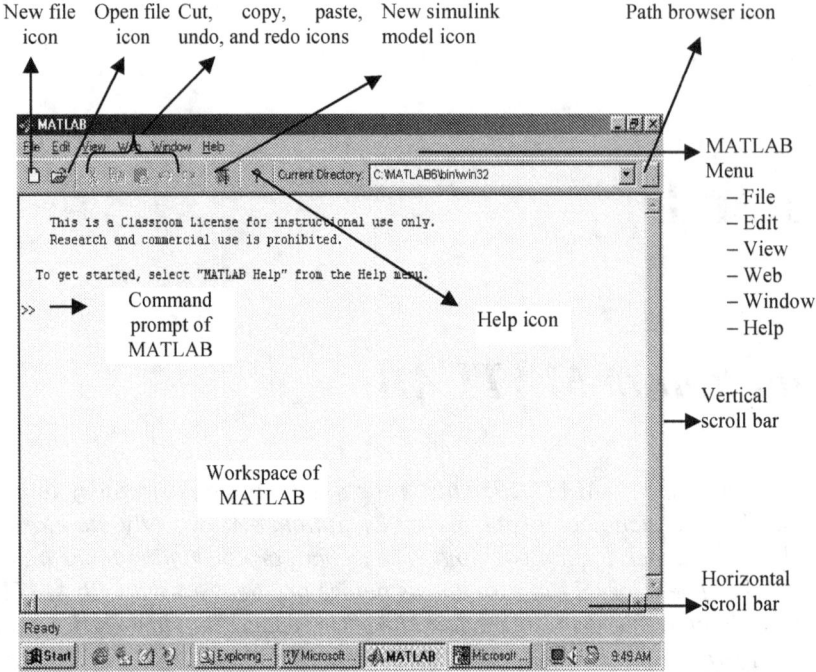

New file icon Open file icon Cut, copy, paste, undo, and redo icons New simulink model icon Path browser icon

MATLAB Menu
– File
– Edit
– View
– Web
– Window
– Help

Command prompt of MATLAB

Help icon

Workspace of MATLAB

Vertical scroll bar

Horizontal scroll bar

Figure 1.1 *Different features of MATLAB command window*

MATLAB responds to user through this prompt. MATLAB cursor will be blinking after >> prompt that says MATLAB is ready to take your commands. To enter any command, type executable MATLAB statements from keyboard and to execute that, press Enter ('↵') key.

⌸ ***MATLAB Menu:*** MATLAB is accompanied with six submenus, namely, File, Edit, View, Web, Window, and Help. Each submenu has its own features. Use the mouse and click different submenus. You will see the pull down menus of figure 1.2. The location of MATLAB menu is shown in figure 1.1. The first submenu is File (figure 1.2.1) that contains opening a new M-file, figure, model, or Graphical User Interface (GUI) layout maker, opening a file which was saved before, loading a saved workspace or importing data from a file, saving the workspace variables, setting the required path to execute a file, printing the workspace, and choosing the command window property. The second one is Edit (figure 1.2.2) that includes cutting, copying, pasting, undoing, and clearing operations. These operations are useful when you type some statements in command prompt. The Edit submenu operations are applicable only for MATLAB workspace not for M-files. An M-file has its own Edit submenu. The third submenu is View (figure 1.2.3) which is accompanied with various window viewing functions such as displaying the workspace variables information, current directory information, command history...etc. The fourth submenu Web (figure 1.2.4) provides the facility to get connected with the MathWorks who is the owner of the software or other related webpages. You may open some graphics window from MATLAB command window or running some M-files. From the fifth submenu window, we can see how many graphics window under MATLAB are open. You can switch from one window to another clicking the mouse to the required window. The last submenu is the Help (figure 1.2.5). MATLAB is affluent in help facilities. Later in this chapter, we mention how one can get help in different ways.

⌸ ***Icons:*** Available icons are shown in the icon bar of figure 1.1. Down the menu bar, icon bar is located. Frequently used operations such as opening a new file, opening an existing file, getting help,...etc are found in the icon menu bar so that user does not have to go through the menu bar over and over. The icons that are displayed are new file, open file, cut, copy, paste, undo, path browser, new simulink model, and help icons. All of these icons' action can be obtained from the menu bar. For example, Help icon activates the same window as Help of View menu does.

⊡ **Workspace:** Workspace is the platform of MATLAB where one executes MATLAB commands. It behaves as the gateway between user and MATLAB. During execution of commands, one may have to deal with some input and output variables. These variables can be one-dimensional array, multi-dimensional array, characters, symbolic objects…etc. Again, to deal with graphics window, we have texts, graphics, or object handles. Workspace holds those variables or handles for you. Not only that, it also exhibits the type or properties of those variables or handles. Like Microsoft Excel, all workspace information can be saved to a file

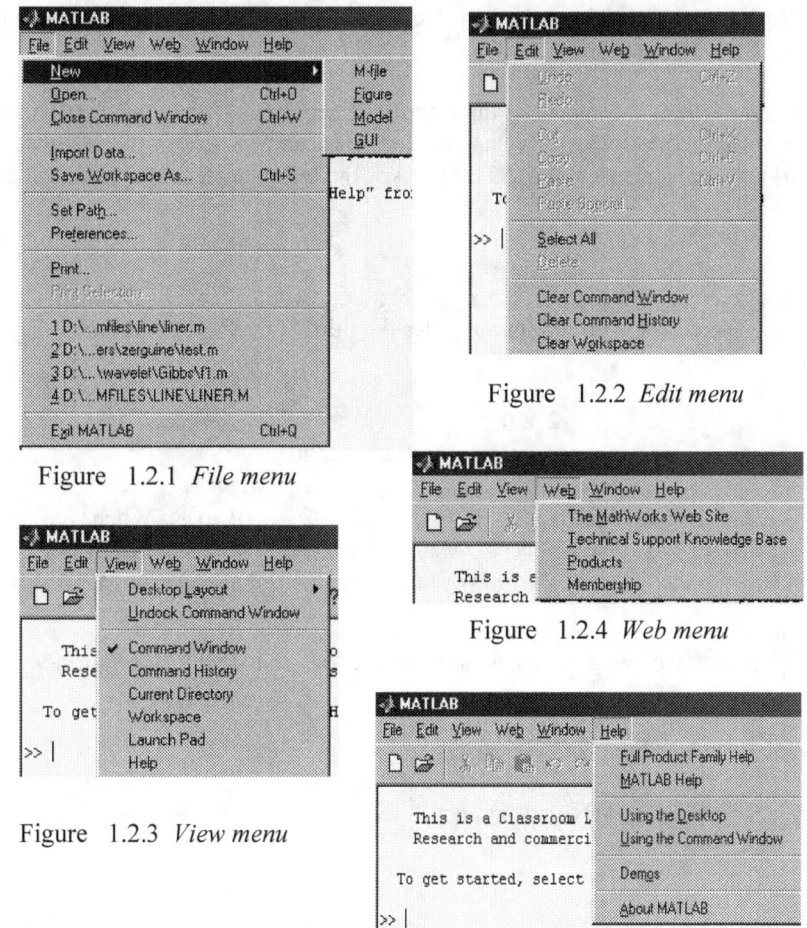

Figure 1.2.1 *File menu*

Figure 1.2.2 *Edit menu*

Figure 1.2.3 *View menu*

Figure 1.2.4 *Web menu*

Figure 1.2.5 *Help menu*

Figure 1.2 *Different pull down menus of MATLAB command window*

once some interaction with MATLAB is conducted.

⊡ **Scroll bars:** Monitor screen is not large enough to display the whole workspace. Scroll bars help us move on to next of the workspace. We have two scroll bars — horizontal and vertical (shown in figure 1.1). When one has to type and edit long string for computation, the use of horizontal scroll bar is obvious. If some matrix or help topic fits out of the monitor screen, we have to use the vertical scroll bar.

1.3 How to get started

New MATLAB users face a common question how one can get started. This tutorial is for the beginners. Here, you will learn the following:

⊡ *How one can enter a vector/matrix*

⊟ *How one can use colon and semicolon operators*
⊟ *How one can call a built-in MATLAB function*
⊟ *How one can open and execute an M–file*
⊟ *How one can plot a graph*

The first thing is user has to be in the command window of MATLAB in some way. In the command window, look for the command prompt '>>'. One can type anything on the command prompt. Widely used key of keyboard is 'Enter' key. Throughout the text, we mention 'pressing the Enter key operation' by symbol '⏎'. Let us proceed with the tutorial.

Entering a matrix: We can have three types of matrices — row, column, and rectangular. Row or column matrices are termed as vectors. Rectangular matrices can be a set of vectors. Enter the row matrix R=[2 3 4 −2 0] into the workspace of MATLAB. Type the following from keyboard on command prompt:

>>R=[2 3 4 -2 0]

Look at the style of typing. There is one space gap between two elements of the matrix R but no space gap at the edge elements. The cursor will be blinking after the third brace (']'). Press 'Enter' key from keyboard and you will see,

    R =
        2   3   4   -2   0
    >>   ← command prompt is ready again

It means that you assigned the matrix [2   3   4   −2   0] to R. Whenever you call R, MATLAB understands the matrix [2   3   4   −2   0]. Matrix R is having five elements, even if R had 100 elements, it would understand the whole matrix that is one of many appreciative features of MATLAB. Next, enter column

matrix C=$\begin{bmatrix} 7 \\ 8 \\ 10 \\ -11 \end{bmatrix}$. After displaying R, command prompt '>>' is ready to take further commands. Type the

following on blinking cursor from the keyboard:

        >>C=[7;8;10;-11] ⏎   you will see,

    C =
             7
             8
            10
           −11
    >>   ← command prompt is ready again

This time you also entered the matrix $\begin{bmatrix} 7 \\ 8 \\ 10 \\ -11 \end{bmatrix}$ to the workspace of MATLAB and assigned that to C. In this style

of typing, there is one semicolon (;) between two consecutive elements of the matrix C but no space gap is necessary. The matrix C could have been entered in another way. Type the following and execute that:

        >>C=[7 8 10 -11]' ⏎   you will see,

    C =
             7
             8
            10
           −11

As if you entered a row matrix but at the end just the transpose operator (') is attached. After that, the

rectangular matrix A=$\begin{bmatrix} 20 & 6 & 7 \\ 5 & 12 & -3 \\ 1 & -1 & 0 \\ 19 & 3 & 2 \end{bmatrix}$ is to be entered:

>>A=[20 6 7;5 12 -3;1 -1 0;19 3 2] ↵   you will see,

A =

```
   20    6    7
    5   12   -3
    1   -1    0
   19    3    2
```

Observe that two consecutive rows of A are separated by semicolons (;) and that consecutive elements in a row are separated by one space gap. Rows of matrix A are typed horizontally. Typing rows vertically one after another is accomplished as follows:

>>A=[20 6 7;   ↵
      5 12 -3;   ↵
      1 -1 0;   ↵
      19 3 2] ↵   you will see,
A =
```
   20    6    7
    5   12   -3
    1   -1    0
   19    3    2
```

So, you finished the first lesson successfully. Congratulations.

Using colon and/or semicolon operator(s): Operators semicolon (';') and colon (':') have special significance in MATLAB. Most numerical computations, MATLAB statements, and M−file programming use these two operators almost in every line. We show some examples of their use. Generation of vectors can easily be performed by colon operator no matter how many elements we need. Carry out the following in command prompt:

>>A=1:4 ↵   you will see,

A =

```
    1    2    3    4
```
We created a vector A (row matrix) where A=[1   2   3   4].

Interact with MATLAB by the following commands:

>>R=1:3:10 ↵   you will see,

R =
```
    1    4    7   10
```
We created a vector R (row matrix) whose elements form an arithmetic progression with first element 1, last element 10, and common difference 3.

Vector with decrement can also be generated:

>>C=[0:−2: −10]' ↵   you will see,

C =
```
    0
   -2
   -4
   -6
```

−8
−10   We created a vector C (column matrix) whose consecutive elements have the decrement 2 with the first element 0 and the last element −10.

MATLAB is also capable of producing vectors whose elements are floating-point numbers. Form a row matrix R whose first element is 3, last element is 6, and increment is 0.5. That is accomplished as follows:

>>R=3:0.5:6 ↵   you will see,

R =
    3.0000   3.5000   4.0000   4.5000   5.0000   5.5000   6.0000

To learn more about colon operator, go through articles 2.22, 2.24, 2.26, and 2.27. Then, what is the use of the semicolon operator? Append a semicolon in the last command and execute that:

>>R=3:0.5:6; ↵   you will see,
>>

Type R in command prompt and press Enter:
    >>R ↵

R =
    3.0000   3.5000   4.0000   4.5000   5.0000   5.5000   6.0000

It indicates that the semicolon operator prevents MATLAB from displaying contents of variable R. We mentioned formation of vectors but in command statements and M−file programming, one may have to compute a long expression and assign that to a variable. If you are sure that typed MATLAB statements are executable, append a semicolon at the end of statement(s) to avoid displaying contents of variables on command window screen.

Calling a MATLAB function from command prompt: In MATLAB, hundreds of M−files are attached. Various toolboxes also enhanced the applicability of the M−file functions. They can be treated as built-in functions. To execute an M−file from the command window, description of the function, numbers of input and output arguments, and the nature of arguments are necessary. The arguments can be real, complex, floating-point, integer numbers, characters, or even symbolic variables depending on the function. Let us start with the simplest example. Say, we want to find $\sin x$ for $x = \frac{3\pi}{2}$, which is −1. MATLAB counterpart of $\sin x$ is 'sin(x)', where x can be any complex number, and it can be a matrix too. Angle $\frac{3\pi}{2}$ is written as '3*pi/2' ('pi' is equivalent to π). Table 3.B provides the correspondence between MATLAB counterparts and symbolic notations. Go to MATLAB command prompt and perform the following:

>>sin(3*pi/2) ↵

ans =
    −1

As another example, factorize the integer 84. We know that 84=2×2×3×7. Subroutine 'factor' can find the factors of integer 84. The implementation is as follows:

>>factor(84) ↵

ans =
    2   2   3   7

The output of function 'factor' is a row matrix. Anyhow, we presented just two examples of calling M−file functions from the command prompt. Explore the other functions that are presented in table 3.B.

Opening and executing an M–file: This is the most important start up for beginners. An M–file can be regarded as a text or script file. A collection of executable MATLAB statements is the contents of an M–file. Before execution of an M–file from command prompt of MATLAB, following steps are necessary:

- Study the problem at hand. Find out what required is from the programming. Figure out how many inputs and outputs there are in the program for which the M–file needs to be written
- Develop a flow chart of your program, i.e., determine the sequence of MATLAB command execution
- Write MATLAB codes of your flow chart in an M–file.

Figure 1.3   *M–file Editor of MATLAB*

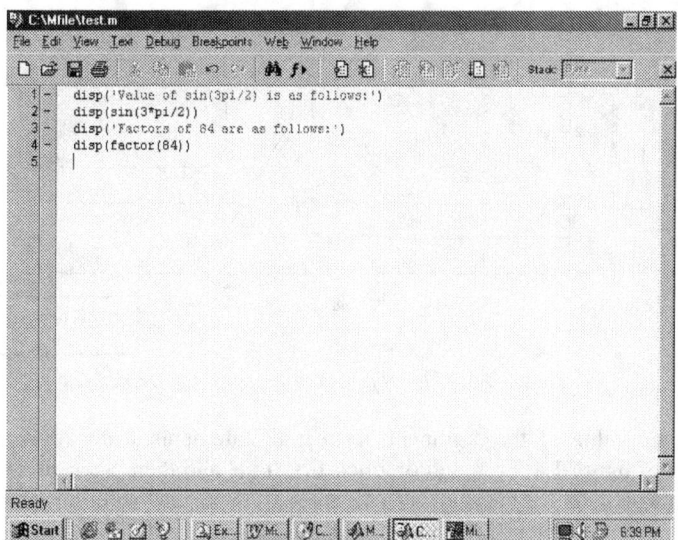

Figure 1.4   *M–file editor containing some executable MATLAB statements*

Before you open an M–file, create a directory or folder for your working path. For this, go to 'Windows Explorer' through 'Start' and 'Program' of Microsoft Windows and select the drive (C:, D:, or E:)

8

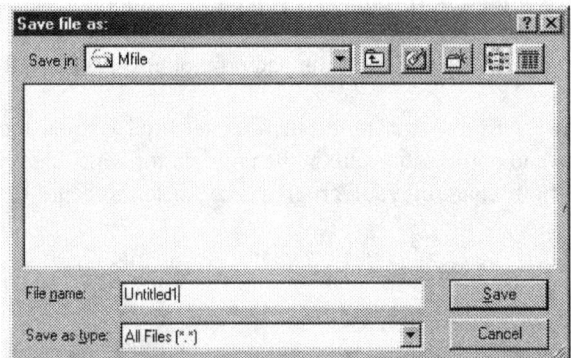

Figure 1.5   *Save dialog box of M-file editor*

where you want to create the directory (suppose, you chose the drive C:). Then, click the 'File' menu of 'Windows Explorer', click the 'New' from pull down menu, and click 'Folder'. You will see 'New Folder' prompts under the drive C. Delete 'New Folder', type 'Mfile' from keyboard, and press enter. So, you created a path for your own by name C:\Mfile. After that, go to MATLAB command prompt. MATLAB does not know that your working path is 'C:\Mfile'. In the command prompt, execute 'cd c:\mfile' to let MATLAB know your path (the same action can be done through the path browser icon as indicated in figure 1.1). Look at figure 1.2.1, M–file is seen under 'New'. Click M–file. MATLAB editor appears before you as shown in figure 1.3. In this editor, one can write any executable MATLAB statements. From previous discussion, you know about the subroutines 'sin' and 'factor'. Let us continue with that. There is a MATLAB function called 'disp', which displays statements under the single quoted comma. If single quoted comma is not there in the argument of

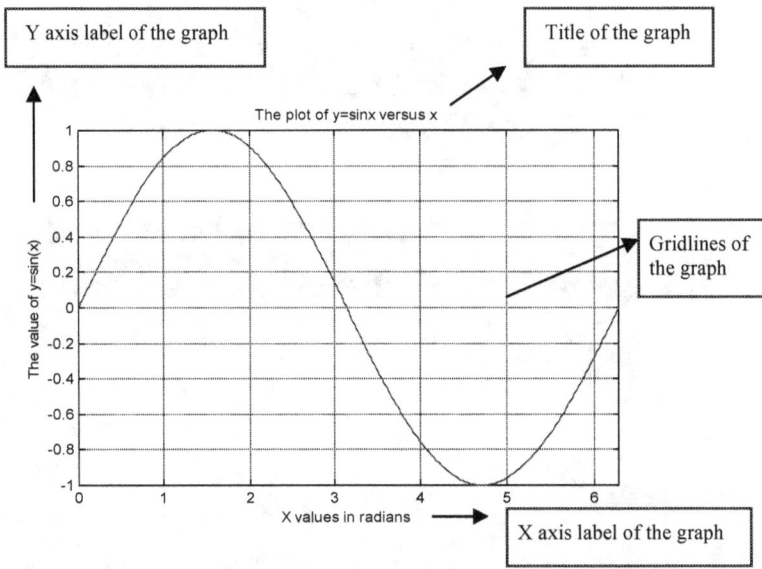

Figure 1.6   *Plot of function $y = \sin x$ for $0 \le x \le 2\pi$*

'disp', it shows the value of the argument. Type the statements from keyboard exactly as they appear in MATLAB editor of figure 1.4. From editor, click first 'File' and then 'Save' to have the save dialog box (can be done by clicking save icon too) as shown in figure 1.5. Type 'test' from keyboard instead of 'Untitled1' in the 'File Name' box and click 'Save'. So, you saved the M–file by the name 'test'. Finally, go to the command prompt and execute the M–file 'test' as follows:

MATLAB Command
>>test ↵

Value of sin(3pi/2) is as follows:
-1

Factors of 84 are as follows:
2   2   3   7

You can see how easy it is. Of coarse, the M–file must be syntactically correct for execution. You finished the lesson of opening and executing an M–file.

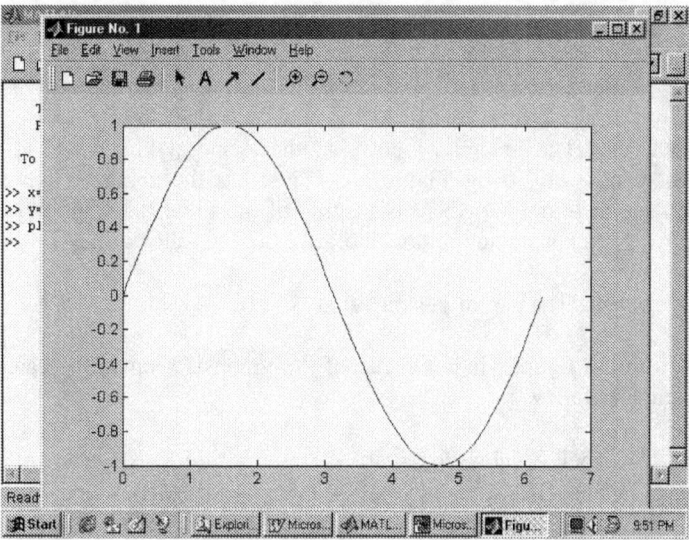

Figure 1.7   *Plot of just the graph of $y = \sin x$ for $0 \le x \le 2\pi$*

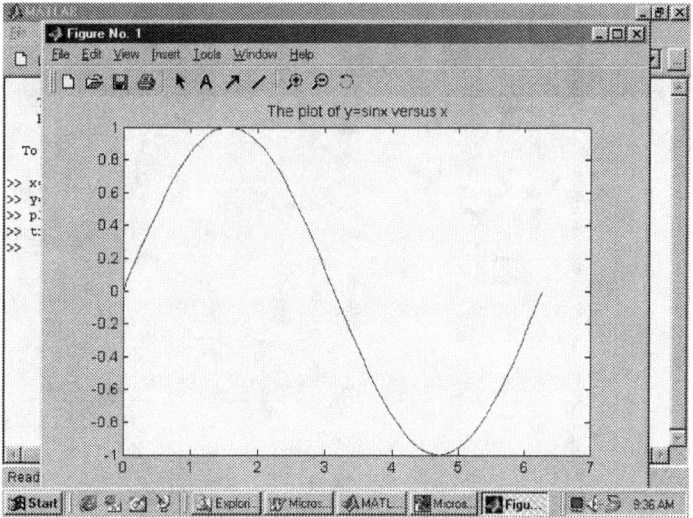

Figure 1.8   *Plot of $y = \sin x$ with title*

Plotting a graph: Even though graphical capability of MATLAB is not discussed in the entire book, just to have you started we are presenting this topic. The best thing is start with an example. Say, we want to plot the function $y = \sin x$ for $0 \le x \le 2\pi$. Plot of this function is shown in figure 1.6. As you see in figure 1.6, the graph has different features such as X-axis label, Y-axis label, Title…etc. Different features of this graph are indicated by arrow marks. Starting from the plot, each feature is put on the graph by one command. We

10

assume that by now you are capable of creating vectors using colon operators. First, we divide the given interval $0 \le x \le 2\pi$ to 100 steps and create a vector x as follows:

>>x=0:2*pi/100:2*pi; ↵

Then, compute $\sin x$ for different elements of vector x by the command 'sin(x)' and assign that to y as follows:

>>y=sin(x); ↵   ← x is a row matrix, so is y with the same length

Function 'plot' can graph $y = \sin x$ taking x and y as arguments, which is presented as follows:

>>plot(x,y) ↵

On top of the MATLAB command window, another window called MATLAB graphics window appears before you. That window titled by 'Figure No 1' just has the graph as depicted in figure 1.7 without any features. As you see in figure 1.7, there is no title of the graph like figure 1.6. Using mouse, switch to the command window, go to command prompt, and carry out the following:

>>title('The plot of y=sinx versus x') ↵

Use mouse to go to the figure window. Title of the graph is seen as indicated in figure 1.8. Add the other features of the graph as follows:

>>xlabel('X values in radians') ↵   ← Add X-axis label
>>ylabel('The value of y=sin(x)') ↵   ← Add Y-axis label
>>grid ↵   ← Add gridlines

See the figure window. Finally, you have the MATLAB output same as figure 1.6. User can also add these features on the plot by clicking Edit or Insert submenu of the graphics window.

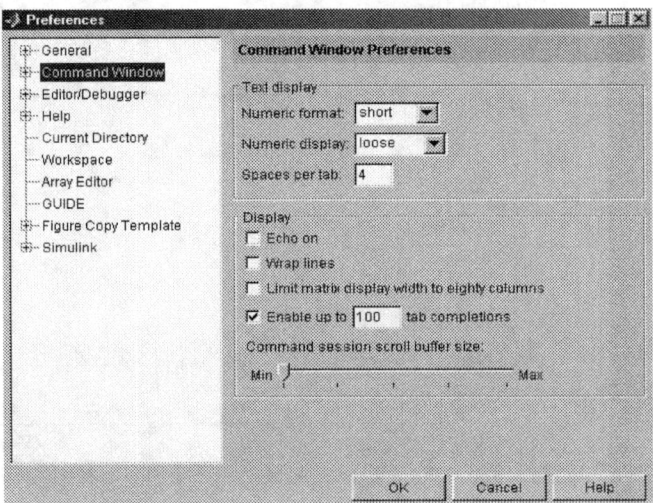

Figure 1.9   *Numeric format dialog box*

1.4 Frequently asked questions about MATLAB environment

New users need to know answers to some questions when they start working in MATLAB. Most of these questions come from MATLAB environment. Some answers are presented as follows:

🔊 *How can I change the numeric format?*

When you perform computation in the command window, output is returned up to four decimal accuracy. This is due to the use of short numeric format, which is the default one. There are other numeric formats also. To reach the numeric format dialog box, clicking operation sequence is MATLAB command window ⇒ File ⇒ Preferences ⇒ Command Window. Figure 1.9 should appear for numeric format dialog box. From the text display format, check the necessary format and click OK.

🔊 *How can I change the font or background color settings?*

Not everybody likes the white background of workspace. One might be interested to change the background color or font color of the command window. That can be carried out through the window shown in figure 1.10. The clicking sequence is MATLAB command window ⇒ File ⇒ Preferences ⇒ Command Window ⇒ Double Click to bring Font & Colors Window. From the displayed 'Text Color' and 'Background color', choose the desired color and click OK.

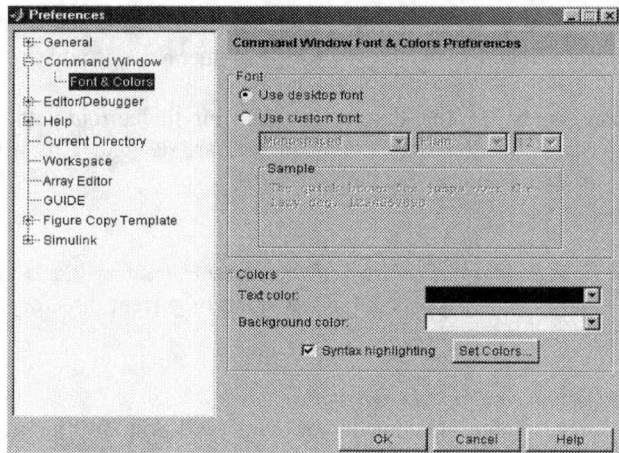

Figure 1.10 *Command window font property dialog box*

🔊 *How can I delete some/all variables from the workspace?*

Using command 'clear' deletes all variables from the workspace. If you want to delete a specific variable, just mention the name beside 'clear'. The command 'who' displays which variables are present in the workspace of MATLAB. Perform the following:

MATLAB Command

>>clear ⏎	← Clear all variables from the workspace
>>who ⏎	← See what variables are present
>>	← It indicates no variable is present
>>A=[1 2 5]; ⏎	← Enter a row matrix A=[1 2 5]
>>B=[4 2 7]; ⏎	← Enter another row matrix B=[4 2 7]
>>who ⏎	

Your variables are:

A   B

>>clear A ⏎	← Delete only matrix A from the workspace
>>who ⏎	

Your variables are:

B   ← As it is expected.

⊟ *How can I delete a file from the current path?*

Detail discussion of saving a file is presented in chapter 11. Type 'delete [file name]' and press 'Enter' to remove a file from the current path. But do not delete necessary file/files from the working path. It is better that you open a file and delete that. Follow the implementation:

MATLAB Command

>>A=[2 3;1 9]; ↵	← Assign matrix $\begin{bmatrix} 2 & 3 \\ 1 & 9 \end{bmatrix}$ to A
>>save test A ↵ 'test.mat'	← Save A in a file by name
>>dir *.mat ↵	← Display how many files with .mat extension are in the current path
test.mat	← This is the file you saved
>>delete test.mat ↵	← Delete the file 'test.mat'
>>dir *.mat ↵	← Just to check if the file 'test.mat' exists
*.mat not found.	← No file by that name is present

Never use the command delete *.*, that deletes all files from the current path. This is a destructive command. If you execute delete *.mat, all files with extension .mat are deleted. This way one can delete the files with extension .m or other located in the current path.

⊟ *How can I know the current path?*

In the upper right position of figure 1.1, the current directory bar is located which indicates in which path the command window is in or type 'cd' (abbreviation of <u>c</u>urrent <u>d</u>irectory) in command prompt and press 'Enter'.

⊟ *How can I clear workspace but not the variables?*

Once you conduct some sessions in the command window, monitor screen keeps all interactive sessions. You can clear the screen contents without removing the variables present in the workspace by the command 'clc' (abbreviation of <u>c</u>lear <u>c</u>ommand window) or performing clicking operation MATLAB command window ⇒ Edit ⇒ Clear command window.

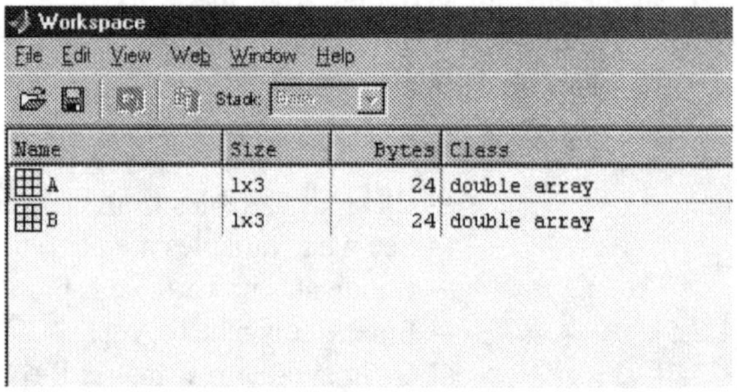

Figure 1.11   *Workspace browser of MATLAB*

⊟ *How can I see different variables in the workspace?*

There are two ways of seeing this – either use the command 'who' or follow the clicking sequence MATLAB command window ⇒ View ⇒ Workspace. Execution of the command 'who' has already been carried out. The command does not return the size of the variable, just provides the names. On the contrary, workspace browser exhibits more information as depicted in figure 1.11. You can even change or edit contents of a variable by double clicking the concern variable situated in workspace browser. Deleting a variable is also possible by right clicking the mouse on a variable in the workspace browser.

⌐ *How can I enter a long command line?*

Sometimes, MATLAB command statements can be too long to fit in one line in the monitor screen. This can be accomplished by the ellipsis (...). We show that considering the example of creating vector x=[1:3:10];. Type it in two lines as follows:

MATLAB Command

>>x=[1:3: ... ↵
10] ↵

x =

    1   4   7   10

Notice that there is one space gap before the ellipsis.

Finally, keep in mind that whatever can be executed in the command window can be executed in an M–file.

1.5 Editing in command prompt

This is advantageous specially for them who work frequently in command window without opening an M–file. Keyboard has different arrow keys marked by ← ↑ → ↓. In addition to arrow keys, control keys are also there. One may type a misspelled command on the command prompt. This causes error message to appear. Instead of retyping the entire line, press appropriate arrow or control key to recall, edit, or reuse MATLAB statements you have typed correctly or wrongly earlier.

		or
Recall a previous line	↑	Cntrl+p
Recall a next line	↓	Cntrl+n
Move back by one character	←	Cntrl+b
Move forward by one character	→	Cntrl+f
Move right by one word	Cntrl+→	Cntrl+r
Move left by one word	Cntrl+←	Cntrl+l
Move to the beginning of a line	Home	Cntrl+a
Move to the end of a line	End	Cntrl+e
Clear a line	Esc	Cntrl+u
Delete the character at cursor	del	Cntrl+d
Delete the character before cursor	backspace	Cntrl+h
Delete to end of a line		Cntrl+k

For example, generate a row vector 1 to 10 with increment 2. Assign the vector to x. The necessary command is 'x=1:2:10'. Mistakenly, you typed 'x+1:2:10'. See the response as follows:

>>x+1:2:10 ↵
??? Undefined function or variable 'x'.

You discovered the mistake and want to correct that. Press ↑ key or cntrl+p to see,
>>x+1:2:10

Edit the command going to + sign using ← or mouse. In the prompt, if you type x and press ↑ again and again, you see the used commands that start with x.

1.6 How to get help

Help facilities of MATLAB are plentiful. One can access to information about a function file in a variety of ways. Command help can find the help of a particular function file. You are familiar with the function 'sin(x)'. Have the online help regarding 'sin(x)' as follows:

MATLAB Command
>>help sin ⏎

SIN   Sine.
   SIN(X) is the sine of the elements of X.

   Overloaded methods
   help sym/sin.m

One disadvantage of this method is that user has to know the exact name of a function. For a novice, this facility may not be appreciative. Casually, you may know a partial name of a function or try to check whether a function exists by that name. Suppose, you want to see whether any function by name 'eye' exists. Execute the following by the intermediacy of command 'lookfor' to see all possible functions bearing the file name 'eye' or having the file name 'eye' partly:

MATLAB Command
>>lookfor eye ⏎
  EYE Identity matrix.
  SPEYE Sparse identity matrix.
  XPHIDE   MATLAB's version of Human Eye Sensitivity towards moving objects.
  commblkeyescatcont.m: % COMMBLKEYESCAT Eye/Scatter Scope dynamic dialog helping function
   COMMBLKEYESCATDISC Eye/Scatter Scope dynamic dialog helping function
  EYEDIASI2 Simulink eye diagram and scatter plot.
   EYESAMPL2 Simulink eye diagram and scatter plot --- sampled time version.
  EYEDIASI Simulink eye diagram and scatter plot.
   EYESAMPL Simulink eye diagram and scatter plot --- sampled time version.
  EYEDIAGRAM Generate an eye diagram.
   ANIMATESCATTEREYE - Does animation of moving offsets for
  scattereyedemo.m: %Demonstration of eye diagram and scatter plot functions
  EYESCAT Produce eye-pattern diagram or scatter plot.
   DSPBLKEYE is the mask function for the DSP Blockset Identity Block
  DMEYINFO Information on "Discrete" Meyer wavelet.
  MEYER Meyer wavelet.
  MEYERAUX Meyer wavelet auxiliary function.
  MEYRINFO Information on Meyer wavelet.

As you see, the return is having all possible matches of functions containing 'eye'. Now command 'help' can be conducted to go through a particular one. Window form help is also there. For this, click different windows of the pull down menu of figire 1.2.5. Make sure you have the full Help CD installed in your system. Any help item preceded by a bullet can be clicked to go inside the item. This help form is better when one navigates MATLAB's capability not looking for a particular function. Besides MATLAB functions, Maple functions are also available. Online information of Maple function is reached by executing mhelp [function name].

Chapter 2

Matrix Fundamentals

This chapter illustrates simple operations on matrices. Matrices are nothing but rectangular arrays of numbers (can be real, integer, or imaginary) or symbolic variables set out in rows and columns. In MATLAB, matrix is a variable on which all manipulations are done. Arithmetic of matrices includes addition, subtraction, multiplication, division, powering, ... etc and manipulation of matrices includes maneuvering of rows and columns, formation of smaller matrices from larger one (and vice versa)...etc. To have manipulative skill, numerical examples are provided for each type of matrix — row, column, and rectangular. Even though row and column matrices are the special cases of rectangular matrices, manipulations of row and column matrices may require different/simpler form of representations. Application of these simple operations is realized when one writes an M–file program describing a real life problem. Matrix size does not affect the concept of procedures, only it does the amount of computing that is involved. We start with the description of a matrix arithmetic or manipulation in symbolic form and continue to the implementation in MATLAB environment.

2.1 Matrix addition

Very basic manipulation of matrices is the addition. See the following examples for addition.

Row matrix addition:

Suppose, $A=[1\ 2\ 3]$ and $B=[9\ 3\ -7]$ are two row matrices, addition of A and B is C, where $C = A + B =[1+9\ 2+3\ 3-7]=[10\ 5\ -4]$.

Column matrix addition:

Addition of column matrices $A=\begin{bmatrix}x\\9\\10\end{bmatrix}$ and $B=\begin{bmatrix}5\\3\\8\end{bmatrix}$ is $C = A + B =\begin{bmatrix}x+5\\12\\18\end{bmatrix}$. We have symbolic variable

x. Declare x as symbolic by the command 'syms'. Transpose operator (') converts a row matrix to column one but operator .' means transpose without conjugate.

Rectangular matrix addition:

$A = \begin{bmatrix} 7 & 4 & 2 \\ 9 & 0 & 1 \\ 10 & 9 & 3 \end{bmatrix}$ and $B = \begin{bmatrix} 5 & 7 & 9 \\ 3 & 1 & 0 \\ 8 & 3 & 0 \end{bmatrix}$ are two rectangular matrices. Their addition is $C = A + B =$

$\begin{bmatrix} 12 & 11 & 11 \\ 12 & 1 & 1 \\ 18 & 12 & 3 \end{bmatrix}$. See all additions as follows:

MATLAB Command

for addition of row matrices,

    >>A=[1 2 3]; ↵
    >>B=[9 3 -7]; ↵
    >>C=A+B ↵

    C=
       10   5   -4

for rectangular matrices,

    >>A=[7 4 2;9 0 1;10 9 3]; ↵
    >>B=[5 7 9;3 1 0;8 3 0]; ↵
    >>C=A+B ↵

    C =
       12 11 11
       12   1   1
       18 12   3

for column matrices,

    >>syms x ↵
    >>A=[x 9 10].'; ↵
    >>B=[5 3 8]'; ↵
    >>C=A+B ↵

    C =

    [x+5]
    [12]
    [18]

Orders of the matrices that are to be added must be same otherwise you will get an error message. Let us sum matrix $A = [2\ 9\ 0]$ and $B = [4\ 5]$ in MATLAB.

MATLAB Command
    >>A=[2 9 0]; ↵
    >>B=[4 5]; ↵
    >>C=A+B ↵
    ??? Error using ==> ±
    Matrix dimensions must agree.

As you see, MATLAB Command Window returns the above error message because order of A is 1×3 and order of B is 1×2, they do not match. This way, addition of floating-point numbers can be carried out also.

Addition of rational numbers:

What if we have matrix elements are rational numbers. We want output matrix to be in rational form too, for example, $A = \begin{bmatrix} \frac{3}{4} & \frac{4}{5} & -\frac{2}{3} \\ \frac{1}{7} & \frac{9}{5} & \frac{5}{9} \end{bmatrix}$ and $B = \begin{bmatrix} \frac{4}{7} & \frac{8}{3} & \frac{2}{3} \\ \frac{3}{4} & \frac{9}{7} & -\frac{1}{9} \end{bmatrix}$, $\therefore$ $A + B = \begin{bmatrix} \frac{3}{4}+\frac{4}{7} & \frac{4}{5}+\frac{8}{3} & -\frac{2}{3}+\frac{2}{3} \\ \frac{1}{7}+\frac{3}{4} & \frac{9}{5}+\frac{9}{7} & \frac{5}{9}-\frac{1}{9} \end{bmatrix} =$

$\begin{bmatrix} \frac{37}{28} & \frac{52}{15} & 0 \\ \frac{25}{28} & \frac{108}{35} & \frac{4}{9} \end{bmatrix}$. This addition is also shown below. Command 'sym' is used to describe the elements inside matrix as rational numbers:

For symbolic addition,
    >>A=sym([3/4 4/5 -2/3;1/7 9/5 5/9]); ↵
    >>B=sym([4/7 8/3 2/3;3/4 9/7 -1/9]); ↵
    >>C=A+B ↵

C =

[37/28,   52/15,   0]
[25/28, 108/35, 4/9]

2.2 Matrix subtraction

Subtraction of matrices is similar to that of addition. Use the same matrices as we chose for article 2.1. Subtraction of B from A should give us $[-8 \quad -1 \quad 10]$, $\begin{bmatrix} x-5 \\ 6 \\ 2 \end{bmatrix}$, and $\begin{bmatrix} 2 & -3 & -7 \\ 6 & -1 & 1 \\ 2 & 6 & 3 \end{bmatrix}$ for row, column,

and rectangular matrices respectively. See the implementation below:

MATLAB Command

for subtraction of row matrices,

>>A=[1 2 3]; ↵
>>B=[9 3 -7]; ↵
>>C=A-B ↵

C=
     -8 -1   10

for rectangular matrices,

>>A=[7 4 2;9 0 1;10 9 3]; ↵
>>B=[5 7 9;3 1 0;8 3 0]; ↵
>>C=A-B ↵

C =
     2 -3 -7
     6 -1   1
     2   6   3

for column matrices,

>>syms x ↵
>>A=[x 9 10].'; ↵
>>B=[5 3 8]'; ↵
>>C=A-B ↵

C =

[x-5]
[6]
[2]

2.3 Matrix multiplication

There are three types of matrix multiplication, namely, scalar, vector, and Kronecker.

2.3.1 Scalar multiplication

Scalar multiplication is performed by the operator .*. Actually, this is point to point multiplication.

Row matrix multiplication:

Say, $A =[1 \quad 2 \quad 3]$ and $B =[9 \quad 3 \quad -7]$ are two row matrices, scalar multiplication of A with B is C, where $C = A .* B =[1 \quad 2 \quad 3].*[9 \quad 3 \quad -7]=[1 \times 9 \quad 2 \times 3 \quad 3 \times (-7)]=[9 \quad 6 \quad -21]$.

Column matrix multiplication:

$A = \begin{bmatrix} x \\ 2y \\ 10 \end{bmatrix}$ and $B = \begin{bmatrix} 5 \\ 3 \\ 8 \end{bmatrix}$ are two column matrices. The scalar multiplication of A with B is $C = A .* B = \begin{bmatrix} 5x \\ 6y \\ 80 \end{bmatrix}$.

Rectangular matrix multiplication:

Scalar multiplication of $A = \begin{bmatrix} 7 & 4 & 2 \\ 9 & 0 & 1 \\ 10 & 9 & 3 \end{bmatrix}$ and $B = \begin{bmatrix} 5 & 7 & 9 \\ 3 & 1 & 0 \\ 8 & 3 & 0 \end{bmatrix}$ is $C = A.*B = \begin{bmatrix} 7\times5 & 4\times7 & 2\times9 \\ 9\times3 & 0\times1 & 1\times0 \\ 10\times8 & 9\times3 & 3\times0 \end{bmatrix} =$

$\begin{bmatrix} 35 & 28 & 18 \\ 27 & 0 & 0 \\ 80 & 27 & 0 \end{bmatrix}$.

Notice the orders of matrices used for scalar multiplication. For all cases, orders of A and B are identical (1×3, 3×1, and 3×3 for row, column, and rectangular matrices respectively). MATLAB implementation for all scalar multiplication is shown below:

MATLAB Command

for scalar multiplication of row matrices,

```
>>A=[1 2 3]; ↵
>>B=[9 3 -7]; ↵
>>C=A.*B ↵
```

C =

     9   6 -21

for rectangular matrices,

```
>>A=[7 4 2;9 0 1;10 9 3]; ↵
>>B=[5 7 9;3 1 0;8 3 0]; ↵
>>C=A.*B ↵
```

C =

    35 28 18
    27   0   0
    80 27   0

for column matrices,

```
>>syms x y ↵
>>A=[x 2*y 10].'; ↵
>>B=[5 3 8]'; ↵
>>C=A.*B ↵
```

C =

[5*x]
[6*y]
[80]

Multiplication of matrix of rational numbers:

If matrix elements are rational numbers and output is wanted in rational numbers, use the command 'sym'. To illustrate this, say, $A = \begin{bmatrix} \frac{3}{4} & \frac{4}{5} & -\frac{2}{3} \\ \frac{1}{7} & \frac{9}{5} & \frac{5}{9} \end{bmatrix}$ and $B = \begin{bmatrix} \frac{4}{7} & \frac{8}{3} & \frac{2}{3} \\ \frac{3}{4} & \frac{9}{7} & -\frac{1}{9} \end{bmatrix}$, scalar product of A and B in rational form is $\begin{bmatrix} \frac{3}{7} & \frac{32}{15} & -\frac{4}{9} \\ \frac{3}{28} & \frac{81}{35} & -\frac{5}{81} \end{bmatrix}$. Its implementation is shown below:

MATLAB Command

```
>>A=sym([3/4 4/5 -2/3;1/7 9/5 5/9]); ↵
>>B=sym([4/7 8/3 2/3;3/4 9/7 -1/9]); ↵
>>C=A.*B ↵
```

C =

[3/7, 32/15, -4/9]
[3/28, 81/35, -5/81]

2.3.2 Vector multiplication

In vector multiplication, matching of matrices' order is more important than it is with the scalar multiplication. The reason is that both matrices have to be of the same order in scalar multiplication but both may not have to be of the same order in the vector multiplication. In general, a matrix A of order $m \times n$ can be multiplied with a matrix B of order $n \times p$. In MATLAB, the vector multiplication is defined by the operator $*$. If vector multiplication of A and B is C, then, order of C is $m \times p$. In MATLAB notation, we can write vector multiplied matrix $C = A*B$ and for scalar multiplication, $A.*B = B.*A$ but for the vector multiplication, $A*B \neq B*A$.

Row matrix multiplication:

A row matrix A has order $1 \times n$, where n can be any integer. A can only be multiplied with a matrix B if B has only n rows no matter how many columns B has. Assume that a matrix B is to be vector multiplied with a row matrix $A = [2 \ -9 \ 7 \ -3]$. Since order of A is 1×4, B must have 4 rows with any number of columns. Take an example of B having three columns, say, $B = \begin{bmatrix} -1 & 1 & -1 \\ -3 & 2 & 0 \\ 7 & 3 & 2 \\ 0 & 4 & 0 \end{bmatrix}$. Vector multiplication of B with A (not A with B) is C, where $C = A*B = [2 \ -9 \ 7 \ -3]*\begin{bmatrix} -1 & 1 & -1 \\ -3 & 2 & 0 \\ 7 & 3 & 2 \\ 0 & 4 & 0 \end{bmatrix} = \begin{bmatrix} [2 \ -9 \ 7 \ -3]*\begin{bmatrix} -1 \\ -3 \\ 7 \\ 0 \end{bmatrix} \end{bmatrix}$

$[2 \ -9 \ 7 \ -3]*\begin{bmatrix} 1 \\ 2 \\ 3 \\ 4 \end{bmatrix} \quad [2 \ -9 \ 7 \ -3]*\begin{bmatrix} -1 \\ 0 \\ 2 \\ 0 \end{bmatrix} = [2(-1)+(-9)(-3)+7\times7+(-3)\times0 \quad 2\times1+(-9)\times2+7\times3+(-3)\times4$

$2\times(-1)+(-9)\times0+7\times2+(-3)\times0] = [74 \ -7 \ 12]$. Orders of A and B are 1×4 and 4×3 ($m=1$, $n=4$, and $p=3$) respectively, so, the vector multiplied matrix C should have the order 1×3.

Column matrix multiplication:

Matrix B is to be vector multiplied with column matrix $A = \begin{bmatrix} -1 \\ -3 \\ 7 \end{bmatrix}$. Since order of A is 3×1, B must have 1 row with any number of columns. Taking $B = [-x \ 2]$, vector multiplication of B with A is $A*B = \begin{bmatrix} \begin{bmatrix} -1 \\ -3 \\ 7 \end{bmatrix}*[-x] & \begin{bmatrix} -1 \\ -3 \\ 7 \end{bmatrix}*[2] \end{bmatrix} = \begin{bmatrix} x & -2 \\ 3x & -6 \\ -7x & 14 \end{bmatrix}$.

Rectangular matrix multiplication:

For rectangular matrix, take the example of a 2×3 matrix, say, $A = \begin{bmatrix} 7 & 4 & 9 \\ 9 & 0 & 2 \end{bmatrix}$. One example of B can be $\begin{bmatrix} 5 & 7 \\ 3 & 1 \\ 2 & 0 \end{bmatrix}$, $\therefore A*B = \begin{bmatrix} 7 & 4 & 9 \\ 9 & 0 & 2 \end{bmatrix}*\begin{bmatrix} 5 & 7 \\ 3 & 1 \\ 2 & 0 \end{bmatrix} = \begin{bmatrix} 7\times5+4\times3+9\times2 & 7\times7+4\times1+9\times0 \\ 9\times5+0\times3+2\times2 & 9\times7+0\times1+2\times0 \end{bmatrix} = \begin{bmatrix} 65 & 53 \\ 49 & 63 \end{bmatrix}$. All types of vector multiplication done in MATLAB are shown as follows:

MATLAB Command

for vector multiplication of row matrix,

```
>>A=[2 -9 7 -3]; ↵
>>B=[ -1 1 -1;-3 2 0;7 3 2;0 4 0]; ↵
>>C=A*B ↵

    C=
```

for column matrix,

```
>>syms x ↵
>>A=[-1 -3 7]'; ↵
>>B=[-x 2]; ↵
>>C=A*B ↵
```

$$\begin{array}{ccc} 74 & -7 & 12 \end{array}$$

for rectangular matrix,

```
>>A=[7 4 9;9 0 2]; ↵
>>B=[5 7;3 1;2 0]; ↵
>>C=A*B ↵
```

C=

$$\begin{array}{cc} 65 & 53 \\ 49 & 63 \end{array}$$

C=

$$\begin{array}{cc} [& x, & -2] \\ [& 3*x, & -6] \\ [& -7*x, & 14] \end{array}$$

Multiplication of the matrix of rational numbers:

Vector multiplication of matrices in rational form is also possible, say, $A = \begin{bmatrix} \frac{3}{4} & \frac{4}{5} \\ \frac{1}{7} & \frac{9}{5} \end{bmatrix}$ and $B = \begin{bmatrix} \frac{4}{7} & \frac{8}{3} \\ \frac{3}{4} & \frac{9}{7} \end{bmatrix}$,

$$\therefore A*B = \begin{bmatrix} \frac{3}{4}\times\frac{4}{7}+\frac{4}{5}\times\frac{3}{4} & \frac{3}{4}\times\frac{8}{3}+\frac{4}{5}\times\frac{9}{7} \\ \frac{1}{7}\times\frac{4}{7}+\frac{9}{5}\times\frac{3}{4} & \frac{1}{7}\times\frac{8}{3}+\frac{9}{5}\times\frac{9}{7} \end{bmatrix} = \begin{bmatrix} \frac{36}{35} & \frac{106}{35} \\ \frac{1403}{980} & \frac{283}{105} \end{bmatrix}:$$

MATLAB Command

```
>>A=sym([3/4 4/5;1/7 9/5]); ↵
>>B=sym([4/7 8/3;3/4 9/7]); ↵
>>C=A*B ↵
```

C =

$$\begin{array}{cc} [& 36/35, & 106/35] \\ [1403/980, & 283/105] \end{array}$$

2.3.3 Kronecker product multiplication

Kronecker product matrix multiplication is useful in some applications. This is, basically, direct product of two matrices. Let matrices A and B be of order $m \times n$ and $p \times q$ respectively. General forms of A and B are $\begin{bmatrix} A_{11} & A_{12} & \cdots & A_{1n} \\ A_{21} & A_{22} & \cdots & A_{2n} \\ & & \vdots & \\ A_{m1} & A_{m2} & \cdots & A_{mn} \end{bmatrix}$ and $\begin{bmatrix} B_{11} & B_{12} & \cdots & B_{1q} \\ B_{21} & B_{22} & \cdots & B_{2q} \\ & & \vdots & \\ B_{p1} & B_{p2} & \cdots & B_{pq} \end{bmatrix}$ respectively. Kronecker product matrix is denoted by $A \otimes B$ and is defined as $\begin{bmatrix} A_{11}B & A_{12}B & \cdots & A_{1n}B \\ A_{21}B & A_{22}B & \cdots & A_{2n}B \\ & & \vdots & \\ A_{m1}B & A_{m2}B & \cdots & A_{mn}B \end{bmatrix}$. Order of $A \otimes B$ is $p\,m \times q\,n$. To illustrate, say,

$$A = \begin{bmatrix} 0 & 1 & 7 \\ -1 & 0 & 2 \end{bmatrix} \text{ and } B = \begin{bmatrix} 1 & 2 \\ 0 & 9 \end{bmatrix}, \quad \therefore \quad A \otimes B = \begin{bmatrix} 0\begin{pmatrix}1&2\\0&9\end{pmatrix} & 1\begin{pmatrix}1&2\\0&9\end{pmatrix} & 7\begin{pmatrix}1&2\\0&9\end{pmatrix} \\ -1\begin{pmatrix}1&2\\0&9\end{pmatrix} & 0\begin{pmatrix}1&2\\0&9\end{pmatrix} & 2\begin{pmatrix}1&2\\0&9\end{pmatrix} \end{bmatrix} =$$

$$\begin{bmatrix} 0 & 0 & 1 & 2 & 7 & 14 \\ 0 & 0 & 0 & 9 & 0 & 63 \\ -1 & -2 & 0 & 0 & 2 & 4 \\ 0 & -9 & 0 & 0 & 0 & 18 \end{bmatrix}.$$ Orders of A and B are 2×3 and 2×2 respectively, hence, order of $A \otimes B$ is

(2×2)×(3×2) or 4×6. See the example below:

MATLAB Command

```
>>A=[0 1 7;-1 0 2]; ↵
>>B=[1 2;0 9]; ↵
>>C=kron(A,B) ↵
```

$$C = \begin{array}{cccccc} 0 & 0 & 1 & 2 & 7 & 14 \\ 0 & 0 & 0 & 9 & 0 & 63 \\ -1 & -2 & 0 & 0 & 2 & 4 \\ 0 & -9 & 0 & 0 & 0 & 18 \end{array}$$

To obtain Kronecker product of two matrices,

MATLAB Steps:
  1. Enter the first matrix A
  2. Enter the second matrix B
  3. Use the command kron(A,B).

2.4 Matrix division

We have mainly two types of division — scalar and vector. For each type of division, again, there are two types of sub-division. One is called left division and the other is right division.

2.4.1 Scalar division

Like multiplication, scalar division is also possible for row, column, or rectangular matrices. Scalar division can be termed as point to point division. Orders of both the divider and dividend matrices must be identical.

Right division:

Right scalar division is computed by the operator ./.

Row matrix division:

$A = [9 \quad 3 \quad -7]$ and $B = [1 \quad 2 \quad 3]$ are two row matrices. Right scalar division of A by B is C, where $C = A./B = [9 \quad 3 \quad -7]./[1 \quad 2 \quad 3] = [\frac{9}{1} \quad \frac{3}{2} \quad \frac{-7}{3}] = [9 \quad 1.5 \quad -2.3333]$.

Column matrix division:

Column matrices $A = \begin{bmatrix} 7 \\ 9 \\ 10 \end{bmatrix}$ and $B = \begin{bmatrix} 5 \\ 3 \\ 8 \end{bmatrix}$ have right scalar division as $A./B = \begin{bmatrix} 7 \\ 9 \\ 10 \end{bmatrix} ./ \begin{bmatrix} 5 \\ 3 \\ 8 \end{bmatrix} = \begin{bmatrix} \frac{7}{5} \\ \frac{9}{3} \\ \frac{10}{8} \end{bmatrix} = \begin{bmatrix} 1.4 \\ 3 \\ 1.25 \end{bmatrix}$.

Rectangular matrix division:

$A = \begin{bmatrix} 7 & 4 \\ 9 & 0 \\ 10 & 9 \end{bmatrix}$ and $B = \begin{bmatrix} 5 & 7 \\ 3 & 1 \\ 8 & 3 \end{bmatrix}$ perform the operation as $C = A./B = \begin{bmatrix} 7 & 4 \\ 9 & 0 \\ 10 & 9 \end{bmatrix} ./ \begin{bmatrix} 5 & 7 \\ 3 & 1 \\ 8 & 3 \end{bmatrix} = \begin{bmatrix} \frac{7}{5} & \frac{4}{7} \\ \frac{9}{3} & \frac{0}{3} \\ \frac{10}{8} & \frac{9}{3} \end{bmatrix} =$

$\begin{bmatrix} 1.4 & 0.5714 \\ 3 & 0 \\ 1.25 & 3 \end{bmatrix}$. See all implementations below:

MATLAB Command

for row matrix,
```
>>A=[9 3 -7]; ↵
>>B=[1 2 3]; ↵
>>C=A./B ↵

C =
        9.0000   1.5000  -2.3333
```
for rectangular matrix,
```
>>A=[7 4;9 0;10 9]; ↵
>>B=[5 7;3 1;8 3]; ↵
```

for column matrix,
```
>>A=[7 9 10]'; ↵
>>B=[5 3 8]'; ↵
>>C=A./B ↵

C =
        1.4000
        3.0000
        1.2500
```

>>C=A./B ↵

C =
```
      1.4000   0.5714
      3.0000        0
      1.2500   3.0000
```

Left division:

The operator .\ calculates left scalar division. Take the same matrices as examples what we have used for the right scalar division. Following left division, one should get $[\frac{1}{9} \quad \frac{2}{3} \quad \frac{-3}{7}]$, $\begin{bmatrix} \frac{5}{7} \\ \frac{1}{3} \\ \frac{4}{5} \end{bmatrix}$, and $\begin{bmatrix} \frac{5}{7} & \frac{7}{4} \\ \frac{1}{3} & \infty \\ \frac{4}{5} & \frac{1}{3} \end{bmatrix}$ for row, column, and rectangular matrices respectively. We wish to see the outputs in symbolic form. See the MATLAB implementation below:

MATLAB Command

for row matrix,
```
>>A=sym([9 3 -7]); ↵
>>B=sym([1 2 3]); ↵
>>C=A.\B ↵

C =

[ 1/9,  2/3,  -3/7]
```

for rectangular matrix,
```
>>A=sym([7 4;9 0;10 9]); ↵
>>B=sym([5 7;3 1;8 3]); ↵
>>C=A.\B ↵

C =

[ 5/7,   7/4]
[ 1/3,   Inf]
[ 4/5,   1/3]
```

for column matrix,
```
>>A=sym([7 9 10]'); ↵
>>B=sym([5 3 8]'); ↵
>>C=A.\B ↵

C =

[ 5/7]
[ 1/3]
[ 4/5]
```

Notice that $\frac{1}{0} = \infty$ is also computed for the rectangular matrix.

2.4.2 Vector division

Vector division is also categorized in two titles – left and right.

Left division:

Left vector division is performed by the operator \. Left vector division of matrix B by matrix A is denoted as $A \backslash B$. If the quotient matrix is X, then, one can write $X = A \backslash B$ and $A\, X = B$. In other words, X is the solution of the matrix equation $A\, X = B$. A matrix B, in general, of order $m \times k$ can be divided by a matrix A of order $m \times n$. The quotient matrix X will be of order $n \times k$. That is, left division is possible if the numbers of rows of A and B are identical. To illustrate this, take $A = \begin{bmatrix} 2 & -2 & 0 & 1 \\ 0 & 1 & 1 & 2 \\ 1 & -4 & 0 & 2 \end{bmatrix}$ and $B = \begin{bmatrix} 4 & 3 \\ -1 & 2 \\ 0 & -4 \end{bmatrix}$.

With this, command $A \backslash B$ should return the solution of the system of equations: $\begin{bmatrix} 2 & -2 & 0 & 1 \\ 0 & 1 & 1 & 2 \\ 1 & -4 & 0 & 2 \end{bmatrix} \begin{bmatrix} x_1 \\ x_2 \\ x_3 \\ x_4 \end{bmatrix} =$

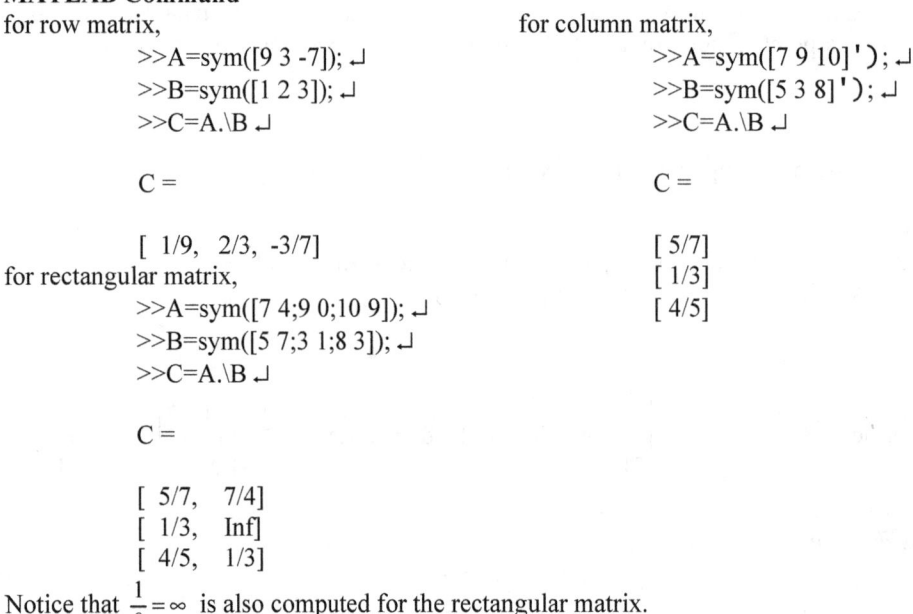

$\begin{bmatrix} 4 & 3 \\ -1 & 2 \\ 0 & -4 \end{bmatrix}$, where $m=3$, $n=4$, and $k=2$. The solution matrix X will be of order 4×2. Rank of matrix A is 3

(see chapter 4 for rank). Matrix A can be partitioned by D and E, so, $A=[D \ E]=\begin{bmatrix} 2 & -2 & 0 & \vdots & 1 \\ 0 & 1 & 1 & \vdots & 2 \\ 1 & -4 & 0 & \vdots & 2 \end{bmatrix}$,

where $D=\begin{bmatrix} 2 & -2 & 0 \\ 0 & 1 & 1 \\ 1 & -4 & 0 \end{bmatrix}$ and $E=\begin{bmatrix} 1 \\ 2 \\ 2 \end{bmatrix}$. Solution of the system of equations is found over each column of B.

First, we consider the first column of B. Nonsingular part of the solution is given by $\begin{bmatrix} x_1 \\ x_2 \\ x_3 \end{bmatrix} = D^{-1}$ [first column of

$B - E \ x_4] = \begin{bmatrix} \frac{2}{3} & 0 & \frac{-1}{3} \\ \frac{1}{6} & 0 & \frac{-1}{3} \\ \frac{-1}{6} & 1 & \frac{1}{3} \end{bmatrix}\begin{bmatrix} \begin{bmatrix} 4 \\ -1 \\ 0 \end{bmatrix} - \begin{bmatrix} 1 \\ 2 \\ 2 \end{bmatrix}[x_4] \end{bmatrix}$ (where D^{-1} is the inverse of the nonsingular square matrix

D)$= \begin{bmatrix} \frac{2}{3} & 0 & \frac{-1}{3} \\ \frac{1}{6} & 0 & \frac{-1}{3} \\ \frac{-1}{6} & 1 & \frac{1}{3} \end{bmatrix}\begin{bmatrix} 4-x_4 \\ -1-2x_4 \\ -2x_4 \end{bmatrix} = \begin{bmatrix} \frac{8}{3} \\ \frac{3x_4+4}{6} \\ \frac{-15x_4-10}{6} \end{bmatrix}$. Therefore, the general solution is given by $\begin{bmatrix} x_1 \\ x_2 \\ x_3 \\ x_4 \end{bmatrix}\begin{bmatrix} \frac{8}{3} \\ \frac{3x_4+4}{6} \\ \frac{-15x_4-10}{6} \\ x_4 \end{bmatrix}$. Using

similar procedure can yield the general solution of the system of equations for the second column of B, which

is $\begin{bmatrix} 3 \\ 2 \\ -4 \end{bmatrix}$, on that account, the general solution is $\begin{bmatrix} x_1 \\ x_2 \\ x_3 \\ x_4 \end{bmatrix} = \begin{bmatrix} \frac{10}{3} \\ \frac{3x_4+11}{6} \\ \frac{1-15x_4}{6} \\ x_4 \end{bmatrix}$. MATLAB solution is shown as follows:

MATLAB Command
>>A=[2 -2 0 1;0 1 1 2;1 -4 0 2]; ↵
>>B=[4 3;-1 2;0 -4]; ↵
>>X=A\B ↵

X =

  2.6667   3.3333
  0.3333   1.8667
    0   0
  -0.6667   0.0667

Insert $x_4 = -\frac{2}{3} = -0.6667$ and $x_4 = \frac{1}{15} = -0.0667$ into the first and second general solutions to have $\begin{bmatrix} x_1 \\ x_2 \\ x_3 \\ x_4 \end{bmatrix} =$

$\begin{bmatrix} \frac{8}{3} \\ \frac{-3\times\frac{2}{3}+4}{6} \\ \frac{15\times\frac{2}{3}-10}{6} \\ -\frac{2}{3} \end{bmatrix} = \begin{bmatrix} 2.6667 \\ 0.3333 \\ 0 \\ -0.6667 \end{bmatrix}$ and $\begin{bmatrix} x_1 \\ x_2 \\ x_3 \\ x_4 \end{bmatrix} = \begin{bmatrix} \frac{10}{3} \\ \frac{3\times\frac{1}{15}+11}{6} \\ \frac{1-15\times\frac{1}{15}}{6} \\ \frac{1}{15} \end{bmatrix} = \begin{bmatrix} 3.3333 \\ 1.8667 \\ 0 \\ 0.0667 \end{bmatrix}$ respectively. So, matrix X is going to be $\begin{bmatrix} 2.6667 \\ 0.3333 \\ 0 \\ -0.6667 \end{bmatrix}$

$\begin{matrix} 3.3333 \\ 1.8667 \\ 0 \\ 0.0667 \end{matrix}$ that is what the return of MATLAB is. $A\backslash B$ returns approximately $A^{-1}B$ for square and

nonsingular A, where A^{-1} is the inverse of A and $A^{-1}B$ is the vector product of matrices A^{-1} and B. Verify this with nonsingular and square $A = \begin{bmatrix} 4 & -1 & 0 \\ 2 & 3 & 1 \\ 0 & 1 & 2 \end{bmatrix}$ and $B = \begin{bmatrix} 3 & -1 \\ 2 & 2 \\ 0 & 4 \end{bmatrix}$. The computation, once again, is

conducted over each column of B. Equivalence of these two commands is shown below:

MATLAB Command for verification:
>>A=[4 -1 0;2 3 1;0 1 2]; ↵
>>B=[3 -1;2 2;0 4]; ↵

Using operator \,
>>A\B ↵

ans =
```
      0.7917  -0.2083
      0.1667   0.1667
     -0.0833   1.9167
```
Using method A⁻¹B,
>>inv(A)*B ↵

ans =
```
      0.7917  -0.2083
      0.1667   0.1667
     -0.0833   1.9167
```

If matrix A is rank deficient [rank deficient means rank of A is less than min (m, n), where A is of order

$m \times n$], the solution will not satisfy $AX = B$. Example of a rank deficient matrix is $A = \begin{bmatrix} 4 & -1 & 0 & 5 \\ 2 & 3 & 1 & -1 \\ 2 & 3 & 1 & -1 \end{bmatrix}$.

Take $B = \begin{bmatrix} 3 & -1 \\ 2 & 2 \\ 0 & 4 \end{bmatrix}$ and perform the operation $A \setminus B$ over these two matrices. Results are shown as follows:

MATLAB Command
>>A=[4 -1 0 5;2 3 1 -1;2 3 1 -1]; ↵
>>B=[3 -1;2 2;0 4]; ↵
>>X=A\B ↵

Warning: Rank deficient, rank = 2 tol = 4.6151e-015.

X =
```
          0        0
     0.5714   1.0000
          0        0
     0.7143        0
```
Check that $AX \neq B$,
>>A*X ↵

ans =
```
     3.0000  -1.0000
     1.0000   3.0000
     1.0000   3.0000
```

Right division:

The operator / accomplishes right vector division. Right vector division X of matrix A by matrix B is denoted as $X = A / B$. X can be defined as $X = (B^T \setminus A^T)^T$, where superscript T indicates transpose of matrix under consideration and \ is the left vector division what we discussed before. A matrix A of order $k \times m$ can be divided by a matrix B of order $n \times m$. The quotient matrix X will be of order $k \times n$. It goes without saying that right vector division is possible if the numbers of columns of A and B are identical. To show by numerical example, take $A = \begin{bmatrix} 2 & -2 \\ -4 & 0 \\ 2 & -3 \end{bmatrix}$ and $B = [4 \quad 3]$. Orders of A and B are 3×2 and 1×2 respectively ($k = 3$, $m = 2$, and $n = 1$). Hence, the quotient matrix X has the order 3×1. Equivalence of methods A / B and $(B^T \setminus A^T)^T$ for the numerical example is shown as follows:

MATLAB Command

using A/B operator,

>>A=[2 -2;-4 0;2 -3]; ↵
>>B=[4 3]; ↵
>>R=A/B ↵

R =

    0.0800
   -0.6400
   -0.0400

Using $(B^T \setminus A^T)^T$ method,

>>R=(B'\A')' ↵

R =

    0.0800
   -0.6400
   -0.0400

For nonsingular and square B, A / B is equivalent to AB^{-1}, where B^{-1} is the inverse of B and AB^{-1} is product of matrices A and B^{-1}. If matrix B is rank deficient, error message appears on execution of command A / B.

Rational form solution:

It may be desirable to have the solution in rational form rather than decimal form. Use the command 'sym(A/B)' to implement that.

2.5 Adding a scalar to each element in a matrix

Assume that 4, $-x$, and 8 are to be added with all elements of row matrix $R = [2 \quad -9 \quad 0 \quad 4 \quad 8]$, column matrix $C = \begin{bmatrix} -3 \\ 14 \\ 15 \end{bmatrix}$, and rectangular matrix $A = \begin{bmatrix} -3 & 0 \\ 14 & 3 \\ 15 & 1 \end{bmatrix}$ respectively. Following addition, matrices are going to be OA=[6 -5 4 8 12], OC=$\begin{bmatrix} -3-x \\ 14-x \\ 15-x \end{bmatrix}$, and OA=$\begin{bmatrix} 5 & 8 \\ 22 & 11 \\ 23 & 9 \end{bmatrix}$ for R, C, and A respectively. In column matrix, we have symbolic variables. To implement that, we need to use command 'sym'. See all additions below:

MATLAB Command

For R,

>>R=[2 –9 0 4 8]; ↵
>>OR=R+4 ↵

OR =

   6 -5 4 8 12

For A,

>>A=[-3 0;14 3;15 1]; ↵
>>OA=A+8 ↵

OA =

For C,

>>C=[-3 14 15]'; ↵
>>syms x ↵
>>OC=C-x ↵

OC =

[-3-x]
[14-x]
[15-x]

$$\begin{matrix} 5 & 8 \\ 22 & 11 \\ 23 & 9 \end{matrix}$$

2.6 Multiplying all elements in a matrix by a scalar

Each element in row matrix $R = [2 \quad 3 \quad 4 \quad 5 \quad 0]$, column matrix $C = \begin{bmatrix} 4 \\ 3 \\ 0 \end{bmatrix}$, and rectangular matrix

$A = \begin{bmatrix} 4 & -9 & 3 \\ 3 & -7 & -1 \end{bmatrix}$ is to be multiplied by -2, 9 and $-3x$ respectively. One should get OR=$[-4 \quad -6 \quad -8 \quad -10$

$0]$, OC=$\begin{bmatrix} 36 \\ 27 \\ 0 \end{bmatrix}$, and OA=$\begin{bmatrix} -12x & 27x & -9x \\ -9x & 21x & 3x \end{bmatrix}$ after multiplication for R, C, and A respectively. See all

multiplication below:

MATLAB Command

for row matrix,

```
>>R=[2 3 4 5 0]; ↵
>>OR=-2*R ↵

 OR =
           -4  -6  -8  -10   0
```

for column matrix,

```
>>C=[4 3 0]'; ↵
>>OC=9*C ↵

OC =
        36
        27
         0
```

for rectangular matrix,

```
>>A=[4 -9 3;3 -7 -1]; ↵
>>syms x ↵
>>OA=-3*x*A ↵

OA =

[ -12*x,  27*x,  -9*x]
[  -9*x,  21*x,   3*x]
```

Order of multiplication is not important, i.e., even if you used commands R*(-2), C*9, and A*(-3*x) for the above mentioned row, column, and rectangular matrices respectively, you would end up with the same results. Try with the following commands:

MATLAB Command

```
>>A=[4 -9 3;3 -7 -1]; ↵
>>syms x ↵
>>O=A*-3*x ↵

O =

[ -12*x,  27*x,  -9*x]
[  -9*x,  21*x,   3*x]
```

Observe that MATLAB is so smart that you do not even need to put the first brace ( ) in the third line of command 'O=A*-3*x'. There is a function 'pretty', which can display output almost in symbolic form. Perform the following:

```
>>pretty(O) ↵
```

```
[-12 x    27 x    -9 x]
[                     ]
[ -9 x    21 x     3 x]
```

2.7 Dividing all elements in a matrix by a scalar

Division of all elements in a matrix is very similar to the multiplication. All you need is use the operator / instead of *. Say, we want to divide each element of column matrix $C = \begin{bmatrix} 4 \\ 3 \\ 0 \end{bmatrix}$ and rectangular matrix

$A = \begin{bmatrix} 4 & -9 & 7 \\ 3 & -7 & 10 \end{bmatrix}$ by 3 and $\frac{7}{9}$ respectively. Resulting matrices should be OC= $\begin{bmatrix} 1.3333 \\ 1 \\ 0 \end{bmatrix}$ and OA=

$\begin{bmatrix} \frac{36}{7} & -\frac{81}{7} & 9 \\ \frac{27}{7} & -9 & \frac{90}{7} \end{bmatrix}$ for C and A respectively. Have the division as follows:

MATLAB Command

for column matrix,
```
>>C=[4 3 0]'; ↲
>>OC=C/3 ↲

OC =
      1.3333
      1.0000
           0
```

for rectangular matrix,
```
>>A=[4 -9 7;3 -7 10]; ↲
>>OA=A/sym(7/9) ↲

OA =

[  36/7,  -81/7,     9]
[  27/7,     -9,  90/7]
```

2.8 Taking reciprocal of all elements in a matrix

Reciprocal of all elements in a matrix can be taken by the operator ./. We wish to find reciprocal of all elements in row matrix $R = [2 \quad 4 \quad 1]$, column matrix $C = \begin{bmatrix} 8 \\ -2 \\ -4 \end{bmatrix}$, and rectangular matrix $A = \begin{bmatrix} 8 & 2 \\ x & 5 \\ 4 & y \end{bmatrix}$. As the

output, the matrices will be OR=$[\frac{1}{2} \quad \frac{1}{4} \quad \frac{1}{1}]$=[0.5 0.25 1], OC=$\begin{bmatrix} \frac{1}{8} \\ -\frac{1}{2} \\ -\frac{1}{4} \end{bmatrix} = \begin{bmatrix} 0.125 \\ -0.5 \\ -0.25 \end{bmatrix}$, and OA=$\begin{bmatrix} \frac{1}{8} & \frac{1}{2} \\ \frac{1}{x} & \frac{1}{5} \\ \frac{1}{4} & \frac{1}{y} \end{bmatrix}$ for R, C,

and A respectively. All manipulations are shown in the following. Matrix elements can be floating-point numbers, integers, and symbolic variables.

MATLAB Command

for row matrix,
```
>>R=[2 4 1]; ↲
>>OR=1./R ↲

OR =
           0.5000   0.2500   1.0000
```
for column matrix,
```
>>C=[8 -2 -4]'; ↲
>>OC=1./C ↲

OC =
           0.1250
```

for rectangular matrix,
```
>>syms x y ↲
>>A=[8 2;x 5;4 y]; ↲
>>OA=1./A ↲

OA =

[  1/8,   1/2]
[  1/x,   1/5]
[  1/4,   1/y]
```

$$-0.5000$$
$$-0.2500$$

2.9 Making elements in a matrix to integers

Subroutine 'fix' can be applied to make floating-point numbers to integers. The subroutine discards fractional parts of the floating-point numbers. Example matrices, whose elements are floating-point numbers, are $R =[1.2578 \quad -9.3445 \quad -8.9999]$ and $A = \begin{bmatrix} 2.5678 & 9.0033 \\ -56.7898 & -5.5555 \\ 6.0989 & 76.3444 \end{bmatrix}$. If you take integer parts of the floating-point numbers of R and A, the resulting matrices are going to be OR=[1 $\quad$ –9 $\quad$ –8] and OA= $\begin{bmatrix} 2 & 9 \\ -56 & -5 \\ 6 & 76 \end{bmatrix}$

for R and A respectively. These conversions are shown below:

MATLAB Command
for row matrix,

```
>>R=[1.2578 -9.3445 -8.9999]; ↵
>>OR=fix(R) ↵
```

```
OR =
         1 -9 -8
```
for rectangular matrix,

```
>>A=[2.5678 9.0033;-56.7898 -5.5555;6.0989 76.3444]; ↵
>>OA=fix(A) ↵
```

```
OA =
         2    9
       -56   -5
         6   76
```

2.10 Rounding elements in a matrix to integers

Conversion of floating-point numbers to rounded integers can be carried out by subroutine 'round'. There is some difference between subroutines 'fix' and 'round'. Rounding means if the fractional part of the floating-point number is greater than or equal to 0.5, it will be taken as 1 and if it is less than 0.5, it will be taken as 0. On the other hand, subroutine 'fix' will discard the fractional part completely regardless of the magnitude. Example matrices we are referring to are $R =[1.5001 \quad -9.5000 \quad -8.4999]$ and $A = \begin{bmatrix} 2.5678 & 9.0033 \\ -56.7898 & -5.5555 \\ 6.0989 & 76.3444 \\ 3.9999 & 2.4567 \end{bmatrix}$. If fractional parts of the elements of R and A are rounded, the resulting matrices are

OR=[2 –10 –8] and OA= $\begin{bmatrix} 3 & 9 \\ -57 & -6 \\ 6 & 76 \\ 4 & 2 \end{bmatrix}$ for R and A respectively. Implementation is shown below:

MATLAB Command
for row matrix,

```
>>R=[1.5001 -9.5000 -8.4999]; ↵
>>OR=round(R) ↵
```

OR =

$$2 \quad -10 \quad -8$$

for rectangular matrix,
>>A=[2.5678 9.0033;-56.7898 -5.5555;6.0989 76.3444;3.9999 2.4567]; ↵
>>OA=round(A) ↵

OA =

$$
\begin{array}{rr}
3 & 9 \\
-57 & -6 \\
6 & 76 \\
4 & 2
\end{array}
$$

2.11 Remainder after integer division

When an integer 3 is divided by integer 2, quotient is 1 and remainder is 1. Unlike floating-point numbers, there are no fractional parts in integer numbers. When 2 is divided by 3, the quotient is 0.6666...., which is a floating-point number but in integer division, the quotient is $\frac{2}{3}=0$ and the remainder is 2–0×3=2.

Remainder after integer division is obtained by subroutine 'rem' (abbreviated form of re<u>mainder</u>). To find the remainder after integer division of all elements in a matrix,

MATLAB Step:

Use command rem(dividend matrix name,divider integer or matrix).

We can divide the problems in two titles – whole matrix by an integer and element to element integer division of two matrices.

Whole matrix by an integer:

Test matrices are $R = [3 \quad 8 \quad 7 \quad 2]$, $C = \begin{bmatrix} 7 \\ -57 \\ 13 \\ 3 \end{bmatrix}$, and $A = \begin{bmatrix} 2 & 9 \\ -56 & -5 \\ 6 & 76 \\ 3 & 2 \end{bmatrix}$. They are to be integer divided

by 3, 4, and –3 for row, column, and rectangular matrices respectively. The quotients are $[\frac{3}{3} \quad \frac{8}{3} \quad \frac{7}{3} \quad \frac{2}{3}] = [1 \quad 2$

$2 \quad 0]$, $\begin{bmatrix} \frac{7}{4} \\ \frac{-57}{4} \\ \frac{13}{4} \\ \frac{3}{4} \end{bmatrix} = \begin{bmatrix} 1 \\ -14 \\ 3 \\ 0 \end{bmatrix}$, and $\begin{bmatrix} \frac{2}{-3} & \frac{9}{-3} \\ \frac{-56}{-3} & \frac{-5}{-3} \\ \frac{6}{-3} & \frac{76}{-3} \\ \frac{3}{-3} & \frac{2}{-3} \end{bmatrix} = \begin{bmatrix} 0 & -3 \\ 18 & 1 \\ -2 & -25 \\ -1 & 0 \end{bmatrix}$ for R, C, and A respectively. Remainders following

integer division are OR=[3–3×1 8–3×2 7–3×2 2–3×0]=[0 2 1 2], OC=$\begin{bmatrix} 7-4\times1 \\ -57-4\times(-14) \\ 13-4\times3 \\ 3-4\times0 \end{bmatrix} = \begin{bmatrix} 3 \\ -1 \\ 1 \\ 3 \end{bmatrix}$, and OA=

$\begin{bmatrix} 2-(-3)\times0 & 9-(-3)\times(-3) \\ -56-(-3)\times18 & -5-(-3)\times1 \\ 6-(-3)\times(-2) & 76-(-3)\times(-25) \\ 3-(-3)\times(-1) & 2-(-3)\times0 \end{bmatrix} = \begin{bmatrix} 2 & 0 \\ -2 & -2 \\ 0 & 1 \\ 0 & 2 \end{bmatrix}$ for R, C, and A respectively. All implementations are shown

as follows:

MATLAB Command

for row matrix,
>>R=[3 8 7 2]; ↵
>>OR=rem(R,3) ↵

OR =

$$0 \quad 2 \quad 1 \quad 2$$

for rectangular matrix,
>>A=[2 9;-56 -5;6 76;3 2]; ↵
>>OA=rem(A,-3) ↵

OA =

$$2 \quad 0$$

for column matrix,
>>C=[7   -57   13   3]'; ↵
>>OC=rem(C,4) ↵

$$\begin{array}{cc} -2 & -2 \\ 0 & 1 \\ 0 & 2 \end{array}$$

OC =
        3
       -1
        1
        3

Element to element integer division of two matrices:

Subroutine 'rem' is equally applicable for element to element integer division. Both divider and dividend matrix have to be of the same order for this purpose. Take the example of dividend matrix as $A = \begin{bmatrix} 2 & 3 & 4 \\ 7 & 9 & 2 \end{bmatrix}$ and divider matrix as $B = \begin{bmatrix} 3 & 4 & 2 \\ -3 & 2 & 3 \end{bmatrix}$. Element to element quotient and remainder matrices following integer division are $\begin{bmatrix} 0 & 0 & 2 \\ -2 & 4 & 0 \end{bmatrix}$ and $\begin{bmatrix} 2 & 3 & 0 \\ 1 & 1 & 2 \end{bmatrix}$ respectively. See the implementation as follows:

MATLAB Command
>>A=[2 3 4;7 9 2]; ↵
>>B=[3 4 2;-3 2 3]; ↵
>>rem(A,B) ↵

ans =
        2   3   0
        1   1   2

2.12 Prime factors of an integer

A prime number is a number that does not have any factors other than unity and itself. Subroutine 'factor' finds the prime numbers of an integer. Output of the subroutine is a row matrix. Input argument of the subroutine can not be a floating-point number. To find the prime factors of an integer,

MATLAB Step:
Use the command factor (integer)

Let us consider the integer 84. Prime factors of 84 are 2, 2, 3, and 7. Input argument of the subroutine 'factor' can not be a matrix too.

MATLAB Command
>>factor(84) ↵

ans=
        2   2   3   7

2.13 Logarithm of elements in a matrix

Frequently used logarithms are logarithm with respect to e, 10, and 2.

2.13.1 Natural logarithm

Subroutine 'log' is used for this purpose. Natural logarithm of any number is taken with respect to base e, where $e = 2.718281828$. To find the natural logarithm,

MATLAB Step:
Use command log(scalar/matrix name).

Natural logarithm of all elements of row matrix $R =[7 \quad 0.8893 \quad 10]$, column matrix $C = \begin{bmatrix} 5 \\ 8 \\ 2 \end{bmatrix}$, and

rectangular matrix $A = \begin{bmatrix} 5 & 45 \\ 8 & 21 \\ 2 & 9 \end{bmatrix}$ is to be taken. Resulting matrices are OR=[ln7 $\quad$ ln0.8893 $\quad$ ln10]=

[1.945910149 $\quad$ −0.117320642 $\quad$ 2.302585093], OC=$\begin{bmatrix} \ln 5 \\ \ln 8 \\ \ln 2 \end{bmatrix} = \begin{bmatrix} 1.609437912 \\ 2.079441542 \\ 0.69314718 \end{bmatrix}$, and OA=$\begin{bmatrix} \ln 5 & \ln 45 \\ \ln 8 & \ln 21 \\ \ln 2 & \ln 9 \end{bmatrix}$=

$\begin{bmatrix} 1.609437912 & 3.80666249 \\ 2.079441542 & 3.044522438 \\ 0.69314718 & 2.197224577 \end{bmatrix}$ for R, C, and A respectively. MATLAB computation is shown below:

MATLAB Command

for row matrix,

        >>R=[7 0.8893 10]; ↵
        >>OR=log(R) ↵

   OR =
        1.9459 -0.1173 2.3026

for column matrix,

        >>C=[5 8 2]'; ↵
        >>OC=log(C) ↵

   OC =
      1.6094
      2.0794
      0.6931

for rectangular matrix,

        >>A=[5 45;8 21;2 9]; ↵
        >>OA=log(A) ↵

   OA =
      1.6094 3.8067
      2.0794 3.0445
      0.6931 2.1972

2.13.2 Common logarithm

Common logarithm of any number is taken with respect to base 10. Subroutine 'log10' is used in this regard. To take common logarithm,

MATLAB Step:
Use the command log10(scalar/matrix name).

Example matrix is $A = \begin{bmatrix} 5 & 45 \\ 8 & 21 \\ 2 & 9 \end{bmatrix}$. Output matrix should be $\begin{bmatrix} \log_{10} 5 & \log_{10} 45 \\ \log_{10} 8 & \log_{10} 21 \\ \log_{10} 2 & \log_{10} 9 \end{bmatrix} = \begin{bmatrix} 0.698970004 & 1.653212514 \\ 0.903089987 & 1.322219295 \\ 0.301029995 & 0.954242509 \end{bmatrix}$.

See the implementation below:

MATLAB Command

        >>A=[5 45;8 21;2 9]; ↵
        >>OA=log10(A) ↵

   OA =
        0.6990 1.6532
        0.9031 1.3222
        0.3010 0.9542

As argument, we can have row, column, or rectangular matrix.

2.13.3 Logarithm w. r. to base 2

Logarithm with respect to base 2 can be taken by the subroutine 'log2'. Test matrix is $A =$ $\begin{bmatrix} 1 & 32 \\ 8 & 0.125 \\ 2 & 64 \end{bmatrix}$. We should have $\begin{bmatrix} \log_2 1 & \log_2 32 \\ \log_2 8 & \log_2 0.125 \\ \log_2 2 & \log_2 64 \end{bmatrix} = \begin{bmatrix} 0 & 5 \\ 3 & -3 \\ 1 & 6 \end{bmatrix}$ after taking the logarithm. That is accomplished as follows:

MATLAB Command

>>A=[1 32;8 0.125;2 64]; ↵
>>OA=log2(A) ↵

OA =

```
0   5
3   -3
1   6
```

Like other two log subroutines, its argument can also be row, column, or rectangular matrix.

2.14 Flipping from left to right/up to down of a matrix

Flipping from left to right:

Flipping from left to right of a row or rectangular matrix is performed by subroutine 'fliplr' (abbreviation of <u>fli</u>pping from <u>l</u>eft to <u>r</u>ight). Suppose, we have row matrix $R = [2 \quad 4 \quad 3 \quad -4 \quad 6 \quad 9 \quad 3 \quad 7 \quad 10]$. If you flip the elements of R from left to right, in the resulting matrix OR, 10 (last element of R) comes first, 7 comes second (second element of R from last), 3 comes third (third element of R from last), and so do the others, hence, OR=[10 $\quad$ 7 $\quad$ 3 9 6 $\quad$ –4 3 4 2]. For a rectangular matrix, flipping operation from left to right will be over each column. That means the first column will be the last column and the second column will be the second from last, and so will be the other columns. Assume that we have rectangular matrix $A =$ $\begin{bmatrix} 4 & 23 & 85 & 34 \\ 5 & 43 & 41 & 87 \\ 8 & 65 & 76 & 71 \end{bmatrix}$. Flipping from left to right of A gives us OA, where OA=$\begin{bmatrix} 34 & 85 & 23 & 4 \\ 87 & 41 & 43 & 5 \\ 71 & 76 & 65 & 8 \end{bmatrix}$.

Since column matrices have only one column, you will not see any change to a column matrix brought about by subroutine 'fliplr'. Have the implementation as shown below:

MATLAB Command

for row matrix,
>>R=[2 4 3 -4 6 9 3 7 10]; ↵
>>OR=fliplr(R) ↵

OR =

```
10   7   3   9   6   -4   3   4   2
```
for rectangular matrix,
>>A=[4 23 85 34;5 43 41 87;8 65 76 71]; ↵
>>OA=fliplr(A) ↵

OA =

```
34   85   23   4
87   41   43   5
71   76   65   8
```

Flipping from up to down:

Subroutine 'flipud' (abbreviation of <u>fli</u>pping from <u>u</u>p to <u>d</u>own) can flip elements of a column or rectangular matrix from up to down. Flip the column matrix $C = \begin{bmatrix} 4 \\ 7 \\ 8 \\ 3 \\ 1 \end{bmatrix}$ from up to down to get the resulting matrix OC=$\begin{bmatrix} 1 \\ 3 \\ 8 \\ 7 \\ 4 \end{bmatrix}$. Flipping operation from up to down of a rectangular matrix will happen over each row.

Assume that we have rectangular matrix $D = \begin{bmatrix} 4 & 23 & 85 \\ 5 & 43 & 41 \\ 8 & 65 & x \\ 3 & 12 & 13 \end{bmatrix}$. Flip D from up to down to have OD=

$\begin{bmatrix} 3 & 12 & 13 \\ 8 & 65 & x \\ 5 & 43 & 41 \\ 4 & 23 & 85 \end{bmatrix}$. No change occurs to a row matrix due to the use of 'flipud' for the reason that row matrices

have only one row. Both examples are implemented as follows:

> **MATLAB Command** for column matrix,
> >>C=[4 7 8 3 1]'; ↵
> >>OC=flipud(C) ↵
>
> OC =
>
>   1
>   3
>   8
>   7
>   4

for rectangular matrix,

> >>syms x ↵
> >>D=[4 23 85;5 43 41;8 65 x;3 12 13]; ↵
> >>OD=flipud(D) ↵
>
> OD =
>
> [3, 12, 13]
> [8, 65, x]
> [5, 43, 41]
> [4, 23, 85]

2.15 Rotation of a matrix by 90^0 or multiple of 90^0s

To have clear idea about the matrix rotation by ninety degrees (90^0) or multiple of ninety degrees ($k\,90^0$, where k is an integer), figure 2.1 is presented in the next page. The figure depicts rotation of a rectangle ABCDEF about the point A in either direction (clockwise and counter clockwise). In clockwise and counter clockwise directions, rotation of a matrix by ninety degrees or multiple of ninety degrees is achieved by subroutine 'rot90' (which is the abbreviation of <u>rot</u>ation by <u>90</u>). Referring to figure 2.1, edge ABC and edge AF of the rectangle can be compared to a row and a column matrix respectively. A row matrix becomes a column matrix and a column matrix becomes a row matrix if they are rotated counter clockwise or clockwise by 90^0. To rotate a matrix by ninety degrees or multiple of ninety degrees,

> **MATLAB Step:**
> *Use command rot90(matrix name, type of rotation).*

MATLAB identifies counter clockwise and clockwise rotation by +ve and −ve signs respectively. Multiplicity of 90^0 is denoted by whole integer. For 90^0 rotation, multiplicity is 1, for 180^0 rotation, multiplicity is 2, and for 270^0 rotation, multiplicity is 3. Let us implement some examples. Row matrix $R = [-1$ -3 2 $0]$ is to be rotated about the first element -1 by 90^0 in clockwise direction. Since the type of rotation

34

Clockwise rotation of the rectangle ABCDEF about A by 90^0:

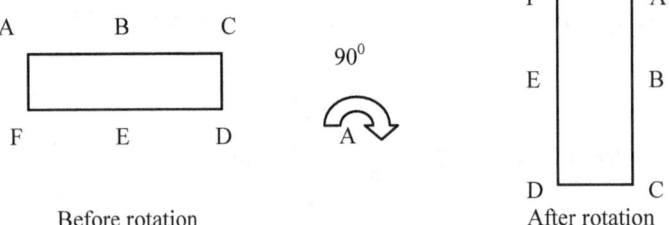

Before rotation

After rotation

Clockwise rotation of the rectangle ABCDEF about A by 180^0:

Before rotation

After rotation

Counter clockwise rotation of the rectangle ABCDEF about A by 270^0:

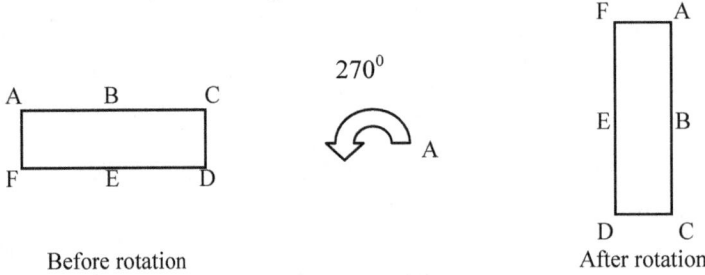

Before rotation

After rotation

Figure 2.1   *Rotation of rectangle ABCDEF about A by different angles*

is clockwise 90^0, -1 recognizes the rotation. Following rotation, the resulting matrix should be OR=$\begin{bmatrix} -1 \\ -3 \\ 2 \\ 0 \end{bmatrix}$. If

the matrix R were to be rotated by 90^0 in counter clockwise direction, the resulting matrix would be OD=$\begin{bmatrix} 0 \\ 2 \\ -3 \\ -1 \end{bmatrix}$. For 90^0 counter clockwise rotation, the type of rotation is +1. See both implementations as follows:

MATLAB Command

for 90^0 clockwise rotation of R,
>>R=[-1 -3 2 0]; ↵
>>OR=rot90(R,-1) ↵

OR =

-1
-3
2
0

for 90^0 counter clockwise rotation of R,
>>OD=rot90(R,1) ↵

OD =

0
2
-3
-1

Next, the column matrix $C=\begin{bmatrix} 9 \\ 8 \\ 7 \\ 1 \end{bmatrix}$ is to be rotated about the first element 9 by 270^0 in clockwise direction,

where the type of rotation is -3. The output matrix should look like OC=[9   8   7   1]. For 270^0 counter clockwise rotation of C about 9, one should use the multiplier +3, where the output is OE=[1   7   8   9].

After that, we rotate rectangular matrix $A = \begin{bmatrix} 2 & 5 & 7 \\ 7 & 0 & 1 \\ 8 & 3 & 6 \\ 6 & 9 & 4 \end{bmatrix}$ about the element 2 by 180^0 in counter clockwise

direction, so, the type of rotation is +2 and the resulting matrix is OA=$\begin{bmatrix} 4 & 9 & 6 \\ 6 & 3 & 8 \\ 1 & 0 & 7 \\ 7 & 5 & 2 \end{bmatrix}$. If the matrix A were to be

rotated by 90^0 in clockwise direction (the type of rotation is −1), the output matrix would be OM= $\begin{bmatrix} 6 & 8 & 7 & 2 \\ 9 & 3 & 0 & 5 \\ 4 & 6 & 1 & 7 \end{bmatrix}$. Rotations of column and rectangular matrices are presented as follows:

MATLAB Command

for 270^0 clockwise rotation of C,

>>C=[9 8 7 1]'; ↵

>>OC=rot90(C,-3) ↵

OC =

9   8   7   1

for 180^0 counter clockwise rotation
of rectangular matrix A,

>>A=[2 5 7;7 0 1;8 3 6;6 9 4]; ↵

>>OA=rot90(A,2) ↵

OA =

4   9   6
6   3   8
1   0   7
7   5   2

for 270^0 counter clockwise rotation
of C,

>>OE=rot90(C,3) ↵

OE =

1   7   8   9

for 90^0 clockwise rotation of A,

>>OM=rot90(A,-1) ↵

OM =

6   8   7   2
9   3   0   5
4   6   1   7

2.16 Raising power to each element of a matrix by a scalar

To raise the power on each element of a matrix by a scalar,

MATLAB Step:

Use the command (matrix name).^scalar.

Power of all elements in row matrix $R = [-1 \quad -3 \quad 2]$, column matrix $C = \begin{bmatrix} -2 \\ x \\ y \end{bmatrix}$, and rectangular

matrix $A = \begin{bmatrix} -2 & -1 \\ 0 & -5 \\ 5 & 6 \end{bmatrix}$ is to be raised by 3, 4, and 4 respectively. After raising the power, the resulting matrices

should be OR=$[(-1)^3 \quad (-3)^3 \quad (2)^3]$=$[-1 \quad -27 \quad 8]$, OC=$\begin{bmatrix} (-2)^4 \\ x^4 \\ y^4 \end{bmatrix} = \begin{bmatrix} 16 \\ x^4 \\ y^4 \end{bmatrix}$, and OA=$\begin{bmatrix} (-2)^4 & (-1)^4 \\ 0^4 & (-5)^4 \\ 5^4 & 6^4 \end{bmatrix} = \begin{bmatrix} 16 & 1 \\ 0 & 625 \\ 625 & 1296 \end{bmatrix}$

for R, C, and A respectively. See all implementations below:

MATLAB Command

for row matrix,

>>R=[-1 -3 2]; ↵

for rectangular matrix,

>>A=[-2 -1;0 -5;5 6]; ↵

>>OR=R.^3 ↵   >>OA=A.^4 ↵

OR =   OA =
      -1 -27   8   16   1

for column matrix,   0   625
>>syms x y ↵   625 1296
>>C=[-2 x y].'; ↵
>>OC=C.^4 ↵

OC =

[16]
[x^4]
[y^4]

2.17 Raising power of 2 as the elements in a matrix

To raise the power of 2 as the elements in a matrix,

MATLAB Step:

Use the command pow2(matrix name).

Power of 2 is to be raised according to row matrix $R=[-1 \ -3 \ 2]$, column matrix $C=\begin{bmatrix} -1 \\ 0 \\ 1 \end{bmatrix}$, and

rectangular matrix $A=\begin{bmatrix} 2 & 3 & 0 \\ 4 & -2 & -1 \end{bmatrix}$, so, consequent matrices are OR=$2^{[-1 \ -3 \ 2]}$=$[2^{-1} \ 2^{-3} \ 2^{2}]$=[0.5 0.125 4],

$OC=2^{\begin{bmatrix} -1 \\ 0 \\ 1 \end{bmatrix}}=\begin{bmatrix} 2^{-1} \\ 2^{0} \\ 2^{1} \end{bmatrix}=\begin{bmatrix} 0.5 \\ 1 \\ 2 \end{bmatrix}$, and $OA=2^{\begin{bmatrix} 2 & 3 & 0 \\ 4 & -2 & -1 \end{bmatrix}}=\begin{bmatrix} 2^{2} & 2^{3} & 2^{0} \\ 2^{4} & 2^{-2} & 2^{-1} \end{bmatrix}=\begin{bmatrix} 4 & 8 & 1 \\ 16 & 0.25 & 0.5 \end{bmatrix}$ for R, C, and A

respectively. Implementations are shown below:

MATLAB Command

for row matrix,   for column matrix,
>>R=[-1 -3 2]; ↵   >>C=[-1 0 1]'; ↵
>>OR=pow2(R) ↵   >>OC=pow2(C) ↵

OR =   OC =
     0.5000   0.1250 4.0000   0.5000

for rectangular matrix,   1.0000
>>A=[2 3 0;4 -2 -1]; ↵   2.0000
>>OA=pow2(A) ↵

OA =
     4.0000   8.0000 1.0000
   16.0000 0.2500 0.5000

The function 'pow2' applies for matrices whose elements are integers and floating-point numbers not symbolic variables.

2.18 Raising power of a scalar as the elements in a matrix

To raise the power of a scalar according to the elements of a matrix,

MATLAB Step:

Use command scalar.^(matrix name).

Power of -3 is to be raised according to row matrix $R = [-1 \ 0 \ 2]$, column matrix $C = \begin{bmatrix} -2 \\ -1 \\ 1 \end{bmatrix}$, and

rectangular matrix $A = \begin{bmatrix} 2 & 3 \\ 4 & -2 \\ 1 & 0 \end{bmatrix}$. Conceivably, the output matrices are OR$=(-3)^{[-1 \ 0 \ 2]}=[(-3)^{-1} \ (-3)^0 \ (-3)^2]=$

$[-0.3333 \ 1 \ 9]$, OC$=(-3)^{\begin{bmatrix} -2 \\ -1 \\ 1 \end{bmatrix}}=\begin{bmatrix} (-3)^{-2} \\ (-3)^{-1} \\ (-3)^1 \end{bmatrix}=\begin{bmatrix} 0.1111 \\ -0.3333 \\ -3 \end{bmatrix}$, and OA$=(-3)^{\begin{bmatrix} 2 & 3 \\ 4 & -2 \\ 1 & 0 \end{bmatrix}}=\begin{bmatrix} (-3)^2 & (-3)^3 \\ (-3)^4 & (-3)^{-2} \\ (-3)^1 & (-3)^0 \end{bmatrix}=\begin{bmatrix} 9 & -27 \\ 81 & 0.1111 \\ -3 & 1 \end{bmatrix}$ for R,

C, and A respectively. See the turnouts as follows:

MATLAB Command

for row matrix,
>>R=[-1 0 2]; ↵
>>OR=(-3).^R ↵

OR =
    -0.3333   1.0000   9.0000

for column matrix,
>>C=[-2 -1 1]'; ↵
>>OC=(-3).^C ↵

OC =
    0.1111
   -0.3333
   -3.0000

for rectangular matrix,
>>A=[2 3;4 -2;1 0]; ↵
>>OA=(-3).^A ↵

OA =
    9.0000   -27.0000
   81.0000   0.1111
   -3.0000   1.0000

2.19 Raising element to element power of two matrices

To raise element to element power of two matrices,
MATLAB Steps:
1. *Enter the base number matrix A*
2. *Enter the power matrix B*
3. *Use the command A.^B.*

Say, the base number matrices are $A = [7 \ 4 \ -5]$, $B = \begin{bmatrix} 3 \\ -x \\ 4 \end{bmatrix}$, and $C = \begin{bmatrix} 3 & -1 \\ -2 & 5 \\ 4 & 7 \end{bmatrix}$. Power matrices are

$D = [2 \ 3 \ -1]$, $E = \begin{bmatrix} 2 \\ 3 \\ 3 \end{bmatrix}$, and $F = \begin{bmatrix} x & -3 \\ 3 & 0 \\ y & 3 \end{bmatrix}$ for row, column, and rectangular matrices respectively. Raising

element to element power should return output matrices as OR$= A.^D = [7^2 \ 4^3 \ (-5)^{-1}]= [49 \ 64 \ -0.2]$,

OC$= B.^E = \begin{bmatrix} 3^2 \\ (-x)^3 \\ 4^3 \end{bmatrix}=\begin{bmatrix} 9 \\ -x^3 \\ 64 \end{bmatrix}$, and OA$=C.^F = \begin{bmatrix} 3^x & (-1)^{-3} \\ (-2)^3 & 5^0 \\ 4^y & 7^3 \end{bmatrix}=\begin{bmatrix} 3^x & -1 \\ -8 & 1 \\ 4^y & 343 \end{bmatrix}$ for row, column, and

rectangular matrices respectively. All outputs are shown in the following:

MATLAB Command

for row matrix,
>>A=[7 4 -5]; ↵
>>D=[2 3 -1]; ↵

for rectangular matrix,
>>C=[3 -1;-2 5;4 7]; ↵
>>syms x y ↵

>>OR=A.^D ↵   >>F=[x -3;3 0;y 3]; ↵
                                    >>OA=C.^F ↵

OR =
            49.0000   64.0000   -0.2000   OA =
for column matrix,
    >>syms x ↵   [3^x,   -1]
    >>B=[3 -x 4].'; ↵   [-8,   1]
    >>E=[2 3 3]'; ↵   [4^y,   343]
    >>OC=B.^E ↵

OC =

    [9]
    [-x^3]
    [64]

Important point is that both base and power matrices have to be of identical order. Matrix elements can be floating-point numbers too.

2.20 The number of non zero elements in a matrix

The number of non-zero elements in a matrix can be found by subroutine 'nnz', which is the abbreviation of number non zero. To find the number of non-zero elements in a matrix,

MATLAB Step:
Use command nnz(matrix name).

Chosen row, column, and rectangular matrices for implementation are $R = [2 \quad 3 \quad 0 \quad 8 \quad -8 \quad 0$

$7 \quad 0 \quad 13]$, $C = \begin{bmatrix} -27 \\ -9 \\ 0 \\ 67 \\ 0 \\ 0 \end{bmatrix}$, and $A = \begin{bmatrix} 0 & 3 & 8 \\ 9 & 0 & 6 \\ 0 & 0 & 1 \end{bmatrix}$ respectively. There are six, three, and five elements, which are

not equal to zero, in R, C, and A respectively. See the outcome as follows:

MATLAB Command for row matrix,   for column matrix,
    >>R=[2 3 0 8 -8 0 7 0 13]; ↵   >>C=[-27 -9 0 67 0 0]'; ↵
    >>nnz(R) ↵   >>nnz(C) ↵

    ans =   ans =
            6   3
for rectangular matrix,
    >>A=[0 3 8;9 0 6;0 0 1]; ↵
    >>nnz(A) ↵

    ans =
            5

2.21 Picking up nonzero elements from a matrix

The subroutine 'nonzeros' can detect the nonzero elements in a matrix. The search is carried out according to columns. The output matrix is a column one regardless of the type of input matrix. To find nonzero elements in a matrix,

MATLAB Step:
Use the command nonzeros(matrix name).

The nonzero elements of row matrix $R = [3 \quad 0 \quad 18 \quad -8 \quad 0 \quad 7 \quad 0 \quad 11]$, column matrix $C = \begin{bmatrix} -3 \\ -7 \\ -1 \\ 0 \\ 7 \\ 0 \end{bmatrix}$,

and rectangular matrix $A = \begin{bmatrix} 0 & 3 & 8 \\ 9 & 0 & 6 \end{bmatrix}$ are $[3 \quad 18 \quad -8 \quad 7 \quad 11]$, $[-3 \quad -7 \quad -1 \quad 7]$, and $[9 \quad 3 \quad 8 \quad 6]$ for R, C,

and A respectively. See the findings below:

MATLAB Command

for row matrix,

    >>R=[3 0 18 -8 0 7 0 11]; ↵

    >>nonzeros(R) ↵

    ans =

        3

       18

       -8

        7

      11

for column matrix,

    >>C=[-3 -7 -1 0 7 0]'; ↵

    >>nonzeros(C) ↵

    ans =

       -3

       -7

       -1

        7

for rectangular matrix,

    >>A=[0 3 8;9 0 6]; ↵

    >>nonzeros(A) ↵

    ans =

        9

        3

        8

        6

2.22 Coloning of matrices

You may have seen TV programs on coloning of sheep, coloning of genes, coloning of computer chips, in MATLAB, nicely you can do coloning of matrices. New matrices can be built from the matrix you have in the workspace of MATLAB. All you need is proper manipulation of colon operator (:). If you are a master of colon operator, you know a lot of MATLAB. To learn about colon operator, execute the following colon operation in MATLAB command window. Assign a row matrix $[2 \quad 4 \quad 3 \quad -10 \quad 0 \quad 9 \quad 73 \quad 29 \quad -31 \quad 50]$ to A:

MATLAB Command

    >>A=[2 4 3 -10 0 9 73 29 -31 50]; ↵

Suppose, we want to form a matrix B, where B will be the second, third, and ninth element of A, i.e., $B = [4 \quad 3 \quad -31]$:

    formation of matrix B,

    >>B=A([2 3 9]) ↵

    B =

      4   3 -31

Then, a matrix C is to be formed from the third through eighth elements of A, i.e., $C = [3 \quad -10 \quad 0 \quad 9 \quad 73 \quad 29]$. You could say, oh we can do it by hand. Of coarse, you can but the thing is that if one has to write a program for a matrix having 200 elements, then, it is not feasible to count the position index of the elements for such a large dimension matrix.

    formation of matrix C,

    >>C=A(3:8) ↵

$$C =$$
$$\begin{array}{cccccc} 3 & -10 & 0 & 9 & 73 & 29 \end{array}$$

What if we have a column matrix $D = \begin{bmatrix} 2 \\ 4 \\ x \\ -10 \\ 0 \\ y \\ 73 \\ a \\ -31 \\ 50 \end{bmatrix}$, enter the matrix into MATLAB workspace:

formation of matrix D,

>>syms x y a ↵
>>D=[2 4 x -10 0 y 73 a -31 50].'; ↵

Next, form matrix E from the tenth and seventh elements of D, i.e., $E = \begin{bmatrix} 50 \\ 73 \end{bmatrix}$ and F from the first five

elements of D, i.e., $F = \begin{bmatrix} 2 \\ 4 \\ x \\ -10 \\ 0 \end{bmatrix}$:

formation of matrix E,   formation of matrix F,

>>E=D([10 7]) ↵   >>F=D(1:5) ↵

E =   F =

[50]   [2]
[73]   [4]
                                                 [x]
                                                 [-10]
                                                 [0]

After that, let us see how coloning of square or rectangular matrices can be accomplished. As an example, say,

$$G = \begin{bmatrix} 8 & 64 & 27 & 56 & 98 & 43 & 4 \\ -64 & 216 & 729 & 40 & 12 & 23 & 568 \\ 678 & -90 & 70 & 61 & 67 & 445 & 3 \\ 1 & 47 & 45 & 72 & 34 & -5 & -7 \\ 3 & 87 & 82 & 29 & 10 & -16 & -59 \end{bmatrix}$$. Input the matrix G into the workspace:

>>G=[8 64 27 56 98 43 4;-64 216 729 40 12 23 568;678 ... ↵
-90 70 61 67 445 3;1 47 45 72 34 -5 -7;3 87 82 29 10 -16 -59] ↵

G =

8	64	27	56	98	43	4
-64	216	729	40	12	23	568
678	-90	70	61	67	445	3
1	47	45	72	34	-5	-7
3	87	82	29	10	-16	-59

In the above MATLAB Command, the last word of the first line is 678. After typing 678, leave one space by pressing spacebar, then type three dots. These three dots mean continuation of MATLAB Command. Press enter key and type other matrix elements of the row which were interrupted. The three dots (...) are called ellipsis. Anyhow, matrix G is now in MATLAB workspace. Perform coloning of G using the following commands. For each case, the required matrix elements are shown by elements inside the dotted box.

Matrix H is to be formed from the second and fourth columns of G :

$$\begin{bmatrix} 8 & 64 & 27 & 56 & 98 & 43 & 4 \\ -64 & 216 & 729 & 40 & 12 & 23 & 568 \\ 678 & -90 & 70 & 61 & 67 & 445 & 3 \\ 1 & 47 & 45 & 72 & 34 & -5 & -7 \\ 3 & 87 & 82 & 29 & 10 & -16 & -59 \end{bmatrix}$$

Matrix K is to be formed from the third and fifth rows of G :

$$\begin{bmatrix} 8 & 64 & 27 & 56 & 98 & 43 & 4 \\ -64 & 216 & 729 & 40 & 12 & 23 & 568 \\ 678 & -90 & 70 & 61 & 67 & 445 & 3 \\ 1 & 47 & 45 & 72 & 34 & -5 & -7 \\ 3 & 87 & 82 & 29 & 10 & -16 & -59 \end{bmatrix}$$

Matrix L is to be formed from the fourth through seventh columns of G :

$$\begin{bmatrix} 8 & 64 & 27 & 56 & 98 & 43 & 4 \\ -64 & 216 & 729 & 40 & 12 & 23 & 568 \\ 678 & -90 & 70 & 61 & 67 & 445 & 3 \\ 1 & 47 & 45 & 72 & 34 & -5 & -7 \\ 3 & 87 & 82 & 29 & 10 & -16 & -59 \end{bmatrix}$$

Matrix M is to be formed from the third through fifth rows of G :

$$\begin{bmatrix} 8 & 64 & 27 & 56 & 98 & 43 & 4 \\ -64 & 216 & 729 & 40 & 12 & 23 & 568 \\ 678 & -90 & 70 & 61 & 67 & 445 & 3 \\ 1 & 47 & 45 & 72 & 34 & -5 & -7 \\ 3 & 87 & 82 & 29 & 10 & -16 & -59 \end{bmatrix}$$

Finally, matrix N is to be formed from the intersection of the third through fifth rows and the fourth through seventh columns of G :

$$\begin{bmatrix} 8 & 64 & 27 & 56 & 98 & 43 & 4 \\ -64 & 216 & 729 & 40 & 12 & 23 & 568 \\ 678 & -90 & 70 & 61 & 67 & 445 & 3 \\ 1 & 47 & 45 & 72 & 34 & -5 & -7 \\ 3 & 87 & 82 & 29 & 10 & -16 & -59 \end{bmatrix}$$

Formation of H , K , L , M , and N is presented as follows:

MATLAB Command

for formation of H,

```
>>H=G(:,[2 4])  ↵
```

H =

```
   64   56
  216   40
  -90   61
   47   72
   87   29
```

for formation of N,

```
>>N=G(3:5,4:7)  ↵
```

N =

```
  61   67  445    3
  72   34   -5   -7
  29   10  -16  -59
```

for formation of M,

```
>>M=G(3:5,:)  ↵
```

M =

for formation of K,

```
>>K=G([3 5],:)  ↵
```

K =

```
  678  -90   70   61   67  445    3
    3   87   82   29   10  -16  -59
```

for formation of L,

```
>>L=G(:,4:7)  ↵
```

L =

```
  56   98   43    4
  40   12   23  568
  61   67  445    3
  72   34   -5   -7
  29   10  -16  -59
```

$$\begin{array}{ccccccc} 678 & -90 & 70 & 61 & 67 & 445 & 3 \\ 1 & 47 & 45 & 72 & 34 & -5 & -7 \\ 3 & 87 & 82 & 29 & 10 & -16 & -59 \end{array}$$

Matrix elements can be floating-point numbers or symbolic variables too. In summary we can say, to pick up

1. *rows command is matrix name (desired row/rows, :),*
2. *columns command is matrix name (:, desired column/columns),*
3. *sub-matrices command is matrix name(desired row/rows, desired column/columns).*

2.23 Appending rows/columns with a matrix

Sometimes, it is necessary that a row or column be appended with an existing matrix in MATLAB workspace. This can be conducted by the following technique:

Appending rows:

Assume that matrix $A = \begin{bmatrix} 1 & 3 & 5 \\ 2 & 6 & 8 \\ 9 & 5 & 0 \\ 4 & 7 & 8 \end{bmatrix}$ is formed by appending two row matrices [9 5   0] and [4   7

8] with the matrix $B = \begin{bmatrix} 1 & 3 & 5 \\ 2 & 6 & 8 \end{bmatrix}$. How would you do that? Enter matrix B into MATLAB and append one

row after another by the command shown below:

MATLAB Command

for entering B,

```
>>B=[1 3 5;2 6 8] ↵

B =
        1    3    5
        2    6    8
```

for appending the second row,

```
>>A =[B;[4 7 8]] ↵

A =
        1    3    5
        2    6    8
        9    5    0
        4    7    8
```

for appending the first row,

```
>>B=[B;[9 5 0]] ↵

B =
        1    3    5
        2    6    8
        9    5    0
```

In the right prompt (>>), command B=[B;[9 5 0]] tells that the row [9 5 0] is appended with existing B (inside the third bracket) and that the result is, again, assigned to B. You can add as many rows as you want to. Important point is that the number of elements in each row that is to be appended must be equal to the number of columns of matrix B. If this is not observed, error message is printed by MATLAB.

Appending columns:

As an example, take the matrix $C = \begin{bmatrix} 1 & 3 & 5 & 9 & 3 \\ 2 & 6 & 8 & 0 & 1 \\ 9 & 5 & 0 & 1 & 9 \end{bmatrix}$, where C is formed by appending two column

matrices $\begin{bmatrix} 9 \\ 0 \\ 1 \end{bmatrix}$ and $\begin{bmatrix} 3 \\ 1 \\ 9 \end{bmatrix}$ with matrix $D = \begin{bmatrix} 1 & 3 & 5 \\ 2 & 6 & 8 \\ 9 & 5 & 0 \end{bmatrix}$. Get matrix D into MATLAB and append one column after

another as follows:

MATLAB Command

for entering D,   for appending the first column,

>>D=[1 3 5;2 6 8;9 5 0] ↵   >>D=[D [9 0 1]'] ↵

D =   D =

$$\begin{matrix} 1 & 3 & 5 \\ 2 & 6 & 8 \\ 9 & 5 & 0 \end{matrix}$$ $$\begin{matrix} 1 & 3 & 5 & 9 \\ 2 & 6 & 8 & 0 \\ 9 & 5 & 0 & 1 \end{matrix}$$

for appending the second column,

>>C =[D [3 1 9]'] ↵

C =

$$\begin{matrix} 1 & 3 & 5 & 9 & 3 \\ 2 & 6 & 8 & 0 & 1 \\ 9 & 5 & 0 & 1 & 9 \end{matrix}$$

Add as many columns as you want just remember that the number of elements in each column that is to be appended must be equal to the number of rows of matrix D. We have shown only integer elements. Appending is also applicable for matrices whose elements are floating-points or symbolic. It is even useful inside a for loop.

2.24 Deleting rows/columns from a matrix

Take a rectangular matrix A as $\begin{bmatrix} 4 & 2 & 0 & 4 & -1 \\ 7 & 5 & 2 & 6 & 0 \\ 5 & 4 & 7 & 8 & 2 \\ 1 & 3 & 4 & 1 & 3 \end{bmatrix}$. Presume that the third row, which is [5 4

7   8   2], is to be deleted from A. After deletion of third row, matrix A should look like $\begin{bmatrix} 4 & 2 & 0 & 4 & -1 \\ 7 & 5 & 2 & 6 & 0 \\ 1 & 3 & 4 & 1 & 3 \end{bmatrix}$ which is implemented as follows:

MATLAB Command for row deletion,

>>A=[4 2 0 4 -1;7 5 2 6 0;5 4 7 8 2;1 3 4 1 3]; ↵

>>A(3,:)=[ ] ↵

A =

$$\begin{matrix} 4 & 2 & 0 & 4 & -1 \\ 7 & 5 & 2 & 6 & 0 \\ 1 & 3 & 4 & 1 & 3 \end{matrix}$$

So, to delete a row from a matrix,

MATLAB Step:

Use the matrix name(row number,:)=[ ].

Again, the second column of A, which is $\begin{bmatrix} 2 \\ 5 \\ 4 \\ 3 \end{bmatrix}$, is to be deleted from A. Once deletion is done, matrix A

becomes $\begin{bmatrix} 4 & 0 & 4 & -1 \\ 7 & 2 & 6 & 0 \\ 5 & 7 & 8 & 2 \\ 1 & 4 & 1 & 3 \end{bmatrix}$. Since we changed matrix A by deleting the third row, we need to reenter A.

See the column deletion below:

MATLAB Command

>>A=[4 2 0 4 -1;7 5 2 6 0;5 4 7 8 2;1 3 4 1 3]; ↵

$$\text{>>A(:,2)=[ ] }\downarrow$$

$$A =$$

$$\begin{array}{rrrr} 4 & 0 & 4 & -1 \\ 7 & 2 & 6 & 0 \\ 5 & 7 & 8 & 2 \\ 1 & 4 & 1 & 3 \end{array}$$

Finally, to delete a column from a matrix,

MATLAB Step:

Use the matrix name(:,column number)=[ ].

2.25 Building a large matrix from smaller matrices

Before you start reading this article, go through the article 2.23. Begin with taking four matrices $A = \begin{bmatrix} 0 & -5 & 6 \\ 7 & 6 & 9 \end{bmatrix}$, $B = \begin{bmatrix} 1 & 7 \\ 9 & 1 \end{bmatrix}$, $C = \begin{bmatrix} 1 \\ 1 \end{bmatrix}$, and $D = \begin{bmatrix} -1 & 0 & 10 & 11 \\ 9 & 7 & 13 & 14 \end{bmatrix}$. We would like to form a large matrix E

from the submatrices A, B, C, and D, where $E = \begin{bmatrix} A & B \\ C & D \end{bmatrix} = \begin{bmatrix} 0 & -5 & 6 & 1 & 7 \\ 7 & 6 & 9 & 9 & 1 \\ 1 & -1 & 0 & 10 & 11 \\ 1 & 9 & 7 & 13 & 14 \end{bmatrix}$. This example is

implemented below:

MATLAB Command

```
>>A=[0 -5 6;7 6 9]; ↵
>>B=[1 7;9 1]; ↵
>>C=[1;1]; ↵
>>D=[-1 0 10 11;9 7 13 14]; ↵
>>E=[A B;C D] ↵
```

$$E =$$

$$\begin{array}{rrrrr} 0 & -5 & 6 & 1 & 7 \\ 7 & 6 & 9 & 9 & 1 \\ 1 & -1 & 0 & 10 & 11 \\ 1 & 9 & 7 & 13 & 14 \end{array}$$

Notice that in the fifth line of MATLAB command, we have E=[A B;C D]. Using the command 'A B' gives us $\begin{bmatrix} 0 & -5 & 6 & 1 & 7 \\ 7 & 6 & 9 & 9 & 1 \end{bmatrix}$. To merge A and B, one should remember that the number of rows of A and B have to be the same (here, it is 2). Similarly, using the command 'C D' merges C and D and provides $\begin{bmatrix} 1 & -1 & 0 & 10 & 11 \\ 1 & 9 & 7 & 13 & 14 \end{bmatrix}$. Matrix formed by [A B] will be placed on the top of the matrix formed by [C D] if the command '[A B;C D]' is used. It is important to mention that the number of columns formed by '[A B]' (here, it is 5) and the number of columns formed by '[C D]' (it is also 5) must be identical. Matrix E can again be merged with some other matrices to form another large dimension matrix.

2.26 Rectangular matrix to row or column matrix

Assume that matrix $A = \begin{bmatrix} 1 & 3 & 5 \\ 2 & 6 & 8 \\ 9 & 5 & 0 \end{bmatrix}$ is to be converted to a column matrix by placing one column after another, i.e., we would like to have a matrix C, where $C = \begin{bmatrix} 1 \\ 2 \\ 9 \\ 3 \\ 6 \\ 5 \\ 5 \\ 8 \\ 0 \end{bmatrix}$. To convert a rectangular matrix A to a

column matrix C,

MATLAB Steps:

    1. Enter the matrix (A) and

    2. Use the command C=A(:).

If another matrix is to be formed by placing one row after another, i.e., our intention is to have a matrix R from A, where R =[1   3   5   2   6   8   9   5   0]. To convert a rectangular matrix A to a row matrix R like this,

MATLAB Steps:

    1. Take the transpose of the given matrix A and assign that to A and

    2. Use the command R=A(:)'.

Both manipulations are shown as follows:

MATLAB Command

for column conversion,

    >>A=[1 3 5;2 6 8;9 5 0]; ↵

    >>C=A(:) ↵

    C =

       1   for row conversion,

       2   >>A=A'; ↵

       9   >>R=A(:)' ↵

       3

       6   R =

       5   1   3   5   2   6   8   9   5   0

       5

       8

       0

The command in the fourth prompt, that is, R=A(:)' has a transpose operator ('). Why is that? Because the command 'A(:)' always puts the columns of A one after another.

2.27 Picking up a row or column from a rectangular matrix

The colon operator is useful in this picking up operation. To pick up a row or column from a rectangular matrix,

MATLAB Step:

    Use the command matrix name (row number,:) for row or matrix name(:,column number) for column matrix.

We illustrate the manipulation on $A = \begin{bmatrix} 2 & 6 & 7 \\ 5 & 7 & 1 \\ 0 & 3 & 4 \\ 6 & 9 & 8 \end{bmatrix}$. We wish to pick up the third row and the second column

from A so that we have R =[0   3   4] and $C = \begin{bmatrix} 6 \\ 7 \\ 3 \\ 9 \end{bmatrix}$. See the manipulation in the next page. It is mention

worthy that matrix A is not changed whichever operation is performed.

MATLAB Command

for picking up a row,
>>A=[2 6 7;5 7 1;0 3 4;6 9 8]; ↵
>>R=A(3,:) ↵

R =
           0   3   4

for picking up a column,
>>C=A(:,2) ↵

C =
                6
                7
                3
                9

2.28 Row or column matrix to a rectangular matrix

A long row or column matrix can be converted to rectangular matrix by subroutine 'reshape'. To perform the conversion,

MATLAB Step:

Use the command reshape (matrix name, number of rows, number of columns).

☞ *Given matrix is a row one*

Consider the row matrix R =[3   14   −9   0   12   11   56   78   9   34   91   30]. There are 12 elements in R. Whatever be the order of the reshaped matrix, the product of the order of the reshaped matrix must be 12. It is evident that we may have 3×4, 4×3, 6×2, or 2×6 matrices from R. When elements are placed consecutively, they can come either in column by column or in row by row.

Column by column:

From R, form a matrix N of order 3×4 in which the first column will be the first three elements of R, the second column will be the second three elements of R, and so will be the others, i.e., R is reshaped as

$$N = \begin{bmatrix} 3 & 0 & 56 & 34 \\ 14 & 12 & 78 & 91 \\ -9 & 11 & 9 & 30 \end{bmatrix}.$$

Row by row:

Then, matrix M of order 3×4 is to be formed from R, in M, the first row will be the first four elements of R, the second row will be the second four elements of R, and so will be the other, i.e., $M = \begin{bmatrix} 3 & 14 & -9 & 0 \\ 12 & 11 & 56 & 78 \\ 9 & 34 & 91 & 30 \end{bmatrix}$. See the formation of M and N as follows:

MATLAB Command
for formation of N,
>>R=[3 14 -9 0 12 11 56 78 9 34 91 30]; ↵
>>N=reshape(R,3,4) ↵

N =
            3   0   56   34
           14   12   78   91
           -9   11   9   30

for formation of M,
>>M=reshape(R,4,3)' ↵

M =
            3   14   -9   0
           12   11   56   78
            9   34   91   30

Observe that the number of rows becomes the number of columns and the number of columns becomes the number of rows in the argument of 'reshape' and that there is a transpose operator (') over the subroutine. The reason for that is the subroutine always operates on column basis.

✍ *Given matrix is a column one*

Test column matrix is $C = \begin{bmatrix} 5 \\ 7 \\ -9 \\ 7 \\ 23 \\ 11 \\ 9 \\ 10 \end{bmatrix}$. There are 8 elements in C. Product of the order of the reshaped

matrix must be 8.

Column by column:

We have two options to reshape C − either 2×4 or 4×2. Say, matrix O of order 2×4 is to be formed from C in which the columns are consecutive elements of C, that is, $O = \begin{bmatrix} 5 & -9 & 23 & 9 \\ 7 & 7 & 11 & 10 \end{bmatrix}$.

Row by row:

Finally, a matrix P of order 2×4 is to be built from C in which the rows are consecutive elements of C and it should look like $P = \begin{bmatrix} 5 & 7 & -9 & 7 \\ 23 & 11 & 9 & 10 \end{bmatrix}$. Just to clarify, the numbers of columns and rows interchange when they are put as argument besides the transpose. The manipulation is presented below:

MATLAB Command

for formation of O,

>>C=[5 7 -9 7 23 11 9 10]';↵
>>O=reshape(C,2,4)↵

O =

    5 -9 23 9
    7   7 11 10

for formation of P,

>>P=reshape(C,4,2)'↵

P =

   5   7 -9   7
  23 11   9 10

2.29 Rectangular matrix from the same row/column/ rectangular matrix

Given a row, column, or rectangular matrix, a rectangular matrix can be formed by placing the given row, column, or rectangular matrix one after another. Subroutine 'repmat' is useful in this regard. This is the abbreviation of <u>rep</u>etition of <u>mat</u>rices. To form a rectangular matrix from the same row, column, or rectangular matrix,

MATLAB Step:

Use the command repmat (row/column/ rectangular matrix, the number of repetition along row, the number of repetition along column).

Matrix $D = \begin{bmatrix} 3 & -1 & 0 \\ 3 & -1 & 0 \\ 3 & -1 & 0 \end{bmatrix}$ is to be formed from $R = \begin{bmatrix} 3 & -1 & 0 \end{bmatrix}$ by placing three R s one over the other. Then,

form $E = \begin{bmatrix} 4 & 4 & 4 & 4 \\ 10 & 10 & 10 & 10 \\ -7 & -7 & -7 & -7 \end{bmatrix}$ placing four column matrices $C = \begin{bmatrix} 4 \\ 10 \\ -7 \end{bmatrix}$ side by side. To operate with

rectangular example, consider $A = \begin{bmatrix} 4 & 10 \\ 7 & 0 \end{bmatrix}$. A large matrix F is to be formed from six A by placing 3 A up

and 3 A down, i.e., $F = \begin{bmatrix} A & A & A \\ A & A & A \end{bmatrix} = \begin{bmatrix} 4 & 10 & 4 & 10 & 4 & 10 \\ 7 & 0 & 7 & 0 & 7 & 0 \\ 4 & 10 & 4 & 10 & 4 & 10 \\ 7 & 0 & 7 & 0 & 7 & 0 \end{bmatrix}$. All examples are shown as follows:

MATLAB Command

from the same row matrix,
>>R=[3 -1 0]; ↵
>>OR=repmat(R,3,1) ↵

OR =
```
        3   -1    0
        3   -1    0
        3   -1    0
```
from the same column matrix,
>>C=[4 10 -7]'; ↵
>>OC=repmat(C,1,4) ↵

OC =
```
        4    4    4    4
       10   10   10   10
       -7   -7   -7   -7
```

from the same rectangular matrix,
>>A=[4 10;7 0]; ↵
>>OA=repmat(A,2,3) ↵

OA =
```
        4   10    4   10    4   10
        7    0    7    0    7    0
        4   10    4   10    4   10
        7    0    7    0    7    0
```

2.30 Summing all elements in a matrix

Using the subroutine 'sum' can add all elements of a row, column, or rectangular matrix. To sum all elements of a matrix,

MATLAB Step:

Use the command sum (matrix name) for row or column matrix or sum (sum (matrix name)) for rectangular matrix.

Example matrices are $R = [1 \ \ -2 \ \ 3 \ \ 9]$, $C = \begin{bmatrix} 23 \\ -20 \\ 30 \\ 8 \end{bmatrix}$, and $A = \begin{bmatrix} 2 & 4 & 7 \\ -2 & 7 & 9 \\ 3 & 8 & -8 \end{bmatrix}$. Sums of all elements are 1−2+3+9

=11, 23−20+30+8=41, and 2+4+7−2+7+9+3+8−8=30 for R, C, and A respectively. See summation of all types of matrices below:

MATLAB Command

for row matrix,
>>R=[1 -2 3 9]; ↵
>>sum(R) ↵

ans =
            11
for rectangular matrix,
>>A=[2 4 7;-2 7 9;3 8 -8]; ↵
>>sum(sum(A)) ↵

ans =
        30

for column matrix,
>>C=[23 -20 30 8]'; ↵
>>sum(C) ↵

ans =
            41

We have two 'sum' subroutines for rectangular matrix. The inner 'sum' performs sum over each column, the result will be a row matrix, and the outer 'sum' provides the sum over the resulting row matrix. For the matrix at hand, sums of the first, second, and third columns are 3, 19, and 8 respectively. That is what is displayed by command 'sum (A)' as follows:

>>sum(A) ⏎

ans =

　　3　19　8

2.31 Cumulative sum of elements in a matrix

Cumulative sum of the elements in a matrix can be obtained by subroutine 'cumsum' (abbreviation of cumulative sum). To take the cumulative sum,

MATLAB Step:
Use the command cumsum(matrix name).

Cumulative sums of elements of row matrix $R = [2 \quad 3 \quad -9 \quad 7]$, column matrix $C = \begin{bmatrix} -2 \\ -7 \\ 9 \\ -10 \end{bmatrix}$, and rectangular

matrix $A = \begin{bmatrix} -2 & 4 & 7 \\ 3 & -4 & -2 \\ 9 & 3 & 9 \end{bmatrix}$ are OR=[2　2+3　2+3–9　2+3–9+7]=[2　5　–4　3], OC= $\begin{bmatrix} -2 \\ -2-7 \\ -2-7+9 \\ -2-7+9-10 \end{bmatrix} = \begin{bmatrix} -2 \\ -9 \\ 0 \\ -10 \end{bmatrix}$,

and OA= $\begin{bmatrix} -2 & 4 & 7 \\ -2+3 & 4-4 & 7-2 \\ -2+3+9 & 4-4+3 & 7-2+9 \end{bmatrix} = \begin{bmatrix} -2 & 4 & 7 \\ 1 & 0 & 5 \\ 10 & 3 & 14 \end{bmatrix}$ for R, C, and A respectively. It is understood

that 'cumsum' operates over each column of rectangular matrix A. All types of cumulative sum examples are shown as follows:

MATLAB Command

for row matrix,

>>R=[2　3　-9　7]; ⏎
>>OR=cumsum(R) ⏎

OR =

　　2　5　-4　3

for column matrix,

>>C=[-2 -7 9 -10]'; ⏎
>>OC=cumsum(C) ⏎

OC =

　　-2
　　-9
　　0
　　-10

for rectangular matrix,

>>A=[-2 4 7;3 -4 -2;9 3 9]; ⏎
>>OA=cumsum(A) ⏎

OA =

　　-2　4　7
　　1　0　5
　　10　3　14

2.32 Producting all elements in a matrix

Subroutine 'prod' can be used (abbreviation of product) for this purpose. To compute product of all elements in a matrix,

MATLAB Step:
Use the command prod (matrix name) for row/column matrix or prod (prod (matrix name)) for rectangular matrix.

As usual, start with the test matrices $R = \begin{bmatrix} 1 & -2 & 3 & 9 \end{bmatrix}$, $C = \begin{bmatrix} 23 \\ -20 \\ 30 \\ 8 \end{bmatrix}$, and $A = \begin{bmatrix} 2 & 4 & 7 \\ -2 & 7 & 9 \\ 3 & 8 & -8 \end{bmatrix}$. Products

of all elements of R, C, and A are PR, PC, and PA respectively, so, PR=1×(−2)×3×9=−54, PC=23×(−20)×30 ×8=−110400, and PA=2×4×7×(−2)×7×9×3×8×(−8)=1354752. Like 'sum', there are two 'prod' subroutines for rectangular matrix. The subroutine functions in a similar way for rectangular matrix as does 'sum'. See all implementations below:

MATLAB Command

for row matrix,

>>R=[1 -2 3 9]; ↵
>>PR=prod(R) ↵

PR =
     -54

for column matrix,

>>C=[23 -20 30 8]'; ↵
>>PC=prod(C) ↵

PC =
     -110400

for rectangular matrix,

>>A=[2 4 7;-2 7 9;3 8 -8]; ↵
>>PA=prod(prod(A)) ↵

PA =
     1354752

for symbolic elements,

>>syms x ↵
>>R=[x x^3 -2*x^3]; ↵
>>prod(R) ↵

ans =

-2*x^7

The function is operational for floating-point numbers and symbolic variables. Product of all elements of $R = \begin{bmatrix} x & x^3 & -2x^3 \end{bmatrix}$ is $-2x^7$, which is presented above as well.

2.33 Cumulative product of elements in a matrix

The subroutine 'cumprod' (abbreviation of <u>cum</u>ulative <u>prod</u>uct) computes the cumulative product of elements in a matrix. To find the cumulative product of elements in a matrix,

    MATLAB Step:

        Use the command cumprod(matrix name).

Cumulative products of the elements of row matrix $R = \begin{bmatrix} 2 & 3 & -9 & 7 \end{bmatrix}$, column matrix $C = \begin{bmatrix} -2 \\ -7 \\ 9 \\ -10 \end{bmatrix}$, and

rectangular matrix $A = \begin{bmatrix} -2 & 4 & 7 \\ 3 & -4 & -2 \\ 9 & 3 & 9 \end{bmatrix}$ are OR $= \begin{bmatrix} 2 & 2\times3 & 2\times3\times(-9) & 2\times3\times(-9)\times7 \end{bmatrix} = \begin{bmatrix} 2 & 6 & -54 & -378 \end{bmatrix}$, OC

$= \begin{bmatrix} -2 \\ -2\times(-7) \\ -2\times(-7)\times9 \\ -2\times(-7)\times9\times(-10) \end{bmatrix} = \begin{bmatrix} -2 \\ 14 \\ 126 \\ -1260 \end{bmatrix}$, and OA $= \begin{bmatrix} -2 & 4 & 7 \\ -2\times3 & 4\times(-4) & 7\times(-2) \\ -2\times3\times9 & 4\times(-4)\times3 & 7\times(-2)\times9 \end{bmatrix} = \begin{bmatrix} -2 & 4 & 7 \\ -6 & -16 & -14 \\ -54 & -48 & -126 \end{bmatrix}$ for R,

C, and A respectively. Like 'cumsum', 'cumprod' operates on column basis for a rectangular matrix. All implementations are shown below:

MATLAB Command

for row matrix,

>>R=[2 3 -9 7]; ↵
>>OR=cumprod(R) ↵

for column matrix,

>>C=[-2 -7 9 -10]'; ↵
>>OC=cumprod(C) ↵

OR =   OC =
      2   6 -54 -378   -2
for rectangular matrix,   14
      >>A=[-2 4 7;3 -4 -2;9 3 9]; ↵   126
      >>OA=cumprod(A) ↵   -1260

OA=
      -2   4   7
      -6 -16 -14
    -54 -48 -126

2.34 Finding the maximum/minimum element from a matrix

Finding the maximum:

To find the maximum element from a matrix,

MATLAB Step:

Use the command max (matrix name) for row/column matrix or use max (max (matrix name)) for rectangular matrix.

Row matrix $R=[1\ \ -2\ \ 3\ \ 9\ \ 78\ \ 90\ \ -90]$, column matrix $C=\begin{bmatrix} 23 \\ -20 \\ 30 \\ 8 \end{bmatrix}$, and rectangular matrix $A=\begin{bmatrix} 2 & 4 & 7 \\ -2 & 7 & 9 \\ 3 & 8 & -8 \end{bmatrix}$ have the maximum values 90, 30, and 9 respectively. How maximum for different matrices can be found is shown as follows:

MATLAB Command

for row matrix,   for column matrix,
      >>R=[1 -2 3 9 78 90 -90]; ↵   >>C=[23 -20 30 8]'; ↵
      >>max(R) ↵   >>max(C) ↵

      ans =   ans =
           90   30
for rectangular matrix,
      >>A=[2 4 7;-2 7 9;3 8 -8]; ↵
      >>max(max(A)) ↵

      ans =
         9

As it happens for rectangular matrix, output of the inner 'max' is a row matrix on finding maximum of each column. Maximum for whole rectangular matrix is found by the outer 'max'. Do not delete R, C, and A, subroutine 'min' is checked with these matrices.

Finding the minimum:

To find the minimum element from a matrix,

MATLAB Step:

Use the command min (matrix name) for row/column matrix or use min(min(matrix name)) for rectangular matrix.

Just mentioned row, column, and rectangular matrices have the minimum values –90, –20, and –8 respectively. Findings of all minimums are shown in the following page:

MATLAB Command

for R,   For C,   For A,
   >>min(R) ↵   >>min(C) ↵   >>min(min(A)) ↵

   ans =   ans =   ans =
          -90   -20   -8

The subroutines are equally applicable for floating-point numbers. They can appear inside any loop statement like 'for' or 'while'. They can be used to find numerically the minimum or maximum between some interval of a function, where the function can be one or two-dimensional.

2.35 Sorting elements in a matrix

Using the subroutine 'sort' can sort the elements of a row or column matrix in ascending order. In the case of a rectangular matrix, the sorting operation will be over each column. To sort the elements in a matrix,

MATLAB Step:
   Use the command sort(matrix name).

In ascending order, sorting of row matrix $R = [1 \quad -2 \quad 3 \quad 9 \quad 0 \quad -5]$ and column matrix $C = \begin{bmatrix} 23 \\ -20 \\ 30 \\ 8 \\ -10 \end{bmatrix}$

yields $OR = [-5 \quad -2 \quad 0 \quad 1 \quad 3 \quad 9]$ and $OC = \begin{bmatrix} -20 \\ -10 \\ 8 \\ 23 \\ 30 \end{bmatrix}$ for R and C respectively. For rectangular matrix $A =$

$\begin{bmatrix} 2 & 4 & 7 \\ -2 & 7 & 9 \\ 3 & 8 & -8 \\ 0 & -3 & -1 \end{bmatrix}$, sorting over each column provides $OA = \begin{bmatrix} -2 & -3 & -8 \\ 0 & 4 & -1 \\ 2 & 7 & 7 \\ 3 & 8 & 9 \end{bmatrix}$. Conduct the subroutine 'sort' as

shown below:

MATLAB Command

for row matrix,   for column matrix,
   >>R=[1 -2 3 9 0 -5]; ↵   >>C=[23 -20 30 8 -10]'; ↵
   >>OR=sort(R) ↵   >>OC=sort(C) ↵

   OR =   OC =
          -5 -2   0   1   3   9   -20
for rectangular matrix,   -10
   >>A=[2 4 7;-2 7 9;3 8 -8;0 -3 -1]; ↵   8
   >>OA=sort(A) ↵   23
                                                    30
   OA =
          -2   -3   -8
           0   4   -1
           2   7   7
           3   8   9

Elements of the matrices can be floating-points too.

2.36 Finding the position indexes of matrix elements with conditions

The subroutine 'find' is useful in this regard. To proceed with, say, $A = \begin{bmatrix} 11 & 10 & 11 & 10 \\ 12 & 10 & -2 & 0 \\ -7 & 17 & 1 & -1 \end{bmatrix}$. We

would like to know what the position indexes of the elements are where the elements are greater than 10. Each position of an element of a matrix needs two indexes two describe it. One is for row and the other is for column. The elements of A, whose position indexes are (1, 1), (2, 1), (3, 2), and (1, 3), are greater than 10. As it happens, MATLAB finds the required index in accordance with column. Since the number of output arguments of the subroutine is two, we have to provide two variables for the return of output. To find the position index with condition,

MATLAB Step:

Use the command [R   C] = find(condition), where index R for row and C for column.

Finding the position indexes where the elements of A are greater than 10 is shown below:

MATLAB Command for rectangular matrix,

where elements >10,

>>A=[11 10 11 10;12 10 -2 0;-7 17 1 -1]; ↵
>>[R C]=find(A>10) ↵

R =

 1
 2
 3
 1

C =

 1
 1
 2
 3

for row matrix,

>>D=[-10 34 1 2 8 4]; ↵
>>R=find(D>=8) ↵

R =

 2   5

for column matrix,

>>E=[-2 8 -2 7]'; ↵
>>C=find(E~=-2) ↵

C =

 2
 4

where elements =10,

>>[R C]=find(A==10) ↵

R =

 1
 2
 1

C =

 2
 2
 4

where elements ≤ 10,

>>[R C]=find(A<=0) ↵

R =

 3
 2
 2
 3

C =

 1
 3
 4
 4

The first position index is (1, 1) which is given by the first elements of R and C, the second position index is (2,1) which is given by the second elements of R and C, and so is the others. To deal with other conditions, in matrix A, what are the position indexes where the elements are equal to 10? The answer is (1, 2), (2, 2), and (1, 4). Again, the position indexes where the elements are less than or equal to zero are (3, 1), (2, 3), (2, 4), and (3, 4). Both implementations are presented in upper right side of this page. As it is presented, the third command prompt >>[R C]=find (A= =10) has '= =' sign. It should be pointed out that '= =' is used for logical comparison and that '=' is used for assigning a vector or matrix to the left variable of sign '='. So far, we considered a rectangular matrix, let us see how the subroutine works for row or column matrix. As an example, take $D = [-10 \quad 34 \quad 1 \quad 2 \quad 8 \quad 4]$, from which, find the position indexes of the elements where they are greater than or equal to 8. Obviously, they are the 2nd and 5th elements. Here, we do not need two arguments for the output of 'find'. To conclude, find the position indexes of the elements of column matrix

54

$E = \begin{bmatrix} -2 \\ 8 \\ -2 \\ 7 \end{bmatrix}$ where the elements are not equal to –2. The 2ⁿᵈ and 4ᵗʰ elements are not equal to –2. The last two

examples are also shown in the previous page. For comparisons, we have used some logical operators such as, greater than or equal to (>=), less than or equal to (<=), not equal to (~=), and equal to (==). One more point is to be spoken of that the output of the subroutine for row matrix is a row one and that for column matrix is a column one.

2.37 Replacing a row/column/submatrix by another

It may be necessary that a row, column or sub matrix of a large rectangular matrix be replaced by another matrix of identical order. Say, that large matrix is $A = \begin{bmatrix} 1 & 2 & 3 & 4 & 5 \\ 7 & 8 & 9 & 0 & 11 \\ 13 & 9 & 2 & 2 & 1 \\ 4 & 7 & 6 & 2 & 90 \end{bmatrix}$. Begin with row

replacement. The second row of A, which is [7 8 9 0 11], is to be replaced by a row matrix [2 2 2 2

2]. After replacement, A should look like $\begin{bmatrix} 1 & 2 & 3 & 4 & 5 \\ 2 & 2 & 2 & 2 & 2 \\ 13 & 9 & 2 & 2 & 1 \\ 4 & 7 & 6 & 2 & 90 \end{bmatrix}$. Then, the forth column of A, which is $\begin{bmatrix} 4 \\ 0 \\ 2 \\ 2 \end{bmatrix}$,

is to be replaced by $\begin{bmatrix} 0 \\ 0 \\ 0 \\ 0 \end{bmatrix}$. The replacement results the new matrix as $\begin{bmatrix} 1 & 2 & 3 & 0 & 5 \\ 7 & 8 & 9 & 0 & 11 \\ 13 & 9 & 2 & 0 & 1 \\ 4 & 7 & 6 & 0 & 90 \end{bmatrix}$. After that, we

proceed with replacement of submatrix. Say, the elements of A inside the dotted marks, which are shown by

$\begin{bmatrix} 1 & 2 & 3 & 4 & 5 \\ 7 & 8 & 9 & 0 & 11 \\ 13 & 9 & 2 & 2 & 1 \\ 4 & 7 & 6 & 2 & 90 \end{bmatrix}$, are to be replaced by a matrix $B = \begin{bmatrix} 6 & 6 \\ 3 & 3 \\ 0 & 0 \end{bmatrix}$. The resulting matrix will be

$\begin{bmatrix} 1 & 2 & 3 & 4 & 5 \\ 7 & 8 & 6 & 6 & 11 \\ 13 & 9 & 3 & 3 & 1 \\ 4 & 7 & 0 & 0 & 90 \end{bmatrix}$. Notice the replacement of the last case. The submatrix is the intersecting elements of

the second through fourth rows with the third through fourth columns of A. Since replacement changes the matrix A, we need to enter matrix A over and over. All replacements are shown below:

MATLAB Command
for row replacement,

>>A=[1 2 3 4 5;7 8 9 0 11;13 9 2 2 1;4 7 6 2 90]; ↵
>>A(2,:) =[2 2 2 2 2] ↵

A =

```
  1   2   3   4   5
  2   2   2   2   2
 13   9   2   2   1
  4   7   6   2  90
```

for column replacement,

>>A=[1 2 3 4 5;7 8 9 0 11;13 9 2 2 1;4 7 6 2 90]; ↵
>>A(:,4)=[0 0 0 0]' ↵

A =

```
  1   2   3   0   5
  7   8   9   0  11
```

$$\begin{matrix} 13 & 9 & 2 & 0 & 1 \\ 4 & 7 & 6 & 0 & 90 \end{matrix}$$

for submatrix replacement,

>>A=[1 2 3 4 5;7 8 9 0 11;13 9 2 2 1;4 7 6 2 90]; ↵
>>B=[6 6;3 3;0 0]; ↵
>>A(2:4,3:4)=B ↵

A =

$$\begin{matrix} 1 & 2 & 3 & 4 & 5 \\ 7 & 8 & 6 & 6 & 11 \\ 13 & 9 & 3 & 3 & 1 \\ 4 & 7 & 0 & 0 & 90 \end{matrix}$$

2.38 Replacing/deleting elements from a matrix

Previous article's replacement includes only row, column, or submatrix. How if the replacement of elements is conducted. For matrix $A = \begin{bmatrix} 1 & 2 & 3 & 7 \\ 7 & 8 & 9 & 6 \\ 13 & 9 & 2 & 1 \end{bmatrix}$, elements 2 in the first row and 1 in the third row are to be replaced by 5 and 9 respectively (depicted in figure 2.2). Position indexes of these two elements are (1, 2) and (3, 4) respectively. The new matrix is going to be $\begin{bmatrix} 1 & 5 & 3 & 7 \\ 7 & 8 & 9 & 6 \\ 13 & 9 & 2 & 9 \end{bmatrix}$ following replacement. The implementation is presented as follows:

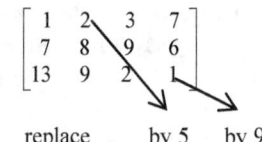
replace   by 5   by 9

Figure 2.2 *Replacement of elements*

MATLAB Command

>>A=[1 2 3 7;7 8 9 6;13 9 2 1]; ↵
>>A(1,2)=5; ↵
>>A(3,4)=9 ↵

A =

$$\begin{matrix} 1 & 5 & 3 & 7 \\ 7 & 8 & 9 & 6 \\ 13 & 9 & 2 & 9 \end{matrix}$$

If you delete some elements from a matrix, the matrix does not exist any more, so, error message appears in MATLAB Command Window due to deletion of some elements.

2.39 Formation of a magic square

A magic square is a square matrix. It has some amazing properties. Summation of any row or column of a magic square gives the same sum. Not only is that, the sum of diagonal elements is also as the sum as the sum of row or column elements. The example of 3×3 magic square, whose common sum is 15, is shown in the following table.

15	15	15	15	15
15	8	1	6	15
15	3	5	7	15
15	4	9	2	15
15	15	15	15	15

Check :

1st   row: 8+1+6 = 15
2nd   row: 3+5+7 = 15

3rd   row: 4+9+2 = 15
1st   column: 8+3+4 = 15
2nd   column: 1+5+9 = 15
3rd   column: 6+7+2 = 15
1st diagonal: 4+5+6 = 15
2nd diagonal: 2+5+8 = 15

This is so easy to implement in MATLAB:

     MATLAB Command for 3×3 magic square,
        >>B=magic(3) ↵

     B =
            8   1   6
            3   5   7
            4   9   2

Finally, to generate a magic square of order $N \times N$,

     MATLAB Step:
         Use the command magic(N) .

2.40 Knowing the order/length of a matrix/vector

If a matrix exists in workspace of MATLAB, order of the matrix can be found using subroutine 'size'. Its output is a two element row vector — the first one for the number of rows and the second one for the number of columns. Necessity of knowing the length or order is understood when an M-file generates a matrix of hundreeds or thousands elements. Order of $A = \begin{bmatrix} 2 & 1 & 0 \\ 4 & 6 & 7 \end{bmatrix}$ is 2×3, hence, 'size(A)' should give us [2   3]. It is possible to know only the number of rows or columns using the command 'size(A,1)' or 'size(A,2)' respectively. Apart from 'size', there is another subroutine called 'length' that returns the number of elements of a vector (row or column matrix). When operated on rectangular matrix, it returns the number of columns.

Conduct 'length' on $R = [1 \quad 2 \quad 0 \quad 3]$ and $C = \begin{bmatrix} 9 \\ 2 \\ 0 \end{bmatrix}$. As output, we should have 4 and 3 for R and C respectively. See all these in the following:

MATLAB Command

for order of A,
     >>A=[2 1 0;4 6 7]; ↵
     >>size(A) ↵

   ans =
       2   3

for the number of column of A,
     >>size(A,2) ↵

   ans =
      3

for the number of elements of R,
     >>R=[1 2 0 3]; ↵
     >>length(R) ↵

   ans =
      4

for the number of row of A,
     >>size(A,1) ↵

   ans =
      2

     >>length(A) ↵

   ans =
      3

for the number of elements of C,
     >>C=[9 2 0]'; ↵
     >>length(C) ↵

   ans =
      3

2.41 Picking up diagonal elements from a matrix

If a square/rectangular matrix is given, we can pick up subdiagonal, diagonal, or superdiagonal elements using the subroutine 'diag'. To pick up any diagonal elements,

MATLAB Step:

Use the command diag(matrix name, type of diagonal).

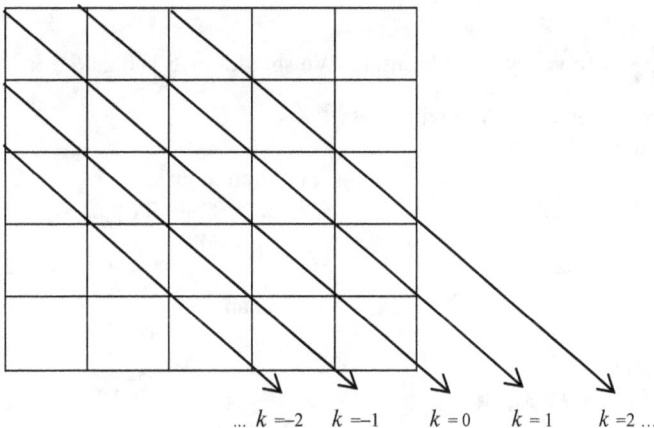

... $k=-2$ $k=-1$ $k=0$ $k=1$ $k=2$ ...

Figure 2.3 *Indexes of different diagonals in a matrix*

Referring to figure 2.3, $k=0$, $k=1$, $k=2$ correspond to the diagonal, the first super diagonal, the second super diagonal respectively, and so on. Again, $k=-1$ and $k=-2$ correspond to the first and second subdiagonals respectively, and so on. To work with diagonal elements, take $A=\begin{bmatrix} 1 & 2 & 3 & 7 \\ 7 & 8 & 9 & 6 \\ 13 & 9 & 2 & 1 \end{bmatrix}$; we want to

pick up diagonal elements from A (shown by the dotted line, i.e., [1 8 2]), where $k=0$. How if can the

second superdiagonal elements $\left(\text{shown by the dotted line in } \begin{bmatrix} 1 & 2 & 3 & 7 \\ 7 & 8 & 9 & 6 \\ 13 & 9 & 2 & 1 \end{bmatrix} \right)$ be? The answer is use the

command 'diag(A,2)'. Use 'diag(A,-1)' to pick up the first subdiagonal elements as shown by dotted mark in

$\begin{bmatrix} 1 & 2 & 3 & 7 \\ 7 & 8 & 9 & 6 \\ 13 & 9 & 2 & 1 \end{bmatrix}$. Regardless of the type of diagonal, output is a column matrix. All examples are shown

below:

MATLAB Command

for the diagonal elements,

    >>A=[1 2 3 7;7 8 9 6;13 9 2 1]; ↵

    >>diag(A,0) ↵

ans =

    1

    8

    2

for the first subdiagonal elements,

    >>diag(A,-1) ↵

ans =

    7

    9

for the second superdiagonal elements,

    >>diag(A,2) ↵

ans =

    3

    6

2.42 Determine whether all elements are nonzero

A situation when we want to determine whether all elements in a matrix are nonzero can take place in M-file programming. This check is performed by the subroutine 'all'. Row or column matrix needs using 'all' once but rectangular matrix needs twice because 'all' operation is performed over columns. If any element is zero, the subroutine returns the output as 0, if not, output is 1. Take the examples of $R = [1 \quad -2 \quad 3]$, $C = \begin{bmatrix} 0 \\ 0 \\ 0 \end{bmatrix}$,

and $A = \begin{bmatrix} 4 & -2 & 9 \\ 9 & 0 & 3 \end{bmatrix}$ to verify the subroutine. We should have 1, 0, and 0 for R, C, and A respectively. These examples are presented in the following:

MATLAB Command

for row matrix,

    >>R=[1 -2 3]; ↵
    >>all(R) ↵

    ans =
        1

for column matrix,

    >>C=[0 0 0]'; ↵
    >>all(C) ↵

    ans =
        0

for rectangular matrix,

    >>A=[4 -2 9;9 0 3]; ↵
    >>all(all(A)) ↵

    ans =
        0

2.43 Determine whether any element is nonzero

Another situation whether any element of a matrix is nonzero can also occur. We can utilize the subroutine 'any' in this regard. If any element is nonzero, the output is 1, if not, the output is 0. So, operation of the subroutine over the matrices R, C, and A of article 2.42 should return 1, 0, and 1 respectively. See the implementation below:

MATLAB Command

for R,

    >>any(R) ↵

    ans =
        0

for C,

    >>any(C) ↵

    ans =
        1

for A,

    >>any(any(A)) ↵

    ans =
        1

2.44 Permutation of matrix elements

All possible combinations of matrix elements can be generated using the subroutine 'perms' (abbreviation of permutations). To have the permutation matrix,

    MATLAB Step:

        Use the command perms (matrix name).

Say, we have three integers 1, 2, and 3. We obtain the permutation matrix from these three elements as follows:

    MATLAB Command

        >>R=[1 2 3]; ↵
        >>perms(R) ↵

        ans =
            3   2   1
            2   3   1
            3   1   2
            1   3   2

$$\begin{matrix} 2 & 1 & 3 \\ 1 & 2 & 3 \end{matrix}$$

The number of rows of permutation matrix is the factorial of the number of elements in the matrix. For example, permutation matrix will be of order $5! \times 5$ for 5 element matrix. The argument matrix can be a rectangular too but the subroutine is applied on the total number of elements of the rectangular matrix. Elements of argument matrix can be floating-point numbers or symbolic variables. If the number of elements is higher, the function takes longer execution time to generate the permutation matrix.

2.45 Multiplying a column by an expression

If a matrix of symbolic variables is given, any column of the matrix can be multiplied by an expression using the subroutine 'mulcol' (abbreviation of multiply column). This subroutine is accompanied with 'linalg' (abbreviation of linear algebra) package, which is conducted by Maple kernel. We want to multiply the second column of $A = \begin{bmatrix} 1 & 5 & 3 \\ 7 & 3 & 2 \\ 3 & 1 & 0 \end{bmatrix}$ by x and the third column by $x^2 - 8$, so, the resulting matrix

should be $B = \begin{bmatrix} 1 & 5x & 3x^2 - 24 \\ 7 & 3x & 2x^2 - 16 \\ 3 & x & 0 \end{bmatrix}$. Multiplication is done one at a time as shown below:

MATLAB Command

```
>>A=[1 5 3;7 3 2;3 1 0]; ↵
>>syms x ↵
>>B=maple('mulcol',A,2,x); ↵    ← Argument 2 indicates the second column. First multiplication
                                    is conducted, then, the output is assigned to B.
>>B=maple('mulcol',B,3,x^2-8) ↵  ← Multiplication of third column with x² - 8
```

B =

```
[    1,    5*x,    3*x^2-24]
[    7,    3*x,    2*x^2-16]
[    3,      x,           0]
```

2.46 Swapping rows or columns of a rectangular matrix

Given a rectangular matrix, we can swap any two rows or columns of the matrix by subroutines 'swaprow' (abbreviation of swap rows) and 'swapcol' (abbreviation of swap columns) respectively. These functions are found in linear algebra package and executed by Maple kernel. To illustrate, take $A = \begin{bmatrix} 1 & 5 & 3 & 4 \\ 7 & 3 & 2 & 9 \\ 3 & 1 & 0 & 0 \end{bmatrix}$, in which the first row is swapped with the third one and the resulting matrix should be

$B = \begin{bmatrix} 3 & 1 & 0 & 0 \\ 7 & 3 & 2 & 9 \\ 1 & 5 & 3 & 4 \end{bmatrix}$. Next, we want to exchange the second and fourth columns of B, the output should be C

$= \begin{bmatrix} 3 & 0 & 0 & 1 \\ 7 & 9 & 2 & 3 \\ 1 & 4 & 3 & 5 \end{bmatrix}$. Implementation is shown below:

MATLAB Command

```
>>A=sym([1 5 3 4;7 3 2 9;3 1 0 0]); ↵
>>B=maple('swaprow',A,1,3) ↵    ← Formation of B
```

B =

```
[ 3,  1,  0,  0]
[ 7,  3,  2,  9]
[ 1,  5,  3,  4]
>>C=maple('swapcol',B,2,4) ↵        ← Formation of C
```

C =

```
[ 3,  0,  0,  1]
[ 7,  9,  2,  3]
[ 1,  4,  3,  5]
```

Maple kernel operates on symbolic elements. What if we have a matrix whose elements are floating point

numbers, say, $R = \begin{bmatrix} 0.33 & 0.5 & 3 \\ 4 & 2 & .7 \\ 2 & -2 & -0.4 \end{bmatrix}$. Exchange the first and third columns of R to have $Q =$

$\begin{bmatrix} 3 & 0.5 & 0.33 \\ 0.7 & 2 & 4 \\ -0.4 & -2 & 2 \end{bmatrix}$. For this, first convert the floating-point numbers to symbolic, then apply 'swapcol',

and finally convert the symbolic values to decimal:

```
>>R=sym([0.33 0.5 3;4 2 .7;2 -2 -.4]); ↵
>>Q=maple('swapcol',R,1,3); ↵              ← Swapping the columns
>>Q=double(Q) ↵                            ← Conversion from symbolic to decimal
```

Q =

```
 3.0000   0.5000   0.3300
 0.7000   2.0000   4.0000
-0.4000  -2.0000   2.0000
```

2.47 Expansion of a matrix by adding some number

Given a matrix of symbolic variables, we can extend the matrix by adding extra rows or columns of some supplied element. The Maple function that does this extension is 'extend'. As usual, consider $A = \begin{bmatrix} 3 & 4 & 2 \\ 0 & 7 & 6 \end{bmatrix}$. Extend A by adding three rows and two columns of −8. Thereupon, we are supposed to have $B =$

$\begin{bmatrix} 3 & 4 & 2 & -8 & -8 \\ 0 & 7 & 6 & -8 & -8 \\ -8 & -8 & -8 & -8 & -8 \\ -8 & -8 & -8 & -8 & -8 \\ -8 & -8 & -8 & -8 & -8 \end{bmatrix}$ as output matrix. Execute the following to have it:

MATLAB Command

```
>>A=sym([3 4 2;0 7 6]); ↵
>>B=maple('extend',A,3,2,-8) ↵     ← Conduct the function 'extend'
```

B =

```
[ 3,  4,  2,  -8,  -8]
[ 0,  7,  6,  -8,  -8]
[ -8, -8, -8, -8,  -8]
[ -8, -8, -8, -8,  -8]
[ -8, -8, -8, -8,  -8]
```

It is evident that inside maple, first argument is the function itself, the second is the number of rows required, the third is the number of columns required, and the fourth one is the element, which is used for filling the matrix.

Chapter 3

Solutions to Algebraic, Trigonometric, and Geometric Problems

This chapter is devised to meet the need of the problems that appear frequently in algebra, trigonometry, geometry, or maybe in some other disciplines. But the nature of problems is mostly selected from polynomial manipulation and geometric object formation. Manipulations like finding roots of a polynomial equation, factorization / multiplication / division / partial fraction of polynomials, representation of geometric objects, two dimensional coordinate geometry problems ...etc are the lay out of this chapter. Attempts have been made to introduce mathematical symbolism and language. Since we pay attention to explanations based on intuitive appeal rather than mathematics, we worked out model problems in detail. A number of examples are devoted to clarify the illustrations.

3.1 Polynomial representation

A polynomial of order n is given by $c_n x^n + c_{n-1} x^{n-1} + c_{n-2} x^{n-2} + \ldots + c_2 x^2 + c_1 x + c_0$. There are $n+1$ coefficients, which are c_n, c_{n-1}, c_{n-2}, ..., c_2, c_1, and c_0, in a polynomial of order n. A row vector in MATLAB represents these $n+1$ coefficients. Suppose we have a polynomial $-17x^3 + 17x^2 - 7x + 1$. Go to MATLAB command window, i.e., command prompt (>>) of MATLAB, type A=[-17 17 -7 1], and press enter, you will see,

    A =

        -17   17   -7   1

The execution says that you assigned the coefficients of polynomial $-17x^3 + 17x^2 - 7x + 1$ as row vector A. To see the powers of x, type poly2str(A,'x') on the command prompt and press enter, you will see,

        ans =

            -17 x^3 + 17 x^2 - 7 x + 1

Here -17 x^3 and +17 x^2 represent $-17x^3$ and $+17x^2$ respectively, and so do others. If you wanted to see the polynomial in terms of z, the command would be poly2str(A,'z'):

>>poly2str(A, 'z') ↵

ans =

$$-17 z^3 + 17 z^2 - 7 z + 1$$

Actually, poly2str is the abbreviation of <u>poly</u>nomial to (2) <u>str</u>ing. This is a subroutine in MATLAB. Symbolic polynomials can also be entered using string form. Any power term like ab^n is written as a*b^n. Utilizing this, we have

the command for symbolic entering of $-17x^3 + 17x^2 - 7x + 1$:

>>syms x ↵   ← Declaration of x as symbolic

>>A=-17*x^3+17*x^2-7*x+1 ↵

A =

-17*x^3+17*x^2-7*x+1   ← A is having the symbolic polynomial $-17x^3 + 17x^2 - 7x + 1$.

3.2 Roots of a polynomial equation

A polynomial equation of degree n has n roots. Depending on the coefficients of a polynomial, the roots can be real or complex. As an illustration, solve polynomial equation $4x^2 + 4x + 1 = 0$. Roots of the quadratic equation are $-\frac{1}{2}$ and $-\frac{1}{2}$ (obtained by using $x = \frac{-b \pm \sqrt{b^2 - 4ac}}{2a}$). The equation can be solved by MATLAB using the subroutine 'roots'. To find the roots of a polynomial equation,

MATLAB Steps:
 1. Enter the polynomial as row vector (A) and
 2. Use the command roots (A).

For the equation $4x^2 + 4x + 1 = 0$, we have the coefficient form A=[4 4 1] and the solution is as follows:

>>A=[4 4 1]; ↵

>>roots(A) ↵

ans =

-0.5000

-0.5000

If the equation were of higher degree than that of 2, for example, $x^5 + 5.5x^4 + 7.5x^3 - 2.5x^2 - 8.5x - 3 = 0$ and whose roots are given by $x = -3, -2, -1, -\frac{1}{2}$, and 1, it could also be solved by the subroutine:

>>A=[1 5.5 7.5 -2.5 -8.5 -3]; ↵

>>roots(A) ↵

ans =

-3.0000

-2.0000

1.0000

-1.0000

-0.5000

An equation may not have real roots always. Such an equation is $x^3 + x^2 + 4x + 4 = 0$. Roots -1, $2i$, and $-2i$ of the equation can be found as follows:

>>A=[1 1 4 4]; ↵

>>roots(A) ↵

ans =

- 0.0000 + 2.0000i

- 0.0000 - 2.0000i

- 1.0000

Very often equations are handled in symbolic form. In that case coefficients of the polynomial are found first by using 'sym2poly' (abbreviation of <u>sym</u>bolic to(<u>2</u>) <u>poly</u>nomial). Verify that with the last equation:

>>syms x ↲
>>y=x^3+x^2+4*x+4; ↲   ← Assigning polynomial to y
>>c=sym2poly(y); ↲   ← Assigning polynomial coefficients to c
>>roots(c) ↲   ← Apply the subroutine

ans =
      -0.0000 + 2.0000i
      -0.0000 - 2.0000i
      -1.0000

3.3 Forming a polynomial equation from its roots

This is a reverse problem to the one as stated in the last article. If roots of a polynomial equation are given, the equation can be obtained from the roots by subroutine 'poly'. Suppose the roots of an equation are -1, 1, 2, and 3. The equation is obtained from the multiplication of $x+1$, $x-1$, $x-2$, and $x-3$ respectively, which is $x^4 - 5x^3 + 5x^2 + 5x - 6 = 0$. MATLAB should return [1 -5 5   5 -6] as the output polynomial. To form a polynomial equation from its roots,

MATLAB Steps:
1. Assign the roots as row vector to x and
2. Use the command poly (x).

For just mentioned four roots, we have x=[-1 1   2   3]. Formation is shown below:

MATLAB Command   To display the polynomial:
   >>x=[-1 1 2 3]; ↲   >>poly2str(poly(x),'x') ↲
   >>poly(x) ↲
                                          ans =
   ans =
       1   -5   5   5   -6   x^4 - 5 x^3 + 5 x^2 + 5 x - 6

Along with the implementation, 'poly2str' displays the polynomial string. If the roots are complex, MATLAB can still form the polynomial. Usually, complex roots are in pairs. Take example of a complex root equation, where the roots are 2, 3, $1-2i$, and $1+2i$, so, root matrix is x=[2   3 $1-2i$ $1+2i$] providing equation $(x-2)$ $(x-3)$ $(x-1+2i)(x-1-2i) = x^4 - 7x^3 + 21x^2 - 37x + 30 = 0$. The output polynomial should be [1 -7 21 -37 30]:

For complex roots: To display the polynomial:
   >>x=[2 3 1-2i 1+2i]; ↲   >>poly2str(poly(x),'x') ↲
   >>poly(x) ↲
                                          ans =
   ans =
       1   -7   21   -37   30   x^4 - 7 x^3 + 21 x^2 - 37 x + 30

3.4 Evaluation of a polynomial at any x

The subroutine 'polyval' (abbreviation of <u>poly</u>nomial <u>val</u>ue at) evaluates a polynomial of x at any x. The subroutine has two input arguments. The first argument takes polynomial coefficients and the second one takes the value of x where we are interested at. To perform the evaluation,

MATLAB Step:
Use the command polyval(polynomial vector, x $-$value).

Suppose we have to evaluate the polynomial $x^4 - 7x^3 + 21x^2 - 37x - 30$ at $x = -3$. The result should be $(-3)^4 - 7(-3)^3 + 21(-3)^2 - 37(-3) - 30 = 540$. The polynomial vector is $[1 \quad -7 \quad 21 \quad -37 \quad -30]$. Implementation is as follows:

MATLAB Command
```
>>y=[1 -7 21 -37 -30]; ↵
>>polyval(y,-3) ↵
```

ans =

540

Another situation is when the polynomial is given in symbolic form. Use the function 'subs' (abbreviation of <u>subs</u>titution) to evaluate the polynomial chosen for 'polyval':

Command for the symbolic polynomial,
```
>>syms x ↵
>>y=x^4-7*x^3+21*x^2-37*x-30; ↵    ← Assigning symbolic polynomial to y
>>subs(y,x,'-3') ↵                 ← Substitution of x by -3 in y
```

ans =

540

3.5 Degree, leading and trailing coefficients of a polynomial

Degree of a polynomial is defined as the power of the highest power term. The polynomial $x^4 - 5x + 7$ has the degree 4. Function 'degree' located in Maple package can find the degree of a polynomial:

MATLAB Command
```
>>syms x ↵               ← Declaration of x as symbolic
>>p=x^4-5*x+7; ↵         ← Assigning polynomial to p
>>maple('degree',p,x) ↵  ← Execute the Maple function
```

ans =

4

The function also applies for product polynomials such as $\left(x^4 - 5x + \dfrac{8}{x^3} \right)(x^3 - 4)$. It does not apply for

indeterminate forms like $f(x) = \dfrac{2x - 2}{x + 9}$ or $(x + 3)(x - 4) - x^2$, which simplifies to $-x - 12$.
```
>>syms x ↵
>>f=(2*x-2)/(x+9); ↵
>>maple('degree',f,x) ↵
```

ans =

FAIL

Lowest degree of a polynomial can be found by function 'ldegree' (abbreviation of <u>l</u>owest <u>degree</u>). For $x^4 - 5x + 7$, the lowest degree is zero. It is determined as follows:
```
>>maple('ldegree',p,x) ↵    ← x^4 - 5x + 7 is assigned to p before
```

ans =

0

Coefficient of the highest power term of a polynomial is called leading coefficient. For example, leading coefficient of polynomial $-9x^5 + 5x^2 - 7x - \dfrac{3}{x} + \dfrac{5}{x^2}$ is -9. MATLAB function used for this purpose is 'lcoeff' (abbreviation of <u>l</u>eading <u>coeff</u>icient), which is implemented as follows:

>>syms x ↵
>>p=-9*x^5+5*x^2-7*x-3/x+5/x^2; ↵
>>maple('lcoeff',p,x) ↵

ans =

-9

On the contrary coefficient of the lowest power term of a polynomial is termed as the trailing coefficient. For the example at hand, the trailing coefficient is 5. Conduct the function 'tcoeff' (abbreviation of <u>t</u>railing <u>coeff</u>icient) for that:

>>maple('tcoeff',p,x) ↵

ans =

5

A list of advanced polynomial functions can be seen on execution of 'mhelp polynom'.

3.6 Multiplication of polynomials

Multiplication of polynomials is an algebraic problem. Multiplication has the other name convolution. Consider the following two polynomials for multiplication,

Polynomial 1: $2x^2 - 3x + 1$ and

Polynomial 2: $3x^2 - 4x + 1$.

Their multiplication is $(2x^2 - 3x + 1)(3x^2 - 4x + 1) = 6x^4 - 17x^3 + 17x^2 - 7x + 1$. To represent $2x^2 - 3x + 1$ and $3x^2 - 4x + 1$, row matrices [2 −3 1] and [3 −4 1] are used. The output should be [6 −17 17 −7 1]. Polynomial multiplication is accomplished by 'conv' (which is the abbreviation of <u>conv</u>olution). To obtain the multiplication of two polynomials,

MATLAB Steps:

1. *Assign the first polynomial to A,*
2. *Assign the second polynomial to B,*
3. *Use the command conv(A,B), and*
4. *Use the subroutine 'poly2str' to display the resulting polynomial.*

Conduct the function as follows:

>>A=[2 -3 1]; ↵   ← Assign $2x^2 - 3x + 1$ to A
>>B=[3 -4 1]; ↵   ← Assign $3x^2 - 4x + 1$ to B
>>R=conv(A,B) ↵   ← Keep the result in R

R =
     6 -17   17   -7   1

when the string required,

>>poly2str(R,'x') ↵

ans =
     6 x^4 - 17 x^3 + 17 x^2 - 7 x + 1

If you had three polynomials to be multiplied, for example,

Polynomial 1: $2x^2 - 3x + 1$,

Polynomial 2: $3x^2 - 4x + 1$, and

Polynomial 3: $9x^2 + 7x + 2$,

where the multiplication is given by $(2x^2 - 3x + 1)(3x^2 - 4x + 1)(9x^2 + 7x + 2) = 54x^6 - 111x^5 + 46x^4 + 22x^3 - 6x^2 - 7x + 2$, there would be two 'conv' subroutines. Inner convolution multiplies polynomial 1 by polynomial 2 and the outer convolution multiplies the resulting one with the third polynomial. See the implementation below:

when coefficients required,

```
>>A=[2 -3 1]; ↵          ← The first polynomial
>>B=[3 -4 1]; ↵          ← The second polynomial
>>C=[9 7 2]; ↵           ← The third polynomial
>>R=conv(conv(A,B),C) ↵       ← Result is in R


R =
          54  -111   46   22   -6   -7   2
```

when string required,

```
>>poly2str(R, 'x') ↵


ans =
          54 x^6 - 111 x^5 + 46 x^4 + 22 x^3 - 6 x^2 - 7 x + 2
```

What if we want to multiply the polynomials in symbolic form. Take three polynomials' example:

```
>>syms x ↵
>>p1=2*x^2-3*x+1; ↵      ← Assign the first polynomial to p1
>>p2=3*x^2-4*x+1; ↵      ← Assign the second polynomial to p2
>>p3=9*x^2+7*x+2; ↵      ← Assign the third polynomial to p3
>>R=expand(p1*p2*p3); ↵  ← Expansion of multiplication is done by the command
                            'expand' and assign the result to R


>>pretty(R) ↵            ← Display the symbolic form


           6       5      4      3      2
       54 x  - 111 x  + 46 x  + 22 x  - 6 x  - 7 x + 2
>>sym2poly(R) ↵          ← To see the polynomial coefficients


ans =
          54  -111   46   22   -6   -7   2
```

3.7 Division of polynomials

In the previous article convolution is applied for multiplication of polynomials. The reverse process of convolution can accomplish the division of polynomials, that is, deconvolution. Its MATLAB counterpart is 'deconv' (abbreviation of deconvolution). Suppose polynomial $x^4 - x + 1$ is to be divided by $x^2 + x + 1$. Perform the long division as follows:

$$
\begin{array}{r}
x^2 - x \\
x^2 + x + 1{\overline{\smash{\big)}\,x^4 - x + 1}} \\
\underline{x^4 + x^3 + x^2} \\
-x^3 - x^2 - x + 1 \\
\underline{-x^3 - x^2 - x} \\
1
\end{array}
$$

From the long division, the polynomials $x^2 + x + 1$, $x^4 - x + 1$, $x^2 - x$, and 1 are called divider, dividend, quotient, and remainder respectively. In MATLAB notation row vectors [1 1 1], [1 0 0 −1 1], [1 −1 0], and [1] stand for divider, dividend, quotient, and remainder respectively. Deconvolution subroutine has two input arguments, namely, dividend and divider polynomials. The outputs are quotient and remainder polynomials. Two matrices are required to assign the two output polynomials. Say, Dv, Dr, R, and Q stand for dividend, divider, remainder and quotient respectively. To perform division using the polynomial coefficients,

MATLAB Steps:

1. *Assign the dividend polynomial to some vector Dv,*
2. *Assign the divider polynomial to some vector Dr, and*
3. *Use the command [Q R]=deconv(Dv,Dr).*

Carry out the division in coefficient form as follows:

MATLAB Command

To display the string,

```
>>Dv=[1 0 0 -1 1]; ↵                    >>poly2str(Q, 'x') ↵
>>Dr=[1 1 1]; ↵
>>[Q R]=deconv(Dv,Dr) ↵                 ans =
                                              x^2 - 1 x
Q =                                     >>poly2str(R,'x') ↵
         1   -1   0
R =                                     ans =
       0   0   0   0   1                       1
```

The symbolic quotient and remainder polynomials are dealt with the functions 'quo' (abbreviation of <u>quo</u>tient) and 'rem' (abbreviation of <u>rem</u>ainder) respectively. These functions are located in Maple package. Verify the last two functions as follows:

```
>>syms x ↵                  ← Declare $x$ as symbolic
>>Dv=x^4-x+1; ↵             ← Assign $x^4 - x + 1$ to Dv
>>Dr=x^2+x+1; ↵             ← Assign $x^2 + x + 1$ to Dr
>>Q=maple('quo',Dv,Dr,x) ↵  ← Calling 'quo' for the quotient polynomial

Q =

x^2-x
>>R=maple('rem',Dv,Dr,x) ↵  ← Calling 'rem' for the remainder polynomial

R =

   1
```

The symbolic form and polynomial coefficients can be seen by exercising the commands 'pretty' and 'sym2poly' on Q and R respectively. Division of two variables' polynomial is also possible by dint of the commands 'quo' and 'rem'. Assume that dividend and divider polynomials are $3x^4 - 8y^4$ and $x - 2y$ respectively. Long division yields quotient and remainder polynomials as $3x^3 + 6x^2y + 12xy^2 + 24y^3$ and $40y^4$ respectively, where x is the prime variable of the given polynomial:

Command for two variables' polynomial division,

```
>>syms x y ↵                ← Declare $x$ and $y$ as symbolic
>>Dv=3*x^4-8*y^4; ↵         ← Assign $3x^4 - 8y^4$ to Dv
>>Dr=x-2*y; ↵               ← Assign $x - 2y$ to Dr
>>Q=maple('quo',Dv,Dr,x); ↵ ← Quotient is assigned to Q
>>pretty(Q) ↵               ← Display symbolic form of Q

         3         2        2        3
      3 x  + 6 y x  + 12 y  x  + 24 y
>>R=maple('rem',Dv,Dr,x); ↵ ← Remainder is assigned to R
>>pretty(R) ↵               ← Display symbolic form of R
         4
      40 y
```

Observe that the last argument 'x' of 'quo' or 'rem' represents the prime variable under consideration.

3.8 Factorization of a polynomial or expression

Factorization is required to express a function or polynomial as the product of linear terms. To factorize a polynomial or expression p,

68

MATLAB Steps:

1. *Enter the variables of p as symbolic,*
2. *Describe the expression or polynomial p, and*
3. *Use the command factor (p).*

Say, we have one variable polynomial $27x^3 - 189x^2 + 441x - 343$. From $(a-b)^3 = a^3 - 3a^2b + 3ab^2 - b^3$, the factored form of $27x^3 - 189x^2 + 441x - 343$ is $(3x-7)^3$. See its implementation below:

MATLAB Command

```
>>syms x ↵
>>p=27*x^3-189*x^2+441*x-343; ↵
>>R=factor(p); ↵                    ← String form of the factor is in R
>>pretty(R) ↵
                    3
          (3 x - 7)
```

By making use of the subroutine two variables' polynomial like $p(x,y) = 5x^3 - 2x^2y - 37xy^2 + 42y^3$ can be factored too. It is given that $p(x,y) = 5x^3 - 2x^2y - 37xy^2 + 42y^3 = (x+3y)(x-2y)(5x-7y)$. Have it as follows:

```
>>syms x y ↵
>>p=5*x^3-2*x^2*y-37*x*y^2+42*y^3; ↵
>>R=factor(p); ↵
>>pretty(R) ↵

          (-7 y + 5 x) (-2 y + x) (3 y + x)
```

Polynomials having rational coefficients are also dealt with the function. One example of that is $a^2 - 2.5a + 1 = \frac{1}{2}(2a-1)(a-2)$:

```
>>syms a ↵
>>p=a^2-2.5*a+1; ↵
>>R=factor(p); ↵
>>pretty(R) ↵

          1/2 (2 a - 1) (a - 2)
```

Factorization involving the symbolic constants like $a^3 \cos x + ya^2 - ay \cos x - y^2 = (a \cos x + y)(a^2 - y)$ is not an exception:

```
>>syms a x y ↵
>>p=a^3*cos(x)+y*a^2-a*y*cos(x)-y^2; ↵
>>R=factor(p); ↵
>>pretty(R) ↵
                    2
          (a  - y ) (cos(x) a + y)
```

3.9 Expansion of a powered polynomial

The function 'expand' expands powered polynomials containing binomial, trinomial, or higher terms. We are going to check $(2x - 3y)^4 = 16x^4 - 96x^3y + 216x^2y^2 - 216xy^3 + 81y^4$:

MATLAB Command

```
>>syms x y ↵
>>R=expand((2*x-3*y)^4); ↵        ← Expansion is stored in R
>>pretty(R) ↵
```

$$16\,x^4\ -\ 96\,x^3\,y + 216\,x^2\,y^2\ -\ 216\,x\,y^3\ +81\,y^4$$

As a trinomial expansion, take example of $(x+y+z)^3 = x^3 + 3x^2y + 3x^2z + 3xy^2 + 6xyz + 3xz^2 + y^3 + 3y^2z + 3yz^2 + z^3$ and implement it as follows:

```
>>syms x y z ↵
>>R=expand((x+y+z)^3); ↵
>>pretty(R) ↵
```

$$x^3 + 3\,x^2\,y + 3\,x^2\,z + 3\,x\,y^2 + 6\,x\,y\,z + 3\,x\,z^2 + y^3 + 3\,y^2\,z + 3\,y\,z^2 + z^3$$

3.10 Horner expansion of a polynomial

Horner representation is the nested form of a polynomial. In some numerical analysis problems, the form is very useful. To obtain the Horner expansion of a polynomial,

MATLAB Steps:
1. *Declare the polynomial variable as symbolic and*
2. *Use the command horner(polynomial string).*

Consider the test polynomial $x^4 - 3x^3 + 2x^2 - x - 7$. We arrange the polynomial as $x^4 - 3x^3 + 2x^2 - x - 7 = -7 - x + 2x^2 - 3x^3 + x^4 = -7 + (-1 + 2x - 3x^2 + x^3)x = -7 + (-1 + (2 - 3x + x^2)x)x = -7 + (-1 + (2 + (-3 + x)x)x)x$, which is the Horner form of $x^4 - 3x^3 + 2x^2 - x - 7$. Obtain it as follows:

MATLAB Command
```
>>syms x ↵
>>H=horner(x^4-3*x^3+2*x^2-x-7); ↵   ← The string form is stored in H
>>pretty(H) ↵
```

$$-7 + (-1 + (2 + (-3 + x)\,x)\,x)\,x$$

Observe the exceptional case for the polynomial $7x^5 - \frac{2}{5}x^2 - x - \frac{6}{7}$. The polynomial has some power terms missing (x^4 and x^3), therefore, $7x^5 - \frac{2}{5}x^2 - x - \frac{6}{7} = -\frac{6}{7} + \left(-1 - \frac{2}{5}x + 7x^4\right)x = -\frac{6}{7} + \left(-1 + \left(-\frac{2}{5} + 7x^3\right)x\right)x$. For this exception:

```
>>H=horner(7*x^5-2/5*x^2-x-6/7); ↵
>>pretty(H) ↵
```

$$- 6/7 + (-1 + (- 2/5 + 7\,x^3)\,x)\,x$$

3.11 Numerator and denominator from an expression

One important use of numerator-denominator separation can be pole-zero analysis from the linear factors or the method can be beneficial to convert sum of rational forms to a single rational form. To obtain the numerator and denominator from an expression,

MATLAB Steps:
1. *Declare the expression variables as symbolic and*
2. *Use the command [n d]=numden(expression in terms of string), where n and d stand for numerator and denominator respectively.*

The subroutine 'numden' is the abbreviated form of <u>num</u>erator and <u>den</u>ominator. Since there are two outputs, numerator and denominator, the number of output arguments of 'numden' is two. Let us see the response of the subroutine for the expression $\dfrac{x^2}{y} + yx + \dfrac{5y}{7x}$, which turns out to $\dfrac{7x^3 + 7x^2 y^2 + 5y^2}{7xy}$ on simplification. Numerator and denominator are $7x^3 + 7x^2 y^2 + 5y^2$ and $7xy$ respectively. There are two independent variables – x and y. MATLAB solution is shown below:

MATLAB Command
>>syms x y ↵
>>[n d]=numden(x^2/y+y*x + 5/7*y/x) ↵

n =

7*x^3+7*y^2*x^2+5*y^2

d =

7*y*x

Command 'pretty' can be used to display numerator and denominator strings in mathematical form. Remember that even though we do not like string form but for the evaluation of an expression in MATLAB, the string form is necessary not the pretty form. As another example, we have $\dfrac{\cos A}{\cos A - 5} - \dfrac{\sin A}{\sin A + 3}$. It is better to split the given function as f1 and f2, hence, f1=$\dfrac{\cos A}{\cos A - 5}$, f2=$\dfrac{\sin A}{\sin A + 3}$, and f1–f2=$\dfrac{\cos A(\sin A + 3) - \sin A(\cos A - 5)}{(\cos A - 5)(\sin A + 3)}$ = $\dfrac{3\cos A + 5\sin A}{(\cos A - 5)(\sin A + 3)}$. Implement that as follows:

>>syms A ↵
>>f1=cos(A)/(cos(A)-5); ↵
>>f2=sin(A)/(sin(A)+3); ↵
>>[n d]=numden(f1-f2); ↵   ← String forms of n and d are returned
>>pretty(n) ↵

    3 cos(A) + 5 sin(A)
>>pretty(d) ↵

    (cos(A) - 5) (sin(A) + 3)

One can transact other manipulations, for example, substitution, multiplication, division...etc on the returned numerator and denominator strings.

3.12 Least common multiplier of integers and functions

Function 'lcm' finds the least common multiplier (LCM) of integers and functions. There can be three possible cases of LCM – two integers, integers of two matrices, or a matrix of integers. To obtain LCM,
MATLAB Step:
Use the command lcm(matrix1,matrix2).
Let us take two integers, 27 and 84. One can find LCM by factoring 27 and 84 (27=3×3×3 and 84=3×2×2×7), which is 3×3×3×2×2×7=756 and is found as follows:
MATLAB Command
>>lcm(27,84) ↵

ans =

756

Find LCM of like positional elements of A=$\begin{bmatrix} 34 \\ 12 \\ 10 \end{bmatrix}$ and B=$\begin{bmatrix} 2 \\ 8 \\ 3 \end{bmatrix}$. LCM of (34,2), (12,8), and (10,3) are 34, 24, and

30 respectively. The output should look like $\begin{bmatrix} 34 \\ 24 \\ 30 \end{bmatrix}$:

>>A=[34 12 10]'; ↵
>>B=[2 8 3]'; ↵
>>lcm(A,B) ↵

ans =

34

24

30

Rectangular matrices can also be the assignees of A and B. Another situation when LCM of all elements of a matrix is to be found can happen. Programming is necessary in this regard. To determine LCM of all elements in a matrix,

MATLAB Steps:

1. *If given matrix A is rectangular, convert it to column one by A=A(:), ignore this step for row or column matrix*
2. *Determine the number of elements N in A by N=length(A)*
3. *Initialize LCM to variable L by L=A(1)*
4. *Use the command for i=1:N, L=lcm(L,A(i)); end*
5. *Display the output stored in L.*

For illustration, LCM of all elements of A=$\begin{bmatrix} 10 & 2 \\ 7 & 9 \\ 18 & 3 \end{bmatrix}$ is 630. Have it as follows:

>>A=[10 2;7 9;18 3]; ↵
>>A=A(:); ↵
>>N=length(A); ↵
>>L=A(1); ↵
>>for i=1:N, L=lcm(L,A(i)); end ↵
>>L ↵

L =

630

Use the Maple procedure to find symbolic LCM. Two expressions $f(x,y) = x^4 - y^4$ and $h(x,y) = x^2 - y^2$ are given to see the implementation. Factorization provides $f(x,y) = x^4 - y^4 = (x^2 + y^2)(x+y)(x-y)$ and $h(x,y) = x^3 - y^3 = (x-y)(x^2 + xy + y^2)$, hence, LCM is $(x^2 + y^2)(x+y)(x-y)\ (x^2 + xy + y^2) = x^6 + x^5 y - x^2 y^4 + x^4 y^2 - xy^5 - y^6$. Exercise it as follows:

>>syms x y ↵
>>f=x^4-y^4; ↵
>>h=x^3-y^3; ↵
>>L=maple('lcm',f,h); ↵
>>pretty(L) ↵

```
      6     5   4 2     2 4     5     6
     x  + y x  - y   x  + y   x  - y   x - y
```

3.13 Greatest common divider of integers and functions

Finding the greatest common divider (GCD) is very similar to the finding of LCM, which is discussed in the previous article. Just substitute 'gcd' instead of 'lcm' for the implementation. Like positional elements of $A=\begin{bmatrix} 10 & 2 \\ 7 & 9 \end{bmatrix}$ and $B=\begin{bmatrix} 7 & 9 \\ 3 & 12 \end{bmatrix}$ have the GCD as matrix $\begin{bmatrix} 1 & 1 \\ 1 & 3 \end{bmatrix}$:

MATLAB Command
```
>>A=[10 2; 7 9];  ↵
>>B=[7 9;3 12];  ↵
>>gcd(A,B)  ↵

ans =
          1    1
          1    3
```

Again, GCD of all elements of column matrix $C=\begin{bmatrix} 45 \\ 12 \\ 18 \\ 75 \end{bmatrix}$ is 3:

```
>>C=[45 12 18 75]';  ↵
>>N=length(C);  ↵
>>G=C(1);  ↵
>>for i=1:N, G=gcd(G,C(i)); end  ↵          ← G contains GCD
>>G  ↵

G =
    3
```

For symbolic GCD examples, choose the same functions as we have for the LCM. By inspection, GCD of $f(x,y)=(x^2+y^2)(x+y)(x-y)$ and $h(x,y)=(x-y)(x^2+xy+y^2)$ is $x-y$:

```
>>syms x y  ↵
>>f=x^4-y^4;  ↵
>>h=x^3-y^3;  ↵
>>maple('gcd',f,h)  ↵

ans =

-y+x
```

3.14 Algebraic substitution of one variable in other

Suppose we have one equation $3x-2y=3$ and one function $e^{2x^2-3xy+5y^2}$. We want to eliminate x from function $e^{2x^2-3xy+5y^2}$, where x is obtained from equation $3x-2y=3$ as $x=\dfrac{2y+3}{3}$. Substitute $x=\dfrac{2y+3}{3}$ in

$e^{2x^2-3xy+5y^2}$ to get $e^{2\left(\frac{2y+3}{3}\right)^2-3\left(\frac{2y+3}{3}\right)y+5y^2}=e^{\frac{35}{9}y^2-\frac{y}{3}+2}$. MATLAB function 'algsubs' (abbreviation of <u>alg</u>ebraic <u>subs</u>titution), which is located in Maple, performs this as follows:

MATLAB Command
```
>>syms x y  ↵             ← Declaration of x and y as symbolic
>>e='3*x-2*y=3';  ↵        ← Assigning 3x-2y=3 to e
>>f='exp(2*x^2-3*x*y+5*y^2)';  ↵ ← Assigning e^{2x²-3xy+5y²} to f
>>R=maple('algsubs',e,f,x);  ↵  ← Elimination of x from f and result is assigned to R
>>pretty(R)  ↵
```

$$\exp(-1/3\,y + 35/9\,y^2 + 2)$$

3.15 Forming complete square of an expression

Assume that the expression $7x^2 - 24xy^2 + 9y^4$ is given. Some algebraic manipulations turn out $7x^2 - 24xy^2 + 9y^4$ to complete square. We carry out that as $7x^2 - 24xy^2 + 9y^4 = 7\left(x^2 - \dfrac{24}{7}xy^2 + \dfrac{9}{7}y^4\right) =$

$7\left[x^2 - 2x\dfrac{12}{7}y^2 + \left(\dfrac{12}{7}y^2\right)^2 + \dfrac{9}{7}y^4 - \left(\dfrac{12}{7}y^2\right)^2\right] = 7\left(x - \dfrac{12y^2}{7}\right)^2 - \dfrac{81y^4}{7}$. The function 'completesquare' of student

package helps us do that:

MATLAB Command

>>maple('with','student'); ↵   ← Activate the student package
>>syms x y ↵
>>e=7*x^2-24*x*y^2+9*y^4; ↵   ← Assign $7x^2 - 24xy^2 + 9y^4$ to e
>>R=maple('completesquare',e,x); ↵   ← Result is assigned to R
>>pretty(R) ↵

$$7\,(x - 12/7\,y^2)^2 - 81/7\,y^4$$

This function is specially useful when one has to express a function containing symbolic coefficients. Such an

example is $ax^2 + bx + c = a\left(x + \dfrac{b}{2a}\right)^2 + c - \dfrac{b^2}{4a}$. The solution is shown below:

>>maple('with','student'); ↵
>>syms a b c x ↵
>>e=a*x^2+b*x+c; ↵
>>R=maple('completesquare',e,x); ↵
>>pretty(R) ↵

$$a\,(x + 1/2\ b/a)^2 - 1/4\ \frac{b^2}{a} + c$$

3.16 Partial fraction in coefficient and symbolic forms

Any function $f(x)$, which is the division of two polynomials like $\dfrac{N(x)}{D(x)}$, where $N(x)$ and $D(x)$ are

numerator and denominator polynomials respectively, can be written as $f(x) = K_m x^m + K_{m-1}x^{m-1} + K_{m-2}x^{m-2}$

$+ \ldots\ldots + K_1 x + K_0 + \dfrac{R_1}{x - P_1} + \dfrac{R_2}{x - P_2} + \dfrac{R_3}{x - P_3} + \ldots\ldots + \dfrac{R_n}{x - P_n}$ if factors of denominator are distinct, that is

$P_1, P_2, P_3, \ldots.$, and P_n are unequal, or as $K_m x^m + K_{m-1}x^{m-1} + K_{m-2}x^{m-2} + \ldots\ldots\ldots + K_1 x + K_0 + \dfrac{R_1}{x - P_0} + \dfrac{R_2}{(x - P_0)^2} +$

$\dfrac{R_3}{(x - P_0)^3} + \ldots. + \dfrac{R_p}{(x - P_0)^p} + \dfrac{R_{p+1}}{x - P_1} + \dfrac{R_{p+2}}{x - P_2} + \ldots. + \dfrac{R_{p+q}}{x - P_q}$ if factors of denominator are repetitive. K's, R's, and P's

are called gains, remnants, and poles of $f(x)$ respectively. Gain polynomial $K_m x^m + K_{m-1}x^{m-1}$

74

$+K_{m-2}x^{m-2}+.........+K_1x+K_0$ does not exist if the degree of numerator is less than that of the denominator. The subroutine used for partial fraction in coefficient form is 'residue' (abbreviation of residual due to polynomial division). It has two input arguments — the first one is the numerator and the other is the denominator polynomial. The number of output arguments is 3 — residuals (R), poles (P), and gains (K, it can be compared to quotient after division). To find partial fraction of $f(x)$ in coefficient form,

MATLAB Steps:

1. *Assign the numerator and denominator polynomials to N and D respectively and*
2. *Use command [R P K]=residue (N, D), where R, P, and K stand for residuals, poles, and gains of f(x) respectively.*

Let us start with partial fraction of $f(x)=\dfrac{N(x)}{D(x)}=\dfrac{x}{(3+2x)(x-1)}$. Setting $D(x)$ to 0 yields $x=-\dfrac{3}{2}$ and 1 — these

are the poles of $f(x)$, namely, $P_1=-\dfrac{3}{2}$ and $P_2=1$. The pole vector is going to be P=[−1.5   1]. Now we have

$f(x)=\dfrac{x}{(3+2x)(x-1)}=\dfrac{\dfrac{x}{2}}{\left(\dfrac{3}{2}+x\right)(x-1)}\equiv\dfrac{A}{x+\dfrac{3}{2}}+\dfrac{B}{x-1}$, where $A=f(x)$ excluding denominator $\left(x+\dfrac{3}{2}\right)\Big|_{x=-\frac{3}{2}}=$

$\dfrac{-\dfrac{3}{2}\times\dfrac{1}{2}}{\left(-\dfrac{3}{2}-1\right)}=\dfrac{3}{10}$ and $B=f(x)$ excluding denominator $(x-1)\Big|_{x=1}=\dfrac{\dfrac{1}{2}}{\left(\dfrac{3}{2}+1\right)}=\dfrac{1}{5}$. The residual vector seems to be R

=[0.3   0.2]. Since the degrees of $N(x)$ and $D(x)$ are 1 and 2 respectively, all K's are zeroes. The gain vector is just K=[ ] that means empty matrix. As input, we have $N(x)=x=[1\quad0]$ and $D(x)=(3+2x)(x-1)=2x^2+x-3=[2\quad1\quad-3]$. To avoid polynomial multiplication, function 'sym2poly' can be used. Working out of this partial fraction along with symbolic form is shown below:

MATLAB Command

in coefficient form,

>>N=[1 0]; ↵
>>D=[2 1 -3]; ↵
>>[R P K]=residue(N,D) ↵

R =
        0.3000
        0.2000
P =
        -1.5000
        1.0000
K =
        [ ]

using the symbolic notation,

>>syms x ↵
>>N=sym2poly(x); ↵
>>D=sym2poly((2*x+3)*(x-1)); ↵
>>[R P K]=residue(N,D) ↵

R =
        0.3000
        0.2000
P =
        -1.5000
        1.0000
K =
        [ ]

Notice that the outputs of 'residue' are in column matrix form.

The second example is the partial fraction of $\dfrac{3x^5-x}{(3x-2)(2x-1)(-x+5)}$. Multiplication of the denominator

and rationalization yield us $\dfrac{-\dfrac{x^5}{2}+\dfrac{x}{6}}{x^3-\dfrac{37}{6}x^2+\dfrac{37}{6}x-\dfrac{5}{3}}$. The highest degrees of numerator and denominator are 5 and

3 respectively. So, the gain polynomial is a polynomial of degree 5−3=2 that means the polynomial is $K_2x^2 + K_1x + K_0$. From the long division, we have

$$-\frac{1}{2}x^2 - \frac{37}{12}x - \frac{1147}{72} \quad \leftarrow \; gain(K)$$

$$x^3 - \frac{37}{6}x^2 + \frac{37}{6}x - \frac{5}{3} \overline{\Big) -\frac{x^5}{2} + \frac{x}{6}}$$

$$-\frac{x^5}{2} + \frac{37}{12}x^4 - \frac{37}{12}x^3 + \frac{5x^2}{6}$$

$$-\frac{37}{12}x^4 + \frac{37}{12}x^3 - \frac{5}{6}x^2 + \frac{x}{6}$$

$$-\frac{37}{12}x^4 + \frac{1369}{72}x^3 - \frac{1369}{72}x^2 + \frac{185}{36}x$$

$$-\frac{1147x^3}{72} + \frac{1309x^2}{72} - \frac{179x}{36}$$

$$-\frac{1147x^3}{72} + \frac{42439x^2}{432} - \frac{42439x}{432} + \frac{5735}{216}$$

$$-\frac{34585x^2}{432} + \frac{40291x}{432} - \frac{5735}{216} \quad \leftarrow \; remainder \; but \; not \; R$$

and $\dfrac{3x^5 - x}{(3x-2)(2x-1)(-x+5)} = \dfrac{-\dfrac{x^5}{2} + \dfrac{x}{6}}{x^3 - \dfrac{37}{6}x^2 + \dfrac{37}{6}x - \dfrac{5}{3}} = K_2x^2 + K_1x + K_0 + \dfrac{-\dfrac{34585x^2}{432} + \dfrac{40291x}{432} - \dfrac{5735}{216}}{\left(x-\dfrac{2}{3}\right)\left(x-\dfrac{1}{2}\right)(x-5)} = -\dfrac{1}{2}x^2 -$

$\dfrac{37}{12}x - \dfrac{1147}{72} + \dfrac{R_1}{x-\dfrac{2}{3}} + \dfrac{R_2}{x-\dfrac{1}{2}} + \dfrac{R_3}{x-5},$ where $K_2 = -\dfrac{1}{2},$ $K_1 = -\dfrac{37}{12},$ $K_0 = -\dfrac{1147}{72},$ $R_1 =$

$\dfrac{-\dfrac{34585x^2}{432} + \dfrac{40291x}{432} - \dfrac{5735}{216}}{\left(x-\dfrac{1}{2}\right)(x-5)}\Bigg|_{x=\frac{2}{3}} = -\dfrac{22}{351} = -0.0627,$ $R_2 = \dfrac{-\dfrac{34585x^2}{432} + \dfrac{40291x}{432} - \dfrac{5735}{216}}{\left(x-\dfrac{2}{3}\right)(x-5)}\Bigg|_{x=\frac{1}{2}} = \dfrac{13}{144} = 0.0903,$ and R_3

$= \dfrac{-\dfrac{34585x^2}{432} + \dfrac{40291x}{432} - \dfrac{5735}{216}}{\left(x-\dfrac{2}{3}\right)\left(x-\dfrac{1}{2}\right)}\Bigg|_{x=5} = -\dfrac{9370}{117} = -80.0855.$ The gain, pole, and residual vectors are K=

$[-\dfrac{1}{2} \quad -\dfrac{37}{12} \quad -\dfrac{1147}{72}]$=[−0.5 −3.0833 −15.9306], P=[5 $\dfrac{2}{3}$ $\dfrac{1}{2}$]=[5 0.6667 0.5], and R=[−80.0855

−0.0627 0.0903] respectively. Notice that corresponding to poles x =5, $\dfrac{2}{3}$, and $\dfrac{1}{2}$, residuals are R_3, R_1, and

R_2 respectively. Following is the implementation:

```
>>syms x ↵
>>N=sym2poly(3*x^5-x); ↵
>>D=sym2poly((3*x-2)*(2*x-1)*(-x+5)); ↵
>>[R P K]=residue(N,D) ↵

R =
        -80.0855
        -0.0627
         0.0903
P =
         5.0000
```

$$K = \begin{matrix} 0.6667 \\ 0.5000 \\ \\ -0.5000 \quad -3.0833 \quad -15.9306 \end{matrix}$$

The third example is the partial fraction of $\dfrac{-2x^7}{(x+1)^4}$ where we have repetitive roots. The gain polynomial is a

polynomial of degree 7−4=3, hence, $\dfrac{-2x^7}{(x+1)^4} = K_3 x^3 + K_2 x^2 + K_1 x + K_0 + \dfrac{R_1}{x+1} + \dfrac{R_2}{(x+1)^2} + \dfrac{R_3}{(x+1)^3} + \dfrac{R_4}{(x+1)^4}$. For

this function, see the long division as follows:

$$
\begin{array}{r}
-2x^3 + 8x^2 - 20x + 40 \leftarrow gain(K) \\
\hline
x^4 + 4x^3 + 6x^2 + 4x + 1 \overline{) -2x^7 } \\
-2x^7 - 8x^6 - 12x^5 - 8x^4 - 2x^3 \\
\hline
+8x^6 + 12x^5 + 8x^4 + 2x^3 \\
+8x^6 + 32x^5 + 48x^4 + 32x^3 + 8x^2 \\
\hline
-20x^5 - 40x^4 - 30x^3 - 8x^2 \\
-20x^5 - 80x^4 - 120x^3 - 80x^2 - 20x \\
\hline
+40x^4 + 90x^3 + 72x^2 + 20x \\
+40x^4 + 160x^3 + 240x^2 + 160x + 40 \\
\hline
-70x^3 - 168x^2 - 140x - 40
\end{array}
$$

In the last division, the divider is $(x+1)^4 = x^4 + 4x^3 + 6x^2 + 4x + 1$. Clearly, the gain and pole polynomials are K=

[−2   8   −20   40] and P=[−1   −1   −1   −1] respectively. Use long division to write $\dfrac{-2x^7}{(x+1)^4} =$

$K_3 x^3 + K_2 x^2 + K_1 x + K_0 + \dfrac{R_1}{x+1} + \dfrac{R_2}{(x+1)^2} + \dfrac{R_3}{(x+1)^3} + \dfrac{R_4}{(x+1)^4}$, where $\dfrac{-70x^3 - 168x^2 - 140x - 40}{(x+1)^4} = \dfrac{R_1}{x+1} + \dfrac{R_2}{(x+1)^2} +$

$\dfrac{R_3}{(x+1)^3} + \dfrac{R_4}{(x+1)^4}$. Heaviside formula states that if $D(x)$ has a root r of multiplicity m, partial fraction

expansion of $\dfrac{N(x)}{D(x)}$ will be of the form $\dfrac{N(x)}{D(x)} = \dfrac{R_1}{x-r} + \dfrac{R_2}{(x-r)^2} + \dfrac{R_3}{(x-r)^3} + ... + \dfrac{R_m}{(x-r)^m}$, provided that the degree

of $N(x)$ is less than that of $D(x)$. On defining $C(x) = (x-r)^m \dfrac{N(x)}{D(x)}$, one can write $R_m = C(r)$, $R_{m-1} = \dfrac{1}{1!} \dfrac{dC(x)}{dx} \bigg|_{x=r}$,

$R_{m-2} = \dfrac{1}{2!} \dfrac{d^2 C(x)}{dx^2} \bigg|_{x=r}$, $R_{m-3} = \dfrac{1}{3!} \dfrac{d^3 C(x)}{dx^3} \bigg|_{x=r}$ .... $R_1 = \dfrac{1}{(m-1)!} \dfrac{d^{m-1} C(x)}{dx^{m-1}} \bigg|_{x=r}$. The problem at hand tells us $C(x) =$

$(x+1)^4 \dfrac{-70x^3 - 168x^2 - 140x - 40}{(x+1)^4} = -70x^3 - 168x^2 - 140x - 40$, where $r = -1$ and $m = 4$, hence, $R_4 =$

$-70x^3 - 168x^2 - 140x - 40 \bigg|_{x=-1} = 2$, $R_3 = \dfrac{1}{1!} \dfrac{d[-70x^3 - 168x^2 - 140x - 40]}{dx} \bigg|_{x=r} = -210x^2 - 336x - 140 \bigg|_{x=-1} = -14$, $R_2 =$

$\dfrac{1}{2!} \dfrac{d^2[-70x^3 - 168x^2 - 140x - 40]}{dx^2} \bigg|_{x=r} = \dfrac{1}{2!} \times [-420x - 336 \bigg|_{x=-1}] = 42$, and $R_1 = \dfrac{1}{3!} \dfrac{d^3[-70x^3 - 168x^2 - 140x - 40]}{dx^3} \bigg|_{x=r} =$

$\dfrac{1}{3!} \times [-420 \bigg|_{x=-1}] = -70$. So, the residual vector is R=[−70   42   −14   2]. Following commands can save this

computation:

```
>>syms x ↵
>>N=sym2poly(-2*x^7); ↵
>>D=sym2poly((x+1)^4); ↵
>>[R P K]=residue(N,D) ↵
```

R =
      -70.0000
       42.0000
      -14.0000
        2.0000
P =
       -1.0000
       -1.0000
       -1.0000
       -1.0000
K =
     -2   8   -20   40

The last example considers poles being complex numbers, for instance, $\dfrac{1}{x^3+1}$. Factorization of x^3+1 gives us

$$(x+1)(x^2-x+1) \quad = \quad (x+1)\left[\left(x-\frac{1}{2}\right)^2+\left(\frac{\sqrt{3}}{2}\right)^2\right] \quad = \quad (x+1)\left(x-\frac{1}{2}+i\frac{\sqrt{3}}{2}\right)\left(x-\frac{1}{2}-i\frac{\sqrt{3}}{2}\right), \quad \therefore \quad \frac{1}{x^3+1} \quad =$$

$$\frac{1}{(x+1)\left(x-\frac{1}{2}+i\frac{\sqrt{3}}{2}\right)\left(x-\frac{1}{2}-i\frac{\sqrt{3}}{2}\right)}=\frac{R_1}{x+1}+\frac{R_2}{x-\frac{1}{2}+i\frac{\sqrt{3}}{2}}+\frac{R_3}{x-\frac{1}{2}-i\frac{\sqrt{3}}{2}}, \quad \text{where} \quad R_1=\frac{1}{x^2-x+1}\Bigg|_{x=-1}=\frac{1}{3}, \quad R_2=$$

$$\frac{1}{(x+1)\left(x-\frac{1}{2}-i\frac{\sqrt{3}}{2}\right)}\Bigg|_{x=\frac{1}{2}-i\frac{\sqrt{3}}{2}}=-\frac{1}{6}+i\frac{\sqrt{3}}{6}, \text{ and } R_3=\frac{1}{(x+1)\left(x-\frac{1}{2}+i\frac{\sqrt{3}}{2}\right)}\Bigg|_{x=\frac{1}{2}+i\frac{\sqrt{3}}{2}}=-\frac{1}{6}-i\frac{\sqrt{3}}{6}. \text{ Finally, the gain,}$$

pole, and residual vectors are K=[ ], P=[−1 $\frac{1}{2}-i\frac{\sqrt{3}}{2}$ $\frac{1}{2}+i\frac{\sqrt{3}}{2}$]=[−1 0.5−i 0.8660   0.5+i 0.8660], and R

=[$\frac{1}{3}$ $-\frac{1}{6}+i\frac{\sqrt{3}}{6}$ $-\frac{1}{6}-i\frac{\sqrt{3}}{6}$]=[0.3333   −0.1667+i 0.2887 −0.1667−i 0.2887] respectively. Notice that the
output is not as our computation order. Our computation looks like the same as the output if we interchange the
first and third outputs. MATLAB finds the solution as follows:

```
>>syms x ↵
>>N=1; ↵
>>D=sym2poly(x^3+1); ↵
>>[R P K]=residue(N,D) ↵
```

R =
       -0.1667 - 0.2887i
       -0.1667 + 0.2887i
        0.3333
P =
        0.5000 + 0.8660i
        0.5000 - 0.8660i
       -1.0000
K =
        [ ]

⌻ *Symbolic form partial fraction*

We have discussed so far only the coefficient forms of partial fraction. Symbolic form partial fraction
is also possible using the Maple procedure. The sub function 'parfrac' (abbreviation of <u>par</u>tial <u>frac</u>tion) located
in 'convert' can be used in this regard. Examples chosen for previous discussion are as follows:

First example:

$$\frac{x}{(3+2x)(x-1)} = \frac{3}{5(2x+3)} + \frac{1}{5(x-1)}$$

Second example:

$$\frac{3x^5 - x}{(3x-2)(2x-1)(-x+5)} = -\frac{x^2}{2} - \frac{37x}{12} - \frac{1147}{72} - \frac{22}{117(3x-2)} + \frac{13}{72(2x-1)} - \frac{9370}{117(x-5)}$$

Third example:

$$\frac{-2x^7}{(x+1)^4} = -2x^3 + 8x^2 - 20x + 40 + \frac{2}{(x+1)^4} - \frac{14}{(x+1)^3} + \frac{42}{(x+1)^2} - \frac{70}{x+1}$$

Fourth example:

$$\frac{1}{x^3+1} = \frac{1}{3(x+1)} - \frac{x-2}{3(x^2-x+1)} \quad \text{(avoiding complex factorization)}$$

Following commands perform straightforward implementation of these examples:

MATLAB Command
for the first example,

```
>>syms x ↵              ← Declaration of symbolic variable
>>N=x; ↵                ← Assign numerator to N
>>D=(3+2*x)*(x-1); ↵    ← Assign denominator to D
>>R=maple('convert',N/D,'parfrac',x); ↵ ← Apply subfunction 'parfrac' and assign  output string to R
>>pretty(R) ↵                           ← Display the symbolic form
```

```
             1            1
      3/5 ----------- + 1/5 ----------
           3 + 2 x          x - 1
```

for the second example,

```
>>syms x ↵
>>N=3*x^5-x; ↵
>>D=(3*x-2)*(2*x-1)*(-x+5); ↵
>>R=maple('convert',N/D,'parfrac',x); ↵
>>pretty(R) ↵
```

```
         2   37       1147   22    1        13    1       9370    1
 - 1/2 x  - ------x - -------- - ------ ---------- + ---- ---------- - ---------- --------
            12         72     117  3 x - 2    72  2 x - 1     117   x - 5
```

for the third example,

```
>>syms x ↵
>>N=-2*x^7; ↵
>>D=(x+1)^4; ↵
>>R=maple('convert',N/D,'parfrac',x); ↵
>>pretty(R) ↵
```

```
        3     2              2         14        42      70
    -2 x  + 8 x  - 20 x + 40 + ----------- - ---------- + --------- - -------
                                   4            3           2        x+1
                               (x +1)        (x+1)       (x+1)
```

for the fourth example,

```
>>syms x ↵
>>F=1/(x^3+1); ↵
```

```
>>R=maple('convert',F,'parfrac',x); ↵
>>pretty(R) ↵
```

$$
\frac{1}{3} \frac{1}{x+1} - \frac{1}{3} \frac{-2+x}{x^2 - x + 1}
$$

3.17 Solutions of different equations

Very often a number of simultaneous linear, algebraic, or trigonometric equations need to be solved. Solutions of such equations are seen by virtue of function 'solve'. To find solution of a set of equations,

MATLAB Step:
Use the command solve('Equation-1','Equation-2',.......so on in string form,'unknowns').

⌦ *Simultaneous linear equations*

One example of simultaneous linear equations involving four unknowns (x, y, z, and u) can be

equation 1: $x - y - 3z + 2u = -8$,

equation 2: $9x + 8y - 7z + u = 5$,

equation 3: $9x + 4y + 2z = 23$, and

equation 4: $-3x + y - 6z + 7u = -12$.

In matrix form, the equations are written as $\begin{bmatrix} 1 & -1 & -3 & 2 \\ 9 & 8 & -7 & 1 \\ 9 & 4 & 2 & 0 \\ -3 & 1 & -6 & 7 \end{bmatrix} \begin{bmatrix} x \\ y \\ z \\ u \end{bmatrix} = \begin{bmatrix} -8 \\ 5 \\ 23 \\ -12 \end{bmatrix}$, wherefrom, the solution is $\begin{bmatrix} x \\ y \\ z \\ u \end{bmatrix}$

$$
= \begin{bmatrix} 1 & -1 & -3 & 2 \\ 9 & 8 & -7 & 1 \\ 9 & 4 & 2 & 0 \\ -3 & 1 & -6 & 7 \end{bmatrix}^{-1} \begin{bmatrix} -8 \\ 5 \\ 23 \\ -12 \end{bmatrix} = \begin{bmatrix} \frac{47}{248} & \frac{-3}{248} & \frac{21}{248} & \frac{-13}{248} \\ \frac{-173}{496} & \frac{107}{1488} & \frac{-5}{1488} & \frac{133}{1488} \\ \frac{-77}{496} & \frac{-133}{1488} & \frac{187}{1488} & \frac{85}{1488} \\ \frac{-1}{496} & \frac{-137}{1488} & \frac{215}{1488} & \frac{233}{1488} \end{bmatrix} \begin{bmatrix} -8 \\ 5 \\ 23 \\ -12 \end{bmatrix} = \begin{bmatrix} 1 \\ 2 \\ 3 \\ 1 \end{bmatrix}
$$. Obtain that from MATLAB as follows:

MATLAB Command

```
>>syms x y z u ↵               ← Declaration of symbolic variables
>>e1='x-y-3*z+2*u=-8'; ↵       ← Assign x - y - 3z + 2u = -8 to e1
>>e2='9*x+8*y-7*z+u=5'; ↵      ← Assign 9x + 8y - 7z + u = 5 to e2
>>e3='9*x+4*y+2*z=23'; ↵       ← Assign 9x + 4y + 2z = 23 to e3
>>e4='-3*x+y-6*z+7*u=-12'; ↵   ←Assign -3x + y - 6z + 7u = -12 to e4
>>d=solve(e1,e2,e3,e4,x,y,z,u) ↵ ← Application of 'solve' on e1, e2, e3, and e4

d =
    u: [1x1 sym]
    x: [1x1 sym]
    y: [1x1 sym]
    z: [1x1 sym]
```

Solution utilizing subroutine 'solve' may be kept as a structured array in workspace of MATLAB. Look at the above example. The output of 'solve' is assigned to d, which is a structured array and each component of the structured array (which are x, y, z, and u) is a 1×1 symbolic matrix. The commands 'd.x', 'd.y', 'd.z', and 'd.u' return contents of x, y, z, and u respectively:

To see the value of x,

>>d.x ↵

ans =

1

Value of y: Value of z: Value of u:

>>d.y ↵   >>d.z ↵   >>d.u ↵

ans =   ans =   ans =

2   3   1

⌗ Several algebraic equations

We present two more examples of the algebraic equations in which power of the variables is greater than 1.

A. $x^2 - y^2 - z^2 = a^2$, $2y + x = 4$, and $z - x = 2$

B. $\cos(x + 2) = y$ and $x = y^2$

Solution of set A:

From the second and third equations, we have $y = \dfrac{4-x}{2}$ and $z = x + 2$. Plugging y and z into the first equation yields $x^2 - \left(\dfrac{4-x}{2}\right)^2 - (x+2)^2 = a^2$. On simplification, it becomes $(x+4)^2 = -16 - 4a^2$, hence, $x = -4 \pm 2\sqrt{-4-a^2}$, whereupon, $y = 4 \mp \sqrt{-4-a^2}$ and $z = -2 \pm 2\sqrt{-4-a^2}$. Finally, solutions of the set are $(x,\ y,\ z) = \left(-4 + 2\sqrt{-4-a^2},\ \ 4 - \sqrt{-4-a^2},\ \ -2 + 2\sqrt{-4-a^2}\right)$ or $\left(-4 - 2\sqrt{-4-a^2},\ \ 4 + \sqrt{-4-a^2},\ \ -2 - 2\sqrt{-4-a^2}\right)$.

Arrange the solution in matrix form to have $(x,\ y,\ z) = \left(\begin{bmatrix} -4 + 2\sqrt{-4-a^2} \\ -4 - 2\sqrt{-4-a^2} \end{bmatrix}, \begin{bmatrix} 4 - \sqrt{-4-a^2} \\ 4 + \sqrt{-4-a^2} \end{bmatrix}, \begin{bmatrix} -2 + 2\sqrt{-4-a^2} \\ -2 - 2\sqrt{-4-a^2} \end{bmatrix}\right.$

$\Big)$. The first, second, and third equations are assigned to e1, e2, and e3 respectively. Instead of e1, e2, and e3, other variables could have been chosen. We have 4 symbolic variables — x, y, z and a. Unknowns under consideration are x, y, and z. Remember that the number of unknowns must be equal to the number of equations otherwise some variables are taken as constants and the other unknowns are solved in terms of the constants. Implementation is shown below:

MATLAB Command

```
>>e1='x^2-y^2-z^2=a^2'; ↵
>>e2='2*y+x=4'; ↵
>>e3='z-x=2'; ↵
>>d=solve(e1,e2,e3,'x','y','z') ↵

d =
      x: [2x1 sym]
      y: [2x1 sym]
      z: [2x1 sym]     ← It indicates that d is a structured array having three components and
                            each one is a 2×1 symbolic matrix
>>pretty(d.x) ↵


        [              2  1/2 ]
        [-4 - 2 (-4 - a  )    ]
        [                     ]
        [              2  1/2 ]
        [-4 + 2 (-4 - a  )    ]
```

>>pretty(d.y) ↵

```
[           2  1/2 ]
[4 + (-4 - a  )    ]
[                  ]
[           2  1/2 ]
[4 - (-4 - a  )    ]
```

>>pretty(d.z) ↵

```
[              2  1/2 ]
[-2 - 2 (-4 - a  )    ]
[                     ]
[              2  1/2 ]
[-2 + 2 (-4 - a  )    ]
```

Notice that second set of solutions comes first as MATLAB output.

Solution of set B:

Not all equations have the symbolic or close form solutions. Some equations are solved numerically. Equations in set B are such examples. Insert the second equation in the first to get $\cos(y^2 + 2) = y$. Half interval search or Newton method can be employed to find the numerical solution. Solution of $\cos(y^2 + 2) = y$ is $y = -0.9855$. Verify this as follows:

MATLAB Command

```
>>e1='cos(x+2)=y'; ↵
>>e2='x=y^2'; ↵
>>d=solve(e1,e2,'x','y') ↵

d =
        x: [1x1 sym]
        y: [1x1 sym]
>>d.x ↵

ans =

cos(2.9712694500986213723672217243464)^2
>>d.y ↵

ans =

cos(2.9712694500986213723672217243464)
```

To have the solution in decimal form, one can use the command 'double'.

Value of x,   Value of y,
   >>double(d.x) ↵   >>double(d.y) ↵

  ans =   ans =
      0.9713   -0.9855

☞ *Solution of trigonometric equations*

Subroutine 'solve' finds another application in solving trigonometric equations. Like algebraic equations, trigonometric equations are also put as strings. Solve the following trigonometric equations.

A . $\sec^2 x + 2\tan x = 0$

B . $3\cos x + 2\sin x = 2$

Solution of equation A:

Using identity $1 + \tan^2 x = \sec^2 x$, the equation can be written as $1 + \tan^2 x + 2\tan x = 0$, which implies $(\tan x + 1)^2 = 0$, or $\tan x = -1$. Within $-\pi \le x \le \pi$, the solution is $\dfrac{3\pi}{4}$ and $-\dfrac{\pi}{4}$. Representation of the equation is 'sec(x)^2+2*tan(x)=0' and the solution is shown below:

MATLAB Command

>>e='sec(x)^2+2*tan(x)=0'; ↵ ← Assign $\sec^2 x + 2\tan x = 0$ to e

>>d=solve(e) ↵

d =

[-1/4*pi]
[3/4*pi]

Solution of equation B:

This is an equation of the form $a\cos x + b\sin x = c$. Dividing both sides by $\sqrt{a^2 + b^2}$ turns out the equation as $\dfrac{3}{\sqrt{3^2 + 2^2}}\cos x + \dfrac{2}{\sqrt{3^2 + 2^2}}\sin x = \dfrac{2}{\sqrt{3^2 + 2^2}}$ or $\dfrac{3}{\sqrt{13}}\cos x + \dfrac{2}{\sqrt{13}}\sin x = \dfrac{2}{\sqrt{13}}$. Assume that $\sin\theta = \dfrac{3}{\sqrt{13}}$, it follows then, $\cos\theta = \dfrac{2}{\sqrt{13}}$. The equation becomes $\sin\theta\cos x + \cos\theta\sin x = \dfrac{2}{\sqrt{13}}$ or $\sin(x + \theta) = \dfrac{2}{\sqrt{13}}$, therefore, we have $x + \theta = \sin^{-1}\dfrac{2}{\sqrt{13}}$ or $\pi - \sin^{-1}\dfrac{2}{\sqrt{13}}$ within $-\pi \le x + \theta \le \pi$, where $\theta = \sin^{-1}\dfrac{3}{\sqrt{13}}$. Finally, the solution is $x = \sin^{-1}\dfrac{2}{\sqrt{13}} - \sin^{-1}\dfrac{3}{\sqrt{13}}$ or $\pi - \sin^{-1}\dfrac{2}{\sqrt{13}} - \sin^{-1}\dfrac{3}{\sqrt{13}}$, which simplifies to $-\tan^{-1}\dfrac{5}{12}$ or $\dfrac{\pi}{2}$. MATLAB solution is as follows:

MATLAB Command

>>e='3*cos(x)+2*sin(x)=2'; ↵

>>d=solve(e) ↵

d =

[1/2*pi]
[-atan(5/12)]

3.18 Expansion of trigonometric functions with multiple angles

From trigonometric identities, we know that the functions such as $\sin 3x$, $\cos 4x$, $\tan 3x$ ... etc can be expanded to smaller angles. To find the expansion of multiple angles in terms of smaller angles,

MATLAB Step:

Use the command expand ('trigonometric function as string').

Let us begin with an elementary identity $\cos 2A = 2\cos^2 A - 1$. See its string and symbolic implementations as follows:

MATLAB Command

In string form,   In symbolic form,

>>syms A ↵   >>pretty(expand(cos(2*A))) ↵

>>expand(cos(2*A)) ↵

ans =
$$2 \cos(A)^2 - 1$$

2*cos(A)^2-1

To have more examples, table 3.A is presented showing only the symbolic computation.

Table 3.A Expansion of trigonometric functions with multiple angles

Trigonometric notation	MATLAB Command in symbolic form
$\sin 2A = 2 \sin A \cos A$	>>syms A ↵ >>pretty(expand(sin(2*A))) ↵ 2 sin(A) cos(A)

$\cos 3A = 4 \cos^3 A - 3 \cos A$	>>syms A ↵
	>>pretty(expand(cos(3*A))) ↵
	             3
	      4 cos(A) - 3 cos(A)

$\tan 3A = \dfrac{3 \tan A - \tan^3 A}{1 - 3 \tan^2 A}$	>>syms A ↵
	>>pretty(expand(tan(3*A))) ↵
	                   3
	      3 tan(A) - tan(A)
	      -------------------- 	
	                   2
	          1 - 3 tan(A)

$\cos(4 \sin^{-1} x) = 8x^4 - 8x^2 + 1$	>>syms x ↵
	>>pretty(expand(cos(4*asin(x)))) ↵
	           2 4
	      1 - 8 x + 8 x

$\tan(5 \tan^{-1} x) = \dfrac{5x - 10x^3 + x^5}{1 - 10x^2 + 5x^4}$	>>syms x ↵
	>>pretty(expand(tan(5*atan(x)))) ↵
	                3 5
	      5 x - 10 x + x
	      -------------------- 	
	                2 4
	      1 - 10 x + 5 x

3.19 Computations of expressions

MATLAB is a very powerful software for computation of expressions. Long and clumsy functions are easily computed in scalar or vector form on its user-friendly platform. Expressions that can be used as the basic of complicated computations are exemplified below. Two types of functions are presented – function of one variable and function of multivariable.

3.19.1 Functions of one variable

Let us consider the following expressions that have to be evaluated for some x :

A. $e^x \cos x$ B. $\dfrac{\log_{10} x}{\cosh x}$ C. $\sin^2 x \cos^2 x$

D. $\dfrac{1}{\sqrt{\tan^2 x + 9x^3}}$ E. $\sum_x -9x^2 \sqrt{\left| \ln \dfrac{x}{2} \right|}$ F. $\prod_x \dfrac{1 + \dfrac{x}{2}}{1 + \dfrac{x}{4}}$

To be specific, expressions A through F have to be computed for $x =1$, 1.5, 3, 3.5, 4, and 2.5. Results computed by a ten-digit calculator are presented in tabular form for expressions A through D in the following page. This will help us get the insight of computational style of MATLAB.

Expression A :	
x	$e^x \cos x$
1	1.46869394
1.5	0.317022143
3	-19.88453084
3.5	-31.01118644
4	-35.68773248
2.5	-9.759927258

Expression B :	
x	$\dfrac{\log_{10} x}{\cosh x}$
1	0
1.5	0.074855696
3	0.047391465
3.5	0.032828926
4	0.02204683
2.5	0.064892567

Expression C :	
x	$\sin^2 x \cos^2 x$
1	0.206705452
1.5	0.004978714
3	0.019518255
3.5	0.107907847
4	0.244707435
2.5	0.229883941

Expression D :	
x	$\dfrac{1}{\sqrt{\tan^2 x + 9x^3}}$
1	0.295843502
1.5	0.066049413
3	0.064147347
3.5	0.05089765
4	0.041618264
2.5	0.084160582

▱ *Expression A*

It is understood that x is in radians. If one has to evaluate expression A for ten or twenty different values of x using calculator, surely he will feel bored. But do not be bored, MATLAB eases your computation. First you have to assign different values of x to x vector, that is, form $x = \begin{bmatrix} 1 \\ 1.5 \\ 3 \\ 3.5 \\ 4 \\ 2.5 \end{bmatrix}$. If you use the command 'cos(x)', MATLAB performs operation $\begin{bmatrix} \cos 1 \\ \cos 1.5 \\ \cos 3 \\ \cos 3.5 \\ \cos 4 \\ \cos 2.5 \end{bmatrix}$. By the way, the subroutine 'cos(x)' of MATLAB takes argument x in radians. Again, using the command 'exp(x)' performs the operation $\begin{bmatrix} e^1 \\ e^{1.5} \\ e^3 \\ e^{3.5} \\ e^4 \\ e^{2.5} \end{bmatrix}$. What we need is the first element of 'cos(x)' is to be multiplied with the first element of 'exp(x)', the second element of 'cos(x)' is to be multiplied with the second element of 'exp(x)', and so is others. This is called scalar or dot multiplication, which is accomplished by the operator .*. Use the command 'exp(x).*cos(x)' to get $\begin{bmatrix} e^1 \cos 1 \\ e^{1.5} \cos 1.5 \\ e^3 \cos 3 \\ e^{3.5} \cos 3.5 \\ e^4 \cos 4 \\ e^{2.5} \cos 2.5 \end{bmatrix}$. That is our objective. If different x values are assigned as row vector, the output is a row vector and if they are as column vector, so is the output. See the computation of expression A in the following page:

MATLAB Command

>>x=[1 1.5 3 3.5 4 2.5]; ↵
>>exp(x).*cos(x) ↵

ans=
    1.4687   0.3170   -19.8845   -31.0112   -35.6877   -9.7599

Compare the output of MATLAB with the computation using a ten-digit calculator. In MATLAB, we entered x vector as a row one. Like expression A, subsequent expressions are explained in terms of column vectors because that is easier to understand but execution of the commands will be in terms of row vectors. Since we are using the short format of floating-point numbers, MATLAB prints floating-point numbers up to 4 decimal places. For example, when $x=1$, the first value obtained by the ten-digit calculator is $e^1 \cos 1 = 1.46869394$. Up to 4 decimal places, the value is 1.4687 that is exactly the MATLAB output.

⌨ *Expression B*

In MATLAB, $\log_{10} x$ and $\cosh x$ are computed by the subroutines 'log10(x)' and 'cosh(x)' respectively.

Use of 'log10(x)' and 'cosh(x)' gives us $\begin{bmatrix} \log_{10} 1 \\ \log_{10} 1.5 \\ \log_{10} 3 \\ \log_{10} 3.5 \\ \log_{10} 4 \\ \log_{10} 2.5 \end{bmatrix}$ and $\begin{bmatrix} \cosh 1 \\ \cosh 1.5 \\ \cosh 3 \\ \cosh 3.5 \\ \cosh 4 \\ \cosh 2.5 \end{bmatrix}$ respectively. Next step is divide the first element

of 'log10(x)' by the first element of 'cosh(x)', the second element of 'log10(x)' by the second element of 'cosh(x)', and so do others. This point to point division operation is achieved by the operator ./. Use the

command 'log10(x)./cosh(x)' to have $\begin{bmatrix} \dfrac{\log_{10} 1}{\cosh 1} \\ \dfrac{\log_{10} 1.5}{\cosh 1.5} \\ \dfrac{\log_{10} 3}{\cosh 3} \\ \dfrac{\log_{10} 3.5}{\cosh 3.5} \\ \dfrac{\log_{10} 4}{\cosh 4} \\ \dfrac{\log_{10} 2.5}{\cosh 2.5} \end{bmatrix}$. As mentioned in expression A, output of MATLAB for

expression B is also rounded up to 4 decimal places. See its implementation below:

MATLAB Command

>>x=[1 1.5 3 3.5 4 2.5]; ↵
>>log10(x)./cosh(x) ↵

ans =
    0   0.0749   0.0474   0.0328   0.0220   0.0649

⌨ *Expression C*

For better understanding, explanation of commands in parts is presented as follows: $\sin(x) = \begin{bmatrix} \sin 1 \\ \sin 1.5 \\ \sin 3 \\ \sin 3.5 \\ \sin 4 \\ \sin 2.5 \end{bmatrix}$,

$$\sin(x).^2 = \begin{bmatrix} \sin^2 1 \\ \sin^2 1.5 \\ \sin^2 3 \\ \sin^2 3.5 \\ \sin^2 4 \\ \sin^2 2.5 \end{bmatrix}, \quad \cos(x) = \begin{bmatrix} \cos 1 \\ \cos 1.5 \\ \cos 3 \\ \cos 3.5 \\ \cos 4 \\ \cos 2.5 \end{bmatrix}, \quad \cos(x).^2 = \begin{bmatrix} \cos^2 1 \\ \cos^2 1.5 \\ \cos^2 3 \\ \cos^2 3.5 \\ \cos^2 4 \\ \cos^2 2.5 \end{bmatrix}, \text{ and } (\sin(x).^2).*(\cos(x).^2) = \begin{bmatrix} \sin^2 1\cos^2 1 \\ \sin^2 1.5\cos^2 1.5 \\ \sin^2 3\cos^2 3 \\ \sin^2 3.5\cos^2 3.5 \\ \sin^2 4\cos^2 4 \\ \sin^2 2.5\cos^2 2.5 \end{bmatrix}. \text{ This}$$

output is also rounded up to 4 decimal places as shown below:

MATLAB Command

      >>x=[1 1.5 3 3.5 4 2.5]; ↵
      >>(sin(x).^2).*(cos(x).^2) ↵

    ans=

      0.2067   0.0050   0.0195   0.1079   0.2447   0.2299

Expression D

In parts, following is the explanation of the commands used: $\tan(x) = \begin{bmatrix} \tan 1 \\ \tan 1.5 \\ \tan 3 \\ \tan 3.5 \\ \tan 4 \\ \tan 2.5 \end{bmatrix}$, $\tan(x).^2 = \begin{bmatrix} \tan^2 1 \\ \tan^2 1.5 \\ \tan^2 3 \\ \tan^2 3.5 \\ \tan^2 4 \\ \tan^2 2.5 \end{bmatrix}$,

$$x.^3 = \begin{bmatrix} 1^3 \\ 1.5^3 \\ 3^3 \\ 3.5^3 \\ 4^3 \\ 2.5^3 \end{bmatrix}, \quad 9*(x.^3) = \begin{bmatrix} 9\times 1^3 \\ 9\times 1.5^3 \\ 9\times 3^3 \\ 9\times 3.5^3 \\ 9\times 4^3 \\ 9\times 2.5^3 \end{bmatrix}, \quad \tan(x).^2 + 9*(x.^3) = \begin{bmatrix} \tan^2 1+9\times 1^3 \\ \tan^2 1.5+9\times 1.5^3 \\ \tan^2 3+9\times 3^3 \\ \tan^2 3.5+9\times 3.5^3 \\ \tan^2 4+9\times 4^3 \\ \tan^2 2.5+9\times 2.5^3 \end{bmatrix}, \quad \text{sqrt}(\tan(x).^2 + 9*(x.^3)) =$$

$$\begin{bmatrix} \sqrt{\tan^2 1+9\times 1^3} \\ \sqrt{\tan^2 1.5+9\times 1.5^3} \\ \sqrt{\tan^2 3+9\times 3^3} \\ \sqrt{\tan^2 3.5+9\times 3.5^3} \\ \sqrt{\tan^2 4+9\times 4^3} \\ \sqrt{\tan^2 2.5+9\times 2.5^3} \end{bmatrix}, \text{ and } 1./\text{sqrt}(\tan(x).^2 + 9*(x.^3)) = \begin{bmatrix} \frac{1}{\sqrt{\tan^2 1+9\times 1^3}} \\ \frac{1}{\sqrt{\tan^2 1.5+9\times 1.5^3}} \\ \frac{1}{\sqrt{\tan^2 3+9\times 3^3}} \\ \frac{1}{\sqrt{\tan^2 3.5+9\times 3.5^3}} \\ \frac{1}{\sqrt{\tan^2 4+9\times 4^3}} \\ \frac{1}{\sqrt{\tan^2 2.5+9\times 2.5^3}} \end{bmatrix}. \text{ Following is the implementation:}$$

MATLAB Command

      >>x=[1 1.5 3 3.5 4 2.5]; ↵
      >>1./sqrt((tan(x).^2)+9*(x.^3)) ↵

    ans=

      0.2958   0.0660   0.0641   0.0509   0.0416   0.0842

Expression E

This expression is the sum form of expression $-9x^2\sqrt{\left|\ln\frac{x}{2}\right|}$ for different x. Expand the summation

$$\sum_x -9x^2\sqrt{\left|\ln\frac{x}{2}\right|} \quad \text{to write} \quad -9\times 1^2\sqrt{\left|\ln\frac{1}{2}\right|} -9\times 1.5^2\sqrt{\left|\ln\frac{1.5}{2}\right|} -9\times 3^2\sqrt{\left|\ln\frac{3}{2}\right|} -9\times 3.5^2\sqrt{\left|\ln\frac{3.5}{2}\right|} -9\times 4^2\sqrt{\left|\ln\frac{4}{2}\right|} -9\times$$

$$2.5^2\sqrt{\left|\ln\frac{2.5}{2}\right|} = -7.4929915-10.861290431-51.577675154-82.475237970-119.887864006-26.571415898 =$$

−298.866474959. Implementation is shown as follows:

MATLAB Command

>>x=[1 1.5 3 3.5 4 2.5]; ↵
>>sum(-9*(x.^2).*sqrt(abs(log(x/2)))) ↵

ans=
-298.8665

The explanatory steps are $x/2=\begin{bmatrix}\frac{1}{2}\\\frac{1.5}{2}\\\frac{3}{2}\\\frac{3.5}{2}\\\frac{4}{2}\\\frac{2.5}{2}\end{bmatrix}$, $\log(x/2)=\begin{bmatrix}\ln\frac{1}{2}\\\ln\frac{1.5}{2}\\\ln\frac{3}{2}\\\ln\frac{3.5}{2}\\\ln\frac{4}{2}\\\ln\frac{2.5}{2}\end{bmatrix}$, $\text{abs}(\log(x/2))=\begin{bmatrix}|\ln\frac{1}{2}|\\|\ln\frac{1.5}{2}|\\|\ln\frac{3}{2}|\\|\ln\frac{3.5}{2}|\\|\ln\frac{4}{2}|\\|\ln\frac{2.5}{2}|\end{bmatrix}$, $\text{sqrt}(\text{abs}(\log(x/2)))=\begin{bmatrix}\sqrt{|\ln\frac{1}{2}|}\\\sqrt{|\ln\frac{1.5}{2}|}\\\sqrt{|\ln\frac{3}{2}|}\\\sqrt{|\ln\frac{3.5}{2}|}\\\sqrt{|\ln\frac{4}{2}|}\\\sqrt{|\ln\frac{2.5}{2}|}\end{bmatrix}$,

$x.{}^\wedge2=\begin{bmatrix}1^2\\1.5^2\\3^2\\3.5^2\\4^2\\2.5^2\end{bmatrix}$, $-9*x.{}^\wedge2=\begin{bmatrix}-9\times1^2\\-9\times1.5^2\\-9\times3^2\\-9\times3.5^2\\-9\times4^2\\-9\times2.5^2\end{bmatrix}$, $-9*x.{}^\wedge2.*\text{sqrt}(\text{abs}(\log(x/2)))=\begin{bmatrix}-9\times1^2\sqrt{|\ln\frac{1}{2}|}\\-9\times1.5^2\sqrt{|\ln\frac{1.5}{2}|}\\-9\times3^2\sqrt{|\ln\frac{3}{2}|}\\-9\times3.5^2\sqrt{|\ln\frac{3.5}{2}|}\\-9\times4^2\sqrt{|\ln\frac{4}{2}|}\\-9\times2.5\sqrt{|\ln\frac{2.5}{2}|}\end{bmatrix}$, and

$\text{sum}(-9*x.{}^\wedge2.*\text{sqrt}(\text{abs}(\log(x/2)))) = -9\times1^2\sqrt{|\ln\frac{1}{2}|}-9\times1.5^2\sqrt{|\ln\frac{1.5}{2}|}-9\times3^2\sqrt{|\ln\frac{3}{2}|}-9\times3.5^2\sqrt{|\ln\frac{3.5}{2}|}-9\times4^2\sqrt{|\ln\frac{4}{2}|}-9\times2.5^2\sqrt{|\ln\frac{2.5}{2}|}$.

⌨ *Expression F*

This expression is a product form of some polynomials that have numerator and denominator polynomials. Let us expand the expression F to write $\prod_x\frac{1+\frac{x}{2}}{1+\frac{x}{4}}=\frac{1+\frac{1}{2}}{1+\frac{1}{4}}\cdot\frac{1+\frac{1.5}{2}}{1+\frac{1.5}{4}}\cdot\frac{1+\frac{3}{2}}{1+\frac{3}{4}}\cdot\frac{1+\frac{3.5}{2}}{1+\frac{3.5}{4}}\cdot\frac{1+\frac{4}{2}}{1+\frac{4}{4}}\cdot\frac{1+\frac{2.5}{2}}{1+\frac{2.5}{4}}=1.2\times$

$1.272727272\times1.428571428\times1.466666666\times1.5\times1.384615384=6.646153833$. Explanation is also attached for

clarification: $x/2=\begin{bmatrix}\frac{1}{2}\\\frac{1.5}{2}\\\frac{3}{2}\\\frac{3.5}{2}\\\frac{4}{2}\\\frac{2.5}{2}\end{bmatrix}$, $1+x/2=\begin{bmatrix}1+\frac{1}{2}\\1+\frac{1.5}{2}\\1+\frac{3}{2}\\1+\frac{3.5}{2}\\1+\frac{4}{2}\\1+\frac{2.5}{2}\end{bmatrix}$, $x/4=\begin{bmatrix}\frac{1}{4}\\\frac{1.5}{4}\\\frac{3}{4}\\\frac{3.5}{4}\\\frac{4}{4}\\\frac{2.5}{4}\end{bmatrix}$, $1+x/4=\begin{bmatrix}1+\frac{1}{4}\\1+\frac{1.5}{4}\\1+\frac{3}{4}\\1+\frac{3.5}{4}\\1+\frac{4}{4}\\1+\frac{2.5}{4}\end{bmatrix}$, $(1+x/2)./(1+x/4)=\begin{bmatrix}\frac{1+\frac{1}{2}}{1+\frac{1}{4}}\\\frac{1+\frac{1.5}{2}}{1+\frac{1.5}{4}}\\\frac{1+\frac{3}{2}}{1+\frac{3}{4}}\\\frac{1+\frac{3.5}{2}}{1+\frac{3.5}{4}}\\\frac{1+\frac{4}{2}}{1+\frac{4}{4}}\\\frac{1+\frac{2.5}{2}}{1+\frac{2.5}{4}}\end{bmatrix}$, and

$\text{prod}((1+x/2)./(1+x/4))=\frac{1+\frac{1}{2}}{1+\frac{1}{4}}\cdot\frac{1+\frac{1.5}{2}}{1+\frac{1.5}{4}}\cdot\frac{1+\frac{3}{2}}{1+\frac{3}{4}}\cdot\frac{1+\frac{3.5}{2}}{1+\frac{3.5}{4}}\cdot\frac{1+\frac{4}{2}}{1+\frac{4}{4}}\cdot\frac{1+\frac{2.5}{2}}{1+\frac{2.5}{4}}$. MATLAB computation is shown as follows:

```
>>x=[1 1.5 3 3.5 4 2.5]; ↵
>>prod((1+x/2)./(1+x/4)) ↵
```

ans=
6.6462

3.19.2 Functions of multivariable

Discussion of article 3.19.1 is applicable when given function is a function of one variable. Situation comes very often when we have to deal with multivariable function. Say, a two variable function is $f(x, y) = -7x^2 + 9y^x + e^{-y^2}$. We wish to evaluate $f(x, y)$ at $x = 0$ and $y = 1$, $\therefore$ $f(0, 1) = 0 + 9 \times 1^0 + e^{-1} = 9.3679$. The string form of $f(x, y)$ is '-7*x^2+9*y^x+exp(-y^2)'. Computation can be carried out by 'eval' (abbreviation of evaluation). Before computation, we need to enter the variables' values that are $x = 0$ and $y = 1$. See the computation as follows:

MATLAB Command
```
>>f='-7*x^2+9*y^x+exp(-y^2)'; ↵
>>x=0; ↵
>>y=1; ↵
>>eval(f) ↵
```

ans =
9.3679

☞ *For several values of* x *and* y :

Another case is when several x and y values are given. We wish to compute $f(x, y) = -7x^2 + 9y^x + e^{-y^2}$ for each set of x and y values, where $x = \begin{bmatrix} 0 \\ 2 \\ -2 \\ -1 \end{bmatrix}$ and $y = \begin{bmatrix} 1 \\ -1 \\ 0 \\ 2 \end{bmatrix}$. Obviously, the output should

be $\begin{bmatrix} f(0, 1) \\ f(2, -1) \\ f(-2, 0) \\ f(-1, 2) \end{bmatrix} = \begin{bmatrix} -7 \times 0^2 + 9 \times 1^0 + e^{-1} \\ -7 \times 2^2 + 9 \times (-1)^2 + e^{-1} \\ -7 \times (-2)^2 + 9 \times 0^{-2} + e^{0} \\ -7 \times (-1)^2 + 9 \times 2^{-1} + e^{-4} \end{bmatrix} = \begin{bmatrix} 9.3679 \\ -18.6321 \\ \infty \\ -2.4817 \end{bmatrix}$. Since x and y are column matrices, point to point

representation of the function is necessary. So, $f(x, y)$ is written as '-7*x.^2+9*y.^x+exp(-y.^2)'. Remember that point to point operation is sensitive for the operators '/', '*', and '^'. Each of these operators dwells with vectors (either row or column matrix) of the same length on either side. Above computation is conducted as follows:

MATLAB Command
```
>>f='-7*x.^2+9*y.^x+exp(-y.^2)'; ↵
>>x=[0 2 -2 -1]'; ↵
>>y=[1 -1 0 2]'; ↵
>>eval(f) ↵
```

ans =
9.3679
-18.6321
Inf
-2.4817

Application of 'eval' can be extended where we have 3, 4, or more variables. Say, we have $h(m, x, y, z) = m\cos(2x + 3yz)$, where the variables under consideration are m, x, y, and z. We want to calculate $h(x, y, z, m)$

for $m =[1 \quad 2 \quad 0 \quad -4]$, $x =[\dfrac{\pi}{2} \quad \dfrac{3\pi}{2} \quad \pi \quad -\dfrac{\pi}{2}]$, $y =[2 \quad 3 \quad -3 \quad -1]$, and $z =[0 \quad -\pi \quad \pi \quad \dfrac{\pi}{4}]$. Assign the output

to O, where O$=[\,h\left(1, \dfrac{\pi}{2}, 2, 0\right) \quad h\left(2, \dfrac{3\pi}{2}, 3, -\pi\right) \quad h(0, \pi, -3, \pi) \quad h\left(-4, -\dfrac{\pi}{2}, -1, \dfrac{\pi}{4}\right)]=[\,\cos\pi \quad 2\cos(3\pi - 9\pi) \quad 0$

$-4\cos\left(-\pi - \dfrac{3\pi}{4}\right)\,]=[-1 \quad 2 \quad 0 \quad -2.8284]$. Carry out the computation as follows:

MATLAB Command
```
>>h='m.*cos(2*x+3*y.*z)';  ↵
>>m=[1 2 0 -4];  ↵
>>x=[pi/2 3*pi/2 pi -pi/2];  ↵
>>y=[2 3 -3 -1];  ↵
>>z=[0 -pi pi pi/4];  ↵
>>O=eval(h)  ↵

O =
      -1.0000    2.0000    0    -2.8284
```

✐ *Computations of advanced multivariable functions*

In image processing problems, computations like $\sum\sum f(xy)$, $\sum\sum f(x+y)$, $\prod\prod f(xy)$, $\prod\prod f(x+y)$, $\sum\prod f(xy)$ …etc take place in different transforms, for example, computation similar to $\sum\sum f(x+y)$ is used in two dimensional discrete cosine transform of an image. Two examples of such computation are presented.

Problem A

Compute $\displaystyle\sum_{y}\sum_{x} \cos(2xy)$ where $x =[0 \quad 2 \quad -1 \quad 1]$ and $y =[\pi \quad -\dfrac{\pi}{4} \quad 0]$.

Solution:

Here the elements of x and y matrices are not consecutive. The double sum $\displaystyle\sum_{y}\sum_{x} \cos(2xy)$ turns to

$$\left\{\begin{array}{l}[\cos(2\times 0\times\pi) + \cos(2\times 2\times\pi) + \cos(2\times(-1)\times\pi) + \cos(2\times 1\times\pi) + \\ \cos(2\times 0\times\left(-\dfrac{\pi}{4}\right)) + \cos(2\times 2\times\left(-\dfrac{\pi}{4}\right)) + \cos(2\times(-1)\times\left(-\dfrac{\pi}{4}\right)) + \cos(2\times 1\times\left(-\dfrac{\pi}{4}\right)) + \\ \cos(2\times 0\times 0) + \cos(2\times 2\times 0) + \cos(2\times(-1)\times 0) + \cos(2\times 1\times 0)]\end{array}\right\} = 8 \text{ (after computation). First, let}$$

us define x as a column matrix and y as a row one so we have $x = \begin{bmatrix} 0 \\ 2 \\ -1 \\ 1 \end{bmatrix}$ and $y =[\pi \quad -\dfrac{\pi}{4} \quad 0]$. Vector

multiplication of x with y is $x\times y = \begin{bmatrix} 0 \\ 2 \\ -1 \\ 1 \end{bmatrix}\times[\pi \quad -\dfrac{\pi}{4} \quad 0]= \begin{bmatrix} 0\times\pi & 0\times\left(-\dfrac{\pi}{4}\right) & 0\times 0 \\ 2\times\pi & 2\times\left(-\dfrac{\pi}{4}\right) & 2\times 0 \\ -1\times\pi & (-1)\times\left(-\dfrac{\pi}{4}\right) & (-1)\times 0 \\ 1\times\pi & 1\times\left(-\dfrac{\pi}{4}\right) & 1\times 0 \end{bmatrix}$ (similar MATLAB

operation is 'x*y'). Multiplying by 2 and taking cosine of each term provide us

90

$$\begin{bmatrix} \cos(2\times0\times\pi) & \cos[\,2\times0\times\left(-\dfrac{\pi}{4}\right)] & \cos(2\times0\times0) \\[2mm] \cos(\,2\times2\times\pi) & \cos[2\times2\times\left(-\dfrac{\pi}{4}\right)] & \cos(2\times2\times0) \\[2mm] \cos(2\times-1\times\pi) & \cos[2\times(-1)\times\left(-\dfrac{\pi}{4}\right)] & \cos[2\times(-1)\times0] \\[2mm] \cos(2\times1\times\pi) & \cos[2\times1\times\left(-\dfrac{\pi}{4}\right)] & \cos(2\times1\times0) \end{bmatrix}$$

[corresponding MATLAB operation is 'cos(2*x*y)'].

Observe that the first, second, and third columns of the last matrix are identical with the first, second, and third rows of $\sum\limits_{y}\sum\limits_{x}\cos(2xy)$ respectively. Then, if we sum over each column, we end up with a row matrix of length 3 [for this, the command is 'sum(cos(2*x*y))']. Finally, resulting row matrix is summed by another 'sum' that is what we are interested in. The whole methodology is as follows:

MATLAB Command
>>x=[0 2 -1 1]'; ↵
>>y=[pi -pi/4 0]; ↵
>>sum(sum(cos(2*x*y))) ↵

ans=
8

Or the other way around, if x be a row matrix and y be a column one, command would be 'sum(sum(cos(2*y*x)))'. Observe that y comes first, then, does x for the multiplication. Just to have a check,

MATLAB Command
>>x=[0 2 -1 1]; ↵
>>y=[pi -pi/4 0]'; ↵
>>sum(sum(cos(2*y*x))) ↵

ans=
8

Problem B

Compute $\sum\limits_{v}\sum\limits_{u}\ln(2u+5v)$ where $u=[4\quad2\quad11\quad3]$ and $v=[\pi\quad2]$.

Solution:

Expanding the double sum, we have $\left\{\begin{array}{l}[\ln(2\times4+5\times\pi)+\ln(2\times2+5\times\pi)+\ln(2\times11+5\times\pi)+\ln(2\times3+5\times\pi)+\\ \ln(2\times4+5\times2)+\ln(2\times2+5\times2)+\ln(2\times11+5\times2)+\ln(2\times3+5\times2)]\end{array}\right\}=$

24.6221 (after computation). Start defining the given values as matrix u=[4 2 11 3] and v=$\begin{bmatrix}\pi\\2\end{bmatrix}$. Form another two matrices x and y. Placing two (the question is why two, because we have two columns in matrix v) u matrices on top of the other forms x. Matrix y is formed by placing four (again, why the number is four, because there are four rows in matrix u) v matrices side by side, hence, x=$\begin{bmatrix}4 & 2 & 11 & 3\\4 & 2 & 11 & 3\end{bmatrix}$ and y=

$\begin{bmatrix}\dfrac{\pi}{2} & \dfrac{\pi}{2} & \dfrac{\pi}{2} & \dfrac{\pi}{2}\end{bmatrix}$. What if '2*x+5*y' is performed. Scale factors 2 and 5 are chosen because coefficients of

u and v are 2 and 5 respectively in given summation $\sum\limits_{v}\sum\limits_{u}\ln(2u+5v)$, therefore, '2*x+5*y'=

$\begin{bmatrix}2\times4+5\times\pi & 2\times2+5\times\pi & 2\times11+5\times\pi & 2\times3+5\times\pi\\ 2\times4+5\times2 & 2\times2+5\times2 & 2\times11+5\times2 & 2\times3+5\times2\end{bmatrix}$ and 'log(2*x+5*y)'=

$$\begin{bmatrix} \ln(2\times4+5\times\pi) & \ln(2\times2+5\times\pi) & \ln(2\times11+5\times\pi) & \ln(2\times3+5\times\pi) \\ \ln(2\times4+5\times2) & \ln(2\times2+5\times2) & \ln(2\times11+5\times2) & \ln(2\times3+5\times2) \end{bmatrix}.$$ Just sum all elements of last matrix

that is what we want to compute. This operation is accomplished by two 'sum' subroutines. Matrices x and y are built from the same u and v matrices by means of the subroutine 'repmat'. Elaborate discussion of 'repmat' has been given in article 2.29. The whole computation is as follows:

MATLAB Command

```
>>u=[4 2 11 3]; ↵
>>v=[pi 2]'; ↵
>>x=repmat(u,2,1); ↵
>>y=repmat(v,1,4); ↵
>>sum(sum(log(2*x+5*y))) ↵
```

ans =

24.6221

We are almost on the verge of computation discussion. Complicated or long functions can be computed straightforwardly and simply so long as one cognizes the representations of functions in MATLAB terminology. Next, the reader needs to be familiar with MATLAB counterparts of commonly used mathematical or calculus functions. Table 3.B is a list of such mathematical functions and their MATLAB counterparts. Writing string for computation is presented in chapter 11.

Table 3.B Some mathematical functions in symbolic forms and their MATLAB counterparts

Mathematical notation	MATLAB notation	Mathematical notation	MATLAB notation
$\sin x$	sin(x)	$\sin^{-1} x$	asin(x)
$\cos x$	cos(x)	$\cos^{-1} x$	acos(x)
$\tan x$	tan(x)	$\tan^{-1} x$	atan(x)
$\cot x$	cot(x)	$\cot^{-1} x$	acot(x)
$\operatorname{cosec} x$	csc(x)	$\sec^{-1} x$	asec(x)
$\sec x$	sec(x)	$\operatorname{cosec}^{-1} x$	acsc(x)
$\sinh x$	sinh(x)	$\sinh^{-1} x$	asinh(x)
$\cosh x$	cosh(x)	$\cosh^{-1} x$	acosh(x)
$\operatorname{sec} h x$	sech(x)	$\sec h^{-1} x$	asech(x)
$\operatorname{cosec} h x$	csch(x)	$\operatorname{cosec} h^{-1} x$	acsch(x)
$\tanh x$	tanh(x)	$\tanh^{-1} x$	atanh(x)
$\coth x$	coth(x)	$\coth^{-1} x$	acoth(x)
π	pi	$\log_{10} x$	log10(x)
A+B	A+B	e^x	exp(x)
A−B	A−B	$\mid x \mid$	abs(x)
A×B	A*B	$\ln x$	log(x)
$\dfrac{A}{B}$	A/B	$\prod$	prod
A^B	A^B	$\sum$	sum
$\int y\,dx$	int(y,x)	$\sqrt{x}$	sqrt(x)
$\dfrac{dy}{dx}$	diff(y,x)		

3.20 Defining a geometric object

Point, line, triangle, square, circle, ellipse, or hyperbola can be the example of a geometric object. We define these objects using Maple procedures, which are located in the geometry package. A geometric object may be entered in MATLAB by several ways. For instance, a triangle can be defined from three points, three sides, and two sides and angle between them. One should activate the geometry package before defining a geometric object. The package does not handle point at infinity or line at infinity. Take the advantage of dos shel key and copy-cut-paste facility to avoid typing nuisance. Since 'D' is the differential operator in Maple, do not use letter 'D' to assign some geometric object. Now we present the defining style of some familiar geometric objects.

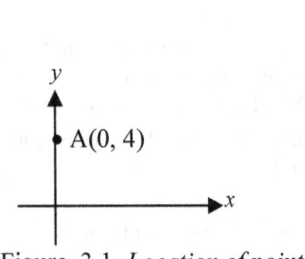

Figure 3.1 *Location of point A*

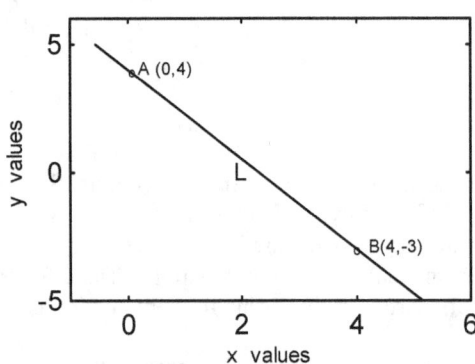

Figure 3.2 *Line L passing through A and B*

3.20.1 Defining a point

A point has no dimensions – length, width, or thickness. It has position only and is represented by a dot. Horizontal and vertical coordinates specify a point in two-dimensional geometry. Say, we have point A(0, 4) (see figure 3.1 for its location). Enter this object as follows:

MATLAB Command
>>maple('with(geometry)'); ↵   ← Activate the geometry package
>>A=maple('point(A,0,4)'); ↵   ← Define point A

3.20.2 Defining a line

A two-dimensional line can be defined from two given points or from its algebraic equation. Say, we have two points whose coordinates are A(0, 4) and B(4, –3) (figure 3.2 shows this). Straight line AB passing through A and B is constructed as follows:

MATLAB Command
>>maple('with(geometry)'); ↵
>>A=maple('point(A,0,4)'); ↵   ← Define point A
>>B=maple('point(B,4,-3)'); ↵   ← Define point B
>>L=maple('line(L,[A,B],[x,y])'); ↵   ←Define line L from A and B

Notice that argument '[x,y]' means the horizontal and vertical axes are x and y respectively. We defined the line AB as L in MATLAB workspace. Equation of a line passing through two points (x_1, y_1) and (x_2, y_2) is given by $\dfrac{x - x_1}{x_1 - x_2} = \dfrac{y - y_1}{y_1 - y_2}$, from which, equation of AB is $7x + 4y - 16 = 0$. Command 'detail' helps us see the details of object L:

>>maple('detail(L)') ↵

ans =

name of the object: L   form of the object: line2d   equation of the line: -16+7*x+4*y = 0

⬚ *Point to remember*

In Command Window, return of subroutine 'detail' is displayed in one line. One may not see all outputs of 'detail' in the monitor screen. Slide the horizontal scroll bar of Command Window to see the rest details of an object.

From the output of 'detail', we can say that name of the object is L, the object is a two-dimensional straight line, and equation of the line is $-16 + 7x + 4y = 0$. Another style of entering a line can be from its algebraic equation. To give an example, enter straight line $7x + 4y - 16 = 0$ in workspace as follows:

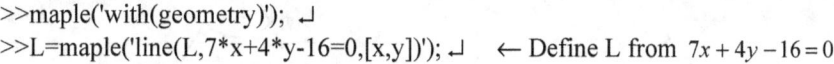

>>maple('with(geometry)'); ↵
>>L=maple('line(L,7*x+4*y-16=0,[x,y])'); ↵   ← Define L from $7x + 4y - 16 = 0$

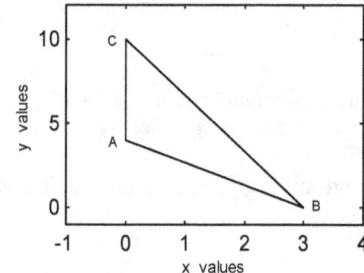

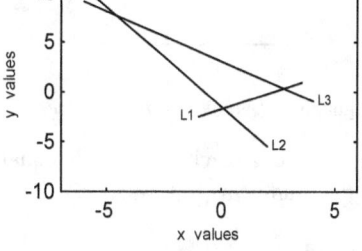

Figure 3.3   *Triangle formed by points A, B, and C*

Figure 3.4 *Triangle formed by lines L1, L2, and L3*

3.20.3 Defining a triangle

A triangle is defined as a polygon that has three sides. Two sides of a triangle meet at a point called vertex. It can be defined in different ways – from three given points, from three lines, from three sides, and two sides and angle between them. We begin with three points A(0, 4), B(3, 0), and C(0, 10) (see figure 3.3 for their location). Triangle T is entered as follows:

MATLAB Command
>>maple('with(geometry)'); ↵
>>A=maple('point(A,0,4)'); ↵   ← Define point A
>>B=maple('point(B,3,0)'); ↵   ← Define point B
>>C=maple('point(C,0,10)'); ↵   ← Define point C
>>T=maple('triangle(T,[A,B,C])'); ↵   ← Define triangle T from points A, B, and C

Then, we have three equations of straight lines, which are given by L1: $3x - 4y = 7$, L2: $4x + 2y = -3$, and L3: $x + y = 3$. Figure 3.4 shows their plots and the triangle formed from their intersections. Construct triangle T from the three lines as follows:

MATLAB Command
>>maple('with(geometry)'); ↵
>>L1=maple('line(L1,3*x-4*y=7,[x,y])'); ↵   ←Define line L1: $3x - 4y = 7$
>>L2=maple('line(L2,4*x+2*y=-3,[x,y])'); ↵   ← Define line L2: $4x + 2y = -3$
>>L3=maple('line(L3,x+y=3,[x,y])'); ↵   ← Define line L3: $x + y = 3$
>>T=maple('triangle(T,[L1,L2,L3],[x,y])'); ↵ ← Define triangle T from L1, L2, and L3
>>maple('detail(T)') ↵   ← To see details of the triangle formed

ans =

name of the object: T   form of the object: triangle2d   method to define the triangle: points   the three vertices: [[1/11, -37/22], [19/7, 2/7], [-9/2, 15/2]]

94

As you see, the command 'detail' can be beneficial for knowing the vertices of the triangle.

Next, three sides 3, 3, and 4 are given, from which, the triangle is to be formed (see figure 3.5 for this).

MATLAB Command
>>maple('with(geometry)'); ↵
>>T=maple('triangle(T,[3,3,4])'); ↵ ←Triangle T from sides

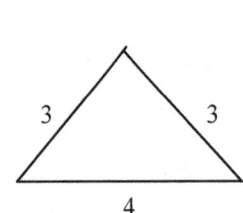

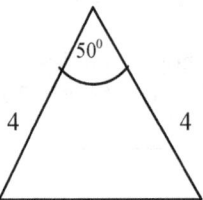

Figure 3.5 *Triangle formed by sides*

Figure 3.6 *Triangle from two sides and angle between them*

Finally, we have two sides, each of them is equal to 4, and angle between them, which is 50^0. Figure 3.6 shows this. Form the triangle as follows:

MATLAB Command
>>maple('with(geometry)'); ↵
>>T=maple('triangle(T,[4,angle=50*pi/180,4])'); ↵

Notice that the angle between the sides is entered in terms of radians.

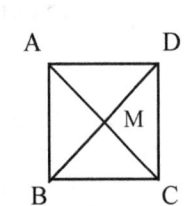

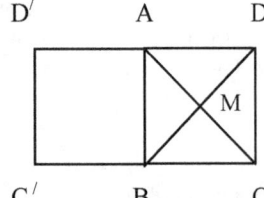

Figure 3.7 *Square ABCD*

Figure 3.8 *Two squares ABCD and ABC'D' can be formed from two adjacent vertices A and B*

3.20.4 Defining a square

A square is a quadrilateral whose sides are equal and angles are right angle. It can be constructed from two opposite vertices of a diagonal, two adjacent vertices, and a vertex and center of the square. Corresponding Maple procedure is 'MakeSquare'. Start with two opposite vertices A(2, 5) and C(9, 0) (referring to figure 3.7).

Our objective is to find other vertices (B and D) and length of diagonal. Midpoint of AC is M$\left(\frac{11}{2}, \frac{5}{2}\right)$. Assume

that coordinates of point B are (u, v). Slopes of AC and BM are $-\frac{5}{7}$ and $\frac{2v-5}{2u-11}$ respectively. Since

AC⊥BM, we have $-\frac{5}{7} \times \frac{2v-5}{2u-11} = -1$, from that cause, $10v - 14u = -52$. Again, distance BM=AM provides

$\left(u - \frac{11}{2}\right)^2 + \left(v - \frac{5}{2}\right)^2 = \left(\frac{7}{2}\right)^2 + \left(\frac{5}{2}\right)^2$. Solve equations involving u and v to write $u = 3$ and 8 and $v = -1$ and 6. So,

coordinates of B and D are (3, −1) and (8, 6) respectively. Diagonal is just the length of AC, which is $\sqrt{74}$. Do

not bother for so much computation as long as MATLAB is there:

MATLAB Command

>>maple('with(geometry)'); ↵
>>A=maple('point(A,2,5)'); ↵   ← Define first corner point A of diagonal
>>C=maple('point(C,9,0)'); ↵   ←Define second corner point C of diagonal
>>S=maple('MakeSquare(S,[A,C,diagonal])'); ↵ ←Form square S from vertices A and C
>>maple('detail(S)') ↵

ans =

>   name of the object: S   form of the object: square2d   the four vertices of the square: [[2, 5], [3, -
>   1], [9, 0], [8, 6]]   the length of the diagonal: 74^(1/2)

Referring to figure 3.7, if A(2, 5) and $M\left(\dfrac{11}{2},\dfrac{5}{2}\right)$ are given, how we find the other coordinates and diagonal.

Carry out that by MATLAB.

MATLAB Command

>>maple('with(geometry)'); ↵
>>A=maple('point(A,2,5)'); ↵   ← Define vertex A of the square
>>M=maple('point(M,11/2,5/2)'); ↵   ← Define center M of the square
>>S=maple('MakeSquare(S,[A,center=M])'); ↵ ← Square S from vertex A and center M
>>maple('detail(S)') ↵

ans =

>   name of the object: S   form of the object: square2d   the four vertices of the square: [[2, 5],
>   [3, -1], [9, 0], [8, 6]]   the length of the diagonal: 74^(1/2)

Finally, if two adjacent vertices A(2, 5) and B(3, −1) of the square ABCD are given, find the other details of the square as follows:

MATLAB Command

>>maple('with(geometry)'); ↵
>>A=maple('point(A,2,5)'); ↵
>>B=maple('point(B,3,-1)'); ↵
>>S=maple('MakeSquare(S,[A,B,adjacent])'); ↵ ← Describe that A and B are two adjacent points of
>   the square
>>maple('detail(S)') ↵

ans =

>   name of the object: S_1   form of the object: square2d   the four vertices of the square: [[2, 5],
>   [3, -1], [-3, -2], [-4, 4]]   the length of the diagonal: 74^(1/2)   name of the object: S_2   form of
>   the object: square2d   the four vertices of the square: [[2, 5], [3, -1], [9, 0], [8, 6]]   the length of
>   the diagonal: 74^(1/2)

Look at figure 3.8. From two adjacent points A and B, we can have two identical squares labeled by ABCD and ABC 'D'. MATLAB return is also consistent with that. Object denoted by S_1 describes one square and S_2 does the other.

3.20.5 Defining a circle

A circle is defined as the set of points in a plane that have the same distance from center. A circle can be defined from three points, from two end points of a diameter, from center and radius, and from an algebraic equation. Examples are presented for each case. Maple procedure 'circle' constructs a circle.

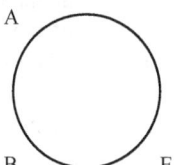

Figure 3.9 *Circle passing through three points A, B, and E*

⌗ From three points

Assume that we have three points A(1, 4), B(12, 8), and E(2, 8). Find a circle passing through these three points (see figure 3.9 for the circle). General equation of a two dimensional circle is $x^2 + y^2 + 2gx + 2fy + c = 0$. Just mentioned three points must satisfy the general equation, hence, we have the following:

for point A => $\quad 17 + 2g + 8f + c = 0$,

for point B => $\quad 208 + 24g + 16f + c = 0$, and

for point E => $\quad 68 + 4g + 16f + c = 0$.

Solving the last three equations relating g, f, and c, we have $g = -7$, $f = -\dfrac{37}{8}$, and $c = 34$, hence, equation

of the circle is $x^2 + y^2 - 14x - \dfrac{37}{4}y + 34 = 0$. Center and radius of the circle are given by $(-g, -f) = \left(7, \dfrac{37}{8}\right)$ and

$\sqrt{g^2 + f^2 - c} = \sqrt{\dfrac{2329}{64}}$ respectively.

MATLAB Command

```
>>maple('with(geometry)');  ↵
>>maple('point(A,1,4),point(B,12,8),point(E,2,8)');  ↵  ← Define the three points together
>>C=maple('circle(C,[A,B,E],[x,y])');  ↵  ← Define the circle C from the given three points
>>maple('detail(C)')  ↵
```

ans =

> name of the object: C   form of the object: circle2d   name of the center: center_C
> coordinates of the center; [7, 37/8]   radius of the circle: 1/64*2329^(1/2)*64^(1/2)   equation
> of the circle: x^2+34+y^2-14*x-37/4*y = 0

⌗ From two end points of a diameter

Suppose, we have two points A(1, 4) and B(5, 10), which are two end points of the diameter of a circle (shown in figure 3.10). Center of the circle is the midpoint of A and B that is M(3, 7). Radius is half of the length AB, where $AB = \sqrt{(1-5)^2 + (4-10)^2} = \sqrt{52}$. Equation is given by $(x-3)^2 + (y-7)^2 = \dfrac{52}{4}$ or $x^2 + y^2 - 6x - 14y + 45 = 0$. Following is the solution:

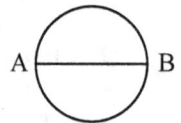

Figure 3.10 *Circle from diameter AB*

MATLAB Command

```
>>maple('with(geometry)');  ↵
>>maple('point(A,1,4),point(B,5,10)');  ↵  ← Define 2 end points of the diameter
>>C=maple('circle(C,[A,B],[x,y])');  ↵
>>maple('detail(C)')  ↵
```

ans =

> name of the object: C   form of the object: circle2d   name of the center: center_C   coordinates
> of the center: [3, 7]   radius of the circle: 1/2*52^(1/2)   equation of the circle: x^2+45+y^2-
> 6*x-14*y = 0

⌗ From center and radius

Assume that the center coordinates are (1, 4) and the radius is $\sqrt{5}$. Equation of the circle is given by $(x-1)^2 + (y-4)^2 = (\sqrt{5})^2$ or $x^2 + y^2 - 2x - 8y + 12 = 0$.

MATLAB Command

```
>>maple('with(geometry)');  ↵
>>A=maple('point(A,1,4)');  ↵                    ← Center is assigned to A
```

>>C=maple('circle(C,[A,sqrt(5)],[x,y])'); ⏎ ← Circle from center and radius
>>maple('detail(C)') ⏎

ans =

name of the object: C   form of the object: circle2d   name of the center: A   coordinates of the center: [1, 4]   radius of the circle: 5^(1/2)   equation of the circle: x^2+12+y^2-2*x-8*y = 0

⌗ *From an equation*

Take the equation of a circle, which is $3x^2 + 3y^2 + 6x - 9y + 1 = 0$. Remember that coefficients of x^2 and y^2 are identical for a circle. Enter it as follows:

MATLAB Command

>>maple('with(geometry)'); ⏎
>>C=maple('circle(C,3*x^2+3*y^2+6*x-9*y+1=0,[x,y])'); ⏎
>>maple('detail(C)') ⏎

ans =

name of the object: C   form of the object: circle2d   name of the center: center_C coordinates of the center: [-1, 3/2]   radius of the circle: 1/12*35^(1/2)*12^(1/2)   equation of the circle: x^2+1/3+y^2+2*x-3*y = 0

3.20.6 Defining a parabola

Parabola is the set of points in a plane that are equidistant from a given line and from a given point (given point is not on the given line). The given line and point are called directrix and focus of the parabola respectively. A parabola is symmetric about a line (which is called axis of the parabola) that passes through the focus at right angles to the directrix. The axis of the parabola intersects the parabola at a point called vertex of the parabola. Figure 3.11 presents all these parameters of a parabola having the equation $y^2 = 4ax$. The name of Maple procedure is 'parabola'. A parabola can be described from five distinct points, from focus and vertex, from directrix and focus, and from a polynomial or algebraic equation. Standard form of a parabola has the equation $y^2 = 4ax$ or $x^2 = 4ay$. Former equation's focus, vertex, axis, and directrix are (a, 0), (0, 0), $y = 0$, and $x = -a$ respectively, and so are (0, a), (0, 0), $x = 0$, and $y = -a$ respectively for the later.

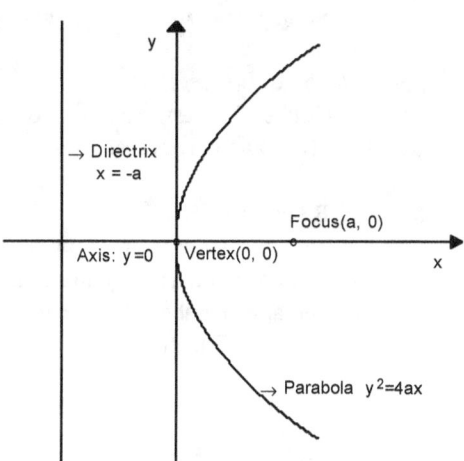

Figure 3.11 *Parabola and its parameters*

From five distinct points:

We take five distinct points, which are A(1, 0), B(0, 1), C(−20, 3), E(−4, −1), and F(−64, 5). Finding equation is carried out from the general equation of a conic, which is $ax^2 + by^2 + cxy + dx + fy + e = 0$. Anyhow, it is provided that the five points lie on parabola $x = 1 + 2y - 3y^2$. Construct that as follows:

MATLAB Command

>>maple('with(geometry)'); ⏎
>>maple('point(A,1,0),point(B,0,1),point(C,-20,3),point(E,-4,-1),point(F,-64,5)'); ⏎
>>P=maple('parabola(P,[A,B,C,E,F],[x,y])'); ⏎ ← Object parabola is assigned to P
>>maple('detail(P)') ⏎

ans =

name of the object: P   form of the object: parabola2d   vertex: [4/3, 1/3]   focus: [5/4, 1/3] directrix: x-17/12 = 0   equation of the parabola: 311040*x-622080*y+933120*y^2-311040 = 0

Observe that output equation of parabola 311040*x-622080*y+933120*y^2-311040 = 0 needs division by 311040 to have it identical with the given equation $x = 1 + 2y - 3y^2$. Parameters of the parabola, vertex, focus, and directrix, are easily seen by dint of command 'detail'.

From focus and vertex:

Test coordinates of focus and vertex are F(1, 2) and V(4, 7) respectively. Equation of the axis is the line passing through focus and vertex, which is A: $3y - 5x - 1 = 0$. If intersection of directrix and axis of the parabola is the point I, then, V is the midpoint of F and I, on that account, coordinates of I are (7, 12). Directrix is perpendicular line to A and passes through I(7, 12), from that cause, equation of the directrix is $3x + 5y - 81 = 0$. Let any point on the parabola be P(x, y). By definition, distance PF is equal to the distance of point P to directrix, hence, $\sqrt{(x-1)^2 + (y-2)^2} = \dfrac{3x + 5y - 81}{\sqrt{3^2 + 5^2}}$ or $25x^2 - 30xy + 9y^2 + 418x + 674y - 6391 = 0$ is the equation of the parabola.

MATLAB Command

>>maple('with(geometry)'); ↵
>>maple('point(F,1,2),point(V,4,7)'); ↵   ← Enter focus F and vertex V
>>P=maple('parabola(P,[focus=F,vertex=V],[x,y])'); ↵ ← Parabola P from F and V
>>maple('Equation(P)') ↵   ← Display the equation of P

ans =

25*x^2-30*x*y+9*y^2+418*x+674*y-6391 = 0

From directrix and focus:

Utilize the last example to have equation of the directrix and coordinates of focus as DX: $3x + 5y - 81 = 0$ and F(1, 2) respectively. Our equation should be $25x^2 - 30xy + 9y^2 + 418x + 674y - 6391 = 0$.

MATLAB Command

>>maple('with(geometry)'); ↵
>>maple('point(F,1,2),line(DX, 3*x+5*y-81=0,[x,y])'); ↵ ← Enter focus F and directrix DX
>>maple('parabola(P,[directrix=DX, focus=F],[x,y])'); ↵ ← Form parabola from F and DX
>>maple('Equation(P)') ↵   ← To verify the equation

ans =

25*x^2-30*x*y+9*y^2+418*x+674*y-6391 = 0

From an equation:

Enter the equation of the parabola $x = 1 + 2y - 3y^2$ as follows:

MATLAB Command

>>maple('with(geometry)'); ↵
>>maple('parabola(P,x=1+2*y-3*y^2,[x,y])'); ↵

Specific parameters of the parabola P are seen by the following commands:

>>maple('coordinates(vertex(P))') ↵ ← To see the coordinates of vertex

ans =

[4/3, 1/3]
>>maple('coordinates(focus(P))') ↵ ← To see the coordinates of focus

ans =

[5/4, 1/3]
>>maple('Equation(directrix(P))') ↵ ← To see the equation of directrix

ans =

x-17/12 = 0

Anyhow, we have outfitted only the construction of circle and parabola with examples. In a similar fashion, one can construct any other conic such as ellipse or hyperbola.

3.21 Area of geometric objects

Two-dimensional closed geometric objects such as triangle, circle, or ellipse form an area. Area of geometrical objects can be found by the subroutine 'area', which is also included in geometry package. Suppose a triangle is formed from three points A(0, 4), B(3, 10), and C(0, 10). Area of this triangle is given by

$\frac{1}{2}\begin{bmatrix} 1 & 1 & 1 \\ 0 & 3 & 0 \\ 4 & 10 & 10 \end{bmatrix}$ =9 square unit. MATLAB helps us find the triangular area as follows:

MATLAB Command

```
>>maple('with(geometry)');  ↵
>>maple('point(A,0,4),point(B,3,10),point(C,0,10)');  ↵  ← Enter points A, B, and C
>>T=maple('triangle(T,[A,B,C])');  ↵                      ← Define triangle T from A, B, and C
>>maple('area(T)')  ↵                                      ← Apply procedure 'area' on T
```

ans =
        9

We defined the triangle from three vertices. If the triangle were defined from three lines or other way, the subroutine 'area' would also be applicable. Next example is a circle whose equation is given by $2x^2 + 2y^2 - x - y - a = 0$, where a is a positive constant. Rearrange the equation as $x^2 + y^2 - \frac{x}{2} - \frac{y}{2} - \frac{a}{2} = 0$ and compare it with the general equation of a circle in two dimensions, $x^2 + y^2 + 2gx + 2fy + c = 0$, to write $g = -\frac{1}{4}$, $f = -\frac{1}{4}$, and $c = -\frac{a}{2}$. Radius of the circle is $\sqrt{g^2 + f^2 - c} = \sqrt{\left(-\frac{1}{4}\right)^2 + \left(-\frac{1}{4}\right)^2 - \left(-\frac{a}{2}\right)} = \sqrt{\frac{4a+1}{8}}$. Area of the circle is given by $\pi r^2 = \pi\left(\sqrt{\frac{4a+1}{8}}\right)^2 = \pi\left(\frac{a}{2} + \frac{1}{8}\right)$. See below how easy it is:

MATLAB Command

```
>>maple('with(geometry)');  ↵
>>maple('assume(a>0)');  ↵                                  ← Clarify the condition of constant a
>>C=maple('circle(C,2*x^2+2*y^2-x-y-a=0,[x,y])');  ↵  ← Define circle C
>>maple('area(C)')  ↵                                       ← Apply the procedure 'area' on C
```

ans =

pi*(1/8+1/2*a)

100

The subroutine 'area' also applies to an ellipse. Say, the example equation is $4x^2 + 9y^2 + 8x - 54y + 49 = 0$.
Rearrange the equation to write $\frac{(x+1)^2}{9} + \frac{(y-3)^2}{4} = 1$. From which, the major and minor axes of the ellipse have lengths 3 and 2 respectively. So, the area of the ellipse is $\pi \times 3 \times 2 = 6\pi$. Obtain that as follows:

MATLAB Command
>>maple('with(geometry)'); ↵
>>E=maple('ellipse(E,4*x^2+9*y^2+8*x-54*y+49=0,[x,y])'); ↵ ← Ellipse is formed from the equation
>>maple('area(E)') ↵

ans =

6*pi

An M−file 'polyarea' (abbreviation of <u>poly</u>gonal <u>area</u>), which is available in Command Window, computes the area of a polygon numerically formed by corner coordinates. To compute the area,

MATLAB Step:
Use the command polyarea (x, y co-ordinates in row/ column matrix form).

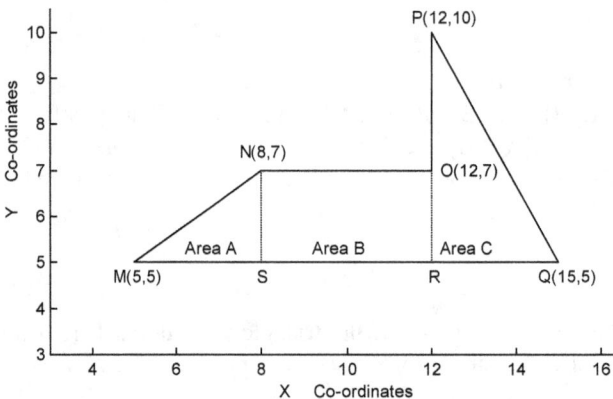

Figure 3.12   *Polygon formed by points M, N, O, P, and Q*

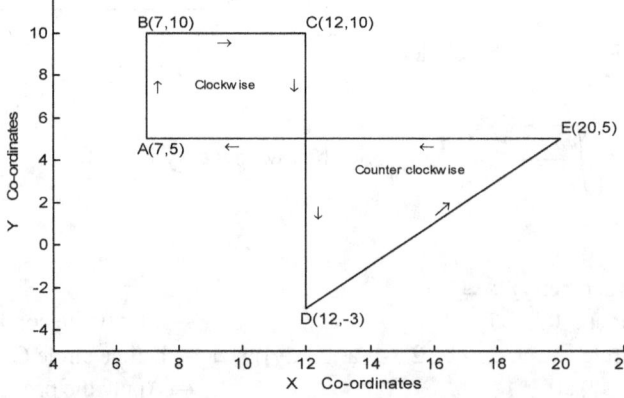

Figure 3.13   *Polygon formed by points A, B, C, D, and E*

As usual, take some examples of points — M(5, 5), N(8, 7), O(12, 7), P(12, 10), and Q(15, 5). Plot of these points is shown in figure 3.12. These five points subtend a polygon MNOPQM. Area of this polygon is equal to sum of area A, B, and C, hence, area A=$\frac{1}{2}$×MS×NS=$\frac{1}{2}$(8–5)(7–5)=3 sq. unit, area B=4×2=8 sq. unit, and

area C=$\frac{1}{2}$×3×5=7.5 sq. unit. Therefore, the total area MNOPQM is 18.5 sq. unit.

MATLAB Command for area MNOPQM,
>>x=[5 8 12 12 15]; ↵
>>y=[5 7 7 10 5]; ↵
>>polyarea(x,y) ↵

ans =
18.5000

During the computation of area, edges of polygon must not intersect. If edge intersection is there, the return is the absolute difference between clockwise and counterclockwise encircled areas. Such an example is indicated by points A(7, 5), B(7, 10), C(12, 10), D(12, –3), and E(20, 5). Subtended area is shown in figure 3.13. As you see, edges AE and CD intersect if one travels according to A → B → C → D → E → A. From figure 3.13, the clockwise area is (12–7)×(10–5)=25 sq. unit and the counterclockwise area is $\frac{1}{2}$(20–12)×(5–(–3))=32 sq. unit. The absolute difference is |25–32|=7 sq. unit. That is what is computed below:

Command for the polygonal area ABCDEA,
>>x=[7 7 12 12 20]; ↵
>>y=[5 10 10 -3 5]; ↵
>>polyarea(x,y) ↵

ans =
7

3.22 Rotation of geometric objects

A geometric object can be rotated in clockwise or counterclockwise direction about some point or line. Needless to say, equation or representation of the geometric object may be required following rotation. Maple procedure 'rotation' is very useful in this regard. We start with a point P whose coordinates are (3, 4)(see figure 3.14). P is rotated by 60^0 counterclockwise about the origin O. New location of P is N. Find the

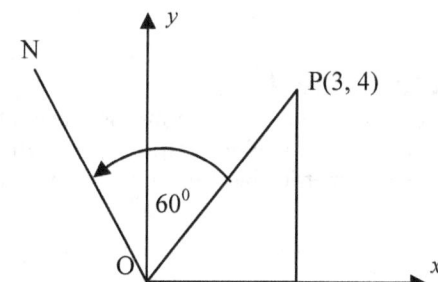

Figure 3.14　*Rotation of point P by 60^0 counterclockwise about the origin*

coordinates of N. Referring to figure 3.14, we have $\angle$ PO$x = \tan^{-1}\frac{4}{3}$, hence, x and y coordinates of N are

$5\cos\left(\tan^{-1}\frac{4}{3}+60^0\right)=\frac{3}{2}-2\sqrt{3}$ and $5\sin\left(\tan^{-1}\frac{4}{3}+60^0\right)=\frac{3\sqrt{3}}{2}+2$ respectively. Have it computed as follows:

102

MATLAB Command

```
>>maple('with(geometry)');  ↵
>>maple('point(P,3,4)');  ↵
>>maple('rotation(N,P,pi/3,counterclockwise)');  ↵
>>maple('coordinates(N)')  ↵
```

ans =

[3/2-2*3^(1/2), 2+3/2*3^(1/2)]

Angle of rotation is inserted in terms of radians.

Next example is the parabola P: $y^2 = 4x$. Parabola P is rotated about the origin by 45^0 clockwise. Figure 3.15 presents both parabolas – before and after the rotation. Our concern is to find the equation of the dotted parabola. As shown in figure 3.15, the new axes are x ' and y '. With the new axes, the equation of the rotated parabola is $y'^2 = 4x'$. The relationship between the original and the rotated points is given by $x' = x\cos\theta + y\sin\theta$ and $y' = y\cos\theta - x\sin\theta$, where counter clockwise θ is taken positive. For the example at hand, we have $\theta = -45^0$, on that cause, $y'^2 = 4x'$ becomes $\dfrac{x^2}{2} + xy + \dfrac{1}{2}y^2 - 2\sqrt{2}\,x + 2\sqrt{2}\,y = 0$. That is what is returned by the procedure 'rotation' as presented below:

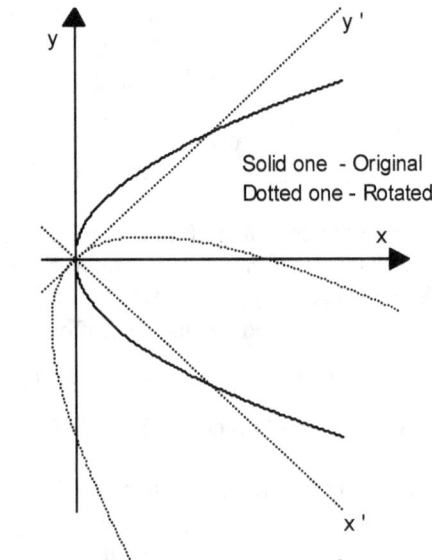

Figure 3.15   *A parabola is rotated 45^0 clockwise about the origin*

MATLAB Command

```
>>maple('with(geometry)');  ↵
>>maple('parabola(P,y^2=4*x,[x,y])');  ↵          ← Form the parabola P: y² = 4x
>>maple('rotation(N,P,pi/4,clockwise)');  ↵       ← N corresponds to the dotted parabola
>>maple('Equation(N)')  ↵                         ← To see the equation of parabola N
```

ans =

1/2*x^2+x*y+1/2*y^2+2*y*2^(1/2)-2*x*2^(1/2) = 0

3.23 Reflection of geometric objects

Reflection of a geometric object is defined as the mirror image of that object with respect to some point or line. Following figures clarify the concept of geometric object's reflection. Maple procedure 'reflection' finds the equation or representation of a geometric object pursuing the reflection.

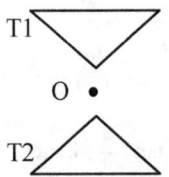

Figure 3.16   *Triangle T1 is the reflection of T2 with respect to point O*

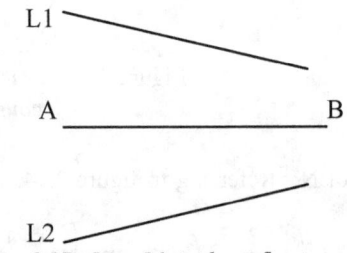

Figure 3.17 *Line L1 is the reflection of L2 with respect to line AB*

Figure 3.16 shows reflection of a triangle T2, which is T1, with respect to point O. Again, figure 3.17 depicts the reelection of line L2 with respect to line AB, which is labeled by line L1. For both illustrations, one can be reflection of the other. Three examples of reflection are mentioned. First, find reflection of point P(2, 5) with respect to point O(3, −2). Reflection of a point is another point. If R is the point following reflection, O is the midpoint of P and R, therefore, coordinates of R are (4, −9). Following is the implementation:

MATLAB Command
>>maple('with(geometry)'); ↵
>>maple('point(P,2,5),point(O,3,-2)'); ↵
>>maple('reflection(R,P,O)'); ↵
>>maple('coordinates(R)') ↵

ans =

[4, -9]

Then, we have an object line L1: $x - y - 5 = 0$ and a reference line AB: $4x - 5y = 8$. We look for image line L2, say, equation of L2 is $y = mx + c$. Slopes of L1, AB, and L2 are 1, $\frac{4}{5}$, and m respectively. It is evident from the slopes that lines L1 and AB are not parallel. Angles between L1 and AB and between AB and L2 are

$\tan^{-1}\dfrac{1 - \frac{4}{5}}{1 + \frac{4}{5}}$ and $\tan^{-1}\dfrac{m - \frac{4}{5}}{1 + \frac{4}{5}m}$ respectively. To fulfill the condition of reflection, both angles are equal, hence,

$\tan^{-1}\dfrac{1 - \frac{4}{5}}{1 + \frac{4}{5}} = \pm\tan^{-1}\dfrac{m - \frac{4}{5}}{1 + \frac{4}{5}m}$. On solving, we have $m = 1$ and $\frac{31}{49}$, but $m \neq 1$ for L1 and AB not being parallel.

Now the equation of L2 becomes $y = \frac{31}{49}x + c$. Take any point that satisfies line AB, we chose (2, 0). Distances

from (2, 0) to L1 and L2 are equal, thereby, $\left|\dfrac{2 - 5}{\sqrt{2}}\right| = \left|\dfrac{\frac{62}{49} + c}{\sqrt{\left(\frac{31}{49}\right)^2 + 1}}\right|$, from which, $c = \frac{61}{49}$. At last the equation of

L2 is $31x - 49y + 61 = 0$. Too much computation is indeed. Avoid that by following:

MATLAB Command
>>maple('with(geometry)'); ↵
>>maple('line(L1,x-y-5=0,[x,y]),line(AB,4*x-5*y=8,[x,y])'); ↵
>>maple('reflection(L2,L1,AB)'); ↵
>>maple('Equation(L2)') ↵

ans =

-31/41*x+49/41*y-61/41 = 0

The last example includes finding the reflection of a circle with respect to a line as indicated in figure 3.18. As usual, assume that the equations are L: $2x - y - 5 = 0$ and C1: $2x^2 + 2y^2 - x - 4y + 1 = 0$. Center and radius of C1 are $\left(\frac{1}{4}, 1\right)$ and $\frac{3}{4}$ respectively. Projection of the center of C1 on line L is point M$\left(\frac{49}{20}, -\frac{1}{10}\right)$. M is the midpoint

104

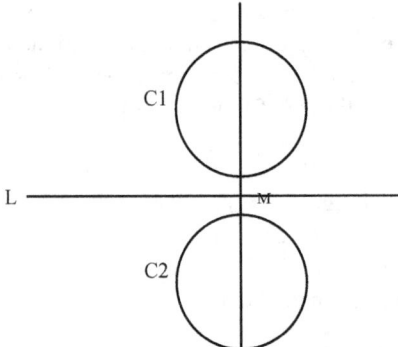

Figure 3.18 *Circle C1 is the reflection of*
C2 with respect to the line L

of the centers of two circles, therefore, the center of C2 is $\left(\dfrac{93}{20}, -\dfrac{6}{5}\right)$. Radius of C2 is the same as that of C1,

so, the equation of C2 is $\left(x - \dfrac{93}{20}\right)^2 + \left(y + \dfrac{6}{5}\right)^2 = \left(\dfrac{3}{4}\right)^2$ or $x^2 + y^2 - \dfrac{93x}{10} + \dfrac{12y}{5} + \dfrac{45}{2} = 0$. See below how simple it

is:

MATLAB Command
>>maple('with(geometry)'); ↵
>>maple('line(L,2*x-y-5=0,[x,y]),circle(C1,2*x^2+2*y^2-x-4*y+1=0,[x,y])'); ↵
>>maple('reflection(C2,C1,L)'); ↵
>>maple('Equation(C2)') ↵

ans =

x^2-93/10*x+y^2+12/5*y+45/2 = 0

3.24 Miscellaneous geometric problems

If we describe the whole geometry package, that requires two or three chapters to describe it. Apart from the previously mentioned tutorials, there are many geometric problems that can be solved conveniently in MATLAB workspace in collaboration with Maple package. More tutorials are presented in the following.

  ⌗ *Perpendicular line to a line passing through a point*

We have a line L: $4x - 7y = 89$ and a point P(0, 4). We wish to find a perpendicular line PL to L that passes through P (see figure 3.19 for illustration). Slope of L is $\dfrac{4}{7}$. Slope of any line

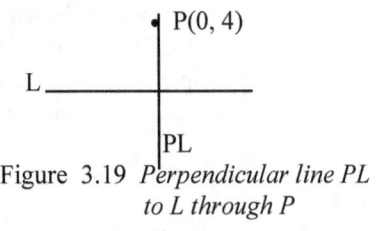

Figure 3.19 *Perpendicular line PL*
to L through P

perpendicular to L is $-\dfrac{7}{4}$. Equation of PL is $y = -\dfrac{7}{4}x + c$, which is

satisfied by P(0, 4). Satisfying by P returns $c = 4$. Finally, PL is given by $4y + 7x = 16$. We use the maple procedure 'PerpendicularLine' for this purpose as follows:

MATLAB Command
>>maple('with(geometry)'); ↵
>>P=maple('point(P,0,4)'); ↵   ← Define point P
>>L=maple('line(L,4*x-7*y=89,[x,y])'); ↵   ← Define line L

>>PL=maple('PerpendicularLine(PL,P,L)'); ↵ ← Line PL from P and L
>>maple('Equation(PL)')   ← Display equation of PL

ans =

16-7*x-4*y = 0

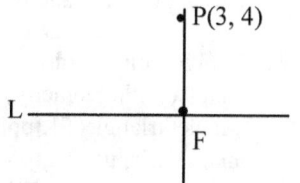

☞ *Projection of a point on a line*

     Assume that equation of the line is L: $4x - 7y = 89$ and the given point is P(3, 4). Find projection of P on L, that is, we have to find coordinates of point F on L, where line L and PF are perpendicular to each other (shown in figure 3.20). Slope of L is $\frac{4}{7}$, hence, slope of PF is $-\frac{7}{4}$.

Figure 3.20   *F is projection of P on L*

Equation of PF is $y - 4 = -\frac{7}{4}(x - 3)$ or $4y + 7x = 37$. Solving L and PF's

equations provides coordinates of F, which are $\left(\frac{123}{13}, -\frac{95}{13}\right)$. Use maple procedure 'projection' in this regard.

MATLAB Command

>>maple('with(geometry)'); ↵
>>maple('point(P,3,4),line(L,4*x-7*y=89,[x,y])'); ↵ ← Enter point and line
>>F=maple('projection(F,P,L)'); ↵   ← Form projection point F
>>maple('coordinates(F)') ↵

ans =

[123/13, -95/13]

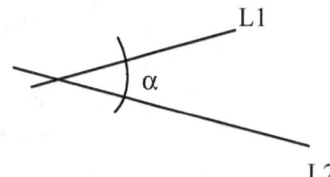

☞ *Finding the angle between two lines*

     Given that m_1 and m_2 are slopes of two straight lines L1 and L2 respectively, angle between them is found by $\alpha = \tan^{-1}\left[\frac{m_1 - m_2}{1 + m_1 m_2}\right]$ (see figure 3.21). Verification can be on lines L1:

Figure 3.21 *Angle between two lines L1 and L2*

$2x - 7y = 4$ and L2: $3x + 4y = 4$. Slopes of the lines are $\frac{2}{7}$ and $-\frac{3}{4}$ respectively, hence, $\alpha = \tan^{-1}\left[\dfrac{\frac{2}{7} - \left(-\frac{3}{4}\right)}{1 + \frac{2}{7}\left(-\frac{3}{4}\right)}\right] =$

$\tan^{-1}\frac{29}{22}$. Maple procedure 'FindAngle' finds the angle between L1 and L2:

MATLAB Command

>>maple('with(geometry)'); ↵
>>maple('line(L1,2*x-7*y=4,[x,y]),line(L2,3*x+4*y=4,[x,y])'); ↵ ← Define the lines L1 and L2
>>maple('FindAngle(L1,L2)') ↵

ans =
    atan(29/22)

☞ *Are two triangles similar?*

     Presume that one triangle is defined from three points – A(0, 0), B(2, 0), and C(2, 3) and the other is from points E(5, 6), F(1, 0), and G(2, −1). We want to test whether they are similar. Two triangles T1 and T2 are said to be similar whose corresponding angles are congruent and whose corresponding sides are in proportion. The function used for this purpose is 'AreSimilar'. Lengths of sides of triangle ABC are 2, 3, and $\sqrt{13}$ and those of triangle EFG are $2\sqrt{13}$, $\sqrt{2}$, and $\sqrt{58}$ respectively, on that account, they are not similar. We name triangles ABC and EFG as T1 and T2 respectively.

MATLAB Command
```
>>maple('with(geometry)');  ↵
>>maple('triangle(T1,[point(A,0,0), point(B,2,0), point(C,2,3)])');  ↵   ← Form the first triangle
>>maple('triangle(T2,[point(E,5,6), point(F,1,0), point(G,2,-1)])');  ↵  ← Form the second triangle
>>maple('AreSimilar(T1,T2)')  ↵                              ← Check similarity of ABC and EFG
```

ans =

false

⌨ *Are three points collinear?*

Three points are said to be collinear if they lie on a straight line. We illustrate that with A(0, 0), B(2, 0), and C(2, 3). Collinearity is satisfied if area of triangle ABC is zero, hence, area ABC=$\frac{1}{2}\begin{bmatrix} 1 & 1 & 1 \\ 0 & 2 & 2 \\ 0 & 0 & 3 \end{bmatrix}$=3. So, A, B, and C are not collinear. Use the Maple procedure 'AreCollinear' for this verification:

MATLAB Command
```
>>maple('with(geometry)');  ↵
>>maple('point(A,0,0),point(B,2,0),point(C,2,3)');  ↵
>>maple('AreCollinear(A,B,C)')  ↵
```

ans =

false

⌨ *Coordinates of a division point*

If the line segment joining two points A (x_1, y_1) and B (x_2, y_2) are internally divided by a ratio k (that is, AE:EB=k :1), coordinates of the division point are given by $\left(\frac{kx_2 + x_1}{k+1}, \frac{ky_2 + y_1}{k+1} \right)$ (see figure 3.22).

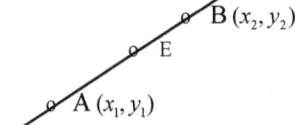

Figure 3.22 *Point E divides AB internally by ratio k*

Example points are A(−2, 5) and B(1, 6) and we wish to find coordinates of E for ratio $\frac{4}{3}$. Insert point's coordinates and ratio to have E$\left(-\frac{2}{7}, \frac{39}{7} \right)$. Conduct the Maple procedure 'OnSegment' for division point:

MATLAB Command
```
>>maple('with(geometry)');  ↵
>>maple('point(A,-2,5),point(B,1,6)');  ↵
>>maple('OnSegment(E,A,B,4/3)');  ↵
>>maple('coordinates(E)')  ↵
```

ans =

[-2/7, 39/7]

☞ *Distance between two points and between a point and a line*

The name of the function is 'distance'. Say, the given points are A(–2, 5) and B(1, 6). Distance AB is $\sqrt{(-2-1)^2+(5-6)^2}=\sqrt{10}$.

MATLAB Command
>>maple('with(geometry)'); ↵
>>maple('point(A,-2,5),point(B,1,6)'); ↵
>>maple('distance(A,B)') ↵

ans =

10^(1/2)

To illustrate the line example, take L: $2x-7y=71$. We wish to find the distance from A to L, which is $\left|\dfrac{2(-2)-7\times5-71}{\sqrt{2^2+7^2}}\right|=\dfrac{110}{\sqrt{53}}$.

MATLAB Command
>>maple('with(geometry)'); ↵
>>maple('point(A,-2,5),line(L,2*x-7*y=71,[x,y])'); ↵
>>maple('distance(A,L)') ↵

ans =

110/53*53^(1/2)

☞ *Is a point on a line or a circle?*

Take example point, line, and circle as A(1, 0), L: $7x-9y-45=0$, and C: $2x^2+2y^2-4x+7y+2=0$ respectively. If A is on L or C, that must satisfy the equation of L or C. Maple procedures 'IsOnLine' and 'IsOnCircle' allows us to check that. For the examples at hand, A does not lie on L but does on C. First check A on L:
MATLAB Command
>>maple('with(geometry)'); ↵
>>maple('point(A,1,0),line(L,7*x-9*y-45=0,[x,y])'); ↵
>>maple('IsOnLine(A,L)') ↵

ans =

false
Then, check whether A is on circle C:
>>maple('circle(C,2*x^2+2*y^2-4*x+7*y+2=0,[x,y])'); ↵
>>maple('IsOnCircle(A,C)') ↵

ans =

true
Sometimes it is necessary to know the condition on which a point is on a line or circle. To accomplish this, we need to append one argument in 'IsOnLine' or 'IsOnCircle'. Say, a point is given as $A\left(a,\dfrac{a}{4}\right)$. Satisfy line L by the last point A to write $7a-\dfrac{9a}{4}-45=0$ or $\dfrac{19a}{4}-45=0$.

MATLAB Command

    >>maple('with(geometry)'); ↵
    >>maple('point(A,a,a/4),line(L,7*x-9*y-45=0,[x,y])'); ↵
    >>maple('IsOnLine(A,L,cond)') ↵

ans =

    IsOnLine: hint: unable to determine if -45+19/4*a is zeroFAIL

Similarly we have the condition $2a^2 + 2\left(\dfrac{a}{4}\right)^2 - 4a + 7\left(\dfrac{a}{4}\right) + 2 = 0$ or $\dfrac{17}{16}a^2 - \dfrac{9}{8}a + 1 = 0$ for the circle C:

MATLAB Command

    >>maple('with(geometry)'); ↵
    >>maple('point(A,a,a/4),circle(C,2*x^2+2*y^2-4*x+7*y+2=0,[x,y])'); ↵
    >>maple('IsOnCircle(A,C,cond)') ↵

ans =

    IsOnCircle: hint: unable to determine if 17/16*a^2+1-9/8*a is zeroFAIL

Median, altitude, and angular bisector from a vertex to a triangle
    Given that three vertices of the triangle ABC are A(2, 7), B(−2, −2), and C(5, 3). Find median, altitude, and angular bisector of the triangle from vertex A. If M is the midpoint of side BC, line AM is the median of the triangle from A (as indicated in figure 3.23).

Coordinates of M are $\left(\dfrac{3}{2}, \dfrac{1}{2}\right)$. Equation of AM is $13x - y - 19 = 0$.

Then, if a perpendicular is drawn from A to BC or to extension of BC, the perpendicular is called altitude of the triangle from A. Equation of BC is $4 - 5x + 7y = 0$. Altitude is the line that is perpendicular to BC

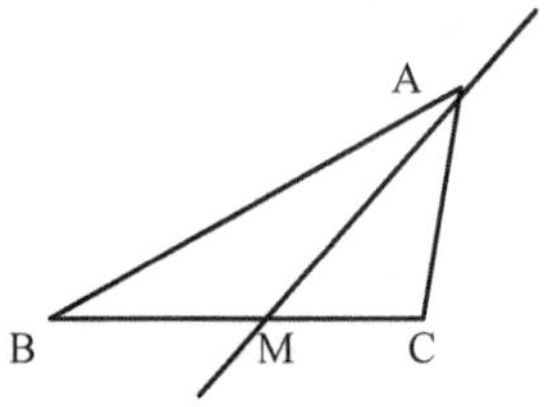

Figure 3.23   *Line AM is the median of triangle ABC from vertex A*

through A, hence, its equation is $7x + 5y - 49 = 0$. Equations of AB and AC are $9x - 4y + 10 = 0$ and $4x + 3y - 29 = 0$ respectively, on that account, equation of the angular bisector of angle BAC is $\dfrac{4x + 3y - 29}{\sqrt{3^2 + 4^2}} = -\dfrac{9x - 4y + 10}{\sqrt{9^2 + 4^2}}$ or $4x\sqrt{97} + 45x - 20y + 3y\sqrt{97} + 50 - 29\sqrt{97} = 0$. Apply Maple functions 'median', 'altitude', and 'bisector' respectively for all these:

MATLAB Command

    >>maple('with(geometry)'); ↵
    >>maple('triangle(T, [point(A,2,7), point(B,-2,-2), point(C,5,3)],[x,y])'); ↵
    >>maple('median(AM,A,T)'); ↵   ← AM represents an equation
    >>maple('Equation(AM)') ↵

ans =

    -19/2+13/2*x-1/2*y = 0

One may be interested to see the end points of the median segment. To do so,

    >>maple('median(AM,A,T,M)'); ↵   ← AM represents line segment
    >>maple('map(coordinates,DefinedAs(AM))') ↵ ← To display end points of AM

ans =

    [[2, 7], [3/2, 1/2]]

Now execute the commands for the altitude,

```
>>maple('altitude(AH,A,T)');          ← We name the altitude as AH
>>maple('Equation(AH)')
```

ans =

-49+7*x+5*y = 0

Like median, you can see the end points of altitude segment AH:

```
>>maple('altitude(AH,A,T,H)');
>>maple('map(coordinates,DefinedAs(AH))')
```

ans =

[[2, 7], [363/74, 217/74]]   ← End points of the altitude segment

Perform the commands for angular bisector too:

```
>>maple('bisector(AS,A,T)');          ← We name the angular bisector as AS
>>maple('simplify(Equation(AS))')  ← See the simplified equation of AS
```

ans =

4*x*97^(1/2)+45*x-20*y+3*y*97^(1/2)+50-29*97^(1/2) = 0

```
>>maple('bisector(AS,A,T,S)');
>>maple('map(coordinates,DefinedAs(AS))')
```

ans =

[[2, 7], [5*(-2+97^(1/2))/(97^(1/2)+5), (-10+3*97^(1/2))/(97^(1/2)+5)]]   ← End points of AS

⌑ Are three straight lines concurrent?

Three straight lines are said to be concurrent if they pass through a single point o as shown in figure 3.24. Test lines are   L1: $2x - 7y = 4$,   L2: $3x + 4y = 4$,   and   L3: $4x - y = 3$. Procedure 'AreConcurrent' can ascertain this. Intersection point of lines L1 and L2 are $\left(\dfrac{44}{29}, -\dfrac{4}{29}\right)$. The coordinates of intersection point do not satisfy L3, hence, the lines are not concurrent. Again, assume that equation of L3 is modified as $4x - by = 3$. To have them concurrent, what is the condition? Satisfy the last equation by intersection point to get $4b + 89 = 0$. Both implementations are shown below:

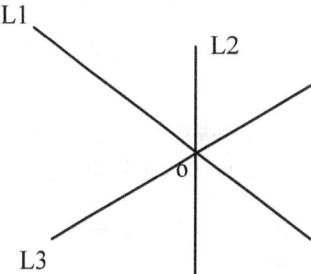

Figure 3.24   *Three concurrent lines L1, L2, and L3 pass through a single point o*

MATLAB Command

```
>>maple('with(geometry)');
>>maple('line(L1,2*x-7*y=4,[x,y]),line(L2,3*x+4*y=4,[x,y]),line(L3,4*x-y=3,[x,y])');
>>maple('AreConcurrent(L1,L2,L3)')
```

ans =

false

Enter the modified equation of L3:

```
>>maple('line(L3,4*x-b*y=3,[x,y])');
>>maple('AreConcurrent(L1,L2,L3,cond)')
```

ans =

AreConcurrent: unble to determine if 89+4*b is zeroFAIL

🖅 *Are four points concyclic?*

Four points are said to be concyclic when they lie on a circle. If circle formed by three points is satisfied by the forth one, the condition is fulfilled. Assume that the given four points are A(2, 8), B(3, 7), C(0, 9), and E(−1, 5). Circle passing through A, B, and C is $x^2 + y^2 + 3x - 7y - 18 = 0$ (see section 3.20.5 for the construction of a circle). Point E does not satisfy the equation of the circle, from that cause, point A, B, C, and E are not concyclic:

MATLAB Command

>>maple('with(geometry)'); ↵
>>maple('point(A,2,8),point(B,3,7),point(C,0,9),point(E,-1,5)'); ↵
>>maple('AreConcyclic(A,B,C,E)') ↵

ans =

false

We are sure that the given four points are not concyclic. Presuppose that you have the flexibility to change the y coordinate of point E. To have four points concyclic, what is the condition then? Find it as follows:

>>maple('point(E,-1,a)'); ↵   ← Just enter point E
>>maple('AreConcyclic(A,B,C,E,cond)'); ↵ ← Find the condition
>>maple('cond') ↵

ans =

-4/166911*(-20+a^2-7*a)/(2+a^2) = 0

From the return, it is obvious that the condition is $-\dfrac{4(a^2 - 7a - 20)}{166911(a^2 + 2)} = 0$ or $a^2 - 7a - 20 = 0$.

🖅 *Is a line tangent to a circle?*

Say, the line is L: $5x - 8y = 3$ and the circle is C: $2x^2 + 2y^2 - 4x + 7y + 2 = 0$. If distance between center of C to line L is equal to the radius of circle C, the condition is satisfied. Center and radius of C are $\left(1, -\dfrac{7}{4}\right)$

and $\dfrac{7}{4}$ respectively. Distance from the center to the line L is $\dfrac{5 \times 1 - 8\left(-\dfrac{7}{4}\right) - 3}{\sqrt{5^2 + (-8)^2}} = \dfrac{16}{\sqrt{89}}$, on this account, L is not

a tangent to C. Verify this by the procedure 'AreTangent':

MATLAB Command

>>maple('with(geometry)'); ↵
>>maple('line(L,5*x-8*y=3,[x,y])'); ↵
>>maple('circle(C,2*x^2+2*y^2-4*x+7*y+2=0,[x,y])'); ↵
>>maple('AreTangent(L,C)') ↵

ans =

false

⌗ *Finding the orthocenter of a triangle*

Perpendiculars drawn from the vertices to the opposite sides of a triangle are concurrent. That concurrent point is called the orthocenter of the triangle. Presume that the triangle ABC is constructed from points A(8, 2), B(2, −1), and C(2, −3). From the vertices A, B, and C, three perpendicular lines AG, BF, and CE are drawn to opposite sides BC, CA, and AB respectively (see attached figure 3.25). The three perpendicular lines meet at orthocenter O. Our objective is to find the coordinates of O. Equations of AG, CE, and BF are $y = 2$, $7 - 6x - 5y = 0$, and $2x + y - 1 = 0$ respectively. Intersection of any two lines is the orthocenter whose coordinates are $\left(-\dfrac{1}{2}, 2\right)$. Conduct the procedure 'orthocenter' for this:

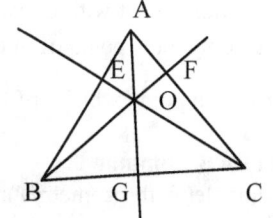

Figure 3.25 *O is the orthocenter of triangle ABC*

MATLAB Command
```
>>maple('with(geometry)');  ↵
>>maple('triangle(T,[point(A,8,2), point(B,2,-1), point(C,2,-3)])');  ↵      ← Form triangle T
>>maple('orthocenter(O,T)');  ↵                                              ← Find orthocenter object O
>>maple('coordinates(O)')  ↵                                                 ← Display coordinates of O
```

ans =

[-1/2, 2]

⌗ *Are two circles orthogonal?*

Two nonconcentric circles are said to be orthogonal if tangents drawn at the point of intersection of the two circles are right angle to each other. According to attached figure 3.26, I is the intersection point and M and N are centers of the two circles. Condition of orthogonality is gratified if the circles follow $MI^2 + NI^2 = MN^2$, where MI and NI are the radius of the circles respectively. Like before, equations of the two circles are given by C1: $x^2 + y^2 - 4x - 4y - 1 = 0$ and C2: $x^2 + y^2 + 6x - 4y - 3 = 0$. Centers and radius of C1 and C2 are M (2, 2) and N (−3, 2) and 3 and 4 respectively. Distance between the centers, MN, is 5, hence, $MI^2 + NI^2 = MN^2$ is satisfied and C1 and C2 are orthogonal. Make certain that as follows:

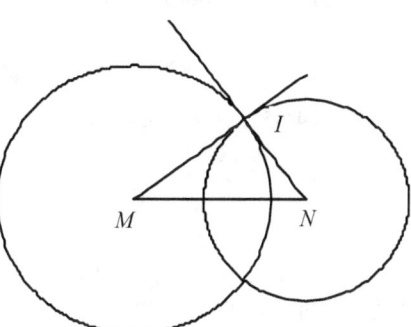

Figure 3.26 *Tangents of orthogonal circles meet at right angle to each other at the point of intersection*

MATLAB Command
```
>>maple('with(geometry)');  ↵
>>maple('circle(C1,x^2+y^2-4*x-4*y-1=0,[x,y])');  ↵
>>maple('circle(C2,x^2+y^2+6*x-4*y-3=0,[x,y])');  ↵
>>maple('AreOrthogonal(C1,C2)')  ↵
```

ans =

true

⌐ *Intersection of some curves*

Geometric objects can intersect. Coordinates of their intersection points are found by function 'intersection'. Start with equations of lines L1: $5x - 8y = 3$ and L2: $7x - 2y = 5$. Both equations are of degree 1, hence, only one point of intersection is there. Solving equations L1 and L2 provides coordinates of the intersection point, which are $\left(\dfrac{17}{23}, \dfrac{2}{23} \right)$.

MATLAB Command
```
>>maple('with(geometry)');  ↵
>>maple('line(L1,5*x-8*y=3,[x,y]),line(L2,7*x-2*y=5,[x,y])');  ↵    ← Enter the lines
>>maple('intersection(I,L1,L2)');  ↵                               ← Intersection object is I
>>maple('coordinates(I)')  ↵
```

ans =

[17/23, 2/23]

Our second example is finding the intersection points of a line and a circle. Assumed equations are L: $x - 8y = 3$ and C: $x^2 + y^2 - x - y - 1 = 0$. Highest degree of the curves are 2, so, there are two intersecting points. Solve the equations of L and C to have the points as $M\left(\dfrac{3}{5} + \dfrac{4\sqrt{221}}{65}, \ -\dfrac{3}{10} + \dfrac{\sqrt{221}}{130} \right)$ and $N\left(\dfrac{3}{5} - \dfrac{4\sqrt{221}}{65}, \ -\dfrac{3}{10} - \dfrac{\sqrt{221}}{130} \right)$.

Have the implementation as follows:
```
>>maple('line(L,x-8*y=3,[x,y]),circle(C,x^2+y^2-x-y-1=0,[x,y])');  ↵
>>maple('intersection(I,L,C,[M,N])');  ↵ ← Object is I and M and N are two components of I
>>M=maple('coordinates(M)')  ↵           ← We assign coordinates of M to M for later use
```

M =

[3/5+4/65*221^(1/2), -3/10+1/130*221^(1/2)]
```
>>N=maple('coordinates(N)')  ↵
```

N =

[3/5-4/65*221^(1/2), -3/10-1/130*221^(1/2)]

One may not appreciate the long string of output. Carry out command 'pretty' to print output in symbolic form:
```
>>pretty(sym(M))  ↵
```

```
         [              1/2                      1/2]
         [3/5 + 4/65  221          - 3/10 + 1/130   221   ]
>>pretty(sym(N))  ↵
```

```
         [              1/2                      1/2 ]
         [3/5 - 4/65  221          - 3/10 - 1/130   221   ]
```

Thus intersection of parabola, ellipse, hyperbola, or any other curves can be obtained by applying the procedure 'intersection'.

⌐ *Circumcircle, incircle, and excircle of a triangle*

Suppose, a triangle is formed from three lines L1: $3x - 4y - 1 = 0$, and L2: $4x - 3y + 5 = 0$, and L3: $8x + 6y - 9 = 0$. We want to find circumcircle, incircle, and excircle of the triangle. Maple functions are also having identical names, as do the circles. Various circles of a triangle are shown in figure 3.27. From

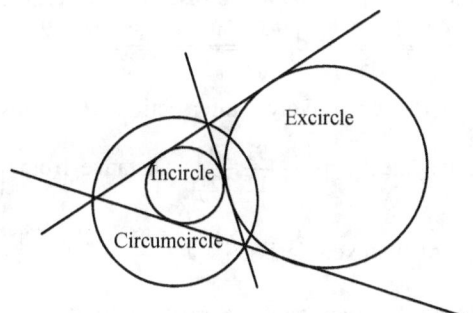

Figure 3.27 *Different circles of a triangle*

intersection of lines, vertices of the triangle are $\left(-\dfrac{23}{7},-\dfrac{19}{7}\right)$, $\left(-\dfrac{1}{16},\dfrac{19}{12}\right)$, and $\left(\dfrac{21}{25},\dfrac{19}{50}\right)$. Circumcircle is the

circle that passes through the vertices of the triangle, thereupon, its equation is given by $x^2+y^2+\dfrac{375}{112}x+\dfrac{95}{84}y$

$-\dfrac{1375}{336}=0$ (see section 3.20.5 for the construction of a circle). Circumcircle is found as follows:

MATLAB Command
```
>>maple('with(geometry)'); ↵
>>maple('line(L1,3*x-4*y-1=0,[x,y]),line(L2,4*x-3*y+5=0,[x,y])'); ↵
>>maple('line(L3,8*x+6*y-9=0,[x,y])'); ↵
>>maple('triangle(T,[L1,L2,L3])'); ↵   ← Form triangle T from L1, L2, and L3
>>maple('circumcircle(C,T)'); ↵        ← Circumcircle is assigned to C
>>maple('Equation(C)') ↵

ans =

x^2-1375/336+y^2+375/112*x+95/84*y = 0
```

Angular bisectors of a triangle are concurrent. The concurrent point is the center of the incircle. Equations of

the angular bisectors of angles between L1 and L2 and L2 and L3 are $\dfrac{3x-4y-1}{\sqrt{3^2+4^2}}=-\dfrac{4x-3y+5}{\sqrt{4^2+3^2}}$ and

$\dfrac{4x-3y+5}{\sqrt{4^2+3^2}}=-\dfrac{8x+6y-9}{\sqrt{8^2+6^2}}$ respectively. On simplification, they become $7x-7y+4=0$ and $16x+1=0$

respectively. Solve the last two equations to get incenter $\left(-\dfrac{1}{16},\dfrac{57}{112}\right)$. Distance from incenter to any of the

given three lines is the radius of incircle $\left(\dfrac{361}{560}\right)$, on that cause, equation of the incircle is

$x^2+y^2+\dfrac{x}{8}-\dfrac{57y}{56}-\dfrac{47871}{313600}=0$. Necessary commands are as follows:

```
>>maple('incircle(I,T)'); ↵        ← Object incircle is assigned to I
>>maple('Equation(I)') ↵

ans =
```

$$x^2 - 47871/313600 + y^2 + 1/8*x - 57/56*y = 0$$

See figure 3.27 for excircle of a triangle. External angle bisectors of vertices formed by L1 and L2 and by L2 and L3 are given by $\dfrac{3x - 4y - 1}{\sqrt{3^2 + 4^2}} = \dfrac{4x - 3y + 5}{\sqrt{4^2 + 3^2}}$ and $\dfrac{4x - 3y + 5}{\sqrt{4^2 + 3^2}} = \dfrac{8x + 6y - 9}{\sqrt{8^2 + 6^2}}$ respectively. Simplify the equations to write $x + y + 6 = 0$ and $12y - 19 = 0$ respectively. Intersection of the last two equations narrates that the center of the excircle touching the line L2 is $\left(-\dfrac{91}{12}, \dfrac{19}{12}\right)$. Distance from $\left(-\dfrac{91}{12}, \dfrac{19}{12}\right)$ to L2 is the radius of excircle $\left(\dfrac{361}{60}\right)$, on that, equation of the excircle is $x^2 + y^2 + \dfrac{91x}{6} - \dfrac{19y}{6} + \dfrac{85729}{3600} = 0$. Figure 3.27 depicts only one excircle of the triangle. There are two more in relation to the other two sides. Avoid the clumsy manipulation by MATLAB as follows:

MATLAB Command

```
>>maple('excircle(E,T)');  ⏎        ← Form the excircle object
>>maple('Equation(E[1])')  ⏎        ← Print the first excircle equation

ans =

x^2+492529/176400+y^2-85/42*x-19/6*y = 0
>>maple('Equation(E[2])')  ⏎        ← Print the second excircle equation.
                                      We found this in preceding discussion.
ans =

x^2+85729/3600+y^2+91/6*x-19/6*y = 0
>>maple('Equation(E[3])')  ⏎        ← Print the third excircle equation

ans =

x^2+95329/6400+y^2+1/8*x+95/8*y = 0
```

Determine medial triangle of a triangle

Medial triangle is a triangle, which is formed by joining the midpoints of the sides of a triangle. Concerning figure 3.28, inner triangle DEF is the medial triangle of the larger triangle ABC. Consider the triangle formed from the three lines of the previous example. Vertices of the triangle for the previous example are $\left(-\dfrac{23}{7}, -\dfrac{19}{7}\right)$, $\left(-\dfrac{1}{16}, \dfrac{19}{12}\right)$, and $\left(\dfrac{21}{25}, \dfrac{19}{50}\right)$. Their midpoints are $\left(-\dfrac{375}{224}, -\dfrac{95}{168}\right)$, $\left(\dfrac{311}{800}, \dfrac{589}{600}\right)$, and $\left(-\dfrac{214}{175}, -\dfrac{817}{700}\right)$ respectively.

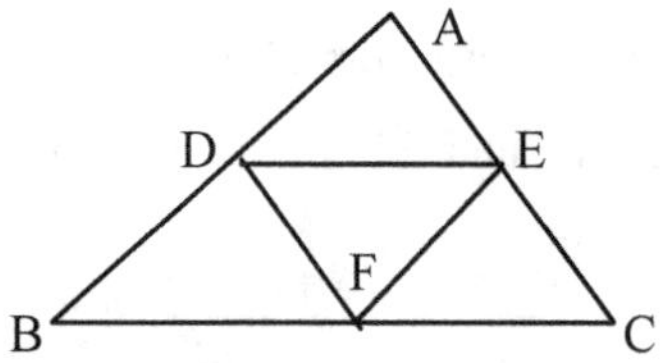

Figure 3.28 *DEF is medial triangle of triangle ABC*

MATLAB Command

```
>>maple('medial(M,T)');  ⏎                          ← Medial triangle object is M
>>O=maple('map(coordinates,DefinedAs(M))')  ⏎       ←Display vertices of M and assign that to O

O =

[[-214/175, -817/700], [-375/224, -95/168], [311/800, 589/600]]
>>pretty(sym(O))  ⏎                                 ← Display symbolic form of the vertices
```

$$\left[\left[\frac{-214}{175}, \frac{-817}{700}\right], \left[\frac{-375}{224}, \frac{-95}{168}\right], \left[\frac{311}{800}, \frac{589}{600}\right]\right]$$

☞ *Centroid of a polygon*

The function used for this purpose is 'centroid'. Say, a quadrilateral is formed from four vertices given by A(2, cos*t*), B(3, 7), C(0, 9), and E(−1, 5) (as depicted in figure 3.29). Coordinates of the centroid are given by $\left(\dfrac{2+3+0-1}{4}, \dfrac{\cos t+7+9+5}{4}\right)$ or $\left(1, \dfrac{21}{4}+\dfrac{\cos t}{4}\right)$.

That is what is returned below:

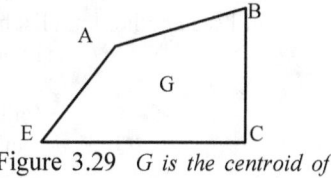

Figure 3.29 *G is the centroid of quadrilateral ABCE*

MATLAB Command
>>maple('with(geometry)'); ↵
>>maple('point(A,2,cos(t)),point(B,3,7),point(C,0,9),point(E,-1,5)'); ↵
>>maple('centroid(G,[A,B,C,E])'); ↵   ← Centroid object is assigned to G
>>maple('coordinates(G)') ↵

ans =

[1, 1/4*cos(t)+21/4]

3.25 Help about the geometry package

Many more functions of the package are yet to be examined. You have seen just a flavor what geometry package can implement. Maybe, you are a smart user of MATLAB packages. Explore comprehensive help of the geometry package typing the following:

MATLAB Command
>>mhelp geometry ↵

Help For: Introduction to the geometry package

Calling Sequence:
  function(args)
  geometry[function](args)

Description:
       ⋮
       ⋮
       ⋮

MATLAB displays a long list help about the available functions in the geometry package. Slide the vertical scroll bar with the help of mouse to see different help paragraphs completely. From the displayed help topic, choose any Maple procedure. Suppose we selected 'circle'. To get the help about circle,

MATLAB Command
>>mhelp circle ↵

Multiple matches found:
    plottools,circle
    geometry,circle

116

Above output indicates that by name circle there are two Maple procedures − one is located in 'plottools' package and the other is in 'geometry' package. Specifically, to display the help of procedure 'circle' of geometry package,

MATLAB Command
>>mhelp geometry[circle] ↵
You will see another long list help regarding circle as follows:
Function: geometry[circle] - define the circles

Calling Sequence:
   circle(c, [A, B, C], n, 'centername'=m)
   circle(c, [A, B], n, 'centername'=m)
   circle(c, [A, rad], n, 'centername'=m)
   circle(c, eqn, n, 'centername'=m)

$\vdots$

In a similar fashion one can explore details of any other procedure of the package. Table 3.C presents some functions found in the geometry package.

Table 3.C   Maple geometry functions

Name of the Maple function	Purpose of the function
EulerLine	find the Euler line of a given triangle
SimsonLine	find the Simson line of a given triangle with respect to a given point on the circumcircle of the triangle
Appolonius	find the Appolonius circles of three given circles
PedalTriangle	find the pedal triangle of a point with respect to a triangle
RadicalCenter	find the radical center of three given circles
RadicalAxis	find the radical axis of two given circles
CircleOfSimilitude	find the circle of similitude of two circles
TangentLine	find the tangents of a point with respect to a circle
dsegment	define a directed segment
AreConjugate	test if two triangles are conjugate for a circle
convexhull	find the convex hull enclosing the given points
similitude	find the insimilitude and outsimilitude of two circles
stretch	find the stretch of a geometric object
inversion	find the inversion of a point, line, or circle with respect to a given circle
AreHarmonic	test if a pair of points is harmonic conjugate to another pair of points
GlideReflection	find the glide-reflection of a geometric object
SpiralRotation	find the spiral-rotation of a geometric object
GergonnePoint	find the Gergonne point of a given triangle
CrossProduct	compute the cross product of two directed segment
Pole	find the pole of a given line with respect to a given conic or a given circle
IsEquilateral	test if a given triangle is equilateral
IsRightTriangle	test if a given triangle is a right triangle
NagelPoint	find the Nagel point of a given triangle
powerpc	compute power of a given point with respect to a given circle
AreParallel	test if two lines are parallel to each other

Chapter 4

Matrix Algebra

Elements arranged in ordered rectangular array are called a matrix. Elements can be symbols, integers, floating-point numbers, or even complex numbers. In MATLAB, the word 'variable' usually refers to a matrix. The primary objective of MATLAB's advent was to facilitate the computation of matrix algebra. But now a days its capability has reached far beyond the objective. Enhanced with Maple package, MATLAB offers a great flexibility of handling various kinds of elements – symbolic or numeric. Rapid use of sophisticated techniques in physical science majors and engineering, especially in control theory, it compels a real understanding of matrix algebra. Before writing the programs or executing MATLAB commands employing matrix algebra, one should have some intuition, which serves as background and helps relate the matrix manipulation such as matrix diagonalization, matrix inversion, matrix decomposition ... etc to develop a complete program.

4.1 Formation of matrix of ones and zeroes

Matrix of ones is necessary in some matrix manipulations. Perform the following commands to form matrix of ones. We form matrices $A = \begin{bmatrix} 1 & 1 & 1 \\ 1 & 1 & 1 \\ 1 & 1 & 1 \\ 1 & 1 & 1 \end{bmatrix}$, $B = \begin{bmatrix} 1 & 1 & 1 \\ 1 & 1 & 1 \\ 1 & 1 & 1 \end{bmatrix}$, and $C = \begin{bmatrix} 1 & 1 & 1 & 1 \\ 1 & 1 & 1 & 1 \end{bmatrix}$. Their orders are 4×3, 3×3, and 2×4 for A, B, and C respectively. To form a matrix of ones,

MATLAB Step:
Use the command ones(row number, column number).

Either the number of rows or the number of columns will do if the matrix is square. For the row and column matrices, the commands would be 'ones(1, column number)' and 'ones(row number, 1)' respectively. Generation of A, B, and C is shown below:

MATLAB Command

for A,   for B,
>>A=ones(4,3) ↵   >>B=ones(3) ↵

A =

1	1	1
1	1	1
1	1	1
1	1	1

B =

1	1	1
1	1	1
1	1	1

for C,

>>C=ones(2,4) ↵

C =

1	1	1	1
1	1	1	1

Formation of matrix of zeroes is quite similar to the formation of matrix of ones. Replacing the subroutine 'ones' by 'zeros' does the formation. To form a matrix of zeroes,

MATLAB Step:
Use the command zeros (row number, column number).

Generate matrices of zeroes like $A = \begin{bmatrix} 0 & 0 & 0 \\ 0 & 0 & 0 \\ 0 & 0 & 0 \\ 0 & 0 & 0 \end{bmatrix}$, $B = \begin{bmatrix} 0 & 0 & 0 \\ 0 & 0 & 0 \\ 0 & 0 & 0 \end{bmatrix}$, and $C = \begin{bmatrix} 0 & 0 & 0 & 0 \\ 0 & 0 & 0 & 0 \end{bmatrix}$ whose orders are 4×3, 3×3,

and 2×4 respectively. Formation of A, B, and C is shown below:

Command for A,

>>A=zeros(4,3) ↵

for B,

>>B=zeros(3) ↵

A =

0	0	0
0	0	0
0	0	0
0	0	0

B =

0	0	0
0	0	0
0	0	0

for C,

>>C=zeros(2,4) ↵

C =

0	0	0	0
0	0	0	0

4.2 Formation of identity matrices

An identity matrix is one whose diagonal elements are unity and off-diagonal elements are zeroes. According to matrix algebra, identity matrices are defined for square matrices, but in MATLAB, you can have identity matrix for a rectangular matrix too. To generate an identity matrix,

MATLAB Step:
Use the command eye(row number, column number).

If the matrix is square, there is no need to write both the column and row numbers, only one will do. Examples

of identity matrices are $A = \begin{bmatrix} 1 & 0 & 0 \\ 0 & 1 & 0 \\ 0 & 0 & 1 \end{bmatrix}$, $B = \begin{bmatrix} 1 & 0 & 0 \\ 0 & 1 & 0 \end{bmatrix}$, and $C = \begin{bmatrix} 1 & 0 & 0 \\ 0 & 1 & 0 \\ 0 & 0 & 1 \\ 0 & 0 & 0 \end{bmatrix}$. Orders of these matrices are

3×3, 2×3, and 4×3 for A, B, and C respectively. How A, B, and C can be formed is shown as follows:

MATLAB Command

for A,
>>A=eye(3) ↵

A =
```
1   0   0
0   1   0
0   0   1
```
for B,
>>B=eye(2,3) ↵

B =
```
1   0   0
0   1   0
```

for C,
>>C=eye(4,3) ↵

C =
```
1   0   0
0   1   0
0   0   1
0   0   0
```

4.3 Formation of diagonal matrices

In a diagonal matrix all elements are zeroes except the diagonal elements (not all elements in the diagonal are zeroes, some elements in the diagonal can be zeroes). To generate a diagonal matrix,

MATLAB Steps:
1. *Assign diagonal elements as row or column matrix to A*
2. *Use the command B=diag(A).*

$B = \begin{bmatrix} 2 & 0 & 0 \\ 0 & 6 & 0 \\ 0 & 0 & 7 \end{bmatrix}$ is the example of a diagonal matrix. Generate the matrix as follows:

MATLAB Command
>>A=[2 6 7]; ↵
>>B=diag(A) ↵

B =
```
2   0   0
0   6   0
0   0   7
```

Elements in matrix A can be floating-point or integer numbers. The diagonal elements can be symbolic too.

Such a matrix is $\begin{bmatrix} 1 & 0 & 0 \\ 0 & x & 0 \\ 0 & 0 & x+5 \end{bmatrix}$. But you need symbolic declaration for this implementation:

MATLAB Command
>>syms x ↵
>>A=[1 x x+5]; ↵
>>B=diag(A) ↵

B =
```
[ 1,   0,    0]
[ 0,   x,    0]
[ 0,   0,  x+5]
```

4.4 Reduced row echelon form of a matrix

Reduced row echelon form of a matrix is required to find the rank of a matrix or linear depenedence of vectors. To obtain the reduced row echelon form of a matrix,

MATLAB Step:

Use the command rref (matrix name).

The subroutine 'rref' is the abbreviation of r̲educed r̲ow e̲chelon f̲orm. See illustrative example for $A =$
$\begin{bmatrix} 6 & 7 & -8 & 8 \\ -1 & -2 & 5 & 3 \\ -3 & 0 & 7 & 5 \end{bmatrix}$. Elementary transformations are applied to obtain the reduced row echelon form of a

matrix. Assume that R_i stands for the ith row of A. The following transformation changes A to reduced row echelon form:

$$\begin{bmatrix} 6 & 7 & -8 & 8 \\ -1 & -2 & 5 & 3 \\ -3 & 0 & 7 & 5 \end{bmatrix} \xrightarrow{\frac{R_1}{6}} \begin{bmatrix} 1 & \frac{7}{6} & -\frac{8}{6} & \frac{8}{6} \\ -1 & -2 & 5 & 3 \\ -3 & 0 & 7 & 5 \end{bmatrix} \xrightarrow{R_1+R_2 \quad and \quad 3R_1+R_3} \begin{bmatrix} 1 & \frac{7}{6} & -\frac{8}{6} & \frac{8}{6} \\ 0 & -\frac{5}{6} & \frac{22}{6} & \frac{26}{6} \\ 0 & \frac{7}{2} & 3 & 9 \end{bmatrix}$$

$$\xrightarrow{\frac{-6R_2}{5}} \begin{bmatrix} 1 & \frac{7}{6} & -\frac{8}{6} & \frac{8}{6} \\ 0 & 1 & -\frac{22}{5} & -\frac{26}{5} \\ 0 & \frac{7}{2} & 3 & 9 \end{bmatrix} \xrightarrow{R_1-\frac{7R_2}{6} \quad and \quad R_3-\frac{7R_2}{2}} \begin{bmatrix} 1 & 0 & \frac{19}{5} & \frac{37}{5} \\ 0 & 1 & -\frac{22}{5} & -\frac{26}{5} \\ 0 & 0 & \frac{92}{5} & \frac{136}{5} \end{bmatrix} \xrightarrow{\frac{5R_3}{92}}$$

$$\begin{bmatrix} 1 & 0 & \frac{19}{5} & \frac{37}{5} \\ 0 & 1 & -\frac{22}{5} & -\frac{26}{5} \\ 0 & 0 & 1 & \frac{34}{23} \end{bmatrix} \xrightarrow{R_2+\frac{22R_3}{5} \quad and \quad R_1-\frac{19R_3}{5}} \begin{bmatrix} 1 & 0 & 0 & \frac{41}{23} \\ 0 & 1 & 0 & \frac{30}{23} \\ 0 & 0 & 1 & \frac{34}{23} \end{bmatrix} = \begin{bmatrix} 1 & 0 & 0 & 1.7826 \\ 0 & 1 & 0 & 1.3043 \\ 0 & 0 & 1 & 1.4783 \end{bmatrix}.$$

Rational form output can be seen by using the command 'sym'. If a matrix has the number of rows greater than the number of columns, the reduced row echelon form of that matrix is a matrix of ones and zeroes. An

example can be $B = \begin{bmatrix} 6 & 7 & -8 \\ -1 & -2 & 5 \\ -3 & 0 & 7 \\ 10 & 13 & 0 \end{bmatrix}$. After elementary transformations, one ends up with $\begin{bmatrix} 1 & 0 & 0 \\ 0 & 1 & 0 \\ 0 & 0 & 1 \\ 0 & 0 & 0 \end{bmatrix}$ for the

row echelon form of B (since the rank of B is 3, not shown here). See both implementations as follows:

MATLAB Command

```
>>A=[6 7 -8 8;-1 -2 5 3;-3 0 7 5];  ↵
>>rref(sym(A)) ↵

ans =

[   1,    0,    0,    41/23]
[   0,    1,    0,    30/23]
[   0,    0,    1,    34/23]
>>rref(A) ↵                          ←For floating-point form

ans =
            1.0000          0         0      1.7826
                 0     1.0000         0      1.3043
                 0          0    1.0000      1.4783
>>B=[6 7 -8;-1 -2 5;-3 0 7;10 13 0];  ↵     ←For matrix B
>>rref(B) ↵
```

ans =

$$\begin{matrix} 1 & 0 & 0 \\ 0 & 1 & 0 \\ 0 & 0 & 1 \\ 0 & 0 & 0 \end{matrix}$$

4.5 Pivoting about a matrix entry

If A denotes a matrix, any element in the matrix is A_{ij}, where i and j correspond to the row and column indexes respectively. Given that A_{ij} is not equal to zero, it is possible to make lower and upper entries of A_{ij} zero applying the elementary transformations. This is called pivoting of matrix A about the entry A_{ij}.

Computation is shown for $A = \begin{bmatrix} 2 & 3 & -4 \\ -1 & -3 & -7 \\ 3 & -7 & -5 \end{bmatrix}$ about A_{23}. $A_{23} = -7$, so, operations $R_1 - (-4)\dfrac{R_2}{-7}$ and

$R_3 - (-5)\dfrac{R_2}{-7}$ provide $\begin{bmatrix} \dfrac{18}{7} & \dfrac{33}{7} & 0 \\ -1 & -3 & -7 \\ \dfrac{26}{7} & -\dfrac{34}{7} & 0 \end{bmatrix}$ (where R_i stands for the ith row of A). Apply the Maple function

'pivot' as follows:

MATLAB Command

```
>>A=sym([2 3 -4;-1 -3 -7;3 -7 -5]); ↵
>>maple('pivot',A,2,3) ↵
```

ans =

```
[ 18/7,    33/7,    0]
[   -1,      -3,   -7]
[ 26/7,   -34/7,    0]
```

4.6 Minor of a matrix

Minor of a matrix A about an entry A_{ij} is defined as the matrix formed by removing the ith row and j^{th} column. To proceed with example, take $A = \begin{bmatrix} 3 & 4 & 2 & 9 \\ 0 & 7 & 6 & 7 \\ 0 & 1 & 5 & 9 \end{bmatrix}$. Element 1 is the entry of A for which $i = 3$ and $j = 2$. So, the minor about A_{32} is given by $B = \begin{bmatrix} 3 & 2 & 9 \\ 0 & 6 & 7 \end{bmatrix}$. Maple function 'minor' picks up the minor from a matrix:

MATLAB Command

```
>>A=sym([3 4 2 9;0 7 6 7;0 1 5 9]); ↵     ←Enter matrix A as symbolic
>>B=maple('minor',A,3,2) ↵                ←Apply the function 'minor'
```

B =

```
[3,  2,  9]
[0,  6,  7]
```

4.7 Adjoint of a square matrix

Adjoint of a square matrix is defined as the transpose of the matrix formed by co-factors of the elements of the square matrix. For citation, say, we have the square matrix $A = \begin{bmatrix} 3 & 4 & 2 \\ 0 & 7 & 6 \\ 1 & 5 & x \end{bmatrix}$, on that, the adjoint

of A is given by $\begin{bmatrix} \begin{pmatrix} 7 & 6 \\ 5 & x \end{pmatrix} & -\begin{pmatrix} 0 & 6 \\ 1 & x \end{pmatrix} & \begin{pmatrix} 0 & 7 \\ 1 & 5 \end{pmatrix} \\ -\begin{pmatrix} 4 & 2 \\ 5 & x \end{pmatrix} & \begin{pmatrix} 3 & 2 \\ 1 & x \end{pmatrix} & -\begin{pmatrix} 3 & 4 \\ 1 & 5 \end{pmatrix} \\ \begin{pmatrix} 4 & 2 \\ 7 & 6 \end{pmatrix} & -\begin{pmatrix} 3 & 2 \\ 0 & 6 \end{pmatrix} & \begin{pmatrix} 3 & 4 \\ 0 & 7 \end{pmatrix} \end{bmatrix}^T = \begin{bmatrix} 7x-30 & 6 & -7 \\ -4x+10 & 3x-2 & -11 \\ 10 & -18 & 21 \end{bmatrix}^T =$

$\begin{bmatrix} 7x-30 & -4x+10 & 10 \\ 6 & 3x-2 & -18 \\ -7 & -11 & 21 \end{bmatrix}$. Maple procedure 'adjoint' or 'adj' situated in package 'linalg' finds the adjoint

of A as shown below:

MATLAB Command

>>syms x ↵   ← Declare x as symbolic
>>A=[3 4 2;0 7 6;1 5 x]; ↵
>>B=maple('adj',A) ↵   ← Assign adjoint matrix to B

B =

```
[ 7*x-30,   -4*x+10,    10]
[      6,     3*x-2,   -18]
[     -7,       -11,    21]
```

4.8 Gaussian elimination of a matrix

Gaussian elimination means making the lower triangle of a matrix zero by performing the elementary row operations. Begin with the example of $A = \begin{bmatrix} x & 1 & 0 \\ 2 & x & -3x \\ 1 & 3 & 5 \end{bmatrix}$. The second row of A is having symbolic expressions. There is no harm in interchanging the rows during gaussian elimination. By doing so, we avoid clumsy expression or zero entry. Interchanging the second and third rows, we have $\begin{bmatrix} x & 1 & 0 \\ 1 & 3 & 5 \\ 2 & x & -3x \end{bmatrix}$. Perform

elementary operations $R_2 - \dfrac{R_1}{x}$ and $R_3 - 2 \times \dfrac{R_1}{x}$ to get $\begin{bmatrix} x & 1 & 0 \\ 0 & \dfrac{3x-1}{x} & 5 \\ 0 & \dfrac{x^2-2}{x} & -3x \end{bmatrix}$, then, $R_3 - \left(\dfrac{x^2-2}{x} \right) \times \dfrac{R_2}{\dfrac{3x-1}{x}}$ to have

$\begin{bmatrix} x & 1 & 0 \\ 0 & \dfrac{3x-1}{x} & 5 \\ 0 & 0 & -\dfrac{14x^2-3x-10}{3x-1} \end{bmatrix}$, which is the required form. Conduct Maple function 'gausselim' to obtain it:

MATLAB Command

```
>>syms x ↵
>>A=[x 1 0;2 x -3*x;1 3 5]; ↵
>>B=maple('gausselim',A); ↵   ← The output matrix is put to B
>>pretty(B) ↵
```

```
[x          1              0          ]
[                                     ]
[        3 x - 1                      ]
[0      -----------      5            ]
[           x                         ]
[                                     ]
[                     2               ]
[                14 x  - 3 x - 10     ]
[0        0     - ---------------------]
[                     3 x - 1         ]
```

Another variant of the function is 'ffgausselim' (abbreviation of fraction free gaussian elimination), which gives

fraction free matrix. The matrix following gaussian elimination is $\begin{bmatrix} x & 1 & 0 \\ 0 & \dfrac{3x-1}{x} & 5 \\ 0 & 0 & -\dfrac{14x^2-3x-10}{3x-1} \end{bmatrix}$. Multiply the

second row with x and the third row with $3x-1$, from that, we have $\begin{bmatrix} x & 1 & 0 \\ 0 & 3x-1 & 5x \\ 0 & 0 & -14x^2+3x+10 \end{bmatrix}$, which is a

fraction free matrix. Apply the function as follows:

```
>>B=maple('ffgausselim',A); ↵
>>pretty(B) ↵
```

```
[ x      1              0          ]
[                                  ]
[ 0    3 x - 1        5 x          ]
[                                  ]
[                      2           ]
[ 0      0        -14 x  + 3 x + 10 ]
```

4.9 Rank of matrices

The number of the linearly independent rows or columns of a matrix A is called the rank of A. If A is of order $M \times N$, then, the rank of $A \leq$ minimum(M , N). By performing the elementary row operations on the rows of a given matrix, a matrix can always be transformed to row echelon form. The number of independent rows of the transformed matrix determines the rank of the matrix. To find the rank of a matrix,

MATLAB Step:

Use the command rank(matrix name).

As an illustration, take the example of $A = \begin{bmatrix} 1 & 2 & 4 & 3 \\ 3 & -1 & 2 & -2 \\ 5 & -4 & 0 & -7 \end{bmatrix}$. Here, the order of A is 3×4, so, M =3 and

N =4. The rank of A must be less than or equal to minimum of 3 and 4, i.e., 3. Denote R_1 , R_2 , and R_3 for the first, second, and third rows respectively. Carrying out $R_2 - 3 \times R_1$ and $R_3 - 5 \times R_1$ on A provides

$$\begin{bmatrix} 1 & 2 & 4 & 3 \\ 0 & -7 & -10 & -11 \\ 0 & -14 & -20 & -22 \end{bmatrix}.$$ Again, perform $R_3 - 2 \times R_2$ on the matrix what we just obtained. This operation

provides us $\begin{bmatrix} 1 & 2 & 4 & 3 \\ 0 & -7 & -10 & -11 \\ 0 & 0 & 0 & 0 \end{bmatrix}$, which is the row echelon form of A. As you see, there are only two

linearly independent rows, therefore, the rank of A is 2, which is less than three. The rank of a row or column matrix is always 1. The elements in the matrix can be floating-point or integer numbers. If the elements are symbolic, even then, function 'rank' is useful provided that variables are declared symbolically. Given that the

rank of $A = \begin{bmatrix} x & x \\ 1 & 1 \end{bmatrix}$ is 1. See all implementations below:

MATLAB Command

for rectangular matrix,
>>A=[1 2 4 3;3 -1 2 -2; 5 -4 0 -7]; ↵
>>rank(A) ↵

ans =
2

for column matrix C = $\begin{bmatrix} 0 \\ -4 \\ 3 \end{bmatrix}$,

>>C=[0 -4 3]'; ↵
>>rank(C) ↵

ans =
1

for row matrix R=[4 –4 5 1 3],
>>R=[4 -4 5 1 3]; ↵
>>rank(R) ↵

ans =
1

for symbolic matrix $A = \begin{bmatrix} x & x \\ 1 & 1 \end{bmatrix}$,

>>syms x ↵
>>A=[x x;1 1]; ↵
>>rank(A) ↵

ans =
1

4.10 Determinant of a square matrix

Determinant of a matrix is possible if the matrix is square. Let us take an example of a square matrix

A, where $A = \begin{bmatrix} 2 & 4 & 3 \\ -4 & 6 & 9 \\ 3 & 7 & 10 \end{bmatrix}$. Determinant of A is denoted by $|A|$ and $|A| = 2 \times \begin{bmatrix} 6 & 9 \\ 7 & 10 \end{bmatrix} -$

$4 \times \begin{bmatrix} -4 & 9 \\ 3 & 10 \end{bmatrix} + 3 \times \begin{bmatrix} -4 & 6 \\ 3 & 7 \end{bmatrix} = 2(60 - 63) - 4(-40 - 27) + 3(-28 - 18) = 124$. To obtain the determinant of a square matrix,

MATLAB Steps:
1. *Enter the square matrix (A)*
2. *Use the command det(A).*

If you enter a rectangular matrix, an error message is seen. What if we have a matrix of symbolic elements,

say, $B = \begin{bmatrix} x & x^2 & y \\ 2x & x & 1 \\ 0 & 2 & 2 \end{bmatrix}$. The determinant of B is $x(2x - 2) - x^2(4x - 0) + y(4x) = 2x^2 - 2x - 4x^3 + 4xy$. There are

two symbolic variables, x and y, in B that must be declared before the matrix B is entered. Both examples are shown in the following page. The determinant of B is assigned to B to apply 'pretty' on B.

MATLAB Command

for numeric elements,   for symbolic elements,

        >>A=[2 4 3;-4 6 9;3 7 10]; ↵   >>syms x y ↵
        >>det(A) ↵   >>B=[x x^2 y;2*x x 1;0 2 2]; ↵
                                            >>B=det(B); ↵
        ans =   >>pretty(B) ↵
               124

$$2 x^2 - 2 x - 4 x^3 + 4 x y$$

4.11 Power of a square matrix

Power of a square matrix is raised by operator '^'. To raise the power of a square matrix,

MATLAB Step:

Use the command (matrix name)^power.

Suppose we have the square matrix $A = \begin{bmatrix} 4 & 5 & 1 \\ 0 & 1 & 3 \\ 8 & 3 & 0 \end{bmatrix}$. Compute A^2. A^2 is vector multiplication of A by A,

therefore, $A^2 = \begin{bmatrix} 4 & 5 & 1 \\ 0 & 1 & 3 \\ 8 & 3 & 0 \end{bmatrix} \times \begin{bmatrix} 4 & 5 & 1 \\ 0 & 1 & 3 \\ 8 & 3 & 0 \end{bmatrix} = \begin{bmatrix} 4\times4+5\times0+1\times8 & 4\times5+5\times1+1\times3 & 4\times1+5\times3+1\times0 \\ 0\times4+1\times0+3\times8 & 0\times5+1\times1+3\times3 & 0\times1+1\times3+3\times0 \\ 8\times4+3\times0+0\times8 & 8\times5+3\times1+0\times3 & 8\times1+3\times3+0\times0 \end{bmatrix} =$

$\begin{bmatrix} 24 & 28 & 19 \\ 24 & 10 & 3 \\ 32 & 43 & 17 \end{bmatrix}$. If the power is 4, then, $A^4 = A^2 \times A^2 = \begin{bmatrix} 24 & 28 & 19 \\ 24 & 10 & 3 \\ 32 & 43 & 17 \end{bmatrix} \times \begin{bmatrix} 24 & 28 & 19 \\ 24 & 10 & 3 \\ 32 & 43 & 17 \end{bmatrix} = \begin{bmatrix} 1856 & 1769 & 863 \\ 912 & 901 & 537 \\ 2344 & 2057 & 1026 \end{bmatrix}$.

If A is a rectangular matrix, an error message is printed by MATLAB. Elements of A can be floating-point

numbers. Matrices of symbolic elements are also computed in a similar fashion. Say, $A = \begin{bmatrix} \frac{1}{3} & \frac{6}{7} \\ \frac{2}{9} & \frac{5}{7} \end{bmatrix}$, hence, A^2

$= \begin{bmatrix} \frac{1}{3} & \frac{6}{7} \\ \frac{2}{9} & \frac{5}{7} \end{bmatrix} \begin{bmatrix} \frac{1}{3} & \frac{6}{7} \\ \frac{2}{9} & \frac{5}{7} \end{bmatrix} = \begin{bmatrix} \frac{19}{63} & \frac{44}{49} \\ \frac{44}{189} & \frac{103}{147} \end{bmatrix}$. All computations are shown below:

MATLAB Command

for A^2,   for A^4,

    >>A=[4 5 1;0 1 3;8 3 0]; ↵   >>A^4 ↵
    >>A^2 ↵

                               ans =

    ans =   1856   1769   863
        24 28 19   912   901   537
        24 10 3   2344 2057 1026
        32 43 17

for rational number,

    >>A=sym([1/3 6/7;2/9 5/7]); ↵
    >>A^2 ↵

    ans =

    [19/63,   44/49]
    [44/189, 103/147]

4.12 Evaluation of a matrix polynomial

Matrix polynomial is similar to the algebraic polynomial with the exception that the variable is a square matrix. To evaluate a matrix polynomial,

MATLAB Steps:

1. *Enter the matrix for polynomial and*
2. *Represent the polynomial in terms of operators $+$, $-$, $*$, and $\wedge$.*

Evaluate the matrix polynomial $P = -3A^3 + 7A^2 + 9I$ for $A = \begin{bmatrix} 1 & 2 \\ 3 & -1 \end{bmatrix}$. We have $A^2 = \begin{bmatrix} 1 & 2 \\ 3 & -1 \end{bmatrix} \times \begin{bmatrix} 1 & 2 \\ 3 & -1 \end{bmatrix} = \begin{bmatrix} 7 & 0 \\ 0 & 7 \end{bmatrix}$, $A^3 = A^2 A = \begin{bmatrix} 7 & 0 \\ 0 & 7 \end{bmatrix} \begin{bmatrix} 1 & 2 \\ 3 & -1 \end{bmatrix} = \begin{bmatrix} 7 & 14 \\ 21 & -7 \end{bmatrix}$, and $I = \begin{bmatrix} 1 & 0 \\ 0 & 1 \end{bmatrix}$, hence, $P = -3 \times \begin{bmatrix} 7 & 14 \\ 21 & -7 \end{bmatrix} + 7 \times \begin{bmatrix} 7 & 0 \\ 0 & 7 \end{bmatrix} + 9 \times \begin{bmatrix} 1 & 0 \\ 0 & 1 \end{bmatrix} = \begin{bmatrix} -21 & -42 \\ -63 & 21 \end{bmatrix} + \begin{bmatrix} 49 & 0 \\ 0 & 49 \end{bmatrix} + \begin{bmatrix} 9 & 0 \\ 0 & 9 \end{bmatrix} = \begin{bmatrix} 37 & -42 \\ -63 & 79 \end{bmatrix}$. The computation is as follows:

MATLAB Command

```
>>A=[1 2;3 -1]; ↵
>>P=-3*A^3+7*A^2+9*eye(2) ↵

P =
        37  -42
       -63   79
```

In the matrix polynomial P, I is the identity matrix of order 2×2, which is generated by the command 'eye'. Order of I is equal to that of A. Elements of matrix A can be floating-points, integer numbers, or even symbolic variables.

4.13 Inverse of a square matrix

Inverse of a square matrix A is defined as $A B = I$, where B is the inverse of A and I is the identity matrix. The order of I depends on the order of A. Make sure that the determinant of A is not zero, then you will not have an inverse of A. A number of methods are available to compute the inverse of a matrix. We present the augment matrix method. To find the inverse of A, a new matrix which is called the augmented matrix is created by appending an identity matrix of the same order with the matrix A, so, the augmented matrix is $[A : I]$. Next row operations that would reduce the matrix A to I are performed. Consequently, I will be replaced by the inverse of A in the augmented matrix. To find the inverse of a matrix A,

MATLAB Steps:

1. *Enter the square matrix (A) and*
2. *Use the command inv(A).*

Take the square matrix example as $A = \begin{bmatrix} 2 & 1 & 5 \\ 0 & 2 & 1 \\ 0 & 0 & 2 \end{bmatrix}$. The augmented matrix is $[A : I] = \begin{bmatrix} 2 & 1 & 5 & \vdots & 1 & 0 & 0 \\ 0 & 2 & 1 & \vdots & 0 & 1 & 0 \\ 0 & 0 & 2 & \vdots & 0 & 0 & 1 \end{bmatrix}$.

Augmented matrix becomes $\begin{bmatrix} 1 & \frac{1}{2} & \frac{5}{2} & \vdots & \frac{1}{2} & 0 & 0 \\ 0 & 1 & \frac{1}{2} & \vdots & 0 & \frac{1}{2} & 0 \\ 0 & 0 & 1 & \vdots & 0 & 0 & \frac{1}{2} \end{bmatrix}$ if elementary row operations $\frac{R_1}{2}$, $\frac{R_2}{2}$, and $\frac{R_3}{2}$ are

performed. Then, $R_1 - \frac{R_2}{2}$ provides $\begin{bmatrix} 1 & 0 & \frac{9}{4} & \vdots & \frac{1}{2} & -\frac{1}{4} & 0 \\ 0 & 1 & \frac{1}{2} & \vdots & 0 & \frac{1}{2} & 0 \\ 0 & 0 & 1 & \vdots & 0 & 0 & \frac{1}{2} \end{bmatrix}$. Finally, operations $R_2 - \frac{R_3}{2}$ and $R_1 - \frac{9R_3}{4}$ convert the last matrix as $\begin{bmatrix} 1 & 0 & 0 & \vdots & \frac{1}{2} & -\frac{1}{4} & -\frac{9}{8} \\ 0 & 1 & 0 & \vdots & 0 & \frac{1}{2} & -\frac{1}{4} \\ 0 & 0 & 1 & \vdots & 0 & 0 & \frac{1}{2} \end{bmatrix}$, therefore, the inverse of A, $B = \begin{bmatrix} \frac{1}{2} & -\frac{1}{4} & -\frac{9}{8} \\ 0 & \frac{1}{2} & -\frac{1}{4} \\ 0 & 0 & \frac{1}{2} \end{bmatrix} =$

$\begin{bmatrix} 0.5 & -0.25 & -1.125 \\ 0 & 0.5 & -0.25 \\ 0 & 0 & 0.5 \end{bmatrix}$. It may be required that the inverse be in rational form. Numeric and rational implementations are presented below:

MATLAB Command for the numeric form,
>>A=[2 1 5;0 2 1;0 0 2]; ↵
>>B=inv(A) ↵

```
        B =
                0.5000      -0.2500      -1.1250
                     0       0.5000      -0.2500
                     0            0       0.5000
```

for rational form,
>>A=sym([2 1 5;0 2 1;0 0 2]); ↵
>>B=inv(A) ↵

```
        B =

        [ 1/2,    -1/4,    -9/8]
        [   0,     1/2,    -1/4]
        [   0,       0,     1/2]
```

Matrices of symbolic variables are easy to deal with 'inv'. Given that the inverse of $\begin{bmatrix} x & y \\ x+y & 1 \end{bmatrix}$ is

$\begin{bmatrix} -\dfrac{1}{y^2 + yx - x} & \dfrac{y}{y^2 + yx - x} \\ \dfrac{x+y}{y^2 + yx - x} & -\dfrac{x}{y^2 + yx - x} \end{bmatrix}$. Obtain that as follows:

MATLAB Command
>>syms x y ↵
>>A=[x y;x+y 1]; ↵
>>B=inv(A); ↵   ← The inverse is stored to B
>>pretty(B) ↵

```
        [          1                      y            ]
        [- ------------------     ------------------   ]
        [            2                        2        ]
        [  -x + x y + y           -x + x y + y         ]
        [                                              ]
        [    x + y                        x            ]
        [-------------------      - ------------------ ]
        [            2                        2        ]
        [ -x + x y + y            -x + x y + y         ]
```

4.14 Characteristic polynomial of a square matrix

Characteristic polynomial of a square matrix A is defined by $|\lambda I - A|$, where I is the identity matrix of the same order as that of A and λ is the eigenvalue of A, which is a scalar. Modulus sign ($|\ldots\ldots|$)

indicates the determinant of the square matrix inside the modulus sign. To evaluate the characteristic polynomial of a square matrix,

MATLAB Step:
Use the command poly(matrix name).

Characteristic polynomial of square matrix $A = \begin{bmatrix} 1 & 2 & -1 \\ -1 & -7 & 2 \\ 0 & 9 & 0 \end{bmatrix}$ is to be evaluated. We have $I = \begin{bmatrix} 1 & 0 & 0 \\ 0 & 1 & 0 \\ 0 & 0 & 1 \end{bmatrix}$,

$\lambda I = \begin{bmatrix} \lambda & 0 & 0 \\ 0 & \lambda & 0 \\ 0 & 0 & \lambda \end{bmatrix}$, $\lambda I - A = \begin{bmatrix} \lambda & 0 & 0 \\ 0 & \lambda & 0 \\ 0 & 0 & \lambda \end{bmatrix} - \begin{bmatrix} 1 & 2 & -1 \\ -1 & -7 & 2 \\ 0 & 9 & 0 \end{bmatrix} = \begin{bmatrix} \lambda-1 & -2 & 1 \\ 1 & \lambda+7 & -2 \\ 0 & -9 & \lambda \end{bmatrix}$, and $| \lambda I - A | = (\lambda-1)[\lambda(\lambda+7)$

$-(-2)(-9)]+2(\lambda-0)+(-9-0)=\lambda^3+6\lambda^2-23\lambda+9$. In MATLAB notation, this polynomial is [1   6   –23   9]. As

another example, we have symbolic matrix, $A = \begin{bmatrix} 1 & x \\ -1 & -7 \end{bmatrix}$, its characteristic polynomial is $\lambda^2 + 6\lambda - 7 + x$.

Both examples are shown below:

MATLAB Command

for coefficient form,   for symbolic variables,
>>A=[1 2 -1;-1 -7 2;0 9 0]; ↵   >>syms x ↵
>>poly(A) ↵   >>A=[1 x;-1 -7]; ↵
                                               >>poly(A) ↵

ans =
        1.0000   6.0000 -23.0000   9.0000   ans =

                                                    t^2+6*t-7+x

Notice that the default output is in terms of t instead of λ. Matrix $\lambda I - A$ is called the characteristic matrix of A. Have that for the first A as follows:

MATLAB Command
>>A=sym([1 2 -1;-1 -7 2;0 9 0]); ↵
>>B=maple('charmat',A,'Lamda') ↵ ←Maple function 'charmat' is applied

B =

[Lamda-1,   -2,   1]
[1,   Lamda+7,   -2]
[0,   -9,   Lamda]

4.15 Basis of some vectors

If S is a set of non-zero vectors such as $S = \{V_1, V_2, V_3, \ldots V_n\}$ spanning the vector space F, then a subset of S is a basis of F. To find the basis, we check the linear dependence of a subset of vectors. If any dependence is there, we exclude the dependent vector to make the subset linearly independent. This way we continue until the whole set of vectors is investigated. Begin with $S = \{V_1, V_2, V_3, V_4\}$, where $V_1 = \begin{bmatrix} 3 \\ 2 \\ 6 \end{bmatrix}$, $V_2 = \begin{bmatrix} 4 \\ 2 \\ 3 \end{bmatrix}$, $V_3 = \begin{bmatrix} 5 \\ 3 \\ 2 \end{bmatrix}$, and $V_4 = \begin{bmatrix} 4 \\ 2 \\ 1 \end{bmatrix}$. First, we check the linear dependence of the subset $\{V_1, V_2, V_3\}$. If the subset $\{V_1, V_2, V_3\}$ has the linear dependence, one can write $V_1 = a\,V_2 + b\,V_3$, where a and b are some constants. In matrix form, they can be written as $\begin{bmatrix} 3 \\ 2 \\ 6 \end{bmatrix} = a \begin{bmatrix} 4 \\ 2 \\ 3 \end{bmatrix} + b \begin{bmatrix} 5 \\ 3 \\ 2 \end{bmatrix}$ or $\begin{cases} 3 = 4a + 5b \\ 2 = 2a + 3b \\ 6 = 3a + 2b \end{cases}$. Solve any two equations of the last set

pertaining to a and b. We chose the first and second ones, thereupon, $a = -\dfrac{1}{2}$ and $b = 1$. The third equation

$6 = 3a + 2b$ is not satisfied by $a = -\dfrac{1}{2}$ and $b = 1$, hence, vectors of the subset $\{V_1, V_2, V_3\}$ are linearly independent. So, we can not exclude neither of them to form the basis. Vectors V_1, V_2, and V_3 are themselves the basis of subset $\{V_1, V_2, V_3\}$. To verify the linear dependence of the whole set, we write $V_4 =$

$a\,V_1 + b\,V_2 + c\,V_3$ or $\begin{cases} 4 = 3a + 4b + 5c \\ 2 = 2a + 2b + 3c \\ 1 = 6a + 3b + 2c \end{cases}$, from which, $a = -\dfrac{4}{11}$, $b = \dfrac{9}{11}$, and $c = \dfrac{4}{11}$. Since V_4 can be expressed

in terms of V_1, V_2, and V_3, it can not be the basis of S. We finished the checking and the basis of S is $\{V_1, V_2, V_3\}$. Maple subroutine 'basis' can help us to find these all:

MATLAB Command

>>maple('v1:=vector([3,2,6]):v2:=vector([4,2,3]):'); ↵ ← Enter V_1 & V_2

>>maple('v3:=vector([5,3,2]):v4:=vector([4,2,1]):'); ↵ ← Enter V_3 & V_4

>>maple('basis({v1,v2,v3,v4})') ↵

ans =

{v1, v2, v3}

In the above execution we assigned the vectors V_1, V_2, V_3, and V_4 to v1, v2, v3, and v4 respectively. The checking is carried out on the order of V_1, V_2, and V_3. When you execute the function, MATLAB may check in other order rather than 1-2-3.

Not necessarily should the vectors be given in individual vector form, they can be in a matrix form too. The last four vectors $\{V_1, V_2, V_3, V_4\}$ can form a matrix, which is $A = \begin{bmatrix} 3 & 4 & 5 & 6 \\ 2 & 2 & 3 & 2 \\ 6 & 3 & 2 & 1 \end{bmatrix}$, where V_1, V_2, V_3,

and V_4 occupy the first, second, third, and fourth columns of A respectively. If we find the basis for the columns of A, that is equivalent to finding the basis of vectors $\{V_1, V_2, V_3, V_4\}$. Check it as follows:

>>maple('A:=matrix([[3,4,5,6],[2,2,3,2],[6,3,2,1]]):'); ↵

>>maple('basis(A,colspace)') ↵

ans =

[VECTOR([3, 2, 6]), VECTOR([4, 2, 3]), VECTOR([5, 3, 2])]

The output corresponds to V_1, V_2, and V_3 respectively. The last argument 'colspace' of the function indicates finding the basis on column space. Rows of matrix A are linearly independent, hence, as row space basis we should get all three rows of A. That is what is returned by the subroutine 'basis':

>>maple('basis(A,rowspace)') ↵

ans =

[VECTOR([3, 4, 5, 6]), VECTOR([2, 2, 3, 2]), VECTOR([6, 3, 2, 1])]

The displayed three rows of A are the basis.

4.16 Eigenvalues and eigenvectors of a square matrix

If A is a square matrix of order $N \times N$, λ is a scalar, and I is an identity matrix of order $N \times N$, then the equation $|\lambda I - A| = 0$ is called the characteristic equation of matrix A. For A of order $N \times N$, the characteristic equation $|\lambda I - A| = 0$ is a polynomial of degree N, and it has N roots designated as λ_1, λ_2, λ_3, ... λ_N. These roots of the characteristic equation are called the eigenvalues of matrix A. To determine the eigenvalues of a square matrix,

MATLAB Step:
Use the command eig(matrix name).

Take an example of a square matrix $A = \begin{bmatrix} 3 & 4 & 5 \\ 4 & -3 & -1 \\ 0 & 0 & 3 \end{bmatrix}$. The characteristic equation of matrix A is $|\lambda I - A| =$

$\lambda \begin{bmatrix} 1 & 0 & 0 \\ 0 & 1 & 0 \\ 0 & 0 & 1 \end{bmatrix} - \begin{bmatrix} 3 & 4 & 5 \\ 4 & -3 & -1 \\ 0 & 0 & 3 \end{bmatrix} = \begin{bmatrix} \lambda-3 & -4 & -5 \\ -4 & \lambda+3 & 1 \\ 0 & 0 & \lambda-3 \end{bmatrix} = (\lambda-3)(\lambda+3)(\lambda-3)+4(-4)(\lambda-3)-5(0) = (\lambda-3)^2(\lambda+3)$

$-16(\lambda-3)=(\lambda-3)(\lambda+5)(\lambda-5)$. The roots of the characteristic equation are $\lambda=3$, 5, and -5, $\therefore$ the eigenvalues of A are 3, 5, and -5. Finding the eigenvalues of A is shown as follows:

MATLAB Command

```
>>A=[3 4 5;4 -3 -1;0 0 3]; ↵
>>eig(A) ↵

ans =
            5
           -5
            3
```

Subroutine 'eig' is also capable of finding the complex eigenvalues of a square matrix. $B = \begin{bmatrix} 4 & 5 \\ -2 & 1 \end{bmatrix}$ is such

an example that has the complex eigenvalues. Characteristic equation of B is $|\lambda I - B| = \lambda \begin{bmatrix} 1 & 0 \\ 0 & 1 \end{bmatrix} -$

$\begin{bmatrix} 4 & 5 \\ -2 & 1 \end{bmatrix} = \begin{bmatrix} \lambda-4 & -5 \\ 2 & \lambda-1 \end{bmatrix} = \lambda^2 - 5\lambda + 14 = 0$ and the roots of the characteristic equation is

$\lambda = \frac{5}{2} \pm j\frac{\sqrt{31}}{2} = 2.5 \pm j2.7839$, which are the eigenvalues of B. The eigenvalues in floating-point and symbolic forms are shown below:

for floating-point eigenvalues,
```
>>B=[4 5;-2 1]; ↵
>>eig(B) ↵

ans =
      2.5000 + 2.7839i
      2.5000 - 2.7839i
```

for eigenvalues in symbolic form,
```
>>B=sym([4 5;-2 1]); ↵
>>pretty(eig(B)) ↵

[                1/2]
[ 5/2 + 1/2 i 31   ]
[                  ]
[                1/2]
[ 5/2 - 1/2 i 31   ]
```

The characteristic polynomial $|\lambda I - A|$ is a polynomial of degree N. N eigenvalues are designated as λ_1, λ_2, λ_3, ... λ_N. For each eigenvalue, which is a scalar, there is a matrix V of order $N \times 1$ which satisfies $AV =$

λV. If this equation is satisfied by the vector V, then V is called an eigenvector of matrix A. Vector multiplication of A and V results the order of AV as $N \times 1$, again, the order of matrix λV is also $N \times 1$. For N eigenvalues, we can have N eigenvectors. The equation $AV = \lambda V$ can be written as $(\lambda I - A)V = 0$. Since $\lambda I - A = 0$ for each eigenvalue, the rank of matrix $A - \lambda I$ must be less than N. The system of equations for an eigenvalue given by the matrix $A - \lambda I$ has infinitely many solutions.

The concept is best conveyed by an example, say, $A = \begin{bmatrix} 2 & 2 & 0 \\ 2 & 1 & 1 \\ -7 & 2 & -3 \end{bmatrix}$. Order of A is 3×3, hence, there are three eigenvalues of A labeled as λ_1, λ_2, and λ_3. $|A - \lambda I| = 0$ provides the eigenvalues, which are $\lambda_1 = 3$, $\lambda_2 = 1$, and $\lambda_3 = -4$. For simplicity, assume that $B = A - \lambda I$. For λ_1, B becomes B_1, $\therefore$ $B_1 = A - \lambda_1 I = \begin{bmatrix} -1 & 2 & 0 \\ 2 & -2 & 1 \\ -7 & 2 & -6 \end{bmatrix}$; rank of B_1 is 2. Similarly, $\lambda_2 = 1$ and $\lambda_3 = -4$ provide $B_2 = A - \lambda_2 I = \begin{bmatrix} 1 & 2 & 0 \\ 2 & 0 & 1 \\ -7 & 2 & -4 \end{bmatrix}$ and $B_3 = A - \lambda_3 I = \begin{bmatrix} 6 & 2 & 0 \\ 2 & 5 & 1 \\ -7 & 2 & 1 \end{bmatrix}$ respectively. The ranks of B_2 and B_3 are 2. Generalized technique is used for the calculation of eigenvectors of A. Any B can be partitioned as $B_k = \begin{bmatrix} R_k & C_k \\ D_k & E_k \end{bmatrix}$, where R_k is nonsingular. The eigenvector V_k for an eigenvalue λ_k can be written as $V_k = \begin{bmatrix} -R_k^{-1} C_k W_k \\ W_k \end{bmatrix}$, where W_k is the vector of variables that comes from the rank of R_k. If B_k has m columns and rank of R_k is k (subscript k of R_k and the last k are not same), the number of variables in the vector W_k is $m - k$. Clearly, inspecting B_1 allows us to write $B_1 = \begin{bmatrix} R_1 & C_1 \\ D_1 & E_1 \end{bmatrix}$, where $R_1 = \begin{bmatrix} -1 & 2 \\ 2 & -2 \end{bmatrix}$, $C_1 = \begin{bmatrix} 0 \\ 1 \end{bmatrix}$, and $W_1 = w$ (since B_1 has 3 columns and the rank of R_1 is 2), on

that, $V_1 = \begin{bmatrix} -R_1^{-1} C_1 W_1 \\ W_1 \end{bmatrix} = \begin{bmatrix} -\begin{bmatrix} -1 & 2 \\ 2 & -2 \end{bmatrix}^{-1} \begin{bmatrix} 0 \\ 1 \end{bmatrix} w \\ w \end{bmatrix} = \begin{bmatrix} -\begin{bmatrix} 1 & 1 \\ 1 & \frac{1}{2} \end{bmatrix} \begin{bmatrix} 0 \\ 1 \end{bmatrix} w \\ w \end{bmatrix} = \begin{bmatrix} -w \\ -\frac{w}{2} \\ w \end{bmatrix}$. Magnitude of vector V_1 is $\sqrt{w^2 + (\frac{w}{2})^2 + w^2}$

$= \frac{3w}{2}$, so, in normalized form, $V_1 = \begin{bmatrix} -w \\ -\frac{w}{2} \\ w \end{bmatrix} / \frac{3w}{2} = \begin{bmatrix} -\frac{2}{3} \\ -\frac{1}{3} \\ \frac{2}{3} \end{bmatrix} = \begin{bmatrix} -0.6667 \\ -0.3333 \\ 0.6667 \end{bmatrix}$. Similarly, we have $B_2 = \begin{bmatrix} 1 & 2 & 0 \\ 2 & 0 & 1 \\ -7 & 2 & -4 \end{bmatrix} =$

$\begin{bmatrix} R_2 & C_2 \\ D_2 & E_2 \end{bmatrix}$, where $R_2 = \begin{bmatrix} 1 & 2 \\ 2 & 0 \end{bmatrix}$ and $C_2 = \begin{bmatrix} 0 \\ 1 \end{bmatrix}$ and $B_3 = \begin{bmatrix} 6 & 2 & 0 \\ 2 & 5 & 1 \\ -7 & 2 & 1 \end{bmatrix} = \begin{bmatrix} R_3 & C_3 \\ D_3 & E_3 \end{bmatrix}$, where $R_3 = \begin{bmatrix} 6 & 2 \\ 2 & 5 \end{bmatrix}$ and

$C_3 = \begin{bmatrix} 0 \\ 1 \end{bmatrix}$ for the other two eigenvalues, therefore, $V_2 = \begin{bmatrix} -R_2^{-1} C_2 W_2 \\ W_2 \end{bmatrix} = \begin{bmatrix} -\begin{bmatrix} 1 & 2 \\ 2 & 0 \end{bmatrix}^{-1} \begin{bmatrix} 0 \\ 1 \end{bmatrix} w \\ w \end{bmatrix} = \begin{bmatrix} -\begin{bmatrix} 0 & \frac{1}{2} \\ \frac{1}{2} & -\frac{1}{4} \end{bmatrix} \begin{bmatrix} 0 \\ 1 \end{bmatrix} w \\ w \end{bmatrix} =$

$\begin{bmatrix} -\frac{w}{2} \\ \frac{w}{4} \\ w \end{bmatrix}$ and $V_3 = \begin{bmatrix} -R_3^{-1} C_3 W_3 \\ W_3 \end{bmatrix} = \begin{bmatrix} -\begin{bmatrix} 6 & 2 \\ 2 & 5 \end{bmatrix}^{-1} \begin{bmatrix} 0 \\ 1 \end{bmatrix} w \\ w \end{bmatrix} = \begin{bmatrix} -\begin{bmatrix} \frac{5}{26} & -\frac{1}{13} \\ -\frac{1}{13} & \frac{3}{13} \end{bmatrix} \begin{bmatrix} 0 \\ 1 \end{bmatrix} w \\ w \end{bmatrix} = \begin{bmatrix} \frac{w}{13} \\ -\frac{3w}{13} \\ w \end{bmatrix}$. Magnitudes of V_2 and V_3 are

$$\sqrt{\frac{w^2}{4}+\frac{w^2}{16}+w^2}=w\sqrt{\frac{21}{16}} \text{ and } \sqrt{\frac{w^2}{169}+\frac{9w^2}{169}+w^2}=w\sqrt{\frac{179}{169}} \text{ respectively. Finally, in normalized form, we have } V_2=$$

$$\begin{bmatrix}-\frac{w}{2}\\\frac{w}{4}\\w\end{bmatrix}/(w\sqrt{\frac{21}{16}})=\begin{bmatrix}-\frac{2}{\sqrt{21}}\\\frac{1}{\sqrt{21}}\\\frac{4}{\sqrt{21}}\end{bmatrix}=\begin{bmatrix}-0.4364\\0.2182\\0.8729\end{bmatrix} \text{ and } V_3=\begin{bmatrix}\frac{w}{13}\\-\frac{3w}{13}\\w\end{bmatrix}/(w\sqrt{\frac{179}{169}})=-\begin{bmatrix}\frac{1}{\sqrt{179}}\\\frac{3}{\sqrt{179}}\\\frac{13}{\sqrt{179}}\end{bmatrix}=\begin{bmatrix}0.0747\\-0.2242\\0.9717\end{bmatrix}. \text{ MATLAB computation of}$$

the eigrenvectors is as follows:

MATLAB Command
```
>>A=[2 2 0;2 1 1;-7 2 -3]; ↵
>>[V E]=eig(A) ↵
```

V =

```
    -0.6667  -0.4364   0.0747
    -0.3333   0.2182  -0.2242
     0.6667   0.8729   0.9717
```

E =

```
     3.0000        0        0
          0   1.0000        0
          0        0  -4.0000
```

Referring to the above output, there are two output arguments of the subroutine 'eig' — V and E. The first argument V is formed by taking the normalized eigenvectors V_1, V_2, and V_3, that is, V=$[V_1 \quad V_2 \quad V_3]$=

$$\begin{bmatrix}\begin{bmatrix}-0.6667\\-0.3333\\0.6667\end{bmatrix}\begin{bmatrix}-0.4364\\0.2182\\0.8729\end{bmatrix}\begin{bmatrix}0.0747\\-0.2242\\0.9717\end{bmatrix}\end{bmatrix}=\begin{bmatrix}-0.6667 & -0.4364 & 0.0747\\-0.3333 & 0.2182 & -0.2242\\0.6667 & 0.8729 & 0.9717\end{bmatrix}. \text{ The second output argument E is a}$$

diagonal matrix. E consists of placing the eigenvalues of A in descending order. One might be interested in viewing the eigenvectors in rational form. Symbolic eigenvectors returned by the subroutine will not be in normalized form instead they will be displayed in greatest common divider (GCD) form. Execution for the symbolic implementation is shown below:

```
>>A=[2 2 0;2 1 1;-7 2 -3]; ↵
>>[V E]= eig(sym(A)) ↵
```

V=

```
    [  2, -2,  1]
    [  1,  1, -3]
    [ -2,  4, 13]
```

E =

```
    [  3,  0,  0]
    [  0,  1,  0]
    [  0,  0, -4]
```

The first output eigenvector (the first column of V) is $\begin{bmatrix}2\\1\\-2\end{bmatrix}$, which is obtained from $V_1=\begin{bmatrix}-\frac{2}{3}\\-\frac{1}{3}\\\frac{2}{3}\end{bmatrix}$ (GCD form is

performed by multiplying -3). In a similar way, the other two eigenvectors, $\begin{bmatrix}-2\\1\\4\end{bmatrix}$ and $\begin{bmatrix}1\\-3\\13\end{bmatrix}$ (the second and third columns of V), are obtained from $V_2 = \begin{bmatrix} -\dfrac{2}{\sqrt{21}} \\ \dfrac{1}{\sqrt{21}} \\ \dfrac{4}{\sqrt{21}} \end{bmatrix}$ and $V_3 = \begin{bmatrix} \dfrac{1}{\sqrt{179}} \\ -\dfrac{3}{\sqrt{179}} \\ \dfrac{13}{\sqrt{179}} \end{bmatrix}$ respectively. GCD forms of the

eigenvectors satisfy the equation $A\,V = \lambda V$ too. As a check, take the first eigenvector V_1, so, $AV_1 =$

$\begin{bmatrix} 2 & 2 & 0 \\ 2 & 1 & 1 \\ -7 & 2 & -3 \end{bmatrix} \begin{bmatrix} 2 \\ 1 \\ -2 \end{bmatrix} = \begin{bmatrix} 6 \\ 3 \\ -6 \end{bmatrix}$ and $\lambda_1 V_1 = 3 \begin{bmatrix} 2 \\ 1 \\ -2 \end{bmatrix} = \begin{bmatrix} 6 \\ 3 \\ -6 \end{bmatrix}$, from which $AV_1 = \lambda_1 V_1$.

4.17 Singular value decomposition

Definitions of the eigenvalues and eigenvectors hold true for square matrices. Spectral decomposition theorem helps any symmetric matrix be decomposed into the product of three matrices. Let A be a matrix of order $M \times N$, A can be expressed as $A = U \times D \times V^T$, where the multiplication sign indicates the matrix multiplication and V^T is the transpose of V. U and V are called the unitary matrices and their determinant should be 1. Orders of U, D, and V are $M \times M$, $M \times N$, and $N \times N$ respectively. This decomposition is termed as the singular value decomposition of A. The decomposition exists for general matrices – square or rectangular. D is a diagonal matrix (can be a rectangular one). Diagonal elements of D are called the singular values of A. Singular values are obtained by taking the positive square roots of the eigenvalues of $A \times A^T$ (A^T is the transpose of A). Matrix $A^T \times A$ or $A \times A^T$ is symmetric and their eigenvalues are nonnegative. Columns of U and V are the eigenvectors of $A \times A^T$ and $A^T \times A$ respectively.

The decomposition is best clarified by the example of $A = \begin{bmatrix} 4 & 1 & 0 & 9 \\ 5 & 7 & -1 & 0 \\ 6 & 9 & 4 & 2 \end{bmatrix}$. Perform the theoretical

computation as $A \times A^T = \begin{bmatrix} 4 & 1 & 0 & 9 \\ 5 & 7 & -1 & 0 \\ 6 & 9 & 4 & 2 \end{bmatrix} \times \begin{bmatrix} 4 & 5 & 6 \\ 1 & 7 & 9 \\ 0 & -1 & 4 \\ 9 & 0 & 2 \end{bmatrix} = \begin{bmatrix} 98 & 27 & 51 \\ 27 & 75 & 89 \\ 51 & 89 & 137 \end{bmatrix}$ and $A^T \times A = \begin{bmatrix} 4 & 5 & 6 \\ 1 & 7 & 9 \\ 0 & -1 & 4 \\ 9 & 0 & 2 \end{bmatrix} \times$

$\begin{bmatrix} 4 & 1 & 0 & 9 \\ 5 & 7 & -1 & 0 \\ 6 & 9 & 4 & 2 \end{bmatrix} = \begin{bmatrix} 77 & 93 & 19 & 48 \\ 93 & 131 & 29 & 27 \\ 19 & 29 & 17 & 8 \\ 48 & 27 & 8 & 85 \end{bmatrix}$. Eigenvalues of $A \times A^T$ are 73.1662, 10.9423, and 225.8915 (see

article 4.16 for the computation of eigenvalues). You can even use MATLAB to find the eigenvalues or eigenvectors:

MATLAB Command for the eigenvalues of $A \times A^T$,
>>A=[4 1 0 9;5 7 -1 0;6 9 4 2]; ↵
>>eig(A*A') ↵

ans =
    73.1662
    10.9423
    225.8915

The singular values are the positive square roots of these eigenvalues, which are 8.5537, 3.3079, and 15.0297 respectively. Diagonal matrix D is formed from these singular values placing them in descending order. If A is rectangular, the rest rows or columns of D are filled by the zeroes. For the matrix A at hand, D should

134

look like
$$\begin{bmatrix} 15.0297 & 0 & 0 & 0 \\ 0 & 8.5537 & 0 & 0 \\ 0 & 0 & 3.3079 & 0 \end{bmatrix}$$
since the order of D is 3× 4. To form U, we need to have the

normalized eigenvectors of $A \times A^T$. The normalized eigenvectors of $A \times A^T$ are $\begin{bmatrix} 0.9062 \\ -0.3269 \\ -0.2682 \end{bmatrix}$, $\begin{bmatrix} -0.1073 \\ -0.7912 \\ 0.6020 \end{bmatrix}$, and

$\begin{bmatrix} 0.4090 \\ 0.5168 \\ 0.7521 \end{bmatrix}$ for the eigenvalues 73.1662, 10.9423, and 225.8915 respectively found from MATLAB as follows

(shown by the output of E1):

Command for the normalized
eigenvectors of A×A^T,
>>[E1 d]=eig(A*A'); ↵
>>E1 ↵

for the determinant of E1,
>>det(E1) ↵

ans =

E1 =
    0.9062   -0.1073   0.4090
   -0.3269   -0.7912   0.5168
   -0.2682   0.6020   0.7521

ans =
            -1

Since the singular values in D, which are 8.5537, 3.3079, and 15.0297 respectively, are in descending order, the eigenvectors just mentioned must be placed in that order. To put the eigenvectors in accordance with the descending order of singular values as in D, interchange the first with the third and then the second with the third columns of E1. Determinant concept tells us to put a negative sign before it if any two columns interchange their positions. As a whole there is no sign change. Placing in the order just mentioned, we have

$$U = \begin{bmatrix} 0.4090 & 0.9062 & -0.1073 \\ 0.5168 & -0.3269 & -0.7912 \\ 0.7521 & -0.2682 & 0.6020 \end{bmatrix}$$
and the determinant of E1 is −1. To make the determinant positive, we

multiply the second column with the negative sign. U should look like $\begin{bmatrix} 0.4090 & -0.9062 & -0.1073 \\ 0.5168 & 0.3269 & -0.7912 \\ 0.7521 & 0.2682 & 0.6020 \end{bmatrix}$. The

negative sign before a determinant is multiplied with any column or row not with all elements (unlike matrix).

Command for the
eigenvalues of A^T×A,
>>eig(A'*A) ↵

for the normalized eigenvectors of A^T×A,
>>[E2 d]=eig(A'*A); ↵
>>E2 ↵

ans =
        10.9423
        -0.0000
        73.1662
       225.8915

E2 =
   -0.2337 -0.7783   0.0445   0.5810
   -0.0688   0.5314 -0.4438   0.7183
    0.9672 -0.1717 -0.0872   0.1658
    0.0721   0.2869   0.8908   0.3450

for the determinant of E2,
>>det(E2) ↵

ans =
        1.0000

Similarly, the normalized eigenvectors of $A^T \times A$ (shown by the output of E2 in the previous page)

are $\begin{bmatrix} -0.2337 \\ -0.0688 \\ 0.9672 \\ 0.0721 \end{bmatrix}$, $\begin{bmatrix} -0.7783 \\ 0.5314 \\ -0.1717 \\ 0.2869 \end{bmatrix}$, $\begin{bmatrix} 0.0445 \\ -0.4438 \\ -0.0872 \\ 0.8908 \end{bmatrix}$, and $\begin{bmatrix} 0.5810 \\ 0.7183 \\ 0.1658 \\ 0.3450 \end{bmatrix}$ for the eigenvalues 10.9423, −0.0000, 73.1662, and

225.8915 respectively. To put the eigenvectors according to the descending order of eigenvalues, interchange the first column with the fourth, then, the second column with the third, next, the third column with the fourth. The determinant of E2 is unity. Three column's interchange results a negative sign to be put before the determinant and this negative sign is inserted in the second column, so, V becomes

$\begin{bmatrix} 0.5810 & -0.0445 & -0.2337 & -0.7783 \\ 0.7183 & 0.4438 & -0.0688 & 0.5314 \\ 0.1658 & 0.0872 & 0.9672 & -0.1717 \\ 0.3450 & -0.8908 & 0.0721 & 0.2869 \end{bmatrix}$. Finally, we have U, D, and V that is what is displayed below

by the outputs of the subroutine 'svd' (abbreviation of <u>s</u>ingular <u>v</u>alue <u>d</u>ecomposition). We should be able to reconstruct A from $U \times D \times V^T$:

Command for the singular value decomposition of A,
>>[U D V]=svd(A) ↵

U =
        0.4090 -0.9062 -0.1073
        0.5168   0.3269 -0.7912
        0.7521   0.2682   0.6020
D =
        15.0297   0   0   0
              0   8.5537   0   0
              0   0   3.3079   0
V =
        0.5810 -0.0445 -0.2337 -0.7783
        0.7183   0.4438 -0.0688   0.5314
        0.1658   0.0872   0.9672 -0.1717
        0.3450 -0.8908   0.0721   0.2869

Command for the proof of A=U×D×V^T,
>>U*D*V' ↵

ans =
        4.0000   1.0000   0.0000   9.0000
        5.0000   7.0000 -1.0000   0.0000
        6.0000   9.0000   4.0000   2.0000

4.18 Matrix norms

Determinant is defined only for the square matrices but it is not defined for the rectangular matrices. Matrix norm is a way of measuring the magnitude of a given matrix by a single number. There are several types of matrix norms such as Frobenius norm, 1-norm (L_1 norm), 2-norm (L_2 norm), ∝-norm (L_∞ norm), etc. Now we discuss how to find the different norms in MATLAB.

Frobenius norm:

The Frobenius norm of any matrix A of order $M \times N$ is defined by $\|A\|_F = \sqrt{\sum_{j=1}^{M}\sum_{i=1}^{N} A_{ij}^2}$. In words, the

Frobenius norm of a matrix is the positive square root of the sum of its squared elements. To compute the Frobenius norm of a matrix,

MATLAB Step:
Use the command norm (matrix name, 'fro').

The Frobenius norms of row matrix $R = [1 \quad -3 \quad 7]$, column matrix $C = \begin{bmatrix} 4 \\ -5 \\ -6 \end{bmatrix}$, and rectangular matrix $A =$

$\begin{bmatrix} 6 & 7 & 8 \\ -1 & -2 & 5 \end{bmatrix}$ are given by $NR = \sqrt{1^2 + (-3)^2 + 7^2} = \sqrt{59} = 7.681145748$, $NC = \sqrt{4^2 + (-5)^2 + (-6)^2} = \sqrt{77} =$

8.774964387, and $NA = \sqrt{6^2 + 7^2 + 8^2 + (-1)^2 + (-2)^2 + 5^2} = \sqrt{179} = 13.37908816$ for R, C, and A respectively. All examples are shown below:

MATLAB Command

for row matrix,

      >>R=[1 -3 7]; ↵

      >>NR=norm(R,'fro') ↵

    NR =

      7.6811

for rectangular matrix,

      >>A=[6 7 8;-1 -2 5]; ↵

      >>NA=norm(A,'fro') ↵

    NA =

      13.3791

for column matrix,

      >>C=[4 -5 -6]'; ↵

      >>NC=norm(C,'fro') ↵

    NC =

      8.7750

L_1 norm:

    To compute L_1 norm of a matrix,

MATLAB Step:
Use the command norm (matrix name, 1).

L_1 norm of a row or column matrix is given by the sum of absolute values of the elements in the matrix. L_1

norms of row matrix $R = [1 \quad -3 \quad 7]$ and column matrix $C = \begin{bmatrix} 4 \\ -5 \\ -6 \end{bmatrix}$ are given by $NR = |1| + |-3| + |7| = 11$ and $NC = |4|$

$+ |-5| + |-6| = 15$ respectively. The norm of rectangular matrix $A = \begin{bmatrix} 6 & 7 & -8 \\ -1 & -2 & 5 \end{bmatrix}$ is given by the largest

column sum of the absolute values of elements in A. In A, the first column sum is $|6| + |-1| = 7$, the second column sum is $|7| + |-2| = 9$, and the third column sum is $|-8| + |5| = 13$. The largest column sum is 13, therefore, L_1 norm of A is $NA = 13$. Implementations of L_1 norms are presented below:

MATLAB Command

for row matrix,

      >>R=[1 -3 7]; ↵

      >>NR=norm(R,1) ↵

    NR =

      11

for rectangular matrix,

      >>A=[6 7 -8;-1 -2 5]; ↵

      >>NA=norm(A,1) ↵

for column matrix,

      >>C=[4 -5 -6]'; ↵

      >>NC=norm(C,1) ↵

    NC =

      15

NA =
13

L_2 norm:

L_2 norm or 2-norm of a matrix A is defined as the largest singular value of A. How singular values can be calculated is discussed in article 4.17. For simplicity, we take help of the subroutine 'svd' to calculate the singular values of a matrix. To compute L_2 norm,

MATLAB Step:
Use the command norm (matrix name,2) or norm (matrix name).

As different matrix examples, take $R=[3 \quad -5 \quad 9]$, $C=\begin{bmatrix} 2 \\ -8 \\ -7 \\ 100 \end{bmatrix}$, and $A=\begin{bmatrix} 6 & 7 & -8 & 0 \\ -1 & -2 & 5 & -4 \\ -3 & 5 & 4 & 5 \end{bmatrix}$. Compute the

singular values as follows:

MATLAB Command

for R,
```
>>R=[3 -5  9]; ↵
>>svd(R) ↵
```

ans =
    10.7238

for C,
```
>>C=[2 -8 -7 100]'; ↵
>>svd(C) ↵
```

ans =
    100.5833

for A,
```
>>A=[6 7 -8 0;-1 -2 5 -4;-3 5 4 5]; ↵
>>svd(A) ↵
```

ans =
    13.3304
    8.7518
    3.9634

The singular values of the different matrices are computed as 10.7238, 100.5833, and 13.3304, 8.7518, and 3.9634 for R, C, and A respectively. The largest singular values are 10.7238, 100.5833, and 13.3304 for R, C, and A respectively. So, the respective L_2 norms are 10.7238, 100.5833, and 13.3304. Conduct the subroutine 'norm' to have these all as follows:

MATLAB Command

for R,
```
>>NR=norm(R,2) ↵
```

NR =
    10.7238

for C,
```
>>NC=norm(C,2) ↵
```

NC =
    100.5833

for A,
```
>>NA=norm(A,2) ↵
```

NA =
    13.3304

L_∞ norm:

To compute L_∞ norm of a matrix,

MATLAB Step:
Use the command norm (matrix name,inf).

L_∞ norm (infinity norm) of a row or column matrix is defined as the largest absolute value of the elements in

the matrix. L_∞ norms of $R=[1 \quad -3 \quad -57 \quad 0]$ and $C=\begin{bmatrix} 14 \\ -50 \\ -62 \\ 13 \end{bmatrix}$ are NR=maximum($|1|$, $|-3|$, $|-57|$, $|0|$)=

maximum(1, 3, 57, 0)=57 and NC=maximum(|14|, |–50|, |–62|, |13|)=maximum(14, 50, 62, 13)=62

respectively. The norm of rectangular matrix $A = \begin{bmatrix} 6 & 7 & -8 \\ -1 & -2 & 5 \\ -3 & 0 & 7 \\ 10 & 13 & 0 \end{bmatrix}$ is the largest row sum of the absolute

values of the elements in A. The first, second, third, and fourth row sums are |6|+|7|+|–8|=21, |–1|+|–2|+|5|=8, |–3|+|0|+|7|=10, and |10|+|13|+|0|=23 respectively. The largest row sum is 23, hence, L_∞ norm of A, NA, is 23. See the findings of L_∞ norms for the different matrices below:

MATLAB Command

for row matrix,

>>R=[1 -3 -57 0]; ↵
>>NR=norm(R,inf) ↵

  NR =
    57

for column matrix,

>>C=[14 -50 -62 13]'; ↵
>>NC=norm(C,inf) ↵

  NC =
    62

for rectangular matrix,

>>A=[6 7 -8;-1 -2 5;-3 0 7;10 13 0]; ↵
>>NA=norm(A,inf) ↵

  NA =
    23

So far what we illustrated is the second argument of 'norm' can be 1, 2, fro, or infinity. The first argument of the subroutine, in general, is a matrix — row, column, or rectangular. There are some other norms that are defined only for the vectors. By vector, what we mean is a row or column matrix.

L_p *norm of a vector:*

The p - norm of a vector X, whose length is N, is $L_p = \left[\sum_{i=1}^{N} |X_i|^p \right]^{\frac{1}{p}}$. To compute L_p norm of a vector,

MATLAB Step:

Use the command norm (matrix name,p).

Test the norm by $R = [9 \ 7 \ -10]$ and $C = \begin{bmatrix} 0 \\ -3 \\ 4 \\ 7 \end{bmatrix}$. There are three and four elements in R and C respectively,

from that, $N = 3$ and $N = 4$ for R and C respectively. Find the 4-norm of R and 5-norm of C, $\therefore$ $p = 4$ and L_4 of $R = [|9|^4 + |7|^4 + |–10|^4]^{1/4} = 11.7347$ and $p = 5$ and L_5 of $C = [|0|^5 + |–3|^5 + |4|^5 + |7|^5]^{1/5} = 7.1025$. See the computations as follows:

MATLAB Command

for row matrix and for L₄,

>>R=[9 7 -10]; ↵
>>L4=norm(R,4) ↵

  L4 =
    11.7347

for column matrix and for L₅,

>>C=[0 -3 4 7]'; ↵
>>L5=norm(C,5) ↵

  L5 =
    7.1025

$L_{-\infty}$ *norm of a vector:*

To compute $L_{-\infty}$ norm of a vector,

MATLAB Step:
Use the command norm (matrix name, –inf).

Minus infinity norm ($L_{-\infty}$) of a vector X is defined as the minimum of the absolute values of the elements in

X . Example matrices are $R =[1 \quad -3 \quad -57 \quad 0]$ and $C = \begin{bmatrix} 14 \\ -50 \\ -62 \\ 13 \end{bmatrix}$. $L_{-\infty}$ of R is NR=minimum(|1|, |–3|, |–57|, |0|)=

minimum(1, 3, 57, 0)=0, and so is NC=minimum(|14|, |–50|, |–62|, |13|)=minimum(14, 50, 62, 13)=13 for $L_{-\infty}$
of C . See the implementation below:

MATLAB Command

for row matrix,
    >>R=[1 -3 -57 0]; ↵
    >>NR=norm(R,-inf) ↵

  NR =
     0

for column matrix,
    >>C=[14 -50 -62 13]'; ↵
    >>NC=norm(C,-inf) ↵

  NC =
     13

4.19 Formation of linearly and logarithmically spaced vectors

Linearly spaced vector elements form an arithmetic progression. If the first element in the vector is a

and common difference of the progression is d , the vector looks like $L = \begin{bmatrix} a \\ a+d \\ a+2d \\ a+3d \\ \vdots \\ a+(N-1)d \end{bmatrix}$, where N is the

number of elements in vector L . Clearly, d is equal to $\dfrac{Last\ element - First\ element}{N-1}$. To form a linearly spaced
vector,

MATLAB Step:
Use command linspace (first element, last element, number of points from first to last).

The subroutine 'linspace' is the abbreviation of <u>lin</u>early <u>spaced</u>. The output of 'linspace' is a row matrix. Say,
we want to form a vector from 3 to 13 with 6 points (inclusive), therefore, $d = \dfrac{13-3}{6-1} = 2$. Output should be [3
5   7   9   11   13]. As a rational example, a column matrix is to be formed from −7 to 3 with 5 points, $\therefore$

$d = \dfrac{3-(-7)}{5-1} = \dfrac{5}{2}$. The vector will be $\begin{bmatrix} -7 \\ -\frac{9}{2} \\ -2 \\ \frac{1}{2} \\ 3 \end{bmatrix}$. Both of them are implemented below:

MATLAB Command

for generation of row vector,
    >>linspace(3,13,6) ↵

  ans =
    3   5   7   9   11   13

for generation of column vector,
    >>sym(linspace(-7,3,5))' ↵

  ans =
    [-7]
    [-9/2]
    [-2]
    [1/2]
    [3]

If the number of points from the first to last element is not specified, 'linspace' generates one hundred points by default.

Logarithmically (base of the logarithm is 10) spaced vector elements form a geometric progression. If the first element in the vector with length N is a and the common ratio of the progression is r, the vector is

given by $L = \begin{bmatrix} a \\ ar \\ ar^2 \\ ar^3 \\ \vdots \\ ar^{N-1} \end{bmatrix}$. To form a logarithmically spaced vector,

MATLAB Step:
Use the command logspace (power of the first element, power of the last element, number of points from the first to last).

The subroutine 'logspace' is the abbreviation of <u>logarithmically</u> <u>spaced</u>. Its output is a row vector. Suppose we wish to form a logarithmically spaced vector where power of the elements will be from 3 to 4 and the number of elements will be 5, so, $a = 10^3$, $N = 5$, and $ar^{N-1} = 10^4$. Plugging a and N allows us to write $r = 10^{\frac{1}{4}}$.

Therefore, the vector is obtained as $L = \begin{bmatrix} 10^3 \\ 10^3 10^{\frac{1}{4}} \\ 10^3 10^{\frac{2}{4}} \\ 10^3 10^{\frac{3}{4}} \\ 10^3 10^{\frac{4}{4}} \end{bmatrix} = \begin{bmatrix} 10^3 \\ 10^{\frac{13}{4}} \\ 10^{\frac{7}{2}} \\ 10^{\frac{15}{4}} \\ 10^4 \end{bmatrix} = \begin{bmatrix} 1000 \\ 1778 \\ 3162 \\ 5623 \\ 10000 \end{bmatrix}$ (neglecting the fractional parts). Following

is the application of 'logspace':

MATLAB Command
>>L=logspace(3,4,5)' ↵

L =
1.0e+004 *

0.1000
0.1778
0.3162
0.5623
1.0000

If the power index of 10 is higher, the elements returned by the 'logspace' will be of higher digits. That is why the return of the subroutine is in exponential form. Let us get clarification of the exponent output. '1.0e+004 *'

means '$1.0 \times 10^4 \times$' and each of the element of $\begin{bmatrix} 0.1000 \\ 0.1778 \\ 0.3162 \\ 0.5623 \\ 1.0000 \end{bmatrix}$ (as shown in output) is multiplied by 10^4. Do not be

confused e+004 with number e, which is equal to 2.7183.

4.20 Trace of a matrix
Trace of a square matrix is defined as the sum of the diagonal elements of the matrix. To calculate the trace of a matrix,

MATLAB Step:
Use the command trace (matrix name).

Trace of square matrix $A = \begin{bmatrix} -2 & 7 & 0 \\ 12 & 4 & 3 \\ 15 & 2 & 0 \end{bmatrix}$ is –2+4+0=2. In accordance with matrix algebra, trace is defined only for the square matrices but in MATLAB, you can find the trace of a rectangular matrix too. The trace of rectangular matrix $R = \begin{bmatrix} 7 & 2 \\ 9 & 5 \\ 0 & -1 \end{bmatrix}$ is 7+5=12. See the implementations below:

MATLAB Command

for square matrix,
```
>>A=[-2 7 0;12 4 3;15 2 0]; ↵
>>T=trace(A) ↵

T =
    2
```

for rectangular matrix,
```
>>R=[7 2;9 5;0 -1]; ↵
>>T=trace(R) ↵

T =
    12
```

Finding the trace of row or column matrices in MATLAB returns the same row or column matrix but the returned matrix is a row one.

4.21 Pseudoinverse of a rectangular matrix

Any non-null matrix A of order $M \times N$ and rank r is equivalent to $P \, A \, Q = \begin{bmatrix} I_r & 0 \\ 0 & 0 \end{bmatrix} = C$, where I_r is the identity matrix of order $r \times r$ and C is called the equivalent canonical form of A. P and Q are non-singular matrices of order $M \times M$ and $N \times N$ respectively. The best thing is start with $A = \begin{bmatrix} 1 & 2 & 3 & -7 \\ 4 & 8 & 12 & -2 \\ 2 & 4 & 6 & -1 \end{bmatrix}$. Elementary row operations are performed on A to obtain the matrix form that is suitable for the determination of rank. $\begin{bmatrix} 1 & 2 & 3 & -7 \\ 4 & 8 & 12 & -2 \\ 2 & 4 & 6 & -1 \end{bmatrix} \xrightarrow{\substack{R_3 - 2\times R_1 \text{ and} \\ R_2 - 4 \times R_1}} \begin{bmatrix} 1 & 2 & 3 & -7 \\ 0 & 0 & 0 & 26 \\ 0 & 0 & 0 & 13 \end{bmatrix}$

$\xrightarrow{R_3 - \frac{R_2}{2}} \begin{bmatrix} 1 & 2 & 3 & -7 \\ 0 & 0 & 0 & 26 \\ 0 & 0 & 0 & 0 \end{bmatrix}$ — this is the required form. Rank of A, r, is 2 (since there are 2 linearly independent rows in the last matrix). P is derived from the identity matrix of order 3×3, that is $I_{3\times3} = \begin{bmatrix} 1 & 0 & 0 \\ 0 & 1 & 0 \\ 0 & 0 & 1 \end{bmatrix}$, using the same sequence of row operations as it is done on A to obtain the suitable matrix for rank determination. The row operations done are $R_3 - 2 \times R_1$, $R_2 - 4 \times R_1$, and $R_3 - \frac{R_2}{2}$. Performing just mentioned operations on $I_{3\times3}$, we have $\begin{bmatrix} 1 & 0 & 0 \\ 0 & 1 & 0 \\ 0 & 0 & 1 \end{bmatrix} \xrightarrow{R_3 - 2 \times R_1 \text{ and } R_2 - 4 \times R_1} \begin{bmatrix} 1 & 0 & 0 \\ -4 & 1 & 0 \\ -2 & 0 & 1 \end{bmatrix} \xrightarrow{R_3 - \frac{R_2}{2}}$

$\begin{bmatrix} 1 & 0 & 0 \\ -4 & 1 & 0 \\ 0 & -\frac{1}{2} & 1 \end{bmatrix}$, which is P. Again, starting from $\begin{bmatrix} 1 & 2 & 3 & -7 \\ 0 & 0 & 0 & 26 \\ 0 & 0 & 0 & 0 \end{bmatrix}$, a series of column operations can obtain the equivalent canonical form of A. Since rank of A is 2, the equivalent canonical form of A should be

$$C = \begin{bmatrix} I_2 & 0 \\ 0 & 0 \end{bmatrix} = \begin{bmatrix} 1 & 0 & 0 & 0 \\ 0 & 1 & 0 & 0 \\ 0 & 0 & 0 & 0 \end{bmatrix}$$ (in $\begin{bmatrix} I_2 & 0 \\ 0 & 0 \end{bmatrix}$, extra zeroes are inserted to make its order to be the same as that of

A). Next, the necessary operations are $\begin{bmatrix} 1 & 2 & 3 & -7 \\ 0 & 0 & 0 & 26 \\ 0 & 0 & 0 & 0 \end{bmatrix} \xrightarrow{\text{swap } C_2 \text{ and } C_4} \begin{bmatrix} 1 & -7 & 3 & 2 \\ 0 & 26 & 0 & 0 \\ 0 & 0 & 0 & 0 \end{bmatrix}$

$\xrightarrow{\frac{C_2}{26}} \begin{bmatrix} 1 & -\frac{7}{26} & 3 & 2 \\ 0 & 1 & 0 & 0 \\ 0 & 0 & 0 & 0 \end{bmatrix} \xrightarrow[\text{and } C_4 - 2 \times C_1]{C_2 + \frac{7}{26} \times C_1, \, C_3 - 3 \times C_1,} \begin{bmatrix} 1 & 0 & 0 & 0 \\ 0 & 1 & 0 & 0 \\ 0 & 0 & 0 & 0 \end{bmatrix}$ what we are after. So, column

operations performed are swap C_2 and C_4, $\frac{C_2}{26}$, $C_2 + \frac{7}{26} \times C_1$, $C_3 - 3 \times C_1$, and $C_4 - 2 \times C_1$. After that, start with

$I_{4 \times 4}$ to obtain Q (since the number of columns of A, N, is 4): $\begin{bmatrix} 1 & 0 & 0 & 0 \\ 0 & 1 & 0 & 0 \\ 0 & 0 & 1 & 0 \\ 0 & 0 & 0 & 1 \end{bmatrix} \xrightarrow{\text{swap } C_2 \text{ and } C_4}$

$\begin{bmatrix} 1 & 0 & 0 & 0 \\ 0 & 0 & 0 & 1 \\ 0 & 0 & 1 & 0 \\ 0 & 1 & 0 & 0 \end{bmatrix} \xrightarrow{\frac{C_2}{26}} \begin{bmatrix} 1 & 0 & 0 & 0 \\ 0 & 0 & 0 & 1 \\ 0 & 0 & 1 & 0 \\ 0 & \frac{1}{26} & 0 & 0 \end{bmatrix} \xrightarrow[\text{and } C_4 - 2 \times C_1]{C_2 + \frac{7}{26} \times C_1, \, C_3 - 3 \times C_1,} \begin{bmatrix} 1 & \frac{7}{26} & -3 & -2 \\ 0 & 0 & 0 & 1 \\ 0 & 0 & 1 & 0 \\ 0 & \frac{1}{26} & 0 & 0 \end{bmatrix}$, which is the matrix

Q. One can check that matrix multiplication of P, A, and Q is C. Since $P A Q = C$, $A = P^{-1} C Q^{-1}$, P and Q are non-singular matrices, and their inverses exist. Using the procedure of article 4.13, inverses of P and Q

are $P^{-1} = \begin{bmatrix} 1 & 0 & 0 \\ 4 & 1 & 0 \\ 2 & \frac{1}{2} & 1 \end{bmatrix}$ and $Q^{-1} = \begin{bmatrix} 1 & 2 & 3 & -7 \\ 0 & 0 & 0 & 26 \\ 0 & 0 & 1 & 0 \\ 0 & 1 & 0 & 0 \end{bmatrix}$ (for simplicity, computation of inverses is avoided).

We have $A = P^{-1} C Q^{-1} = \begin{bmatrix} 1 & 0 & 0 \\ 4 & 1 & 0 \\ 2 & \frac{1}{2} & 1 \end{bmatrix} \begin{bmatrix} 1 & 0 & 0 & 0 \\ 0 & 1 & 0 & 0 \\ 0 & 0 & 0 & 0 \end{bmatrix} \begin{bmatrix} 1 & 2 & 3 & -7 \\ 0 & 0 & 0 & 26 \\ 0 & 0 & 1 & 0 \\ 0 & 1 & 0 & 0 \end{bmatrix}$. Since P^{-1} is non-singular, its

columns must be linearly independent and P^{-1} can be partitioned as $P^{-1} = [D_{M \times r} \quad Z_{M \times (M-r)}]$. Again, nonsingularity of Q^{-1} manifests that rows of Q^{-1} are linearly independent and they can be partitioned as $Q^{-1} = $

$\begin{bmatrix} W_{r \times N} \\ H_{(N-r) \times N} \end{bmatrix}$. One can write $A = P^{-1} C Q^{-1} = [D_{M \times r} \quad Z_{M \times (M-r)}] \begin{bmatrix} I_r & 0 \\ 0 & 0 \end{bmatrix} \begin{bmatrix} W_{r \times N} \\ H_{(N-r) \times N} \end{bmatrix} = [D_{M \times r} \quad Z_{M \times (M-r)}] \begin{bmatrix} W_{r \times N} \\ 0 \end{bmatrix} = $

$D_{M \times r} W_{r \times N}$ following matrix multiplication. Thus factoring A as $A_{M \times N} = D_{M \times r} W_{r \times N}$ is called the full rank factorization. For the given numerical example of A, write $P^{-1} = [D_{M \times r} \quad Z_{M \times (M-r)}] = [D_{3 \times 2} \quad Z_{3 \times 1}] = $

$\left[\begin{bmatrix} 1 & 0 \\ 4 & 1 \\ 2 & \frac{1}{2} \end{bmatrix} \begin{bmatrix} 0 \\ 0 \\ 1 \end{bmatrix} \right]$, where $D_{3 \times 2} = \begin{bmatrix} 1 & 0 \\ 4 & 1 \\ 2 & \frac{1}{2} \end{bmatrix}$ and $Z_{3 \times 1} = \begin{bmatrix} 0 \\ 0 \\ 1 \end{bmatrix}$ and $Q^{-1} = \begin{bmatrix} W_{r \times N} \\ H_{(N-r) \times N} \end{bmatrix} = \begin{bmatrix} W_{2 \times 4} \\ H_{2 \times 4} \end{bmatrix} = \begin{bmatrix} \begin{bmatrix} 1 & 2 & 3 & -7 \\ 0 & 0 & 0 & 26 \end{bmatrix} \\ \begin{bmatrix} 0 & 0 & 1 & 0 \\ 0 & 1 & 0 & 0 \end{bmatrix} \end{bmatrix}$,

where $W_{2 \times 4} = \begin{bmatrix} 1 & 2 & 3 & -7 \\ 0 & 0 & 0 & 26 \end{bmatrix}$ and $H_{2 \times 4} = \begin{bmatrix} 0 & 0 & 1 & 0 \\ 0 & 1 & 0 & 0 \end{bmatrix}$. With the fact that $A_{M \times N} = D_{M \times r} W_{r \times N}$,

pseudoinverse $G_{N \times M}$ of $A_{M \times N}$ is given by $G_{N \times M} = W_{r \times N}^T \left(D_{M \times r}^T A_{M \times N} W_{r \times N}^T \right)^{-1} D_{M \times r}^T$, where $W_{r \times N}^T$ is the transpose of matrix $W_{r \times N}$, and so is $D_{M \times r}^T$ of $D_{M \times r}$. Matrix $G_{N \times M}$ has some amazing properties such as $\begin{cases} AGA = A \\ GAG = G \\ GA \ is \ symmetric \\ AG \ is \ symmetric \end{cases}$.

Matrices P and Q are not unique but $G_{N \times M}$ is unique. This is also called the Moore–Penrose inverse of A. For

the example at hand, we obtain $G_{N \times M} = G_{4 \times 3} = W_{2 \times 4}^T \left(D_{3 \times 2}^T A_{3 \times 4} W_{2 \times 4}^T \right)^{-1} D_{3 \times 2}^T = \begin{bmatrix} 1 & 2 & 3 & -7 \\ 0 & 0 & 0 & 26 \end{bmatrix}^T \begin{bmatrix} 1 & 0 \\ 4 & 1 \\ 2 & \frac{1}{2} \end{bmatrix}^T \cdot$

$\begin{bmatrix} 1 & 2 & 3 & -7 \\ 4 & 8 & 12 & -2 \\ 2 & 4 & 6 & -1 \end{bmatrix} \begin{bmatrix} 1 & 2 & 3 & -7 \\ 0 & 0 & 0 & 26 \end{bmatrix}^T \end{bmatrix}^{-1} \begin{bmatrix} 1 & 0 \\ 4 & 1 \\ 2 & \frac{1}{2} \end{bmatrix}^T = \begin{bmatrix} 1 & 2 & 3 & -7 \\ 0 & 0 & 0 & 26 \end{bmatrix}^T \begin{bmatrix} 413 & -442 \\ \frac{175}{2} & -65 \end{bmatrix}^{-1} \begin{bmatrix} 1 & 0 \\ 4 & 1 \\ 2 & \frac{1}{2} \end{bmatrix}^T =$

$\begin{bmatrix} 1 & 0 \\ 2 & 0 \\ 3 & 0 \\ -7 & 26 \end{bmatrix} \begin{bmatrix} -\frac{1}{182} & \frac{17}{455} \\ -\frac{5}{676} & \frac{59}{1690} \end{bmatrix} \begin{bmatrix} 1 & 4 & 2 \\ 0 & 1 & \frac{1}{2} \end{bmatrix} = \begin{bmatrix} -\frac{1}{182} & \frac{1}{65} & \frac{1}{130} \\ -\frac{1}{91} & \frac{2}{65} & \frac{1}{65} \\ -\frac{3}{182} & \frac{3}{65} & \frac{3}{130} \\ -\frac{2}{13} & \frac{2}{65} & \frac{1}{65} \end{bmatrix} = \begin{bmatrix} -0.0055 & 0.0154 & 0.0077 \\ -0.0110 & 0.0308 & 0.0154 \\ -0.0165 & 0.0462 & 0.0231 \\ -0.1538 & 0.0308 & 0.0154 \end{bmatrix}$. Have the

pseudoinverse of A computed by the agency of 'pinv' as follows:

MATLAB Command
>>A=[1 2 3 -7;4 8 12 -2;2 4 6 -1]; ↵
>>G=pinv(A) ↵

ans =

-0.0055	0.0154	0.0077
-0.0110	0.0308	0.0154
-0.0165	0.0462	0.0231
-0.1538	0.0308	0.0154

4.22 Bilinear and quadratic forms

If we define X and Y as row and column vectors of length M and N respectively and A is a matrix

of order $M \times N$, then XAY is called a bilinear form. For example, $X = [x_1 \quad x_2]$, $Y = \begin{bmatrix} y_1 \\ y_2 \\ y_3 \end{bmatrix}$, and $A =$

$\begin{bmatrix} 8 & 3 & -2 \\ 9 & 7 & 5 \end{bmatrix}$ express XAY as $[x_1 \quad x_2] \times \begin{bmatrix} 8 & 3 & -2 \\ 9 & 7 & 5 \end{bmatrix} \times \begin{bmatrix} y_1 \\ y_2 \\ y_3 \end{bmatrix} = 8x_1y_1 + 9y_1x_2 + 3y_2x_1 + 7y_2x_x - 2x_1y_3 + 5x_2y_3$ – that is

what the bilinear form is. See the implementation below:

MATLAB Command
>>syms x1 x2 y1 y2 y3 ↵ ← Declare all elements of X and Y as symbolic
>>X=[x1 x2]; ↵ ← Define row vector X
>>Y=[y1 y2 y3].'; ↵ ← Define column vector Y
>>A=[8 3 -2;9 7 5]; ↵ ← Enter matrix A
>>O=simplify(X*A*Y); ↵ ← Simplified output is assigned to O
>>pretty(O) ↵

8 y1 x1 + 9 y1 x2 + 3 y2 x1 + 7 y2 x2 - 2 y3 x1 + 5 y3 x2

In bilinear form, if X is equal to Y and A is a square matrix, then XAX is called a quadratic form. Quadratic form is actually the sum of squares, which can be written as $\sum_{i=1}^{M}(x_i - \bar{x})^2 = XAX$. The quadratic expression has uses in the analysis of variance in statistics. In quadratic form, A is a matrix of order $M \times M$ and X is a vector of length M. Finally, to show with an example, say, we have $X = [x_1 \quad x_2 \quad x_3]$ and $A = \begin{bmatrix} 2 & 3 & -4 \\ 1 & 9 & 7 \\ 2 & 3 & -6 \end{bmatrix}$,

hence, the form is $[x_1 \quad x_2 \quad x_3] \begin{bmatrix} 2 & 3 & -4 \\ 1 & 9 & 7 \\ 2 & 3 & -6 \end{bmatrix} \begin{bmatrix} x_1 \\ x_2 \\ x_3 \end{bmatrix} = 2x_1^2 + 9x_2^2 - 6x_3^2 + 4x_1x_2 - 2x_1x_3 + 10x_2x_3$, that is what is shown below:

MATLAB Command

```
>>syms x1 x2 x3 ↵
>>X=[x1 x2 x3]; ↵
>>A=[2 3 -4;1 9 7;2 3 -6]; ↵
>>O=simplify(X*A*X.'); ↵
>>pretty(O) ↵
             2                        2          2
     2 x1  + 4 x1 x2 - 2 x1 x3 + 9 x2  + 10 x2 x3 - 6 x3
```

Here X.' means the transpose without conjugate of X. One should take this into account when a row matrix of symbolic variables is converted to a column one (or vice versa).

4.23 Orthonormalization of a matrix

A matrix Q is said to be an orthonormal basis for the range of A if Q satisfies $Q^T Q = I$, where I is the identity matrix of order $r \times r$ and r is the rank of A. We begin with example of $A = \begin{bmatrix} 2 & 2 & -3 \\ 2 & 2 & 2 \\ -2 & -2 & 0 \\ 0 & 0 & 2 \end{bmatrix}$.

The rank of A is 2. Once Q is found, $Q^T Q$ should provide us identity matrix $\begin{bmatrix} 1 & 0 \\ 0 & 1 \end{bmatrix}$. The question is how

one can find the orthonormal matrix of A. First, we find $A \times A^T$, which is $\begin{bmatrix} 2 & 2 & -3 \\ 2 & 2 & 2 \\ -2 & -2 & 0 \\ 0 & 0 & 2 \end{bmatrix} \times$

$\begin{bmatrix} 2 & 2 & -2 & 0 \\ 2 & 2 & -2 & 0 \\ -3 & 2 & 0 & 2 \end{bmatrix} = \begin{bmatrix} 17 & 2 & -8 & -6 \\ 2 & 12 & -8 & 4 \\ -8 & -8 & 8 & 0 \\ -6 & 4 & 0 & 4 \end{bmatrix}$, call the last matrix B. Denote characteristic matrix of B

as C, therefore, C is given by $C = \lambda I - B = \begin{bmatrix} \lambda - 17 & -2 & 8 & 6 \\ -2 & \lambda - 12 & 8 & -4 \\ 8 & 8 & \lambda - 8 & 0 \\ 6 & -4 & 0 & \lambda - 4 \end{bmatrix}$, on that, eigenvalues of B are

$\lambda = 0$, 0, 16, and 25. Nonzero eigenvalues 16 and 25 provide characteristic matrices as $C_1 = \begin{bmatrix} -1 & -2 & 8 & 6 \\ -2 & 4 & 8 & -4 \\ 8 & 8 & 8 & 0 \\ 6 & -4 & 0 & 12 \end{bmatrix}$ and $C_2 = \begin{bmatrix} 8 & -2 & 8 & 6 \\ -2 & 13 & 8 & -4 \\ 8 & 8 & 17 & 0 \\ 6 & -4 & 0 & 21 \end{bmatrix}$. Ranks of C_1 and C_2 are 3. We look for the generalized eigenvectors of B for eigenvalues 16 and 25 respectively (see article 4.16 for the eigenvectors).

Required partitioned matrices are $C_1 = \begin{bmatrix} -1 & -2 & 8 & \vdots & 6 \\ -2 & 4 & 8 & \vdots & -4 \\ 8 & 8 & 8 & \vdots & 0 \\ \cdots & \cdots & \cdots & & \\ 6 & -4 & 0 & & 12 \end{bmatrix}$ and $C_2 = \begin{bmatrix} 8 & -2 & 8 & \vdots & 6 \\ -2 & 13 & 8 & \vdots & -4 \\ 8 & 8 & 17 & \vdots & 0 \\ \cdots & \cdots & \cdots & & \\ 6 & -4 & 0 & & 21 \end{bmatrix}$ respectively.

Generalized eigenvectors of B are $V_1 = \begin{bmatrix} -\begin{bmatrix} -1 & -2 & 8 \\ -2 & 4 & 8 \\ 8 & 8 & 8 \end{bmatrix}^{-1} \begin{bmatrix} 6 \\ -4 \\ 0 \end{bmatrix} w_1 \\ w_1 \end{bmatrix} = \begin{bmatrix} -w_1 \\ \dfrac{3}{2}w_1 \\ -\dfrac{1}{2}w_1 \\ w_1 \end{bmatrix}$ and $V_2 =$

$\begin{bmatrix} -\begin{bmatrix} 8 & -2 & 8 \\ -2 & 13 & 8 \\ 8 & 8 & 17 \end{bmatrix}^{-1} \begin{bmatrix} 6 \\ -4 \\ 0 \end{bmatrix} w_2 \\ w_2 \end{bmatrix} = \begin{bmatrix} -\dfrac{11}{2}w_2 \\ -3w_2 \\ 4w_2 \\ w_2 \end{bmatrix}$ corresponding to $\lambda = 16$ and $\lambda = 25$ respectively. Magnitudes of

V_1 and V_2 are $\dfrac{3}{\sqrt{2}}w_1$ and $\dfrac{15}{2}w_2$ respectively. Normalized eigenvectors are $\begin{bmatrix} -w_1 \\ \dfrac{3}{2}w_1 \\ -\dfrac{1}{2}w_1 \\ w_1 \end{bmatrix} / (\dfrac{3}{\sqrt{2}}w_1) = \begin{bmatrix} -\dfrac{\sqrt{2}}{3} \\ \dfrac{1}{\sqrt{2}} \\ -\dfrac{\sqrt{2}}{6} \\ \dfrac{\sqrt{2}}{3} \end{bmatrix} =$

$\begin{bmatrix} -0.4714 \\ 0.7071 \\ -0.2357 \\ 0.4714 \end{bmatrix}$ and $\begin{bmatrix} -\dfrac{11}{2}w_2 \\ -3w_2 \\ 4w_2 \\ w_2 \end{bmatrix} / (\dfrac{15}{2}w_2) = \begin{bmatrix} -\dfrac{11}{15} \\ -\dfrac{2}{5} \\ \dfrac{8}{15} \\ \dfrac{2}{15} \end{bmatrix} = \begin{bmatrix} -0.7333 \\ -0.4 \\ 0.5333 \\ 0.1333 \end{bmatrix}$ for $\lambda = 16$ and $\lambda = 25$ respectively. Multiplying an

eigenvector with a negative sign does not make any difference. Eigenvectors are written as the descending

order of eigenvalues. That is how the orthonormal basis of A is $Q = \begin{bmatrix} 0.7333 & 0.4714 \\ 0.4 & -0.7071 \\ -0.5333 & 0.2357 \\ -0.1333 & -0.4714 \end{bmatrix}$. Implement that as

follows:

MATLAB Command

for the orthonormal matrix of A,
```
>>A=[2 2 -3;2 2 2;-2 -2 0;0 0 2]; ↵
>>Q=orth(A) ↵
```

Q =
```
      0.7333    0.4714
      0.4000   -0.7071
     -0.5333    0.2357
     -0.1333   -0.4714
```

for finding the rank of A,
```
>>rank(A) ↵
```

ans =

2

for the proof of $Q^T Q = I$,
```
>> Q'*Q ↵
```

ans =
```
      1.0000   -0.0000
     -0.0000    1.0000
```

There is another variant in this regard, which is located in the Maple package. The name of the function is 'orthog' (applicable for square matrix only). A square matrix A is said to be orthogonal if it satisfies $AA^T = A^T A = I$, where A^T is the transpose of A and I is the identity matrix. Output of 'orthog' is either true or false depending on the orthogonality of A. Assume that the given square matrix is $A = \begin{bmatrix} \frac{2}{3} & -\frac{2}{3} & \frac{1}{3} \\ \frac{1}{3} & \frac{2}{3} & \frac{2}{3} \\ \frac{2}{3} & \frac{1}{3} & -\frac{2}{3} \end{bmatrix}$.

We want to check the orthogonality of A. Now we have $A \times A^T = \begin{bmatrix} \frac{2}{3} & -\frac{2}{3} & \frac{1}{3} \\ \frac{1}{3} & \frac{2}{3} & \frac{2}{3} \\ \frac{2}{3} & \frac{1}{3} & -\frac{2}{3} \end{bmatrix} \times \begin{bmatrix} \frac{2}{3} & \frac{1}{3} & \frac{2}{3} \\ -\frac{2}{3} & \frac{2}{3} & \frac{1}{3} \\ \frac{1}{3} & \frac{2}{3} & -\frac{2}{3} \end{bmatrix} = \begin{bmatrix} 1 & 0 & 0 \\ 0 & 1 & 0 \\ 0 & 0 & 1 \end{bmatrix}$ and $A^T \times A = \begin{bmatrix} \frac{2}{3} & \frac{1}{3} & \frac{2}{3} \\ -\frac{2}{3} & \frac{2}{3} & \frac{1}{3} \\ \frac{1}{3} & \frac{2}{3} & -\frac{2}{3} \end{bmatrix} \times \begin{bmatrix} \frac{2}{3} & -\frac{2}{3} & \frac{1}{3} \\ \frac{1}{3} & \frac{2}{3} & \frac{2}{3} \\ \frac{2}{3} & \frac{1}{3} & -\frac{2}{3} \end{bmatrix} = \begin{bmatrix} 1 & 0 & 0 \\ 0 & 1 & 0 \\ 0 & 0 & 1 \end{bmatrix}$. In either multiplication, we get the identity matrix, hence, A is orthogonal. Test it as follows:

MATLAB Command

```
>>A=sym([2/3 -2/3 1/3;1/3 2/3 2/3;2/3 1/3 -2/3]); ↵
>>maple('orthog',A) ↵
```

ans =

true

4.24 Differentiation and integration of matrix functions

If $A(t) = \begin{bmatrix} A_{11}(t) & A_{12}(t) & A_{13}(t) & ..... & A_{1n}(t) \\ A_{21}(t) & A_{22}(t) & A_{23}(t) & ..... & A_{2n}(t) \\ & & \vdots & & \\ A_{m1}(t) & A_{m2}(t) & A_{m3}(t) & ..... & A_{mn}(t) \end{bmatrix}$ is a matrix function of t, where the order of $A(t)$ is $m \times n$, its derivative is defined as $\frac{d[A(t)]}{dt} = \begin{bmatrix} \frac{d[A_{11}(t)]}{dt} & \frac{d[A_{12}(t)]}{dt} & \frac{d[A_{13}(t)]}{dt} & ..... & \frac{d[A_{1n}(t)]}{dt} \\ \frac{d[A_{21}(t)]}{dt} & \frac{d[A_{22}(t)]}{dt} & \frac{d[A_{23}(t)]}{dt} & ..... & \frac{d[A_{2n}(t)]}{dt} \\ & & \vdots & & \\ \frac{d[A_{m1}(t)]}{dt} & \frac{d[A_{m2}(t)]}{dt} & \frac{d[A_{m3}(t)]}{dt} & ..... & \frac{d[A_{mn}(t)]}{dt} \end{bmatrix}$. In a similar fashion, integration of $A(t)$ is defined as $\int A(t)dt = \begin{bmatrix} \int A_{11}(t)dt & \int A_{12}(t)dt & \int A_{13}(t)dt & ..... & \int A_{1n}dt \\ \int A_{21}(t)dt & \int A_{22}(t)dt & \int A_{23}(t)dt & ..... & \int A_{2n}dt \\ & & \vdots & & \\ \int A_{m1}(t)dt & \int A_{m2}(t)dt & \int A_{m3}(t)dt & ..... & \int A_{mn}dt \end{bmatrix}$. Say, $A(t) =$

$\begin{bmatrix} \sin t & e^{-t} & t^2 \\ 2t & \ln t & 8 \end{bmatrix}$, $\therefore \dfrac{d[A(t)]}{dt} = \begin{bmatrix} \cos t & -e^{-t} & 2t \\ 2 & \dfrac{1}{t} & 0 \end{bmatrix}$, again, $\dfrac{d^2[A(t)]}{dt^2} = \begin{bmatrix} -\sin t & e^{-t} & 2 \\ 0 & -\dfrac{1}{t^2} & 0 \end{bmatrix}$. Integration of the same

matrix function is $\int A(t)dt = \begin{bmatrix} -\cos t & -e^{-t} & \frac{1}{3}t^3 \\ t^2 & t\ln t - t & 8t \end{bmatrix}$. String representation is necessary to describe the element

functions, constant of integration is avoided, and declaration of symbolic variable is required. Computation of

matrix function may be required following differentiation or integration. We wish to compute $\dfrac{d^2[A(t)]}{dt^2}$ at $t=1$,

$\therefore \dfrac{d^2[A(t)]}{dt^2} = \begin{bmatrix} -\sin t & e^{-t} & 2 \\ 0 & -\dfrac{1}{t^2} & 0 \end{bmatrix} = \begin{bmatrix} -\sin 1 & e^{-1} & 2 \\ 0 & -\dfrac{1}{1^2} & 0 \end{bmatrix} = \begin{bmatrix} -0.8415 & 0.3679 & 2 \\ 0 & -1 & 0 \end{bmatrix}$. These all are implemented as

follows:

MATLAB Command for $\dfrac{d[A(t)]}{dt}$,

```
>>syms t ↵
>>A=[sin(t) exp(-t) t^2;2*t log(t) 8]; ↵
>>diff(A) ↵

ans =

[  cos(t),   -exp(-t),    2*t]
[     2,        1/t,        0]
```

for $\dfrac{d^2[A(t)]}{dt^2}$,

```
>>diff(A,2) ↵

ans =

[ -sin(t),    exp(-t),     2]
[     0,     -1/t^2,       0]
```

for $\int A(t)\,dt$,

```
>>int(A) ↵

ans =

[ -cos(t),       -exp(-t),    1/3*t^3]
[    t^2,       t*log(t)-t,       8*t]
```

for $\dfrac{d^2[A(t)]}{dt^2}$ at $t=1$,

```
>>d=diff(A,2); ↵          ← d²[A(t)]/dt² is stored to d

>>t=1; ↵
>>subs(d) ↵              ← Substitute t=1 by subroutine 'subs'

ans =
        -0.8415      0.3679     2.0000
             0      -1.0000          0
```

4.25 Minimal polynomial of a square matrix

Cayley-Hamilton theorem says that any square matrix A satisfies the characteristic polynomial $C(\lambda) = |\lambda I - A|$, where λ is a scalar and I is the identity matrix of the same order as that of A. The characteristic polynomial $C(\lambda)$ is not necessarily the polynomial of lowest degree such that $C(A) = 0$. If f is any polynomial such that $f(A) = 0$, then f is called an annihilating polynomial of A. Both the characteristic and the minimal polynomials are the annihilating polynomials. There is an algorithm for evaluating the minimal polynomial of a square matrix A. The algorithm is presented in figure 4.1. For numerical illustration, we find minimal polynomial of $A = \begin{bmatrix} 1 & 0 & 0 \\ -1 & 0 & 2 \\ 1 & 1 & -1 \end{bmatrix}$. First, check $A = K_0 I$, where $I = \begin{bmatrix} 1 & 0 & 0 \\ 0 & 1 & 0 \\ 0 & 0 & 1 \end{bmatrix}$. Clearly, there is no K_0 such that $A = K_0 I$, so, we go for next checking $A^2 = K_1 A + K_0 I$. For that, we need A^2, which is

$\begin{bmatrix} 1 & 0 & 0 \\ -1 & 0 & 2 \\ 1 & 1 & -1 \end{bmatrix} \times \begin{bmatrix} 1 & 0 & 0 \\ -1 & 0 & 2 \\ 1 & 1 & -1 \end{bmatrix} = \begin{bmatrix} 1 & 0 & 0 \\ 1 & 2 & -2 \\ -1 & -1 & 3 \end{bmatrix}$. Matrix equation $A^2 = K_1 A + K_0 I$ can be written as

$\begin{bmatrix} 1 & 0 & 0 \\ 1 & 2 & -2 \\ -1 & -1 & 3 \end{bmatrix} = K_1 \begin{bmatrix} 1 & 0 & 0 \\ -1 & 0 & 2 \\ 1 & 1 & -1 \end{bmatrix} + K_0 \begin{bmatrix} 1 & 0 & 0 \\ 0 & 1 & 0 \\ 0 & 0 & 1 \end{bmatrix}$. Entries indexed by (1,1) and (2,1) of the last matrix

equation say that $\begin{cases} 1 = K_1 + K_0 \\ 1 = -K_1 \end{cases}$, from which, $\begin{cases} K_0 = 2 \\ K_1 = -1 \end{cases}$. Having found $\begin{cases} K_0 = 2 \\ K_1 = -1 \end{cases}$, $K_1 A + K_0 I$ is equal to

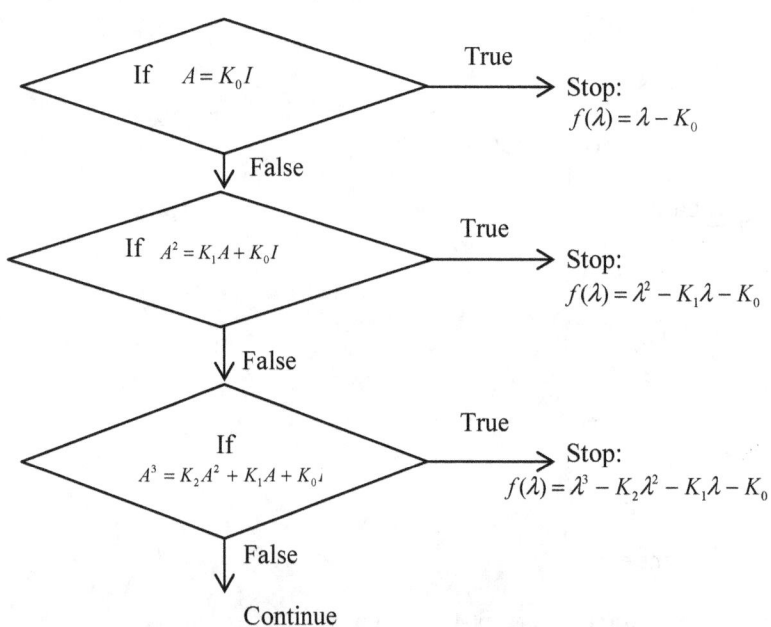

Figure 4.1 *Algorithm for calculating the minimal polynomial*

$\begin{bmatrix} -1 & 0 & 0 \\ 1 & 0 & -2 \\ -1 & -1 & 1 \end{bmatrix} + \begin{bmatrix} 2 & 0 & 0 \\ 0 & 2 & 0 \\ 0 & 0 & 2 \end{bmatrix} = \begin{bmatrix} 1 & 0 & 0 \\ 1 & 2 & -2 \\ -1 & -1 & 3 \end{bmatrix}$, which is equal to A^2. We have to stop here according to the

algorithm, therefore, the minimal polynomial of A is $f(A) = A^2 - K_1 A - K_0 = A^2 + A - 2$. With the help of Maple function 'minpoly', one can find the minimal polynomial conveniently as follows:

MATLAB Command

>>A=sym([1,0,0;-1,0,2;1,1,-1]); ↵

>>maple('minpoly',A,'A') ↵

ans =

-2+A+A^2   ← As we found $f(A) = A^2 + A - 2$

Do not be confused with two A arguments of 'minpoly'. The first A means the given matrix and the second one is the variable of the output polynomial. One can use other variables like x or t.

4.26 Generating equations from coefficient matrices or vice versa

Suppose we have a coefficient matrix $A = \begin{bmatrix} 2 & 3 & -4 & 5 \\ -1 & -3 & -7 & 3 \\ 3 & -7 & 0 & 2 \end{bmatrix}$ and a column vector $C = \begin{bmatrix} a \\ b \\ 3 \end{bmatrix}$.

Equations from the matrix A and the column vector C are to be formed in terms of variables x, y, z, and

u. The equations can be obtained from matrix equation $\begin{bmatrix} 2 & 3 & -4 & 5 \\ -1 & -3 & -7 & 3 \\ 3 & -7 & 0 & 2 \end{bmatrix} \begin{bmatrix} x \\ y \\ z \\ u \end{bmatrix} = \begin{bmatrix} a \\ b \\ 3 \end{bmatrix}$. Following matrix

multiplication, we have $\begin{cases} 2x + 3y - 4z + 5u = a \\ -x - 3y - 7z + 3u = b \\ 3x - 7y + 2u = 3 \end{cases}$. Maple function 'geneqns' is used to generate the equations:

MATLAB Command

>>maple('A:=matrix([[2,3,-4,5],[-1,-3,-7,3],[3,-7,0,2]])'); ↵

>>maple('C:=vector([a,b,3])'); ↵

>>maple('geneqns(A,[x,y,z,u],C)') ↵

          ans =

{2*x+3*y-4*z+5*u = a, -x-3*y-7*z+3*u = b, 3*x-7*y+2*u = 3}

To display the output in pretty form,

>>maple('o:=geneqns(A,[x,y,z,u],C):'); ↵   ←Assign output to o

>>maple('interface(prettyprint=1):o'); ↵

{2 x + 3 y - 4 z + 5 u = a, -x - 3 y - 7 z + 3 u = b, 3 x - 7 y + 2 u = 3}

The reverse operation that is from equations to matrix is performed by Maple subroutine 'genmatrix'. Assume that the given equations are $x + 2z = a - 8y$, $3x - 5y = 6 - z$, and $z = -x - 5y + 23$. Organize the equations to write

$\begin{cases} x + 8y + 2z = a \\ 3x - 5y + z = 6 \\ x + 5y + z = 23 \end{cases}$, from which, we have $\begin{bmatrix} 1 & 8 & 2 \\ 3 & -5 & 1 \\ 1 & 5 & 1 \end{bmatrix} \begin{bmatrix} x \\ y \\ z \end{bmatrix} = \begin{bmatrix} a \\ 6 \\ 23 \end{bmatrix}$. Coefficient matrix and column vector are

$\begin{bmatrix} 1 & 8 & 2 \\ 3 & -5 & 1 \\ 1 & 5 & 1 \end{bmatrix}$ and $\begin{bmatrix} a \\ 6 \\ 23 \end{bmatrix}$ respectively.

MATLAB Command

>>maple('eqns:={x+2*z=a-8*y,3*x-5*y=6-z,z=-x-5*y+23}:'); ↵

>>A=maple('genmatrix(eqns,[x,y,z],flag)'); ↵

>>sym(A) ↵

ans =

```
[ 1,   8,   2,    a]
[ 3,  -5,   1,    6]
[ 1,   5,   1,   23]
```

In output matrix, the first three columns are the coefficient matrix and the fourth column is the column vector. Also notice that one does not need to rearrange an equation when it is entered into MATLAB. Viewing coefficient matrix and column vector differently is executed as follows:

Command for viewing
the coefficient matrix,   for viewing the column vector,
>>A=maple('genmatrix(eqns,[x,y,z],C)'); ↵   >>sym(maple('print(C)')).' ↵
>>sym(A) ↵

                                                     ans =
ans =
                                                     [a]
```
[ 1,   8,   2]                                       [  6]
[ 3,  -5,   1]                                       [ 23]
[ 1,   5,   1]
```

Operator .' is used to convert the row matrix output to a column one without conjugate. Usefulness of the subroutine is understood when we have 6 or 7 variables' equations.

4.27 Matrix exponential

For every square matrix A, e^A can be expressed as $I+A+\dfrac{A^2}{2!}+\dfrac{A^3}{3!}+\dfrac{A^4}{4!}+\dfrac{A^5}{5!}+\dots$. Utilizing the annihilating polynomial of A, higher powers of A can be reduced to the lower ones. Given that $A=\begin{bmatrix} 4 & 0 & 0 \\ -4 & 0 & 0 \\ 4 & 4 & 4 \end{bmatrix}$, compute e^{At}. Use the concept of article 4.25 to get the minimal polynomial of A, which is $f(A)=A^2-4A$. We know that A satisfies the minimal polynomial equation $A^2-4A=0$, from which, $A^2=4A$. Now, the expansion of e^{At} is $I+At+\dfrac{(At)^2}{2!}+\dfrac{(At)^3}{3!}+\dfrac{(At)^4}{4!}+\dfrac{(At)^5}{5!}+\dots$. As we mentioned, the higher order terms (say, A^3) of A are reduced as $A^3=A^2A=4AA=4A^2=4^2A$. In a similar way, we have $A^4=4^3A$, $A^5=4^4A$, $A^6=4^5A$ ... etc. So, the series $I+At+\dfrac{(At)^2}{2!}+\dfrac{(At)^3}{3!}+\dfrac{(At)^4}{4!}+\dfrac{(At)^5}{5!}+\dots$ becomes $I+At+\dfrac{4At^2}{2!}+$

$\dfrac{4^2At^3}{3!}+\dfrac{4^3At^4}{4!}+\dfrac{4^4At^5}{5!}+\dots \quad = \quad I+A\left[t+\dfrac{4t^2}{2!}+\dfrac{4^2t^3}{3!}+\dfrac{4^3t^4}{4!}+\dfrac{4^4t^5}{5!}+\dots\right] \quad = \quad I+\dfrac{A}{4}\left[4t+\dfrac{(4t)^2}{2!}+\dfrac{(4t)^3}{3!}+\dfrac{(4t)^4}{4!}+\right.$

$\left.\dfrac{(4t)^5}{5!}+\dots\right]=I+\dfrac{A}{4}[e^{4t}-1]$. Finally, the matrix exponential of A is $e^{At}=I+\dfrac{A}{4}[e^{4t}-1]=\begin{bmatrix} 1 & 0 & 0 \\ 0 & 1 & 0 \\ 0 & 0 & 1 \end{bmatrix}+\dfrac{e^{4t}-1}{4}\times$

$\begin{bmatrix} 4 & 0 & 0 \\ -4 & 0 & 0 \\ 4 & 4 & 4 \end{bmatrix}=\begin{bmatrix} 1 & 0 & 0 \\ 0 & 1 & 0 \\ 0 & 0 & 1 \end{bmatrix}+\begin{bmatrix} e^{4t}-1 & 0 & 0 \\ 1-e^{4t} & 0 & 0 \\ e^{4t}-1 & e^{4t}-1 & e^{4t}-1 \end{bmatrix}=\begin{bmatrix} e^{4t} & 0 & 0 \\ 1-e^{4t} & 1 & 0 \\ e^{4t}-1 & e^{4t}-1 & e^{4t} \end{bmatrix}$. MATLAB counterpart

for the matrix exponential is 'expm'. Apply that as follows:

MATLAB Command

```
>>syms t ↵
>>A=sym([4 0 0;-4 0 0;4 4 4]); ↵
>>y=expm(A*t) ↵

y =

[    exp(4*t),            0,            0]
[ -exp(4*t)+1,           1,            0]
[  exp(4*t)-1,    exp(4*t)-1,    exp(4*t)]
```

If necessary, use the command 'pretty' to display the symbolic form. There are other methods to find the matrix exponential. One of them uses Jordan form, which is discussed in the subsequent section.

4.28 Normalization of a vector

Sometimes a vector needs normalization. For example, we have the vector $A = \begin{bmatrix} a \\ 3 \\ 4 \end{bmatrix}$. Magnitude of A

is $\sqrt{a^2 + 3^3 + 4^2} = \sqrt{a^2 + 25}$. Having known the magnitude, normalized A is given by $\begin{bmatrix} \dfrac{a}{\sqrt{a^2+25}} \\ \dfrac{3}{\sqrt{a^2+25}} \\ \dfrac{4}{\sqrt{a^2+25}} \end{bmatrix}$. Maple

function 'normalize' can be conducted on this:

MATLAB Command

```
>>maple('A:=vector([a,3,4])'); ↵   ← Define vector A
>>N=maple('normalize(A)'); ↵       ← Assign the output of 'normalize' to N
>>pretty(sym(N)) ↵
```

```
[        a                    3                    4           ]
[-------------------    -----------------    ------------------]
[     2    1/2             2    1/2             2    1/2        ]
[  (| a | + 25)         (| a | + 25)         (| a | + 25)      ]
```

Usually, the variables are complex that is why a appears as $|a|$. Notice that the output is a row matrix. There is another alternative if you use the command 'syms' as follows:

```
>>syms a ↵
>>A=[a 3 4]; ↵
>>N=A./sqrt(sum(A.^2)); ↵
>>pretty(N) ↵
```

```
[       a                   3                   4          ]
[------------------    ----------------    -----------------]
[    2    1/2             2    1/2            2    1/2       ]
[  (a + 25)            (a + 25)            (a + 25)         ]
```

In addition to the symbolic elements, numerical normalized vector is also found for $A = \begin{bmatrix} 5 \\ 3 \\ 4 \end{bmatrix}$, where the

normalized vector is $A = \begin{bmatrix} 5 \\ 3 \\ 4 \end{bmatrix} / \sqrt{50} = \begin{bmatrix} 0.7071 \\ 0.4243 \\ 0.5657 \end{bmatrix}$. Carry out the following command:

>>A=[5 3 4]; ↵
>>A./sqrt(sum(A.^2)) ↵

ans =
     0.7071   0.4243   0.5657

4.29 LU triangular factorization of a matrix

In the triangular factorization, a square matrix is expressed as the product of two triangular matrices – a lower and an upper triangular matrices. Examples of the lower and upper triangular matrices are $\begin{bmatrix} 1 & 0 & 0 \\ 7 & 0 & 0 \\ 2 & 0 & 5 \end{bmatrix}$ (super diagonal elements are zeroes) and $\begin{bmatrix} 1 & 2 & 3 \\ 0 & 0 & 3 \\ 0 & 0 & 5 \end{bmatrix}$ (subdiagonal elements are zeroes) respectively. If A is a square matrix, A can be written as $A = L \times U$, where L and U correspond to the lower and upper triangular matrices respectively of the same order as that of A. The factorization is not a unique one. One may end up with different L and U but the product of L and U is A that is for sure. We perform Gaussian elimination with partial pivoting until the given matrix A is transformed to the upper triangular one and store the multipliers of different elementary transformations. Say, $A = \begin{bmatrix} 2 & 3 & 4 & 7 \\ 2 & 0 & 1 & 3 \\ 9 & 1 & 3 & 7 \\ 1 & 8 & 2 & 4 \end{bmatrix}$, operations $R_2 - R_1$, $R_3 - \dfrac{9}{2}R_1$,

and $R_4 - \dfrac{1}{2}R_1$ transform A to $\begin{bmatrix} 2 & 3 & 4 & 7 \\ 0 & -3 & -3 & -4 \\ 0 & -\dfrac{25}{2} & -15 & -\dfrac{49}{2} \\ 0 & \dfrac{13}{2} & 0 & \dfrac{1}{2} \end{bmatrix}$ and we have the multipliers $\begin{bmatrix} 1 \\ \dfrac{9}{2} \\ \dfrac{1}{2} \end{bmatrix}$. Next, $R_3 - \dfrac{25}{6}R_2$

and $R_4 + \dfrac{13}{6}R_2$ give us $\begin{bmatrix} 2 & 3 & 4 & 7 \\ 0 & -3 & -3 & -4 \\ 0 & 0 & -\dfrac{5}{2} & -\dfrac{47}{6} \\ 0 & 0 & -\dfrac{13}{2} & -\dfrac{49}{6} \end{bmatrix}$ along with the accumulation of multipliers $\begin{bmatrix} 1 & \\ \dfrac{9}{2} & \dfrac{25}{6} \\ \dfrac{1}{2} & -\dfrac{13}{6} \end{bmatrix}$.

Perform $R_4 - \dfrac{13}{5}R_3$ on the last matrix to get the required upper triangular form $U = \begin{bmatrix} 2 & 3 & 4 & 7 \\ 0 & -3 & -3 & -4 \\ 0 & 0 & -\dfrac{5}{2} & -\dfrac{47}{6} \\ 0 & 0 & 0 & \dfrac{61}{5} \end{bmatrix}$

and the last multiplier accumulations $\begin{bmatrix} 1 & & \\ \dfrac{9}{2} & \dfrac{25}{6} & \\ \dfrac{1}{2} & -\dfrac{13}{6} & \dfrac{13}{5} \end{bmatrix}$. For the lower triangular matrix, we have to have an

identity matrix of order 4×4 (same as that of A), which is $\begin{bmatrix} 1 & 0 & 0 & 0 \\ 0 & 1 & 0 & 0 \\ 0 & 0 & 1 & 0 \\ 0 & 0 & 0 & 1 \end{bmatrix}$. Relpace the lower triangular

zeroes of the identity matrix with the accumulated multipliers, hence, we have the lower triangular matrix $L =$

$\begin{bmatrix} 1 & 0 & 0 & 0 \\ 1 & 1 & 0 & 0 \\ \dfrac{9}{2} & \dfrac{25}{6} & 1 & 0 \\ \dfrac{1}{2} & -\dfrac{13}{6} & \dfrac{13}{5} & 1 \end{bmatrix}$. One can check that $\begin{bmatrix} 2 & 3 & 4 & 7 \\ 2 & 0 & 1 & 3 \\ 9 & 1 & 3 & 7 \\ 1 & 8 & 2 & 4 \end{bmatrix} = \begin{bmatrix} 1 & 0 & 0 & 0 \\ 1 & 1 & 0 & 0 \\ \dfrac{9}{2} & \dfrac{25}{6} & 1 & 0 \\ \dfrac{1}{2} & -\dfrac{13}{6} & \dfrac{13}{5} & 1 \end{bmatrix} \times$

$\begin{bmatrix} 2 & 3 & 4 & 7 \\ 0 & -3 & -3 & -4 \\ 0 & 0 & -\dfrac{5}{2} & -\dfrac{47}{6} \\ 0 & 0 & 0 & \dfrac{61}{5} \end{bmatrix}$. Maple function 'LUdecomp', abbreviation of Lower Upper decomposition, can

have the symbolic output as follows:

MATLAB Command

```
>>maple('A:=matrix(4,4,[2,3,4,7,2,0,1,3,9,1,3,7,1,8,2,4])');↵   ← Enter A
>>maple('LUdecomp(A,L=L,U=U)'); ↵                              ← Apply 'LUdecomp'
>>L=sym(maple('print(L)')) ↵                                   ← Display L


L =


[   1,      0,      0,     0]
[   1,      1,      0,     0]
[ 9/2,   25/6,      1,     0]
[ 1/2,  -13/6,   13/5,     1]
>>U=sym(maple('print(U)')) ↵                                   ← Display U


U =


[   2,      3,      4,       7]
[   0,     -3,     -3,      -4]
[   0,      0,   -5/2,   -47/6]
[   0,      0,      0,    61/5]
>>L*U ↵                                                        ← To check A = L × U


ans =


[ 2,   3,   4,   7]
[ 2,   0,   1,   3]
[ 9,   1,   3,   7]
[ 1,   8,   2,   4]                                            ← Same as A
```

You can know more about the subroutine executing 'mhelp LUdecomp' in the command window. There is one more subroutine by name 'lu', which provides numerical factorization of a matrix. Execute 'help lu' in this regard.

4.30 QR decomposition of a matrix

Before going to the details of QR decomposition, we demonstrate the idea underlying the decomposition. Consider the two-dimensional $x-y$ geometry plane depicted in figure 4.2. Referring to the figure, unit vectors $\begin{bmatrix} 1 \\ 0 \end{bmatrix}$ and $\begin{bmatrix} 0 \\ 1 \end{bmatrix}$ direct along the x and y axes respectively. Any point on $x-y$ plane can be represented by the vector $\begin{bmatrix} x \\ y \end{bmatrix}$. Suppose, $x-y$ axes are rotated counterclockwise by an angle θ, where new

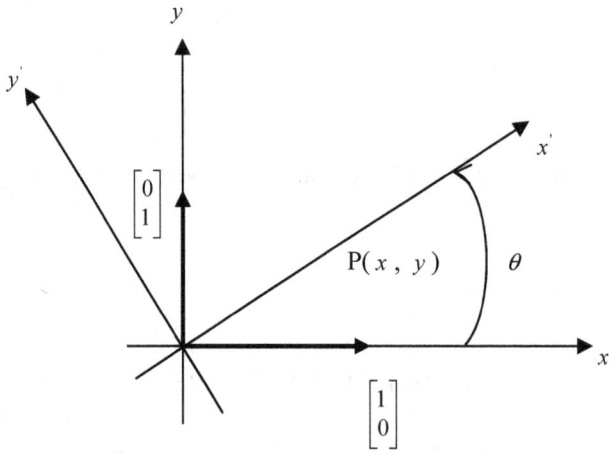

Figure 4.2 *Rotation of $x-y$ axes by an angle θ*

axes are labeled by x' and y'. With reference to the new axes system, components of the vectors $\begin{bmatrix} 1 \\ 0 \end{bmatrix}$ and $\begin{bmatrix} 0 \\ 1 \end{bmatrix}$ are $\begin{bmatrix} \cos\theta \\ -\sin\theta \end{bmatrix}$ and $\begin{bmatrix} \sin\theta \\ \cos\theta \end{bmatrix}$ respectively. Taking these as basis, form matrix $Q = \begin{bmatrix} \cos\theta & \sin\theta \\ -\sin\theta & \cos\theta \end{bmatrix}$. Choosing proper θ, where $\theta = \tan^{-1}\dfrac{y}{x}$, makes any point $\begin{bmatrix} x \\ y \end{bmatrix}$ other than origin fall on x' axis, where the same point has the coordinate vector $\begin{bmatrix} x' \\ 0 \end{bmatrix}$ in x'-y' system. Now get down the topic. Any vector $\begin{bmatrix} x \\ y \end{bmatrix}$ can be transformed to the upper triangular form $\begin{bmatrix} x' \\ 0 \end{bmatrix}$ applying matrix Q, that is, $\begin{bmatrix} \cos\theta & \sin\theta \\ -\sin\theta & \cos\theta \end{bmatrix}\begin{bmatrix} x \\ y \end{bmatrix} = \begin{bmatrix} x' \\ 0 \end{bmatrix}$. Matrix Q is orgthogonal because $|Q|=1$ and $QQ^T = I$. Utilize the orthogonality to write $\begin{bmatrix} x \\ y \end{bmatrix} = \begin{bmatrix} \cos\theta & -\sin\theta \\ \sin\theta & \cos\theta \end{bmatrix}\begin{bmatrix} x' \\ 0 \end{bmatrix}$.

The last equation says that any non-zero vector (A) can be expressed as the product of two matrices, i.e, $A = Q \times R$, where Q is orthogonal and R is upper triangular (more appropriately, $A = Q \times R$ should be written as $A = Q^T \times R$. As long as Q is orthogonal, it does not matter whether we write Q or Q^T). This is the basic concept of QR decomposition. Ongoing discussion dealt with a vector of two elements now we concentrate on the situation when the number of elements is more than two. One transformation is used, called

Householder transformation, in this regard. The transformation is given by $X = \begin{bmatrix} x_1 \\ x_2 \\ \vdots \\ x_N \end{bmatrix}$, $Y = \begin{bmatrix} |X| \\ 0 \\ \vdots \\ 0 \end{bmatrix}$, $U = X - Y$,

and $Q = I - \dfrac{UU^T}{|X|(|X|-x_1)}$, where orders of X, Y, U, I, and Q are $N \times 1$, $N \times 1$, $N \times 1$, $N \times N$, and $N \times N$

respectively. Do not be perplexed with the notations. We take $X = \begin{bmatrix} 2 \\ 4 \\ -2 \\ 1 \end{bmatrix}$ for finding the orthogonal matrix Q,

hence, $|X| = \sqrt{2^2 + 4^2 + (-2)^2 + 1} = 5$, $x_1 = 2$, $Y = \begin{bmatrix} 5 \\ 0 \\ 0 \\ 0 \end{bmatrix}$, $U = X - Y = \begin{bmatrix} -3 \\ 4 \\ -2 \\ 1 \end{bmatrix}$, $I = \begin{bmatrix} 1 & 0 & 0 & 0 \\ 0 & 1 & 0 & 0 \\ 0 & 0 & 1 & 0 \\ 0 & 0 & 0 & 1 \end{bmatrix}$, and $Q =$

$\begin{bmatrix} 1 & 0 & 0 & 0 \\ 0 & 1 & 0 & 0 \\ 0 & 0 & 1 & 0 \\ 0 & 0 & 0 & 1 \end{bmatrix} - \dfrac{1}{5(5-2)} \begin{bmatrix} -3 \\ 4 \\ -2 \\ 1 \end{bmatrix} [-3 \quad 4 \quad -2 \quad 1] = \begin{bmatrix} \frac{2}{5} & \frac{4}{5} & -\frac{2}{5} & \frac{1}{5} \\ \frac{4}{5} & -\frac{1}{15} & \frac{8}{15} & -\frac{4}{15} \\ -\frac{2}{5} & \frac{8}{15} & \frac{11}{15} & \frac{2}{15} \\ \frac{1}{5} & -\frac{4}{15} & \frac{2}{15} & \frac{14}{15} \end{bmatrix}$. With $R = Y$, one can check that $X =$

$Q \times R$ – that is what we need from the decomposition.

The method is applied to a rectangular matrix with little digression. For rectangular matrix, we will have one Q for each column until the transformed form is the upper triangular. The overall Q is obtained from

the multiplication of these Q's. To illustrate this, take $A = \begin{bmatrix} 3 & 3 & 3 \\ 0 & -3 & -3 \\ 0 & -4 & -4 \\ 4 & 4 & 4 \end{bmatrix}$. Apply Householder transformation

on the first column $\begin{bmatrix} 3 \\ 0 \\ 0 \\ 4 \end{bmatrix}$ of A. The necessary computations are $X = \begin{bmatrix} 3 \\ 0 \\ 0 \\ 4 \end{bmatrix}$, $|X| = 5$, $Y = \begin{bmatrix} 5 \\ 0 \\ 0 \\ 0 \end{bmatrix}$, $U = X - Y = \begin{bmatrix} -2 \\ 0 \\ 0 \\ 4 \end{bmatrix}$,

Q_1 (call for the first column) $= \begin{bmatrix} 1 & 0 & 0 & 0 \\ 0 & 1 & 0 & 0 \\ 0 & 0 & 1 & 0 \\ 0 & 0 & 0 & 1 \end{bmatrix} - \dfrac{1}{5(5-3)} \begin{bmatrix} -2 \\ 0 \\ 0 \\ 4 \end{bmatrix} \times [-2 \quad 0 \quad 0 \quad 4] = \begin{bmatrix} \frac{3}{5} & 0 & 0 & \frac{4}{5} \\ 0 & 1 & 0 & 0 \\ 0 & 0 & 1 & 0 \\ \frac{4}{5} & 0 & 0 & -\frac{3}{5} \end{bmatrix}$, and $Q_1 \times A =$

$\begin{bmatrix} \frac{3}{5} & 0 & 0 & \frac{4}{5} \\ 0 & 1 & 0 & 0 \\ 0 & 0 & 1 & 0 \\ \frac{4}{5} & 0 & 0 & -\frac{3}{5} \end{bmatrix} \times \begin{bmatrix} 3 & 3 & 3 \\ 0 & -3 & -3 \\ 0 & -4 & -4 \\ 4 & 4 & 4 \end{bmatrix} = \begin{bmatrix} 5 & 5 & 5 \\ 0 & -3 & -3 \\ 0 & -4 & -4 \\ 0 & 0 & 0 \end{bmatrix}$. Observe the matrix inside the dotted mark of the last

matrix – that is $\begin{bmatrix} -3 & -3 \\ -4 & -4 \\ 0 & 0 \end{bmatrix}$. Corresponding to the second column of $Q_1 \times A$, we have Q_2 but Q_2 acquires the

form $\begin{bmatrix} 1 & 0 & 0 & 0 \\ 0 & & & \\ 0 & & \hat{Q}_2 & \\ 0 & & & \end{bmatrix}$, where $\hat{Q}_2$ is of order 3×3 (since the number of rows of $\begin{bmatrix} -3 & -3 \\ -4 & -4 \\ 0 & 0 \end{bmatrix}$ is 3) and it is again obtained from the last matrix conducting Householder transfornation. Computations are $X = \begin{bmatrix} -3 \\ -4 \\ 0 \end{bmatrix}$, $|X| = 5$, U

$$= \begin{bmatrix} -8 \\ -4 \\ 0 \end{bmatrix}, \quad \hat{Q}_2 = \begin{bmatrix} 1 & 0 & 0 \\ 0 & 1 & 0 \\ 0 & 0 & 1 \end{bmatrix} - \frac{1}{5(5+3)} \begin{bmatrix} -8 \\ -4 \\ 0 \end{bmatrix} \begin{bmatrix} -8 & -4 & 0 \end{bmatrix} = \begin{bmatrix} -\frac{3}{5} & -\frac{4}{5} & 0 \\ -\frac{4}{5} & \frac{3}{5} & 0 \\ 0 & 0 & 1 \end{bmatrix}, \text{ on that, } Q_2 = \begin{bmatrix} 1 & 0 & 0 & 0 \\ 0 & -\frac{3}{5} & -\frac{4}{5} & 0 \\ 0 & -\frac{4}{5} & \frac{3}{5} & 0 \\ 0 & 0 & 0 & 1 \end{bmatrix}.$$

After that, we have $\hat{Q}_2 \times \begin{bmatrix} -3 & -3 \\ -4 & -4 \\ 0 & 0 \end{bmatrix} = \begin{bmatrix} 5 & 5 \\ 0 & 0 \\ 0 & 0 \end{bmatrix}$, which is completely the upper triangular form, so, we stop

finding Q_3. Insert $\begin{bmatrix} 5 & 5 \\ 0 & 0 \\ 0 & 0 \end{bmatrix}$ back to $\begin{bmatrix} 5 & 5 & 5 \\ 0 & -3 & -3 \\ 0 & -4 & -4 \\ 0 & 0 & 0 \end{bmatrix}$ on the dotted elements to get the entire upper triangular

matrix $R = \begin{bmatrix} 5 & 5 & 5 \\ 0 & 5 & 5 \\ 0 & 0 & 0 \\ 0 & 0 & 0 \end{bmatrix}$. Then, the overall Q is $Q_1 \times Q_2 = \begin{bmatrix} \frac{3}{5} & 0 & 0 & \frac{4}{5} \\ 0 & 1 & 0 & 0 \\ 0 & 0 & 1 & 0 \\ \frac{4}{5} & 0 & 0 & -\frac{3}{5} \end{bmatrix} \times \begin{bmatrix} 1 & 0 & 0 & 0 \\ 0 & -\frac{3}{5} & -\frac{4}{5} & 0 \\ 0 & -\frac{4}{5} & \frac{3}{5} & 0 \\ 0 & 0 & 0 & 1 \end{bmatrix} =$

$\begin{bmatrix} \frac{3}{5} & 0 & 0 & \frac{4}{5} \\ 0 & -\frac{3}{5} & -\frac{4}{5} & 0 \\ 0 & -\frac{4}{5} & \frac{3}{5} & 0 \\ \frac{4}{5} & 0 & 0 & -\frac{3}{5} \end{bmatrix}$. Finally, we have the decomposition as $\begin{bmatrix} 3 & 3 & 3 \\ 0 & -3 & -3 \\ 0 & -4 & -4 \\ 4 & 4 & 4 \end{bmatrix} = \begin{bmatrix} \frac{3}{5} & 0 & 0 & \frac{4}{5} \\ 0 & -\frac{3}{5} & -\frac{4}{5} & 0 \\ 0 & -\frac{4}{5} & \frac{3}{5} & 0 \\ \frac{4}{5} & 0 & 0 & -\frac{3}{5} \end{bmatrix} \times$

$\begin{bmatrix} 5 & 5 & 5 \\ 0 & 5 & 5 \\ 0 & 0 & 0 \\ 0 & 0 & 0 \end{bmatrix}$. However, the matrix Q is not unique. Output of MATLAB can yield different Q but anyhow

$A = Q \times R$ is observed. See below how easy it is in MATLAB by dint of Maple function 'QRdecomp':

MATLAB Command

```
>>maple('A:=matrix(4,3,[3,3,3,0,-3,-3,0,-4,-4,4,4,4])');  ↵      ← Enter A
>>R=maple('QRdecomp(A,Q=Q)');  ↵                                 ← Apply function 'QRdecomp'
>>Q=sym(maple('print(Q)'))  ↵                                    ← Display Q
```

Q =

```
[ 3/5,    0,   4/5,     0]
[   0, -3/5,     0,   4/5]
[   0, -4/5,     0,  -3/5]
[ 4/5,    0,  -3/5,     0]
>>sym(R)  ↵                                                      ← Display R
```

ans =

```
[ 5,  5,  5]
[ 0,  5,  5]
[ 0,  0,  0]
[ 0,  0,  0]
```

As we mentioned, the output Q is not the same as found before. The numerical counterpart of QR decomposition is 'qr'. To see details of 'qr', execute 'help qr' in Command Window.

4.31 Basis for the null space of a matrix

Recall that the eigenvector V of a square matrix A comes from the characteristic equation $|A - \lambda I| = 0$ and it satisfies the equation $AV = \lambda V$. A basis vector Y for the null space of matrix A (does not have to be square) is obtained from the equation $AY = 0$. All rules and methods of the article 4.16 are applicable here but that will be on A in lieu of B. Nullity v of a matrix A is defined as the difference between the number of columns of A and rank of A. If order of A is $M \times N$ and nullity of A is v, the basis matrix for the null space of A will be of order $N \times v$. We explicate the concept with $A =$

$$\begin{bmatrix} 2 & 3 & 1 & 2 & 3 & 1 \\ 8 & 8 & 8 & -3 & -3 & -3 \end{bmatrix}.$$ The rank of matrix A is 2 since both rows are linearly independent.

Nullity of A is $v = 6-2=4$, hence, the null space of A has four basis vectors, which form a matrix of order

$N \times v = 6 \times 4$. Matrix A can be partitioned as $\begin{bmatrix} 2 & 3 & \vdots & 1 & 2 & 3 & 1 \\ 8 & 8 & \vdots & 8 & -3 & -3 & -3 \end{bmatrix}$, where $\begin{bmatrix} 2 & 3 \\ 8 & 8 \end{bmatrix}$ is non

singular. Utilize the notion of generalized eigenvectors of article 4.16. Since the nullity is 4, four unknowns (independent variables) are required to find the generalized eigenvectors. Say, the variables are w_1, w_2, w_3,

and w_4. The basis vectors are obtained from $Y = \begin{bmatrix} -\begin{bmatrix} 2 & 3 \\ 8 & 8 \end{bmatrix}^{-1} \begin{bmatrix} 1 & 2 & 3 & 1 \\ 8 & -3 & -3 & -3 \end{bmatrix} \begin{bmatrix} w_1 \\ w_2 \\ w_3 \\ w_4 \end{bmatrix} \\ \begin{bmatrix} w_1 \\ w_2 \\ w_3 \\ w_4 \end{bmatrix} \end{bmatrix} =$

$\begin{bmatrix} -2w_1 + \frac{25}{8}w_2 + \frac{33}{8}w_3 + \frac{17}{8}w_4 \\ w_1 - \frac{11}{4}w_2 - \frac{15}{4}w_3 - \frac{7}{4}w_4 \\ w_1 \\ w_2 \\ w_3 \\ w_4 \end{bmatrix}$ (following matrix multiplication). Choose the arbitrary w's to have the basis. For

$\begin{Bmatrix} w_1 = 1 \\ w_2 = 0 \\ w_3 = 0 \\ w_4 = 0 \end{Bmatrix}$, $\begin{Bmatrix} w_1 = 0 \\ w_2 = 1 \\ w_3 = 0 \\ w_4 = 0 \end{Bmatrix}$, $\begin{Bmatrix} w_1 = 0 \\ w_2 = 0 \\ w_3 = 1 \\ w_4 = 0 \end{Bmatrix}$, and $\begin{Bmatrix} w_1 = 0 \\ w_2 = 0 \\ w_3 = 0 \\ w_4 = 1 \end{Bmatrix}$, we have $\begin{bmatrix} -2 \\ 1 \\ 1 \\ 0 \\ 0 \\ 0 \end{bmatrix}$, $\begin{bmatrix} \frac{25}{8} \\ -\frac{11}{4} \\ 0 \\ 1 \\ 0 \\ 0 \end{bmatrix}$, $\begin{bmatrix} \frac{33}{8} \\ -\frac{15}{4} \\ 0 \\ 0 \\ 1 \\ 0 \end{bmatrix}$, and $\begin{bmatrix} \frac{17}{8} \\ -\frac{7}{4} \\ 0 \\ 0 \\ 0 \\ 1 \end{bmatrix}$ respectively. Basis for the

null space of A is $Y = \begin{bmatrix} -2 & \frac{25}{8} & \frac{33}{8} & \frac{17}{8} \\ 1 & -\frac{11}{4} & -\frac{15}{4} & -\frac{7}{4} \\ 1 & 0 & 0 & 0 \\ 0 & 1 & 0 & 0 \\ 0 & 0 & 1 & 0 \\ 0 & 0 & 0 & 1 \end{bmatrix}$. One can verify that $\begin{bmatrix} 2 & 3 & 1 & 2 & 3 & 1 \\ 8 & 8 & 8 & -3 & -3 & -3 \end{bmatrix} \times$

$\begin{bmatrix} -2 & \frac{25}{8} & \frac{33}{8} & \frac{17}{8} \\ 1 & -\frac{11}{4} & -\frac{15}{4} & -\frac{7}{4} \\ 1 & 0 & 0 & 0 \\ 0 & 1 & 0 & 0 \\ 0 & 0 & 1 & 0 \\ 0 & 0 & 0 & 1 \end{bmatrix} = \begin{bmatrix} 0 & 0 & 0 & 0 \\ 0 & 0 & 0 & 0 \end{bmatrix}$, hence, $A \times Y = 0$ is satisfied but the point is Y is not unique.

Conduct the function 'null' to find the basis of null space of A as follows:

MATLAB Command

```
>>A=[2 3 1 2 3 1;8 8 8 -3 -3 -3]; ↵
>>Y=null(sym(A)) ↵
```

Y =

$$
\begin{bmatrix}
-27/8, & -19/8, & -2, & -11/8 \\
0, & 0, & 1, & 0 \\
15/4, & 11/4, & 1, & 7/4 \\
0, & 1, & 0, & 0 \\
1, & 0, & 0, & 0 \\
0, & 0, & 0, & 1
\end{bmatrix}
$$

The output Y is different from the computed one as we mentioned. The columns of Y (that is, basis of the null space of A) are not normalized (norm of the first column is not unity). For conceptual perception, symbolic form is better but orthonormal basis is necessary for the numerical computation. Numerical basis is obtained from the singular value decomposition. To get the orthonormalized basis numerically, execute 'null(A)' instead of 'null(sym(A))'.

4.32 Jordan form decomposition of a square matrix

If the eigenvalues of a square matrix A are all distinct, then, by choosing the set of eigenvectors as a basis, matrix A has a diagonal representation with the eigenvalues on the diagonal. Matrix A may not have distinct eigenvalues always. For repeated eigenvalues, it is not always possible to find a diagonal matrix representation. However, it is possible to find some basis vectors so that the new representation is almost a diagonal form, called a Jordan canonical form or Jordan form. The form has the eigenvalues of A in the diagonal and either 0 or 1 in the superdiagonal or subdiagonal. For example, if A has an eigenvalue λ with multiplicity 4, then Jordan form (assuming superdiagonal) takes one of the following forms –
$$
\begin{bmatrix}
\lambda & 0 & 0 & 0 \\
0 & \lambda & 0 & 0 \\
0 & 0 & \lambda & 0 \\
0 & 0 & 0 & \lambda
\end{bmatrix},
$$

$$
\begin{bmatrix}
\lambda & 1 & 0 & 0 \\
0 & \lambda & 0 & 0 \\
0 & 0 & \lambda & 0 \\
0 & 0 & 0 & \lambda
\end{bmatrix},
\begin{bmatrix}
\lambda & 1 & 0 & 0 \\
0 & \lambda & 0 & 0 \\
0 & 0 & \lambda & 1 \\
0 & 0 & 0 & \lambda
\end{bmatrix},
\begin{bmatrix}
\lambda & 1 & 0 & 0 \\
0 & \lambda & 1 & 0 \\
0 & 0 & \lambda & 1 \\
0 & 0 & 0 & \lambda
\end{bmatrix},
\text{ or }
\begin{bmatrix}
\lambda & 1 & 0 & 0 \\
0 & \lambda & 1 & 0 \\
0 & 0 & \lambda & 0 \\
0 & 0 & 0 & \lambda
\end{bmatrix}.
$$
Which Jordan form the matrix A assumes depends on the nature of A. A square matrix of type
$$
\begin{bmatrix}
\lambda & 1 & 0 & 0 \\
0 & \lambda & 1 & 0 \\
0 & 0 & \lambda & 1 \\
0 & 0 & 0 & \lambda
\end{bmatrix}
$$
is called a Jordan block. We have shown only the upper Jordan block, there can be lower Jordan form also, for example, $\begin{bmatrix} \lambda & 0 \\ 1 & \lambda \end{bmatrix}$,
$$
\begin{bmatrix}
\lambda & 0 & 0 \\
1 & \lambda & 0 \\
0 & 1 & \lambda
\end{bmatrix}, \text{ or }
\begin{bmatrix}
\lambda & 0 & 0 & 0 \\
1 & \lambda & 0 & 0 \\
0 & 1 & \lambda & 0 \\
0 & 0 & 1 & \lambda
\end{bmatrix}.
$$

Consider a square matrix A of order $N \times N$. One eigenvalue provides one eigenvector. For repeated eigenvalues, say λ, of multiplicity m, a procedure is adopted to find m linearly independent eigenvectors. A vector V is said to be a generalized eigenvector of A of grade k associated with λ, if $B^k V = 0$ and $B^{k-1} V \neq 0$, where $B = A - \lambda I$. Compute ranks of B^i, where $i = 0, 1, 2, 3 \ldots$ until $N - m$. Define $V_k = V$, $V_{k-1} = BV$, $V_{k-2} = B^2 V$, $V_{k-3} = B^3 V$ ..... $V_1 = B^{k-1} V$, then the set of vectors $\{V_1, V_2, V_3, ..V_k\}$ is called the generalized eigenvectors associated with λ. Once all basis vectors are found, one can write $Q^{-1}AQ = J$, where J is called the Jordan canonical form consisted of the Jordan blocks. It is noteworthy that matrix Q is not unique. But with a particular order of eigenvalues J is unique. Let S_i denote the null space of B^i, that is, S_i consists of all X such that $B^i X = 0$. It is evident that if Y is in S_i, then Y is also in S_{i+1}. Therefore, S_i is a subspacce of S_{i+1} and denoted by $S_i \subset S_{i+1}$. In general, it can be said that $S_0 \subset S_1 \subset S_2 \subset S_3 \subset \dots\dots$ and $0 = v_0 \le v_1 \le v_2 \le v_3 \dots v_k = m$, where v_i is the nullity of B^i. The generalized eigenvectors just defined follow then V_i is in S_i but not in S_{i-1}. Do not be overwhelmed by notations. We take the numerical example of $A =$

$$\begin{bmatrix} 3 & -1 & 1 & 1 & 0 & 0 \\ 1 & 1 & -1 & -1 & 0 & 0 \\ 0 & 0 & 2 & 0 & 1 & 1 \\ 0 & 0 & 0 & 2 & -2 & -1 \\ 0 & 0 & 0 & 0 & 2 & 2 \\ 0 & 0 & 0 & 0 & 2 & 2 \end{bmatrix}$$
. Having acquainted with the previous articles, one can use MATLAB to

inquest the different quantities. The eigenvalues of A are 0, 2 (of multiplicity 4), and 4. Carry out the following to have the eigenvalues of A :

```
>>A=sym([3 -1 1 1 0 0;1 1 -1 -1 0 0;0 0 2 0 1 1;0 0 0 2 -2 -1;0 0 0 0 2 2;0 0 0 0 2 2]); ↵
>>eig(A) ↵
```

ans =

[0]
[4]
[2]
[2]
[2]
[2]

Pertaining to the eigenvalue 2, we have $N = 6$, $m = 4$, $N - m = 2$, $B = A - \lambda I = \begin{bmatrix} 1 & -1 & 1 & 1 & 0 & 0 \\ 1 & -1 & -1 & -1 & 0 & 0 \\ 0 & 0 & 0 & 0 & 1 & 1 \\ 0 & 0 & 0 & 0 & -2 & -1 \\ 0 & 0 & 0 & 0 & 0 & 2 \\ 0 & 0 & 0 & 0 & 2 & 0 \end{bmatrix}$, rank

of $B = 4$, $B^2 = \begin{bmatrix} 0 & 0 & 2 & 2 & -1 & 0 \\ 0 & 0 & 2 & 2 & 1 & 0 \\ 0 & 0 & 0 & 0 & 2 & 2 \\ 0 & 0 & 0 & 0 & -2 & -4 \\ 0 & 0 & 0 & 0 & 4 & 0 \\ 0 & 0 & 0 & 0 & 0 & 4 \end{bmatrix}$, rank of $B^2 = 3$, $B^3 = \begin{bmatrix} 0 & 0 & 0 & 0 & -2 & -2 \\ 0 & 0 & 0 & 0 & -2 & 2 \\ 0 & 0 & 0 & 0 & 4 & 4 \\ 0 & 0 & 0 & 0 & -8 & -4 \\ 0 & 0 & 0 & 0 & 0 & 8 \\ 0 & 0 & 0 & 0 & 8 & 0 \end{bmatrix}$, and rank of $B^3 = $

2, hence, we have to stop here with grade $k = 3$. We summarize the different ranks, nullities, and null space in tabular form as follows:

$v_3 - v_2 = 1$	$v_2 - v_1 = 1$	$v_1 - v_0 = 2$	$\leftarrow$ Number of the independent vectors in S_i but not in S_{i-1}
V_1	$V_2 = BV_1$	$V_3 = B^2V_1$ V_4	$\leftarrow V_1, V_2, V_3,$ and V_4 are found below

$\xleftarrow{\qquad\qquad S_1 \qquad\qquad}$

$\xleftarrow{\qquad\qquad\qquad S_2 \qquad\qquad\qquad}$

$\xleftarrow{\qquad\qquad\qquad\qquad S_3 \qquad\qquad\qquad\qquad}$

160

Rank of B^0 (since B^0 is an identity matrix) is 6 and nullity of B^0 is $v_0 = 6-6=0$,

Rank of B^1 is 4 and nullity of B^1 is $v_1 = 6-4=2$,

Rank of B^2 is 3 and nullity of B^2 is $v_2 = 6-3=3$, and

Rank of B^3 is 2 and nullity of B^3 is $v_3 = 6-2=4$.

Since the multiplicity is 4, we are supposed to have 4 linearly independent vectors for $\lambda = 2$. So, there

is a vector V, for which, $B^3 V = 0$ but $B^2 V \neq 0$. Such a V can be $\begin{bmatrix} 1 \\ 1 \\ 1 \\ 1 \\ 0 \\ 0 \end{bmatrix}$. Denote that as $V_1 = \begin{bmatrix} 1 \\ 1 \\ 1 \\ 1 \\ 0 \\ 0 \end{bmatrix}$, on that, $V_2 = BV_1$

$= \begin{bmatrix} 2 \\ -2 \\ 0 \\ 0 \\ 0 \\ 0 \end{bmatrix}$, and $V_3 = B^2 V_1 = \begin{bmatrix} 4 \\ 4 \\ 0 \\ 0 \\ 0 \\ 0 \end{bmatrix}$. From the above table and arrows, the second and third space can have two basis

vectors. We found them labelled by V_1 and V_2. The first space needs two vectors. Out of two, we found V_3.

For V_4, we search one such that $V_4 \neq 0$ but $BV_4 = 0$. One choice can be $V_4 = \begin{bmatrix} 0 \\ 0 \\ 1 \\ -1 \\ 0 \\ 0 \end{bmatrix}$. We completed all searches

for eigenvectors associated with $\lambda = 2$. Then, we go for $\lambda = 0$, for which, $B = \begin{bmatrix} 3 & -1 & 1 & 1 & 0 & 0 \\ 1 & 1 & -1 & -1 & 0 & 0 \\ 0 & 0 & 2 & 0 & 1 & 1 \\ 0 & 0 & 0 & 2 & -2 & -1 \\ 0 & 0 & 0 & 0 & 2 & 2 \\ 0 & 0 & 0 & 0 & 2 & 2 \end{bmatrix}$, rank of B is 5, nullity of B is 1 (hence, one unknown variable), and

generalized eigenvector is $\begin{bmatrix} 3 & -1 & 1 & 1 & 0 \\ 1 & 1 & -1 & -1 & 0 \\ 0 & 0 & 2 & 0 & 1 \\ 0 & 0 & 0 & 2 & -2 \\ 0 & 0 & 0 & 0 & 2 \\ & & & w & \end{bmatrix}^{-1} \begin{bmatrix} 0 \\ 0 \\ 1 \\ -1 \\ 2 \end{bmatrix} w = \begin{bmatrix} 0 \\ -\frac{w}{2} \\ 0 \\ -\frac{w}{2} \\ -w \\ w \end{bmatrix}$. Choose $w = 1$ to get the basis $V_5 =$

$\begin{bmatrix} 0 \\ -\frac{1}{2} \\ 0 \\ -\frac{1}{2} \\ -1 \\ 1 \end{bmatrix}$ associated with $\lambda = 0$. After that, we have $B = \begin{bmatrix} -1 & -1 & 1 & 1 & 0 & 0 \\ 1 & -3 & -1 & -1 & 0 & 0 \\ 0 & 0 & -2 & 0 & 1 & 1 \\ 0 & 0 & 0 & -2 & -2 & -1 \\ 0 & 0 & 0 & 0 & -2 & 2 \\ 0 & 0 & 0 & 0 & 2 & -2 \end{bmatrix}$ (for $\lambda = 4$) whose

rank is 5 and the generalized eigenvector is $\begin{bmatrix} -1 & -1 & 1 & 1 & 0 \\ 1 & -3 & -1 & -1 & 0 \\ 0 & 0 & -2 & 0 & 1 \\ 0 & 0 & 0 & -2 & -2 \\ 0 & 0 & 0 & 0 & -2 \\ & & & w & \end{bmatrix}^{-1} \begin{bmatrix} 0 \\ 0 \\ 1 \\ -1 \\ 2 \end{bmatrix} w = \begin{bmatrix} -\frac{w}{2} \\ 0 \\ w \\ -\frac{3w}{2} \\ w \\ w \end{bmatrix}$. Thereupon, the last basis vector is $V_6 = \begin{bmatrix} -\frac{1}{2} \\ 0 \\ 1 \\ -\frac{3}{2} \\ 1 \\ 1 \end{bmatrix}$ (for $w = 1$). With basis $\{V_1,\ V_2,\ V_3,\ V_4,\ V_5,\ V_6\}$, matrix Q is obtained as

$$\begin{bmatrix} 1 & 2 & 4 & 0 & 0 & -\frac{1}{2} \\ 1 & -2 & 4 & 0 & -\frac{1}{2} & 0 \\ 1 & 0 & 0 & 1 & 0 & 1 \\ 1 & 0 & 0 & -1 & -\frac{1}{2} & -\frac{3}{2} \\ 0 & 0 & 0 & 0 & -1 & 1 \\ 0 & 0 & 0 & 0 & 1 & 1 \end{bmatrix}, \text{ therefore, } Q^{-1} = \begin{bmatrix} 0 & 0 & \frac{1}{2} & \frac{1}{2} & 0 & \frac{1}{4} \\ \frac{1}{4} & -\frac{1}{4} & 0 & 0 & \frac{1}{8} & 0 \\ \frac{1}{8} & \frac{1}{8} & -\frac{1}{8} & -\frac{1}{8} & 0 & 0 \\ 0 & 0 & \frac{1}{2} & -\frac{1}{2} & -\frac{1}{2} & -\frac{3}{4} \\ 0 & 0 & 0 & 0 & -\frac{1}{2} & \frac{1}{2} \\ 0 & 0 & 0 & 0 & \frac{1}{2} & \frac{1}{2} \end{bmatrix}, \text{ and } Q^{-1}AQ =$$

$$\begin{bmatrix} 2 & 0 & 0 & 0 & 0 & 0 \\ 1 & 2 & 0 & 0 & 0 & 0 \\ 0 & 1 & 2 & 0 & 0 & 0 \\ 0 & 0 & 0 & 2 & 0 & 0 \\ 0 & 0 & 0 & 0 & 0 & 0 \\ 0 & 0 & 0 & 0 & 0 & 4 \end{bmatrix}.$$ Put aside all these computations utilizing the function 'jordan' as follows:

MATLAB Command

```
>>[Q J]=jordan(A); ↵
>>pretty(Q) ↵
```

```
[ -1/4     0     -1    -1/2     0    1/4 ]
[                                        ]
[  0      1/4    -1     1/2     0    1/4 ]
[                                        ]
[ 1/2      0      0      0    -5/2  -3/2 ]
[                                        ]
[ -3/4    1/4     0      0      2    3/2 ]
[                                        ]
[ 1/2     1/2     0      0      0     0  ]
[                                        ]
[ 1/2    -1/2     0      0      0     0  ]
```

```
>>pretty(J) ↵
```

```
[4   0   0   0   0   0]
[                     ]
[0   0   0   0   0   0]
[                     ]
[0   0   2   1   0   0]
[                     ]
[0   0   0   2   1   0]
[                     ]
[0   0   0   0   2   0]
[                     ]
[0   0   0   0   0   2]
```

Notice that the returned Jordan canonical form and the computed one are not identical. Why is that? This is due to the choice of order of the basis vectors. Instead of $\{V_1,\ V_2,\ V_3,\ V_4,\ V_5,\ V_6\}$, choose the basis vectors' order

$\{V_6, V_5, V_3, V_2, V_1, V_4\}$ to write $Q = \begin{bmatrix} -\frac{1}{2} & 0 & 4 & 2 & 1 & 0 \\ 0 & -\frac{1}{2} & 4 & -2 & 1 & 0 \\ 1 & 0 & 0 & 0 & 1 & 1 \\ -\frac{3}{2} & -\frac{1}{2} & 0 & 0 & 1 & -1 \\ 1 & -1 & 0 & 0 & 0 & 0 \\ 1 & 1 & 0 & 0 & 0 & 0 \end{bmatrix}$. If you do so, $Q^{-1}AQ$ yields

$\begin{bmatrix} 4 & 0 & 0 & 0 & 0 & 0 \\ 0 & 0 & 0 & 0 & 0 & 0 \\ 0 & 0 & 2 & 1 & 0 & 0 \\ 0 & 0 & 0 & 2 & 1 & 0 \\ 0 & 0 & 0 & 0 & 2 & 0 \\ 0 & 0 & 0 & 0 & 0 & 2 \end{bmatrix}$, which is identical with the MATLAB return.

☞ Formation of a Jordan block

Now the reader is familiar with the Jordan blocks. There is a Maple function called 'JordanBlock', which can build upper Jordan block matrix. Say, we want to build a 5×5 upper Jordan block for repeated eigenvalue a. To have the matrix, implement the following:

MATLAB Command

```
>>syms a ↵
>>maple('JordanBlock',a,5) ↵

ans =

[ a,  1,  0,  0,  0]
[ 0,  a,  1,  0,  0]
[ 0,  0,  a,  1,  0]
[ 0,  0,  0,  a,  1]
[ 0,  0,  0,  0,  a]
```

4.33 Cholesky decomposition of a square matrix

Any positive definite matrix A can be decomposed as $A = C \times C^T$, where C is the lower triangular and has positive entries on the main diagonal and C^T represents the transpose of C. C is termed as the Cholesky factor of A. To explain the compuation, consider a 4×4 general matrix, which is $A =$

$\begin{bmatrix} A_{11} & A_{12} & A_{13} & A_{14} \\ A_{21} & A_{22} & A_{23} & A_{24} \\ A_{31} & A_{32} & A_{33} & A_{34} \\ A_{41} & A_{42} & A_{43} & A_{44} \end{bmatrix}$. C should look like $\begin{bmatrix} C_{11} & 0 & 0 & 0 \\ C_{21} & C_{22} & 0 & 0 \\ C_{31} & C_{32} & C_{33} & 0 \\ C_{41} & C_{42} & C_{43} & C_{44} \end{bmatrix}$, hence, $C \times C^T = \begin{bmatrix} C_{11} & 0 & 0 & 0 \\ C_{21} & C_{22} & 0 & 0 \\ C_{31} & C_{32} & C_{33} & 0 \\ C_{41} & C_{42} & C_{43} & C_{44} \end{bmatrix}$

$\times \begin{bmatrix} C_{11} & C_{21} & C_{31} & C_{41} \\ 0 & C_{22} & C_{32} & C_{42} \\ 0 & 0 & C_{33} & C_{43} \\ 0 & 0 & 0 & C_{44} \end{bmatrix} = \begin{bmatrix} C_{11}^2 & C_{21}C_{11} & C_{31}C_{11} & C_{11}C_{41} \\ C_{11}C_{21} & C_{21}^2 + C_{22}^2 & C_{31}C_{21} + C_{32}C_{22} & C_{21}C_{41} + C_{22}C_{42} \\ C_{11}C_{31} & C_{21}C_{31} + C_{22}C_{32} & C_{31}^2 + C_{32}^2 + C_{33}^2 & C_{31}C_{41} + C_{32}C_{42} + C_{33}C_{43} \\ C_{11}C_{41} & C_{21}C_{41} + C_{22}C_{42} & C_{31}C_{41} + C_{32}C_{42} + C_{33}C_{43} & C_{41}^2 + C_{42}^2 + C_{43}^2 + C_{44}^2 \end{bmatrix}$.

Equate different entries to write $C_{11}^2 = A_{11}$, $C_{11}C_{21} = A_{21}$, $C_{11}C_{31} = A_{31}$, $C_{11}C_{41} = A_{41}$, $C_{21}^2 + C_{22}^2 = A_{22}$, $C_{21}C_{31} + C_{22}C_{32} = A_{32}$, $C_{21}C_{41} + C_{22}C_{42} = A_{42}$, $C_{31}^2 + C_{32}^2 + C_{33}^2 = A_{33}$, $C_{31}C_{41} + C_{32}C_{42} + C_{33}C_{43} = A_{43}$, and $C_{41}^2 + C_{42}^2 + C_{43}^2 + C_{44}^2 = A_{44}$. Since the matrix is symmetric, equate only the lower triangle. A chain of substitution can uncover different C's. As numerical example, we have $A = \begin{bmatrix} 16 & 4 & 8 & 4 \\ 4 & 10 & 8 & 4 \\ 8 & 8 & 12 & 10 \\ 4 & 4 & 10 & 12 \end{bmatrix}$, from that,

$C_{11} = \sqrt{A_{11}} = \sqrt{16} = 4$, $\quad C_{21} = \dfrac{A_{21}}{C_{11}} = \dfrac{4}{4} = 1$, $\quad C_{31} = \dfrac{A_{31}}{C_{11}} = \dfrac{8}{4} = 2$, $\quad C_{41} = \dfrac{A_{41}}{C_{11}} = \dfrac{4}{4} = 1$, $\quad C_{22} = \sqrt{A_{22} - C_{21}^2} = \sqrt{10 - 1^2} = 3$,

$C_{32} = \dfrac{A_{32} - C_{21}C_{31}}{C_{22}} = \dfrac{8 - 1 \times 2}{3} = 2$, $\quad C_{42} = \dfrac{A_{42} - C_{21}C_{41}}{C_{22}} = \dfrac{4 - 1 \times 1}{3} = 1$, $\quad C_{33} = \sqrt{A_{33} - C_{31}^2 - C_{32}^2} = \sqrt{12 - 2^2 - 2^2} = 2$,

$C_{43} = \dfrac{A_{43} - C_{31}C_{41} - C_{32}C_{42}}{C_{33}} = \dfrac{10 - 2 \times 1 - 2 \times 1}{2} = 3$, and $\quad C_{44} = \sqrt{A_{44} - C_{41}^2 - C_{42}^2 - C_{43}^2} = \sqrt{12 - 1^2 - 1^2 - 3^2} = 1$. Having

found all C's, one can write the Cholesky factor $C = \begin{bmatrix} 4 & 0 & 0 & 0 \\ 1 & 3 & 0 & 0 \\ 2 & 2 & 2 & 0 \\ 1 & 1 & 3 & 1 \end{bmatrix}$. Have the factor by the Maple

function 'cholesky' as follows:

MATLAB Command
>>A=sym([16 4 8 4;4 10 8 4;8 8 12 10;4 4 10 12]); ↵
>>maple('cholesky',A) ↵

ans =

[4, 0, 0, 0]
[1, 3, 0, 0]
[2, 2, 2, 0]
[1, 1, 3, 1]

4.34 Gram Schmidt orthogonal vectors from independent vectors

Suppose $\{ E_1, E_2, E_3, \ldots, E_m \}$ is a set of linearly independent vectors. A set of orthogonal or orthonormal vectors can be obtained from the set $\{ E_1, E_2, E_3, \ldots, E_m \}$. Let us say the orthogonal set is $\{ U_1, U_2, U_3, U_4, \ldots, U_m \}$. Then, the orthonormal vectors from the desired set is $\left\{ u_1 = \dfrac{U_1}{\| U_1 \|}, \ u_2 = \dfrac{U_2}{\| U_2 \|}, \right.$ $u_3 = \dfrac{U_3}{\| U_3 \|}, \ \ldots\ldots, \ u_m = \dfrac{U_m}{\| U_m \|} \left. \right\}$. The Gram Schmidt computation procedure is as follows:

$U_1 = E_1$

$U_2 = E_2 - \langle u_1, E_2 \rangle u_1$

$U_3 = E_3 - \langle u_1, E_3 \rangle u_1 - \langle u_2, E_3 \rangle u_2$ $\quad U_4 = E_4 - \langle u_1, E_4 \rangle u_1 - \langle u_2, E_4 \rangle u_2 - \langle u_3, E_4 \rangle u_3$

………………………………so on.

The notation $\langle A, B \rangle$ is called the inner product or scalar product of vectors A and B. Consider that the vectors

are given by $A = \begin{bmatrix} A_1 \\ A_2 \\ A_3 \end{bmatrix}$ and $B = \begin{bmatrix} B_1 \\ B_2 \\ B_3 \end{bmatrix}$, then, the inner product is $\langle A, B \rangle = A_1 B_1 + A_2 B_2 + A_3 B_3$. In matrix notation,

$\langle A, B \rangle$ can be written as $[A_1 \quad A_2 \quad A_3] \begin{bmatrix} B_1 \\ B_2 \\ B_3 \end{bmatrix}$. Two vectors A and B are said to be orthogonal if and only if

$\langle A, B \rangle = 0$. Again, a vector A is said to be normalized if and only if $\langle A, A \rangle = 1$. A set of basis vectors $\{ U_1, U_2, U_3, U_4, \ldots, U_m \}$ is said to be an orthonormal basis if and only if $\langle U_i, U_j \rangle = \begin{cases} 0 & i \neq j \\ 1 & i = j \end{cases}$.

We compute the Gram Schmidt orthogonal vectors considering the set $\left\{ E_1 = \begin{bmatrix} 4 \\ 5 \\ 3 \end{bmatrix}, \ E_2 = \begin{bmatrix} 9 \\ 2 \\ 1 \end{bmatrix}, \ E_3 = \begin{bmatrix} 0 \\ 3 \\ 4 \end{bmatrix}, \right.$

$\left. E_4 = \begin{bmatrix} 5 \\ 2 \\ 0 \end{bmatrix} \right\}$. First, verify whether the given vectors are linearly independent. Apply the concept of article 4.15.

If you do so, you will end up with $E_4 = \frac{47}{79}E_1 + \frac{23}{79}E_2 - \frac{41}{79}E_3$. It indicates that E_1, E_2, and E_3 are linearly independent. For orthogonalization, we concentrate only on independent vectors, therefore, required

computations are $U_1 = E_1 = \begin{bmatrix} 4 \\ 5 \\ 3 \end{bmatrix}$, $\|U_1\| = \sqrt{4^2 + 5^2 + 3^2} = 5\sqrt{2}$, $u_1 = \frac{U_1}{\|U_1\|} = \frac{1}{5\sqrt{2}} \begin{bmatrix} 4 \\ 5 \\ 3 \end{bmatrix} = \begin{bmatrix} \frac{4}{5\sqrt{2}} \\ \frac{1}{\sqrt{2}} \\ \frac{3}{5\sqrt{2}} \end{bmatrix}$, $U_2 = E_2 - \langle u_1, E_2 \rangle u_1 =$

$\begin{bmatrix} 9 \\ 2 \\ 1 \end{bmatrix} - \left\{ \begin{bmatrix} \frac{4}{5\sqrt{2}} & \frac{1}{\sqrt{2}} & \frac{3}{5\sqrt{2}} \end{bmatrix} \begin{bmatrix} 9 \\ 2 \\ 1 \end{bmatrix} \right\} \begin{bmatrix} \frac{4}{5\sqrt{2}} \\ \frac{1}{\sqrt{2}} \\ \frac{3}{5\sqrt{2}} \end{bmatrix} = \begin{bmatrix} 9 \\ 2 \\ 1 \end{bmatrix} - \frac{49}{5\sqrt{2}} \begin{bmatrix} \frac{4}{5\sqrt{2}} \\ \frac{1}{\sqrt{2}} \\ \frac{3}{5\sqrt{2}} \end{bmatrix} = \begin{bmatrix} \frac{127}{25} \\ -\frac{29}{10} \\ -\frac{97}{50} \end{bmatrix}$, $\qquad \|U_2\| = \sqrt{\left(\frac{127}{25}\right)^2 + \left(-\frac{29}{10}\right)^2 + \left(-\frac{97}{50}\right)^2} = \frac{3\sqrt{422}}{10}$,

$u_2 = \frac{U_2}{\|U_2\|} = \frac{1}{\frac{3\sqrt{422}}{10}} \begin{bmatrix} \frac{127}{25} \\ -\frac{29}{10} \\ -\frac{97}{50} \end{bmatrix} = \begin{bmatrix} \frac{127\sqrt{422}}{3165} \\ -\frac{29\sqrt{422}}{1266} \\ -\frac{97\sqrt{422}}{6330} \end{bmatrix}$, and $U_3 = E_3 - \langle u_1, E_3 \rangle u_1 - \langle u_2, E_3 \rangle u_2 = \begin{bmatrix} 0 \\ 3 \\ 4 \end{bmatrix} - \left\{ \begin{bmatrix} \frac{4}{5\sqrt{2}} & \frac{1}{\sqrt{2}} & \frac{3}{5\sqrt{2}} \end{bmatrix} \begin{bmatrix} 0 \\ 3 \\ 4 \end{bmatrix} \right\}$

$\begin{bmatrix} \frac{4}{5\sqrt{2}} \\ \frac{1}{\sqrt{2}} \\ \frac{3}{5\sqrt{2}} \end{bmatrix} - \left\{ \begin{bmatrix} \frac{127\sqrt{422}}{3165} & -\frac{29\sqrt{422}}{1266} & -\frac{97\sqrt{422}}{6330} \end{bmatrix} \begin{bmatrix} 0 \\ 3 \\ 4 \end{bmatrix} \right\} \begin{bmatrix} \frac{127\sqrt{422}}{3165} \\ -\frac{29\sqrt{422}}{1266} \\ -\frac{97\sqrt{422}}{6330} \end{bmatrix} = \begin{bmatrix} \frac{79}{1899} \\ -\frac{1817}{1899} \\ \frac{2923}{1899} \end{bmatrix}$. Since E_4 is linearly dependent, we stop here.

Finally, we have the orthogonal set as $\{ U_1, U_2, U_3 \} = \left\{ \begin{bmatrix} 4 \\ 5 \\ 3 \end{bmatrix}, \begin{bmatrix} \frac{127}{25} \\ -\frac{29}{10} \\ -\frac{97}{50} \end{bmatrix}, \begin{bmatrix} \frac{79}{1899} \\ -\frac{1817}{1899} \\ \frac{2923}{1899} \end{bmatrix} \right\}$. Avoid so much computational

hassle by applying the Maple function 'GramSchmidt' as follows:

MATLAB Command
>>maple('E1:=vector([4,5,3]):E2:=vector([9,2,1]):E3:=vector([0,3,4]):E4:=vector([5,2,0]):'); ↵
>>maple('GramSchmidt([E1,E2,E3,E4])') ↵

ans =

[[4, 5, 3], [127/25, -29/10, -97/50], [79/1899, -1817/1899, 2923/1899]]

4.35 Condition number of a matrix

The condition number of a matrix is defined as the ratio of its largest to smallest singular values. A large condition number indicates contiguous linear dependency among the columns of the matrix or in other words, the matrix is nearly a singular one. The function that determines condition number is 'cond'. For

convenience, we find the singular values of the matrix $A = \begin{bmatrix} 2 & 3 & 4 \\ 8 & -1 & 2 \\ 2 & 2 & 0 \\ 9 & 1 & 3 \end{bmatrix}$ implementing the subroutine 'svd'

(see article 4.17 for the singular value decomposition):

MATLAB Command

>>A=[2 3 4;8 -1 2;2 2 0;9 1 3]; ↵
>>svd(A) ↵

ans =

13.1401
4.5586
1.8858

From the MATLAB return, the largest singular value is 13.1401 and the smallest one is 1.8858, hence, the condition number of A is $\dfrac{13.1401}{1.8858}$=6.9679.

>>cond(A) ↵   ← Apply the function

ans =

6.9679   ← As it is expected

There are other variants of function 'cond', for example, 'condest' and 'condeig'. Have the online help executing 'help condest' and 'help condeig' respectively.

4.36 Special matrices

A number of matrices can be generated in MATLAB. We mention some of them. Depending on the purpose, the generated matrix can be in floating-point form, in integer form, or even in symbolic form. Some other matrices are presented in the following.

4.36.1 Formation of a Hilbert matrix

A Hilbert matrix is one in which each entry of the matrix is formed by $\dfrac{1}{i+j-1}$, where i is the row

index and j is the column index. The matrix is a symmetric one. A 2×2 Hilbert matrix is generated as $H_{2\times2}=$

$\begin{bmatrix} H_{11} & H_{12} \\ H_{21} & H_{22} \end{bmatrix} = \begin{bmatrix} \frac{1}{1+1-1} & \frac{1}{1+2-1} \\ \frac{1}{2+1-1} & \frac{1}{2+2-1} \end{bmatrix} = \begin{bmatrix} 1 & \frac{1}{2} \\ \frac{1}{2} & \frac{1}{3} \end{bmatrix} = \begin{bmatrix} 1 & 0.5 \\ 0.5 & 0.3333 \end{bmatrix}$. As another example, say Hilbert matrix of

order 4×4 is to be formed, so, $H_{4\times4}=\begin{bmatrix} H_{11} & H_{12} & H_{13} & H_{14} \\ H_{21} & H_{22} & H_{23} & H_{24} \\ H_{31} & H_{32} & H_{33} & H_{34} \\ H_{41} & H_{42} & H_{43} & H_{44} \end{bmatrix} = \begin{bmatrix} \frac{1}{1+1-1} & \frac{1}{1+2-1} & \frac{1}{1+3-1} & \frac{1}{1+4-1} \\ \frac{1}{2+1-1} & \frac{1}{2+2-1} & \frac{1}{2+3-1} & \frac{1}{2+4-1} \\ \frac{1}{3+1-1} & \frac{1}{3+2-1} & \frac{1}{3+3-1} & \frac{1}{3+4-1} \\ \frac{1}{4+1-1} & \frac{1}{4+2-1} & \frac{1}{4+3-1} & \frac{1}{4+4-1} \end{bmatrix} =$

$\begin{bmatrix} 1 & \frac{1}{2} & \frac{1}{3} & \frac{1}{4} \\ \frac{1}{2} & \frac{1}{3} & \frac{1}{4} & \frac{1}{5} \\ \frac{1}{3} & \frac{1}{4} & \frac{1}{5} & \frac{1}{6} \\ \frac{1}{4} & \frac{1}{5} & \frac{1}{6} & \frac{1}{7} \end{bmatrix} = \begin{bmatrix} 1 & 0.5 & 0.3333 & 0.25 \\ 0.5 & 0.3333 & 0.25 & 0.2 \\ 0.3333 & 0.25 & 0.2 & 0.1667 \\ 0.25 & 0.2 & 0.1667 & 0.1429 \end{bmatrix}$. Corresponding MATLAB subroutine is 'hilb',

which is the abbreviation of <u>hilb</u>ert. Rational form can be seen using command 'sym'.

MATLAB Command

for matrix of order 2×2,   for matrix of order 4×4,
>>H=hilb(2) ↵   >>H=hilb(4) ↵

H =   H =
    1.0000   0.5000   1.0000 0.5000 0.3333 0.2500
    0.5000   0.3333   0.5000 0.3333 0.2500 0.2000
in rational form, of order 2×2,   0.3333 0.2500 0.2000 0.1667
>>H=sym(hilb(2)) ↵   0.2500 0.2000 0.1667 0.1429
    in rational form, of order 4×4,
H =   >>H=sym(hilb(4)) ↵

[1,   1/2]   H =
[1/2, 1/3]

$$\begin{bmatrix} 1, & 1/2, & 1/3, & 1/4 \\ 1/2, & 1/3, & 1/4, & 1/5 \\ 1/3, & 1/4, & 1/5, & 1/6 \\ 1/4, & 1/5, & 1/6, & 1/7 \end{bmatrix}$$

However, to form a Hilbert matrix of order $N \times N$,

MATLAB Step:
Use the command hilb(N).

4.36.2 Formation of a Hadamard matrix

The number of rows or columns of a Hadamard matrix, N, must be an integer which is 2^m, where $m = 1, 2, 3 \ldots$etc. The lowest order Hadamard matrix is of order 2×2, where $N = 2$ and $m = 1$ and it is given by $H_{2\times2} = \begin{bmatrix} 1 & 1 \\ 1 & -1 \end{bmatrix}$. Advantage of recursive relationship is taken to form the Hadamard matrices of other orders.

Letting H_N represent the Hadamard matrix of order $N \times N$, the recursive relationship is given by $H_N =$

$\begin{bmatrix} H_{N/2} & H_{N/2} \\ H_{N/2} & -H_{N/2} \end{bmatrix}$, so, $H_{4\times4} = H_4 = \begin{bmatrix} H_2 & H_2 \\ H_2 & -H_2 \end{bmatrix} = \begin{bmatrix} \begin{bmatrix} 1 & 1 \\ 1 & -1 \end{bmatrix} & \begin{bmatrix} 1 & 1 \\ 1 & -1 \end{bmatrix} \\ \begin{bmatrix} 1 & 1 \\ 1 & -1 \end{bmatrix} & -\begin{bmatrix} 1 & 1 \\ 1 & -1 \end{bmatrix} \end{bmatrix} = \begin{bmatrix} 1 & 1 & 1 & 1 \\ 1 & -1 & 1 & -1 \\ 1 & 1 & -1 & -1 \\ 1 & -1 & -1 & 1 \end{bmatrix}$. Similarly,

Hadamard matrix of order 8×8 can be generated by $\begin{bmatrix} H_4 & H_4 \\ H_4 & -H_4 \end{bmatrix}$. The matrices are used in image processing problems. To generate Hadamard matrix of order $N \times N$,

MATLAB Step:
Use the command hadamard(N).

Formation of different Hadamard matrices is shown below exercising the subroutine 'hadamard':
MATLAB Command
for the matrix of order 2×2,   for the matrix of order 4×4,
>>H=hadamard(2) ↵   >>H=hadamard(4) ↵

H =   H =
  1   1   1   1   1   1
  1 -1   1 -1   1 -1
for the matrix of order 8×8,   1   1 -1 -1
>>H=hadamard(8) ↵   1 -1 -1   1

H =
   1   1   1   1   1   1   1   1
   1 -1   1 -1   1 -1   1 -1
   1   1 -1 -1   1   1 -1 -1
   1 -1 -1   1   1 -1 -1   1
   1   1   1   1 -1 -1 -1 -1
   1 -1   1 -1 -1   1 -1   1
   1   1 -1 -1 -1 -1   1   1
   1 -1 -1   1 -1   1   1 -1

4.36.3 Companion matrix of a polynomial

If λ is an eigenvalue of the square matrix A, the characteristic polynomial of A is defined by $|\lambda I - A|$. In general, the characteristic polynomial is given by $G(\lambda) = \alpha_n \lambda^n + \alpha_{n-1} \lambda^{n-1} + \alpha_{n-2} \lambda^{n-2} + \alpha_{n-3} \lambda^{n-3} +$

..... $+ \alpha_1 \lambda + \alpha_0$. Order of the characteristic polynomial is n and there are $n+1$ polynomial coefficients in $G(\lambda)$. Given that the polynomial coefficients of the characteristic polynomial $G(\lambda)$, the companion matrix can be

generated as $\begin{bmatrix} \frac{-\alpha_{n-1}}{\alpha_n} & \frac{-\alpha_{n-2}}{\alpha_n} & \cdots & \frac{-\alpha_1}{\alpha_n} & \frac{-\alpha_0}{\alpha_n} \\ 1 & 0 & \cdots & 0 & 0 \\ 0 & 1 & \vdots & 0 & 0 \\ \vdots & \vdots & \ddots & \vdots & \vdots \\ 0 & 0 & \cdots & 1 & 0 \end{bmatrix}$. To obtain a companion matrix from G(λ),

MATLAB Step:
Use the command compan (polynomial name).

Consider a characteristic polynomial $G(\lambda) = -6\lambda^3 + 3\lambda^2 + \lambda + 7$. In matrix form, $G(\lambda)$ is written as [−6　3　1　7]. Order of the polynomial is 3. There are four polynomial coefficients that can be written as α_3, α_2, α_1, and α_0. Compare the given polynomial with the general polynomial to write $\alpha_3 = -6$, $\alpha_2 = 3$, $\alpha_1 = 1$, and $\alpha_0 = 7$, so,

the companion matrix is obtained as $\begin{bmatrix} \frac{-\alpha_2}{\alpha_3} & \frac{-\alpha_1}{\alpha_3} & \frac{-\alpha_0}{\alpha_3} \\ 1 & 0 & 0 \\ 0 & 1 & 0 \end{bmatrix} = \begin{bmatrix} \frac{-3}{-6} & \frac{-1}{-6} & \frac{-7}{-6} \\ 1 & 0 & 0 \\ 0 & 1 & 0 \end{bmatrix} = \begin{bmatrix} \frac{1}{2} & \frac{1}{6} & \frac{7}{6} \\ 1 & 0 & 0 \\ 0 & 1 & 0 \end{bmatrix} =$

$\begin{bmatrix} 0.5 & 0.1667 & 1.1667 \\ 1 & 0 & 0 \\ 0 & 1 & 0 \end{bmatrix}$. Formation of the matrix, along with the symbolic form, from $G(\lambda)$ is presented as

follows:

MATLAB Command
for numeric form,
>>P=[-6 3 1 7]; ↵
>>compan(P) ↵

ans =
```
        0.5000    0.1667    1.1667
        1.0000         0         0
             0    1.0000         0
```

for symbolic form,
>>sym(compan(P)) ↵

ans =
```
[ 1/2, 1/6, 7/6]
[   1,   0,   0]
[   0,   1,   0]
```

4.36.4 Vandermonde matrix of a column vector

If $C = \begin{bmatrix} C_1 \\ C_2 \\ C_3 \\ \vdots \\ C_N \end{bmatrix}$ is a column matrix, then Vandermonde matrix is defined as $V =$

$\begin{bmatrix} C_1^{N-1} & C_1^{N-2} & C_1^{N-3} & C_1^1 & & 1 \\ C_2^{N-1} & C_2^{N-2} & C_2^{N-3} & C_2^1 & \cdots & 1 \\ C_3^{N-1} & C_3^{N-2} & C_3^{N-3} & C_3^1 & & 1 \\ & & \vdots & & & \\ C_N^{N-1} & C_N^{N-2} & C_N^{N-3} & C_N^1 & & 1 \end{bmatrix}$. MATLAB counterpart in this regard is 'vander' that comes from

Vandermonde. Input matrix of the subroutine can be a row or column one. Input is taken as columnwise regardless of the type of input matrices. Elements in the matrix can be floating-points or even complex numbers. The subroutine does not work for rectangular matrix. Test matrices are $R = [2 \quad 3 \quad 5 \quad -1]$ and $C =$

$\begin{bmatrix} 1 \\ 4 \\ 6 \end{bmatrix}$. There are 4 and 3 elements in R and C, so, $N=4$ and $N=3$ for R and C respectively. Vandermonde

matrices of R and C are VR=$\begin{bmatrix} 2^3 & 2^2 & 2^1 & 1 \\ 3^3 & 3^2 & 3^1 & 1 \\ 5^3 & 5^2 & 5^1 & 1 \\ (-1)^3 & (-1)^2 & (-1)^1 & 1 \end{bmatrix} = \begin{bmatrix} 8 & 4 & 2 & 1 \\ 27 & 9 & 3 & 1 \\ 125 & 25 & 5 & 1 \\ -1 & 1 & -1 & 1 \end{bmatrix}$ and VC=$\begin{bmatrix} 1^2 & 1 & 1 \\ 4^2 & 4 & 1 \\ 6^2 & 6 & 1 \end{bmatrix} =$

$\begin{bmatrix} 1 & 1 & 1 \\ 16 & 4 & 1 \\ 36 & 6 & 1 \end{bmatrix}$ respectively. Implementation is shown below:

MATLAB Command

for row matrix,
>>R=[2 3 5 -1]; ↵
>>vander(R) ↵

VR =

          8   4   2   1
         27   9   3   1
        125   25   5   1
         -1   1   -1   1

for column matrix,
>>C=[1 4 6]'; ↵
>>vander(C) ↵

VC =

          1   1   1
         16   4   1
         36   6   1

Vandermonde matrix is used in approximation and polynomial interpolation problems.

4.36.5 Dissimilarity matrix of some points

Suppose we have two points in $x-y$ plane given by (x_1, y_1) and (x_2, y_2). Let d_{12} denote the Euclidean or straight-line distance between these two points from the first point, whence, $d_{12} = \sqrt{(x_2 - x_1)^2 + (y_2 - y_1)^2}$. If we have N points, the straight-line distance between the ith and j^{th} points from the ith point is $d_{ij} = \sqrt{(x_j - x_i)^2 + (y_j - y_i)^2}$. A matrix can be formed from different i's and j's like

$D = \begin{bmatrix} d_{11} & d_{12} & d_{13} & \cdots d_{1N} \\ d_{21} & d_{22} & d_{23} & \cdots d_{2N} \\ d_{31} & d_{32} & d_{33} & \cdots d_{3N} \\ & & \vdots & \\ d_{N1} & d_{N2} & d_{N3} & \cdots d_{NN} \end{bmatrix}$. The matrix D is termed as the dissimilarity matrix (because the elements

measure how far separated two points are), understandably, $d_{ij} = d_{ji}$ and $d_{ii} = 0$ since distance of a point from itself is zero. Take an example of four points — $(x_1, y_1) = (1, 0)$, $(x_2, y_2) = (2, -6)$, $(x_3, y_3) = (-5, -2)$, and

$(x_4, y_4) = (6, 1)$. The dissimilarity matrix will have the form $D = \begin{bmatrix} d_{11} & d_{12} & d_{13} & d_{14} \\ d_{21} & d_{22} & d_{23} & d_{24} \\ d_{31} & d_{32} & d_{33} & d_{34} \\ d_{41} & d_{42} & d_{43} & d_{44} \end{bmatrix}$. Computations of the

various distances are $d_{11} = d_{22} = d_{33} = d_{44} = 0$, $d_{12} = d_{21} = \sqrt{(2-1)^2 + (-6-0)^2} = 6.0828$, $d_{13} = d_{31} = \sqrt{(-5-1)^2 + (-2-0)^2} = 6.3246$, $d_{14} = d_{41} = \sqrt{(6-1)^2 + (1-0)^2} = 5.0990$, $d_{23} = d_{32} = \sqrt{(-5-2)^2 + (-2+6)^2} = 8.0623$, $d_{24} = d_{42} = \sqrt{(6-2)^2 + (1+6)^2} = 8.0623$, and $d_{34} = d_{43} = \sqrt{(6+5)^2 + (1+2)^2} = 11.4018$. Form the matrix $D =$

$$\begin{bmatrix} 0 & 6.0828 & 6.3246 & 5.0990 \\ 6.0828 & 0 & 8.0623 & 8.0623 \\ 6.3246 & 8.0623 & 0 & 11.4018 \\ 5.0990 & 8.0623 & 11.4018 & 0 \end{bmatrix}$$ from the computed distances. Formation of D is shown as

follows:

MATLAB Command

>>x=[1 2 -5 6]'; ↵
>>y=[0 -6 -2 1]'; ↵
>>xx=repmat(x,1,4)-repmat(x',4,1); ↵
>>yy=repmat(y,1,4)-repmat(y',4,1); ↵
>>D=sqrt(xx.^2+yy.^2) ↵

$D =$

0	6.0828	6.3246	5.0990
6.0828	0	8.0623	8.0623
6.3246	8.0623	0	11.4018
5.0990	8.0623	11.4018	0

As shown in the output, subroutine 'repmat' is used for the formation of repetitive matrices (see article 2.29 for

'repmat'). The given coordinates can be written as $x = \begin{bmatrix} x_1 \\ x_2 \\ x_3 \\ x_4 \end{bmatrix} = \begin{bmatrix} 1 \\ 2 \\ -5 \\ 6 \end{bmatrix}$ and $y = \begin{bmatrix} y_1 \\ y_2 \\ y_3 \\ y_4 \end{bmatrix} = \begin{bmatrix} 0 \\ -6 \\ -2 \\ 1 \end{bmatrix}$. Following is the

explanation of the commands used: $x=[1 \quad 2 \quad -5 \quad 6]' \rightarrow \begin{bmatrix} 1 \\ 2 \\ -5 \\ 6 \end{bmatrix}$, $x' \rightarrow [1 \ 2 \ -5 \ 6]$, repmat(x,1,4) $\rightarrow \begin{bmatrix} 1 & 1 & 1 \\ 2 & 2 & 2 \\ -5 & -5 & -5 \\ 6 & 6 & 6 \end{bmatrix}$

$\begin{bmatrix} 1 \\ 2 \\ -5 \\ 6 \end{bmatrix}$, repmat(x',4,1) $\rightarrow \begin{bmatrix} 1 & 2 & -5 & 6 \\ 1 & 2 & -5 & 6 \\ 1 & 2 & -5 & 6 \\ 1 & 2 & -5 & 6 \end{bmatrix}$, xx=repmat(x,1,4)-repmat(x',4,1) $\rightarrow \begin{bmatrix} 1-1 & 1-2 & 1+5 \\ 2-1 & 2-2 & 2+5 \\ -5-1 & -5-2 & -5+5 \\ 6-1 & 6-2 & 6+5 \end{bmatrix}$

$\begin{bmatrix} 1-6 \\ 2-6 \\ -5-6 \\ 6-6 \end{bmatrix}$, xx.^2 $\rightarrow \begin{bmatrix} (1-1)^2 & (1-2)^2 & (1+5)^2 & (1-6)^2 \\ (2-1)^2 & (2-2)^2 & (2+5)^2 & (2-6)^2 \\ (-5-1)^2 & (-5-2)^2 & (-5+5)^2 & (-5-6)^2 \\ (6-1)^2 & (6-2)^2 & (6+5)^2 & (6-6)^2 \end{bmatrix}$, in a similar fashion, yy.^2 $\rightarrow \begin{bmatrix} (0-0)^2 \\ (-6-0)^2 \\ (-2-0)^2 \\ (1-0)^2 \end{bmatrix}$

$\begin{bmatrix} (0+6)^2 & (0+2)^2 & (0-1)^2 \\ (-6+6)^2 & (-6+2)^2 & (-6-1)^2 \\ (-2+6)^2 & (-2+2)^2 & (-2-1)^2 \\ (1+6)^2 & (1+2)^2 & (1-1)^2 \end{bmatrix}$, and finally, sqrt(xx.^2+yy.^2) $\rightarrow \begin{bmatrix} \sqrt{(1-1)^2+(0-0)^2} & \sqrt{(1-2)^2+(0+6)^2} \\ \sqrt{(2-1)^2+(-6-0)^2} & \sqrt{(2-2)^2+(-6+6)^2} \\ \sqrt{(-5-1)^2+(-2-0)^2} & \sqrt{(-5-2)^2+(-2+6)^2} \\ \sqrt{(6-1)^2+(1-0)^2} & \sqrt{(6-2)^2+(1+6)^2} \end{bmatrix}$

$\begin{bmatrix} \sqrt{(1+5)^2+(0+2)^2} & \sqrt{(1-6)^2+(0-1)^2} \\ \sqrt{(2+5)^2+(-6+2)^2} & \sqrt{(2-6)^2+(-6-1)^2} \\ \sqrt{(-5+5)^2+(-2+2)^2} & \sqrt{(-5-6)^2+(-2-1)^2} \\ \sqrt{(6+5)^2+(1+2)^2} & \sqrt{(6-6)^2+(1-1)^2} \end{bmatrix}$ — that is what the required D is. Dissimilarity matrices are used

in the cluster analysis.

4.36.6 Formation of a Pascal matrix

The function $(1+x)^m$ can be written in a series as $(1+x)^m = 1 + mx + \dfrac{m(m-1)x^2}{2!} + \dfrac{m(m-1)(m-2)x^3}{3!} + \dots\dots$ $+x^m$. The coefficients of the series for different values of m are given as follows:

$$
\begin{array}{llcccccccc}
m=0 & 1 & & & & 1 & & & & \\
m=1 & 1+x & & & & 1 & 1 & & & \\
m=2 & 1+2x+x^2 & & & 1 & 2 & 1 & & & \\
m=3 & 1+3x+3x^2+x^3 & & 1 & 3 & 3 & 1 & & \\
m=4 & 1+4x+6x^2+4x^3+x^4 & & 1 & 4 & 6 & 4 & 1 & \\
m=5 & 1+5x+10x^2+10x^3+5x^4+x^5 & 1 & 5 & 10 & 10 & 5 & 1 & \\
m=6 & 1+6x+15x^2+20x^3+15x^4+6x^5+x^6 & 1 & 6 & 15 & 20 & 15 & 6 & 1
\end{array}
$$

$\dots\dots\dots$ and so on.

Pascal matrices can be formed by picking up the elements in a diamond ($\Diamond$) from the different coefficients of x according to figure 4.3. To form a Pascal matrix of order $N \times N$,

MATLAB Step:

Use the command pascal (N).

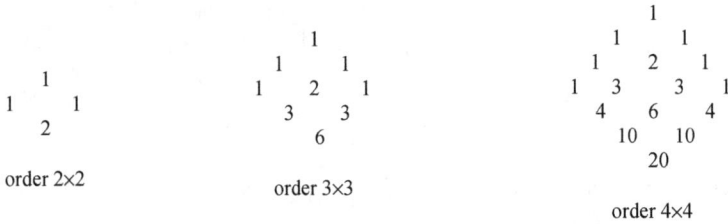

order 2×2   order 3×3   order 4×4

Figure 4.3   *Pascal matrices of different orders*

Formation of a Pascal matrix P of order 4×4 is shown below:
MATLAB Command

>>P=pascal(4) ↵

$$
P = \begin{bmatrix}
1 & 1 & 1 & 1 \\
1 & 2 & 3 & 4 \\
1 & 3 & 6 & 10 \\
1 & 4 & 10 & 20
\end{bmatrix}
$$

4.36.7 Formation of a toeplitz matrix

A toeplitz matrix is one whose entries are constant along each diagonal. For example, a 4×4 and a 3×5 toeplitz matrix will take the form as $T_{4\times4} = \begin{bmatrix} d & c & e & f \\ c & d & c & e \\ e & c & d & c \\ f & e & c & d \end{bmatrix}$ and $T_{3\times5} = \begin{bmatrix} d & c & e & f & b \\ c & d & c & e & f \\ e & c & d & c & e \end{bmatrix}$, where a, b, c, d, e, and f are real numbers. Toeplitz is a symmetric matrix. To form a toeplitz matrix,

MATLAB Step:

Use the command toeplitz (row or column for square toeplitz) or toeplitz (super diagonal elements, sub diagonal elements in row or column matrix for rectangular toeplitz).

Generate a square toeplitz of order 5×5 from $C = \begin{bmatrix} 5 \\ 0 \\ -4 \\ 9 \\ 1 \end{bmatrix}$, so, $T_{5 \times 5} = \begin{bmatrix} 5 & 0 & -4 & 9 & 1 \\ 0 & 5 & 0 & -4 & 9 \\ -4 & 0 & 5 & 0 & -4 \\ 9 & -4 & 0 & 5 & 0 \\ 1 & 9 & -4 & 0 & 5 \end{bmatrix}$.

As another example, a toeplitz of order 3×5 is to be formed where the subdiagonal elements are $R = [10 \ -9 \ 7]$ and the superdiagonal elements are $C = [10 \quad 55 \quad 5 \quad 40 \quad 6]$. The matrix should look like $\begin{bmatrix} 10 & 55 & 5 & 40 & 6 \\ -9 & 10 & 55 & 5 & 40 \\ 7 & -9 & 10 & 55 & 5 \end{bmatrix}$. Both examples are presented below:

MATLAB Command

for 5×5 toeplitz matrix,

      >>C=[5 0 -4 9 1]'; ↵

      >>T=toeplitz(C) ↵

for 3×5 toeplitz matrix,

      >>R=[10 -9 7]; ↵

      >>C=[10 55 5 40 6]; ↵

      >>T=toeplitz(R,C) ↵

T =

```
    5   0  -4   9   1
    0   5   0  -4   9
   -4   0   5   0  -4
    9  -4   0   5   0
    1   9  -4   0   5
```

T =

```
   10  55   5  40   6
   -9  10  55   5  40
    7  -9  10  55   5
```

The first elements of R and C must be identical when one rectangular toeplitz is formed otherwise error message is displayed by MATLAB. Toeplitz matrix is used in digital signal processing, for instance, in linear prediction of speech.

4.36.8 Formation of a Hankel matrix

Hankel matrix can be square or rectangular. Elements in the square Hankel matrix are zeroes below the first anti-diagonal and the matrix is symmetric. Suppose a square Hankel matrix HC is to be formed from $C = \begin{bmatrix} 6 \\ -3 \\ 9 \end{bmatrix}$. HC should look like $\begin{bmatrix} 6 & -3 & 9 \\ -3 & 9 & 0 \\ 9 & 0 & 0 \end{bmatrix}$. A rectangular Hankel matrix HA is to be formed in which the super anti-diagonal and sub anti-diagonal elements are $A = [6 \quad 7 \quad -1]$ and $B = [-1 \quad 9 \quad 2 \quad 4]$ respectively, so, HA should be $\begin{bmatrix} 6 & 7 & -1 & 9 \\ 7 & -1 & 9 & 2 \\ -1 & 9 & 2 & 4 \end{bmatrix}$. The last element of A must be equal to the first element of B. See their formation in the following:

MATLAB Command

for the square Hankel,

      >>C=[6 -3 9]'; ↵

      >>HC=hankel(C) ↵

for the rectangular Hankel,

      >>A=[6 7 -1]; ↵

      >>B=[-1 9 2 4]; ↵

      >>HA=hankel(A,B) ↵

HC =

```
    6  -3   9
   -3   9   0
    9   0   0
```

HA =

```
    6   7  -1   9
    7  -1   9   2
   -1   9   2   4
```

However, to form a Hankel matrix,

MATLAB Step:

      Use the command hankel (row or column matrix for square hankel) or hankel (super anti-diagonal elements, sub anti-diagonal elements in row or column matrix form for rectangular hankel).

172

4.36.9 Formation of a spiral matrix

Shown figure 4.4 is the way by which a rectangular spiral is formed. Starting from 1, as the element emerges horizontally or vertically, it is increased by 1. The spiral matrix is always a square one. To form a spiral matrix of order $N \times N$,

MATLAB Step:

Use the command spiral (N).

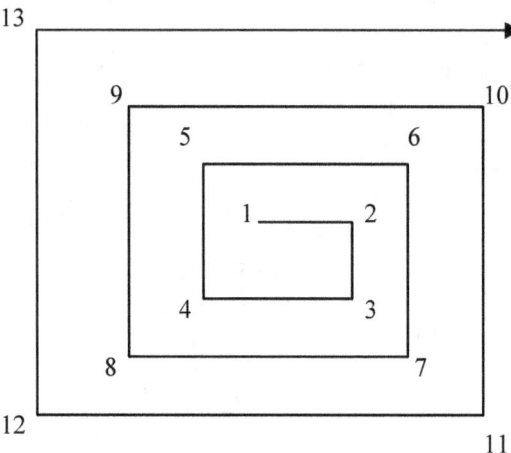

Figure 4.4   *Formation of a rectangular spiral*

If N is odd, the location of 1 will be $\left(\dfrac{N+1}{2}, \dfrac{N+1}{2}\right)$ and if N is even, the location of 1 will be $\left(\dfrac{N}{2}, \dfrac{N}{2}\right)$.

Then, a 4×4 or 5×5 spiral matrix will be $\begin{bmatrix} 7 & 8 & 9 & 10 \\ 6 & 1 & 2 & 11 \\ 5 & 4 & 3 & 12 \\ 16 & 15 & 14 & 13 \end{bmatrix}$ or $\begin{bmatrix} 21 & 22 & 23 & 24 & 25 \\ 20 & 7 & 8 & 9 & 10 \\ 19 & 6 & 1 & 2 & 11 \\ 18 & 5 & 4 & 3 & 12 \\ 17 & 16 & 15 & 14 & 13 \end{bmatrix}$

respectively. For a $N \times N$ spiral, the elements vary from 1 to N^2. See the formation of square spiral matrices as follows:

MATLAB Command

for the square spiral of 4×4,

    >>P=spiral(4) ↵

P =

```
     7    8    9   10
     6    1    2   11
     5    4    3   12
    16   15   14   13
```

for the square spiral of 5×5,

    >>P=spiral(5) ↵

P =

```
    21   22   23   24   25
    20    7    8    9   10
    19    6    1    2   11
    18    5    4    3   12
    17   16   15   14   13
```

4.36.10 Formation of a band matrix

Elements of a band matrix are zeroes eveywhere except along a band or strip that runs diagonally through the martix, usually but not necessarily, about the principal diagonal. An example of a band matrix can

be $B = \begin{bmatrix} 1 & 9 & 2 & 0 & 0 & 0 \\ -2 & 1 & 9 & 2 & 0 & 0 \\ a & -2 & 1 & 9 & 2 & 0 \\ 0 & a & -2 & 1 & 9 & 2 \\ 0 & 0 & a & -2 & 1 & 9 \\ 0 & 0 & 0 & a & -2 & 1 \end{bmatrix}$, where the band elements are $[a \quad -2 \quad 1 \quad 9 \quad 2]$. In general, the width of the band can be any value up to a number that would include the entire matrix within the band. We have the Maple function 'band' that can generate the band matrix. See the implementation of B as follows:

MATLAB Command

>>maple('A:=vector([a,-2,1,9,2])'); ↵   ← Assign the band elements to vector A
>>B=maple('band(A,6)'); ↵
>>sym(B) ↵

ans =

```
[ 1,   9,   2,   0,   0,   0]
[-2,   1,   9,   2,   0,   0]
[ a,  -2,   1,   9,   2,   0]
[ 0,   a,  -2,   1,   9,   2]
[ 0,   0,   a,  -2,   1,   9]
[ 0,   0,   0,   a,  -2,   1]
```

The number of elements in the vector A must be an odd number and it must not be greater than N, where $N \times N$ is the order of the band matrix. If N is greater than the number of elements in A, the remaining super and sub diagonal elements are set to zero. Since a tridiagonal matrix is primarily a band one, it can be created utilizing the subroutine. Have a tridiagonal matrix for the vector $[a \quad -2 \quad 1]$ as follows:

Command for the tridiagonal matrix,

>>maple('A:=vector([a,-2,1])'); ↵ ← Assign the tridiagonal elements to vector A
>>B=maple('band(A,6)'); ↵
>>sym(B) ↵

ans =

```
[-2,   1,   0,   0,   0,   0]       ← All we have is three diagonals - main, sub, and super
[ a,  -2,   1,   0,   0,   0]
[ 0,   a,  -2,   1,   0,   0]
[ 0,   0,   a,  -2,   1,   0]
[ 0,   0,   0,   a,  -2,   1]
[ 0,   0,   0,   0,   a,  -2]
```

4.36.11 Sylvester matrix from two polynomials

If two polynomials are given as $A(x) = a_m x^m + a_{m-1} x^{m-1} + a_{m-2} x^{m-2} + \ldots\ldots\ldots + a_2 x^2 + a_1 x + a_0$ and $B(x) = b_n x^n + b_{n-1} x^{n-1} + b_{n-2} x^{n-2} + \ldots\ldots\ldots + b_2 x^2 + b_1 x + b_0$, a matrix can be generated from the coefficients of the polynomials, which is

$$S = \begin{bmatrix} a_m & a_{m-1} & a_{m-2} & \cdots & a_2 & a_1 & a_0 & 0 & 0 & \cdots & 0 & 0 \\ 0 & a_m & a_{m-1} & \cdots & a_3 & a_2 & a_1 & a_0 & 0 & \cdots & 0 & 0 \\ 0 & 0 & a_m & \cdots & a_4 & a_3 & a_2 & a_1 & a_0 & \cdots & 0 & 0 \\ & & & \vdots & & & & & & & & \\ 0 & 0 & 0 & & & & & & & & a_1 & a_0 \\ \hdashline b_n & b_{n-1} & b_{n-2} & \cdots & b_2 & b_1 & b_0 & 0 & 0 & \cdots & 0 & 0 \\ 0 & b_n & b_{n-1} & \cdots & b_3 & b_2 & b_1 & b_0 & 0 & \cdots & 0 & 0 \\ 0 & 0 & b_n & \cdots & b_4 & b_3 & b_2 & b_1 & b_0 & \cdots & 0 & 0 \\ & & & \vdots & & & & & & & & \\ 0 & 0 & 0 & & & & & & & & b_1 & b_0 \end{bmatrix}$$

This matrix is called the Sylvester matrix. It is a square matrix of order $(m+n) \times (m+n)$, where the first m rows come from $A(x)$ and the second n rows come from $B(x)$. S is also called the resultant of $A(x)$ and $B(x)$.

174

To show by numerical example, assume that $A(x) = -9 + 8x - 6x^3$ and $B(x) = 4x^2 - 3x - 1$. Degrees of $A(x)$ and $B(x)$ are $m = 3$ and $n = 2$ respectively. Order of S should be 5×5 (since $m + n = 5$). On comparsion, we have $a_3 = -6$, $a_2 = 0$, $a_1 = 8$, $a_0 = -9$, $b_2 = 4$, $b_1 = -3$, and $b_0 = -1$. Therefore, the matrix S is given by

$$\begin{bmatrix} -6 & 0 & 8 & -9 & 0 \\ 0 & -6 & 0 & 8 & -9 \\ 4 & -3 & -1 & 0 & 0 \\ 0 & 4 & -3 & -1 & 0 \\ 0 & 0 & 4 & -3 & -1 \end{bmatrix}$$. Conduct Maple function 'sylvester' to have the matrix as follows:

MATLAB Command

```
>>syms x ↵
>>A=-9+8*x-6*x^3; ↵          ← Assign polynomial A(x) to A
>>B=4*x^2-3*x-1; ↵           ← Assign polynomial B(x) to B
>>S=maple('sylvester',A,B,x) ↵  ← Apply the Maple function 'sylvester'

S =

[ -6,  0,  8, -9,  0]
[  0, -6,  0,  8, -9]
[  4, -3, -1,  0,  0]
[  0,  4, -3, -1,  0]
[  0,  0,  4, -3, -1]
```

We did not include all aspects of the matrix theory due to the limitation of length. Accessible functions can come into view in two titles – M-file functions and Maple functions. Usually, Maple functions are suitable for symbolic implementation while on the contrary M-file functions can be applied for numerical computation. Readers can have online assistance about the other available subroutines executing 'help matfun' and 'mhelp linalg' in command window. It is important to mention that searching assistance for M-file functions and Maple functions starts with help and mhelp respectively.

Chapter 5

Problems of Differential Calculus

Differential calculus finds wide spectrum of uses in many branches of pure and applied mathematics. In a broad sense, problems pertaining to differentials are the contents of this chapter. Differentials fall into two major categories – ordinary (i.e., use of $\frac{d}{dx}$ or $\frac{d^n}{dx^n}$) and partial (use of $\frac{\partial}{\partial x}$ or other derivatives). Now a days problems of differentials are having an articulate mixing up with those of other disciplines. Ordinary differentials are prevalently put to use in analytic geometry, for example, in curve tracing and polynomial approximation of some functions. Application of partial derivatives is implied to vector calculus problems. MATLAB, in association with Maple package, has the capability of maneuvering the differentials both in numeric and in symbolic forms. Specimen problems are provided to demonstrate the utility of MATLAB in this facilitation.

5.1 Limit of a function

Limit of a function is the rudimentary problem of differential calculus. It is found by the subroutine 'limit'. To determine the limit,

MATLAB Steps:
1. *Declare variables of the limit function as symbolic and*
2. *Use the command limit (function as string, variable of the limit, limiting value).*

Begin with the example $\displaystyle\lim_{x \to 2} \frac{10\sin^2(x-2)}{(x-2)^2}$. Substitution of $x-2=u$ provides $\displaystyle\lim_{u \to 0} \frac{10\sin^2 u}{u^2}$. We know that $\displaystyle\lim_{\theta \to 0} \frac{\sin\theta}{\theta}=1$, thereupon, $\displaystyle\lim_{u \to 0} \frac{10\sin^2 u}{u^2}=10$. The string representation of $\frac{10\sin^2(x-2)}{(x-2)^2}$ is 10*sin(x-2)^2/(x-2)^2. The limiting value is $x \to 2$. The solution is as follows:

MATLAB Command

>>syms x ↵
>>limit(10*sin(x-2)^2/(x-2)^2,x,2) ↵

ans =

10

Supplementary exercise:

Determine the limits of the following functions:

A. $\underset{x \to a}{Lt} \dfrac{x^4 - a^4}{\sqrt{x} - \sqrt{a}}$

B. $\underset{x \to \infty}{Lt} \dfrac{2x^3 - x}{3x^3 - 7x - 8}$

C. $\underset{\theta \to 0}{Lt} \dfrac{\cos^2\left(\dfrac{\pi}{2}\cos\theta\right)}{\sin^2\theta}$

D. $\underset{k \to 2}{Lt} \underset{x \to k}{Lt} \dfrac{(k^2 - 4)\tan 2(x - k)}{(k - 2)(x - k)}$

E. $\underset{n \to 0}{Lt} \underset{m \to 0}{Lt} \dfrac{(1 - e^{\frac{Mm}{\pi}})(1 - e^{\frac{Nn}{\pi}})}{(1 - e^m)(1 - e^n)}$

⎙ Problem A

The given function is $\dfrac{x^4 - a^4}{\sqrt{x} - \sqrt{a}}$. Variables involved in this function are x and a. It is better if we type numerator and denominator separately. String forms of the numerator and denominator are x^4-a^4 and sqrt(x)-sqrt(a) respectively. We are going to assign them to n and d respectively. The given function is formed from n/d. In doing so, long functions can be entered without complicity. When $x \to a$, $\dfrac{x^4 - a^4}{\sqrt{x} - \sqrt{a}}$ takes $\dfrac{0}{0}$ form. Differentiating numerator and denominator w. r. to x provides $\dfrac{4x^3}{\dfrac{1}{2\sqrt{x}}} = 8x^{\frac{7}{2}}$. So, the limit of the function

is $8a^{\frac{7}{2}}$.

⎙ Problem B

The function $\dfrac{2x^3 - x}{3x^3 - 7x - 8}$ takes $\dfrac{\infty}{\infty}$ form when $x \to \infty$. Rationalization by x^3 provides $\dfrac{2x^3 - x}{3x^3 - 7x - 8} = \dfrac{2 - \dfrac{1}{x^2}}{3 - \dfrac{7}{x^2} - \dfrac{8}{x^3}}$. Setting $x \to \infty$ gives us the limit of the function, which is $\dfrac{2}{3}$. Numerator and denominator strings are 2*x^3-3 and 3*x^3-7*x-8 respectively.

⎙ Problem C

Very often, this type of limit occurs to compute the radiation intensity of an antenna. In this problem, $\dfrac{\cos^2\left(\dfrac{\pi}{2}\cos\theta\right)}{\sin^2\theta}$ takes $\dfrac{0}{0}$ form as $\theta \to 0$. Variable of the function is θ. Using the identities $\sin^2 A = \dfrac{1 - \cos 2A}{2}$ and $\cos^2 A = \dfrac{1 + \cos 2A}{2}$ turns out the function to be $\dfrac{1 + \cos(\pi \cos\theta)}{1 - \cos 2\theta}$, still this is $\dfrac{0}{0}$ form. Differentiating numerator and denominator of the last function w. r. to θ gives us $\dfrac{\pi \sin\theta \sin(\pi \cos\theta)}{2\sin 2\theta}$. By dint of identity $\sin 2A = 2\sin A \cos A$, we have $\dfrac{\pi \sin\theta \sin(\pi \cos\theta)}{2\sin 2\theta} = \dfrac{\pi \sin(\pi \cos\theta)}{4\cos\theta}$. Finally, substitution of $\theta = 0$ provides the limit

0. Write t instead of θ (because of its unavailability). String forms of the numerator (n) and denominator (d) with the variable t are cos(pi/2*cos(t))^2 and sin(t)^2 respectively.

Problem D

This is a function of double limits. One is $k \to 2$ and the other is $x \to k$. The inner limit is computed first. For $x \to k$, $\frac{(k^2-4)\tan 2(x-k)}{(k-2)(x-k)}$ takes $\frac{0}{0}$ form. Differentiate the numerator and denominator w. r. to x to have $\frac{(k^2-4)\tan 2(x-k)}{(k-2)(x-k)} = \frac{2(k^2-4)\sec^2 2(x-k)}{k-2}$ (the differentiation is partial). The last expression becomes $\frac{2(k^2-4)}{k-2}$ for $x \to k$. So, $\underset{k\to 2}{Lt}\ \underset{x\to k}{Lt}\ \frac{(k^2-4)\tan 2(x-k)}{(k-2)(x-k)} = \underset{k\to 2}{Lt}\ \frac{2(k^2-4)}{(k-2)}$, which takes $\frac{0}{0}$ form again for $k \to 2$. Factorization and simplification provide $\underset{k\to 2}{Lt}\ \frac{2(k^2-4)}{(k-2)} = \underset{k\to 2}{Lt}\ 2(k+2)$ =8. Required strings are n=(k^2-4)* tan(2*(x-k)) and d=(k-2)*(x-k) respectively. After computation of the first limit resulting output is assigned to f1, then, f1 is computed w.r.to k.

Problem E

This is another example of double limit. This kind of limit is seen in computation of array factor of a two dimensional array in antenna engineering. Because of $\frac{0}{0}$ form for $m \to 0$, partial differentiation w.r.to m is carried out on the numerator and denominator, on that cause, we have $\frac{Me^{\frac{Mm}{\pi}}(1-e^{\frac{Nn}{\pi}})}{\pi e^m(1-e^n)}$. Setting $m \to 0$ gives forth $\underset{n\to 0}{Lt}\ \frac{M(1-e^{\frac{Nn}{\pi}})}{\pi(1-e^n)}$, which is again $\frac{0}{0}$ form for $n \to 0$, hence, $\underset{n\to 0}{Lt}\ \frac{M(1-e^{\frac{Nn}{\pi}})}{\pi(1-e^n)} = \underset{n\to 0}{Lt}\ \frac{MNe^{\frac{Nn}{\pi}}}{\pi^2 e^n}$ (partial differentiation of the numerator and denominator w.r.to n)= $\frac{MN}{\pi^2}$. This time we have four variables in the given function, namely, M, N, m, and n (we can not write n for numerator). The numerator and denominator strings are p=(1-exp(-M*m/pi))*(1-exp(-N*n/pi)) and q=(1-exp(m))*(1-exp(n)) respectively.

Implementation of the limits from A through E is presented as follows:

MATLAB Command

for problem A,
```
>>syms x a ↵
>>n=x^4-a^4; ↵
>>d=sqrt(x)-sqrt(a); ↵
>>limit(n/d,x,a) ↵

ans =

8*a^(7/2)
```
for problem B,
```
>>syms x ↵
>>n=2*x^3-x; ↵
>>d=3*x^3-7*x-8; ↵
>>limit(n/d,x,inf) ↵

ans =

2/3
```
for problem C,
```
>>syms t ↵
>>n=cos(pi/2*cos(t))^2; ↵
>>d=sin(t)^2; ↵
>>limit(n/d,t,0) ↵

ans =

0
```
for problem D,
```
>>syms k x ↵
>>n=(k^2-4)*tan(2*(x-k)); ↵
>>d=(k-2)*(x-k); ↵
>>f=n/d; ↵
>>f1=limit(f,x,k); ↵
>>limit(f1,k,2) ↵

ans =
```

178

for problem E,
```
>>syms M N m n ↵                                    8
>>p=(1-exp(M*m/pi))*(1-exp(N*n/pi)); ↵
>>q=(1-exp(m))*(1-exp(n)); ↵
>>f1=limit(p/q,m,0); ↵
>>limit(f1,n,0) ↵
```

ans =

N/pi^2*M

Limit of a vector function:

Sometimes we may have a function, which is given in a matrix form. $F(x) = \begin{bmatrix} \dfrac{1-\cos x}{\sin x} & \dfrac{x^2-4}{x-2} \\ \dfrac{\tan 2x}{x} & \dfrac{-7x^4}{x^4+2} \end{bmatrix}$

is such an example. We wish to find $\underset{x \to 0}{Lt}\ F(x)$, that is, $\underset{x \to 0}{Lt}\ F(x) = \begin{bmatrix} \underset{x \to 0}{Lt}\ \dfrac{1-\cos x}{\sin x} & \underset{x \to 0}{Lt}\ \dfrac{x^2-4}{x-2} \\ \underset{x \to 0}{Lt}\ \dfrac{\tan 2x}{x} & \underset{x \to 0}{Lt}\ \dfrac{-7x^4}{x^4+2} \end{bmatrix}$, for

which $\underset{x \to 0}{Lt}\ \dfrac{1-\cos x}{\sin x}\ \left(\dfrac{0}{0}\ \text{form} \right) = \underset{x \to 0}{Lt}\ \dfrac{\sin x}{\cos x} = 0$, $\underset{x \to 0}{Lt}\ \dfrac{x^2-4}{x-2} = 2$, $\underset{x \to 0}{Lt}\ \dfrac{\tan 2x}{x} = 2\ \underset{x \to 0}{Lt}\ \dfrac{\tan 2x}{2x} = 2\left(\text{since,} \right.$

$\underset{\theta \to 0}{Lt}\ \dfrac{\tan \theta}{\theta} = 1 \left. \right)$, and $\underset{x \to 0}{Lt}\ \dfrac{-7x^4}{x^4+2} = 0$. Collecting all, we have $\underset{x \to 0}{Lt}\ F(x) = \begin{bmatrix} 0 & 2 \\ 2 & 0 \end{bmatrix}$. In order for easy entering,

the first row strings for $\dfrac{1-\cos x}{\sin x}$ and $\dfrac{x^2-4}{x-2}$ and the second row strings for $\dfrac{\tan 2x}{x}$ and $\dfrac{-7x^4}{x^4+2}$ are assigned to

f1 and f2 and f3 and f4 respectively, then, $F(x) = \begin{bmatrix} \text{f1} & \text{f2} \\ \text{f3} & \text{f4} \end{bmatrix}$. The limiting value of the variable can only be a

scalar, that is, for the example at hand, x can only be zero not a matrix of order 2×2. Implement the vector function as follows:

Command for the matrix function,
```
        >>syms x ↵
        >>f1=(1-cos(x))/sin(x); ↵
        >>f2=(x^2-4)/(x-2); ↵
        >>f3=tan(2*x)/x; ↵
        >>f4=-7*x^4/(x^4+2); ↵
        >>F=[f1 f2;f3 f4]; ↵
        >>limit(F,x,0) ↵

        ans =

        [ 0,  2]
        [ 2,  0]
```

Use of the definition of derivative:

By definition, we know that $\dfrac{dy}{dt} = \underset{h \to 0}{Lt} \dfrac{y(t+h) - y(t)}{h}$, where $y(t)$ is a function of t and that $y(t)$ is continuous, say, $y(t) = \sin t$, $\therefore y(t+h) = \sin(t+h)$ and $\dfrac{d(\sin t)}{dt} = \underset{h \to 0}{Lt} \dfrac{\sin(t+h) - \sin t}{h}$. The evaluation of the limit must be $\cos t$, which is the derivative of $\sin t$. Let us verify that,

MATLAB Command

```
>>syms t h ↵
>>limit((sin(t+h)-sin(t))/h,h,0) ↵

ans =

cos(t)
```

Left and right hand limits:

To have the idea about the left and right hand limits, let us consider the function $f(x) = \dfrac{1}{x(x-2)}$. Plot of the function for $-2 \le x \le 4$ is depicted in figure 5.1. As it is shown, the function $f(x)$ is not defined at $x = 0$ and $x = 2$. The functional values of $f(x)$ slightly before $x = 0$ and after $x = 0$ are termed as the left hand and right hand limits of $f(x)$ at $x = 0$ and denoted by $\underset{x \to 0_-}{Lt} f(x)$ and $\underset{x \to 0_+}{Lt} f(x)$ respectively. From the function, we observe that $f(x) = \infty$ at $x = 0_-$ and $f(x) = -\infty$ at $x = 0_+$, hence, $\underset{x \to 0_-}{Lt} f(x) = \infty$ and $\underset{x \to 0_+}{Lt} f(x) = -\infty$. Similarly, another transition of $f(x)$ is depicted in figure 5.1 at $x = 2$, where $\underset{x \to 2_-}{Lt} f(x) = -\infty$ and $\underset{x \to 2_+}{Lt} f(x) = \infty$.

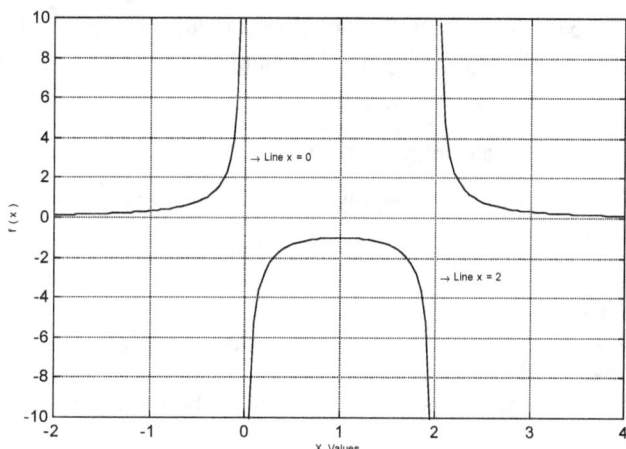

Figure 5.1 *Plot of function* $f(x) = \dfrac{1}{x(x-2)}$ *for interval* $-2 \le x \le 4$

The choice of finding the left and right hand limits is also incorporated in the subroutine 'limit'. See all these as follows:

Finding the left hand limit at $x = 0$,

```
>>syms x ↵
>>f=1/x/(x-2); ↵
>>limit(f,x,0,'left') ↵
```

Finding the right hand limit at $x = 0$,

```
>>syms x ↵
>>f=1/x/(x-2); ↵
>>limit(f,x,0,'right') ↵
```

ans =	ans =
inf	-inf

Finding the left hand limit at $x = 2$,

```
>>syms x ↵
>>f=1/x/(x-2); ↵
>>limit(f,x,2,'left') ↵
```

ans =

-inf

Finding the right hand limit at $x = 2$,

```
>>syms x ↵
>>f=1/x/(x-2); ↵
>>limit(f,x,2,'right') ↵
```

ans =

inf

The concept of limit is the very basic of calculus. Some other basic concepts are continuity and discontinuity of a function. Maple procedures are also avialble for these decision making functions. Get Maple help from the Command Window of MATLAB by executing 'mhelp iscont' and 'mhelp discont' respectively.

5.2 Derivatives of the polynomials using coefficients

As stated in chapter 3, polynomials can be represented by the coefficients. Derivatives of different orders of a polynomial can be taken on these coefficients by using the subroutine 'polyder' (abbreviation of polynomial derivative). You can say, 'polyder' of MATLAB simulates the differential operator $\frac{d}{dx}$ of calculus. The subroutine is applicable only for the polynomial coefficients not for the function strings.

5.2.1 Derivatives of different orders

⌗ *First derivative*

Consider the polynomial $9x^5 - 8x^4 + 3x^3 - 2x + 7$. What is the first derivative of this polynomial? The first derivative of this polynomial is given by $\frac{d}{dx}[9x^5 - 8x^4 + 3x^3 - 2x + 7] = 9 \times 5x^{5-1} - 8 \times 4x^{4-1} + 3 \times 3x^{3-1} - 2x^{1-1} + 0$

$= 45x^4 - 32x^3 + 9x^2 - 2$. The coefficient vectors for the given and the derivative polynomials are [9 –8 3 0 –2 7] and [45 –32 9 0 –2] respectively. Implementation of the first derivative is shown below:

MATLAB Command

for derivative coefficients,

```
>>A=[9 -8 3 0 -2 7]; ↵
>>polyder(A) ↵
```

ans =

      45 -32   9   0 -2

to display the derivative string,

```
>>B=polyder(A); ↵   ←Assign dy/dx to B
>>poly2str(B,'x') ↵
```

ans =

      45 x^4 - 32 x^3 + 9 x^2 - 2

The string of the derivative polynomial is displayed by the subroutine 'poly2str'. However, to take the first derivative of a polynomial,

MATLAB Steps:

1. Assign the polynomial coefficients to some row or column matrix A and
2. Use the command polyder(A).

⌗ *Second/third/other derivatives*

The second and third derivatives require using the subroutine polyder two and three times respectively, i.e., $\frac{d^2y}{dx^2} \equiv$ polyder(polyder(y)) and $\frac{d^3y}{dx^3} \equiv$ polyder(polyder(polyder(y))). Assume that the second and third derivatives of the polynomial $-7x^5 + 2x^4 + 1$ are to be found. The second derivative is given by

$$\frac{d^2(-7x^5 + 2x^4 + 1)}{dx^2} = \frac{d}{dx}\left(\frac{d}{dx}(-7x^5 + 2x^4 + 1)\right) = \frac{d(-35x^4 + 8x^3)}{dx} = -140x^3 + 24x^2.$$ The third derivative is

$$\frac{d^3(-7x^5 + 2x^4 + 1)}{dx^3} = \frac{d(-140x^3 + 24x^2)}{dx} = -420x^2 + 48x.$$ In MATLAB notation, the given, second, and third polynomials are [−7 2 0 0 0 1], [−140 24 0 0], and [−420 48 0] respectively. Implementation of the second and third derivatives is shown as follows:

MATLAB Command

for the second derivative coefficients,

```
>>A=[-7 2 0 0 0 1]; ↵
>>polyder(polyder(A)) ↵
```

ans =

    -140   24   0   0

to display the second derivative string,

```
>>B=polyder(polyder(A)); ↵
>>poly2str(B,'x') ↵
```

ans =

    -140 x^3 + 24 x^2

for the third derivative coefficients,

```
>>polyder(polyder(polyder(A))) ↵
```

ans =

    -420   48   0

to display the third derivative string,

```
>>B=polyder(polyder(polyder(A))); ↵
>>poly2str(B,'x') ↵
```

ans =

    -420 x^2 + 48 x

The number of 'polyder' is the same as the order of derivative. If there were fourth derivative, the number of 'polyder' would be four. We assigned the output of the subroutine to B, if only derivative coefficients are necessary, B will have these coefficients. Repetitive writing of 'polyder' can be avoided by using a for loop. To find the Nth order derivative of a polynomial A using for loop,

MATLAB Steps:

1. Assign the polynomial coefficients to some row/column matrix A
2. Initialize the resulting derivative matrix B using command B=A
3. Use for loop counter index as the order of the derivative as follows:
    for k=1:N B=polyder(B); end
4. Display derivative polynomial B using command poly2str (B,'x').

Let us execute the just mentioned third order derivative of the polynomial using the for loop:

```
>>A=[-7 2 0 0 0 1]; ↵
>>B= A; ↵
>>for k=1:3 B=polyder(B); end ↵
>>poly2str(B,'x') ↵
```

ans =

    -420 x^2 + 48 x

5.2.2 Derivatives of the product polynomials

Suppose we want to find the first derivative of the product of polynomials $7x^2 + 2x - 9$ and $3x^3 + 4x + 1$. The derivative is found by using the rule $\frac{d}{dx}[uv] = u\frac{dv}{dx} + v\frac{du}{dx}$, therefore, $\frac{d}{dx}[(7x^2 + 2x - 9) \times (3x^3 + 4x + 1)] = (7x^2 + 2x - 9)\frac{d}{dx}(3x^3 + 4x + 1) + (3x^3 + 4x + 1)\frac{d}{dx}(7x^2 + 2x - 9) = 105x^4 + 24x^3 + 3x^2 + 30x - 34$ (after simplification). You can still use 'polyder' to evaluate the first derivative of the product polynomial but the number of input arguments in the subroutine is two. To find the derivative of the product of two polynomials,

MATLAB Steps:

1. Assign the first and second polynomial coefficients to matrices A and B respectively,
2. Use the command C=polyder(A,B), where output derivative is assigned to C, and

3. Display the derivative polynomial C using poly2str(C,'x') .

Implement the example as follows:

MATLAB Command

>>A=[7 2 -9]; ↵
>>B=[3 0 4 1]; ↵
>>C=polyder(A,B); ↵
>>poly2str(C,'x') ↵

ans =
$$105 \text{ x}^4 + 24 \text{ x}^3 + 3 \text{ x}^2 + 30 \text{ x} - 34$$

5.2.3 Derivatives of the division of two polynomials

As an example, find $\dfrac{d}{dx}\left[\dfrac{x^2+x+1}{x^3-x+1}\right]$. For the given division polynomial, say, N stands for the numerator and D stands for the denominator. In MATLAB notation, we have N= $x^2 +x+1$ =[1   1   1] and D= $x^3 -x+1$ =[1   0   −1   1]. Using the rule $\dfrac{d}{dx}\left[\dfrac{u}{v}\right]=\dfrac{v\dfrac{du}{dx}-u\dfrac{dv}{dx}}{v^2}$ can find the differentiation of division of two polynomials, $\therefore$ $\dfrac{d}{dx}\left[\dfrac{x^2+x+1}{x^3-x+1}\right]$ = $\dfrac{(x^3-x+1)\dfrac{d}{dx}(x^2+x+1)-(x^2+x+1)\dfrac{d}{dx}(x^3-x+1)}{(x^3-x+1)^2}$ =

$\dfrac{(x^3-x+1)(2x+1)-(x^2+x+1)(3x^2-1)}{(x^3-x+1)^2}=\dfrac{-x^4-2x^3-4x^2+2x+2}{x^6-2x^4+2x^3+x^2-2x+1}$ (after simplification)= $\dfrac{NO}{DO}$, where NO stands for the output numerator and DO does for the output denominator of the polynomial after the derivative is taken. As regards, MATLAB should yield NO= $-x^4 - 2x^3 - 4x^2 + 2x + 2$ =[−1   −2   −4   2   2] and DO= $x^6 - 2x^4 + 2x^3 + x^2 - 2x + 1$ =[1   0   −2   2   1   −2   1]. 'Polyder' can, again, be used for the evaluation of the first derivative of division of two polynomials but the number of output arguments in the subroutine is two. One is for numerator and the other is for denominator. As argument, numerator comes first then does the denominator. Finally, to take the derivative,

MATLAB Steps:
1. *Assign the given numerator and denominator polynomial coefficients to row matrix N and D respectively,*
2. *Use the command [NO DO]=polyder(N,D), and*
3. *Use poly2str to display the output.*

Above-mentioned example is shown below:
MATLAB Command

>>N=[1 1 1]; ↵
>>D=[1 0 -1 1]; ↵
>>[NO DO]=polyder(N,D) ↵

NO =
        -1 -2 -4 2 2
DO =
        1 0 -2 2 1 -2 1

displaying the numerator and denominator strings,

>>poly2str(NO,'x') ↵

ans =
$$-1 \text{ x}^4 - 2 \text{ x}^3 - 4 \text{ x}^2 + 2 \text{ x} + 2$$

>>poly2str(DO,'x') ↵

ans =
$$x^6 - 2 x^4 + 2 x^3 + x^2 - 2 x + 1$$

5.2.4 Value of the derivatives of a polynomial at some x
Evaluate the following:

A. $\dfrac{d^3}{dx^3}\left(-7x^5 + 2x^4 + 1\right)$ at $x = -1$

B. $\dfrac{d}{dx}[(7x^2 + 2x - 9) \times (3x^3 + 4x + 1)]$ at $x = 0$

C. $\dfrac{d}{dx}\left[\dfrac{x^2 + x + 1}{x^3 - x + 1}\right]$ at $x = -2$

Computations of the articles 5.2.1, 5.2.2, and 5.2.3 provide

A. At $x = -1$, $\dfrac{d^3}{dx^3}\left(-7x^5 + 2x^4 + 1\right) = -420x^2 + 48x = -420 \times (-1)^2 + 48 \times \quad (-1) = -468.$

B. At $x = 0$, $\dfrac{d}{dx}[(7x^2 + 2x - 9) \times (3x^3 + 4x + 1)] = 105x^4 + 24x^3 + 3x^2 + 30x - 34 = -34.$

C. At $x = -2$, $\dfrac{d}{dx}\left[\dfrac{x^2 + x + 1}{x^3 - x + 1}\right] = \dfrac{-x^4 - 2x^3 - 4x^2 + 2x + 2}{x^6 - 2x^4 + 2x^3 + x^2 - 2x + 1} = \dfrac{-(-2)^4 - 2(-2)^3 - 4(-2)^2 + 2(-2) + 2}{(-2)^6 - 2(-2)^4 + 2(-2)^3 + (-2)^2 - 2(-2) + 1} = -\dfrac{18}{25}.$

The subroutine 'polyval' can evaluate a polynomial at some x. Use it after the derivative is taken. All examples are shown as follows:

MATLAB Command

for problem A,
```
>>A=[-7 2 0 0 0 1]; ↵
>>B=polyder(polyder(polyder(A))); ↵
>>polyval(B,-1) ↵

ans =
    -468
```

for problem C,
```
>>N=[1 1 1]; ↵
>>D=[1 0 -1 1]; ↵
>>[NO DO]=polyder(N,D); ↵
>>sym(polyval(NO,-2)/polyval(DO,-2)) ↵

ans =

-18/25
```

for problem B,
```
>>A=[7 2 -9]; ↵
>>B=[3 0 4 1]; ↵
>>C=polyder(A,B); ↵
>>polyval(C,0) ↵

ans =

    -34
```

Regarding problem C, the commands 'polyval(NO,-2)' and 'polyval(DO,-2)' compute the numerator and denominator polynomials at $x = -2$ respectively. The command 'sym' is used on the division of numerator and denominator to have the rational form output. Without the command 'sym', the output of problem C would be a floating-point number.

5.3 Symbolic differentiation of functions
Most problems of the last section are associated with the coefficients of the polynoimials. Symbolic differentiation of functions can be conveniently carried out in MATLAB by the subroutine 'diff', which is the

abbreviation of <u>differ</u>entiation. In other words, 'diff' of MATLAB performs the same operation as does $\frac{d}{dx}$ of calculus. For this purpose, functions must be written in string form according to MATLAB notation. How a function is represented as string is illustrated in chapter 11.

5.3.1 Standard functions

In this section, we concentrate on the differentiation of standard functions. To find the derivative of a standard function,

MATLAB Step:
Use the command diff('function as string').

There are two ways to carry out the differentiation – with and without the command 'sym'. Let us take the example of $\sin^{-1} x$. According to MATLAB notation, $\sin^{-1} x$ is represented as 'asin(x)'. Differentiation of $\sin^{-1} x$, $\frac{d}{dx}(\sin^{-1} x)$, is $\frac{1}{\sqrt{1-x^2}}$. To differentiate $\sin^{-1} x$ without 'sym',

MATLAB Command
>>diff('asin(x)').↵

ans =
1/(1-x^2)^(1/2)

MATLAB returned the result of $\frac{d}{dx}(\sin^{-1} x)$ as 1/(1-x^2)^(1/2). Now x^2, 1-x^2, (1-x^2)^(1/2), 1/(1-x^2)^(1/2) mean x^2, $1-x^2$, $\sqrt{1-x^2}$, and $\frac{1}{\sqrt{1-x^2}}$ respectively, which is equivalent to the standard result. One disadvantage of this method is that the output can not be seen in symbolic form. To do so, use command 'pretty':

MATLAB Command
>>pretty(diff('asin(x)')) ↵

```
            1
    ------------------
           2  1/2
      ( 1 - x   )
```

The other way is declare the variable under consideration as symbolic from the beginning of differentiation. To show this,

MATLAB Command
>>syms x ↵
>>pretty(diff(asin(x))) ↵

```
           1
    --------------------
           2   1/2
       ( 1 - x   )
```

Pretty form is visually attractive but the mathematical maneuvering requires the string form. Table 5.A provides the differentiation of other standard functions in MATLAB environment. MATLAB codes of the mathemetical functions are seen in table 3.C.

Table 5.A Differentiation of some standard functions

Mathematical Notation	Equivalent MATLAB Notation
$\frac{d}{dx}(\sin x) = \cos x$	pretty(diff('sin(x)')) $\Rightarrow$ cos(x)
$\frac{d}{dx}(\cos x) = -\sin x$	pretty(diff('cos(x)')) $\Rightarrow$ -sin(x)
$\frac{d}{dx}(\tan x) = \sec^2 x = 1 + \tan^2 x$	pretty(diff('tan(x)')) $\Rightarrow$ 1 + tan(x)²
$\frac{d}{dx}(\cot x) = -\cos ec^2 x$ $= -(1 + \cot^2 x)$	pretty(diff('cot(x)')) $\Rightarrow$ -1 - cot(x)²
$\frac{d}{dx}(\sec x) = \sec x \tan x$	pretty(diff('sec(x)')) $\Rightarrow$ sec(x) tan(x)
$\frac{d}{dx}(\cos ecx) = -\cos ecx \cot x$	pretty(diff('csc(x)')) $\Rightarrow$ -csc(x) cot(x)
$\frac{d}{dx}(\cos^{-1} x) = -\frac{1}{\sqrt{1-x^2}}$	pretty(diff('acos(x)')) $\Rightarrow$ $-\dfrac{1}{(1-x^2)^{1/2}}$
$\frac{d}{dx}(\tan^{-1} x) = \frac{1}{1+x^2}$	pretty(diff('atan(x)')) $\Rightarrow$ $\dfrac{1}{1+x^2}$
$\frac{d}{dx}(\cot^{-1} x) = -\frac{1}{1+x^2}$	pretty(diff('acot(x)')) $\Rightarrow$ $-\dfrac{1}{1+x^2}$
* $\frac{d}{dx}(\sec^{-1} x) = \frac{1}{x\sqrt{x^2-1}}$	pretty(diff('asec(x)')) $\Rightarrow$ $\dfrac{1}{x^2\left(1-\dfrac{1}{x^2}\right)^{1/2}}$
* $\frac{d}{dx}(\cos ec^{-1}x) = -\frac{1}{x\sqrt{x^2-1}}$	pretty(diff('acsc(x)')) $\Rightarrow$ $-\dfrac{1}{x^2\left(1-\dfrac{1}{x^2}\right)^{1/2}}$
$\frac{d}{dx}(\sinh x) = \cosh x$	pretty(diff('sinh(x)')) $\Rightarrow$ cosh(x)
$\frac{d}{dx}(\cosh x) = \sinh x$	pretty(diff('cosh(x)')) $\Rightarrow$ sinh(x)
$\frac{d}{dx}(\tanh x) = 1 - \tanh^2 x$	pretty(diff('tanh(x)')) $\Rightarrow$ 1 - tanh(x)²
$\frac{d}{dx}(\coth x) = 1 - \coth^2 x$	pretty(diff('coth(x)')) $\Rightarrow$ 1 - coth(x)²
$\frac{d}{dx}(\sec hx) = -\sec hx \tanh x$	pretty(diff('sech(x)')) $\Rightarrow$ -sech(x) tanh(x)
$\frac{d}{dx}(\cos echx) = -\cos echx \coth x$	pretty(diff('csch(x)')) $\Rightarrow$ -csch(x) coth(x)
$\frac{d}{dx}(\sinh^{-1} x) = \frac{1}{\sqrt{1+x^2}}$	pretty(diff('asinh(x)')) $\Rightarrow$ $\dfrac{1}{(1+x^2)^{1/2}}$
* $\frac{d}{dx}(\cosh^{-1} x) = \pm\frac{1}{\sqrt{x^2-1}}$	pretty(diff('acosh(x)')) $\Rightarrow$ $\dfrac{1}{(x-1)^{1/2}(x+1)^{1/2}}$
$\frac{d}{dx}(\tanh^{-1} x) = \frac{1}{1-x^2}$	pretty(diff('atanh(x)')) $\Rightarrow$ $\dfrac{1}{1-x^2}$

186

Continuation of the previous table:

Mathematical Notation	Equivalent MATLAB Notation

$\frac{d}{dx}(\coth^{-1}x) = -\dfrac{1}{x^2-1}$	``` pretty(diff('acoth(x)')) ⇒``` `- - ------------`
	 `                2`
	 `             x  - 1`

* $\frac{d}{dx}(\sec h^{-1}x) = \pm\dfrac{1}{x\sqrt{1-x^2}}$	``` pretty(diff('asech(x)')) ⇒```
	 `                    1`
	 ` - ------------------------------------`
	 `       2        1/2        1/2`
	 `      x  (1/x - 1)   (1/x + 1)`

$\frac{d}{dx}(\cos ech^{-1}x) = -\dfrac{1}{x\sqrt{1+x^2}}$	``` pretty(diff('acsch(x)')) ⇒``` `            1` ` - --------------------------` `        2 /      1  \ 1/2` ` x	1 + ------|`   `	2 |`   ` \ x /`
$\frac{d}{dx}(e^x) = e^x$	``` pretty(diff('exp(x)')) ⇒ exp(x)```
$\frac{d}{dx}(\log_e x) = \frac{1}{x}$	``` pretty(diff('log(x)')) ⇒ 1/x```

$\frac{d}{dx}(x^n) = nx^{n-1}$	`            n`
	 `         x   n`
	 ``` pretty(diff('x^n')) ⇒``` `  ---------`
	 `            x`

$\frac{d}{dx}(a^x) = a^x \ln a$	`            x` ``` pretty(diff('a^x')) ⇒ a log(a)```

* Returns for $\sec^{-1}x$ and $\cos ec^{-1}x$ do not look the standard results. Simple manipulation turns out that they are

equivalent, for example, pretty(diff('asec(x)'))= $\dfrac{1}{x^2\sqrt{1-\frac{1}{x^2}}} = \dfrac{1}{x^2\sqrt{\frac{x^2-1}{x^2}}} = \dfrac{1}{x^2\frac{\sqrt{x^2-1}}{x}} = \dfrac{1}{x\sqrt{x^2-1}}$, which is the same

as standard result.
* Notice that the differentiation of $\cosh^{-1}x$ and $\sec h^{-1}x$ can have positive or negative differential coefficients. But in MATLAB, they are defined only for one type of sign. If it is $\cosh^{-1}x$, the differential coefficient is

$+\dfrac{1}{\sqrt{x^2-1}}$ and if it is $\sec h^{-1}x$, the differential coefficient is $-\dfrac{1}{x\sqrt{1-x^2}}$.

5.3.2 Composite functions

Composite function means function of a function. MATLAB is also capable of differentiating composite functions. All you need is put the composite function inside 'diff' in string form. We assign the string output of 'diff' to variable d for the following composite functions and exercise the command 'pretty' to print the symbolic form.

Find the derivatives of the following composite functions:

A. $x^3\tan^3 x$ B. $(12x^4 - 4x + 1)(9x^3 - 5)$ C. $\dfrac{7x-3}{x^2\sqrt{2x^5-13}}$

D. $\sqrt[4]{[3\sin^3(3x-1) + \ln x]^3}$ E. $x^4 J_3(x)$ F. $\tan\left(\dfrac{u-5}{2u+5}\right)$

⌨ *Function A*

The function $x^3\tan^3 x$ is the product of x^3 and $\tan^3 x$, here, x^3 is a function of x and $\tan^3 x$ is a

function of $\tan x$, again, $\tan x$ is a function of x . Using the rule $\dfrac{d}{dx}[uv] = u\dfrac{dv}{dx} + v\dfrac{du}{dx}$ does the differentiation of

$x^3 \tan^3 x$, $\therefore \dfrac{d}{dx}[x^3 \tan^3 x] = 3x^2 \tan^3 x + 3x^3 \tan^2 x \sec^2 x$. The string of $x^3 \tan^3 x$ is x^3*tan(x)^3. To solve this,

MATLAB Command

```
>>d=diff('x^3*tan(x)^3'); ↵
>>pretty(d) ↵
```

$$3 x^2 \tan(x)^3 + 3 x^3 \tan(x)^2 (1 + \tan(x)^2)$$

Observe that $\sec^2 x$ is returned as $1 + \tan^2 x$.

☞ Function B

$(12x^4 - 4x + 1)(9x^3 - 5)$ is the product of two polynomials, hence, the rule $\dfrac{d}{dx}[u\,v] = u\dfrac{dv}{dx} + v\dfrac{du}{dx}$ is also

applicable here, on that, $\dfrac{d}{dx}[(12x^4 - 4x + 1)(9x^3 - 5)] = (12x^4 - 4x + 1)\,27x^2 + (9x^3 - 5)(48x^3 - 4)$. In string form, the function is (12*x^4-4*x+1)*(9*x^3-5). The solution is as follows:

MATLAB Command

```
>>d=diff('(12*x^4-4*x+1)*(9*x^3-5)'); ↵
>>pretty(d) ↵
```

$$(48 x^3 - 4)(9 x^3 - 5) + 27 (12 x^4 - 4 x + 1) x^2$$

Simplified polynomial is seen by the command 'expand':

```
>>pretty(expand(d)) ↵
```

$$756 x^6 - 384 x^3 + 20 + 27 x^2$$

☞ Function C

This example is the division of two functions. The numerator is $7x - 3$ and the denominator is

$x^2\sqrt{2x^5 - 13}$, which is, again, product of two functions. Utilize $\dfrac{d}{dx}[u\,v] = u\dfrac{dv}{dx} + v\dfrac{du}{dx}$ and $\dfrac{d}{dx}\left[\dfrac{u}{v}\right] = \dfrac{v\dfrac{du}{dx} - u\dfrac{dv}{dx}}{v^2}$

for the multiplication and division respectively, $\therefore \dfrac{d}{dx}\left[\dfrac{7x - 3}{x^2\sqrt{2x^5 - 13}}\right] = \dfrac{(x^2\sqrt{2x^5 - 13})7 - (7x - 3)\dfrac{d}{dx}(x^2\sqrt{2x^5 - 13})}{(x^2\sqrt{2x^5 - 13})^2} =$

$\dfrac{7x^2\sqrt{2x^5 - 13} - (7x - 3)\left[2x\sqrt{2x^5 - 13} + x^2\dfrac{1}{2\sqrt{2x^5 - 13}}.10x^4\right]}{x^4(2x^5 - 13)} = \dfrac{7}{x^2\sqrt{2x^5 - 13}} - \dfrac{2(7x - 3)}{x^3\sqrt{2x^5 - 13}} - \dfrac{5(7x - 3)x^2}{\sqrt{(2x^5 - 13)^3}}$. The numerator and

denominator strings are 7*x-3 and x^2*sqrt(2* x^5-13) respectively. See the implementation below:

MATLAB Command

```
>>d=diff('(7*x-3)/(x^2*sqrt(2*x^5-13))'); ↵
>>pretty(d) ↵
```

$$\dfrac{7}{x^2 (2 x^5 - 13)^{1/2}} - 2 \dfrac{7 x - 3}{x^3 (2 x^5 - 13)^{1/2}} - 5 \dfrac{(7 x - 3) x^2}{(2 x^5 - 13)^{3/2}}$$

⚘ *Function D*

The function $\sqrt[4]{[3\sin^3(3x-1)+\ln x]^3}$ can be written in power index form as $[3\sin^3(3x-1)+\ln x]^{\frac{3}{4}}$ whose string is given by $(3*\sin(3*x-1)^3+\log(x))^{(3/4)}$, $\therefore \dfrac{d}{dx}\left[\sqrt[4]{[3\sin^3(3x-1)+\ln x]^3}\right] =$

$\dfrac{3}{4}[3\sin^3(3x-1)+\ln x]^{\frac{3}{4}-1}\,[9\sin^2(3x-1)\cos(3x-1).3+\dfrac{1}{x}] = \dfrac{3}{4}[3\sin^3(3x-1)+\ln x]^{-\frac{1}{4}}\{27\sin^2(3x-1)\cos(3x-1)+\dfrac{1}{x}\} =$

$\dfrac{3\{27\sin^2(3x-1)\cos(3x-1)+\dfrac{1}{x}\}}{4\{3\sin^3(3x-1)+\ln x\}^{\frac{1}{4}}}$.

MATLAB Command

```
>>d=diff('(3*sin(3*x-1)^3+log(x))^(3/4)'); ↵
>>pretty(d) ↵
                        2
            27 sin(3 x - 1)   cos(3 x - 1) + 1/x
      3/4 ------------------------------------------
                    3              1/4
          ( 3 sin(3 x - 1)   + log(x) )
```

⚘ *Function E*

$J_3(x)$ is a special type of function. $J_3(x)$ is called the Bessel function of the first kind of order 3. We know the recursion formula of $J_v(x)$, which is, for any real number v, $x\dfrac{d}{dx}[J_v(x)] = -vJ_v(x)+xJ_{v-1}(x)$. $x^4J_3(x)$ is the product of x^4 and $J_3(x)$, $\therefore \dfrac{d}{dx}[x^4J_3(x)] = x^4\dfrac{d}{dx}[J_3(x)]+J_3(x)\dfrac{d}{dx}[x^4] = x^4\,[-\dfrac{3}{x}J_3(x)+J_2(x)]+J_3(x)\,4x^3$. The string of the function is x^4*besselj(3,x).

MATLAB Command

```
>>d=diff('x^4*besselj(3,x)'); ↵
>>pretty(d) ↵

       3                4 /                    besselj(3, x)  \
    4 x   besselj(3, x) + x  | besselj(2, x) - 3 ----------------- |
                             \                         x        /
```

⚘ *Function F*

Evidently, the independent variable is u . Write $y=\tan\left(\dfrac{u-5}{2u+5}\right)=\tan z$, where $z=\dfrac{u-5}{2u+5}$, then, apply

$\dfrac{dy}{du}=\dfrac{dy}{dz}\dfrac{dz}{du}$ to have $\dfrac{dy}{du}=\left[\dfrac{1}{2u+5}-\dfrac{2(u-5)}{(2u+5)^2}\right]\sec^2\left(\dfrac{u-5}{2u+5}\right)$. MATLAB code of the function is tan((u-5)/(2*u+5)).

MATLAB Command

```
>>d=diff('tan((u-5)/(2*u+5))'); ↵
>>pretty(d) ↵

      /          u - 5      2 \  /      1             u - 5       \
      | 1 +  tan( ------------ )  |  |  ----------- - 2 ---------------  |
      \          2 u + 5       /  |  2 u + 5              2       |
                                  \                    (2 u + 5)    /
```

As you see, $\sec^2 x$ is returned as $1+\tan^2 x$ and display of parenthesis ($\ldots$) for the bulk expression is accomplished by characters '|', '/', and '\'.

5.4 Successive differentiation of functions

Successive differentiation means finding the higher order derivatives of a differentiable function $y = f(x)$. According to calculus notation, $\dfrac{d^n y}{dx^n}$ is the nth order derivative of y or $\dfrac{d^n y}{dx^n}$ is the successive differentiation of y for n times. For instance, $\dfrac{d^5 y}{dx^5}$ is the successive differentiation of y for 5 times. MATLAB comforts in performing successive differentiation of any order. To evaluate the successive derivative of any order,

MATLAB Step:

Use the command diff('function as string', the order of the differentiation).

Consider the successive differentiation of $y = \cos(2x-5)$ for five times, i.e., evaluate $\dfrac{d^5 y}{dx^5}$. Different derivatives are $\dfrac{dy}{dx} = -2\sin(2x-5)$, $\dfrac{d^2 y}{dx^2} = \dfrac{d}{dx}[-2\sin(2x-5)] = -4\cos(2x-5)$, $\dfrac{d^3 y}{dx^3} = \dfrac{d}{dx}[-4\cos(2x-5)] = 8\sin(2x-5)$, $\dfrac{d^4 y}{dx^4} = \dfrac{d}{dx}[8\sin(2x-5)] = 16\cos(2x-5)$, and $\dfrac{d^5 y}{dx^5} = \dfrac{d}{dx}[16\cos(2x-5)] = -32\sin(2x-5)$. In string form, $\cos(2x-5)$ and the fifth derivative $\dfrac{d^5 y}{dx^5}$ are represented as cos(2*x-5) and -32*sin(2*x-5) respectively. To get the successive differentiation done,

MATLAB Command

>>diff('cos(2*x-5)',5) ↵

ans =

-32*sin(2*x-5)

More examples on the successive differentiation are furnished below:

⊟ Example A

Evaluate $\dfrac{d^4}{dx^4}[(1-7x+x^3)^3]$. The order of the successive differentiation is 4. Differentiate the function to write $\dfrac{d}{dx}[(1-7x+x^3)^3] = 3(1-7x+x^3)^2(3x^2-7)$, $\dfrac{d^2}{dx^2}[(1-7x+x^3)^3] = \dfrac{d}{dx}[3(1-7x+x^3)^2(3x^2-7)] = 3[(1-7x+x^3)^2 \times 6x + (3x^2-7)2(1-7x+x^3)(3x^2-7)] = 18x(1-7x+x^3)^2 + 6(3x^2-7)^2(1-7x+x^3)$, $\dfrac{d^3}{dx^3}[(1-7x+x^3)^3] = \dfrac{d}{dx}[18x(1-7x+x^3)^2 + 6(3x^2-7)^2(1-7x+x^3)] = 18(1-7x+x^3)^2 + 108x(1-7x+x^3)(3x^2-7) + 6(3x^2-7)^3$, and $\dfrac{d^4}{dx^4}[(1-7x+x^3)^3] = \dfrac{d}{dx}\left[\dfrac{d^3}{dx^3}[(1-7x+x^3)^3]\right] = 216x(3x^2-7)^2 + 648x^2(1-7x+x^3) + 144(3x^2-7)(1-7x+x^3)$ (on

simplification). Implement it as follows:

MATLAB Command

>>d=diff('(1-7*x+x^3)^3',4); ↵
>>pretty(d) ↵

```
                2  2                3  2                3            2
   216 (-7 + 3 x )  x  + 648 (1 - 7 x + x )x  + 144 (1 - 7 x + x )(-7 + 3 x )
```

Have the simplified form using the command 'expand':

>>pretty(expand(d)) ↵

$$17640\ x^3 - 17640\ x^5 + 3024\ x^2 + 1080\ x\ -\ 1008$$

⊟ Example B

Evaluate $7\dfrac{d^4y}{dx^4} + \dfrac{d^2y}{dx^2} - 20y$ for $y = e^{-2x}\cos(3x-4)$. Carry out the various differentiations to have $\dfrac{dy}{dx}$

$= -2e^{-2x}\cos(3x-4) - 3e^{-2x}\sin(3x-4)$, $\dfrac{d^2y}{dx^2} = -5e^{-2x}\cos(3x-4) + 12e^{-2x}\sin(3x-4)$, $\dfrac{d^3y}{dx^3} = 46e^{-2x}\cos(3x-4) -$

$9e^{-2x}\sin(3x-4)$, and $\dfrac{d^4y}{dx^4} = -119e^{-2x}\cos(3x-4) - 120e^{-2x}\sin(3x-4)$, hence, $7\dfrac{d^4y}{dx^4} + \dfrac{d^2y}{dx^2} - 20y = 7[-119e^{-2x}\cos(3x-4)$

$-120\ e^{-2x}\sin(3x\ -4)] + [-5e^{-2x}\cos(3x-4) + 12\ e^{-2x}\sin(3x-4)] - 20e^{-2x}\cos(3x-4) = -858e^{-2x}\cos(3x-4) - 828e^{-2x}\sin(3x-$

4) . So much computation is easily bypassed as follows:

MATLAB Command

>>syms x ↵   ← Declare the variable x as symbolic

>>y=exp(-2*x)*cos(3*x-4); ↵   ← Assign $e^{-2x}\cos(3x-4)$ to y

>>d=7*diff(y,4)+diff(y,2)-20*y; ↵   ← Assign $7\dfrac{d^4y}{dx^4} + \dfrac{d^2y}{dx^2} - 20y$ to d

>>pretty(d) ↵

-858 exp(-2 x) cos(3 x - 4) - 828 exp(-2 x) sin(3 x - 4)

⊟ Example C

Evaluate $\dfrac{d^4}{dx^4}[(1-7x+x^3)^3]$ at $x = -1$. From example A, we have $\dfrac{d^4}{dx^4}[(1-7x+x^3)^3] = 216x(3x^2-7)^2 +$

$648x^2(1-7x+x^3) + 144\ (3x^2-7)(1-7x+x^3)$. Plugging $x = -1$ yields $(216)(-1)(3-7)^2 + 648\{1-7(-1)-1\} +$

$144(3-7)\{1-7(-1)-1\} = -2952$. String evaluation is accomplished by the subroutine 'eval' and -1 should be

assigned to x before the evaluation of the string assigned to d is executed. See the implementation as follows:

MATLAB Command

>>d=diff('(1-7*x+x^3)^3',4); ↵

>>x=-1; ↵

>>eval(d) ↵

ans =

-2952

⊟ Example D

Evaluate $7\dfrac{d^4y}{dx^4} + \dfrac{d^2y}{dx^2} - 20y$ at $x = \pi$ for $y = e^{-2x}\cos(3x-4)$. Example B tells us that $7\dfrac{d^4y}{dx^4} + \dfrac{d^2y}{dx^2} - 20y$

$= -858e^{-2x}\cos(3x-4) - 828e^{-2x}\sin(3x-4) = -858e^{-2\pi}\cos(3\pi-4) - 828e^{-2\pi}\sin(3\pi-4) = 858e^{-2\pi}\cos 4 - 828e^{-2\pi}\sin 4 =$

0.1229. See the computation below:

>>syms x ↵

>>y=exp(-2*x)*cos(3*x-4); ↵

>>d=7*diff(y,4)+diff(y,2)-20*y; ↵

>>x=pi; ↵

>>eval(d) ↵

ans =

0.1229

⊟ Example E

Not all functions have the independent variable x. Compute $\dfrac{d^2}{dz^2}\left[\dfrac{z-4}{z+3}\right]$ at $z=-5$. It is found that

$\dfrac{d^2}{dz^2}\left[\dfrac{z-4}{z+3}\right]=\dfrac{-14}{(z+3)^3}$, therefore, at $z=-5$, $\dfrac{d^2}{dz^2}\left[\dfrac{z-4}{z+3}\right]$ becomes $\dfrac{-14}{(-5+3)^3}=\dfrac{7}{4}$. This implementation is also shown below:

```
>>syms z ↵
>>y=(z-4)/(z+3); ↵
>>d=diff(y,2); ↵
>>z=-5; ↵
>>sym(eval(d)) ↵   ← The output in rational form exercising 'sym'

ans =

7/4
```

5.5 Partial differentiation of functions

To analytically find out the partial derivative of a function of several variables, one variable is taken as variable and the rest are assumed to be constants. It is evident that the total number of partial derivatives is equal to the number of independent variables. Subroutine 'diff' is also applicable to compute the partial derivative of a multivariable function. According to calculus, partial derivative of f with respect to x is denoted by $\dfrac{\partial f}{\partial x}$. In MATLAB jargon, $\dfrac{\partial f}{\partial x}$ is equivalent to diff(f,'x'). Example of a two variables' function is

$U = f(x,y) = x^2 + y^3 - 7x^2 y^5$. Several partial derivatives of U are $\dfrac{\partial U}{\partial x} = 2x - 14xy^5$, $\dfrac{\partial U}{\partial y} = 3y^2 - 35x^2 y^4$,

$\dfrac{\partial^2 U}{\partial y \partial x} = -70xy^4$, $\dfrac{\partial^2 U}{\partial y^2} = 6y - 140x^2 y^3$, and $\dfrac{\partial^3 U}{\partial y^3} = 6 - 420x^2 y^2$. Implementations of all these are shown in the next page:

MATLAB Command for $\dfrac{\partial U}{\partial x}$,

```
>>U='x^2+y^3-7*x^2*y^5'; ↵
>>Ux=diff(U,'x') ↵

Ux =

2*x-14*x*y^5
```

for $\dfrac{\partial^2 U}{\partial y \partial x}$,

```
>>Uyx=diff(diff(U,'y'),'x') ↵

Uyx =

-70*x*y^4
```

for $\dfrac{\partial^2 U}{\partial y^2}$,

```
>>Uyy=diff(U,'y',2) ↵

Uyy =
```

for $\dfrac{\partial U}{\partial y}$,

```
>>Uy=diff(U,'y') ↵

Uy =

3*y^2-35*x^2*y^4
```

another way of finding $\dfrac{\partial^2 U}{\partial y \partial x}$,

```
>>Uy=diff(U,'y'); ↵
>>Uyx=diff(Uy,'x') ↵

Uyx =

-70*x*y^4
```

for $\dfrac{\partial^3 U}{\partial y^3}$,

```
>>Uyyy=diff(U,'y',3) ↵

Uyyy =
```

6*y-140*x^2*y^3   6-420*x^2*y^2

Referring to the above output, we assigned the different partial derivatives $\frac{\partial U}{\partial x}$, $\frac{\partial U}{\partial y}$, $\frac{\partial^2 U}{\partial y \partial x}$, $\frac{\partial^2 U}{\partial y^2}$, and $\frac{\partial^3 U}{\partial y^3}$ to Ux, Uy, Uyx, Uyy, and Uyyy respectively. For Ux, first type the capital U and then do small x that is how Ux is written. This is just to follow the convention of partial derivatives. One can write any variable he wants to. Another way of finding $\frac{\partial^2 U}{\partial y \partial x}$ can be in steps. For this, first find $\frac{\partial U}{\partial y}$, then, assign the result to some variable Uy, and after that, differentiate Uy w.r.to. x. This technique is useful if one has to find different types of partial derivatives from a single one. The example presented is a two-variable one, it can be a function of any number of variables like 3, 4, or 5. If one needs to see the symbolic form, command 'pretty' can be used. What if we need to evaluate $\frac{\partial^3 U}{\partial y^3}$ at $x=1$ and $y=3$. The answer is $\frac{\partial^3 U}{\partial y^3} = 6 - 420 x^2 y^2 = 6-420 \times 1 \times 9 = -3774$. Evaluation is shown below:

Command for the evaluation of $\frac{\partial^3 U}{\partial y^3}$ at $x=1$ and $y=3$,

>>x=1; y=3; ↵
>>eval(Uyyy) ↵

ans =
        -3774

5.6 Derivatives of the parametric equations

Equations like $x = \sin t \cos^2 t$ and $y = \cos^3 t$ are called the parametric equations. Both x and y are the functions of parameter t. Our objective is to find different derivatives such as $\frac{dy}{dx}$, $\frac{d^2 y}{dx^2}$, or $\frac{d^3 y}{dx^3}$ in terms of parameter t. Differentiate x and y to get $\frac{dx}{dt} = -2\sin^2 t \cos t + \cos^3 t$ and $\frac{dy}{dt} = -3\cos^2 t \sin t$ respectively, from which, $\frac{dy}{dx} = \frac{dy}{dt} \bigg/ \frac{dx}{dt} = \frac{-3\cos^2 t \sin t}{-2\sin^2 t \cos t + \cos^3 t} = \frac{-3\sin t \cos t}{-2\sin^2 t + \cos^2 t}$. Obtain $\frac{dy}{dx}$ as follows:

MATLAB Command

>>syms t ↵
>>x=sin(t)*cos(t)^2; ↵   ← Enter $x = \sin t \cos^2 t$
>>y=cos(t)^3; ↵   ← Enter $y = \cos^3 t$

>>d=diff(y,t)/diff(x,t); ↵   ← Apply $\frac{dy}{dx} = \frac{dy}{dt} \bigg/ \frac{dx}{dt}$ and assign

output to d
>>pretty(d) ↵

```
                             2
                  sin(t) cos(t)
       -3 -----------------------------------
                  3            2
              cos(t) - 2 sin(t)  cos(t)
```
It goes without saying that simplification is necessary:

>>pretty(simplify(d)) ↵

```
                  sin(t) cos(t)
       -3 ---------------------
                      2
              3 cos(t)  - 2
```

Notice that trigonometric identity $\sin^2 t = 1 - \cos^2 t$ is used in the denominator. For the second derivative, one

can write $\dfrac{d^2 y}{dx^2} = \dfrac{d}{dx}\left(\dfrac{dy}{dx}\right) = \dfrac{d}{dt}\left(\dfrac{dy}{dx}\right)\bigg/ \dfrac{dx}{dt} = \dfrac{d}{dt}\left[\dfrac{-3\sin t\cos t}{-2\sin^2 t + \cos^2 t}\right]\bigg/ (-2\sin^2 t\cos t + \cos^3 t) =$

$\dfrac{3(\cos^2 t - 2)}{9\cos^4 t - 12\cos^2 t + 4}\bigg/ (-2\sin^2 t\cos t + \cos^3 t) = \dfrac{3(\cos^2 t - 2)}{\cos t(27\cos^6 t - 54\cos^4 t + 36\cos^2 t - 8)}$. Implement it as follows:

MATLAB Command

```
>>syms t ↵
>>x=sin(t)*cos(t)^2; ↵
>>y=cos(t)^3; ↵
>>d=diff(y,t)/diff(x,t); ↵          ← Form dy/dx and assign that to d
>>d2=diff(d,t)/diff(x,t); ↵          ← Form d²y/dx² and assign it to d2
>>pretty(simplify(d2)) ↵
```

$$3\ \frac{\cos(t)^2 - 2}{\cos(t)\ (27\cos(t)^6 - 54\cos(t)^4 + 36\cos(t)^2 - 8\ )}$$

In a similar way, the third order derivative $\dfrac{d^3 y}{dx^3}$ can be computed from $\dfrac{d}{dt}\left(\dfrac{d^2 y}{dx^2}\right)\bigg/ \dfrac{dx}{dt}$. At some point,

evaluation of the derivative may be necessary for specific parameter. Suppose we want to find $\dfrac{d^2 y}{dx^2}$ at $t = \pi$.

Insert $t = \pi$ to $\dfrac{3(\cos^2 t - 2)}{\cos t(27\cos^6 t - 54\cos^4 t + 36\cos^2 t - 8)}$ to have $\dfrac{3(1-2)}{-(27 - 54 + 36 - 8)} = 3$. Evidently, the last assignee

d2 has the expression of $\dfrac{d^2 y}{dx^2}$ in terms of t . Substitute $t = \pi$ in d2 as follows:

```
>>t=pi; ↵
>>subs(d2) ↵

ans =
        3
```

5.7 Derivatives of the implicit functions

An equation can be a function of both the independent and dependent variables. This type of function is termed as the implicit function. One illustration of implicit function is $y^2 + 2xy = 0$, where $y = f(x)$. Independent and dependent variables of the implicit function are x and y respectively. Differentiate both

sides of the equation with respect to x to write $2y\dfrac{dy}{dx} + 2x\dfrac{dy}{dx} + 2y = 0$, from which, $\dfrac{dy}{dx} = -\dfrac{y}{x+y}$. The

resembling Maple function is 'implicitdiff'. As argument insert equation, dependent variable, and independent variable respectively. See the implementation below:

MATLAB Command

```
>>maple('e:=y^2+2*y*x=0'); ↵       ← Assign y² + 2xy = 0 to e
>>d=maple('implicitdiff(e,y,x)'); ↵  ← Assign Maple output to d
>>pretty(sym(d)) ↵                   ← Conversion of d to symbolic
```

$$- \frac{y}{y + x}$$

The second derivative $\frac{d^2 y}{dx^2}$ is also computed from $y^2 + 2xy = 0$. We obtained that $\frac{dy}{dx} = -\frac{y}{x+y}$, so, $\frac{d^2 y}{dx^2} =$

$$-\frac{(x+y)\frac{dy}{dx} - y\left(1 + \frac{dy}{dx}\right)}{(x+y)^2} = \frac{y^2 + 2xy}{x^3 + 3x^2 y + 3xy^2 + y^3}$$ (on insertion of $\frac{dy}{dx} = -\frac{y}{x+y}$). Its implementation is as follows:

MATLAB Command

```
>>maple('e:=y^2+2*y*x=0'); ↵        ← Assign y² + 2xy = 0 to e
>>d=maple('implicitdiff(e,y,x,x)'); ↵ ← Assign Maple output to d
>>pretty(sym(d)) ↵                   ← Conversion of d to symbolic
```

```
        y (y + 2 x)
  ---------------------------------
   3     2       2     3
   y + 3 y x + 3 y x + x
```

Observe that 'implicitdiff(e,y,x)' is equivalent to $\frac{dy}{dx}$ and that 'implicitdiff(e,y,x, x)' and 'implicitdiff(e,y,x,x,x)'

are to $\frac{d^2 y}{dx^2}$ and $\frac{d^3 y}{dx^3}$ respectively, where e refers to the equation pertaining to x and y.

What if we wish to find the partial derivatives from a set of equations. The best thing is start with two

variables' example. Two equations are given by $f: x^2 - e^z = y$ and $g: y^3 - 2xz = 5$. We intend to find $\frac{\partial y}{\partial x}$ and

$\frac{\partial z}{\partial x}$, where y and z are functions of x (which is the independent variable). Of coarse, f and g are the

functions of dependent variables y and z. Partial differentiation on both sides of f and g provides

$2x - e^z \frac{\partial z}{\partial x} = \frac{\partial y}{\partial x}$ and $3y^2 \frac{\partial y}{\partial x} - 2x \frac{\partial z}{\partial x} - 2z = 0$ respectively. From the last two equations, solve for $\frac{\partial y}{\partial x}$ and $\frac{\partial z}{\partial x}$ to

write $\frac{\partial y}{\partial x} = 2 \frac{2x^2 + ze^z}{3y^2 e^z + 2x}$ and $\frac{\partial z}{\partial x} = 2 \frac{3y^2 x - z}{3y^2 e^z + 2x}$ respectively. Following commands can have $\frac{\partial y}{\partial x}$ and $\frac{\partial z}{\partial x}$ very

simply:

MATLAB Command

```
>>maple('f:=x^2-exp(z)=y'); ↵         ← Assign x² - eᶻ = y to f
>>maple('g:=y^3-2*x*z=5'); ↵          ← Assign y³ - 2xz = 5 to g
>>maple('o:=implicitdiff({f,g},{y,z},{y,z},x)'); ↵  ← Apply the subroutine and assign output to o
>>maple('interface(prettyprint=1):o'); ↵            ← Symbolic display of contents of o
```

```
                  2                          2
               3 y x - z                   2 x + exp(z) z
   {D(z) = 2 ----------------,   D(y) = 2 ------------------}
                  2                          2
               3 y exp(z) + 2 x            3 y exp(z) + 2 x
```

It is evident that $D(z) \equiv \frac{\partial z}{\partial x}$ and $D(y) \equiv \frac{\partial y}{\partial x}$. The function 'interface' communicates between Maple and user

interface (which is called Iris). Have online Maple help about the function executing 'mhelp interface' in the

Command Window of MATLAB. The function can take many arguments such as 'echo', 'errorbreak', 'labelling', 'screenwidth'...etc. One of them is 'prettyprint'. It is similar to the command 'pretty' of symbolic toolbox. The value prettyprint=1 tells the interface to display output in accordance with two dimensional character based formatter. For simplicity, we assigned output of 'implicitdiff' to variable o, which is preceded by interface.

5.8 Taylor series expansion of a function

Taylor series approximates a function by a polynomial utilizing the derivatives of the function. If any function $f(x)$ and its derivatives are differentiable, then Taylor series expansion of the function is given by

$$f(x) = f(0) + xf'(0) + \frac{x^2}{2!}f''(0) + \frac{x^3}{3!}f'''(0) + \frac{x^4}{4!}f^{iv}(0) + \frac{x^5}{5!}f^{v}(0) + \cdots + \frac{x^n}{n!}f^{n}(0) + \cdots to \infty \qquad ...(5-1).$$

In equation (5-1) $f(0)$, $f'(0)$, and $f''(0)$ mean the values of $f(x)$, $\frac{d}{dx}[f(x)]$, $\frac{d^2}{dx^2}[f(x)]$ at $x=0$ respectively, and so on. The higher is the number of terms of the polynomial, the better is the approximation of the function $f(x)$. To obtain Taylor series approximation of a function,

MATLAB Step:
Use the command taylor('function as string').

Let us take the example of $\sin x$, where $f(x) = \sin x$, $f'(x) = \cos x$, $f''(x) = -\sin x$, $f'''(x) = -\cos x$, $f^{iv}(x) = \sin x$, $f^{v}(x) = \cos x$, $\cdots$ etc and $f(0) = 0$, $f'(0) = 1$, $f''(0) = 0$, $f'''(0) = -1$, $f^{iv}(0) = 0$, $f^{v}(0) = 1$, ... etc, therefore, $\sin x = x - \frac{x^3}{3!} + \frac{x^5}{5!} - \cdots = x - \frac{x^3}{6} + \frac{x^5}{120} - \cdots$. Implementation of Taylor series of $\sin x$ is shown below:

MATLAB Command

>>syms x ⏎   ← State variable of $\sin x$ as symbolic
>>S=taylor(sin(x)) ⏎   ← Taylor series is assigned to S

S =

x-1/6*x^3+1/120*x^5

Display the contents of S in symbolic form using the command 'pretty':
>>pretty(S) ⏎

$$x - 1/6\ x^3 + 1/120\ x^5$$

☞ *Taylor series of some standard functions*

Taylor series of some known functions is furnished. Verify MATLAB validity of the series.
Function: e^{-x}

Taylor series: $e^{-x} = 1 - x + \frac{x^2}{2!} - \frac{x^3}{3!} + \frac{x^4}{4!} - \frac{x^5}{5!} + \cdots$

$$\Rightarrow 1 - x + \frac{x^2}{2} - \frac{x^3}{6} + \frac{x^4}{24} - \frac{x^5}{120} + \cdots$$

MATLAB Command

>>syms x ⏎
>>S=taylor(exp(-x)); ⏎
>>pretty(S) ⏎

$$1 - x + 1/2\ x^2 - 1/6\ x^3 + 1/24\ x^4 - 1/120\ x^5$$

Function: $\ln(1-x) = \log_e(1-x)$

Taylor series: $-x - \dfrac{x^2}{2} - \dfrac{x^3}{3} - \dfrac{x^4}{4} - \dfrac{x^5}{5} - \dfrac{x^6}{6} - \cdots$

MATLAB Command

```
>>syms x ↵
>>S=taylor(log(1-x)); ↵
>>pretty(S) ↵
                    2        3        4        5
        -x - 1/2 x  - 1/3 x  - 1/4 x  - 1/5 x
```

Function: $\ln x = \log_e x$

Taylor series: Does not exist since $\ln x$ at $x = 0$ is not defined. Error message is printed by MATLAB.

MATLAB Command

```
>>syms x ↵
>>S=taylor(log(x)) ↵
??? Error using ==> sym/taylor
Error, does not have a taylor expansion, try series()
```

⌗ *Taylor series of a composite function*

Not only the subroutine expands standard functions to series but also it looks for the approximation of composite functions. Such an example is $\sin x \cos x$. Its Taylor series is also obtained from the equation (5–1), which is given by $x - \dfrac{2}{3}x^3 + \dfrac{2}{15}x^5 - \dfrac{4}{315}x^7 \ldots$. Accomplishment is shown below:

MATLAB Command

```
>>syms x ↵
>>S=taylor(sin(x)*cos(x)); ↵
>>pretty(S) ↵
                   3        5
        x - 2/3 x  + 2/15 x
```

⌗ *Taylor series for the required order*

Subroutine 'taylor' attempts to expand argument function *up to the fifth order that is the default order.* A situation when the order more than 5 is necessary can occur in higher order approximation of a function. To expand a function up to certain order, say, n,

MATLAB Step:

    Use the command taylor('function as string', $n + 1$).

Notice that the required order is exclusive, i.e., degree of the polynomial is one less than the order argument of 'taylor'. Consider the example of $\sin x$, up to the 8th order, expansion of $\sin x$ is $x - \dfrac{x^3}{3!} + \dfrac{x^5}{5!} - \dfrac{x^7}{7!} + \cdots = x - \dfrac{x^3}{6} + \dfrac{x^5}{120} - \dfrac{x^7}{5040} + \cdots$. It is shown as follows:

MATLAB Command

```
>>syms x ↵
>>S=taylor(sin(x),9); ↵
>>pretty(S) ↵
                 3         5          7
        x - 1/6 x  + 1/120 x  - 1/5040 x
```

⌗ *Taylor series about a certain point*

Taylor series expansion of a function $f(x)$ about $x = a$ is given by

$$f(x) = f(a) + (x-a)f'(a) + \frac{(x-a)^2}{2!}f''(a) + \frac{(x-a)^3}{3!}f'''(a) + \cdots + \frac{(x-a)^n}{n!}f^n(a) + \cdots to \ \infty. \qquad \ldots(5\text{--}2).$$

To obtain an approximation of $f(x)$ about $x = a$ up to the order n,

MATLAB Step:
    Use the command taylor('function as string', n +1,a).

Say, the expansion of $f(x) = \sin x$ about $x = \frac{\pi}{2}$ up to the 4th order is to be found. Utilize different derivatives of

$\sin x$ to write $f'\left(\frac{\pi}{2}\right) = \cos\frac{\pi}{2} = 0$, $\ f''\left(\frac{\pi}{2}\right) = -\sin\left(\frac{\pi}{2}\right) = -1$, $\ f'''\left(\frac{\pi}{2}\right) = -\cos\left(\frac{\pi}{2}\right) = 0$, and $\ f^{iv}\left(\frac{\pi}{2}\right) = \sin\frac{\pi}{2} = 1$, hence,

from equation (5–2), $\sin x \big|_{at\ x=\frac{\pi}{2}} = 1 - \frac{(x-\frac{\pi}{2})^2}{2} + \frac{(x-\frac{\pi}{2})^4}{24} - \ldots$. How if $\sin x$ is to be expanded about a variable u,

evidently, up to the second order, $\sin x \big|_{at\ x=u} = \sin u + (x-u)\cos u - \frac{1}{2}\sin u(x-u)^2$. Carry out the expansion as follows:

MATLAB Command
```
>>syms x ↵
>>S=taylor(sin(x),5,pi/2); ↵
>>pretty(S) ↵
```
$$\qquad\qquad 1 - 1/2\ (x - 1/2\ pi)^2 + 1/24\ (x - 1/2\ pi)^4$$

To expand about u,
```
>>syms x u ↵
>>S=taylor(sin(x),3,u); ↵
>>pretty(S) ↵
```
$$\sin(u) + \cos(u)\ (x - u) - 1/2\ \sin(u)\ (x - u)^2$$

Computation of the series after expansion

  Assume that $\sin x \big|_{at\ x=\frac{\pi}{2}}$ up to the 4th order is to be evaluated at $x = \frac{3\pi}{4}$. Use the last expansion to get

computation as $1 - \dfrac{\left(\dfrac{3\pi}{4} - \dfrac{\pi}{2}\right)^2}{2} + \dfrac{\left(\dfrac{3\pi}{4} - \dfrac{\pi}{2}\right)^4}{24} = 1 - \dfrac{\pi^2}{32} + \dfrac{\pi^4}{6144}$. Have the computation as follows:

MATLAB Command
```
>>syms x pi ↵
>>S=taylor(sin(x),5,pi/2); ↵
>>x=3*pi/4; ↵
>>pretty(eval(S)) ↵  ← Computation by subroutine 'eval'
```
$$1 - 1/32\ pi^2 + 1/6144\ pi^4$$

Decimal form is seen by virtue of the command 'double':
```
>>double(eval(S)) ↵

ans =
        0.7074
```

Taylor series of a multivariable function

  Shown illustrations so far deal only with the functions of one variable. Another nice feature of 'taylor'
is that it is well adapted to the case when a function is described by two or more independent variables.

General expansion of Taylor series of a multivariable function $f = f(x_1, x_2, x_3, \ldots, x_m)$ at $x_m = a$ up to the nth

order is given by $f(x_1, x_2, x_3, \ldots, x_m) \Big|_{at\ x_m = a} = f(x_1, x_2, x_3, \ldots, x_m = a) + \frac{(x_m - a)}{1!} \frac{\partial f}{\partial x_m} \Big|_{x_m = a} + \frac{(x_m - a)^2}{2!} \frac{\partial^2 f}{\partial x^2_m} \Big|_{x_m = a} +$

$\frac{(x_m - a)^3}{3!} \frac{\partial^3 f}{\partial x^3_m} \Big|_{x_m = a} + \ldots + \frac{(x_m - a)^n}{n!} \frac{\partial^n f}{\partial x^n_m} \Big|_{x_m = a}$. To obtain the Taylor series of a multivariable function up to the

order n about some point a,

MATLAB Steps:
    1. Declare the independent variables as symbolic and
    2. Use the command taylor('function as string', independent variable for variation,
    n+1, a).

We illustrate the implementation on $f(x, y, z, v) = e^{2x^2 + 3y^2 - 7z + 8v}$ — say, $f(x, y, z, v)$ is to be expanded about $x = 2$

up to the *3rd* order, obviously, $x_1 = x$, $x_2 = y$, $x_3 = z$, and $x_4 = v$. The number of independent variables, m, is 4.

We have $\frac{\partial f}{\partial x} = 4x\ e^{2x^2 + 3y^2 - 7z + 8v}$, $\frac{\partial^2 f}{\partial x^2} = (4 + 16x^2)\ e^{2x^2 + 3y^2 - 7z + 8v}$, $\frac{\partial^3 f}{\partial x^3} = (48x + 64x^3)\ e^{2x^2 + 3y^2 - 7z + 8v}$, therefore, $f \Big|_{at\ x = 2}$

$= e^{3y^2 - 7z + 8v + 8}$, $\frac{\partial f}{\partial x} \Big|_{at\ x = 2} = 8\ e^{3y^2 - 7z + 8v + 8}$, $\frac{\partial^2 f}{\partial x^2} \Big|_{at\ x = 2} = 68\ e^{3y^2 - 7z + 8v + 8}$, and $\frac{\partial^3 f}{\partial x^3} \Big|_{at\ x = 2} = 608\ e^{3y^2 - 7z + 8v + 8}$. From the general

series, the expansion is $e^{2x^2 + 3y^2 - 7z + 8v} \Big|_{at\ x = 2} = e^{3y^2 - 7z + 8v + 8} + 8(x - 2)e^{3y^2 - 7z + 8v + 8} + 34(x - 2)^2 e^{3y^2 - 7z + 8v + 8} + \frac{304}{3}(x -$

$2)^3 e^{3y^2 - 7z + 8v + 8}$. The string representation of $e^{2x^2 + 3y^2 - 7z + 8v}$ is 'exp(2*x^2+3*y^2-7*z+8*v)'. The execution is as

follows:

MATLAB Command
```
>>syms x y z v ↵
>>S=taylor(exp(2*x^2+3*y^2-7*z+8*v),x,4,2); ↵
>>pretty(S) ↵
```

$$\%1 + 8\ \%1\ (x - 2) + 34\ \%1\ (x - 2)^2 + 304/3\ \%1\ (x - 2)^3$$

$$\%1 := \exp(8 + 8\ v + 3\ y^2 - 7\ z)$$

The last line indicates that the expression $\exp(8 + 8\ v + 3\ y^2 - 7\ z)$ is equivalent to %1. The exponent $e^{3y^2 - 7z + 8v + 8}$
will be substituted where %1 is found. This type of shortcut is seen for the repetitive long expressions.

5.9 Jacobian matrix of some function

Suppose a vector function F is formed by m functions and given by $F = \begin{bmatrix} f_1 \\ f_2 \\ f_3 \\ \vdots \\ f_m \end{bmatrix}$. Any row function of

F is denoted by f_i, where i can be from 1 to m. Assume that f_i is a function of n variables, where the
variables are x_1, x_2, x_3, ... x_n. A matrix of partial derivatives can be formed as

$$\begin{bmatrix} \dfrac{\partial f_1}{\partial x_1} & \dfrac{\partial f_1}{\partial x_2} & \dfrac{\partial f_1}{\partial x_3} & \cdots & \dfrac{\partial f_1}{\partial x_n} \\[2mm] \dfrac{\partial f_2}{\partial x_1} & \dfrac{\partial f_2}{\partial x_2} & \dfrac{\partial f_2}{\partial x_3} & \cdots & \dfrac{\partial f_2}{\partial x_n} \\[2mm] \dfrac{\partial f_3}{\partial x_1} & \dfrac{\partial f_3}{\partial x_2} & \dfrac{\partial f_3}{\partial x_3} & \cdots & \dfrac{\partial f_3}{\partial x_n} \\[2mm] & & \vdots & & \\[2mm] \dfrac{\partial f_m}{\partial x_1} & \dfrac{\partial f_m}{\partial x_2} & \dfrac{\partial f_m}{\partial x_3} & \cdots & \dfrac{\partial f_m}{\partial x_n} \end{bmatrix}$$. The matrix of partial derivatives is known as the jacobian matrix of F and

mathematically denoted by $J_F(X)$, where X is the vector formed by $x_1, x_2, x_3, \ldots x_n$. Sometimes jacobian

is called $F : R^n \to R^m$ mapping. Let the mapping $F : R^2 \to R^3$ be defined by $F(x,y) = \begin{bmatrix} x^3 + y^2 x \\ e^{xy} \\ \sin(2x - 3y) \end{bmatrix}$. Find

$J_F(x,y)$. Each row of F is a function of 2 variables, $\therefore n = 2$. There are 3 rows in F, $\therefore m = 3$. The order of

the jacobian matrix $J_F(x,y)$ will be $m \times n$ or 3×2. We have $f_1 = x^3 + y^2 x$, $f_2 = e^{xy}$, $f_3 = \sin(2x - 3y)$, $x_1 = x$,

and $x_2 = y$ and $J_F(x,y) = \begin{bmatrix} \dfrac{\partial f_1}{\partial x_1} & \dfrac{\partial f_1}{\partial x_2} \\[2mm] \dfrac{\partial f_2}{\partial x_1} & \dfrac{\partial f_2}{\partial x_2} \\[2mm] \dfrac{\partial f_3}{\partial x_1} & \dfrac{\partial f_3}{\partial x_2} \end{bmatrix}$, where $\dfrac{\partial f_1}{\partial x_1} = 3x^2 + y^2$, $\dfrac{\partial f_1}{\partial x_2} = 2xy$, $\dfrac{\partial f_2}{\partial x_1} = ye^{xy}$, $\dfrac{\partial f_2}{\partial x_2} = xe^{xy}$,

$\dfrac{\partial f_3}{\partial x_1} = 2\cos(2x - 3y)$, and $\dfrac{\partial f_3}{\partial x_2} = -3\cos(2x - 3y)$, on that, $J_F(x,y) = \begin{bmatrix} 3x^2 + y^2 & 2xy \\ ye^{xy} & xe^{xy} \\ 2\cos(2x - 3y) & -3\cos(2x - 3y) \end{bmatrix}$. To obtain

the jacobian of a vector function,

MATLAB Steps:

1. *Declare the independent variables as symbolic,*
2. *In string form, assign the first row of the vector function to some variable f1, the second row to some variable f2, ...and so do other rows, and*
3. *Use the command jacobian ([form the vector function matrix from f1, f2, f3,....], independent variables in the vector function).*

For the example at hand,
MATLAB Command

```
>>syms x y ↵
>>f1=x^3+y^2*x; ↵                ← Assign x³ + y²x to f1
>>f2=exp(x*y); ↵                 ← Assign eˣʸ to f2
>>f3=sin(2*x-3*y); ↵             ← Assign sin(2x − 3y) to f3
>>J=jacobian([f1;f2;f3],[x y]); ↵  ← Assign jacobian to J
>>pretty(J) ↵
```

```
[       2    2                        ]
[    3 x  + y            2 x y         ]
[                                     ]
[   y exp(x y)         x exp(x y)     ]
[                                     ]
[2 cos(2 x - 3 y)     -3 cos(2 x - 3 y)]
```

At some point, say, we need to compute the jacobian for $x = \frac{\pi}{2}$ and $y = 0$, i.e., $J_F\left(\frac{\pi}{2}, 0\right) = ?$ Output should be

$$\begin{bmatrix} \frac{3\pi^2}{4} & 0 \\ 0 & \frac{\pi}{2} \\ -2 & 3 \end{bmatrix} = \begin{bmatrix} 7.4022 & 0 \\ 0 & 1.5708 \\ -2 & 3 \end{bmatrix}.$$ It is obvious that assignee J of the last command contains the jacobian.

Perform the substitution as follows:

>>x=sym(pi/2); ↵   ← Define $\frac{\pi}{2}$ as symbolic

>>y=sym(0); ↵
>>subs(J) ↵   ← Substitution by the subroutine 'subs'

ans =

[3/4*pi^2,   0]
[0,   1/2*pi]
[-2,   3]
>>double(subs(J)) ↵   ← Display the result in decimal form

ans =

    7.4022   0
        0   1.5708
   -2.0000   3.0000

Take one more example of F, where $F(r, \theta, \varphi) = \begin{bmatrix} r^2 \sin\theta \\ \cos(\theta + \varphi) \end{bmatrix}$. It follows then, $f_1 = r^2 \sin\theta$, $f_2 = \cos(\theta + \varphi)$,

$x_1 = r$, $x_2 = \theta$, $x_3 = \varphi$, and $J_F(r, \theta, \varphi) = \begin{bmatrix} \frac{\partial f_1}{\partial x_1} & \frac{\partial f_1}{\partial x_2} & \frac{\partial f_1}{\partial x_3} \\ \frac{\partial f_2}{\partial x_1} & \frac{\partial f_2}{\partial x_2} & \frac{\partial f_2}{\partial x_3} \end{bmatrix} = \begin{bmatrix} 2r\sin\theta & r^2\cos\theta & 0 \\ 0 & -\sin(\theta + \varphi) & -\sin(\theta + \varphi) \end{bmatrix}$. Letters t

and p are used instead of θ and φ. See the jacobian of the last example as follows:

MATLAB Command
>>syms r t p ↵
>>f1=r^2*sin(t); ↵
>>f2=cos(t+p); ↵
>>J=jacobian([f1;f2],[r t p]); ↵
>>pretty(J) ↵

```
[                  2                    ]
[   2 r sin(t)    r  cos(t)      0       ]
[                                        ]
[      0        -sin(t + p)   -sin(t + p) ]
```

5.10 Hessian matrix of a function

Let f be a function of n variables denoted by $x_1, x_2, x_3, \ldots x_n$ and that f is twice continuously

differentiable, then Hessian of f, $H_f(X)$, is the matrix $H_f(X) = \begin{bmatrix} \frac{\partial^2 f}{\partial x_i \partial x_j} \end{bmatrix}$, where $X = [x_1, x_2, x_3, \ldots x_n]$ and

i, j vary from 1 to n. Hessian is associated with the quadratic form of a matrix. We illustrate the formation of Hessian matrix considering the function $f = x^3 y + z^4$. We have three variables – $x_1 = x$, $x_2 = y$, and $x_3 = z$.

Hessian matrix is given by $H_f(x, y, z) = \begin{bmatrix} \dfrac{\partial^2 f}{\partial x^2} & \dfrac{\partial^2 f}{\partial x \partial y} & \dfrac{\partial^2 f}{\partial x \partial z} \\ \dfrac{\partial^2 f}{\partial y \partial x} & \dfrac{\partial^2 f}{\partial y^2} & \dfrac{\partial^2 f}{\partial y \partial z} \\ \dfrac{\partial^2 f}{\partial z \partial x} & \dfrac{\partial^2 f}{\partial z \partial y} & \dfrac{\partial^2 f}{\partial z^2} \end{bmatrix}$. Different second order partial derivatives are

$\dfrac{\partial^2 f}{\partial x^2} = 6xy$, $\dfrac{\partial^2 f}{\partial y^2} = 0$, $\dfrac{\partial^2 f}{\partial z^2} = 12z^2$, $\dfrac{\partial^2 f}{\partial x \partial y} = 3x^2$, $\dfrac{\partial^2 f}{\partial x \partial z} = 0$, $\dfrac{\partial^2 f}{\partial y \partial x} = 3x^2$, $\dfrac{\partial^2 f}{\partial y \partial z} = 0$, $\dfrac{\partial^2 f}{\partial z \partial x} = 0$, and $\dfrac{\partial^2 f}{\partial z \partial y} = 0$. Put them

together to form the Hessian matrix as $H_f(x, y, z) = \begin{bmatrix} 6xy & 3x^2 & 0 \\ 3x^2 & 0 & 0 \\ 0 & 0 & 12z^2 \end{bmatrix}$. Conduct the command 'hessian' for this:

MATLAB Command

>>maple('f:=x^3*y+z^4'); ↵   ← Assign $x^3 y + z^4$ to f

>>H=maple('hessian(f,[x,y,z])'); ↵ ← Apply 'hessian' on f and assign output to H

>>pretty(sym(H)) ↵   ← Convert contents of H to symbolic

```
        [              2          ]
        [6 x y      3 x        0   ]
        [                          ]
        [     2                    ]
        [3 x        0          0   ]
        [                          ]
        [                        2]
        [ 0         0        12 z  ]
```

5.11 Gradient, divergence, and curl of different fields

Very often problems related to the gradient of scalar field and the divergence and curl of vector fields are seen in vector calculus. Linear algebra package has the facility to implement these three operations. Gradient, divergence, and curl expressions are dependent on coordinate system. Geometry of three basic coordinate systems, rectangular or Cartesian, cylindrical, and spherical, are presented in figure 10.11. Any point in space is described by (x, y, z), (ρ, φ, z), and (r, θ, φ) for rectangular, cylindrical, and spherical coordinate systems respectively.

5.11.1 Gradient of a scalar field

Gradient of any scalar field S is denoted by ∇S. Field S is a scalar on the contrary ∇S is a vector. Operator ∇ (nabla) is different for different coordinate system. Apply Maple procedure 'grad' to find the gradient of a scalar field. For each example, we assign the output of 'grad' to 'g'. Return of 'grad' is a three-element row matrix, which corresponds consecutive components of the output vector. Command 'sym' is used to convert Maple output to symbolic form. Expressions for the gradients in different coordinate systems are presented below:

In Cartesian system: $\nabla S = \dfrac{\partial S}{\partial x}\bar{a}_x + \dfrac{\partial S}{\partial y}\bar{a}_y + \dfrac{\partial S}{\partial z}\bar{a}_z$

In cylindrical system: $\nabla S = \dfrac{\partial S}{\partial \rho}\bar{a}_\rho + \dfrac{1}{\rho}\dfrac{\partial S}{\partial \varphi}\bar{a}_\varphi + \dfrac{\partial S}{\partial z}\bar{a}_z$

In spherical system: $\nabla S = \dfrac{\partial S}{\partial r}\bar{a}_r + \dfrac{1}{r}\dfrac{\partial S}{\partial \theta}\bar{a}_\theta + \dfrac{1}{r\sin\theta}\dfrac{\partial S}{\partial \varphi}\bar{a}_\varphi$

Examples of finding the gradient:

⊞ In rectangular system

Suppose $S = (x^2 + 2y)z^2$, the gradient is $\nabla S = \frac{\partial S}{\partial x}\bar{a}_x + \frac{\partial S}{\partial y}\bar{a}_y + \frac{\partial S}{\partial z}\bar{a}_z$. Partial derivatives are $\frac{\partial S}{\partial x} = 2xz^2$, $\frac{\partial S}{\partial y} = 2z^2$, and $\frac{\partial S}{\partial z} = 2z(x^2 + 2y)$. Summing all, the gradient ∇S is equal to $2xz^2\bar{a}_x + 2z^2\bar{a}_y + 2z(x^2 + 2y)\bar{a}_z$. The string form of $(x^2 + 2y)z^2$ is (x^2+2*y)*z^2. Entering order of the variables is x, y, and z respectively. MATLAB implementation is as follows:

MATLAB Command

>>maple('S:=(x^2+2*y)*z^2'); ↵   ← Assign $(x^2 + 2y)z^2$ to S
>>g=maple('grad(S,[x,y,z])'); ↵   ← Apply the procedure 'grad' on S
>>pretty(sym(g)) ↵

```
        [   2       2         2      ]
        [2 x z    2 z     2 (x + 2 y) z ]
```

⊞ In cylindrical system

The scalar field is $A = (\rho + 2)^2 z\cos 3\varphi$. From gradient expression, we have $\nabla A = \frac{\partial A}{\partial \rho}\bar{a}_\rho + \frac{1}{\rho}\frac{\partial A}{\partial \varphi}\bar{a}_\varphi + \frac{\partial A}{\partial z}\bar{a}_z$, where partial derivatives are $\frac{\partial A}{\partial \rho} = 2(\rho + 2)z\cos 3\varphi$, $\frac{\partial A}{\partial \varphi} = -3(\rho + 2)^2 z\sin 3\varphi$, and $\frac{\partial A}{\partial z} = (\rho + 2)^2\cos 3\varphi$. Substitute all partial derivatives to write $\nabla A = 2(\rho + 2)z\cos 3\varphi\bar{a}_\rho - \frac{3(\rho + 2)^2 z\sin 3\varphi}{\rho}\bar{a}_\varphi + (\rho + 2)^2\cos 3\varphi\bar{a}_z$. Use r and p instead of ρ and φ respectively. String for the scalar function A is (r+2)^2*z*cos(3*p). The solution is shown below:

MATLAB Command

>>maple('A:=(r+2)^2*z*cos(3*p)'); ↵
>>g=maple('grad(A,[r,p,z], coords=cylindrical)'); ↵
>>pretty(sym(g)) ↵

```
   [                                    2                              ]
   [                           (r + 2)  z sin(3 p)            2        ]
   [2 (r + 2) z cos(3 p)   -3 ------------------------   (r + 2) cos(3 p) ]
   [                                    r                              ]
```

⊞ In spherical system

The field is $R = 3r^2\sin\varphi\cos\theta$. Applying expression $\nabla R = \frac{\partial R}{\partial r}\bar{a}_r + \frac{1}{r}\frac{\partial R}{\partial \theta}\bar{a}_\theta + \frac{1}{r\sin\theta}\frac{\partial R}{\partial \varphi}\bar{a}_\varphi$ yields $\nabla R = 6r\sin\varphi\cos\theta\bar{a}_r - 3r\sin\varphi\sin\theta\bar{a}_\theta + \frac{3r\cos\varphi\cos\theta}{\sin\theta}\bar{a}_\varphi$. The string of R is 3*r^2*sin(p)*cos(t). Unavailability of the symbols θ and φ in the command window makes us use t and p instead of them. Implement it as follows:

MATLAB Command

>>maple('R:=3*r^2*sin(p)*cos(t)'); ↵
>>g=maple('grad(R,[r,t,p],coords=spherical)'); ↵
>>pretty(sym(g)) ↵

```
   [                                      r cos(p) cos(t)  ]
   [6 r sin(p) cos(t)   -3 r sin(p) sin(t)   3 ----------------]
   [                                          sin(t)       ]
```

5.11.2 Divergence of a vector field

Divergence of a vector field $\overline{A}$ is symbolized by $\nabla \bullet \overline{A}$. $\overline{A}$ is a vector function but $\nabla \bullet \overline{A}$ is a scalar. The vector $\overline{A}$ is entered as three-element row matrix. Different components of the vector field are put in order, for example, $A_r \to A_\theta \to A_\varphi$ for spherical system. A string represents each component. The resembling function of Maple is 'diverge'. Output of 'diverge' is assigned to d for convenience. Divergence expressions for different coordinate systems are as follows:

In Cartesian system: $\nabla \bullet \overline{A} = \dfrac{\partial A_x}{\partial x} + \dfrac{\partial A_y}{\partial y} + \dfrac{\partial A_z}{\partial z}$

In cylindrical system: $\nabla \bullet \overline{A} = \dfrac{1}{\rho} \dfrac{\partial(\rho A_\rho)}{\partial \rho} + \dfrac{1}{\rho} \dfrac{\partial A_\varphi}{\partial \varphi} + \dfrac{\partial A_z}{\partial z}$

In spherical system: $\nabla \bullet \overline{A} = \dfrac{1}{r^2} \dfrac{\partial(r^2 A_r)}{\partial r} + \dfrac{1}{r \sin\theta} \dfrac{\partial(A_\theta \sin\theta)}{\partial \theta} + \dfrac{1}{r \sin\theta} \dfrac{\partial A_\varphi}{\partial \varphi}$

⊟ In rectangular system

Example vector is $\overline{A} = \ln(x+y)\overline{a}_x + \cos y\, \overline{a}_y + e^{7z}\, \overline{a}_z$. Divergence of $\overline{A}$ is simple and straightforward, $\nabla \bullet \overline{A} = \dfrac{\partial A_x}{\partial x} + \dfrac{\partial A_y}{\partial y} + \dfrac{\partial A_z}{\partial z} = \dfrac{\partial[\ln(x+y)]}{\partial x} + \dfrac{\partial(\cos y)}{\partial y} + \dfrac{\partial(e^{7z})}{\partial z} = \dfrac{1}{x+y} - \sin y + 7e^{7z}$. The string for $\overline{A}$ is [log(x+y),cos(y),exp(7*z)]. Find $\nabla \bullet \overline{A}$ as follows:

MATLAB Command

```
>>maple('A:=[log(x+y),cos(y),exp(7*z)]'); ↵
>>d=maple('diverge(A,[x,y,z])'); ↵
>>pretty(sym(d)) ↵
              1
          ---------- -  sin(y)  +  7 exp(7 z)
            x + y
```

⊟ In cylindrical system

Given vector is $\overline{A} = \rho^2 z\overline{a}_\rho + \sin\varphi\, \overline{a}_\varphi + \rho z \cos\varphi\, \overline{a}_z$. Necessary expression is $\nabla \bullet \overline{A} = \dfrac{1}{\rho} \dfrac{\partial(\rho A_\rho)}{\partial \rho} + \dfrac{1}{\rho} \dfrac{\partial A_\varphi}{\partial \varphi} + \dfrac{\partial A_z}{\partial z}$, hence, $\dfrac{\partial(\rho A_\rho)}{\partial \rho} = \dfrac{\partial(\rho\rho^2 z)}{\partial \rho} = 3\rho^2 z$, $\dfrac{\partial A_\varphi}{\partial \varphi} = \dfrac{\partial(\sin\varphi)}{\partial \varphi} = \cos\varphi$, and $\dfrac{\partial A_z}{\partial z} = \dfrac{\partial(\rho z \cos\varphi)}{\partial z} = \rho\cos\varphi$. Collecting all, $\nabla \bullet \overline{A} = 3\rho z + \dfrac{\cos\varphi}{\rho} + \rho\cos\varphi = \dfrac{3\rho^2 z + \cos\varphi + \rho^2 \cos\varphi}{\rho}$. Different components are entered in order as $A_\rho \to A_\varphi \to A_z$ (r for ρ and p for φ). In string, vector $\overline{A}$ is written as [r^2*z,sin(p),r*z*cos(p)].

MATLAB Command

```
>>maple('A:=[r^2*z,sin(p),r*z*cos(p)]'); ↵
>>d=maple('diverge(A,[r,p,z],coords=cylindrical)'); ↵
>>pretty(sym(d)) ↵
            2                 2
         3 r  z + cos(p) + r  cos(p)
        -------------------------------
                      r
```

⊟ In spherical system

Chosen vector for this coordinate system is $\overline{A} = r^2 \overline{a}_r + \sin\theta\cos\varphi\, \overline{a}_\theta + r\cos\varphi\, \overline{a}_\varphi$. Use the general expression to write $\nabla \bullet \overline{A} = \dfrac{1}{r^2} \dfrac{\partial(r^2 A_r)}{\partial r} + \dfrac{1}{r \sin\theta} \dfrac{\partial(A_\theta \sin\theta)}{\partial \theta} + \dfrac{1}{r \sin\theta} \dfrac{\partial A_\varphi}{\partial \varphi} = \dfrac{1}{r^2} \dfrac{\partial(r^2 r^2)}{\partial r} + \dfrac{1}{r \sin\theta} \dfrac{\partial(\sin^2\theta\cos\varphi)}{\partial \theta} +$

$$\frac{1}{r\sin\theta}\frac{\partial(r\cos\varphi)}{\partial\varphi}=\frac{4r^2\sin\theta+2\cos\theta\sin\theta\cos\varphi-r\sin\varphi}{r\sin\theta}$$. As symbolic variables, t and p are used instead of θ and φ.

MATLAB Command

```
>>maple('A:=[r^2,sin(t)*cos(p),r*cos(p)]');  ↵
>>d=maple('diverge(A,[r,t,p],coords=spherical)');  ↵
>>pretty(sym(d))  ↵
```

```
                 2
        4 r  sin(t) + 2 cos(p) cos(t) sin(t) - r sin(p)
        -------------------------------------------------
                          r sin(t)
```

5.11.3 Curl of a vector field

Symbolic notation for the curl of a vector field $\overline{A}$ is $\nabla\times\overline{A}$. $\overline{A}$ is a vector, so is $\nabla\times\overline{A}$. The name of MATLAB counterpart is also 'curl'. Style of implementation of curl is similar to that of divergence. Different curl expressions are presented below:

In Cartesian system: $\nabla\times\overline{A}=\begin{vmatrix}\overline{a}_x & \overline{a}_y & \overline{a}_z \\ \dfrac{\partial}{\partial x} & \dfrac{\partial}{\partial y} & \dfrac{\partial}{\partial z} \\ A_x & A_y & A_z\end{vmatrix}=\left[\dfrac{\partial A_z}{\partial y}-\dfrac{\partial A_y}{\partial z}\right]\overline{a}_x+\left[\dfrac{\partial A_x}{\partial z}-\dfrac{\partial A_z}{\partial x}\right]\overline{a}_y+\left[\dfrac{\partial A_y}{\partial x}-\dfrac{\partial A_x}{\partial y}\right]\overline{a}_z$

In cylindrical system: $\nabla\times\overline{A}=\dfrac{1}{\rho}\begin{vmatrix}\overline{a}_\rho & \rho\overline{a}_\varphi & \overline{a}_z \\ \dfrac{\partial}{\partial\rho} & \dfrac{\partial}{\partial\varphi} & \dfrac{\partial}{\partial z} \\ A_\rho & \rho A_\varphi & A_z\end{vmatrix}=\left[\dfrac{1}{\rho}\dfrac{\partial A_z}{\partial\varphi}-\dfrac{\partial A_\varphi}{\partial z}\right]\overline{a}_\rho+\left[\dfrac{\partial A_\rho}{\partial z}-\dfrac{\partial A_z}{\partial\rho}\right]\overline{a}_\varphi+\dfrac{1}{\rho}\left[\dfrac{\partial(\rho A_\varphi)}{\partial\rho}-\dfrac{\partial A_\rho}{\partial\varphi}\right]\overline{a}_z$

In spherical system: $\nabla\times\overline{A}=\dfrac{1}{r^2\sin\theta}\begin{vmatrix}\overline{a}_r & r\overline{a}_\theta & r\sin\theta\,\overline{a}_\varphi \\ \dfrac{\partial}{\partial r} & \dfrac{\partial}{\partial\theta} & \dfrac{\partial}{\partial\varphi} \\ A_r & rA_\theta & r\sin\theta\,A_\varphi\end{vmatrix}=\dfrac{1}{r\sin\theta}\left[\dfrac{\partial(A_\varphi\sin\theta)}{\partial\theta}-\dfrac{\partial A_\theta}{\partial\varphi}\right]\overline{a}_r$

$$+\dfrac{1}{r}\left[\dfrac{1}{\sin\theta}\dfrac{\partial A_r}{\partial\varphi}-\dfrac{\partial(rA_\varphi)}{\partial r}\right]\overline{a}_\theta+\dfrac{1}{r}\left[\dfrac{\partial(rA_\theta)}{\partial r}-\dfrac{\partial A_r}{\partial\theta}\right]\overline{a}_\varphi$$

Vectors employed for the illustration of divergence in different coordinate systems are used for finding the curls too. Assignee c contains the return of 'curl' in the following implementation.

⬚ In rectangular system

$$\nabla\times\overline{A}=\left[\dfrac{\partial A_z}{\partial y}-\dfrac{\partial A_y}{\partial z}\right]\overline{a}_x+\left[\dfrac{\partial A_x}{\partial z}-\dfrac{\partial A_z}{\partial x}\right]\overline{a}_y+\left[\dfrac{\partial A_y}{\partial x}-\dfrac{\partial A_x}{\partial y}\right]\overline{a}_z,\qquad \overline{A}=\ln(x+y)\overline{a}_x+\cos y\,\overline{a}_y+e^{7z}\,\overline{a}_z,$$

$\dfrac{\partial A_z}{\partial y}=\dfrac{\partial(e^{7z})}{\partial y}=0$, $\dfrac{\partial A_y}{\partial z}=\dfrac{\partial(\cos y)}{\partial z}=0$, $\dfrac{\partial A_x}{\partial z}=\dfrac{\partial[\ln(x+y)]}{\partial z}=0$, $\dfrac{\partial A_z}{\partial x}=\dfrac{\partial(e^{7z})}{\partial x}=0$, $\dfrac{\partial A_y}{\partial x}=\dfrac{\partial(\cos y)}{\partial x}=0$, and

$\dfrac{\partial A_x}{\partial y}=\dfrac{\partial[\ln(x+y)]}{\partial y}=\dfrac{1}{x+y}$. Inserting all partial derivatives, we have $\nabla\times\overline{A}=-\dfrac{\overline{a}_z}{x+y}$. A matrix can be formed

from the three components of $\nabla\times\overline{A}$, which is $\left[0 \quad 0 \quad -\dfrac{1}{x+y}\right]$. That is what is found below:

MATLAB Command

```
>>maple('A:=[log(x+y),cos(y),exp(7*z)]'); ↵
>>c=maple('curl(A,[x,y,z])'); ↵
>>pretty(sym(c)) ↵
```

$$
\begin{bmatrix}
 & & 1 & \\
0 & 0 & -\text{-----------} \\
 & & x+y &
\end{bmatrix}
$$

⊟ **In cylindrical system**

$$\overline{A} = \rho^2 z \overline{a}_\rho + \sin\varphi\, \overline{a}_\varphi + \rho z \cos\varphi\, \overline{a}_z, \quad \nabla \times \overline{A} = \left[\frac{1}{\rho}\frac{\partial A_z}{\partial \varphi} - \frac{\partial A_\varphi}{\partial z}\right]\overline{a}_\rho + \left[\frac{\partial A_\rho}{\partial z} - \frac{\partial A_z}{\partial \rho}\right]\overline{a}_\varphi + \frac{1}{\rho}\left[\frac{\partial(\rho A_\varphi)}{\partial \rho} - \frac{\partial A_\rho}{\partial \varphi}\right]\overline{a}_z, \text{ partial}$$

derivatives are $\dfrac{\partial A_z}{\partial \varphi} = -\rho z \sin\varphi$, $\dfrac{\partial A_\varphi}{\partial z} = 0$, $\dfrac{\partial A_\rho}{\partial z} = \rho^2$, $\dfrac{\partial A_z}{\partial \rho} = z\cos\varphi$, $\dfrac{\partial(\rho A_\varphi)}{\partial \rho} = \dfrac{\partial(\rho \sin\varphi)}{\partial \rho} = \sin\varphi$, and $\dfrac{\partial A_\rho}{\partial \varphi} = 0$.

Plugging all partial derivatives, we have $\nabla \times \overline{A} = -z\sin\varphi\, \overline{a}_\rho + (\rho^2 - z\cos\varphi)\overline{a}_\varphi + \dfrac{\sin\varphi}{\rho}\overline{a}_z$. The curl vector in matrix

form is written as $\left[-z\sin\varphi \quad \rho^2 - z\cos\varphi \quad \dfrac{\sin\varphi}{\rho}\right]$.

MATLAB Command

```
>>maple('A:=[r^2*z,sin(p),r*z*cos(p)]'); ↵
>>c=maple('curl(A,[r,p,z],coords=cylindrical)'); ↵
>>pretty(sym(c)) ↵
```

$$
\begin{bmatrix}
 & 2 & \sin(p) \\
-z\sin(p) & r - z\cos(p) & \text{----------} \\
 & & r
\end{bmatrix}
$$

⊟ **In spherical system**

$$\overline{A} = r^2 \overline{a}_r + \sin\theta\cos\varphi\, \overline{a}_\theta + r\cos\varphi\, \overline{a}_\varphi, \qquad \nabla \times \overline{A} = \frac{1}{r\sin\theta}\left[\frac{\partial(A_\varphi \sin\theta)}{\partial\theta} - \frac{\partial A_\theta}{\partial\varphi}\right]\overline{a}_r + \frac{1}{r}\left[\frac{1}{\sin\theta}\frac{\partial A_r}{\partial\varphi} - \frac{\partial(rA_\varphi)}{\partial r}\right]\overline{a}_\theta +$$

$\dfrac{1}{r}\left[\dfrac{\partial(rA_\theta)}{\partial r} - \dfrac{\partial A_r}{\partial\theta}\right]\overline{a}_\varphi$, $\dfrac{\partial(A_\varphi \sin\theta)}{\partial\theta} = \dfrac{\partial(r\cos\varphi\sin\theta)}{\partial\theta} = r\cos\varphi\cos\theta$, $\dfrac{\partial A_\theta}{\partial\varphi} = \dfrac{\partial(\sin\theta\cos\varphi)}{\partial\varphi} = -\sin\theta\sin\varphi$, $\dfrac{\partial A_r}{\partial\varphi} = \dfrac{\partial(r^2)}{\partial\varphi} = 0$,

$\dfrac{\partial(rA_\varphi)}{\partial r} = \dfrac{\partial(r^2\cos\varphi)}{\partial r} = 2r\cos\varphi$, $\dfrac{\partial(rA_\theta)}{\partial r} = \dfrac{\partial(r\sin\theta\cos\varphi)}{\partial r} = \sin\theta\cos\varphi$, and $\dfrac{\partial A_r}{\partial\theta} = \dfrac{\partial(r^2)}{\partial\theta} = 0$. After manipulation we

have $\quad \nabla \times \overline{A} = \dfrac{r\cos\theta\cos\varphi + \sin\theta\sin\varphi}{r\sin\theta}\overline{a}_r - 2\cos\varphi\, \overline{a}_\theta + \dfrac{\sin\theta\cos\varphi}{r}\overline{a}_\varphi = \left[\dfrac{r\cos\theta\cos\varphi + \sin\theta\sin\varphi}{r\sin\theta} \quad -2\cos\varphi \quad \dfrac{\sin\theta\cos\varphi}{r}\right]$.

See below how easy it is:

MATLAB Command

```
>>maple('A:=[r^2,sin(t)*cos(p),r*cos(p)]'); ↵
>>c=maple('curl(A,[r,t,p],coords=spherical)'); ↵
>>pretty(sym(c)) ↵
```

$$
\begin{bmatrix}
2 & & \\
r\cos(t)\cos(p) + r\sin(t)\sin(p) & & \sin(t)\cos(p) \\
\text{--} & -2\cos(p) & \text{------------------} \\
2 & & r \\
r\sin(t) & &
\end{bmatrix}
$$

5.11.4 Composite manipulation involving gradient, divergence, and curl

There are instances where scalar or vector manipulations like $\nabla \bullet (\overline{A} \times \overline{B})$, $\nabla \bullet (\nabla \times \overline{A})$, $\nabla \times (\nabla \times \overline{A})$, ... etc are necessary. Few examples are cited regarding composite manipulation involving gradient, divergence, and curl.

⊟ **Example of $\nabla \times (\nabla \times \overline{A})$**

Compute $\nabla \times (\nabla \times \overline{A})$ for $\overline{A} = \rho^2 z \overline{a}_\rho + \sin \varphi \, \overline{a}_\varphi + \rho z \cos \varphi \, \overline{a}_z$.

Solution:

From article 5.11.3, we have $\nabla \times \overline{A} = -z \sin \varphi \, \overline{a}_\rho + (\rho^2 - z \cos \varphi) \overline{a}_\varphi + \dfrac{\sin \varphi}{\rho} \overline{a}_z$. Say $\overline{B} = \nabla \times \overline{A}$ and perform

another curl operation on $\overline{B}$ using $\left[\dfrac{1}{\rho} \dfrac{\partial B_z}{\partial \varphi} - \dfrac{\partial B_\varphi}{\partial z} \right] \overline{a}_\rho + \left[\dfrac{\partial B_\rho}{\partial z} - \dfrac{\partial B_z}{\partial \rho} \right] \overline{a}_\varphi + \dfrac{1}{\rho} \left[\dfrac{\partial (\rho B_\varphi)}{\partial \rho} - \dfrac{\partial B_\rho}{\partial \varphi} \right] \overline{a}_z$, where $B_\rho = -z \sin \varphi$,

$B_\varphi = \rho^2 - z \cos \varphi$, and $B_z = \dfrac{\sin \varphi}{\rho}$. Thus we obtain $\nabla \times (\nabla \times \overline{A}) = \nabla \times \overline{B} = \dfrac{1 + \rho^2}{\rho^2} \cos \varphi \, \overline{a}_\rho + \dfrac{1 - \rho^2}{\rho^2} \sin \varphi \, \overline{a}_\varphi + 3\rho \, \overline{a}_z$.

MATLAB Command

```
>>maple('A:=[r^2*z,sin(p),r*z*cos(p)]');  ↵
>>maple('B:=curl(A,[r,p,z],coords=cylindrical)');  ↵ ← ∇×A̅  is assigned to B
>>O=maple('curl(B,[r,p,z],coords=cylindrical)');  ↵ ← The output of ∇×B̅  is assigned to O
>>pretty(sym(O))  ↵
```

```
[   cos(p)                                              ]
[ ---------- + r cos(p)                                 ]
[    r                              sin(p)              ]
[ ------------------------    - sin(p) + ---------   3 r]
[           r                              2            ]
[                                          r            ]
```

Taking the help of the command 'simplify' can carry out further simplification:

```
>>pretty(simplify(sym(O)))  ↵
```

```
[             2                    2                ]
[cos(p) (1 + r )        sin(p) (r  - 1)             ]
[------------------   - ------------------      3 r ]
[        2                      2                   ]
[        r                      r                   ]
```

⊟ **Example of $\nabla \times (\overline{A} \times \overline{B})$**

If $\overline{A} = x^2 \overline{a}_x + y \overline{a}_y + (x + z) \overline{a}_z$ and $\overline{B} = zx \overline{a}_x + z^3 \overline{a}_z$, find $\nabla \times (\overline{A} \times \overline{B})$.

Solution:

First we need to find $\overline{A} \times \overline{B}$, which is $\overline{A} \times \overline{B} = \begin{vmatrix} \overline{a}_x & \overline{a}_y & \overline{a}_z \\ x^2 & y & x + z \\ zx & 0 & z^3 \end{vmatrix} = yz^3 \, \overline{a}_x + [zx(x + z) - x^2 z^3] \overline{a}_y - xyz \, \overline{a}_z$.

Say $\overline{C} = \overline{A} \times \overline{B}$ to write $\nabla \times (\overline{A} \times \overline{B}) = \nabla \times \overline{C} = \left[\dfrac{\partial C_z}{\partial y} - \dfrac{\partial C_y}{\partial z} \right] \overline{a}_x + \left[\dfrac{\partial C_x}{\partial z} - \dfrac{\partial C_z}{\partial x} \right] \overline{a}_y + \left[\dfrac{\partial C_y}{\partial x} - \dfrac{\partial C_x}{\partial y} \right] \overline{a}_z$, where $C_x = yz^3$,

$C_y = zx(x + z) - x^2 z^3$, and $C_z = -xyz$, finally, $\nabla \times \overline{C} = (3z^2 x^2 - x^2 - 3zx) \overline{a}_x + (3yz^2 + yz) \overline{a}_y + (2zx + z^2 - 2xz^3 - z^3) \overline{a}_z$.

Cross product of two vectors is conducted by subroutine 'crossprod'. The computation is so simple as shown below:

MATLAB Command

```
>>maple('A:=[x^2,y,x+z]');  ↵
>>maple('B:=[z*x,0,z^3]');  ↵
>>maple('C:=crossprod(A,B)');  ↵          ← $\overline{A} \times \overline{B}$ is assigned to C
>>O=maple('curl(C,[x,y,z])');  ↵          ← $\nabla \times \overline{C}$ is assigned to O
>>pretty(simplify(sym(O)))  ↵             ← Print the simplified contents of output O
```

$$[\quad\quad 2\quad 2\,2\quad\quad\quad 2\quad\quad\quad\quad\quad 2\quad\quad 3\quad 3\;]$$
$$[-3\;z\;x - x\; + 3\;x\;z\quad\quad 3\;y\;z\; + y\;z\quad\quad 2\;z\;x + z\; - 2\;x\;z\; - z\;]$$

⌗ **Example of** $\nabla \bullet (\overline{A} \times \overline{B})$

Choose $\overline{A}$ and $\overline{B}$ from cylindrical system, where $\overline{A} = \rho^2 z \overline{a}_\rho + \rho z \overline{a}_z$ and $\overline{B} = z \overline{a}_\varphi + \sin \varphi \overline{a}_\rho$.

Solution:

The cross product of two vectors $\overline{A}$ and $\overline{B}$ in cylindrical system is similar to that of rectangular

system, so, $\overline{C} = \overline{A} \times \overline{B} = \begin{bmatrix} \overline{a}_\rho & \overline{a}_\varphi & \overline{a}_z \\ \rho^2 z & 0 & \rho z \\ \sin\varphi & z & 0 \end{bmatrix} = -\rho z^2 \overline{a}_\rho + \rho z \sin\varphi \overline{a}_\varphi + \rho^2 z^2 \overline{a}_z$, thereupon, $\nabla \bullet (\overline{A} \times \overline{B}) = \nabla \bullet \overline{C} =$

$\dfrac{1}{\rho}\dfrac{\partial(\rho C_\rho)}{\partial\rho} + \dfrac{1}{\rho}\dfrac{\partial C_\varphi}{\partial\varphi} + \dfrac{\partial C_z}{\partial z} = z(-2z + \cos\varphi + 2\rho^2)$. Obtain all these very simply:

MATLAB Command

```
>>maple('A:=[r^2*z,0,r*z]');  ↵
>>maple('B:=[sin(p),z,0]');  ↵
>>maple('C:=crossprod(A,B)');  ↵          ← No need to specify the coordinate system
>>O=maple('diverge(C,[r,p,z],coords=cylindrical)');  ↵
>>pretty(sym(O))  ↵
```

$$\quad\quad\quad\quad\quad\quad\quad 2$$
$$z\;(-2\;z + \cos(p) + 2\;r\;)$$

5.11.5 Laplacian of a scalar field

Laplacian of a scalar function V is denoted by $\nabla^2 V$. Laplacian is the composite of gradient and divergence operators. Actually, Laplacian of V is defined as the divergence of the gradient of V. Function V is said to be harmonic if $\nabla^2 V = 0$ whose MATLAB cognate is 'laplacian'. Assignee of the function will be L for the subsequent implementations. Expressions for the laplacians in various coordinate systems are $\nabla^2 V =$

$\dfrac{\partial^2 V}{\partial x^2} + \dfrac{\partial^2 V}{\partial y^2} + \dfrac{\partial^2 V}{\partial z^2}$, $\quad \nabla^2 V = \dfrac{1}{\rho}\dfrac{\partial}{\partial\rho}\left(\rho\dfrac{\partial V}{\partial\rho}\right) + \dfrac{1}{\rho^2}\dfrac{\partial^2 V}{\partial\varphi^2} + \dfrac{\partial^2 V}{\partial z^2}$, and $\quad \nabla^2 V = \dfrac{1}{r^2}\dfrac{\partial}{\partial r}\left(r^2\dfrac{\partial V}{\partial r}\right) + \dfrac{1}{r^2\sin\theta}\dfrac{\partial}{\partial\theta}\left(\sin\theta\dfrac{\partial V}{\partial\theta}\right) +$

$\dfrac{1}{r^2\sin^2\theta}\dfrac{\partial^2 V}{\partial\varphi^2}$ for rectangular, cylindrical, and spherical systems respectively. Start with the function $V = \dfrac{x^2 y^3}{z}$

given in rectangular system, on partial differentiation, $\dfrac{\partial^2 V}{\partial x^2} = \dfrac{2y^3}{z}$, $\dfrac{\partial^2 V}{\partial y^2} = \dfrac{6yx^2}{z}$, and $\dfrac{\partial^2 V}{\partial z^2} = \dfrac{2x^2 y^3}{z^3}$, hence,

$\nabla^2 V = \dfrac{2y^3}{z} + \dfrac{6yx^2}{z} + \dfrac{2x^2 y^3}{z^3}$.

MATLAB Command

```
>>maple('V:=x^2*y^3/z');  ↵
>>L=maple('laplacian(V,[x,y,z])');  ↵
>>pretty(sym(L))  ↵
```

$$\quad\quad\quad\quad 3\quad\quad\quad\quad\quad 2\quad\quad\quad\quad\quad 2\;3$$
$$\quad\quad\quad\quad y\quad\quad\quad\quad x\;y\quad\quad\quad\quad x\;y$$
$$2\;\text{---------} + 6\;\text{-----------} + 2\;\text{--------------}$$
$$\quad\quad\quad\quad z\quad\quad\quad\quad z\quad\quad\quad\quad\quad 3$$
$$\quad\quad\quad\quad\quad\quad\quad\quad\quad\quad\quad\quad\quad\quad\quad\quad z$$

Next, we have $V = z\rho\cos^2\varphi$ in cylindrical system, therefore, $\nabla^2 V = \dfrac{1}{\rho}\dfrac{\partial}{\partial\rho}\left(\rho\dfrac{\partial(z\rho\cos^2\varphi)}{\partial\rho}\right) + \dfrac{1}{\rho^2}\dfrac{\partial^2(z\rho\cos^2\varphi)}{\partial\varphi^2} +$

$\dfrac{\partial^2(z\rho\cos^2\varphi)}{\partial z^2} = \dfrac{z\cos^2\varphi}{\rho} - \dfrac{2z\cos 2\varphi}{\rho} = \dfrac{-z\cos^2\varphi + 2z\sin^2\varphi}{\rho}$ (on simplification).

MATLAB Command

```
>>maple('V:=z*r*cos(p)^2');  ↵
>>L=maple('laplacian(V,[r,p,z],coords=cylindrical)');  ↵
>>pretty(sym(L))  ↵
                      2            2
        -z cos(p)  + 2 z sin(p)
        ---------------------------------
                      r
```

The last implementation is on $V = r\sin\theta\cos^2\varphi$, which is in spherical system, clearly, $\nabla^2 V = \dfrac{1}{r^2}\dfrac{\partial}{\partial r}\left(r^2\dfrac{\partial V}{\partial r}\right) +$

$\dfrac{1}{r^2\sin\theta}\dfrac{\partial}{\partial\theta}\left(\sin\theta\dfrac{\partial V}{\partial\theta}\right) + \dfrac{1}{r^2\sin^2\theta}\dfrac{\partial^2 V}{\partial\varphi^2} = \dfrac{1}{r^2}\times\dfrac{\partial}{\partial r}\left(r^2\dfrac{\partial[r\sin\theta\cos^2\varphi]}{\partial r}\right) + \dfrac{1}{r^2\sin\theta}\dfrac{\partial}{\partial\theta}\left(\sin\theta\dfrac{\partial[r\sin\theta\cos^2\varphi]}{\partial\theta}\right) + \dfrac{1}{r^2\sin^2\theta}\quad\times$

$\dfrac{\partial^2[r\sin\theta\cos^2\varphi]}{\partial\varphi^2} = \dfrac{2\sin\theta\cos^2\varphi}{r} + \dfrac{\cos 2\theta\cos^2\varphi}{r\sin\theta} - \dfrac{2\cos 2\varphi}{r\sin\theta} = \dfrac{2 - 3\cos^2\varphi}{r\sin\theta}$ (after trigonometric manipulation). Obtain

that simply executing the following:

MATLAB Command

```
>>maple('V:=r*sin(t)*cos(p)^2');  ↵
>>L=maple('laplacian(V,[r,t,p],coords=spherical)');  ↵
>>pretty(simplify(sym(L)))  ↵


                      2
          3 cos(p)  - 2
        - ----------------------
             r sin(t)
```

5.11.6 Computation of the net flux crossing a surface

Calculate the total outward flux of vector $\overline{A} = \rho^3\sin\varphi\,\overline{a}_\rho + z\cos^2\varphi\,\overline{a}_\varphi + \rho z^2\,\overline{a}_z$ through the hollow cylinder defined by $1 \le \rho \le 2$ and $0 \le z \le 4$.

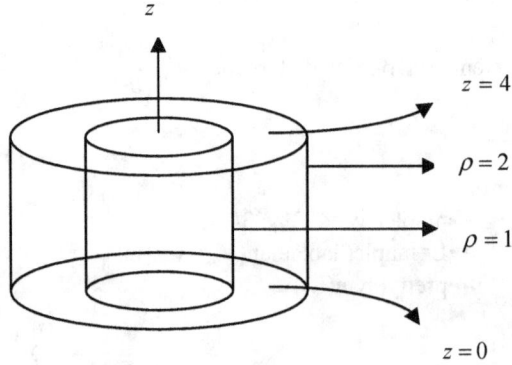

Figure 5.2 *The surface enclosed by* $1 \le \rho \le 2$ *and* $0 \le z \le 4$

Solution:

Figure 5.2 depicts the hollow cylinder defined by $1 \leq \rho \leq 2$ and $0 \leq z \leq 4$. The total flux due to the vector $\overline{A}$ coming out of the above hollow cylindrical surface is given by $\int\limits_{v} (\nabla \bullet \overline{A}) dv$, where $\nabla \bullet \overline{A} =$

$$\frac{1}{\rho}\frac{\partial(\rho A_{\rho})}{\partial \rho} + \frac{1}{\rho}\frac{\partial A_{\varphi}}{\partial \varphi} + \frac{\partial A_{z}}{\partial z} = \frac{1}{\rho}\frac{\partial(\rho^{4}\sin\varphi)}{\partial \rho} + \frac{1}{\rho}\frac{\partial(z\cos^{2}\varphi)}{\partial \varphi} + \frac{\partial(\rho z^{2})}{\partial z} = 4\rho^{2}\sin\varphi - \frac{2z\cos\varphi\sin\varphi}{\rho} + 2z\rho$$ and the

elementary volume dv in cylindrical system is equal to $\rho\, d\rho\, d\varphi\, dz$. For the enclosed surface different limits are $\rho \to 1 \sim 2$, $\varphi \to 0 \sim 2\pi$, and $z \to 0 \sim 4$. Having known the divergence and limits, the total flux is computed

as $\int\limits_{z=0}^{z=4}\int\limits_{\varphi=0}^{\varphi=2\pi}\int\limits_{\rho=1}^{\rho=2}\left[4\rho^{2}\sin\varphi - \frac{2z\cos\varphi\sin\varphi}{\rho} + 2z\rho\right]\rho\, d\rho\, d\varphi\, dz = \int\limits_{z=0}^{z=4}\int\limits_{\varphi=0}^{\varphi=2\pi}\int\limits_{\rho=1}^{\rho=2}2z\rho\,\rho\, d\rho\, d\varphi\, dz$ (since $\int\limits_{\varphi=0}^{\varphi=2\pi}\sin\varphi\, d\varphi$ or $\int\limits_{\varphi=0}^{\varphi=2\pi}\cos\varphi\, d\varphi$

is 0)$= 4\pi\int\limits_{z=0}^{z=4}\int\limits_{\rho=1}^{\rho=2}z\rho^{2}\, d\rho\, dz = \frac{224\pi}{3}$. The whole computation is as follows:

MATLAB Command

```
>>maple('A:=[r^3*sin(p),z*cos(p)^2,r*z^2]'); ↵
>>maple('B:=diverge(A,[r,p,z],coords=cylindrical)'); ↵        ← ∇ • Ā is assigned to B
>>maple('int(int(int(B*r,r=1..2),p=0..2*pi),z=0..4)') ↵       ← Triple integration of ( ∇ • Ā ) ρ
```

ans =

224/3*pi

Volumetric integral is basically a triple one that is why we need subroutine 'int' three times. Elaborate discussion of 'int' is presented in next chapter. In the volumetric integrand, $\nabla \bullet \overline{A}$ is multiplied by ρ, which is accomplished by the command B*r (integrand of innermost integration). Just to show the difference between the entering style of integration in Symbolic Math Toolbox and Maple package, $\int\limits_{x=a}^{x=b} y\, dx$ is written as int(y, x, a, b) and int(y, $x = a .. b$) in former and later respectively.

We presented few model computations regarding symbolic maneuvering of dot/cross product, gradient, divergence, curl, or composite operations. Above educational examples demonstrate how problems of differential vector calculus are implemented conveniently. The concept can be applied to electromagnetics problems utilizing Maple package in collaboration with Symbolic Toolbox of MATLAB.

5.12 Curvature and radius of curvature of a curve

The rate at which the tangent line to a curve changes direction is called the curvature of the curve. Curavture of a straight line is zero at all points on the line. Curvature is denoted by ρ. Symbolically, the definition of curvature is given by $\rho = \left|\frac{dT}{ds}\right|$, where T is the unit tangent at any point on the curve and s is the arc length. If a curve has equation $y = f(x)$, its curvature is derived as $\rho = \dfrac{\left|\dfrac{d^{2}y}{dx^{2}}\right|}{\left[1 + \left(\dfrac{dy}{dx}\right)^{2}\right]^{\frac{3}{2}}}$. Find the curvature of

circle $x^{2} + y^{2} = a^{2}$. The equation is an implicit one. The first and second order derivatives are $\dfrac{dy}{dx} = -\dfrac{x}{y}$ and

$\dfrac{d^2y}{dx^2} = -\dfrac{x^2 + y^2}{y^3}$ respectively (see article 5.7 for implicit derivative), from which, $\rho = \dfrac{\left| -\dfrac{x^2 + y^2}{y^3} \right|}{\left[1 + \dfrac{x^2}{y^2} \right]^{\frac{3}{2}}} = \dfrac{1}{a}$ (on

insertion of $x^2 + y^2 = a^2$). Have the computation as follows:

MATLAB Command

>>maple('c:=y^2+x^2=a^2'); ↵   ← Assign $x^2 + y^2 = a^2$ to c

>>maple('d1:=implicitdiff(c,y,x)'); ↵   ← Assign $\dfrac{dy}{dx}$ to d1

>>maple('d2:=implicitdiff(c,y,x,x)'); ↵   ← Assign $\dfrac{d^2y}{dx^2}$ to d2

>>maple('r:=d2/(1+d1^2)^(3/2)'); ↵   ← Compute ρ and assign that to r

>>r=maple('algsubs(x^2+y^2=a^2,r)'); ↵   ← Algebraic substitution of $x^2 + y^2 = a^2$

>>r=simple(sym(r)) ↵   ← Take the simplified form of r

r =

-1/a

The negative sign appears because we did not take the absolute value. We know that the radius of curvature is the reciprocal of its curvature, which is a for the circle. Equation of a curve can be in other form. When the

equation is $x = f(y)$, $\rho = \dfrac{\left| \dfrac{d^2x}{dy^2} \right|}{\left[1 + \left(\dfrac{dx}{dy} \right)^2 \right]^{\frac{3}{2}}}$, and so are $\rho = \dfrac{\left| \dfrac{dx}{dt} \dfrac{d^2y}{dt^2} - \dfrac{d^2x}{dt^2} \dfrac{dy}{dt} \right|}{\left[\left(\dfrac{dx}{dt} \right)^2 + \left(\dfrac{dy}{dt} \right)^2 \right]^{\frac{3}{2}}}$ and $\rho = \dfrac{\left| r^2 + 2\left(\dfrac{dr}{d\theta} \right)^2 - r\dfrac{d^2r}{d\theta^2} \right|}{\left[r^2 + \left(\dfrac{dr}{d\theta} \right)^2 \right]^{\frac{3}{2}}}$ for

parametric curve $\begin{cases} x = x(t) \\ y = y(t) \end{cases}$ and polar curve $r = f(\theta)$ respectively. We present one more example.

⌼ Compute the unit tangent vector, unit normal vector, curvature, and center of curvature for the parabola $\begin{cases} x = t^2 \\ y = 2t \end{cases}$ at any point (x, y) .

Solution:

Position vector of the curve is $\bar{r}(t) = t^2 \bar{a}_x + 2t\bar{a}_y$, where $\bar{a}_x$ and $\bar{a}_y$ correspond to unit vectors along x

and y directions respectively. Unit tangent vector $\bar{T}$ is defined as $\bar{T} = \dfrac{d\bar{r}(t)}{ds}$. For the given parabola, we have

$\bar{T} = \dfrac{\dfrac{d\bar{r}(t)}{dt}}{\dfrac{ds}{dt}} = \dfrac{2t\bar{a}_x + 2\bar{a}_y}{\sqrt{\left(\dfrac{dx}{dt} \right)^2 + \left(\dfrac{dy}{dt} \right)^2}} = \dfrac{2t\bar{a}_x + 2\bar{a}_y}{\sqrt{4t^2 + 4}} = \dfrac{t\bar{a}_x + \bar{a}_y}{\sqrt{t^2 + 1}}$. As a matrix, $\bar{T}$ is written as $\left[\dfrac{t}{\sqrt{t^2 + 1}} \quad \dfrac{1}{\sqrt{t^2 + 1}} \right]$. Obtain

this as follows:

MATLAB Command

>>syms t ↵   ← Declare parameter t as symbolic

>>x=t^2; y=2*t; ↵   ← Define $x = x(t)$ and $y = y(t)$

```
>>r=[x y]; dr=diff(r); ↵      ← r is the position vector r̄(t) and dr= dr̄(t)/dt
```

```
>>ds=sqrt(sum(dr.^2)); ↵  ← ds= |ds/dt|
```

```
>>T=dr/ds; ↵                  ← T is the unit tangent vector
>>pretty(T) ↵                 ← Print T
```

```
[       t                      1          ]
[---------------      ---------------- ]
[    2    1/2            2    1/2    ]
[ (t  + 1)              (t  + 1)       ]
```

Then, we look for the normal vector $\overline{N}$, which is defined as $\overline{N} = \dfrac{d\overline{T}}{ds}$, from that, $\overline{N} = \dfrac{\frac{d\overline{T}}{dt}}{\frac{ds}{dt}} = \dfrac{d}{dt}\left[\dfrac{t\overline{a}_x + \overline{a}_y}{\sqrt{t^2+1}}\right] \Big/$

$\sqrt{4t^2+4} = \dfrac{\overline{a}_x - t\overline{a}_y}{2(t^2+1)^2}$ (on differentiation and simplification). Modulus of $\overline{N}$ is $M = \left|\dfrac{\overline{a}_x - t\overline{a}_y}{2(t^2+1)^2}\right| = \dfrac{1}{2(t^2+1)^{\frac{3}{2}}}$ and the

unit normal vector $\overline{n} = \dfrac{\overline{N}}{M} = \dfrac{\overline{a}_x - t\overline{a}_y}{\sqrt{t^2+1}} = \begin{bmatrix} \dfrac{1}{\sqrt{t^2+1}} & \dfrac{-t}{\sqrt{t^2+1}} \end{bmatrix}$ (in matrix form). Accomplish that as follows:

MATLAB Command
```
>>N=diff(T)/ds; ↵          ← N is normal vector
>>M=sqrt(sum(N.^2)); ↵  ← M is modulus M
>>n=simple(N/M); ↵        ← n is simplified unit normal vector n̄
>>pretty(n) ↵               ← Print n̄
```

```
[       1                      t          ]
[---------------      - ---------------- ]
[    2    1/2            2    1/2    ]
[ (t  + 1)              (t  + 1)       ]
```

Curvature ρ is nothing but the modulus of normal vector $\overline{N}$, which is $M = \dfrac{1}{2(t^2+1)^{\frac{3}{2}}}$. Conduct the following

to have ρ:

```
>>p=simple(M); ↵          ← Compute simplified M, which is ρ and assign that to p
>>pretty(p) ↵              ← Display ρ
```

```
              1
    1/2 ---------------
          2    3/2
        (t  + 1)
```

The last computation can be found by observing that $\overline{C} = \overline{r}(t) + R\,\overline{n}$, where $\overline{C}$ and R are called the center of

curvature vector and radius of curvature $\left(R = \dfrac{1}{\rho}\right)$ respectively, then promptly, $\overline{C} = (t^2\overline{a}_x + 2t\overline{a}_y) + \dfrac{1}{\rho}\dfrac{\overline{a}_x - t\overline{a}_y}{\sqrt{t^2+1}} =$

$(3t^2+2)\overline{a}_x - 2t^3\,\overline{a}_y = [3t^2+2 \quad -2t^3]$ (as a row matrix):

```
>>C=r+n/p; ↵                      ← Form C̄
>>pretty(simple(C)) ↵             ← Display simplified C
```

$$\begin{bmatrix} 2 & 3 \\ 3t+2 & -2t \end{bmatrix}$$

5.13 Tangential and normal components of acceleration

Given that $\bar{r}(t) = x(t)\,\bar{a}_x + y(t)\,\bar{a}_y$ is the position vector of a particle in the curvilinear motion, the velocity vector $\bar{v}(t)$ is defined as $\bar{v}(t) = \dfrac{d\bar{r}(t)}{dt}$. Vector $\bar{v}(t)$ is always directed along the tangent line to the curve whose position vector is $\bar{r}(t)$ and has magnitude $\dfrac{ds}{dt}$, where $\dfrac{ds}{dt} = \sqrt{\left(\dfrac{dx}{dt}\right)^2 + \left(\dfrac{dy}{dt}\right)^2}$. Component of $\bar{v}(t)$ in the direction of the tangent is $\bar{v}(t)$ itself and the normal component is zero, therefore, $\bar{v}(t) = \dfrac{d\bar{r}(t)}{dt} = \dfrac{d\bar{r}(t)}{ds}\dfrac{ds}{dt} = \bar{T}\dfrac{ds}{dt}$, where $\bar{T}$ is the unit tangent vector. That is not the incident with acceleration. Acceleration has tangential as well as normal component. Differentiating the equation $\bar{v}(t) = \bar{T}\dfrac{ds}{dt}$, we have $\dfrac{d[\bar{v}(t)]}{dt} = \dfrac{d\bar{T}}{dt}\dfrac{ds}{dt} + \bar{T}\dfrac{d^2s}{dt^2}$ or $\bar{a}(t) = \dfrac{d\bar{T}}{dt}\dfrac{ds}{dt} + \bar{T}\dfrac{d^2s}{dt^2}$, where $\bar{a}(t)$ is the acceleration vector. Manipulate as shown in last article to write $\bar{a}(t) = \bar{T}\dfrac{d^2s}{dt^2} + \rho\bar{n}\left(\dfrac{ds}{dt}\right)^2$. So, the tangential and normal components of acceleration are $\dfrac{d^2s}{dt^2}$ and $\rho\left(\dfrac{ds}{dt}\right)^2$ respectively.

Problem

A particle moves around an ellipse according to equation $\begin{cases} x = a\cos t \\ y = b\sin t \end{cases}$. Find the tangential and normal components of acceleration.

Solution:

The computation is straightforward — $\dfrac{dx}{dt} = -a\sin t$, $\dfrac{dy}{dt} = b\cos t$, $\dfrac{d^2x}{dt^2} = -a\cos t$, $\dfrac{d^2y}{dt^2} = -b\sin t$, $\dfrac{ds}{dt} = \sqrt{a^2\sin^2 t + b^2\cos^2 t}$, $\dfrac{d^2s}{dt^2} = \dfrac{(a^2-b^2)\sin t\cos t}{\sqrt{a^2\sin^2 t + b^2\cos^2 t}}$, and $\rho = \dfrac{\left|\begin{matrix} \dfrac{dx}{dt} & \dfrac{d^2y}{dt^2} \\ \dfrac{d^2x}{dt^2} & \dfrac{dy}{dt} \end{matrix}\right|}{\left[\left(\dfrac{dx}{dt}\right)^2 + \left(\dfrac{dy}{dt}\right)^2\right]^{\frac{3}{2}}} = \dfrac{ab}{[a^2\sin^2 t + b^2\cos^2 t]^{\frac{3}{2}}}$, on that, the tangential and normal components of acceleration are $\dfrac{d^2s}{dt^2} = \dfrac{(a^2-b^2)\sin t\cos t}{\sqrt{a^2\sin^2 t + b^2\cos^2 t}}$ and $\rho\left(\dfrac{ds}{dt}\right)^2 = \dfrac{ab}{\sqrt{a^2\sin^2 t + b^2\cos^2 t}}$ respectively.

MATLAB Command

```
>>syms a b t ↵          ← Declare the symbolic variables
>>x=a*cos(t); y=b*sin(t); ↵    ← Define x = a cos t and y = b sin t

>>s=sqrt(diff(x)^2+diff(y)^2); ↵   ← s ⇒ √((dx/dt)² + (dy/dt)²)
```

>>T1=simple(diff(s)); ↵   ← T1 contains $\dfrac{d^2s}{dt^2}$

>>pretty(T1) ↵

$$\frac{\sin(t)\,\cos(t)\,(a^2 - b^2)}{(a^2\,\sin(t)^2 + b^2\,\cos(t)^2)^{1/2}}$$

>>r=(diff(x)*diff(y,2)-diff(y)*diff(x,2))/s^3; ↵   ← r contains the curvature ρ

>>N=simple(r*s^2); ↵   ← N is $\rho\left(\dfrac{ds}{dt}\right)^2$

>>pretty(N) ↵

$$\frac{a\,b}{(a^2 - a^2\,\cos(t)^2 + b^2\,\cos(t)^2)^{1/2}}$$

5.14 Tangent and normal planes

If a space curve has the parametric equation $x = f(t)$, $y = g(t)$, and $z = h(t)$, its tangent line is given by direction numbers $\dfrac{\partial f}{\partial t} : \dfrac{\partial g}{\partial t} : \dfrac{\partial h}{\partial t}$ and the equation of normal plane at any point (x_1, y_1, z_1) is given by $\dfrac{\partial f}{\partial t}(x - x_1)$ $+\dfrac{\partial g}{\partial t}(y - y_1) + \dfrac{\partial h}{\partial t}(z - z_1)$=0. Compute the direction ratios of the tangent line and the equation of the normal plane for the space curve $x = 2t$, $y = 3t^2$, and $z = -t$ at point $t = 3$. Clearly at $t = 3$ the direction ratios are $[\dfrac{\partial f}{\partial t} : \dfrac{\partial g}{\partial t} : \dfrac{\partial h}{\partial t}]$=[2:6$t$:−1]=[2:18:−1] and $(x, y, z) = (6, 27, -3)$ and the equation of the normal plane is $2(x - 6)$ $+18(y - 27) - (z - (-3))$=0 or $2x + 18y - z - 501 = 0$. Implement it as follows:

MATLAB Command

```
>>syms t x y z ↵
>>r=[2*t 3*t^2 -t]; ↵
>>dr=subs(diff(r),t,3) ↵                  ← Print the direction ratios

dr =
      2   18   -1
>>sum(dr.*([x,y,z]-subs(r,t,3))) ↵     ← Find the equation

ans =

2*x-501+18*y-z
```

A surface, which is described by $f(x, y, z) = 0$, has the tangent plane at any point (x_1, y_1, z_1) having equation $\dfrac{\partial f}{\partial x}(x - x_1) + \dfrac{\partial f}{\partial y}(y - y_1) + \dfrac{\partial f}{\partial z}(z - z_1)$=0 and the normal line having direction ratios $\dfrac{\partial f}{\partial x} : \dfrac{\partial f}{\partial y} : \dfrac{\partial f}{\partial z}$. Find the direction ratios of the normal line and equation of the tangent plane to sphere $x^2 + y^2 + z^2 = 125$ at (3, −4, 10). For the spherical surface, we have $f = x^2 + y^2 + z^2 - 125$, $\dfrac{\partial f}{\partial x} = 2x$, $\dfrac{\partial f}{\partial y} = 2y$, $\dfrac{\partial f}{\partial z} = 2z$, hence, the direction ratios of the normal line are 6:−8:20 and equation of the tangent plane is $6(x - 3) - 8(y + 4) + 20(z - 10)$=0 or $6x - 8y + 20z - 250 = 0$. MATLAB solution is as follows:

MATLAB Command

```
>>syms x y z ↵
>>f=x^2+y^2+z^2-125; ↵
>>d=[diff(f,x) diff(f,y) diff(f,z)]; ↵        ← Form vector d=[ ∂F/∂x  ∂F/∂y  ∂F/∂z ]
>>dr=subs(d,{x,y,z},{3,-4,10}) ↵              ← dr is the direction ratio vector

dr =
       6  -8   20
>>e=sum(dr.*[[x y z]-[3 -4 10]]); ↵           ← e is the equation
>>pretty(e) ↵

       6 x - 250 - 8 y + 20 z
```

5.15 Determining the existence of scalar and vector potentials

Let $\overline{F} = (F_x, F_y, F_z)$ be a vector field in 3-D rectangular space such that F_x, F_y, and F_z have continuous partial derivatives, $\overline{F}$ is said to have a scalar potential function if the components of $\overline{F}$ satisfy $\frac{\partial F_y}{\partial x} = \frac{\partial F_x}{\partial y}$, $\frac{\partial F_z}{\partial x} = \frac{\partial F_x}{\partial z}$, and $\frac{\partial F_z}{\partial y} = \frac{\partial F_y}{\partial z}$. Satisfying the partial derivative's equality implies that $\nabla V = \overline{F}$, where $V(x, y, z)$ is a scalar potential function. The corresponding Maple function is 'potential'.

⌸ Problem

Test whether the scalar potential exists for the vector field $\overline{F} = xy\,\overline{a}_x + yz\,\overline{a}_y + (x + z)\overline{a}_z$.

Solution:

From the given field various field components are $F_x = xy$, $F_y = yz$, and $F_z = x + z$. Partial derivatives provide $\frac{\partial F_y}{\partial x} = 0$, $\frac{\partial F_x}{\partial y} = x$, $\frac{\partial F_z}{\partial x} = 1$, $\frac{\partial F_x}{\partial z} = 0$, $\frac{\partial F_z}{\partial y} = 0$, and $\frac{\partial F_y}{\partial z} = y$. It is evident that $\frac{\partial F_y}{\partial x} \neq \frac{\partial F_x}{\partial y}$, $\frac{\partial F_z}{\partial x} \neq \frac{\partial F_x}{\partial z}$, and $\frac{\partial F_z}{\partial y} \neq \frac{\partial F_y}{\partial z}$, hence, scalar potential of $\overline{F}$ does not exist. The MATLAB proof is as follows:

MATLAB Command

```
>>maple('F:=[x*y,y*z,x+z]'); ↵
>>maple('potential(F,[x,y,z],V)') ↵

ans =

false
```

As another example, take the vector field $\overline{F} = yz\,\overline{a}_x + (xz + z^3)\,\overline{a}_y + (xy + 3yz^2)\overline{a}_z$. It follows then, $\frac{\partial F_y}{\partial x} = \frac{\partial F_x}{\partial y} = z$, $\frac{\partial F_z}{\partial x} = \frac{\partial F_x}{\partial z} = y$, and $\frac{\partial F_z}{\partial y} = \frac{\partial F_y}{\partial z} = x + 3z^2$. Needless to say, the scalar potential exists for this $\overline{F}$, as it does, how one can find the scalar potential V. Go for back integration from any partial derivative, so, $\nabla V = \overline{F} \Rightarrow \frac{\partial V}{\partial x} = F_x = yz$, from which, $V = xyz + f(y, z)$ (integration w.r.to. x). Then, apply $\frac{\partial V}{\partial y} = F_y$ to have $xz + \frac{\partial [f(y,z)]}{\partial y} = xz + z^3$. Integrate the last equation w.r.to. y to get $f(y, z) = z^3 y + g(z)$, therefore, $V = xyz + z^3 y + g(z)$, after that, $\frac{\partial V}{\partial z} = F_z$ gives $xy + 3z^2 y + \frac{d[g(z)]}{dz} = xy + 3z^2 y$, which implies $g(z)$ is a constant. Make the constant 0, so, the scalar potential becomes $V(x, y, z) = xyz + z^3 y$ that is what is shown in the following:

MATLAB Command

```
>>maple('F:=[y*z,x*z+z^3,x*y+3*y*z^2]'); ↵
>>maple('potential(F,[x,y,z],V)')  ↵        ← The return is assigned to V

ans =

true
>>maple('V')  ↵

ans =

x*y*z+y*z^3
```

There are three input arguments of the function 'potential'. The first, second, and third arguments represent the vector function, variables under consideration, and scalar potential respectively. So far we discussed only the existence of scalar potential but there can exist a vector potential $\overline{A}$ for a given vector field $\overline{F}$. Existence of vector potential $\overline{A}$ necessitates that $\nabla \times \overline{A} = \overline{F}$ and which insinuates $\nabla \bullet \overline{F} = 0$. Maple function 'vecpotent' can perform the checking and find the vector potential as well. As usual, take a vector field $\overline{F} = yz\overline{a}_x + 3xz\,\overline{a}_y + (x^2y + 3yz^2)\overline{a}_z$. Its divergence is $\nabla \bullet \overline{F} = \dfrac{\partial[yz]}{\partial x} + \dfrac{\partial[3xz]}{\partial y} + \dfrac{\partial[x^2y + 3yz^2]}{\partial z} = 6y$, which is not equal to zero, on that, vector field of $\overline{F}$ does not exist:

MATLAB Command

```
>>maple('F:=[y*z,3*x*z,x^2*y+3*y*z^2]'); ↵
>>maple('vecpotent(F,[x,y,z],A)')  ↵

ans =

false
```

To show the fulfillment of the condition, assume that $\overline{F} = -3y^2z\overline{a}_x + (3x - 1)\overline{a}_z$. Conduct the following:

```
>>maple('F:=[-3*y^2*z,0,3*x-1]'); ↵
>>maple('vecpotent(F,[x,y,z],A)')  ↵        ← A contains the vector potential

ans =

true                                        ← It indicates existence of A
>>maple('print(A)')  ↵                      ← To display vector potential

ans =

VECTOR([-3*y*x+y, 3/2*z^2*y^2, 0])          ← A is a three element row matrix.
```

From the output we can say that $\overline{A} = (-3yx + y)\overline{a}_x + \dfrac{3}{2}z^2y^2\overline{a}_y$. Just to check, $\nabla \times \overline{A} =$

$$\begin{bmatrix} \overline{a}_x & \overline{a}_y & \overline{a}_z \\ \dfrac{\partial}{\partial x} & \dfrac{\partial}{\partial y} & \dfrac{\partial}{\partial z} \\ -3yx + y & \dfrac{3}{2}z^2y^2 & 0 \end{bmatrix} = -3y^2z\overline{a}_x + (3x - 1)\overline{a}_z,$$ which is exactly the same as the given $\overline{F}$.

5.16 Minimum and maximum of a function

Minimum and maximum are often required to trace or plot a curve. They can be absolute or relative. We classify the finding of minimum or maximum by the title one-dimensional and two-dimensional functions. Computation required can be in numerical or symbolic form. A number of problems are discussed pertaining to the maximum and minimum of a function. Before you go for computation, decide whether a minimum or maximum exists.

5.16.1 Minimum/maximum of a one-dimensional function

If continuous function $y = f(x)$ has a relative maximum or minimum at $x = a$ and $\frac{dy}{dx}$ exists, then,

$\frac{dy}{dx} = 0$ at $x = a$. This is the necessary condition. For sufficient condition, we have to have the second order

derivative. It states that $\left.\frac{d^2 y}{dx^2}\right|_{x=a} > 0$ for minimum and

$\left.\frac{d^2 y}{dx^2}\right|_{x=a} < 0$ for maximum. The test fails when $\frac{d^2 y}{dx^2} = 0$, which indicates the possibility of existing inflection points. That is not the sufficient condition of having point of inflection. If $\frac{d^2 y}{dx^2} = 0$ and if $\frac{d^2 y}{dx^2}$ has the same sign for values immediately preceding $x = a$ and immediately following $x = a$, then, $x = a$ does not yield a point of inflection, and otherwise $x = a$ is a point of inflection. At the point of inflection, the

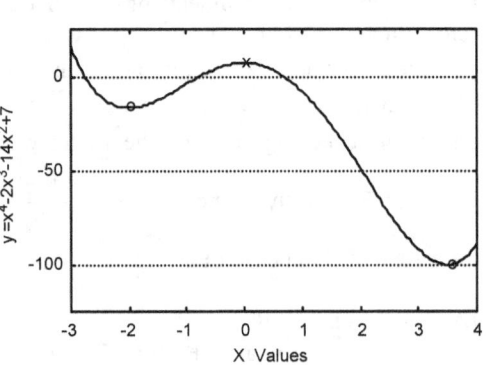

Figure 5.3 Plot of y=x⁴-2x³-14x²+7 for -3 ≤x≤ 4

turning tendency of a curve changes from concave downward to concave upward or vice versa.

We explain the concept with two examples. The first one is $y = f(x) = x^4 - 2x^3 - 14x^2 + 7$. The function is graphed in figure 5.3 for $-3 \leq x \leq 4$, where marks 'o' and 'x' indicate the relative minimum and maximum respectively. For the function at hand, we have $\frac{dy}{dx} = 4x^3 - 6x^2 - 28x$. Set $\frac{dy}{dx}$ to 0 to get the critical points $x = -2$, 0, and $\frac{7}{2}$. The double derivative of y is $\frac{d^2 y}{dx^2} = 12x^2 - 12x - 28$. Plugging critical points says that $\left.\frac{d^2 y}{dx^2}\right|_{x=-2} = 44$ (positive $\Rightarrow$ minimum), $\left.\frac{d^2 y}{dx^2}\right|_{x=0} = -28$ (negative $\Rightarrow$ maximum), and $\left.\frac{d^2 y}{dx^2}\right|_{x=\frac{7}{2}} = 77$ (positive $\Rightarrow$

minimum). That is what is shown in figure 5.3. Finally, the functional values at the critical points are -17, 7, and $-\frac{1603}{16}$ for $x = -2$, 0, and $\frac{7}{2}$ respectively. Implementation can be conducted as follows:

MATLAB Command

```
>>syms x ↵                          ← Declare x as symbolic
>>y=x^4-2*x^3-14*x^2+7; ↵           ← Enter y = x⁴ − 2x³ − 14x² + 7
>>d1=diff(y); ↵                     ← dy/dx is put to d1
>>r=solve(d1) ↵                     ←Find the critical points by setting dy/dx =0,
                                      where r contains the critical points.

r =

[   0]
[  -2]
[ 7/2]
```

```
>>d2=diff(d1); ↵        ← d²y/dx² is assigned to d2

>>subs(d2,x,r) ↵        ← Substitute the critical values of r to d²y/dx²


ans =

[ -28]          ← − ve  indicates relative maximum
[  44]          ← + ve  indicates relative minimum
[  77]          ← + ve  indicates relative minimum
```

Functional values can be seen by,

```
>>subs(y,x,r) ↵         ← Substitute the critical values to function y


ans =

[        7]
[      -17]
[ -1603/16]
```

The next example is $y = x^4 - 12x^3$. See figure 5.4 for the plot of the function. The first derivative $\frac{dy}{dx}$ is $4x^3 - 36x^2$. From $\frac{dy}{dx} = 0$, the critical values are $x = 0, 0,$ and 9. The second derivative of the function is $\frac{d^2y}{dx^2} = 12x^2 - 72x$. Now $\left.\frac{d^2y}{dx^2}\right|_{x=0} = 0$ tells us that there may be

a point of inflection at $x = 0$. To verify that we take two x values slightly before and after $x = 0$, say, the x values are $x = -\frac{1}{10}$ and $x = \frac{1}{10}$, it follows then,

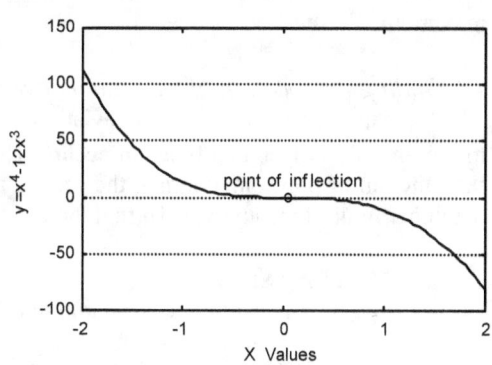

Figure 5.4   Plot of $y = x^4 - 12x^3$ for $-2 \le x \le 2$

$\left.\frac{d^2y}{dx^2}\right|_{x=-\frac{1}{10}} = \frac{183}{25}$ (positive) and $\left.\frac{d^2y}{dx^2}\right|_{x=\frac{1}{10}} = -\frac{177}{25}$ (negative). Since sign change of $\frac{d^2y}{dx^2}$ is there immediately

preceding and following $x = 0$, $x = 0$ is a point of inflection for the curve $y = x^4 - 12x^3$ as shown in figure 5.4 (the inflection is convcave upward to concave downward). Implement it as follows:

MATLAB Command

```
>>syms x ↵
>>y=x^4-12*x^3; ↵
>>d1=diff(y); ↵
>>r=solve(d1) ↵

r =

[ 0]
[ 0]
[ 9]
>>d2=diff(d1); ↵
>>subs(d2,x,r) ↵

ans =
```

[0]   ← 0 indicates the possibility of point of inflection
[0]
[324]

>>x=sym(-1/10); ↵   ← Assign $-\dfrac{1}{10}$ to x

>>subs(d2,x) ↵

ans =

183/25   ← Value of $\dfrac{d^2y}{dx^2}$ at $x=-\dfrac{1}{10}$, which is +ve

>>x=sym(1/10); ↵   ← Assign $\dfrac{1}{10}$ to x

>>subs(d2,x) ↵

ans =

-177/25   ← Value of $\dfrac{d^2y}{dx^2}$ at $x=\dfrac{1}{10}$, which is −ve

In addition to the analytical solution numerical subroutines are also available to find the minimum or maximum of a function.

☞ Finding the location of a minimum over some interval

Subroutine 'fminbnd' (abbreviation of functional minimum at) is used to find the value of x at which minimum of a polynomial/function occurs. The subroutine has three input arguments — the first argument is the function or polynomial string, the second argument is the lower limit of the intereval, and the third one is the upper limit of the interval. To find the location of a minimum of a function of x,

MATLAB Step:
Use the command fminbnd('function as string', lower limit of x, upper limit of x).

As an example, take $f(x)=7x^2-9x+7$. Analytically one can find the location of minimum for $f(x)$ by differentiating $f(x)=7x^2-9x+7$, setting the first derivative to zero, and solving for x, hence, $\dfrac{d}{dx}(7x^2-9x+7)$ $=14x-9=0$ and $x=\dfrac{9}{14}=0.642857142$. Since the second derivative is +ve (7), there is no need to check for the maximum or minimum. We can choose any interval, say, we selected the interval from $x=-4$ to 3. Have the location of minimum as follows:

MATLAB Command
>>fminbnd('7*x^2-9*x+7',-4,3) ↵

ans=
0.6429

Finding the first derivative may not be feasible for all functions it suggests then investigate the function numerically over certain interval of x by 'fminbnd'. Unfortunately, the subroutine determines only the first minimum so if we have mutiple minima, checking from the interval to interval is required.

☞ Finding the location of a maximum over some interval

By program 'fminbnd' can deal only with the finding of a minimum. A situation where $f(x)$ has a maximum can occur very often. If a function of x has a relative maximum between some interval, multiplying the function by −ve sign changes the function have minimum over the same interval of x. This artifice can be applied to determine the maximum of $f(x)$. Say, $f(x) = \sin 2x$, it is given that the function has a maximum at $x = \dfrac{\pi}{4} = 0.7854$. We search the maximum from 0 to π as follows:

MATLAB Command
>>fminbnd('-sin(2*x)',0,pi) ↵

ans=
0.7854

☞ *Location of the minimum/maximum of a composite function*
    Not only 'fminbnd' determines the location of the minimum of a standard function but also it looks for the location of minimum of a composite function. Function $h(x)$ $= \dfrac{e^{-2(x-1)}\cos^3(x-1)}{x} + (x-1)^2$ is such an example. Its plot is presnted in figure 5.5. From the figure one can say that the minimum of $h(x)$ occurs somewhere between $x = 1$ and $x = 2$ but we are not sure about the exact location of the minimum. We can compute it as follows:

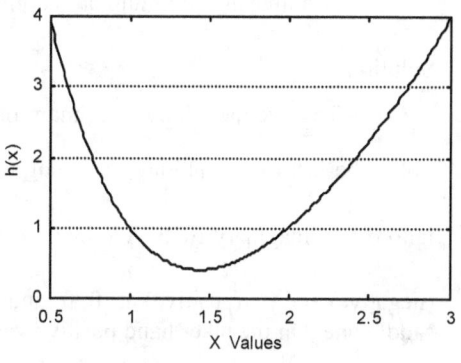

Figure 5.5   Plot of h(x) for $0.5 \leq x \leq 3$

MATLAB Command
>>h='exp(-2*(x-1))*cos(x-1)^3/x+(x-1)^2 '; ↵
>>m=fminbnd(h,1,2) ↵

m =
1.4392

From the MATLAB return the exact location of the minimum is at $x = 1.4392$. What if you are interested to find the functional value of $h(x)$ at $x = 1.4392$. Obviously, variable m contains the location. The computation is as follows:

>>x=m;   ← Assign contents of m to x
>>eval(h) ↵   ← Computation of $h(x)$ at location of minimum

ans =
0.4069

That is what is seemed the value of $h(x)$ according to figure 5.5. In a similar fashion, the location of a maximum can be found too preceding a negative sign for a composite function. One should keep the independent variable as x for the function string. *Independent variable other than x is not supported by 'fmin'.*

5.16.2 Minimum/maximum of a two-dimensional function

    General form of a two-dimensional function is $z = f(x, y)$, where x and y are two independent variables. If $z = f(x, y)$ has a relative maximum or minimum at $P(x_1, y_1)$, where $P(x_1, y_1)$ is an interior point of a region R in the x-y plane and if $\dfrac{\partial f(x,y)}{\partial x}$ and $\dfrac{\partial f(x,y)}{\partial y}$ exist at $P(x_1, y_1)$, then, it is necessary that $\dfrac{\partial f(x,y)}{\partial x} = 0$ and $\dfrac{\partial f(x,y)}{\partial y} = 0$ at $P(x_1, y_1)$. For sufficient conditions, we define a quantity D as $\left(\dfrac{\partial^2 f(x,y)}{\partial x^2}\right)\left(\dfrac{\partial^2 f(x,y)}{\partial y^2}\right) - \left(\dfrac{\partial^2 f(x,y)}{\partial y \partial x}\right)^2$ and D must be greater than 0 to exist a relative minimum or maximum. Over that region R at

$P(x_1, y_1)$, $z = f(x, y)$ has a relative maximum if $\dfrac{\partial^2 f(x, y)}{\partial x^2}$ or $\dfrac{\partial^2 f(x, y)}{\partial y^2}$ <0 and a relative minimum if $\dfrac{\partial^2 f(x, y)}{\partial x^2}$ or $\dfrac{\partial^2 f(x, y)}{\partial y^2}$ >0. If D <0, there is neither a minimum nor a maximum at $P(x_1, y_1)$, which indicates $\dfrac{\partial^2 f(x, y)}{\partial x^2}$ and $\dfrac{\partial^2 f(x, y)}{\partial y^2}$ are of opposite signs and the point $P(x_1, y_1)$ is called a saddle point or minimax. Plot of $z = f(x, y)$, in general, is a surface.

⊟ Problem

Examine the minimum or maximum for the surface $z = x^3 - 3xy + y^3 + 7$.

Solution:

To have the relative minimum or maximum, $\dfrac{\partial z}{\partial x} = 0$ and $\dfrac{\partial z}{\partial y} = 0$, hence, $3x^2 - 3y = 0$ and $-3x + 3y^2 = 0$.
Solve the last two equations pertaining to x and y to get $(x, y) = (0,0)$, $(1,1)$, and two imaginary roots. We ignore the imaginary roots. We have $D = \left(\dfrac{\partial^2 z}{\partial x^2}\right)\left(\dfrac{\partial^2 z}{\partial y^2}\right) - \left(\dfrac{\partial^2 z}{\partial y \partial x}\right)^2 = 6x \cdot 6y - (-3)^2 = 36\, xy - 9$ and D becomes -9

(negative) and 27 (positive) at (0, 0) and (1, 1) respectively. The negative value says that the point (0, 0) is a saddle one. On the other hand positive value says that there may be a relative minimum or maximum at (1, 1), for which, $\dfrac{\partial^2 z}{\partial x^2} = 6x = 6$ (positive), so, (1, 1) is the point corresponding to a relative minimum, and the functional value of z at (1, 1) is 6. Implementation is as follows:

MATLAB Command

>>syms x y ↵	← Declare x and y as symbolic
>>z=x^3-3*x*y+y^3+7; ↵	← Define $z = x^3 - 3xy + y^3 + 7$
>>e1=diff(z,x); e2=diff(z,y); ↵	← $\dfrac{\partial z}{\partial x}$ and $\dfrac{\partial z}{\partial y}$ are assigned to e1 and e2
>>S=solve(e1,e2) ↵	← Solve $\dfrac{\partial z}{\partial x} = 0$ and $\dfrac{\partial z}{\partial y} = 0$ and assign output to S

S =
    x: [4x1 sym]
    y: [4x1 sym]   ← It indicates that output S is a structured array having four components

>>[S.x S.y] ↵   ← Print the critical points

ans =

```
[               0,                0]
[               1,                1]
[ -1/2-1/2*i*3^(1/2),   -1/2+1/2*i*3^(1/2)]
[ -1/2+1/2*i*3^(1/2),   -1/2-1/2*i*3^(1/2)]
```
>>D=diff(z,x,2)*diff(z,y,2)-(diff(diff(z,y),x))^2; ↵   ←Form D
>>subs(D,{x,y},{0,0}) ↵   ← Compute D at (0, 0)

ans =
      −9   ← −ve value indicates saddle point at (0, 0)
>>subs(D,{x,y},{1,1}) ↵   ← Compute D at (1, 1)

ans =
      27   ← +ve value indicates no saddle point at (4, 4)

>>subs(diff(z,x,2),{x,y},{1,1}) ↵   ← Check $\dfrac{\partial^2 z}{\partial x^2}$ at (1, 1)

ans =
         6   ← +ve value says that relative minimum at (1, 1)
>>subs(z,{x,y},{1,1}) ↵   ← Minimum value of z at (1, 1)

ans =
         6

You can bypass all these steps by taking the help of Maple function 'maximize' or 'minimize'.

MATLAB Command

```
>>syms x y ↵
>>z=x^3-3*x*y+y^3+7; ↵
>>maple('minimize',z) ↵
```

ans =

6

Functions 'maximize' and 'minimize' can take different kind of arguments. Get Maple help from the MATLAB Command Window executing 'mhelp maximize'. The reader might ask why we need so much computation before applying the function 'minimize'. The reason for this is initially we do not know whether the critical points are minimum, maximum, or saddle ones. This is understood by the following:

>>maple('maximize',z) ↵   ← Try to find the maximum of z

ans =
         7   ←This is not the maximum of z but saddle point value

The subroutines are applicable for one dimension too. There is another function by the name 'fmins', which is the two-dimensional counterpart of 'fmin' (presented in article 5.16.1). Explore the style of implementation for 'fmins' which is left as an exercise for the reader.

Chapter 6

Problems of Integral Calculus

Problems of integral calculus are the subject matter of this chapter. Integration is the reverse process of differentiation. With precision, integration enables us to calculate analytically/numerically areas, volumes, surface areas, and centroids bounded or contained by curved lines or surfaces and many other quantities which it is necessary to find in both pure and applied mathematics. It will be instructive to the reader that simpler commands turn out clumsy and tedious integration or problems derived from integration to be easily performed. If the reader is well equipped with those powerful weapons, he can proceed to accomplish more fully the complicated uses to which the implementation may be applied. In doing so, one would appreciate the usefulness of subroutine 'int' that is the MATLAB's counterpart of $\int$ .... dx .

6.1 Symbolic integration of standard functions

In this section, you will learn how to integrate standard indefinite integrals in MATLAB environment. Like differentiation, integration of functions can be evaluated in MATLAB by entering the functions in terms of strings or symbolic expressions. There are several types of integration, namely, single, double...etc. We discuss different integration one by one. Substantial examples are furnished expecting you to have good training in applying the subroutine 'int', which is the abbreviation of <u>int</u>egration. You can say integration symbol $\int$ .... dx of calculus $\equiv$ 'int' of MATLAB. Indefinite integration of a function is accompanied by a constant. That constant is avoided in MATLAB. To evaluate a standard integrand,

MATLAB Step:
       Use the command int('function as string').

Let us consider the integrand $\cos x$. We know that $\int \cos x\, dx = \sin x$. To integrate $\cos x$, we can use two styles as follows:

MATLAB Command
for string form,
        >>int('cos(x)') ↵

for symbolic form,
        >>syms x ↵
        >>int(cos(x)) ↵

    ans =

ans =

sin(x)

sin(x)

Integrations of several standard functions and their MATLAB implementation are provided in the following tables.

Table 6.A   Integrations of trigonometric functions and their MATLAB counterparts

Mathematical Notation	Equivalent MATLAB Notation
$\int \sin x\,dx = -\cos x$	int('sin(x)') $\Rightarrow$ -cos(x)
$\int \cos x\,dx = \sin x$	int('cos(x)') $\Rightarrow$ sin(x)
$\int \tan x\,dx = -\ln \cos x$	int('tan(x)') $\Rightarrow$ -log(cos(x))
$\int \cot x\,dx = \ln \sin x$	int('cot(x)') $\Rightarrow$ log(sin(x))
$\int \sec x\,dx = \ln(\sec x + \tan x)$	int('sec(x)') $\Rightarrow$ log(sec(x)+tan(x))
$\int \cos ecx\,dx = \ln(\cos ecx - \cot x)$	int('csc(x)') $\Rightarrow$ log(csc(x)-cot(x))
$\int \sin^{-1} x\,dx = x\sin^{-1} x + \sqrt{1-x^2}$	int('asin(x)') $\Rightarrow$ x*asin(x)+(1-x^2)^(1/2)
$\int \cos^{-1} x\,dx = x\cos^{-1} x - \sqrt{1-x^2}$	int('acos(x)') $\Rightarrow$ x*acos(x)-(1-x^2)^(1/2)
$\int \tan^{-1} x\,dx = x\tan^{-1} x - \dfrac{1}{2}\ln(1+x^2)$	int('atan(x)') $\Rightarrow$ x*atan(x)-1/2*log(1+x^2)
$\int \cot^{-1} x\,dx = x\cot^{-1} x + \dfrac{1}{2}\ln(1+x^2)$	int('acot(x)') $\Rightarrow$ x*acot(x)+1/2*log(1+x^2)
$\int \sec^{-1} x\,dx = x\sec^{-1} x - \ln(x+\sqrt{x^2 -1})$	int('asec(x)') $\Rightarrow$ x*asec(x)-log((1+sqrt(1-1/x^2))*x)
$\int \cos ec^{-1} x\,dx = x\cos ec^{-1} x + \cosh^{-1} x$	int('acsc(x)') $\Rightarrow$ x*acsc(x)+log((1+sqrt(1-1/x^2))*x)

* It seems that the string of int('asec(x)') does not look like its mathematical form. Convert the string to

mathematical form, which turns to $x\sec^{-1} x - \ln\left[\left(1+\sqrt{1-\dfrac{1}{x^2}}\right)x\right]$. One can write $\sqrt{1-\dfrac{1}{x^2}} = \sqrt{\dfrac{x^2 -1}{x^2}} = \dfrac{\sqrt{x^2 -1}}{x}$, $\therefore$

$\ln\left[\left(1+\sqrt{1-\dfrac{1}{x^2}}\right)x\right] = \ln\left[\left(1+\dfrac{\sqrt{x^2 -1}}{x}\right)x\right] = \ln(x+\sqrt{x^2 -1})$ so the output of int('asec(x)') resembles to the

mathematical form. Similar circumstance also arises with the output string of $\int \cos ec^{-1} x\,dx$. Here you have to use the identity $\cosh^{-1} x = \ln(x+\sqrt{x^2 -1})$. Tables 6.B and 6.C show more integration.

Table 6.B   Integrations of hyperbolic functions and their MATLAB counterparts

Mathematical Notation	Equivalent MATLAB Notation
$\int \sinh x\,dx = \cosh x$	int('sinh(x)') $\Rightarrow$ cosh(x)
$\int \cosh x\,dx = \sinh x$	int('cosh(x)') $\Rightarrow$ sinh(x)
$\int \tanh x\,dx = \ln \cosh x$	int('tanh(x)') $\Rightarrow$ log(cosh(x))
$\int \coth x\,dx = \ln \sinh x$	int('coth(x)') $\Rightarrow$ log(sinh(x))
$\int \sec hx\,dx = \tan^{-1}(\sinh x)$	int('sech(x)') $\Rightarrow$ atan(sinh(x))

Continuation of the previous table:

Mathematical Notation	Equivalent MATLAB Notation
$\int \cos ech x dx = \ln \tanh \dfrac{x}{2}$	int('csch(x)') $\Rightarrow$ log(tanh(1/2*x))
$\int \sinh^{-1} x dx = x \sinh^{-1} x - \sqrt{1+x^2}$	int('asinh(x)') $\Rightarrow$ x*asinh(x)-(1+x^2)^(1/2)
$\int \cosh^{-1} x dx = x \cosh^{-1} x - \sqrt{x^2-1}$	int('acosh(x)') $\Rightarrow$ x*acosh(x)-(-1+x)^(1/2) *(1+x)^(1/2)
$\int \tanh^{-1} x dx = x \tanh^{-1} x + \dfrac{1}{2}\ln(1-x^2)$	int('atanh(x)') $\Rightarrow$ x*atanh(x)+1/2*log(1-x^2)
$\int \coth^{-1} x dx = x \coth^{-1} x + \dfrac{1}{2}\ln(x^2-1)$	int('acoth(x)') $\Rightarrow$ x*acoth(x)+1/2*log(x^2-1)
$\int \sec h^{-1} x dx = x \sec h^{-1} x - \tan^{-1}\left(\dfrac{\sqrt{1-x^2}}{x}\right)$	int('asech(x)') $\Rightarrow$ x*asech(x)-atan((1/x-1)^(1/2) *(1/x+1)^(1/2))
$\int \cos ech^{-1} x dx = x \cos ech^{-1} x + \ln(x+\sqrt{1+x^2})$	int('acsch(x)') $\Rightarrow$ x*acsch(x)+ log((1+sqrt(1+1/x^2))*x)

Table 6.C Integrations of other functions and their MATLAB counterparts

Mathematical Notation	Equivalent MATLAB Notation
$\int x^n dx = \dfrac{x^n+1}{n+1}$	int('x^n') $\Rightarrow$ x^(n+1)/(n+1)
$\int e^x dx = e^x$	int('exp(x)') $\Rightarrow$ exp(x)
$\int \ln x dx = x \ln x - x$	int('log(x)') $\Rightarrow$ x*log(x)-x
$\int a^x dx = \dfrac{a^x}{\ln a}$	int('a^x') $\Rightarrow$ 1/log(a)*a^x

6.1.1 Single integration of a composite function

Integrations of composite functions are evaluated using the standard results and some other techniques such as substitution of variables, part by part integration...etc. The function 'int' also performs the single integration of composite function expressed as a string (presented in chapter 11).

Evaluate the following single indefinite integrals in MATLAB:

A . $\int \dfrac{(\ln x)^4}{x} dx$ $\qquad$ B . $\int \dfrac{1}{x(x^2+1)^2} dx$ $\qquad$ C . $\int \sin^3 x \cos^3 x dx$

D . $\int e^{-7x} \cos(3x-34) dx$ $\qquad$ E . $\int \dfrac{1}{\sqrt{3u^2-3u+1}} du$ $\qquad$ F . $\int \log_{10} x dx$

⌧ Integrand A

Method of substitution is used. Substitute $\ln x = u$, it follows then, $du = \dfrac{1}{x} dx$ and $\int \dfrac{(\ln x)^4}{x} dx = \int u^4 du = \dfrac{1}{5} u^5 = \dfrac{1}{5}(\ln x)^5$. To evaluate integrand A,

MATLAB Command

```
>>I=int('(log(x)^4)/x');  ↵          ← Integration is assigned to I
>>pretty(I) ↵                         ← Display the symbolic form of output
```

```
          5
1/5 log(x)
```

Integrand B

For this integral, first partial fraction of the integrand $\frac{1}{x(x^2+1)^2}$ is done. Forming partial fraction of the integrand provides $\frac{1}{x(x^2+1)^2}=\frac{1}{x}-\frac{x}{x^2+1}-\frac{x}{(x^2+1)^2}$, $\therefore \int\frac{1}{x(x^2+1)^2}dx=\int\frac{1}{x}dx-\int\frac{x}{x^2+1}dx-\int\frac{x}{(x^2+1)^2}dx=$ $\ln x-\frac{1}{2}\ln(x^2+1)+\frac{1}{2(x^2+1)}$. To evaluate the integrand,

MATLAB Command

```
>>I=int('1/(x*((x^2+1)^2))'); ↵
>>pretty(I) ↵
```

```
                        2                  1
log(x) - 1/2 log(x  + 1) + 1/2  -----------
                                       2
                                      x  + 1
```

Integrand C

Out of many ways to integrating the problem, the method of substitution is used. Either the substitution of $\sin x = z$ or $\cos x = z$ will do. Say, $\cos x = z$, hence, we have $dz = -\sin x dx$ and $\int \sin^3 x\cos^3 x dx =$ $\int\sin^2 x\sin x\cos^3 x dx=\int(1-\cos^2 x)\sin x\cos^3 x dx=-\int(1-z^2)z^3 dz=\int(z^5-z^3)dz=\frac{1}{6}z^6-\frac{1}{4}z^4=\frac{1}{6}\cos^6 x-\frac{1}{4}\cos^4 x$. To integrate this function,

MATLAB Command

```
>>I=int('sin(x)^3*cos(x)^3'); ↵
>>pretty(I) ↵
```

```
           2       4            4
- 1/6 sin(x)  cos(x)  - 1/12 cos(x)
```

It is important to mention that MATLAB output can yield different form of results. An illustration of this is easily seen by comparing the integration done by the method of substitution $\left(\frac{1}{6}\cos^6 x-\frac{1}{4}\cos^4 x\right)$ with the output returned by MATLAB $\left(-\frac{1}{6}\sin^2 x\cos^4 x-\frac{1}{12}\cos^4 x\right)$. Simple manipulation would remove such ambiguity:

$$\frac{1}{6}\cos^6 x-\frac{1}{4}\cos^4 x=\cos^4 x\left(\frac{\cos^2 x}{6}-\frac{1}{4}\right)=\cos^4 x\left(\frac{1-\sin^2 x}{6}-\frac{1}{4}\right)=\cos^4 x\left(\frac{-\sin^2 x}{6}-\frac{1}{12}\right)=-\frac{1}{6}\sin^2 x\cos^4 x-\frac{1}{12}\cos^4 x.$$

Integrand D

This is the product form of two functions, e^{-7x} and $\cos(3x-34)$. Rule of integration by parts should be used for this integrand, which is given by $\int[uv]\,dx=u\int vdx-\int\left\{\frac{du}{dx}\int vdx\right\}dx$. Say, I=$\int e^{-7x}\cos(3x-34)\,dx$, one can write I=$\frac{1}{3}e^{-7x}\sin(3x-34)+\frac{7}{3}\int e^{-7x}\sin(3x-34)dx=\frac{1}{3}e^{-7x}\sin(3x-34)+\frac{7}{3}\left[-\frac{1}{3}e^{-7x}\cos(3x-34)-\frac{7}{3}\int e^{-7x}\cos(3x-34)\,dx\right]$ $=\frac{1}{3}e^{-7x}\sin(3x-34)-\frac{7}{9}e^{-7x}\cos(3x-34)-\frac{49}{9}$I. Solving for I yields I=$-\frac{7}{58}e^{-7x}\cos(3x-34)+\frac{3}{58}e^{-7x}\sin(3x-34)$. The integration is as follows:

MATLAB Command
>>I=int('exp(-7*x)*cos(3*x-34)'); ↵
>>pretty(I) ↵

- 7/58 exp(-7 x) cos(3 x - 34) + 3/58 exp(-7 x) sin(3 x - 34)

⊡ Integrand E

In this integration, the integrand is arranged similar to the standard form, $\therefore \int \dfrac{1}{\sqrt{3u^2 - 3u + 1}} du =$

$\int \dfrac{1}{\sqrt{3\left(u^2 - u + \dfrac{1}{3}\right)}} du = \int \dfrac{1}{\sqrt{3\left[\left(u - \dfrac{1}{2}\right)^2 + \dfrac{1}{3} - \dfrac{1}{4}\right]}} du = \int \dfrac{1}{\sqrt{\left\{\sqrt{3}\left(u - \dfrac{1}{2}\right)\right\}^2 + \left(\dfrac{1}{2}\right)^2}} du$. Substitute $\sqrt{3}\left(u - \dfrac{1}{2}\right) = z$ to get $\sqrt{3}du$

$= dz$. So, the integration becomes $\dfrac{1}{\sqrt{3}} \int \dfrac{1}{\sqrt{z^2 + \left(\dfrac{1}{2}\right)^2}} dz$. We know from the standard integration that $\int \dfrac{dx}{\sqrt{x^2 + a^2}} =$

$\sinh^{-1}\dfrac{x}{a}$. Therefore, we obtain $\dfrac{1}{\sqrt{3}} \int \dfrac{1}{\sqrt{z^2 + \left(\dfrac{1}{2}\right)^2}} dz = \dfrac{1}{\sqrt{3}} \sinh^{-1}(2z) = \dfrac{1}{\sqrt{3}} \sinh^{-1}\left\{2\sqrt{3}\left(u - \dfrac{1}{2}\right)\right\}$. For integrand E,

MATLAB Command
>>I=int('1/sqrt(3*u^2-3*u+1)'); ↵
>>pretty(I) ↵
```
          1/2          1/2
1/3    3      asinh(2  3     (u - 1/2))
```

⊡ Integrand F

All integrands can not be integrated by MATLAB, such an example is $\log_{10} x$. Try with this,

MATLAB Command
>>int('log10(x)') ↵

Warning: Explicit integral could not be found.
  > In C:\MATLAB\toolbox\symbolic\@sym\int.m at line 58
    In C:\MATLAB\toolbox\symbolic\@char\int.m at line 9

ans =

int(log10(x),x)

Above error message shows that $\log_{10} x$ is not defined in symbolic toolbox, so, integration of $\log_{10} x$ is not

possible. Some manipulations can be used to carry out this integration. We know that $\log_{10} x = \dfrac{\ln x}{\ln 10}$, therefore,

$\int \log_{10} x \, dx = \int \dfrac{\ln x}{\ln 10} dx = \dfrac{x \ln x - x}{\ln 10}$.

MATLAB Command
>>pretty(int('log(x)/log(10) ')) ↵

```
  x log(x)       x
----------  -  ---------
  log(10)       log(10)
```

6.1.2 Double integration of a composite function

Two 'int' subroutines are adopted to find the double integration of the composite function of two variables. General form of a double integration is $\iint f(x,y)dxdy$. There are two independent variables, x and y, in this integration. When the inner integration is performed, y is kept constant. To perform a double integration,

MATLAB Step:
> *Use the command int(int('function as string of two independent variables', 'first independent variable'), 'second independent variable').*

Evaluate the following double indefinite integrals using MATLAB:

$$A.\ \iint xye^{x^2+y^2}dxdy \qquad B.\ \iint\frac{dudv}{u+v^2} \qquad C.\ \iint r^2\sin^2\theta\cos\theta\,drd\theta$$

⊟ Integrand A

For inner integration, y is kept constant, $\therefore\ \iint xye^{x^2+y^2}dxdy=\frac{1}{2}\iint y(2x)e^{x^2+y^2}dxdy=\frac{1}{2}\iint ye^{x^2+y^2}d(x^2)dy=$

$\frac{1}{2}\int y\left(\int e^{x^2+y^2}d(x^2)\right)dy=\frac{1}{2}\int ye^{x^2+y^2}dy=\frac{1}{4}\int 2ye^{x^2+y^2}dy=\frac{1}{4}\int e^{x^2+y^2}d(y^2)$, x is kept constant$=\frac{1}{4}e^{x^2+y^2}$.

MATLAB Command

```
>>I=int(int('x*y*exp(x^2+y^2)','x'),'y'); ↵        ← The output is assigned to I
>>pretty(I) ↵                                        ← Show the output
                2   2
        1/4 exp( x  + y  )
```

⊟ Integrand B

$$\iint\frac{dudv}{u+v^2}=\int\left(\int\frac{du}{u+v^2}\right)dv=\int\ln(u+v^2)\,dv=v\ln(u+v^2)-\int v.\frac{1}{u+v^2}.2v\,dv\text{, rule of integration by parts is used}$$

$$=v\ln(u+v^2)-\int\frac{2v^2dv}{u+v^2}=v\ln(u+v^2)-\int\frac{2(u+v^2-u)dv}{u+v^2}=v\ln(u+v^2)-\int 2dv+2u\int\frac{dv}{u+v^2}=v\ln(u+v^2)-2v+2u\int\frac{dv}{v^2+\left(\sqrt{u}\right)^2}$$

$$=v\ln(u+v^2)-2v+2u.\frac{1}{\sqrt{u}}\tan^{-1}\frac{v}{\sqrt{u}}=v\ln(u+v^2)-2v+2\sqrt{u}\tan^{-1}\frac{v}{\sqrt{u}}.$$

MATLAB Command

```
>>I=int(int('1/(u+v^2)','u'),'v'); ↵
>>pretty(I) ↵


              2            1/2            v
  v log( u + v  ) - 2 v  + 2 u     atan(----------)
                                           1/2
                                          u
```

⊟ Integrand C

There is no symbol like θ available in workspace of MATLAB except LaTex command, which is used in the graphics. Writing t instead of θ, we have $\iint r^2\sin^2\theta\cos\theta\,drd\theta=\int\sin^2\theta\cos\theta\left(\int r^2dr\right)d\theta=$

$\frac{1}{3}r^3\int\sin^2\theta\cos\theta\,d\theta=\frac{1}{3}r^3\left(\frac{1}{3}\sin^3\theta\right)$ (substitution of $z=\sin\theta$ is used)$=\frac{1}{9}r^3\sin^3\theta$.

MATLAB Command

```
>>I=int(int('r^2*sin(t)^2*cos(t)','r'),'t'); ↵
>>pretty(I) ↵
```

$$\frac{1}{9}\ \sin(t)^3\ r^3$$

There is some other toolbox called the student package, which is run by the Maple package. Symbolic form is necessary to run the toolbox. Function of the package that performs double integration is 'Doubleint'. Step by step procedure for the example C is presented as follows:

MATLAB Command

```
>>maple('with','student'); ↵        ← Activate the student package
>>syms r t ↵                        ← Declare the variables of ∬r² sin²θ cosθ drdθ , r (r) and θ (t), as symbolic
>>f=r^2*sin(t)^2*cos(t); ↵          ← Define integrand and assign that to f
>>I=maple('Doubleint',f,r,t); ↵     ← Integration of f is assigned to I
>>R=maple('map','value',I); ↵       ← Evaluate integration and assign the output to R
>>pretty(R) ↵
```

$$\frac{1}{9}\ \sin(t)^3\ r^3$$

6.1.3 Triple integration of a composite function

In triple integration, just you need to employ the subroutine 'int' three times nestedly. To perform a triple integration,

MATLAB Step:

Use the command int(int(int('function as string of three independent variables', 'first independent variable'), 'second independent variable'),'third independent variable').

Evaluate the following triple indefinite integrals using MATLAB:

$$A\ .\ \iiint(z^3-1)xye^{x^2+y^2}dxdydz \qquad\qquad B\ .\ \iiint\frac{v}{w+1}\ln u\,dudvdw$$

$$C\ .\ \iiint r^{\frac{3}{2}}\sin\theta\cos^2\varphi\,dr\,d\theta\,d\varphi$$

⌗ Integrand A

$$\iiint(z^3-1)xye^{x^2+y^2}dxdydz=\iint(z^3-1)y\left(\int xe^{x^2+y^2}dx\right)dydz\ =\ \iint(z^3-1)y\left(\frac{1}{2}\int e^{x^2+y^2}d(x^2)\right)dy\,dz\ =\ \frac{1}{2}\iint(z^3-1)y\ \times$$

$$e^{x^2+y^2}dydz=\frac{1}{4}\int(z^3-1)\left(\int 2ye^{x^2+y^2}dy\right)dz=\frac{1}{4}\int(z^3-1)\left(\int e^{x^2+y^2}d(y^2)\right)dz=\frac{1}{4}e^{x^2+y^2}\int(z^3-1)dz=\frac{1}{4}e^{x^2+y^2}\left(\frac{1}{4}z^4-z\right).$$

MATLAB Command

```
>>I=int(int(int('(z^3-1)*x*y*exp(x^2+y^2)','x'),'y'),'z'); ↵
>>pretty(I) ↵
```

$$1/4\ \exp(x^2\ +\ y^2\ )\ (1/4\ z^4\ -\ z)$$

⌗ Integrand B

$$\iiint\frac{v}{w+1}\ln u\,dudvdw=\iint\frac{v}{w+1}\left(\int\ln u\,du\right)dvdw=\iint\frac{v}{w+1}(u\ln u-u)\,dvdw=(u\ln u-u)\int\frac{1}{w+1}\left(\int v\,dv\right)dw=(u\ln u-$$

$$u)\int\frac{1}{w+1}\left(\frac{1}{2}v^2\right)dw=\frac{1}{2}(u\ln u-u)v^2\int\frac{1}{w+1}\,dw=\frac{1}{2}\ (u\ln u-u)\ v^2\ln(w+1)=\frac{1}{2}u\ln u\ v^2\ln(w+1)-\frac{1}{2}uv^2\ \ln(w+1).$$

MATLAB Command

```
>>I=int(int(int('log(u)*v/(w+1)','u'),'v'),'w'); ↵
>>pretty(I) ↵
```

$$\frac{1}{2}\ \log(w+1)\ v^{2}\ \ u\ \log(u)\ -\ 1/2\ \ \log(w+1)\ v^{2}\ \ u$$

⊟ Integrand C

Use t and p instead of θ and φ respectively. We have $\iiint r^{\frac{3}{2}}\sin\theta\cos^{2}\varphi\,dr\,d\theta\,d\varphi =$

$\iint\sin\theta\cos^{2}\varphi\left(\int r^{\frac{3}{2}}dr\right)d\theta\,d\varphi=\iint\sin\theta\cos^{2}\varphi\left(\frac{2}{5}r^{\frac{5}{2}}\right)d\theta\,d\varphi=\frac{2}{5}r^{\frac{5}{2}}\iint\sin\theta\cos^{2}\varphi\,d\theta\,d\varphi=\frac{2}{5}r^{\frac{5}{2}}\int\cos^{2}\varphi\left(\int\sin\theta\,d\theta\right)d\varphi \qquad =$

$\frac{2}{5}r^{\frac{5}{2}}\int\cos^{2}\varphi\ (-\cos\theta)\,d\varphi=-\frac{2}{5}r^{\frac{5}{2}}\cos\theta\int\cos^{2}\varphi\,d\varphi$ and that $\int\cos^{2}\varphi\,d\varphi=\frac{1}{2}\int(1+\cos2\varphi)\,d\varphi=\frac{1}{2}\left(\varphi+\frac{1}{2}\sin2\varphi\right)=$

$\frac{\varphi}{2}+\frac{2\sin\varphi\cos\varphi}{4}=\frac{\varphi}{2}+\frac{\sin\varphi\cos\varphi}{2}$. The integration becomes $-\frac{2}{5}\cos\theta\ r^{\frac{5}{2}}\left(\frac{\varphi}{2}+\frac{\sin\varphi\cos\varphi}{2}\right)$.

MATLAB Command
```
>>I=int(int(int('r^(3/2)*sin(t)*cos(p)^2','r'),'t'),'p'); ↵
>>pretty(I) ↵
```
$$\qquad\qquad\qquad\qquad\qquad 5/2$$
$$- 2/5\ (1/2\ \cos(p)\ \sin(p) + 1/2\ p)\ \cos(t)\ r$$

What if we solve the integrand C by the student package:
```
>>maple('with','student'); ↵
>>syms r t p ↵              ← Declare variables r (r), θ (t), and φ (p) as symbolic
>>f= r^(3/2)*sin(t)*cos(p)^2; ↵   ← Define integrand and assign that to f
>>I=maple('Tripleint',f,r,t,p); ↵  ← Function 'Tripleint' performs triple integration
>>R=maple('map','value',I); ↵
>>pretty(R) ↵
```
$$\qquad\qquad\qquad\qquad\qquad 5/2$$
$$- 2/5\ (1/2\ \cos(p)\ \sin(p) + 1/2\ p)\ \cos(t)\ r$$

6.2 Definite integration

Definite integrals include lower and upper limits to the indefinite integrals. General forms of a single, double, and triple definite integration are $\int_{x=a}^{x=b}f(x)dx$, $\int_{y=c}^{y=d}\int_{x=a}^{x=b}f(x,y)dxdy$, and $\int_{z=e}^{z=f}\int_{y=c}^{y=d}\int_{x=a}^{x=b}f(x,y,z)dxdydz$ respectively.

The subroutine 'int' with different number of input arguments is put to use for the definite integration. Another two arguments are appended with 'int' — one for lower limit and the other for upper limit.

6.2.1 Single definite integration

To evaluate a single definite integral of the form $\int_{x=a}^{x=b}f(x)\,dx$,

MATLAB Step:

Use the command int(f(x) as string, x , a , b).

Writing the independent variable inside 'int' is optional. If the lower and upper limits have symbolic constants, those symbolic constants can not be the independent variables of the function string.

Evaluate the following single definite integrals using MATLAB:

We consider the same integrands as we discussed in article 6.1.1. Some lower and upper limits are included in the integration.

$A.\ \int\limits_{x=1}^{x=4}\dfrac{(\ln x)^4}{x}\,dx$ $\qquad B.\ \int\limits_{x=3}^{x=9}\dfrac{1}{x(x^2+1)^2}\,dx$ $\qquad C.\ \int\limits_{x=-\frac{\pi}{2}}^{x=\frac{\pi}{2}}\sin^3 x\cos^3 x\,dx$

$D.\ \int\limits_{x=0}^{x=3\pi}e^{-7x}\cos(3x-34)\,dx$ $\qquad E.\ \int\limits_{u=0}^{u=4}\dfrac{1}{\sqrt{3u^2-3u+1}}\,du$

⌗ **Integrand A**

$$\int\limits_{x=1}^{x=4}\frac{(\ln x)^4}{x}\,dx=\left[\tfrac{1}{5}(\ln x)^5\right]_{x=1}^{x=4}=\tfrac{1}{5}(\ln 4)^5-\tfrac{1}{5}(\ln 1)^5=\tfrac{1}{5}(2\ln 2)^5-0=\tfrac{32}{5}(\ln 2)^5,\ \ \because\ \ln 1=0.$$

MATLAB Command

```
>>R=int('log(x)^4/x',1,4); ↵
>>pretty(R) ↵
                    5
          32/5  log(2)
```

What if this integrand had lower and upper limits, which are symbolic constants. Such an example is $\int\limits_{x=a^2}^{x=v^2}\dfrac{(\ln x)^4}{x}\,dx$, $\therefore\ \int\limits_{x=a^2}^{x=v^2}\dfrac{(\ln x)^4}{x}\,dx=\left[\tfrac{1}{5}(\ln x)^5\right]_{x=a^2}^{x=v^2}=\dfrac{32}{5}(\ln v)^5-\dfrac{32}{5}(\ln a)^5$. The required command for that would be,

```
>>R=int('log(x)^4/x','x','a^2','v^2'); ↵
>>pretty(R) ↵
               5              5
     32/5 log(v)   -  32/5 log(a)
```

⌗ **Integrand B**

$$\int\limits_{x=3}^{x=9}\frac{1}{x(x^2+1)^2}\,dx=\left[\ln x-\frac{1}{2}\ln(x^2+1)+\frac{1}{2(x^2+1)}\right]_{x=3}^{x=9}=\left[\ln 9-\frac{1}{2}\ln(9^2+1)+\frac{1}{2(9^2+1)}\right]-\left[\ln 3-\frac{1}{2}\ln(3^2+1)+\right.$$

$$\left.\frac{1}{2(3^2+1)}\right]=\ln 9-\frac{1}{2}\ln 82+\frac{1}{164}-\left(\ln 3-\frac{1}{2}\ln 10+\frac{1}{20}\right)=\ln\frac{9}{3}-\frac{1}{2}\ln\frac{82}{10}+\frac{1}{164}-\frac{1}{20}=\ln 3-\frac{1}{2}\ln\frac{41}{5}-\frac{9}{205}.$$

MATLAB Command

```
>>R=int('1/(x*((x^2+1)^2))',3,9); ↵
>>pretty(R) ↵

     log(3) - 1/2 log(41) - 9/205 + 1/2 log(5)
```

⌗ **Integrand C**

$$\int\limits_{x=-\frac{\pi}{2}}^{x=\frac{\pi}{2}}\sin^3 x\cos^3 x\,dx=\left[-\frac{1}{6}\sin^2 x\cos^4 x-\frac{1}{12}\cos^4 x\right]_{x=-\frac{\pi}{2}}^{x=\frac{\pi}{2}}=\left[-\frac{1}{6}\sin^2\frac{\pi}{2}\cos^4\frac{\pi}{2}-\frac{1}{12}\cos^4\frac{\pi}{2}\right]-\left[-\frac{1}{6}\sin^2\left(-\frac{\pi}{2}\right)\ \times\right.$$

$$\left.\cos^4\left(-\frac{\pi}{2}\right)-\frac{1}{12}\cos^4\left(-\frac{\pi}{2}\right)\right]=0,\ \because\ \cos\frac{\pi}{2}=0.$$

MATLAB Command

```
>>int('sin(x)^3*cos(x)^3',-pi/2,pi/2) ↵

ans =

     0
```

⊟ Integrand D

$$\int_{x=0}^{x=3\pi} e^{-7x} \cos(3x-34)dx = \left[-\frac{7}{58}e^{-7x}\cos(3x-34)+\frac{3}{58}e^{-7x}\sin(3x-34)\right]_{x=0}^{x=3\pi} = \left[-\frac{7}{58}e^{-21\pi}\cos(9\pi-34)+\frac{3}{58}e^{-21\pi}\times\right.$$

$$\left.\sin(9\pi-34)\right] - \left[-\frac{7}{58}\cos(-34)+\frac{3}{58}\sin(-34)\right] = e^{-21\pi}\left[\frac{7}{58}\cos(34)+\frac{3}{58}\sin(34)\right]+\frac{7}{58}\cos(34)+\frac{3}{58}\sin(34).$$

MATLAB Command

```
>>R=int('exp(-7*x)*cos(3*x-34)',0,3*pi); ↵
>>pretty(R) ↵
```

    (7/58 cos(34) + 3/58 sin(34)) exp(-21 pi) + 7/58 cos(34) + 3/58 sin(34)

⊟ Integrand E

$$\int_{u=0}^{u=4}\frac{du}{\sqrt{3u^2-3u+1}} = \left[\frac{1}{\sqrt{3}}\sinh^{-1}\left\{2\sqrt{3}\left(u-\frac{1}{2}\right)\right\}\right]_{u=0}^{u=4} = \frac{1}{\sqrt{3}}\sinh^{-1}\left\{2\sqrt{3}\left(\frac{7}{2}\right)\right\}-\frac{1}{\sqrt{3}}\sinh^{-1}\left\{2\sqrt{3}\left(\frac{-1}{2}\right)\right\} =$$

$$\frac{1}{\sqrt{3}}\sinh^{-1}(7\sqrt{3}) - \frac{1}{\sqrt{3}}\sinh^{-1}(-\sqrt{3}). \text{ We know that } \sinh^{-1}x = \ln(x+\sqrt{x^2+1}), \therefore \sinh^{-1}(7\sqrt{3}) = \ln[7\sqrt{3}+\sqrt{(7\sqrt{3})^2+1}]$$

$$= \ln(7\sqrt{3}+2\sqrt{37}) \text{ and } \sinh^{-1}(-\sqrt{3}) = \ln[-\sqrt{3}+\sqrt{(-\sqrt{3})^2+1}] = \ln(2-\sqrt{3}). \text{ So, the definite integral E becomes}$$

$$\frac{1}{\sqrt{3}}\ln(7\sqrt{3}+2\sqrt{37})-\frac{1}{\sqrt{3}}\ln(2-\sqrt{3}).$$

MATLAB Command

```
>>R=int('1/sqrt(3*u^2-3*u+1)',0,4); ↵
>>pretty(R) ↵
```

                  1/2   1/2   1/2   1/2   1/2
      1/3 log(7 3   + 2 37)3   - 1/3 log(2 - 3   )3

                                                                    1/2
Confusion may arise looking at the last pretty form of the integration. The representation 7 3   means $7\sqrt{3}$

                                                      1/2
not $\sqrt{73}$. If it were $\sqrt{73}$, the expression would look like 73   and there would not be any space gap between 7 and 3.

6.2.2 Double definite integration

To perform a double definite integral of the form $\int_{y=c}^{y=d}\int_{x=a}^{x=b}f(x,y)\,dxdy$,

MATLAB Step:
    Use the command int(int($f(x,y)$ as string, x, a, b), y, c, d).

If the lower and upper limits are the functions of some variables involved in integration, symbolic declaration is needed before the integration. Above-mentioned step is applicable as far as the upper and lower limits are constant numbers.

Evaluate the following double definite integrals:

$$A. \int_{v=3}^{v=6}\int_{u=0}^{u=v}\frac{dudv}{u^2+v^2} \qquad\qquad B. \int_{y=1}^{y=2}\int_{x=0}^{x=1}xye^{x^2+y^2}dxdy$$

$$C. \int_{\theta=\frac{\pi}{6}}^{\theta=\frac{\pi}{4}}\int_{r=0}^{r=4\sin 2\theta}r\sin\theta\cos\theta drd\theta \qquad\qquad D. \int_{z=2}^{z=4}\int_{x=-\infty}^{x=\infty}\frac{3zdxdz}{x^2+z^4}$$

◻ Integrand A

$$\int_{v=3}^{v=6}\int_{u=0}^{u=v}\frac{du\,dv}{u^2+v^2}=\int_{v=3}^{v=6}\left(\int_{u=0}^{u=v}\frac{du}{u^2+v^2}\right)dv=\int_{v=3}^{v=6}\left[\frac{1}{v}\tan^{-1}\frac{u}{v}\right]_{u=0}^{u=v}dv=\int_{v=3}^{v=6}\left[\frac{1}{v}\tan^{-1}\frac{v}{v}-0\right]dv=\frac{\pi}{4}\int_{v=3}^{v=6}\frac{1}{v}dv=\frac{\pi}{4}\left[\ln v\right]_{v=3}^{v=6}=\frac{\pi}{4}[\ln 6-$$

$$\ln 3]=\frac{\pi}{4}\ln\frac{6}{3}=\frac{\pi}{4}\ln 2.$$

MATLAB Command

```
>>syms u v ↵
>>R=int(int('1/(u^2+v^2)',u,0,'v'),v,3,6); ↵
>>pretty(R)
        1/4 (log(2) + log(3)) pi - 1/4 log(3) pi
```

For the inner integration, the upper limit is a function of the second independent variable. The output returned by MATLAB is $\frac{1}{4}(\ln 2+\ln 3)\pi-\frac{1}{4}(\ln 3)\pi$. It becomes $\frac{\pi}{4}\ln 2$ after simplification as shown below:

```
>>simplify(R) ↵

ans =

1/4*pi*log(2)
```

◻ Integrand B

$$\int_{y=1}^{y=2}\int_{x=0}^{x=1}xye^{x^2+y^2}dxdy=\int_{y=1}^{y=2}\frac{1}{2}y\left(\int_{x=0}^{x=1}2xe^{x^2+y^2}dx\right)dy=\frac{1}{2}\int_{y=1}^{y=2}y\left[e^{x^2+y^2}\right]_{x=0}^{x=1}dy=\frac{1}{2}\int_{y=1}^{y=2}y\left[e^{y^2+1}-e^{y^2}\right]dy=\frac{1}{2}\int_{y=1}^{y=2}ye^{y^2+1}dy-$$

$$\frac{1}{2}\int_{y=1}^{y=2}ye^{y^2}dy=\frac{1}{4}\left[e^{y^2+1}\right]_{y=1}^{y=2}-\frac{1}{4}\left[e^{y^2}\right]_{y=1}^{y=2}=\frac{1}{4}(e^5-e^2)-\frac{1}{4}(e^4-e).$$

MATLAB Command

```
>>R=int(int('x*y*exp(x^2+y^2)','x',0,1),'y',1,2) ↵
>>pretty(R)

    1/4 exp(5) - 1/4 exp(4) - 1/4 exp(2) + 1/4 exp(1)
```

◻ Integrand C

$$\int_{\theta=\frac{\pi}{6}}^{\theta=\frac{\pi}{4}}\int_{r=0}^{r=4\sin 2\theta}r\sin\theta\cos\theta\,dr\,d\theta=\int_{\theta=\frac{\pi}{6}}^{\theta=\frac{\pi}{4}}\sin\theta\cos\theta\left(\int_{r=0}^{r=4\sin 2\theta}r\,dr\right)d\theta=\int_{\theta=\frac{\pi}{6}}^{\theta=\frac{\pi}{4}}\sin\theta\cos\theta\frac{1}{2}\left[r^2\right]_{r=0}^{r=4\sin 2\theta}d\theta=\int_{\theta=\frac{\pi}{6}}^{\theta=\frac{\pi}{4}}\sin\theta\cos\theta\times$$

$$\left(\frac{16\sin^2 2\theta}{2}\right)d\theta=32\int_{\theta=\frac{\pi}{6}}^{\theta=\frac{\pi}{4}}\sin^3\theta\cos^3\theta\,d\theta\ \text{(identity }\sin 2\theta\equiv 2\sin\theta\cos\theta\text{ is used)}=32\int_{\frac{1}{2}}^{\frac{1}{\sqrt{2}}}z^3(1-z^2)\,dz\ \text{(substitution of}$$

$$z=\sin\theta\text{)}=32\left[\frac{z^4}{4}-\frac{z^6}{6}\right]_{z=\frac{1}{2}}^{z=\frac{1}{\sqrt{2}}}=32\left[\frac{1}{4}\cdot\frac{1}{4}-\frac{1}{6}\cdot\frac{1}{8}-\frac{1}{4}\cdot\frac{1}{16}+\frac{1}{6}\cdot\frac{1}{64}\right]=\frac{11}{12}.\ \text{Due to unavailability of symbol, t will be used}$$

instead of θ. Like integrand A, symbolic declaration is necessary because the upper limit of the inner integration is a function of the outer variable θ.

MATLAB Command

```
>>syms r t ↵
>>int(int('r*sin(t)*cos(t)',r,0,'4*sin(2*t)'),t,pi/6,pi/4) ↵

ans =
```

234

11/12

⊟ Integrand D

This integration helps us implement integration involving infinity sign (∞). We have $\int\limits_{z=2}^{z=4}\int\limits_{x=-\infty}^{x=\infty}\dfrac{3zdxdz}{x^2+z^4}$

$=\int\limits_{z=2}^{z=4}3z\left(\int\limits_{x=-\infty}^{x=\infty}\dfrac{dx}{x^2+(z^2)^2}\right)dz=\int\limits_{z=2}^{z=4}3z\left[\dfrac{1}{z^2}\tan^{-1}\left(\dfrac{x}{z^2}\right)\right]_{x=-\infty}^{x=\infty}dz=\int\limits_{z=2}^{z=4}3z.\dfrac{1}{z^2}[\tan^{-1}(\infty)-\tan^{-1}(-\infty)]dz$. Since $\tan(\infty)=\dfrac{\pi}{2}$ and

$\tan(-\infty)=-\dfrac{\pi}{2}$, the integration becomes $3\pi\int\limits_{z=2}^{z=4}\dfrac{1}{z}dz=3\pi[\ln z]_{z=2}^{z=4}=3\pi\ln 2$.

MATLAB Command
>>int(int('3*z/(x^2+z^4)','x',-inf,inf),'z',2,4) ⏎

ans =

3*log(2)*pi

6.2.3 Triple definite integration

To evaluate a triple definite integral of the form $\int\limits_{z=e}^{z=f}\int\limits_{y=c}^{y=d}\int\limits_{x=a}^{x=b}f(x,y,z)\,dxdydz$,

MATLAB Step:
Use the command int(int(int($f(x,y,z)$ as string, x , a , b) , y , c , d), z , e , f).

Evaluate the following triple definite integrals:

$A.\ \int\limits_{x=0}^{x=a^2}\int\limits_{y=0}^{y=x}\int\limits_{z=0}^{z=y}x^2y^2z\,dzdydx$ $\qquad B.\ \int\limits_{\varphi=0}^{\varphi=\pi}\int\limits_{\theta=0}^{\theta=\frac{\pi}{2}}\int\limits_{r=0}^{r=2\cos\varphi}r^2\sin\theta\sin\varphi\,dr\,d\theta\,d\varphi$

$C.\ \int\limits_{u=0}^{u=\sqrt{5}}\int\limits_{v=0}^{v=\sqrt{9-u^2}}\int\limits_{w=0}^{w=\sqrt[3]{(9-u^2-v^2)^2}}u\ \sqrt[3]{9-u^2-v^2}\ dw\,dv\,du$

⊟ Integrand A

$\int\limits_{x=0}^{x=a^2}\int\limits_{y=0}^{y=x}\int\limits_{z=0}^{z=y}x^2y^2z\,dzdydx=\int\limits_{x=0}^{x=a^2}\int\limits_{y=0}^{y=x}x^2y^2\left[\int\limits_{z=0}^{z=y}zdz\right]dydx=\dfrac{1}{2}\int\limits_{x=0}^{x=a^2}\int\limits_{y=0}^{y=x}x^2y^2[z^2]_{z=0}^{z=y}\,dydx=\dfrac{1}{2}\int\limits_{x=0}^{x=a^2}\int\limits_{y=0}^{y=x}x^2y^2y^2dydx$ $=$

$\dfrac{1}{2}\int\limits_{x=0}^{x=a^2}x^2\left[\int\limits_{y=0}^{y=x}y^4dy\right]dx=\dfrac{1}{2}\int\limits_{x=0}^{x=a^2}x^2\dfrac{1}{5}[y^5]_{y=0}^{y=x}\,dx=\dfrac{1}{10}\int\limits_{x=0}^{x=a^2}x^2x^5\,dx=\dfrac{1}{10}\int\limits_{x=0}^{x=a^2}x^7dx=\dfrac{1}{80}[x^8]_{x=0}^{x=a^2}=\dfrac{1}{80}(a^2)^8=\dfrac{a^{16}}{80}$.

MATLAB Command
>>syms x y z ⏎
>>R=int(int(int('x^2*y^2*z',z,0,'y'),y,0,'x'),x,0,'a^2'); ⏎
>>pretty(R) ⏎

$\qquad\qquad\qquad$ 16
$\qquad\qquad$ 1/80 a

⊟ Integrand B

$\int\limits_{\varphi=0}^{\varphi=\pi}\int\limits_{\theta=0}^{\theta=\frac{\pi}{2}}\int\limits_{r=0}^{r=2\cos\varphi}r^2\sin\theta\sin\varphi\,dr\,d\theta\,d\varphi=\int\limits_{\varphi=0}^{\varphi=\pi}\int\limits_{\theta=0}^{\theta=\frac{\pi}{2}}\sin\theta\sin\varphi\left[\int\limits_{r=0}^{r=2\cos\varphi}r^2\,dr\right]d\theta\,d\varphi=\int\limits_{\varphi=0}^{\varphi=\pi}\int\limits_{\theta=0}^{\theta=\frac{\pi}{2}}\sin\theta\ \sin\varphi\dfrac{1}{3}[r^3]_{r=0}^{r=2\cos\varphi}\,d\theta$

$d\varphi=\dfrac{1}{3}\int\limits_{\varphi=0}^{\varphi=\pi}\int\limits_{\theta=0}^{\theta=\frac{\pi}{2}}\sin\theta\sin\varphi[2\cos\varphi]^3d\theta\,d\varphi=\dfrac{8}{3}\int\limits_{\varphi=0}^{\varphi=\pi}\sin\varphi\cos^3\varphi\left(\int\limits_{\theta=0}^{\theta=\frac{\pi}{2}}\sin\theta\,d\theta\right)d\varphi=\dfrac{8}{3}\int\limits_{\varphi=0}^{\varphi=\pi}\sin\varphi\cos^3\varphi(1)\,d\varphi=\dfrac{8}{3}\int\limits_{\varphi=0}^{\varphi=\pi}\sin\varphi\times$

$\cos^3\varphi d\varphi$ =0. Once again, t and p will be used for θ and φ respectively.

MATLAB Command

>>syms r t p ↵
>>int(int(int('r^2*sin(t)*sin(p)',r,0,'2*cos(p)'),t,0,pi/2),p,0,pi) ↵

ans =

0

⊟ Integrand C

$$\int_{u=0}^{u=\sqrt{5}}\int_{v=0}^{v=\sqrt{9-u^2}}\int_{w=0}^{w=\sqrt[3]{(9-u^2-v^2)^2}} u\sqrt[3]{9-u^2-v^2}\ dw\,dv\,du \quad = \quad \int_{u=0}^{u=\sqrt{5}}\int_{v=0}^{v=\sqrt{9-u^2}} u\sqrt[3]{9-u^2-v^2}\int_{w=0}^{w=\sqrt[3]{(9-u^2-v^2)^2}} dw\,dv\,du \quad =$$

$$\int_{u=0}^{u=\sqrt{5}}\int_{v=0}^{v=\sqrt{9-u^2}} u\sqrt[3]{9-u^2-v^2}\ \sqrt[3]{(9-u^2-v^2)^2}\,dv\,du = \int_{u=0}^{u=\sqrt{5}}\int_{v=0}^{v=\sqrt{9-u^2}} u(9-u^2-v^2)\,dv\,du \quad = \quad \int_{u=0}^{u=\sqrt{5}} u\left(\int_{v=0}^{v=\sqrt{9-u^2}}(9-u^2-v^2)\,dv\right)du \quad =$$

$$\int_{u=0}^{u=\sqrt{5}} u\left[(9-u^2)v-\frac{1}{3}v^3\right]_{v=0}^{v=\sqrt{9-u^2}}du = \int_{u=0}^{u=\sqrt{5}} u\left[(9-u^2)^{\frac{3}{2}}-\frac{1}{3}(9-u^2)^{\frac{3}{2}}\right]du = \frac{2}{3}\int_{u=0}^{u=\sqrt{5}}u(9-u^2)^{\frac{3}{2}}du \quad = \quad -\frac{1}{3}\left[\frac{2}{5}(9-u^2)^{\frac{5}{2}}\right]_{u=0}^{u=\sqrt{5}} \quad =$$

$$-\frac{2}{15}\left[(9-5)^{\frac{5}{2}}-9^{\frac{5}{2}}\right]=-\frac{2}{15}[32-243]=\frac{422}{15}.$$

MATLAB Command

>>syms u v w ↵
>>int(int(int('u*(9-u^2-v^2)^(1/3)',w,0,'(9-u^2-v^2)^(2/3)'),v,0,'(9-u^2)^(1/2)'),u,0,sqrt(5)) ↵

ans =

422/15

You may feel cumbersome to write the long string for multiple integration. Perform the integration step by step. How it can be done is mentioned for this example. The inner integration $\int_{w=0}^{w=\sqrt[3]{(9-u^2-v^2)^2}} u\sqrt[3]{9-u^2-v^2}\ dw$ is performed first by the following command:

>>syms w ↵
>>y1=int('u*(9-u^2-v^2)^(1/3)',w,0,'(9-u^2-v^2)^(2/3)'); ↵

So, you performed the inner integration and stored the result in variable y1. Then, the given integration becomes $\int_{u=0}^{u=\sqrt{5}}\int_{v=0}^{v=\sqrt{9-u^2}} y1\ dv\,du$. Integrate it w. r. to v by the following:

>>syms v ↵
>>y2=int(y1,v,0,'(9-u^2)^(1/2)'); ↵

Again, integration w. r. to v is performed and the output is stored in variable y2. Next, the integration takes the form $\int_{u=0}^{u=\sqrt{5}} y2\ du$. Finally, get the last integration done using

>>int(y2,'u',0,sqrt(5)) ↵

ans =

422/15

6.3 Numerical integration

All types of integrands do not have the close form analytical integration. Numerical methods are employed in that case. Usually, numerical integrations are carried out by two ways:

1. Trapezoidal rule and
2. Simpson rule.

Both methods apply to the definite integration of different orders. Each rule, trapezoidal or Simpson's, has the rigorous analytical treatment how it computes integration numerically. For simplicity, we avoided the theoretical background of numerical integration. We describe just the implementation of numerical integration in MATLAB environment.

6.3.1 Single integration

We compute the integration of the type $\int_{x=a}^{x=b} f(x)\,dx$ numerically employing the above two methods.

🖘 *Trapezoidal rule*

In MATLAB there is a subroutine called 'trapz', which is the abbreviation of trapezoidal. It can have two input arguments. The first argument is the x vector, which takes care of the x limits and the other argument is the y vector, which takes care of the integrand. Suppose we wish to evaluate the definite integral $\int_{x=0}^{x=2} x^3\,dx$ using the trapezoidal rule. Here the lower limit is $a=0$ and the upper limit is $b=2$. Many steps can be taken between these two limits. The higher is the number of steps, the better is the accuracy of the numerical integration. Take the step size as 0.05. The analytical integration of $\int_{x=0}^{x=2} x^3\,dx$ is $\left[\frac{1}{4}x^4\right]_{x=0}^{x=2} = 4$. To evaluate an integrand using the trapezoidal rule,

MATLAB Steps:
1. *Create the x vector as [lower limit : increment : upper limit],*
2. *Evaluate the integrand vector y for vector x, and*
3. *Use the subroutine trapz (x, y).*

For the above integration, the x vector is [0:0.05:2] and the y vector is computed for each element of x. The evaluation is as follows:

MATLAB Command
```
>>x=[0:0.05:2]; ↵
>>y=(x.*x).*x; ↵
>>trapz(x,y) ↵

ans =
        4.0025
```

Maybe you are disappointed with the output returned by MATLAB because the exact value of the integration is 4 while the output of MATLAB is 4.0025. For the same integration, making the step size 0.01 returns the output as follows:
```
>>x=[0:0.01:2]; ↵
>>y=(x.*x).*x; ↵
>>trapz(x,y) ↵

ans =
        4.0001
```
Again, choosing the step size 0.005 instead of 0.01 yields
```
>>x=[0:0.005:2]; ↵
```

```
>>y=(x.*x).*x; ↵
>>trapz(x,y) ↵
```

ans =
4.0000

That is what is expected.

Supplementary exercise:

Evaluate the following integrals numerically using the trapezoidal rule:

$$A. \int_{x=-\frac{\pi}{3}}^{x=\frac{\pi}{2}} \sin^3 x \cos^3 x\, dx \qquad B. \int_{x=3}^{x=9} \frac{1}{x(x^2+1)^2}\, dx \qquad C. \int_{x=0}^{x=3\pi} e^{-7x} \cos(3x-34)\, dx$$

⊡ Integral A

Analytical evaluation of the integral is presented in article 6.1.1, which is $\frac{1}{6}\cos^6 x - \frac{1}{4}\cos^4 x$, ∴

$$\int_{x=-\frac{\pi}{3}}^{x=\frac{\pi}{2}} \sin^3 x \cos^3 x\, dx = \left[\frac{1}{6}\cos^6 x - \frac{1}{4}\cos^4 x\right]_{x=-\frac{\pi}{3}}^{x=\frac{\pi}{2}} = 0 - \left(\frac{1}{6}\cos^6 \frac{\pi}{3} - \frac{1}{4}\cos^4 \frac{\pi}{3}\right) = \frac{1}{4\times 2^4} - \frac{1}{6\times 2^6} = \frac{5}{384} = 0.0130.$$ The lower and

upper limits of the integration are $-\frac{\pi}{3}$ and $\frac{\pi}{2}$ respectively. Let us take 100 steps between $-\frac{\pi}{3}$ and $\frac{\pi}{2}$. So,

the step increment is $\dfrac{\frac{\pi}{2}-\left(-\frac{\pi}{3}\right)}{100} = \dfrac{\pi}{120}$, the x vector is formed from $-\frac{\pi}{3}$ to $\frac{\pi}{2}$ with step $\frac{\pi}{120}$, and the y will be computed for each element of the x vector from $\sin^3 x \cos^3 x$. See the implementation as follows:

MATLAB Command
```
>>x=[-pi/3:pi/120:pi/2]; ↵
>>y=(sin(x).^3).*(cos(x).^3); ↵
>>trapz(x,y) ↵
```

ans =
0.0130

⊡ Integral B

From the article 6.1.1, we have $\int_{x=3}^{x=9} \dfrac{1}{x(x^2+1)^2}\, dx = \ln 3 - \dfrac{1}{2}\ln\dfrac{41}{5} - \dfrac{9}{205} = 0.0026.$ The lower limit, upper

limit, and step increment (assume 100 steps) are 3, 9, and $\frac{9-3}{100} = 0.06$ respectively. See the solution below:

```
>>x=[3:.06:9]; ↵
>>y=1./(x.*(x.^2+1).^2); ↵
>>trapz(x,y) ↵
```

ans =
0.0026

⊡ Integral C

From the same article, we get $\int_{x=0}^{x=3\pi} e^{-7x}\cos(3x-34)\, dx = e^{-21\pi}\left[\dfrac{7}{58}\cos(34) + \dfrac{3}{58}\sin(34)\right] + \dfrac{7}{58}\cos(34) +$

$\dfrac{3}{58}\sin(34) = -0.0750.$ The lower and upper limits are 0 and 3π respectively. Choose 300 steps so the step

increment is $\frac{3\pi-0}{300} = \frac{\pi}{100}$. Computation is presented as follows:

```
>>x=[0:pi/100:3*pi]; ↵
>>y=exp(-7*x).*cos(3*x-34); ↵
>>trapz(x,y) ↵
```

ans =
-0.0757

Compare the analytical and the trapezoidal outputs. They differ in the fourth digit. If one takes 4800 steps, then the output becomes −0.0750. That is a limitation of the trapezoidal rule. Remember that the vector y is computed for each element of x that is why commands '.*' or './' are used instead of '*' or '/' respectively, which are applicable for the strings.

⌗ Simpson's rule

So many steps are essential to have the better accuracy using the trapezoidal rule when the integrand has some stiff or abrupt curvature. This is just because that the trapezoidal rule assumes a straight line between the consecutive points. In Simpson's rule the assumed curve is a parabola (the curve of degree 2) or a curve of the higher degree more than 2 between the consecutive points. Better accuracy is achieved with fewer number of points using the Simpson's rule.

There is a subroutine called 'quad' in MATLAB. Idea behind the subroutine is the application of adaptive recursive Simpson's rule. The default relative error for the subroutine is 10^{-6}. To evaluate an integrand using the rule,

MATLAB Steps:
    1. Describe integrand by subroutine inline and assign that to some variable f and
    2. Use the command quad (f, lower limit, upper limit).

Compared to the trapezoidal rule, there is no number of steps in the argument of the subroutine 'quad'. The subroutine adaptively decides the steps within a relative error 10^{-6}. The subroutine 'inline' helps us construct an online (in the command window) equation or expression. More importantly, the function description must be in point to point string form (see chapter 11). If the reader is not satisfied with the relative error 10^{-6}, choice for inserting the relative error is there in the subroutine. We choose the same definite integrals as we have shown for the trapezoidal rule's in the supplementary exercise. Point to point computation of the integrand A, $\sin^3 x \cos^3 x$, for different x is written as 'sin(x).^3.*cos(x).^3' and the variable under consideration is x. So, the online representation of the integrand is inline('sin(x).^3.*cos(x).^3','x'). Let us execute this:

MATLAB Command
```
>>f=inline('sin(x).^3.*cos(x).^3','x'); ↵
>>quad(f,-pi/3,pi/2) ↵
```

ans =
0.0130

The output 0.0130 is equal to the analytical value of the integration. One more argument is appended with the subroutine if one wishes to insert the relative error. To perform the integration with the relative error,

MATLAB Steps:
    1. Describe integrand by subroutine inline and assign that to some variable f and
    2. Use the command quad (f, lower limit, upper limit, relative error).

We know from the previous discussion that the default relative error of 'quad' is 10^{-6}. What if we try with another relative error, say, 0.0001. Execution for the integrand A with this relative error is as follows:

```
>>quad(f,-pi/3,pi/2,0.0001) ↵
```

ans =

0.0130

Even though the chosen relative error (0.0001) is more than the default error (10^{-6}), the output is the same as the analytical computation (0.0130). Lower tolerance can be entered using the exponential form, for example, 0.000001 is written as 1e−6. Now execute 'quad' for the integrands B and C. The point to point representations of the integrands $\dfrac{1}{x(x^2+1)^2}$ and $e^{-7x}\cos(3x-34)$ are 1./(x.*(x.^2+1).^2) and exp(-7*x).*cos(3*x-34) respectively. In both integrands variables to be considered are x. Limits of the integrand B are 3 and 9. See the implementation below:

MATLAB Command

```
>>f=inline('1./(x.*(x.^2+1).^2)','x'); ↵
>>quad(f,3,9) ↵
```

ans =
0.0026

Command for the integrand C,
```
>>f=inline('exp(-7*x).*cos(3*x-34)','x'); ↵
>>quad(f,0,3*pi) ↵
```

ans =
-0.0750

The output is more accurate than that of the trapezoidal one (same as analytical, −0.0750).

%M – file Program for double integration of $\int_{y=c}^{y=d}\int_{x=a}^{x=b} f(x, y)\,dxdy$ using the trapezoidal rule

```
I = [ ];                %Opening an empty matrix I
for   v = c : (d−c)/N : d
        u = [a : (b−a)/M : b];    % u is a row matrix/vector
        compute f(x, y), where x = u and y = v    % v is a scalar and row
                                                   matrix u is assigned to x
        so, the computed f(x, y) will be a row matrix, call this matrix w,
        where w = f(u, v), actually, w will have the value of f(x, y) for the same
        y but for different x
        I = [I  trapz(u, w)];    % Computation of the inner integration and
                                  storing that to I one after another for different v
end
z = [c : (d−c)/N : d];    % Creating a row matrix z between lower and upper
                           limits of y with N steps

O = trapz(z, I);    % Computation of the outer integral and assign that to
                     O that is the output
```

Figure 6.1 *Program algorithms of double integration using the trapezoidal rule*

6.3.2 Double integration

Numerical double integration is also possible using the trapezoidal or Simpson's rule.

☞ *Using the trapezoidal rule*

The trapezoidal rule may not be feasible for double or triple integration because it needs M–file programming and it is less accurate than the Simpson's rule for the stiff curves. However, if it is necessary to perform the double integration by the trapezoidal rule, M–file procedure can be adopted. One or two lines command statement (s) can be executed in the Command Window directly but if many line statements are there, an editor is needed where we put all commands that is why M–file programming is required. Details of M–file's opening and execution are discussed in chapters 1 and 11. Only M–file program statements of the double integration are presented here.

The general form of the double integration is $\int_{y=c}^{y=d}\int_{x=a}^{x=b} f(x, y)\,dxdy$. The x and y limits are a and b and c and d respectively. Difference between a and b and c and d can be divided into M and N steps respectively. So, the respective step increments/decrements are $\frac{b-a}{M}$ and $\frac{d-c}{N}$. Two 'trapz' subroutines are required — one for the inner and the other for the outer integration. Algorithms for the numerical computation of $\int_{y=c}^{y=d}\int_{x=a}^{x=b} f(x, y)\,dxdy$ with M and N steps for the inner and the outer integration respectively are given in figure 6.1.

Let us implement the M–file program for the definite integral $\int_{y=1}^{y=2}\int_{x=0}^{x=1} xye^{x^2+y^2}\,dxdy$. Evaluation of the integrand is shown in article 6.2.2, which is $\int_{y=1}^{y=2}\int_{x=0}^{x=1} xye^{x^2+y^2}\,dxdy = \frac{1}{4}(e^5 - e^2) - \frac{1}{4}(e^4 - e) = 22.2860$. We choose 300 (M=300) and 400 (N=400) as the steps of x and y limits respectively. Increments for x and y are $\frac{1-0}{300} = \frac{1}{300}$ and $\frac{2-1}{400} = \frac{1}{400}$ respectively. Computation of the integrand can be performed using the subroutine 'eval'.

M–file for $\int_{y=1}^{y=2}\int_{x=0}^{x=1} xye^{x^2+y^2}\,dxdy$ will have the following:

```
        I=[];
     for v=1:1/400:2
        u=[0:1/300:1];
        f='x.*y.*exp(x.^2+y.^2)';
        x=u;
        y=v;
        w=eval(f);
        I=[I trapz(u,w)];
     end
        z=[1:1/400:2];
        O=trapz(z,I)
```

Write the above program statements in an M–file editor from the Command Window and save that with the name, say, trpdbl.m. Go to the MATLAB prompt and set the path where the file trpdbl.m is. Execution of trpdbl.m returns:

```
>>trpdbl ↵

O =
        22.2864
```

Notice that we do not type the file extension .m while the M–file is executed nor do we put ';' after the statement O=trapz(z,I). In doing so, the numerical integration value (that is assigned to O) is displayed in MATLAB Command Window. Numerically computed value differs in the 4th decimal place.

Supplementary exercise:

Evaluate the following double integrals numerically using the trapezoidal rule:

$$A. \int_{y=0}^{y=\pi} \int_{x=0}^{x=\frac{\pi}{2}} x\cos(x^2 + y)\, dx\, dy \qquad\qquad B. \int_{v=2}^{v=3} \int_{u=0}^{u=2} u\sqrt{16 - v^2}\, du\, dv$$

⊟ **Integrand** *A*

$$\int_{y=0}^{y=\pi} \int_{x=0}^{x=\frac{\pi}{2}} x\cos(x^2 + y)\, dx\, dy \quad=\quad \frac{1}{2}\int_{y=0}^{y=\pi} [\sin(x^2 + y)]_{x=0}^{x=\frac{\pi}{2}}\, dy \quad=\quad \frac{1}{2}\int_{y=0}^{y=\pi} [\sin\left(\frac{\pi^2}{4} + y\right) - \sin y]\, dy \quad=$$

$$\frac{1}{2}\left[\cos y - \cos\left(\frac{\pi^2}{4} + y\right)\right]_{y=0}^{y=\pi} = \frac{1}{2}\left[\cos\pi - \cos\left(\frac{\pi^2}{4} + \pi\right) - \left\{1 - \cos\left(\frac{\pi^2}{4}\right)\right\}\right] = \frac{1}{2}\left[-2 + 2\cos\left(\frac{\pi^2}{4}\right)\right] = -1.7812.$$

M–file for $\displaystyle\int_{y=0}^{y=\pi} \int_{x=0}^{x=\frac{\pi}{2}} x\cos(x^2 + y)\, dx\, dy$:

```
        I=[];
    for v=0:pi/500:pi
        u=[0:pi/600:pi/2];
        f='x.*cos(x.^2+y)';
        x=u;
        y=v;
        w=eval(f);
        I=[I trapz(u,w)];
    end
        z=[0:pi/500:pi];
        O=trapz(z,I)
```

After running the M–file as we mentioned for the previous problem,

$$O =$$
$$-1.7812$$

Here the x and y limits are 0 and $\dfrac{\pi}{2}$ and 0 and π respectively. We have 300 and 500 steps for x and y respectively. The respective increments are $\dfrac{\pi}{600}$ and $\dfrac{\pi}{500}$ for x and y respectively. This integrand returns accurate result (same even in the 4th decimal place).

⊟ **Integrand** *B*

$$\int_{v=2}^{v=3} \int_{u=0}^{u=2} u\sqrt{16 - v^2}\, du\, dv \;=\; \frac{1}{2}\int_{v=2}^{v=3} [u^2]_{u=0}^{u=2} \sqrt{16 - v^2}\, dv \;=\; 2\int_{v=2}^{v=3} \sqrt{16 - v^2}\, dv \;=\; 2\left[\frac{v\sqrt{16 - v^2}}{2} + 8\sin^{-1}\frac{v}{4}\right]_{v=2}^{v=3} \;=$$

$$2\left[\frac{3\sqrt{7} - 2\sqrt{12}}{2} + 8\left(\sin^{-1}\frac{3}{4} - \sin^{-1}\frac{1}{2}\right)\right] = 6.2005.$$ The variables involved in the integrand are u and v. This may create a mixing up with the M–file's u and v. It is better if we use dummy variables, that is, $\displaystyle\int_{v=2}^{v=3} \int_{u=0}^{u=2} u\sqrt{16 - v^2}\, du\, dv = \int_{y=2}^{y=3} \int_{x=0}^{x=2} x\sqrt{16 - y^2}\, dx\, dy$. The program statements for $\displaystyle\int_{v=2}^{v=3} \int_{u=0}^{u=2} u\sqrt{16 - v^2}\, du\, dv$ are as follows:

```
        I=[];
    for v=2:1/400:3
        u=[0:1/400:2];
        f='x.*sqrt(16-y.^2)';
        x=u;
        y=v;
```

```
        w=eval(f);
        I=[I trapz(u,w)];
    end
        z=[2:1/400:3];
        O=trapz(z,I)
```
Execution of the M–file with 400 steps for both u and v returns,

O =

6.2005

⌗ *Using the subroutine dblquad (Simpson's rule)*

There is another subroutine called 'dblquad' (abbreviation of double quadrature or in other words, double integration) in MATLAB for the double numerical integration. The subroutine is an extension of the Simpson's rule in two dimensions. To evaluate an integration of the type $\int_{y=c}^{y=d}\int_{x=a}^{x=b} f(x, y)\,dxdy$ using 'dblquad',

MATLAB Steps:

1. *Describe the function $f(x, y)$ as point to point form using subroutine inline and assign that to f and*
2. *Use the command dblquad (f, lower limit of x, upper limit of x, lower limit of y, upper limit of y) if default relative error is used, or use command dblquad (f, lower limit of x, upper limit of x, lower limit of y, upper limit of y, relative error) if specified relative error is required.*

Evaluate the following double integrals numerically using the subroutine 'dblquad':

$$A.\ \int_{v=-5}^{v=3}\int_{u=3}^{u=7}(u^2+v)\sqrt{u}\,dudv \qquad\qquad B.\ \int_{\theta=-\frac{\pi}{3}}^{\theta=\frac{\pi}{2}}\int_{r=2}^{r=5}r^3(\cos^2\theta+\sin^4\theta)\,dr\,d\theta$$

$$C.\ \int_{z=-4}^{z=6}\int_{w=1}^{w=2}\frac{z^3}{w^2+z^4}\,dw\,dz \qquad\qquad D.\ \int_{y=0}^{y=\pi}\int_{x=0}^{x=\frac{\pi}{2}}x\cos(x^2+y)\,dxdy$$

⌗ Integral A

$$\int_{v=-5}^{v=3}\int_{u=3}^{u=7}(u^2+v)\sqrt{u}\,dudv \ = \ \int_{v=-5}^{v=3}\left[\frac{2}{7}u^{\frac{7}{2}}+\frac{2}{3}vu^{\frac{3}{2}}\right]_{u=3}^{u=7}dv \ = \ \int_{v=-5}^{v=3}\left[\frac{2}{7}(7^{\frac{7}{2}}-3^{\frac{7}{2}})+\frac{2}{3}v(7^{\frac{3}{2}}-3^{\frac{3}{2}})\right]dv \ = \ \frac{2}{7}(7^{\frac{7}{2}}-3^{\frac{7}{2}})8+$$

$\frac{1}{3}(7^{\frac{3}{2}}-3^{\frac{3}{2}})$ $(9-25)=1896.3148$. The point to point form of $(u^2+v)\sqrt{u}$ is (u.^2+v).*sqrt(u). The lower and upper limits of u are 3 and 7 and the same for v are −5 and 3 respectively. The variables u and v can be thought as x and y respectively. See the integration below:

MATLAB Command

>>f=inline('(u.^2+v).*sqrt(u)','u','v'); ↵
>>dblquad(f,3,7,-5,3) ↵

ans =

1.8963e+003

The output is in the exponential form, which is 1.8963e+003=1896.3.

⌗ Integral B

$$\int_{\theta=-\frac{\pi}{3}}^{\theta=\frac{\pi}{2}}\int_{r=2}^{r=5}r^3(\cos^2\theta+\sin^4\theta)\,dr\,d\theta=\frac{1}{4}[5^4-2^4]\int_{\theta=-\frac{\pi}{3}}^{\theta=\frac{\pi}{2}}(\cos^2\theta+\sin^4\theta)d\theta=\frac{609}{4}\left[\frac{1}{8}\sin\theta\cos\theta+\frac{7}{8}\theta-\frac{1}{4}\sin^3\theta\cos\theta\right.$$

$$\left.\right]_{\theta=-\frac{\pi}{3}}^{\theta=\frac{\pi}{2}}=\frac{609}{4}\left[\frac{7\pi}{16}-\left\{-\frac{1}{8}\sin\frac{\pi}{3}\cos\frac{\pi}{3}-\frac{7\pi}{24}+\frac{1}{4}\times\sin^3\frac{\pi}{3}\cos\frac{\pi}{3}\right\}\right]=344.6455.$$ The variables of this integrand are

r and θ but t is used instead of θ, hence, the point to point form is r.^3.*(cos(t).^2+sin(t).^4). The lower and upper limits of r and θ are 2 and 5 and $-\dfrac{\pi}{3}$ and $\dfrac{\pi}{2}$ respectively. Computation of this integral is shown as follows:

>>f=inline('r.^3.*(cos(t).^2+sin(t).^4)','r','t'); ↵
>>dblquad(f,2,5,-pi/3,pi/2) ↵

ans =
　　344.6455

⊟ Integral C

$$\int_{z=-4}^{z=6}\int_{w=1}^{w=2}\frac{z^3}{w^2+z^4}\,dw\,dz = \int_{z=-4}^{z=6}\left[\frac{z^3}{z^2}\tan^{-1}\frac{w}{z^2}\right]_{w=1}^{w=2}dz \;\left(\text{using}\;\int\frac{1}{x^2+a^2}dx=\frac{1}{a}\tan^{-1}\frac{x}{a}\right)=\int_{z=-4}^{z=6}\left[z\tan^{-1}\frac{2}{z^2}-z\tan^{-1}\frac{1}{z^2}\right.$$

$$\left.\right]dz = \int_{z=-4}^{z=6} z\tan^{-1}\frac{z^2}{z^4+2}\,dz \;\left(\text{using}\;\tan^{-1}a-\tan^{-1}b=\tan^{-1}\frac{a-b}{1+ab}\right)=\frac{1}{2}\int_{p=16}^{p=36}\tan^{-1}\frac{p}{p^2+2}\,dp \;(\text{substitution of }z^2\text{ by }p)$$

$$=\left[\frac{p}{2}\tan^{-1}\frac{p}{p^2+2}\right]_{p=16}^{p=36}-\frac{1}{2}\int_{p=16}^{p=36}p\frac{d}{dp}\left[\tan^{-1}\frac{p}{p^2+2}\right]dp \;\;(\text{integration by parts})\;=18\tan^{-1}\frac{18}{649}-8\tan^{-1}\frac{8}{129}-\frac{1}{2}\int_{p=16}^{p=36}p\times$$

$$\frac{1}{1+\left(\dfrac{p}{p^2+2}\right)^2}\times\frac{d}{dp}\left[\frac{p}{p^2+2}\right]dp=18\tan^{-1}\frac{18}{649}-8\tan^{-1}\frac{8}{129}-\frac{1}{2}\int_{p=16}^{p=36}p\times\frac{1}{1+\left(\dfrac{p}{p^2+2}\right)^2}\times\frac{p^2+2-2p^2}{(p^2+2)^2}dp\;=\;18\tan^{-1}\frac{18}{649}-$$

$$8\tan^{-1}\frac{8}{129}+\frac{1}{2}\int_{p=16}^{p=36}\frac{p^3-2p}{p^4+5p^2+4}dp=18\tan^{-1}\frac{18}{649}-8\tan^{-1}\frac{8}{129}+\frac{1}{4}\int_{w=256}^{w=1296}\frac{w-2}{(w+4)(w+1)}dw\;(\text{substitution of }p^2\text{ by }w)=$$

$$18\tan^{-1}\frac{18}{649}-8\tan^{-1}\frac{8}{129}+\frac{1}{4}\int_{w=256}^{w=1296}\left[\frac{2}{w+4}-\frac{1}{w+1}\right]dw=18\tan^{-1}\frac{18}{649}-8\tan^{-1}\frac{8}{129}+\frac{1}{4}[2\ln(w+4)-\ln(w+1)]_{w=256}^{w=1296}\;\;=$$

$$18\tan^{-1}\frac{18}{649}-8\tan^{-1}\frac{8}{129}+\frac{1}{4}\left[2\ln 5-\ln\left(\frac{1297}{257}\right)\right]=0.4036.$$ The integrand has the variables w and z. In point to point form, the integrand is z.^3./(w.^2+z.^4). The inner and outer limits are 1 and 2 and −4 and 6 respectively. Implementation is shown as follows:

MATLAB Command
>>f=inline('z.^3./(w.^2+z.^4)','w','z'); ↵
>>dblquad(f,1,2,-4,6) ↵

ans =
　　0.4036

It is indeed amazing that few lines of command do so much computation.

⊟ Integral D

　　Analytical computation of the integration is shown earlier in this article, which is −1.7812. Compute it as follows:

MATLAB Command
>>f=inline('x.*cos(x.^2+y)','x','y'); ↵
>>dblquad(f,0,pi/2,0,pi) ↵

ans =
　　-1.7812

Until now we did not cite any use of the relative error for 'dblquad'. The default relative error of 'dblquad' is also 10^{-6}. As it is mentioned in the MATLAB steps, one argument is appended with the 'dblquad' beside the

%General M – file program for triple integration of $\int\limits_{z=e}^{z=f} \int\limits_{y=c}^{y=d} \int\limits_{x=a}^{x=b} f(x, y, z)\,dxdydz$

using the trapezoidal rule

Iz = [];   %Opening an empty matrix Iz for outermost integration

$x = [a : \dfrac{b-a}{M} : b];$ %Creating a row matrix x according to x limits with M steps

$vy = [c : \dfrac{d-c}{N} : d];$ %Creating a row matrix vy according to y limits with N steps

Define the function string in point to point form and assign that to f

$for \ w = e : \dfrac{f-e}{P} : f$ %First for loop comes according to z limits

considering P steps

      Iy = [];   %Opening an empty matrix Iy for the middle integration

$\qquad for \ \ v = c : \dfrac{d-c}{N} : d$ %Second for loop comes considering y limits with N steps

Assign the for loop counter index of the middle and outermost integration to y and z respectively

Compute f(x, y, z), which is a row matrix for row matrix x and scalars y and z and assign the computed value to some variable I by I = eval(f);

Compute the innermost integration by trapz(x, I) and assign that integrated value one by one to Iy by Iy = [Iy trapz(x, I)];

      end

Compute the middle integration by trapz(vy, Iy) and assign that integrated value one by one to Iz by Iz = [Iz trapz(vy, Iy)];

end

$vz = [e : \dfrac{f-e}{P} : f];$ % Creating a row matrix vz between lower and upper limits of z with P steps

O = trapz(vz, Iz) % Computation of the outermost integral and assigned that to O that is the output

Figure 6.2 *Program algorithms of the triple integration using the three trapezoidal subroutines*

last upper limit when the use of relative error is necessary. Just to show by illustration, we wish to find

$\int\limits_{y=0}^{y=\pi} \int\limits_{x=0}^{x=\frac{\pi}{2}} x\cos(x^2 + y)\,dxdy$ within the relative error 0.01. The command is as follows:

MATLAB Command

```
>>f=inline('x.*cos(x.^2+y)','x','y'); ↵
>>dblquad(f,0,pi/2,0,pi,0.01) ↵
```

ans =

$$-1.7812$$

The output is still accurate even with the less relative error. Whatsoever numerical integration subroutine is used, an M−file can be written to obtain the specified precision with changing relative error if it is necessary.

6.3.3 Triple integration

The numerical integration of the triple integrals is introduced in this section. The trapezoidal and Simpson's methods can be extended to compute triple integrals. Error propagation of the triple integration is higher than that of single or double one. $\int_{z=e}^{z=f} \int_{y=c}^{y=d} \int_{x=a}^{x=b} f(x, y, z)\,dx\,dy\,dz$ is the general form of a triple integral. It is evident that depending on the function's stiffness, choice of the step size of trapezoidal rule and the relative error of Simpson's rule (or in other words, subroutines 'quad' or 'dblquad') matter computational accuracy of the triple integration. Computation can be carried out in several ways using 'trapz', 'quad', and/or 'dblquad'. Two methods of computation are mentioned. The first one is the direct use of three trapezoidal subroutines and the second one is the use of one 'trapz' and one 'dblquad' subroutines. Program algorithms of both methods are shown in figures 6.2 and 6.3 respectively.

%General M − file program for triple integration of $\int_{z=e}^{z=f} \int_{y=c}^{y=d} \int_{x=a}^{x=b} f(x, y, z)\,dx\,dy\,dz$

using one trapezoidal and one dblquad subroutines

I = [];   %Opening an empty matrix I for storing dblquad output

Split the point to point string form of function f(x, y, z) excluding the function associated with z and assign them to s1, s2, s3, ...etc whatever is necessary

for $z = e : \dfrac{f - e}{P} : f$ %For loop counter index comes according to z limits with P steps

Compute the z part of f(x, y, z) what was excluded in earlier splitting

Convert the computed z part of f(x, y, z) to string form by num2str subroutine and assign that to sz

Form the complete function string using command strcat from s1, s2, s3,.....etc and sz and assign the complete string to g

Declare g as function of x and y by inline subroutine and assign that to f

I = [I dblquad(f, a, b, c, d)];   %Computation of the innermost and middle integration and storing that to I one after another for different z

end

$z = [e : \dfrac{f - e}{P} : f]$;   % Creating a row matrix z between lower and upper limits of z with P steps

O = trapz(z, I)   % Computation of the outermost integral using trapezoidal rule and assign that to O that is the output

Figure 6.3   *Program algorithms of the triple integration using one trapezoidal and one dblquad subroutines*

246

As usual, we proceed with the numerical computation of $\int_{z=0}^{z=\pi}\int_{y=\frac{\pi}{2}}^{y=\pi}\int_{x=0}^{x=\frac{\pi}{2}} \sin(2x-3y-z)\,dx\,dy\,dz$. The close

form value of the integral can be evaluated very easily, which is $\int_{z=0}^{z=\pi}\int_{y=\frac{\pi}{2}}^{y=\pi}\int_{x=0}^{x=\frac{\pi}{2}} \sin(2x-3y-z)\,dx\,dy\,dz =$

$-\frac{1}{2}\int_{z=0}^{z=\pi}\int_{y=\frac{\pi}{2}}^{y=\pi}[\cos(2x-3y-z)]_{x=0}^{x=\frac{\pi}{2}}\,dy\,dz = \int_{z=0}^{z=\pi}\int_{y=\frac{\pi}{2}}^{y=\pi}\cos(3y+z)\,dy\,dz = \frac{1}{3}\int_{z=0}^{z=\pi}[\sin(3y+z)]_{y=\frac{\pi}{2}}^{y=\pi}\,dz = \frac{1}{3}\int_{z=0}^{z=\pi}[\cos z - \sin z]\,dz =$

$\frac{1}{3}[\sin z + \cos z]_{z=0}^{z=\pi} = -\frac{2}{3}$. Let us continue with the algorithm of the three trapezoidal rules. From the limits of x,

y, and z, $a=0$, $b=\frac{\pi}{2}$, $c=\frac{\pi}{2}$, $d=\pi$, $e=0$, and $f=\pi$. Choose 150, 150, and 200 steps for x, y, and z

limits respectively, therefore, M=150, N=150, and P=200. Different row matrices for this program are x=

[a: $\frac{b-a}{M}$:b]=[0: $\frac{\pi}{300}$: $\frac{\pi}{2}$], vy=[c: $\frac{d-c}{N}$:d]=[$\frac{\pi}{2}$: $\frac{\pi}{300}$: π], and vz=[e: $\frac{f-e}{P}$:f]=[0: $\frac{\pi}{200}$: π]. The applicable string

form of the integrand is sin(2*x-3*y-z). The program is as follows:

M–file program for the computation of $\int_{z=0}^{z=\pi}\int_{y=\frac{\pi}{2}}^{y=\pi}\int_{x=0}^{x=\frac{\pi}{2}} \sin(2x-3y-z)\,dx\,dy\,dz$ using the three trapezoidal

subroutines:

```
Iz=[];
x=[0:pi/300:pi/2];
vy=[pi/2:pi/300:pi];
f='sin(2*x-3*y-z)';
for w=0:pi/200:pi
        Iy=[];
        for v=pi/2:pi/300:pi
        y=v;
        z=w;
        I=eval(f);
        Iy=[Iy trapz(x,I)];
    end
        Iz=[Iz trapz(vy,Iy)];
end
        vz=[0:pi/200:pi];
        O=trapz(vz,Iz)
```
Execution of the above M–file returns,

$$O =$$

-0.6666

Execution of the triple integration takes considerable time. The execution time is a function of the processor speed you have. Reason for this is that there are three loops and for each loop computations of hundreds of vectors are being carried out. Our system was a Pentium III, 400 MHz, 64 Mbytes RAM, and Windows 98 operated. Execution time for the above program was 43.34 seconds.

The next agenda is how we utilize one trapezoidal and one dblquad subroutines to compute the triple integration. Observe the integrand $\sin(2x-3y-z)$. For fixed value of z, the integration turns out to be a double one so we can apply 'dblquad' to compute the double integration. There is some complicity with the first argument of 'dblquad', which has to be point to point string form. The point to point string form of the integrand is sin(2*x-3*y-z). We can not compute sin(2*x-3*y-z) for fixed z without inserting z in the argument of sine. A string is a set of characters and operators (see chap. 11). It must be correct syntactically before its computation. Definition of the first argument of the 'inline' will be different for different z. The numerical value of any computed function of z needs converting to string form before the variable string for different z is evaluated. This is accomplished by the subroutine 'num2str' (abbreviation of <u>num</u>ber to (<u>2</u>) <u>str</u>ing). The string 'sin(2*x-3*y-z)' is divided in three parts — 'sin(2*x-3*y-', 'z', and ')'. We assign them to s1, sz, and s2 respectively (as mentioned in program algorithms of figure 6.3) before the variable integrand $\sin(2x - 3y - z)$ is defined. The different strings are combined to form the integrand using the subroutine 'strcat' (abbreviation of <u>str</u>ing's <u>cat</u>enation). The following program statements summarize the whole jargon:

M−file for $\int\limits_{z=0}^{z=\pi} \int\limits_{y=\frac{\pi}{2}}^{y=\pi} \int\limits_{x=0}^{x=\frac{\pi}{2}} \sin(2x - 3y - z) \, dx \, dy \, dz$ using one trpaz and one dblquad subroutines:

```
            I=[];
            s1='sin(2*x-3*y-';
            s2=')';
     for z=0:pi/70:pi
            sz=num2str(z);
            g=strcat(s1,sz,s2);
            f=inline(g,'x','y');
            I=[I dblquad(f,0,pi/2,pi/2,pi)];
     end
            z=[0:pi/70:pi];
            O=trapz(z,I)
```

After execution, we have,

$$O =$$
$$-0.6666$$

For the outermost trapezoidal integration, we chose the number of steps of the z limits as 70, hence, P=70. Specific relative error can be inserted in 'dblquad' if it is necessary. The time required to run the program was 72.23 seconds. The execution time of the later is more than that of the former.

Supplementary exercise:

Evaluate the following triple integrals numerically using both methods:

A . $\int\limits_{z=2}^{z=4} \int\limits_{y=1}^{y=2} \int\limits_{x=0}^{x=1} zx \ln(x^2 + y) \, dx \, dy \, dz$

B . $\int\limits_{\varphi=0}^{\phi=2\pi} \int\limits_{\theta=-\frac{\pi}{2}}^{\theta=\frac{\pi}{2}} \int\limits_{r=0}^{r=2} r^2(\cos^2\theta + \cos^2\varphi) \, dr \, d\theta \, d\varphi$

▱ Integral A

$\int\limits_{z=2}^{z=4} \int\limits_{y=1}^{y=2} \int\limits_{x=0}^{x=1} zx \ln(x^2 + y) \, dx \, dy \, dz$ $= \dfrac{1}{2}\int\limits_{z=2}^{z=4} \int\limits_{y=1}^{y=2} [z[(x^2 + y)\ln(x^2 + y) - (x^2 + y)]_{x=0}^{x=1} \, dy \, dz$ $= \dfrac{1}{2}\int\limits_{z=2}^{z=4} \int\limits_{y=1}^{y=2} z \times$

$[(y + 1)\ln(y + 1) - (y + 1) - y\ln y + y] \, dy \, dz$ $= \dfrac{1}{2}\int\limits_{z=2}^{z=4} \int\limits_{y=1}^{y=2} z[(y + 1)\ln(y + 1) - 1 - y\ln y] \, dy \, dz$ $= \dfrac{1}{2}\int\limits_{z=2}^{z=4} z\left[\dfrac{1}{2}(y + 1)^2\ln(y + 1)\right.$

$\left. -\dfrac{1}{4}(y + 1)^2 - y - \dfrac{1}{2}y^2\ln y + \dfrac{1}{4}y^2\right]_{y=1}^{y=2} dz$ $= \dfrac{1}{2}\int\limits_{z=2}^{z=4} z\left[\dfrac{3^2\ln 3}{2} - \dfrac{3^2}{4} - 2 - \dfrac{2^2\ln 2}{2} + \dfrac{2^2}{4} - \dfrac{2^2\ln 2}{2} + \dfrac{2^2}{4} + 1 - \dfrac{1}{4}\right] dz$ $=$

$\left[\dfrac{9\ln 3}{2} - 4\ln 2 - \dfrac{3}{2}\right]\dfrac{1}{4}[z^2]_{z=2}^{z=4} = \dfrac{27}{2}\ln 3 - \dfrac{9}{2} - 12\ln 2 = 2.0135.$

M−file for $\int\limits_{z=2}^{z=4} \int\limits_{y=1}^{y=2} \int\limits_{x=0}^{x=1} zx \ln(x^2 + y) \, dx \, dy \, dz$ using the three trapezoidal subroutines:

In point to point string form the integrand $zx \ln(x^2 + y)$ becomes z.*x.*log(x.^2+y). The chosen number of steps are M=100, N=100, and P=100 for x, y, and z respectively. The different limits are $a = 0$,

$b=1$, $c=1$, $d=2$, $e=2$, and $f=4$. Various vectors are x=[a:$\dfrac{b-a}{M}$:b]=[0:$\dfrac{1}{100}$:1], vy=[c:$\dfrac{d-c}{N}$:d]= [1:$\dfrac{1}{100}$:2], and vz=[e:$\dfrac{f-e}{P}$:f]=[2:$\dfrac{1}{50}$:4]. The complete program is as follows:

```
Iz=[];
x=[0:1/100:1];
vy=[1:1/100:2];
f='z.*x.*log(x.^2+y)';
for w=2:1/50:4
        Iy=[];
    for v=1:1/100:2
        y=v;
        z=w;
        I=eval(f);
        Iy=[Iy trapz(x,I)];
    end
        Iz=[Iz trapz(vy,Iy)];
end
        vz=[2:1/50:4];
        O=trapz(vz,Iz)
```

This M–file returns,
$\qquad$ O =
$\qquad\qquad$ 2.0136

Execution time of the above program was 29.17 seconds.

M–file for $\int\limits_{z=2}^{z=4}\int\limits_{y=1}^{y=2}\int\limits_{x=0}^{x=1} zx\ln(x^2+y)\,dx\,dy\,dz$ using one trapz and one dblquad subroutines:

$\qquad$ Variable z is appearing as the first character of the integrand string 'z.*x.*log(x.^2+y)'. Excluding that z, the rest of the string is defined as s1='.*x.*log(x.^2+y)'. You can check the different limits and the number of steps required for this method in the last trapezoidal integration. Following is the whole program:

```
I=[];
s1='.*x.*log(x.^2+y)';
for z=2:1/50:4
        sz=num2str(z);
        g=strcat(sz,s1);
        f=inline(g,'x','y');
        I=[I dblquad(f,0,1,1,2)];
end
        z=[2:1/50:4];
        O=trapz(z,I)
```

The output is returned as
$\qquad$ O =
$\qquad\qquad$ 2.0135

Execution time: 40.81 seconds

Integral B

$\qquad$ Not to be puzzled with different variables, it is suggested that you use the variables x, y, and z as the independent variables of the integration (for r, θ, and φ respectively). Using the dummy variables allows

us to write $\int\limits_{\varphi=0}^{\varphi=2\pi}\int\limits_{\theta=-\frac{\pi}{2}}^{\theta=\frac{\pi}{2}}\int\limits_{r=0}^{r=2} r^2(\cos^2\theta+\cos^2\varphi)\,dr\,d\theta\,d\varphi = \int\limits_{z=0}^{z=2\pi}\int\limits_{y=-\frac{\pi}{2}}^{y=\frac{\pi}{2}}\int\limits_{x=0}^{x=2} x^2(\cos^2 y+\cos^2 z)\,dx\,dy\,dz$. First we compute the

analytical value of the definite integral as $\int\limits_{z=0}^{z=2\pi}\int\limits_{y=-\frac{\pi}{2}}^{y=\frac{\pi}{2}}\int\limits_{x=0}^{x=2} x^2(\cos^2 y+\cos^2 z)\,dx\,dy\,dz = \int\limits_{z=0}^{z=2\pi}\int\limits_{y=-\frac{\pi}{2}}^{y=\frac{\pi}{2}} (\cos^2 y+\cos^2 z)\dfrac{1}{3}$

$$[x^3]_{x=0}^{x=2}\,dy\,dz = \frac{8}{3}\int_{z=0}^{z=2\pi}\int_{y=-\frac{\pi}{2}}^{y=\frac{\pi}{2}}(\cos^2 y + \cos^2 z\,)\,dy\,dz = \frac{4}{3}\int_{z=0}^{z=2\pi}\int_{y=-\frac{\pi}{2}}^{y=\frac{\pi}{2}}(1+\cos 2y + 2\cos^2 z\,)\,dy\,dz = \frac{4}{3}\int_{z=0}^{z=2\pi}[y + \frac{1}{2}\sin 2y +$$

$$2y\cos^2 z]_{y=-\frac{\pi}{2}}^{y=\frac{\pi}{2}}\,dz = \frac{4}{3}\int_{z=0}^{z=2\pi}[\pi + 2\pi\cos^2 z]\,dz = \frac{4}{3}\int_{z=0}^{z=2\pi}[\pi + \pi(1+\cos 2z)]\,dz = \frac{4\pi}{3}\int_{z=0}^{z=2\pi}(2+\cos 2z)\,dz = \frac{4\pi}{3}[2z + \frac{1}{2}\sin 2z]_{z=0}^{z=2\pi} =$$

$$\frac{16\pi^2}{3} = 52.6379.$$

M–file for $\int_{z=0}^{z=2\pi}\int_{y=-\frac{\pi}{2}}^{y=\frac{\pi}{2}}\int_{x=0}^{x=2} x^2(\cos^2 y + \cos^2 z\,)\,dx\,dy\,dz$ using the three trapezoidal subroutines:

For this integration, the different limits and the various vectors are $a=0$, $b=2$, $c=-\frac{\pi}{2}$, $d=\frac{\pi}{2}$, $e=0$, and $f=2\pi$ and x=[a:$\frac{b-a}{M}$:b]=[0:$\frac{1}{50}$:2], vy=[c:$\frac{d-c}{N}$:d]=[$-\frac{\pi}{2}$:$\frac{\pi}{100}$:$\frac{\pi}{2}$], and vz= [e:$\frac{f-e}{P}$:f]=[0:$\frac{\pi}{50}$:2π] respectively for the steps M=100, N=100, and P=100. The MATLAB codes are as follows:

```
                  Iz=[];
                  x=[0:1/50:2];
                  vy=[-pi/2:pi/100:pi/2];
                  f='x.^2.*(cos(y).^2+cos(z).^2)';
            for w=0:pi/50:2*pi
                  Iy=[];
                for v=-pi/2:pi/100:pi/2
                  y=v;
                  z=w;
                  I=eval(f);
                  Iy=[Iy trapz(x,I)];
                end
                  Iz=[Iz trapz(vy,Iy)];
            end
                  vz=[0:pi/50:2*pi];
                  O=trapz(vz,Iz)
```

After running the above M–file, we find that
$$O =$$
$$52.6405$$
Execution time: 28.07 seconds

M–file for $\int_{z=0}^{z=2\pi}\int_{y=-\frac{\pi}{2}}^{y=\frac{\pi}{2}}\int_{x=0}^{x=2} x^2(\cos^2 y + \cos^2 z\,)\,dx\,dy\,dz$ using one trapz and one dblquad subroutines:

The integrand string 'x.^2.*(cos(y).^2+cos(z).^2)' is split in three parts — 'x.^2.*(cos(y).^2+cos(', 'z', and ').^2)'. The complete codes are as follows:

```
                  I=[];
                  s1='x.^2.*(cos(y).^2+cos(';
                  s2=').^2)';
            for z=0:pi/50:2*pi
                  sz=num2str(z);
                  g=strcat(s1,sz,s2);
                  f=inline(g,'x','y');
                  I=[I dblquad(f,0,2,-pi/2,pi/2)];
            end
                  z=[0:pi/50:2*pi];
```

250

```
O=trapz(z,I)
```
The numerical output is returned as
```
                O =
```
52.6379

Execution time: 25.38 seconds.

6.4 Arc length of different curves

If function $y = f(x)$ is continuous and single valued from $x = a$ to $x = b$, as shown in figure 6.4, arc

length ABC between these two lines $x = a$ and $x = b$ is given by $s = \int\limits_{x=a}^{x=b} \sqrt{1 + \left(\dfrac{dy}{dx}\right)^2}\, dx$.

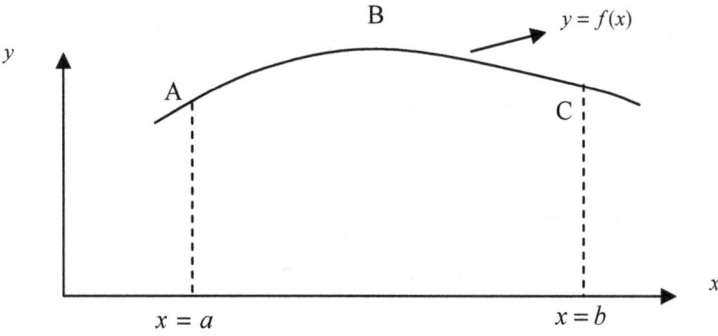

Figure 6.4   *Illustration of arc length of a curve when its*
equation is $y = f(x)$

⊟ Example

Compute the arc length of the curve $y = x^2 - 3x + 4$ from $x = -2$ to $x = 3$.

Solution:

Plot of $y = f(x) = x^2 - 3x + 4$ is shown in figure 6.5. The required arc length is shown by the thick

curve in the same figure, which necessitates the variation of x from -2 to 3, hence, $\dfrac{dy}{dx} = 2x - 3$ and the

arc   length   is $s = \int\limits_{x=-2}^{x=3} \sqrt{1 + (2x-3)^2}\, dx$. Compare the integration with $\int \sqrt{x^2 + a^2}\, dx = \dfrac{x\sqrt{x^2 + a^2}}{2} +$

$\dfrac{a^2 \ln(x + \sqrt{x^2 + a^2}\ )}{2}$ to   write $s = \dfrac{1}{2}\left[\dfrac{(2x-3)\sqrt{4x^2 - 12x + 10}}{2} + \dfrac{1}{2}\ln\left(2x - 3 + \sqrt{4x^2 - 12x + 10}\ \right) \right]_{x=-2}^{x=3} = \dfrac{1}{2}\left[\dfrac{3\sqrt{10}}{2} + \right.$

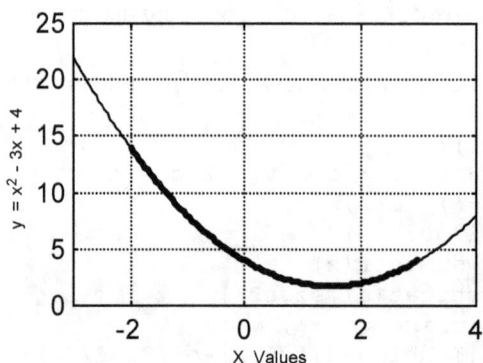

Figure 6.5   *Plot of $y = x^2 - 3x + 4$ for $-3 \le x \le 4$*

$$\frac{1}{2}\ln(3+\sqrt{10})-\left(\frac{-7\sqrt{50}}{2}+\frac{1}{2}\ln(-7+\sqrt{50})\right)\right]=\frac{3\sqrt{10}}{4}+\frac{\ln(3+\sqrt{10})}{4}+\frac{35\sqrt{2}}{4}-\frac{\ln(-7+5\sqrt{2})}{4}$$ (on simplification) =

$$\frac{3\sqrt{10}}{4}-\frac{\ln(-3+\sqrt{10})}{4}+\frac{35\sqrt{2}}{4}-\frac{\ln(-7+5\sqrt{2})}{4}$$ (rationalization of the first logarithmic argument)=15.861718 units.

Computations are as follows:

MATLAB Command

```
>>syms x ↵
>>y=x^2-3*x+4; ↵         ← Define y = x² − 3x + 4

>>d=diff(y,x); ↵          ← Perform dy/dx

>>I=sqrt(1+d^2); ↵        ← Form I = √(1+(dy/dx)²)

>>L=int(I,-2,3); ↵        ← Apply 'int' on I
>>pretty(L) ↵
```

$$3/4\ 10^{1/2} - 1/4\ \log(-3 + 10^{1/2}) + 35/4\ 2^{1/2} - 1/4\ \log(-7 + 5\ 2^{1/2})$$

Decimal form is seen by,

```
>>double(L) ↵

    ans =
            15.8617
```

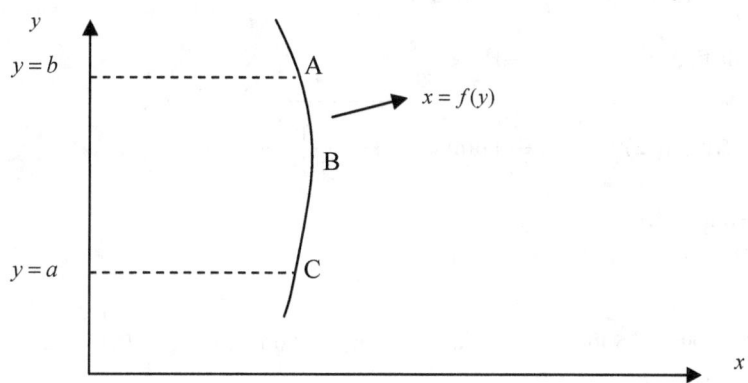

Figure 6.7   *Illustration of arc length of a curve when its equation is $x = f(y)$*

Not necessarily the given equation will be in $y = f(x)$ form, we may have an equation in the form $x = f(y)$. Illustration of this type of curve is shown in figure 6.7. For this curve, the arc length is given by

$$s = \int_{y=a}^{y=b}\sqrt{1+\left(\frac{dx}{dy}\right)^2}\ dy\ .$$

⊟ Example

Calculate the arc length of the curve $y^2 = 100x$ from $y = 5$ to $y = 15$.

252

Solution:

Figure 6.6 depicts the plot of $y^2 = 100x$ for $0 \le y \le 20$. The thick curve of the figure refers to the required arc length. We have $x = \dfrac{y^2}{100}$, $\dfrac{dx}{dy} = \dfrac{y}{50}$, and the arc length $s = \displaystyle\int_{y=5}^{y=15} \sqrt{1 + \left(\dfrac{y}{50}\right)^2}\, dy =$

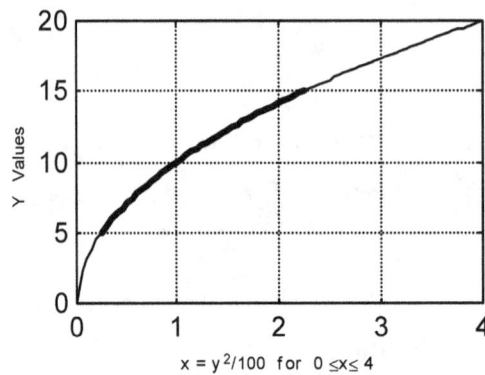

Figure 6.6 *Plot of $y^2 = 100x$ for $0 \le y \le 20$*

$$50 \left[\frac{y}{100}\sqrt{1 + \frac{y^2}{2500}} + \frac{1}{2}\ln\left(\frac{y}{50} + \sqrt{\frac{y^2}{2500} + 1} \right) \right]_{y=5}^{y=15} = \frac{3}{4}\sqrt{109} - 25\ln(-3 + \sqrt{109}) + 50\ln 10 - \frac{\sqrt{101}}{4} - 25\ln(1 + \sqrt{101})$$

(after simplification)=10.21373519 units. Have it done as follows:

MATLAB Command

>>syms y ↵

>>x=y^2/100; ↵   ← Define $x = \dfrac{y^2}{100}$

>>d=diff(x,y); ↵   ← Have $\dfrac{dx}{dy}$

>>I=sqrt(1+d^2); ↵   ← Form I $= \sqrt{1 + \left(\dfrac{dx}{dy}\right)^2}$

>>L=int(I,y,5,15); ↵
>>pretty(L) ↵

$$\qquad\quad 1/2 \qquad\qquad\quad 1/2 \qquad\qquad\qquad\qquad 1/2$$
$$3/4\ 109\quad - 25\ \log(-3 + 109\ \ ) + 50\ \log(2) + 50\ \log(5) - 1/4\ 101$$

$$\qquad\qquad\qquad\qquad\qquad 1/2$$
$$- 25\ \log(1 + 101\ \ )$$

In decimal form,

>>double(L) ↵

ans =
           10.2137

We may have other form of curve, for example, polar or parametric. The arc length in parametric form is given by $s = \displaystyle\int_{t=t_1}^{t=t_2} \sqrt{\left(\dfrac{dx}{dt}\right)^2 + \left(\dfrac{dy}{dt}\right)^2}\, dt$.

⊟ Example

Compute the arc length of a curve for $\dfrac{\pi}{3} \le t \le \dfrac{2\pi}{3}$ whose parametric equations are $x = 3\cos t$ and $y = 2\sin t$.

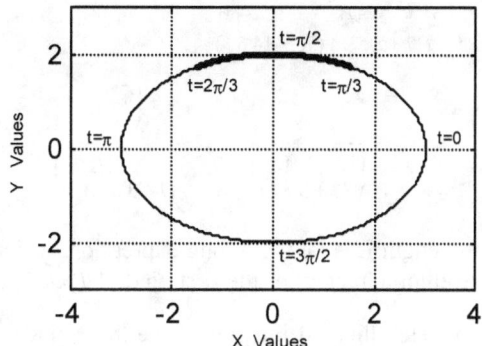

Figure 6.8   *Plot of* $x = 3\cos t$ *and* $y = 2\sin t$
for $0 \le t \le 2\pi$

Solution:

The parametric curve is graphed in figure 6.8. It traces an ellipse as shown in figure 6.8, hence,

$\dfrac{dx}{dt} = -3\sin t$, $\dfrac{dy}{dt} = 2\cos t$, and the arc length (marked by the thick curve) $s = \displaystyle\int_{t=\frac{\pi}{3}}^{t=\frac{2\pi}{3}} \sqrt{9\sin^2 t + 4\cos^2 t} \; dt = 3.0644$

(by 200 steps trapezoidal integration). Try with the following:

MATLAB Command

```
>>syms t ↵
>>x=3*cos(t); ↵
>>y=2*sin(t); ↵
>>I=sqrt(diff(x,t)^2+diff(y,t)^2); ↵
>>L=int(I,t,pi/3,2*pi/3); ↵
>>pretty(L) ↵
```

```
          1/2  1/2
      1/3 31    3

                                               1/2      1/2
       /    4                 8          \    3   (7 + 3 5  )
     + |----------- + -----------------------| EllipticF(--------------------------, %1)
       |    1/2         1/2        1/2    |                1/2 1/2
       \ 3 + 5      (3 + 5  ) (7 + 3 5  ) /        (296 + 132 5   )

          +

                                                            1/2
       /    4                 8          \        7 + 3 5
     |- ------------- - -----------------------| EllipticF(-------------------------, %1)
       |     1/2         1/2        1/2    |                1/2 1/2
       \   3 + 5      (3 + 5  ) (7 + 3 5  ) /        (136 + 60 5   )


                              1/2          1/2
                  1/2       3   (7 + 3 5  )
          (15 + 7 5  ) EllipticPi(---------------------------, 1, %1)
                                        1/2  1/2
                              (296 + 132 5   )
       - 16 -------------------------------------------------------------
                      1/2        1/2            1/2
                (3 + 5  ) (7 + 3 5  ) (30 + 14 5  )
```

$$\begin{array}{c} \qquad\qquad 1/2 \\ \qquad 1/2 \qquad\qquad\quad 7+3\ 5 \\ (15+7\ 5\quad)\ \text{EllipticPi}(\text{-----------------------},\ 1,\ \%1) \\ \qquad\qquad\qquad\qquad\quad 1/2\ \ 1/2 \\ \qquad\qquad\qquad\quad (136+60\ 5\quad) \\ +\ 16\ \text{---} \\ \qquad\quad 1/2 \qquad\quad 1/2 \qquad\qquad 1/2 \\ \qquad (3+5\quad)\ (7+3\ 5\quad)\ (30+14\ 5\quad) \end{array}$$

$$\%1\ :=\left(\frac{90}{(7+3\ 5^{1/2})^{2}}+42\ \frac{5^{1/2}}{(7+3\ 5^{1/2})^{2}}\right)^{1/2}$$

As you see, the close form arc length L is taking a long expression, which makes us familiarize with two new functions, namely, incomplete elliptic integral of the first kind, *EllipticF*(x,k), whose definition is *EllipticF*(x,k) $=\int_{t=0}^{t=x}\dfrac{dt}{\sqrt{1-t^{2}}\sqrt{1-k^{2}t^{2}}}$ and incomplete elliptical integral of the third kind, *EllipticPi*(x, v, k), whose definition is

EllipticPi$(x, v, k) = \int_{t=0}^{t=x}\dfrac{dt}{(1-v\ t^{2})\sqrt{1-t^{2}}\sqrt{1-k^{2}t^{2}}}$. Notice the style of presenting a long expression in MATLAB.

The last numeric paragraph of the output tells us that %1 is equivalent to $\sqrt{\dfrac{90}{(7+3\sqrt{5})^{2}}+42\dfrac{\sqrt{5}}{(7+3\sqrt{5})^{2}}}$. However,

we can have the arc length in decimal by

>>double(L) ↵

ans =
            3.0644

Challenging case:

   If the derivative $\dfrac{dy}{dx}$ of $y=f(x)$ curve is not defined (i.e., value of $\dfrac{dy}{dx}$ becomes infinity), the arc length can not be found using the above formula. Such a situation can be when one tries to find the arc length of the ellipse from $t=\dfrac{\pi}{2}$ to π . Observe that $\dfrac{dy}{dx}=-\infty$ but $\dfrac{dx}{dy}=0$ at $t=\pi$ and Cartesian form of the ellipse is $\dfrac{x^{2}}{9}+\dfrac{y^{2}}{4}=1$. We can divide the arc from $t=\dfrac{\pi}{2}$ to π in two parts — one is from $t=\dfrac{\pi}{2}$ to $\dfrac{3\pi}{4}$ and the other from $t=\dfrac{3\pi}{4}$ to π . Use $\dfrac{dy}{dx}$ (where the curve is $y=f(x)=\dfrac{2}{3}\sqrt{9-x^{2}}$) for the first part and $\dfrac{dx}{dy}$ (where the curve is $x=f(y)=\dfrac{3}{2}\sqrt{4-y^{2}}$) for the other and add the two lengths to find the total arc length otherwise use the numerical integration. This is left as an exercise for the reader.

   Another type of arc length computation can come into view is polar form's.

Arc length in polar form: $\quad s=\int_{\theta=\theta_{1}}^{\theta=\theta_{2}}\sqrt{r^{2}+\left(\dfrac{dr}{d\theta}\right)^{2}}\ d\theta$ when $r=f(\theta)$

$$s=\int_{r=r_{1}}^{r=r_{2}}\sqrt{1+r^{2}\left(\dfrac{d\theta}{dr}\right)^{2}}\ dr \text{ when } \theta=f(r)$$

⊟ Example

Compute the arc length of a curve for $\frac{\pi}{2} \le \theta \le \pi$ whose polar equation is given by $r = 1 + \cos\theta$.

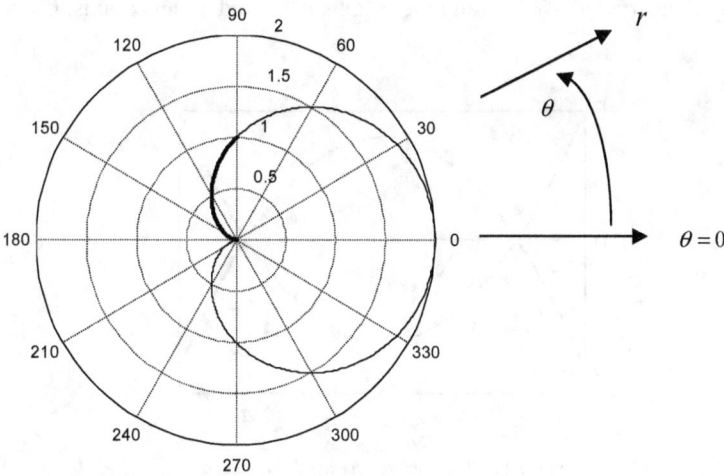

Figure 6.9 *Plot of* $r = 1 + \cos\theta$ *for* $0 \le \theta \le 2\pi$

Solution:

Plot of the polar curve and the required arc length are shown in figure 6.9. The length is

$$s = \int_{\theta=\frac{\pi}{2}}^{\theta=\pi} \sqrt{r^2 + \left(\frac{dr}{d\theta}\right)^2}\ d\theta = \int_{\theta=\frac{\pi}{2}}^{\theta=\pi} \sqrt{(1+\cos\theta)^2 + \sin^2\theta}\ d\theta = \int_{\theta=\frac{\pi}{2}}^{\theta=\pi} 2\cos\frac{\theta}{2}d\theta = 4 - 2\sqrt{2}.$$ The computation is shown as

follows:

MATLAB Command

>>syms t ↵   ← Use t instead of θ
>>r=1+cos(t); ↵   ← Define $r = 1 + \cos\theta$

>>I=sqrt(r^2+diff(r,t)^2); ↵   ← Form $I = \sqrt{r^2 + \left(\dfrac{dr}{d\theta}\right)^2}$

>>L=int(I,t,pi/2,pi); ↵   ← Apply 'int' on I
>>pretty(L) ↵

```
                 1/2      1/2   1/2
              2 4    -  2      4
```

Simplify to have the equivalent symbolic form:

>>simplify(L) ↵

ans =

4-2*2^(1/2)

6.5 Volume of a solid of revolution

If the area bounded by the single valued curve $y = f(x)$ from $x = a$ to $x = b$ is rotated about x axis, volume of the solid formed is given by $V = \pi \int_{x=a}^{x=b} [f(x)]^2 dx$. Or, if the area bounded by the single valued curve x

256

$= f(y)$ from $y = a$ to $y = b$ is rotated about y axis, volume of the solid formed is given by $V = \pi \int\limits_{y=a}^{y=b} [f(y)]^2 dy$.

⊟ Example

Area formed by the curve $y = 4x - x^2$ and the x axis is rotated about x axis. Calculate the volume of the solid of revolution.

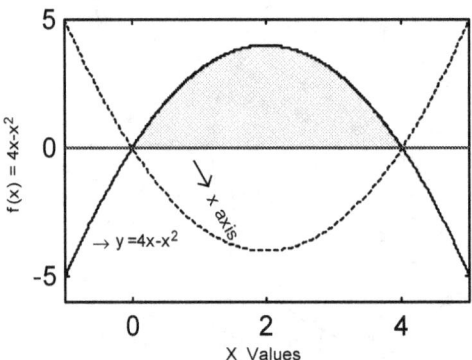

Figure 6.10 *Bounded area for solid of revolution*

Solution:

The curve $y = 4x - x^2$ and the x axis intersect at points (0, 0) and (0, 4). The volume of solid due to rotation of the shaded area (shown in figure 6.10) about the x axis is given by $V = \pi \int\limits_{x=a}^{x=b}[f(x)]^2 dx = \pi \int\limits_{x=0}^{x=4}(4x - x^2)^2 dx = \frac{512\pi}{15}$. Implementation is simple and straightforward.

MATLAB Command

>>syms x ↵

>>y=4*x-x^2; ↵   ← Define $y = 4x - x^2$

>>V=pi*int(y^2,x,0,4) ↵   ← Apply $V = \pi \int\limits_{x=a}^{x=b}[f(x)]^2 dx$

V =

512/15*pi

Axis of rotation can be other than x or y axis. The problem is somewhat clumsy. Once you know the calculus behind the problem, implementation becomes easier. We provide one example of that.

⊟ Example

Area bounded by the parabola P: $y^2 = 4ax$ and the straight line L: $y = 2x$ is rotated about L. Find the volume of the solid of revolution.

Solution:

Referring to figure 6.11, coordinates of any point C on the parabola P is (x, y). Parabola P and the line L intersect at points A and B, whose coordinates are (0,0) and (a, $2a$) respectively. Variable distance between any point on ACB to line AB, CH, is

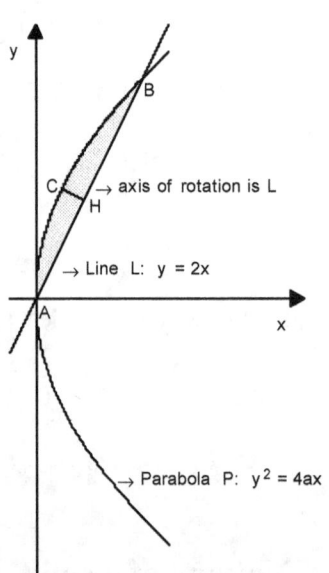

Figure 6.11 *Shaded area is rotated about AB to form solid of revolution*

given by $\dfrac{y-2x}{\sqrt{1^2+(-2)^2}}$ or $\dfrac{y-2x}{\sqrt{5}}$. The distance AC is $\sqrt{x^2+y^2}$. In AC and CH, y must be substituted from the

parabola expression, which is $y=2\sqrt{ax}$. Since AH⊥CH, AH$=\sqrt{AC^2-CH^2}$. If we label AH$=h$, the required

volume is $V=\pi\int_A^B CH^2\,dh$, where $\dfrac{dh}{dx}=\dfrac{d}{dx}\left[\sqrt{AC^2-CH^2}\right]=\dfrac{d}{dx}\left[\sqrt{x^2+y^2-\left(\dfrac{y-2x}{\sqrt{5}}\right)^2}\right]=\dfrac{d}{dx}\left[\dfrac{x+2y}{\sqrt{5}}\right]=$

$\dfrac{1}{\sqrt{5}}\left(1+2\sqrt{\dfrac{a}{x}}\right)$ (inserting $y=2\sqrt{ax}$). The final expression of the volume is $V=\pi\int_{x=0}^{x=a}\left(\dfrac{2\sqrt{ax}-2x}{\sqrt{5}}\right)^2\dfrac{1}{\sqrt{5}}\left(1+2\sqrt{\dfrac{a}{x}}\right)dx$

$=\dfrac{2\pi\,a^3}{15\sqrt{5}}$ cubic unit (on integration and simplification). You can bypass this non lenient computation by the

following:

MATLAB Command

>>syms x a y real ↵	← Declare variables as real symbolic
>>CH=(y-2*x)/sqrt(5); ↵	← Define CH$=\dfrac{y-2x}{\sqrt{5}}$
>>AC=sqrt(x^2+y^2); ↵	← Define AC$=\sqrt{x^2+y^2}$
>>h=simple(sqrt(AC^2-CH^2)); ↵	← Find $h=\sqrt{AC^2-CH^2}$ in simplified form
>>y=2*sqrt(a*x); ↵	← Define $y=2\sqrt{ax}$
>>h=subs(h); ↵	← Substitute $y=2\sqrt{ax}$ in h
>>CH=subs(CH); ↵	← Substitute $y=2\sqrt{ax}$ in CH
>>I=diff(h)*CH^2; ↵	← Form the integrand of V
>>V=simple(pi*int(I,x,0,a)); ↵	← Simplified volume is stored in V
>>pretty(V) ↵	

```
            1/2   3
   2/75 pi 5     a~
```

There are two commands for simplification – 'simple' and 'simplify'. We used the first one. Referring to the output, 'a~' means real a. One may raise the question why we have to declare the variables as real. The answer is that MATLAB subroutines are usually operational for the complex functions. Simplification of real expressions is easier than that of the complex counterpart. Non lenient expressions are easy to handle in MATLAB provided that one applies appropriate sequence and subroutine. In this particular problem, do not substitute $y=2\sqrt{ax}$ before the simplification. That may turn the MATLAB subroutines non operational.

6.6 Surface of revolution

Surface of revolution due to the rotation about x axis is given by $S_x=2\pi\int_{x=a}^{x=b}y\sqrt{1+\left(\dfrac{dy}{dx}\right)^2}\,dx$, where

$y=f(x)$ and the same for the rotation about y axis is given by $S_y=2\pi\int_{y=a}^{y=b}y\sqrt{1+\left(\dfrac{dx}{dy}\right)^2}\,dy$, where $x=f(y)$. We

begin the computation of surface area considering the first example of article 6.5. The surface of revolution due to the rotation of the shaded area about x axis of figure 6.10 is $S_x = 2\pi \int\limits_{x=0}^{x=4} (4x - x^2)\sqrt{1 + (4 - 2x)^2}\ dx$. Carry out a

long and clumsy integration to have $S_x = 2\pi \left[\dfrac{31\sqrt{17}}{8} - \dfrac{65}{32}\ln(\sqrt{17} - 4) \right]$ square unit. Following is the procedure:

MATLAB Command

```
>>syms x ↵
>>y=4*x-x^2; ↵                          ← Define y = 4x - x²
```

```
>>I=y*sqrt(1+diff(y)^2); ↵              ← Form the integrand y√(1+(dy/dx)²)
```

```
>>S=2*pi*int(I,x,0,4); ↵                ← Apply the formula of Sₓ
>>pretty(S) ↵
```

$$
2\ pi\ \left|31/8\ 17^{1/2}\ -\ \frac{65}{32}\ \log(-4 + 17^{1/2})\right|
$$

Pertaining to the second example of the last article, finding the surface of revolution due to the rotation of the shaded area about the line AB is slightly challenging. The surface area is given by $S_h = 2\pi \int\limits_A^B CH\sqrt{1 + \left[\dfrac{d(CH)}{dh}\right]^2}\ dh$. Convert the axis of rotation h in terms of x, hence, $S_h =$

$2\pi\int\limits_A^B CH\sqrt{1 + \left[\dfrac{d(CH)}{dx}\Big/\dfrac{dh}{dx}\right]^2}\ \dfrac{dh}{dx}\ dx = 2\pi \int\limits_{x=0}^{x=a} \dfrac{2\sqrt{ax} - 2x}{\sqrt{5}} \sqrt{1 + \left[\dfrac{d}{dx}\left[\dfrac{2\sqrt{ax} - 2x}{\sqrt{5}}\right]\Big/\dfrac{1}{\sqrt{5}}\left(1 + 2\sqrt{\dfrac{a}{x}}\right)\right]^2}\ \dfrac{1}{\sqrt{5}}\left(1 + 2\sqrt{\dfrac{a}{x}}\right)\ dx$. I do

not think anyone would be interested to carry out such a painstaking integration. Let us take MATLAB's help:

MATLAB Command

```
>>syms x a y real ↵                     ← Declare variables as real symbolic
>>CH=(y-2*x)/sqrt(5); ↵                 ← Define CH = (y-2x)/√5
```

```
>>AC=sqrt(x^2+y^2); ↵                   ← Define AC = √(x²+y²)
>>h=simple(sqrt(AC^2-CH^2)); ↵          ← Find h = √(AC²-CH²) in simplified form
>>y=2*sqrt(a*x); ↵                      ← Define y = 2√(ax)
>>h=subs(h); ↵                          ← Substitute y = 2√(ax) in h
>>CH=subs(CH); ↵                        ← Substitute y = 2√(ax) in CH
```

```
>>I=CH*sqrt(1+(diff(CH)/diff(h))^2)*diff(h); ↵ ← Form the integrand CH√(1+[d(CH)/dx / dh/dx]²) dh/dx
```

```
>>I=simple(I); ↵                        ← Simplified integrand is taken before integration
>>S=simple(2*pi*int(I,x,0,a)); ↵        ← Integration is also simplified
>>pretty(S) ↵
```

$$
1/30\ pi\ 5^{1/2}\ a\sim^2\ (14\ 2^{1/2} + 3 \log(3\ a\sim + 2\ 2^{1/2}\ (a\sim^2)^{1/2}) - 16 - 3 \log(a\sim))
$$

If we write the mathematical code, the close from integration is $\dfrac{\pi a^2 \sqrt{5}}{30}\left[\,14\sqrt{2}+3\ln(3a+2\sqrt{2}\,a)-16-3\ln a\,\right]$.

All integrands may not yield the close form integration in this regard one can use the numerical computation techniques as mentioned in article 6.3. Anyhow we have shown the style of computation for few integral calculus problems. This type of computation can occur in calculating area included by some curves, moment of inertia about some axis, centroid of an object…etc. Techniques so illustrated can be employed to solve those problems as well.

6.7 Summation of a series

A series can be finite or infinite. Take the example of series $1+2+3+4+\ldots\ldots$. The common term of the series is k, where k varies from 1 to infinity. Certainly this is an infinite series. Again, $3+7+11+15+19+23$ is a finite series. The common term of the series is $4n-1$, where n takes the value from 1 to 6. Another form of series representation can be a summation form, for example, $\displaystyle\sum_{n=1}^{n=4}\dfrac{1}{2n}=\dfrac{1}{2}+\dfrac{1}{4}+\dfrac{1}{6}+\dfrac{1}{8}$, which is a finite series (obviously, the increment of n is consecutive, if not, we need to make it consecutive for MATLAB implementation). Our concern is how to represent a series in MATLAB. A series is represented in MATLAB in terms of the common term. Along with the common term, the starting and final values corresponding to the common term variable must be presented. Subroutine 'symsum' (abbreviation of <u>sym</u>bolic <u>sum</u>mation) is capable of summing a series. To determine the summation of a series, whose common term is a function of n (n is an integer),

MATLAB Steps:
1. *Declare the common term variable n as symbolic and*
2. *Use the command symsum(common term as string, starting value of n, final value of n).*

The finite series $3+7+11+15+19+23$ has the common term $4n-1$ and sum 78. The string form of $4n-1$ is 4*n-1. The starting and final n's are 1 and 6 respectively. See the solution as follows:

MATLAB Command

>>syms n ↵
>>symsum(4*n-1,1,6) ↵

ans =

78

Supplementary exercise:

Determine the summation of the following series:

A. $\dfrac{1}{1^2}+\dfrac{1}{2^2}+\dfrac{1}{3^2}+\dfrac{1}{4^2}+\dfrac{1}{5^2}+\ldots\ldots\ldots\ldots$

B. $\dfrac{1}{2.4}+\dfrac{1}{3.5}+\dfrac{1}{4.6}+\dfrac{1}{5.7}+\dfrac{1}{6.8}+\dfrac{1}{7.9}+\dfrac{1}{8.10}$

C. $\dfrac{100}{6+3^2}+\dfrac{100}{6+5^2}+\dfrac{100}{6+7^2}+\dfrac{100}{6+9^2}$

D. $\dfrac{3}{2!}+\dfrac{3}{3!}+\dfrac{3}{4!}+\dfrac{3}{5!}+\dfrac{3}{6!}+\dfrac{3}{7!}+\ldots\ldots\ldots+\infty$

E. $\displaystyle\sum_{n=1,3,}^{n=11}\dfrac{1}{2n}$

F. Generate the expression $\ 1+2\cos 2x+3\cos 4x+$
$4\cos 6x+5\cos 8x+6\cos 10x+7\cos 12x$

Summations of the series A through E in MATLAB are presented as follows:

MATLAB Command

for the series A,

```
>>syms n ↵
>>symsum(1/n^2,1,inf) ↵
```

ans =

1/6*pi^2

for the series C,

```
>>syms n ↵
>>symsum(100/(6+(2*n-1)^2),2,5) ↵
```

ans =

381520/29667

for the series B,

```
>>syms n ↵
>>symsum(1/((n+1)*(n+3)),1,7) ↵
```

ans =

14/45

for the series D,

```
>>syms n ↵
>>symsum(3/sym('(n+1)!'),1,inf) ↵
```

ans =

3*exp(1)*(1-2*exp(-1))

simplification of the output for the series D,

```
>>simplify(symsum(3/sym('(n+1)!'),1,inf)) ↵
```

ans =

3*exp(1)-6

for the series E,

```
>>syms p ↵
>>symsum(1/2/(2*p-1),1,6) ↵
```

ans =

3254/3465

⊟ Series A

The common term is $\frac{1}{n^2}$ and the series is infinite. The starting and final values of n are 1 and ∞ respectively. This is a well-known series. The close form summation of the series is $\frac{\pi^2}{6}$. String representation of $\frac{1}{n^2}$ is 1/n^2.

⊟ Series B

The denominator of each term is in the product form. The common term is $\frac{1}{(n+1)(n+3)}$, where n takes the value from 1 to 7. Summation of this series is $\frac{14}{45}$. The string of $\frac{1}{(n+1)(n+3)}$ is 1/((n+1)*(n+3)).

⊟ Series C

Little intuition shows that the common term is $\frac{100}{6+(2n-1)^2}$, where n changes from 2 to 5 and the sum is $\frac{381520}{29667}$. In string form the common term becomes 100/(6+(2*n-1)^2).

⊟ Series D

This series has the factorial terms. The factorial of n is represented by sym('n!'). The common term is $\frac{3}{(n+1)!}$, where n varies from 1 to ∞. Actually, the series is derived from the series of $e = 1 + \frac{1}{1!} + \frac{1}{2!} + \frac{1}{3!} +$

$\frac{1}{4!}+\frac{1}{5!}+\ldots\ldots\ldots+\infty$. Using the series of e helps us write $\frac{3}{2!}+\frac{3}{3!}+\frac{3}{4!}+\frac{3}{5!}+\frac{3}{6!}+\frac{3}{7!}+\ldots\ldots\ldots+\infty=3e-6$. The factorial of $n+1$, i.e., $(n+1)!$ can be entered to MATLAB as sym('(n+1)!') so the string form of the common term is 3/sym('(n+1)!'). The output is returned as 3*exp(1)*(1-2*exp(-1)). It can be simplified further with the command 'simplify' as shown in the previous page.

⊟ Series E

The series is already given in summation form but the problem is the variation of n is not consecutive. Change n by $2p-1$, on that account, $\sum\limits_{n=1,3,}^{n=11}\frac{1}{2n}=\sum\limits_{p=1}^{p=6}\frac{1}{2(2p-1)}=\frac{3254}{3465}$.

⊟ Series F

Sometimes it is easier to generate a long expression by 'symsum'. We can avoid typing nuisance by doing so. By inspection, the common term is $(n+1)\cos 2nx$ and the variation of n is from 0 to 6. Generate the expression as follows:

MATLAB Command
>>syms x n ↵
>>symsum((n+1)*cos(2*n*x),n,0,6) ↵

ans =

1+2*cos(2*x)+3*cos(4*x)+4*cos(6*x)+5*cos(8*x)+6*cos(10*x)+7*cos(12*x)

6.8 Some integral functions

Different kinds of integral functions are available in MATLAB. The functions may be dealt with numerically or symbolically. Many functions, which handle symbolic input/output, are located in the symbolic or Maple package. M–file functions are also available in the Command Window. We present some of them.

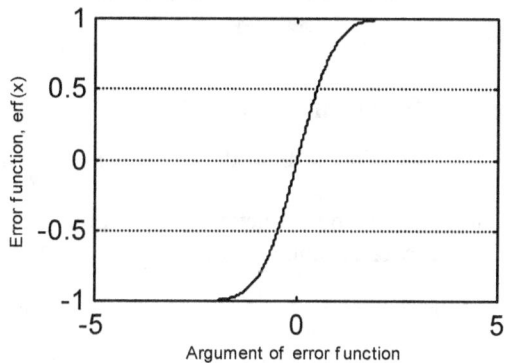

Figure 6.12   Plot of error function for interval $-5 \leq x \leq 5$

6.8.1 Error function

In MATLAB the error function is defined as $erf(x)=\frac{2}{\sqrt{\pi}}\int\limits_0^x e^{-t^2}dt$. This is an odd function and argument x of the function can be any real number (i.e., $-\infty \leq x \leq \infty$). The range of the function lies between -1 and 1 (i.e., $-1\leq erf(x)\leq 1$). The function is graphed in figure 6.12. For example, we wish to compute $erf(2)$ $=\frac{2}{\sqrt{\pi}}\int\limits_0^2 e^{-t^2}dt$. Numerically using the trapezoidal or Simpson's rule can compute $erf(2)$. Without going to the details, it is given that the trapezoidal rule of numerical integration provides $erf(2)=0.995322265$. Argument of the function can be a matrix (row, column, or rectangular) too. Let us say, $A=\begin{bmatrix}0 & -1 & 0.4\\ 2 & 1.3 & 3\end{bmatrix}$, it follows then,

262

$\text{erf(A)} = \begin{bmatrix} erf(0) & erf(-1) & erf(0.4) \\ erf(2) & erf(1.3) & erf(3) \end{bmatrix} = \begin{bmatrix} 0 & -0.8427 & 0.4284 \\ 0.995 & 0.9340 & 1 \end{bmatrix}$. Implementation for the both examples is shown as follows:

MATLAB Command

for scalar argument,

>>erf(2) ↵

ans =

0.9953

for matrix argument,

>>A=[0 -1 0.4;2 1.3 3]; ↵
>>O=erf(A) ↵

O =

| 0 | -0.8427 | 0.4284 |
| 0.9953 | 0.9340 | 1.0000 |

In summary, to evaluate the error function at any real number or matrix,

MATLAB Step:
Use the command erf(scalar/matrix name).

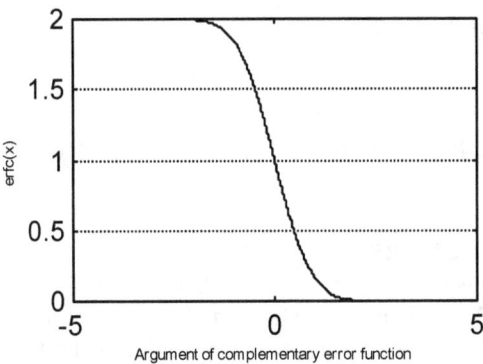

Figure 6.13   *Plot of complementary error function*
for interval -5 ≤ x ≤ 5

6.8.2 Complementary error function

The complementary error function is defined as $erfc\ (x) = \dfrac{2}{\sqrt{\pi}} \int\limits_{x}^{\infty} e^{-t^2} dt$. Unlike error function the complementary error function is not an odd function (see figure 6.13). The complementary relationship between the error and complementary error function is $erf(x) + erfc(x) = 1$. For the function, we have $-\infty \le x \le \infty$ and $0 \le erfc(x) \le 2$. It is given that $erfc(1) = \dfrac{2}{\sqrt{\pi}} \int\limits_{1}^{\infty} e^{-t^2} dt = 0.157299207$ [numerically, $erfc(1)$ can be computed by taking the fact that $erfc(1) = 1 - erf(1)$]. Like error function the argument can be a scalar, row, column or rectangular matrix. To show the scalar computation,

MATLAB Command

>>erfc(1) ↵

ans =

0.1573

6.8.3 Gamma function

The gamma function is given by $\Gamma x = \int\limits_{t=0}^{t=\infty} t^{x-1} e^{-t} dt$. Γx is not defined at $x = 0$ or negative integers but it is defined between any two consecutive negative integers. The range of gamma function includes positive and negative real numbers. Figure 6.14 depicts the plot of gamma function for some interval. To evaluate the gamma function,

MATLAB Step:
Use the command gamma (scalar/matrix name).

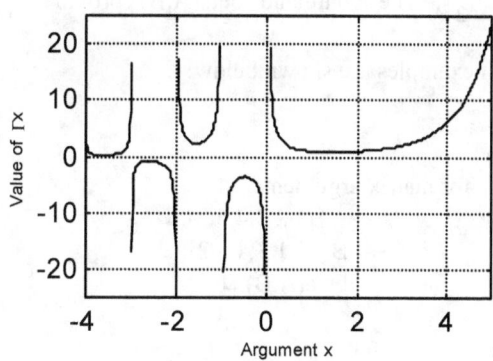

Figure 6.14   *Plot of gamma function for interval* $-4 \leq x \leq 5$

As example, find $\Gamma 4$, $\therefore$ $\Gamma 4 = \int_0^\infty t^3 e^{-t} dt = [-t^3 e^{-t} - 3t^2 e^{-t} - 6te^{-t} - 6e^{-t}]_{t=0}^{t=\infty} = [0-(-6)] = 6$. Test the function with

matrix argument. Say, $A = \begin{bmatrix} 0 & 0.5 & 1.5 \\ -0.4 & -1 & -3.2 \end{bmatrix}$, then, gamma(A) performs $\begin{bmatrix} \Gamma 0 & \Gamma(0.5) & \Gamma(1.5) \\ \Gamma(-0.2) & \Gamma(-1) & \Gamma(-3.2) \end{bmatrix} =$

$\begin{bmatrix} undefined & 1.7725 & 0.8862 \\ -5.8211 & undefined & 0.6891 \end{bmatrix}$. See the response of MATLAB for these computations as follows:

MATLAB Command
>>gamma(4) ↵

    ans =

       6

for matrix argument,
>>A=[0 0.5 1.5;-0.4 -1 -3.2]; ↵
>>gamma(A) ↵

    ans =

       Inf   1.7725   0.8862
   -3.7230   Inf   0.6891

6.8.4 Beta function

The definition of beta function is given by $B(z,w) = \int_{t=0}^{t=1} t^{z-1}(1-t)^{w-1} dt$. The beta function has two input arguments, z and w. Both arguments must be real and non-negative (i.e., $0 < z$ or $w < \infty$, z or w can not be 0). The function can be expressed in terms of the gamma function, which is $B(z,w) = \dfrac{\Gamma z \; \Gamma w}{\Gamma(z+w)}$. One important property of the function is that $B(z,w) = B(w,z)$. To evaluate the beta function at some z and w,

MATLAB Step:
Use the command beta(the first argument–z, the second argument–w).

We present a scalar example first. Evaluate $B(2, 3)$. We have $B(2, 3) = \int_{t=0}^{t=1} t(1-t)^2 \, dt = \left[\frac{t^2}{2} - \frac{2}{3}t^3 + \frac{1}{4}t^4\right]_{t=0}^{t=1} = \frac{1}{2} -$

$\frac{2}{3} + \frac{1}{4} = \frac{1}{12} = 0.0833$. The arguments of beta function can be matrices too. Say, the matrix arguments are given

as $A = \begin{bmatrix} 1 & 2 & 3 \\ 4 & 5 & 3 \end{bmatrix}$ and $B = \begin{bmatrix} 3 & 1 & 3 \\ 4 & 5 & 2 \end{bmatrix}$. The command beta(A,B) provides $\begin{bmatrix} beta(1,3) & beta(2,1) & beta(3,3) \\ beta(4,4) & beta(5,5) & beta(3,2) \end{bmatrix} =$

$\begin{bmatrix} 0.3333 & 0.5 & 0.0333 \\ 0.0071 & 0.0016 & 0.0833 \end{bmatrix}$. Both examples are shown below:

MATLAB Command

for scalar argument,
>>beta(2,3) ↵

ans =
    0.0833

for matrix arguments,
>>A=[1 2 3;4 5 3]; ↵
>>B=[3 1 3;4 5 2]; ↵
>>beta(A,B) ↵

ans =
    0.3333   0.5000   0.0333
    0.0071   0.0016   0.0833

6.8.5 Sine integral

The sine integral is defined as $Si(x) = \int_{t=0}^{t=x} \frac{\sin t}{t} \, dt$. $Si(x)$ is an odd function. The plot of sine integral is shown in figure 6.15. Its MATLAB counterpart is 'sinint' (abbreviation of <u>sin</u>e <u>int</u>egral). To evaluate $Si(x)$,

MATLAB Step:
Use the command sinint(scalar or row matrix name).

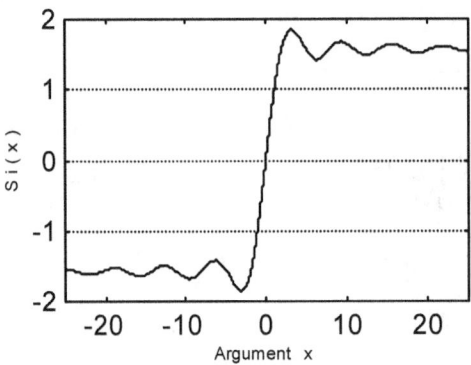

Figure 6.15   *Plot of sine integral for interval* $-25 \leq x \leq 25$

Elements of the argument matrices can be real or complex. For illustration, we wish to compute sine integral for each element of A=[0 2.1] so the output should be $[\, Si(0) \quad Si(2.1) \,] = [\, 0 \quad 1.6487 \,]$. Foregoing matrix elements are obtained by using the trapezoidal rule integration. The differentiation of sine integral can be done too, clearly, $\frac{d}{dx}[Si(x)] = \frac{\sin x}{x}$. Both examples are shown as follows:

MATLAB Command

for computation of sine integral,
>>A=[0 2.1]; ↵
>>sinint(A) ↵

ans =

for differentiation of sine integral,
>>syms x ↵
>>diff(sinint(x)) ↵

ans =

$$0 \quad 1.6487$$

$$\sin(x)/x$$

6.8.6 Cosine integral

The definition of the cosine integral is $Ci(x) = \int_{t=\infty}^{t=x} \frac{\cos t}{t} dt$. Some manipulations allow us to write $Ci(x)$ as $\gamma + \ln x - \int_{t=0}^{t=x} \frac{1 - \cos t}{t} dt$, where $\gamma = 0.577215664$ is called the Euler's constant. $Ci(x)$ is graphed for interval $0 \le x \le 25$ in figure 6.16. MATLAB counterpart of $Ci(x)$ is 'cosint' (abbreviation of <u>cos</u>ine <u>int</u>egral). To evaluate $Ci(x)$,

MATLAB Step

Use the command cosint(scalar or row matrix name).

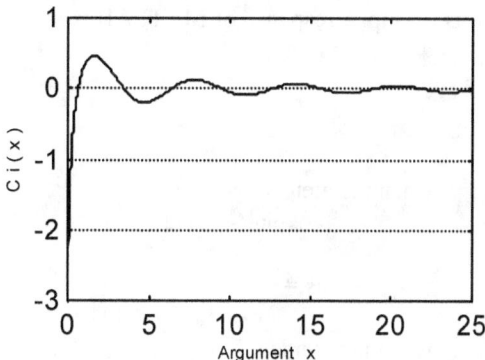

Figure 6.16   *Plot of cosine integral for interval $0 \le x \le 25$*

For illustration, cosine integrals of the elements of A =[0   4] are to be evaluated so output should look like $[\, Ci(0) \quad Ci(4)\,] = [\infty \quad -0.1410]$ (on numerical computation). If any element of scalar / row matrix argument is negative number, the output is a complex number. See the following for implementations:

MATLAB Command

for computation of cosine integral,
        >>A=[0 4]; ↵
        >>cosint(A) ↵

ans =
      Inf   -0.1410

for −ve elements,
        >>cosint(-2.2) ↵

ans =
      0.3751 + 3.1416i

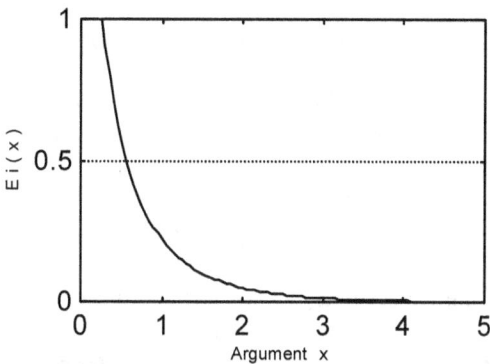

Figure 6.17   *Plot of exponential integral for interval $0 \le x \le 5$*

6.8.7 Exponential integral

Expression for the exponential integral is given by $Ei(x) = \int_{t=x}^{t=\infty} \frac{e^{-t}}{t} dt$. Shown figure 6.17 is the plot of exponential integral for interval $0 \leq x \leq 5$. The counterpart is 'expint' (abbreviation of <u>exp</u>onential <u>int</u>egral). To compute $Ei(x)$,

MATLAB Step:

Use the command expint (scalar or matrix).

Any scalar or matrix can be the arguments of 'expint'. To show by an example, say, $C = \begin{bmatrix} 3.2 \\ 4 \\ 2 \end{bmatrix}$, expint(C) returns

$\begin{bmatrix} Ei(3.2) \\ Ei(4) \\ Ei(2) \end{bmatrix} = \begin{bmatrix} 0.0101 \\ 0.0038 \\ 0.0489 \end{bmatrix}$. Including this example, response of MATLAB is shown below when arguments' element is −ve, 0, or ∞.

MATLAB Command

for column matrix C,
```
>>C=[3.2 4 2]'; ↵
>>expint(C) ↵

ans =

        0.0101
        0.0038
        0.0489
```
when the element is 0,
```
>>expint(0) ↵
```

Warning: Log of zero.
> In C:\MATLAB\toolbox\matlab\specfun\expint.m at line 53

```
ans =
        Inf
```

when the element is −ve,
```
>>expint(-2.1) ↵

ans =
        -5.3332 - 3.1416i
```
when the element is ∞,
```
>>expint(inf) ↵

ans =
        NaN
```

6.8.8 Bessel's equation and functions

The differential equation $x^2 \frac{d^2y}{dx^2} + x \frac{dy}{dx} + (x^2 - \alpha^2)y = 0$ is called the Bessel's equation of order α, where $\alpha \geq 0$. Basically, the equation is a second order differential equation but traditionally α is referred to as the order of the Bessel's equation. One of the power series solutions of the equation is the Bessel's function $J_\alpha(x)$,

where $J_\alpha(x) = \sum_{n=0}^{\infty} \frac{(-1)^n}{2^{2n+\alpha} \Gamma(n+1)\Gamma(n+\alpha+1)} x^{2n+\alpha}$. $J_\alpha(x)$ is called the Bessel function of the first kind of order α.

This series converges for all positive x (i.e., $x \geq 0$). Γn is the gamma function, which is defined in article 6.8.3. Bessel function of the first kind is graphed in figure 6.18. To evaluate Bessel function of the first kind, $J_\alpha(x)$,

MATLAB Step:

Use the command besselj (order, x−value).

To clarify by examples, value of the Bessel function of the first kind of order 1 at $x = 3$ [$J_1(3)$] is required.

Using the expression of $J_\alpha(x)$, one can write that $J_1(3) = \sum_{n=0}^{\infty} \frac{(-1)^n}{\Gamma(n+1)\Gamma(n+2)} \left(\frac{3}{2}\right)^{2n+1} = \left(\frac{3}{2}\right) \frac{1}{\Gamma 1 \Gamma 2} -$

$$\left(\frac{3}{2}\right)^{3}\frac{1}{\Gamma2\Gamma3}+\left(\frac{3}{2}\right)^{5}\frac{1}{\Gamma3\Gamma4}-\left(\frac{3}{2}\right)^{7}\frac{1}{\Gamma4\Gamma5}+\left(\frac{3}{2}\right)^{9}\frac{1}{\Gamma5\Gamma6}-\left(\frac{3}{2}\right)^{11}\frac{1}{\Gamma6\Gamma7}+\left(\frac{3}{2}\right)^{13}\frac{1}{\Gamma7\Gamma8}-...=1.5-1.6875+0.6328-0.1187+$$

$0.0133-0.0010+0.0000-...$(up to 4 decimal accuracy)$=0.3389$. To have the computation,

MATLAB Command
>>besselj(1,3) ↵

ans =
      0.3391

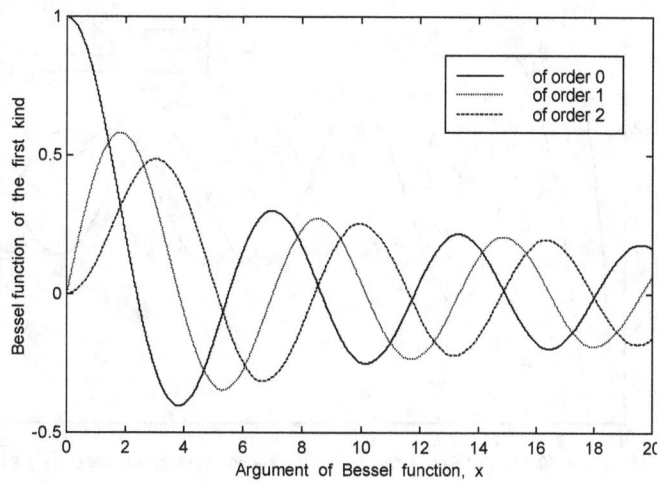

Figure 6.18 *Plot of Bessel function of the first kind for interval $0 \le x \le 20$*

The arguments of 'besselj' can be like positional elements of two matrices. What we mean is if $A=\begin{bmatrix}0 & 1\\ 2 & 3\end{bmatrix}$ and

$B=\begin{bmatrix}0 & 5.5\\ \infty & 1\end{bmatrix}$, then the command besselj(A,B) should return $\begin{bmatrix}J_0(0) & J_1(5.5)\\ J_2(\infty) & J_3(1)\end{bmatrix}=\begin{bmatrix}1 & -0.3414\\ 0 & 0.0196\end{bmatrix}$. To show this,

MATLAB Command
>>A=[0 1;2 3]; ↵
>>B=[0 5.5;inf 1]; ↵
>>besselj(A,B) ↵

ans =
      1.0000   -0.3414
      NaN   0.0196

Symbolic differentiation and integration of Bessel's functions can be carried out too. We know that $\frac{d}{dx}[J_n(x)]=\frac{n}{x}J_n(x)-J_{n+1}(x)$ and $\int x^{n+1}J_n(x)\,dx=x^{n+1}J_{n+1}(x)$. Do not be confused, n is used instead of α for order. Following is the differentiation and integration:

MATLAB Command
>>syms x n ↵   ← Declare n and x as symbolic
>>y=besselj(n,x); ↵   ← Define $y=J_n(x)$

>>pretty(diff(y)) ↵   ← Perform $\frac{d}{dx}[J_n(x)]$

$$-\text{besselj}(n+1, x) \;+\; \frac{n\ \text{besselj}(n, x)}{x}$$

>>pretty(int(x^(n+1)*y)) ↵   ← Carry out $\int x^{n+1} J_n(x)\,dx$

$$x^{(n+1)}\ \text{besselj}(n+1, x)$$

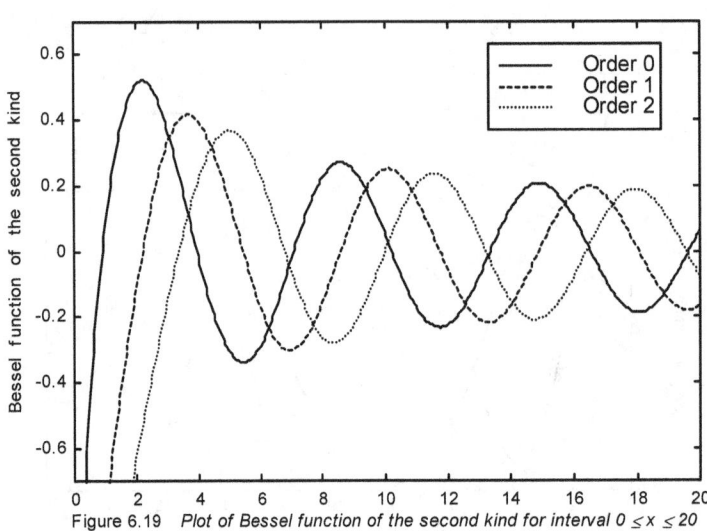

Figure 6.19   Plot of Bessel function of the second kind for interval $0 \le x \le 20$

When α of $J_\alpha(x)$ is positive or negative odd multiple of $\frac{1}{2}$, one can express $J_\alpha(x)$ in terms of the sine and cosine functions. To show an example, we present the implementation of $J_{-\frac{3}{2}}(x) = \sqrt{\dfrac{2}{x\pi}}\left[-\sin x - \dfrac{\cos x}{x} \right]$. Maple procedure 'convert' helps us do that.

MATLAB Command

>>y=maple('convert(besselj(-3/2,x),sincos)'); ↵
>>pretty(sym(y)) ↵

$$-\ \frac{2^{1/2}\ (\sin(x)\,x + \cos(x))}{\text{pi}^{1/2}\ x^{3/2}}$$

Notice that the argument 'sincos' of 'convert' tells the procedure to display the expression in terms of the sine and cosine terms. The output of 'convert' is stored to some variable y. Contents of y are converted to symbolic variable to display the output in symbolic form by making the use of command 'pretty'.

The second family of the Bessel's function is given by $Y_\alpha(x) = \dfrac{J_\alpha(x)\cos(\alpha\pi) - J_{-\alpha}(x)}{\sin(\alpha\pi)}$. $Y_\alpha(x)$ is called the Bessel's function of the second kind of order α, where α is not an integer (if it is an integer, limiting value is taken). $Y_\alpha(x)$ is a linear combination of $J_\alpha(x)$ and $J_{-\alpha}(x)$. Figure 6.19 depicts the plot of Bessel's function of the second kind for some interval. General solution of the Bessel's equation of order α is given by $y(x) = C_1 J_\alpha(x) + C_2 Y_\alpha(x)$. However, to evaluate the Bessel's function of the second kind, $Y_\alpha(x)$,

MATLAB Step:
Use the command bessely(order,x−value).

Compute $Y_{\frac{3}{2}}(4)$. We defined $Y_\alpha(x)$ previously, on that account, $Y_{\frac{3}{2}}(4) = \dfrac{J_{\frac{3}{2}}(4)\cos\left(\dfrac{3}{2}\pi\right) - J_{-\frac{3}{2}}(4)}{\sin\left(\dfrac{3}{2}\pi\right)} = 0.3671$.

Computation is shown as follows:

MATLAB Command
```
>>bessely(3/2,4) ↵

ans =
        0.3671
```

The argument of the Bessel's function mentioned so far is real number. There are circumstances when it is indispensable to use complex number arguments, for example, skin effect analysis of some material due to the high frequency current. Then, $J_\alpha(x)$ becomes $J_\alpha(ix)$ which is denoted by $I_\alpha(x)$ and called the modified Bessel function of the first kind. The MATLAB generic is 'besseli'. We encounter another function with complex argument for $Y_\alpha(x)$. That is called the modified Bessel's function of the second kind, which is denoted by $K_\alpha(x)$. The resembling MATLAB function is 'besselk'. Writing input arguments of 'besseli' and 'besselk' follows the same style as do 'besselj' and 'bessely'. That is not all. The last member is the Bessel's function of the third kind. It has other acronym − Hankel's function. Hankel function is of two types − the first kind [$H_\alpha^1(x)$] and the second kind [$H_\alpha^2(x)$]. They are defined by $H_\alpha^1(x) = J_\alpha(x) + jY_\alpha(x)$ and $H_\alpha^2(x) = J_\alpha(x) - jY_\alpha(x)$ respectively. Corresponding MATLAB functions are 'besselh(α ,1, x)' and 'besselh(α ,2, x)' respectively. We summarize all Bessel functions in table 6.D.

Table 6.D   Bessel functions and their MATLAB counterparts

Name	Symbolic notation	MATLAB counterpart
Bessel function of the first kind	$J_\alpha(x)$	besselj(α , x)
Bessel function of the second kind	$Y_\alpha(x)$	bessely(α , x)
Modified Bessel function of the first kind	$I_\alpha(x)$	besseli(α , x)
Modified Bessel function of the second kind	$K_\alpha(x)$	besselk(α , x)
Hankel function of the first kind	$H_\alpha^1(x)$	besselh(α ,1, x)
Hankel function of the second kind	$H_\alpha^2(x)$	besselh(α ,2, x)

Chapter 7

Problems of Complex Variables and Differential Equations

There are two parts in this chapter as the name implies. The first part highlights the basic implementation of complex variables' problems. In general, most subroutines (appropriately, built-in M-files or Maple functions) are written for complex numbers. We deposited the common problems of complex variables to grasp the concept of implementation and pay unified attention. Presence of symbolic toolbox enhances the beauty of computation. We devoted to the solutions of differential equations in the second part. Differential equations are equations that comprise from the derivatives of some unknown functions. Very often, the differential equations come into view in mathematical models that endeavor to describe the real-life problems. Discrete version of the differential equation, which is difference equation, is addressed in chapter 8. Broadly speaking, solving differential equations falls into two categories – one is the analytical or symbolic solution and the other is numerical solution. Pedagogically, analytical one is preferable, on the contrary, major simulation enigma demands the numerical solution.

7.1 Complex variable analysis

Collected in the first section are the basic manipulations from complex algebra. Arithmetic of complex numbers can be different from that of real numbers arising from the fact that a real number is scalar in nature whereas a complex number is a vector one. Basic properties of complex numbers lay the foundations of the analysis of complex functions. One typical use of complex functions can be in electrical circuits. A number of MATLAB subroutines accept complex arguments. We outline few of them.

7.1.1 Representation of complex numbers

In complex variable analysis, the basic imaginary number is defined by $\sqrt{-1}$. Symbolically, it is denoted by i or j. MATLAB allows us to represent a complex number with the same imaginary unit. As an example, complex number $4 + 5i$ can be entered into MATLAB by any of the expressions 4+5i, 4+5*i, 4+i*5, 4+5*j, or 4+5*sqrt(-1). The imaginary number unit i or j can also be a variable or loop counter index because they are written as functions. Be aware that if i or j is used as a variable, any complex number like

4+5*i assumes the variable value that is why it is better not to use i or j as complex number and loop counter index in a program where one has to deal with the both. From the definition of complex number, one can write $i^2 = -1$, $i^3 = i^2 i = -i$, $i^4 = i^2 i^2 = 1$, $i^5 = i^4 i = i$ .... etc.

7.1.2 Entering matrices of complex numbers

Matrix of complex numbers follows the similar entering style to that of integer or real numbers' with little difference in conjugateness. Enter the complex number matrices $R = [3-i \quad 4i \quad -4]$, $C = \begin{bmatrix} 7i \\ -4+5i \\ 8i \end{bmatrix}$, and

$A = \begin{bmatrix} 2 & 5-i & 9i \\ 7i & 2+i & 11i \end{bmatrix}$ into MATLAB as follows:

MATLAB Command

for R,   for C,
>>R=[3-i   4i -4] ↵   >>C=[7i -4+5i 8i].' ↵

R =   C =
      3.0000 - 1.0000i   0 + 4.0000i -4.0000   0 + 7.0000i
for A,   -4.0000 + 5.0000i
>>A=[2 5-i 9i;7i 2+i 11i] ↵   0 + 8.0000i

A =
      2.0000   5.0000 - 1.0000i   0 + 9.0000i
           0 + 7.0000i   2.0000 + 1.0000i   0 +11.0000i

Notice that we have the operator .' at the end when the column matrix is entered. For real matrix, it is just '. There is a difference between the two operators − .' means the transpose without conjugate but ' means the transpose with conjugate. Use of operator ' would assign $\begin{bmatrix} -7i \\ -4-5i \\ -8i \end{bmatrix}$ to C. Then, how do we enter the matrix of rational numbers? The answer is just use the command 'sym'. To show this, enter the matrix $R = \begin{bmatrix} 2+i\dfrac{3}{7} & \dfrac{5}{8}i \end{bmatrix}$ as follows:

>>R=sym([2+i*3/7 5*i/8]) ↵

R =

[(2)+(3/7)*i, (0)+(5/8)*i]

Another singular feature of MATLAB is that we can enter a matrix of symbolic variables too. One can use the command 'syms' for that. As usual, take $A = \begin{bmatrix} a+ib & \cos x + i \sin x \\ 2+4i & -4i \end{bmatrix}$ and have A in MATLAB as follows:

>>syms a b x ↵   ← Declare all variables symbolically
>>A=[a+i*b cos(x)+i*sin(x);2+4i -4i] ↵

A =

[a+i*b,   cos(x)+i*sin(x)]
[(2)+(4)*i,   (0)-(4)*i]

7.1.3 Addition/subtraction of complex numbers

Two matrices of complex numbers $A = \begin{bmatrix} 1+2i & 4+3i \\ 7-5i & 5+9i \end{bmatrix}$ and $B = \begin{bmatrix} 6i & -4 \\ 2+77i & 2-8i \end{bmatrix}$ are added to get $C =$

$A + B = \begin{bmatrix} 1+8i & 3i \\ 9+72i & 7+i \end{bmatrix}$. Have it done as follows:

MATLAB Command

>>A=[1+2i 4+3i;7-5i 5+9i]; ↵
>>B=[6i -4;2+77i 2-8i]; ↵
>>C=A+B ↵

C =

        1.0000 + 8.0000i   0 + 3.0000i
        9.0000 +72.0000i   7.0000 + 1.0000i

For the same matrices, the subtraction is performed by $A - B$. They can be a row, column, or rectangular matrix containing integers, floating-point numbers,
or symbolic variables.

7.1.4 Multiplication of complex numbers

Like real numbers, the multiplication of complex numbers can also be of two types — scalar and vector (see the details of the scalar and vector multiplications in article 2.3). The scalar multiplication of $A = \begin{bmatrix} 1+2i & 4+3i & 6i \\ 7-5i & 5+9i & 3-5i \end{bmatrix}$ and $B = \begin{bmatrix} 1+2i & 4-3i & 0 \\ 7-5i & 5-7i & 4i \end{bmatrix}$ is $C = \begin{bmatrix} 1+2i & 4+3i & 6i \\ 7-5i & 5+9i & 3-5i \end{bmatrix} .* \begin{bmatrix} 1+2i & 4-3i & 0 \\ 7-5i & 5-7i & 4i \end{bmatrix} =$

$\begin{bmatrix} (1+2i)^2 & (4+3i)(4-3i) & 6i \times 0 \\ (7-5i)^2 & (5+9i)(5-7i) & (3-5i)4i \end{bmatrix} = \begin{bmatrix} 1+4i+4i^2 & 16-9i^2 & 0 \\ 49-70i+25i^2 & 25+10i-63i^2 & 12i-20i^2 \end{bmatrix} =$

$\begin{bmatrix} -3+4i & 25 & 0 \\ 24-70i & 88+10i & 12i+20 \end{bmatrix}$ (on substitution of $i^2 = -1$). Have the multiplication as follows:

MATLAB Command

>>A=[1+2i 4+3i 6i;7-5i 5+9i 3-5i]; ↵
>>B=[1+2i 4-3i 0;7-5i 5-7i 4i]; ↵
>>C=A.*B ↵

C =

        -3.0000 + 4.0000i   25.0000   0
        24.0000 -70.0000i   88.0000 +10.0000i   20.0000 +12.0000i

As an example of the rational number matrices, consider $A = \begin{bmatrix} \frac{9}{2}-\frac{4}{5}i \\ -\frac{6}{7}+7i \end{bmatrix}$ and $B = \begin{bmatrix} \frac{19}{9}-\frac{4}{25}i \\ \frac{1}{4}+5i \end{bmatrix}$, where $A.*B =$

$\begin{bmatrix} (\frac{9}{2}-\frac{4}{5}i)(\frac{19}{9}-\frac{4}{25}i) \\ (-\frac{6}{7}+7i)(\frac{1}{4}+5i) \end{bmatrix} = \begin{bmatrix} \frac{19}{2}-\frac{542}{225}i+\frac{16}{125}i^2 \\ -\frac{6}{28}-\frac{71}{28}i+35i^2 \end{bmatrix} = \begin{bmatrix} \frac{19}{2}-\frac{542}{225}i-\frac{16}{125} \\ -\frac{6}{28}-\frac{71}{28}i-35 \end{bmatrix} = \begin{bmatrix} \frac{2343}{250}-\frac{542}{225}i \\ -\frac{493}{14}-\frac{71}{28}i \end{bmatrix}$. See the implementation as follows:

MATLAB Command

>>A=sym([9/2-4/5*i -6/7+7i].'); ↵
>>B=sym([19/9-4/25*i 1/4+5i].'); ↵
>>A.*B ↵

ans =

[2343/250-542/225*i]
[-493/14-71/28*i]

One sample example is presented for the vector multiplication taking $A = \begin{bmatrix} 1+i & 2-i & 7+i \\ i & 2i & -3i \end{bmatrix}$ and $B = \begin{bmatrix} 7i \\ 9-i \\ 3i \end{bmatrix}$.

The vector multiplication of A with B is $\begin{bmatrix} (1+i)7i + (2-i)(9-i) + (7+i)3i \\ i(7i) + 2i(9-i) - 3i(3i) \end{bmatrix} = \begin{bmatrix} 7i - 7 + 18 - 11i - 1 + 21i - 3 \\ -7 + 18i + 2 + 9 \end{bmatrix} =$

$\begin{bmatrix} 7+17i \\ 4+18i \end{bmatrix}$ as shown below:

MATLAB Command
```
>>A=[1+i 2-i 7+i;i 2i -3i]; ↵
>>B=[7i 9-i 3i].'; ↵
>>A*B ↵
```

ans =
                7.0000 +17.0000i
                4.0000 +18.0000i

7.1.5 Division of complex numbers

The general form of division of two complex numbers is $\dfrac{c+id}{a+ib}$. Rationalization yields $\dfrac{(c+id)(a-ib)}{(a+ib)(a-ib)}$

$= \dfrac{ca + ida - ibc - i^2bd}{a^2 - i^2b^2} = \dfrac{ca + bd + ida - ibc}{a^2 + b^2}$. Considering $A = 3 + 4i$ and $B = 7 - 8i$, division of A by B is $\dfrac{A}{B} =$

$\dfrac{3+4i}{7-8i} = \dfrac{3 \times 7 + 4 \times (-8) + i[4 \times 7 - (-8) \times 3)]}{7^2 + 8^2} = \dfrac{-11 + i52}{113} = -0.0973 + i\,0.4602$. Carry out that as follows:

MATLAB Command
```
>>A=3+4i; ↵
>>B=7-8i; ↵
>>A/B ↵
```

ans =
                -0.0973 + 0.4602i

As we pointed out in article 2.4, division is also of two types — scalar and vector. One example of scalar

division is presented taking into account $A = \begin{bmatrix} 3-5i & 5 \\ 7+7i & -7i \\ 2+6i & 8+i \end{bmatrix}$ and $B = \begin{bmatrix} 9i & 3 \\ 2+5i & 2-i \\ -7i & 4i \end{bmatrix}$, the division is $D =$

$A./B = \begin{bmatrix} \dfrac{3-5i}{9i} & \dfrac{5}{3} \\ \dfrac{7+7i}{2+5i} & \dfrac{-7i}{2-i} \\ \dfrac{2+6i}{-7i} & \dfrac{8+i}{4i} \end{bmatrix} = \begin{bmatrix} -\dfrac{5}{9} - \dfrac{1}{3}i & \dfrac{5}{3} \\ \dfrac{49}{29} - \dfrac{21}{29}i & \dfrac{7}{5} - \dfrac{14}{5}i \\ -\dfrac{6}{7} + \dfrac{2}{7}i & \dfrac{1}{4} - 2i \end{bmatrix} = \begin{bmatrix} -0.5556 - 0.3333i & 1.6667 \\ 1.6897 - 0.7241i & 1.4 - 2.8i \\ -0.8571 + 0.2857i & 0.25 - 2i \end{bmatrix}$. Ascertain the division as

follows:

```
>>A=sym([3-5i 5;7+7i -7i;2+6i 8+i]); ↵
>>B=sym([9i 3;2+5i 2-i;-7i 4i]); ↵
>>D=A./B ↵
```

D =

```
[      -5/9-1/3*i,            5/3]
[ 49/29-21/29*i,      7/5-14/5*i]
[      -6/7+2/7*i,          1/4-2*i]
```

The operators displayed in the output may be confusing. By sequence the division is done first, then, is multiplication, so, –6/7+2/7*i should be read as $-\dfrac{6}{7}+i\dfrac{2}{7}$ not $-\dfrac{6}{7}+\dfrac{2}{7i}$. You can even have the pretty form as follows:

>>pretty(D) ↵

```
[   - 5/9 - 1/3 i            5/3        ]
[                                       ]
[   49       21                         ]
[  ----- - ------ i    7/5 - 14/5 i ]
[   29       29                         ]
[                                       ]
[ - 6/7 + 2/7 i          1/4 - 2 i   ]
```

To have the result in decimal form, use the command 'double':

>>double(D) ↵

ans =

```
-0.5556 - 0.3333i   1.6667
 1.6897 - 0.7241i   1.4000 - 2.8000i
-0.8571 + 0.2857i   0.2500 - 2.0000i
```

Verification of the vector division is left as an exercise for the reader.

7.1.6 Modulus and argument of a complex number

Modulus or absolute value of a complex number $A+iB$ is given by $\sqrt{A^2+B^2}$. To take the modulus,

MATLAB Step:
Use the command abs(complex scalar or matrix name).

Subroutine 'abs' is the abbreviation of <u>abs</u>olute value. Modulus of $4+i3$ is 5. The subroutine applies to matrices too and the argument matrix can be row, column, or rectangular. The modulus of each element of R =[$12+i5$ $-4-i3$ $-8+i6$] is [$\sqrt{12^2+5^2}$ $\sqrt{(-4)^2+(-3)^2}$ $\sqrt{(-8)^2+6^2}$]=[13   5   10]. Both examples are shown as follows:

MATLAB Command
for the single number,   for the row matrix R,
>>abs(4+3i) ↵   >>R=[12+5i -4-3i -8+6i]; ↵
                                         >>abs(R) ↵

   ans =   ans =
        5   13   5   10

Argument of a complex number $A+iB$ is given by $\tan^{-1}\dfrac{B}{A}$. To find the argument,

MATLAB Step:
Use the command angle (complex scalar or matrix name).

Actually, 'angle' in the abbreviation of phase <u>angle</u>. The argument $\tan^{-1}\dfrac{B}{A}$ can take any value from $-\pi$ to π.

Arguments of $4+i3$ and each element of $R=[12+i5 \quad -4-i3 \quad -8+i6]$ are $\tan^{-1}\dfrac{3}{4}=0.6435^c$ and $[\tan^{-1}\dfrac{5}{12}$

$\tan^{-1}\dfrac{-3}{-4} \quad \tan^{-1}\dfrac{6}{-8}]=[0.3948^c \quad -2.4981^c \quad 2.4981^c]$ respectively. The arctan is the four quadrant inverse

tangent. Findings of these arguments are presented as follows:

MATLAB Command

for a single number,

>>angle(4+3i) ↵

ans =

0.6435

for the row matrix R,

>>R=[12+5i -4-3i -8+6i]; ↵

>>angle(R) ↵

ans =

0.3948 -2.4981   2.4981

7.1.7 Conjugate of a complex number

The conjugate of a complex number $A+iB$ is given by $A-iB$. To find the conjugate,

MATLAB Step:

Use the command conj(complex scalar or matrix name).

The subroutine 'conj' is named from <u>conj</u>ugate. Its argument can be row, column, or rectangular matrix.

Conjugates of $4+i3$ and all elements of $\begin{bmatrix} 12+5i \\ -4-3i \\ -8+6i \end{bmatrix}$ are $4-i3$ and $\begin{bmatrix} 12-5i \\ -4+3i \\ -8-6i \end{bmatrix}$ respectively. Implementations are

shown below:

MATLAB Command

for the single number,

>>conj(4+3i) ↵

ans =

4.0000 - 3.0000i

for the column matrix C,

>>C=[12+5i -4-3i -8+6i].'; ↵

>>conj(C) ↵

ans =

12.0000 - 5.0000i

-4.0000 + 3.0000i

-8.0000 - 6.0000i

7.1.8 Real and imaginary parts of a complex number

A complex number $A+iB$ has the real part A and the imaginary part B. To find the real or imaginary part of complex number(s),

MATLAB Step:

Use the commands real (complex scalar or matrix name) for the real part and imag (complex scalar or matrix name) for the imaginary part.

The real part of $4+i3$ is 4 and the imaginary parts of each element of $C=\begin{bmatrix} x+i5 \\ -4-iy \\ -8+i3y \end{bmatrix}$ are $\begin{bmatrix} 5 \\ -y \\ 3y \end{bmatrix}$. See their

findings in the following:

MATLAB Command

for the single number,

>>real(4+3i) ↵

for the imaginary part of C,

>>syms x y real ↵

>>C=[x+i*5 -4-i*y -8+i*3*y].'; ↵
>>imag(C) ↵

ans =

4

ans =

```
[  5 ]
[ -y ]
[ 3*y]
```

Observe that the declaration of symbolic variables x and y needs to include 'real' otherwise MATLAB assumes these variables to be complex. For long expressions the real and imaginary parts are computed in a different way. If $Z = x + iy$ is a complex number, its conjugate is $\bar{Z} = x - iy$ so the real and imaginary parts of Z are given by $x = \dfrac{Z + \bar{Z}}{2}$ and $y = \dfrac{Z - \bar{Z}}{2i}$ respectively. To show the anomaly, let us see the response of MATLAB for $\dfrac{\cos(x + iy)}{\sin(x - iy)}$.

MATLAB Command

>>syms x y ↵

>>f=cos(x+i*y)/sin(x-i*y); ↵   ← Assign $\dfrac{\cos(x + iy)}{\sin(x - iy)}$ to f

>>R=real(f); ↵   ← Assign the real part of f to R
>>pretty(R) ↵
```
       cos(x + i y)              cos(x + i y)
 1/2 ---------------- + 1/2 conj(----------------)
       sin(x - i y)              sin(x - i y)
```
>>I=imag(f); ↵   ← Assign imaginary part of f to I
>>pretty(I) ↵
```
        / cos(x + i y)         cos(x + i y) \
 - 1/2 i|---------------- - conj(----------------)|
        \ sin(x - i y)         sin(x - i y) /
```

As you see, the real denominator may not be obtained for the real and imaginary parts of a long expression. But the form can serve the computational purpose. Calculate the real and imaginary parts of $\dfrac{\cos(x + iy)}{\sin(x - iy)}$ for $x = \pi$ and $y = 3$. For $x = \pi$ and $y = 3$, $\dfrac{\cos(x + iy)}{\sin(x - iy)}$ becomes $\dfrac{\cos(\pi + i3)}{\sin(\pi - i3)} = \dfrac{-\cos(i3)}{\sin(i3)} = \dfrac{-\cosh 3}{i \sinh 3} = \dfrac{i \cosh 3}{\sinh 3}$, therefore, the real part is zero and the imaginary part is $\dfrac{\cosh 3}{\sinh 3}$. The assignees R and I are having the real and imaginary parts. Carry out the substitution as follows:

>>subs(R,{x,y},{'pi','3'}) ↵   ← Substitute for the real part

ans =

0

>>subs(I,{x,y},{'pi','3'}) ↵   ← Substitute for the imaginary part

ans =

cosh(3)/sinh(3)

The reverse problem can also be seen, that is, the real and imaginary parts are given, from which, a complex matrix or expression is to be formed. As usual, form the complex matrix from $R = \begin{bmatrix} \cos x \\ 2 \end{bmatrix}$ and $I = \begin{bmatrix} \sin x \\ 3 \end{bmatrix}$, so,

the output matrix should be $\begin{bmatrix} \cos x + i \sin x \\ 2 + i3 \end{bmatrix}$. Have it as follows:

```
>>syms x ↵
>>R=[cos(x) 2].'; ↵          ← Enter the real part matrix
>>I=[sin(x) 3].'; ↵          ← Enter the imaginary part matrix
>>R+i*I ↵                    ← Form the complex matrix

ans =

[ cos(x)+i*sin(x)]
[         2+3*i]
```

7.1.9 Evaluation of a complex number polynomial

If $f(z)$ is a complex number polynomial of variable z, the polynomial can be computed at any z by the subroutine 'polyval'. As an illustration, compute $f(z) = z^4 + 2iz^3 - 7z + 6$ at $z = 2 - 3i$. The computation is $f(2-3i) = (2-3i)^4 + 2i(2-3i)^3 - 7(2-3i) + 6 = 16 - 96i + 216i^2 - 216i^3 + 81i^4 + 2i(8 - 36i + 54i^2 - 27i^3) - 14 + 21i + 6 = -109 + 49i$. In MATLAB notation, the polynomial is $[1 \quad 2i \quad 0 \quad -7 \quad 6]$. See the computation below:

MATLAB Command
```
>>y=[1 2i 0 -7 6]; ↵
>>polyval(y,2-3i) ↵

ans =
        -1.0900e+002   +4.9000e+001i
```

It is evident that $-1.0900e+002$ and $+4.9000e+001i$ displayed in exponential form mean -109 and $49i$ respectively. The function is not defined for the symbolic variables.

7.1.10 Logarithms of a complex number

Logarithm of a complex number is another complex number. The modulus and argument of a complex number $A + iB$ are given by $\sqrt{A^2 + B^2}$ and $\tan^{-1}\frac{B}{A}$ respectively. In exponential form, one can write $A + iB = \sqrt{A^2 + B^2}\, e^{i\tan^{-1}\frac{B}{A}}$, from which, the logarithm of $A + iB$ with respect to e is $\ln(A + iB) = \ln\left(\sqrt{A^2 + B^2}\, e^{i\tan^{-1}\frac{B}{A}}\right) = \ln\left(\sqrt{A^2 + B^2}\right) + \ln\left(e^{i\tan^{-1}\frac{B}{A}}\right) = \frac{1}{2}\ln(A^2 + B^2) + i\tan^{-1}\frac{B}{A}$. Take a numerical example of $A + iB$ as $5 + i4$, $\therefore \ln(5 + i4) = \frac{1}{2}\ln(5^2 + 4^2) + i\tan^{-1}\frac{4}{5} = 1.8568 + j0.6747$, $\tan^{-1}\frac{B}{A}$ must be in radians. Similarly, the logarithm of $A + iB$ with respect to base 10 can also be taken, which is $\log_{10}(A + iB) = \log_{10}\left(\sqrt{A^2 + B^2}\, e^{i\tan^{-1}\frac{B}{A}}\right) = \log_{10}\left(\sqrt{A^2 + B^2}\right) + \log_{10}\left(e^{i\tan^{-1}\frac{B}{A}}\right) = \frac{1}{2}\log_{10}(A^2 + B^2) + i\tan^{-1}\frac{B}{A}\log_{10}e$. Computation of $\log_{10}(5 + i4)$ is $\frac{1}{2}\log_{10}(5^2 + 4^2) + i\tan^{-1}\frac{4}{5}\log_{10}e = 0.8064 + i0.2930$. There is another logarithm subroutine that computes the logarithm with respect to base 2. Argument of any logarithm subroutine can be a matrix too. To illustrate, assume that $C = \begin{bmatrix} 4+7i \\ -5i \\ -6+10i \end{bmatrix}$, $\therefore$ $\log_2 C = \begin{bmatrix} \log_2(4+7i) \\ \log_2(-5i) \\ \log_2(-6+10i) \end{bmatrix} = \begin{bmatrix} \frac{1}{2}\log_2(4^2+7^2) + i\tan^{-1}\frac{7}{4}\log_2 e \\ \frac{1}{2}\log_2(0^2+(-5)^2) + i\tan^{-1}\frac{-5}{0}\log_2 e \\ \frac{1}{2}\log_2(10^2+(-6)^2) + i\tan^{-1}\frac{10}{-6}\log_2 e \end{bmatrix} =$

$\begin{bmatrix} 3.0112 + i1.5172 \\ 2.3219 - i2.2662 \\ 3.5437 + i3.0458 \end{bmatrix}$. All computations are shown below:

MATLAB Command

for the natural logarithm,

>>A=5+4i; ↵
>>log(A) ↵

ans =

1.8568 + 0.6747i

for the common logarithm,

>>log10(A) ↵

ans =

0.8064 + 0.2930i

for the logarithm of the column matrix C w.r.to base 2,

>>C=[4+7i -5i -6+10i].'; ↵
>>log2(C) ↵

ans =

3.0112 + 1.5172i
2.3219 - 2.2662i
3.5437 + 3.0458i

Finally, to take the logarithms of complex number(s),

MATLAB Step:

Use the commands log, log10, or log2 (complex scalar or matrix name) for natural, common, or base 2 logarithm respectively.

7.1.11 A complex number with complex power

The general form of a complex number with complex power is $(A+iB)^{C+iD}$. Little manipulation is carried out until we separate $(A+iB)^{C+iD}$ into the real and imaginary parts, for that, $(A+iB)^{C+iD} =$

$\left(\sqrt{A^2+B^2}\; e^{i\tan^{-1}\frac{B}{A}}\right)^{C+iD} = \left(\sqrt{A^2+B^2}\right)^{C+iD} e^{i(C+iD)\tan^{-1}\frac{B}{A}} = \left(\sqrt{A^2+B^2}\right)^{C}\left(\sqrt{A^2+B^2}\right)^{iD} e^{-D\tan^{-1}\frac{B}{A}+iC\tan^{-1}\frac{B}{A}} = \left(\sqrt{A^2+B^2}\right)^{C}$

$e^{-D\tan^{-1}\frac{B}{A}}e^{iC\tan^{-1}\frac{B}{A}}\left(\sqrt{A^2+B^2}\right)^{iD} = \left(\sqrt{A^2+B^2}\right)^{C} e^{-D\tan^{-1}\frac{B}{A}}e^{iC\tan^{-1}\frac{B}{A}}e^{iD\ln\sqrt{A^2+B^2}} = \left(\sqrt{A^2+B^2}\right)^{C} e^{-D\tan^{-1}\frac{B}{A}}e^{i\left(C\tan^{-1}\frac{B}{A}+D\ln\sqrt{A^2+B^2}\right)}$. The

real and imaginary parts are $\left(\sqrt{A^2+B^2}\right)^{C} e^{-D\tan^{-1}\frac{B}{A}}\cos\left(C\tan^{-1}\frac{B}{A}+D\ln\sqrt{A^2+B^2}\right)$ and

$\left(\sqrt{A^2+B^2}\right)^{C} e^{-D\tan^{-1}\frac{B}{A}}\sin\left(C\tan^{-1}\frac{B}{A}+D\ln\sqrt{A^2+B^2}\right)$ respectively. For the numerical example, say, $(2-i4)^{3+i5}$ is

to be calculated. On comparison, we have $A=2$, $B=-4$, $C=3$, and $D=5$, therefore, the real and imaginary

parts are $\left(\sqrt{2^2+(-4)^2}\right)^{3} e^{-5\tan^{-1}\frac{-4}{2}}\cos\left(3\tan^{-1}\frac{-4}{2}+5\ln\sqrt{2^2+(-4)^2}\right) = -11749.3230$ and

$\left(\sqrt{2^2+(-4)^2}\right)^{3} e^{-5\tan^{-1}\frac{-4}{2}}\sin\left(3\tan^{-1}\frac{-4}{2}+5\ln\sqrt{2^2+(-4)^2}\right) = -19402.1141$ respectively. The computation in

exponential form is shown as follows:

MATLAB Command

>>(2-4i)^(3+5i) ↵

ans =

$$-1.1749e+004 \ -1.9402e+004i$$

7.1.12 Rectangular to polar conversion or vice versa

Given a complex number in rectangular form $A+iB$, the polar form of the number is $re^{i\theta}$, where $r=\sqrt{A^2+B^2}$ and $\theta=\tan^{-1}\dfrac{B}{A}$. Its MATLAB counterpart is 'cart2pol' (abbreviation of Cartesian to(2) polar, *use command cart2pol (A, B)*). Again, given the polar form of a number is $re^{i\theta}$, the reverse conversion is $A=r\cos\theta$ and $B=r\sin\theta$. The resembling MATLAB function is 'pol2cart' (abbreviation of polar to(2) Cartesian, *use command pol2cart (θ in radians, r)*). Unavailability of θ makes us write t instead of θ. Say, a number $5+i4$ is in rectangular form. Its polar form is $(r, \theta)=\left(\sqrt{5^2+4^2}, \tan^{-1}\dfrac{4}{5}\right)=(6.4031, 0.6747^c)$. From the last polar form $6.4031\,e^{i0.6747^c}$, the rectangular form is $(A, B)=(6.4031\cos 0.6747^c, 6.4031\sin 0.6747^c)=$ (5.0001, 3.998). Both conversions are presented as follows:

MATLAB Command

for rectangular to polar conversion,
>>[t r]=cart2pol(5,4) ⏎

for polar to rectangular conversion,
>>[A B]=pol2cart(0.6747,6.4031) ⏎

t =
    0.6747
r =
    6.4031

A =
    5.0001
B =
    3.9998

Referring to the return, the discrepancy is due to the round off error. Several conversions in the form of a matrix are also handled by each of the subroutines.

7.1.13 The nth root of a complex number

The problem statement is to find the nth root of a complex number $A+iB$. If the nth root is x, then, we have $x=\sqrt[n]{A+iB}$, or $x^n=A+iB$, or $x^n-A-iB=0$. Basically, finding the nth root of $A+iB$ is nothing but solving $x^n-A-iB=0$. It is easily solved by the subroutine 'solve'. Find the 4th root of $(4+3i)^2$. Simplification turns $(4+3i)^2$ to $7+i24$. In polar form, one can write $7+i24=\sqrt{7^2+24^2}\,e^{i\tan^{-1}\frac{24}{7}}=25\,e^{i\tan^{-1}\frac{24}{7}}$. The roots are obtained from $25^{\frac{1}{4}}\,e^{i\frac{2\pi n+\tan^{-1}\frac{24}{7}}{4}}$, where $n=0$, 1, 2, and 3. They are given by $\sqrt{5}\,e^{i\frac{1}{4}\tan^{-1}\frac{24}{7}}$, $\sqrt{5}\,e^{i\frac{2\pi+\tan^{-1}\frac{24}{7}}{4}}$, $\sqrt{5}\,e^{i\frac{4\pi+\tan^{-1}\frac{24}{7}}{4}}$, and $\sqrt{5}\,e^{i\frac{6\pi n+\tan^{-1}\frac{24}{7}}{4}}$ respectively. Apply the conversion from the polar to rectangular to write $\sqrt{5}\,e^{i\frac{1}{4}\tan^{-1}\frac{24}{7}}=\sqrt{5}\,\cos\left(\dfrac{1}{4}\tan^{-1}\dfrac{24}{7}\right)+i\sqrt{5}\sin\left(\dfrac{1}{4}\tan^{-1}\dfrac{24}{7}\right)$. The trigonometric identities allow us to write

$\cos x=\sqrt{\dfrac{1+\cos 2x}{2}}$ and $\sin x=\sqrt{\dfrac{1-\cos 2x}{2}}$, hence, $\sqrt{5}\,\cos\left(\dfrac{1}{4}\tan^{-1}\dfrac{24}{7}\right)=\sqrt{5}\,\sqrt{\dfrac{1}{2}\left[1+\cos\left(\dfrac{1}{2}\tan^{-1}\dfrac{24}{7}\right)\right]}$, again,

$\cos\left(\dfrac{1}{2}\tan^{-1}\dfrac{24}{7}\right)=\sqrt{\dfrac{1}{2}\left[1+\cos\left(\tan^{-1}\dfrac{24}{7}\right)\right]}=\sqrt{\dfrac{1}{2}\left[1+\dfrac{7}{25}\right]}=\dfrac{4}{5}$, on that, $\sqrt{5}\,\cos\left(\dfrac{1}{4}\tan^{-1}\dfrac{24}{7}\right)=\sqrt{5}\,\sqrt{\dfrac{1}{2}\left[1+\dfrac{4}{5}\right]}=\dfrac{3}{\sqrt{2}}$. In a

similar manipulation, $+i\sqrt{5}\sin\left(\dfrac{1}{4}\tan^{-1}\dfrac{24}{7}\right)$ turns out to be $+i\dfrac{1}{\sqrt{2}}$. So, the 4th root corresponding to $n=0$ is

$\dfrac{3}{\sqrt{2}}+i\dfrac{1}{\sqrt{2}}$. Similar simplification provides $\sqrt{5}\,e^{i\frac{2\pi+\tan^{-1}\frac{24}{7}}{4}}=-\dfrac{1}{\sqrt{2}}+i\dfrac{3}{\sqrt{2}}$, $\sqrt{5}\,e^{i\frac{4\pi+\tan^{-1}\frac{24}{7}}{4}}=-\dfrac{3}{\sqrt{2}}-i\dfrac{1}{\sqrt{2}}$, and

$\sqrt{5}\,e^{i\frac{6\pi+\tan^{-1}\frac{24}{7}}{4}}=\dfrac{1}{\sqrt{2}}-i\dfrac{3}{\sqrt{2}}$ for $n=1$, 2, and 3 respectively. In matrix form, the roots are arranged as

$$\begin{bmatrix} \dfrac{3}{\sqrt{2}}+i\dfrac{1}{\sqrt{2}} \\ -\dfrac{1}{\sqrt{2}}+i\dfrac{3}{\sqrt{2}} \\ -\dfrac{3}{\sqrt{2}}-i\dfrac{1}{\sqrt{2}} \\ \dfrac{1}{\sqrt{2}}-i\dfrac{3}{\sqrt{2}} \end{bmatrix}$$. Our concern is to solve $x^4-(4+3i)^2=0$. See the solution as follows:

MATLAB Command

>>S=solve('x^4-(4+3*i)^2'); ↵ ← We assigned the output of 'solve' to S
>>pretty(S) ↵

```
[      1/2          1/2  ]
[ 1/2 2   - 3/2  i 2     ]
[                        ]
[      1/2          1/2  ]
[- 1/2 2   + 3/2  i 2    ]
[                        ]
[      1/2          1/2  ]
[ 3/2 2   + 1/2  i 2     ]
[                        ]
[      1/2          1/2  ]
[- 3/2 2   -  1/2 i 2    ]
```

If you are interested in decimal form, the command 'double' can make that happen:
>>double(S) ↵

ans =
    0.7071 - 2.1213i
   -0.7071 + 2.1213i
    2.1213 + 0.7071i
   -2.1213 - 0.7071i

The surds $\dfrac{3}{\sqrt{2}}$ and $\dfrac{1}{\sqrt{2}}$ are returned as $\dfrac{3\sqrt{2}}{2}$ and $\dfrac{\sqrt{2}}{2}$. The roots are not returned according to the order of $n=$ 0, 1, 2, and 3. Order is not so important here. While typing the argument of 'solve', the complex number should be entered as the string. The subroutine does not work if one types the argument as 'x^4-(4+3i)^2'; 3i is applicable for the complex number but for string entering, it should be 3*i. If higher order roots are required, surely, one has to appreciate the root finding capability of 'solve'.

7.1.14 Expansion of complex trigonometric, hyperbolic, and other functions

Sometimes the arguments of trigonometric, hyperbolic, and other functions have complex number expressions. If the argument of a trigonometric function is complex, that may turn out to be real in hyperbolic form or vice versa. One example can be $\cos ix=\cosh x$. It is written from De Moivre's theorem $\cos x=\dfrac{e^{ix}+e^{-ix}}{2}$, in which, $\cos ix = \dfrac{e^{iix} + e^{-iix}}{2} = \dfrac{e^{-x} + e^{x}}{2} = \cosh x$. Expansion of this type is also possible in MATLAB using the command 'expand'. See the execution below:

MATLAB Command
>>syms x ↵
>>expand(cos(i*x)) ↵

ans =

cosh(x)

More examples that involve the complex expansions are also accomplished as follows:

MATLAB Command for $\sin(x + iy) = \sin x \cosh y + i \cos x \sinh y$,
>>syms x y ↵
>>expand(sin(x+i*y)) ↵

ans =

sin(x)*cosh(y)+i*cos(x)*sinh(y)

for $(2x + iy)^3 = 8x^3 + i12x^2 y - 6xy^2 - iy^3$,
>>syms x y ↵
>>expand((2*x+i*y)^3) ↵

ans =

8*x^3+12*i*x^2*y-6*x*y^2-i*y^3

for $\tanh i2y = \dfrac{2i \tan y}{1 - \tan^2 y}$,
>>syms y ↵
>>S=expand(tanh(i*2*y)); ↵
>>pretty(S) ↵

```
                i tan(y)
       2 --------------------
                   2
            1 - tan(y)
```

7.1.15 Mean of complex numbers

The mean of complex numbers is defined as the means of the real and the imaginary parts. For example, mean of all elements of $R = [5 + 4i \quad 7 - i \quad -6 + i]$ is $\dfrac{5 + 7 - 6}{3} + i\dfrac{4 - 1 + 1}{3} = 2 + i\dfrac{4}{3} = 2 + i1.3333$. See this as follows:

MATLAB Command
>>R=[5+4i 7-i -6+i]; ↵
>>mean(R) ↵

ans =
2.0000 + 1.3333i

The argument of 'mean' can be row, column or rectangular matrix of complex numbers. For the rectangular matrix, means are computed over columns.

7.1.16 Sum or product of all elements of a complex matrix

Descriptions of the subroutines 'sum' and 'prod' are presented in the articles 2.30 and 2.32 respectively. Sum and product of all elements of $A = \begin{bmatrix} 2+i & 3i & 4+8i \\ xi & x-i & 1 \end{bmatrix}$ are $2+i+3i+4+8i+xi+x-i+1 = 7+x+i(x+11)$ and $(2+i)(3i)(4+8i)(xi)(x-i)(1) = -60x - i60x^2$ respectively. Execution of the two subroutines are presented below:

MATLAB Command
```
>>syms x  ↵
>>A=[2+i 3i 4+8i;x*i x-i 1];  ↵
>>sum(sum(A)) ↵            ← For sum

ans =

7+11*i+i*x+x
>>prod(prod(A)) ↵          ← For product

ans =

-60*i*x*(x-i)            ← The output is not separated in real and imaginary parts
>>expand(prod(prod(A))) ↵   ← Conduct the command 'expand' to separate them
ans =

-60*i*x^2-60*x
```

7.1.17 Geometric mean of complex numbers

If we have n complex numbers like $A_1 + iB_1$, $A_2 + iB_2$, $A_3 + iB_3$, ..., and $A_n + iB_n$, the geometric mean of these numbers is given by $\sqrt[n]{(A_1 + iB_1)(A_2 + iB_2)(A_3 + iB_3)...........(A_n + iB_n)}$. Calculate the geometric mean of the complex numbers $-4i$, $3-4i$, $1+i2$, and $2i$. Their product is $-4i(3-i4)(1+i2)2i = 88+16i$. There are four numbers, $\therefore$ $n = 4$. The geometric mean is $\sqrt[4]{88+16i} = \left[\sqrt{88^2 + 16^2}\ e^{i\tan^{-1}\frac{2}{11}} \right]^{\frac{1}{4}} = \left(\sqrt{8000} \right)^{\frac{1}{4}} \left(e^{i\tan^{-1}\frac{2}{11}} \right)^{\frac{1}{4}} = 8000^{\frac{1}{8}} \times \left[\cos\left(\frac{1}{4}\tan^{-1}\frac{2}{11} \right) + i\sin\left(\frac{1}{4}\tan^{-1}\frac{2}{11} \right) \right] = 3.0722 + 0.1382\,i$. First, assign all numbers to matrix R, then, use the command 'prod(R)^(1/4)' to have the geometric mean:

MATLAB Command
```
>>R=[-4*i 3-4*i 1+i*2 2*i];  ↵
>>prod(R)^(1/4) ↵

ans =
        3.0722 + 0.1382i
```

7.1.18 Powering a complex matrix

Assume that power of each element of $C = \begin{bmatrix} 2+ix \\ 1+i \\ 3i \end{bmatrix}$ is to be raised by 2. After powering, we should have $\begin{bmatrix} (2+ix)^2 \\ (1+i)^2 \\ (3i)^2 \end{bmatrix} = \begin{bmatrix} 4 - x^2 + 4ix \\ 2i \\ -9 \end{bmatrix}$. Have it as follows:

284

MATLAB Command

```
>>syms x  ↵
>>C=[2+i*x 1+i 3i].';  ↵
>>C.^2 ↵
```

ans =

```
[ (2+i*x)^2]
[       2*i]
[        -9]
```

It seems that the first element of output is not expanded. Obtain it by the command 'expand':

```
>>expand(C.^2) ↵
```

ans =

```
[ 4+4*i*x-x^2]
[         2*i]
[          -9]
```

That is what we need from the expansion. Power of a complex square matrix can also be computed. Given that

$A = \begin{bmatrix} 2+i & 3-i \\ 2i & 1 \end{bmatrix}$, calculate A^3. This is actually vector multiplication $A \times A \times A$. On squaring and taking cube,

we can write $A^2 = A \times A = \begin{bmatrix} 2+i & 3-i \\ 2i & 1 \end{bmatrix} \times \begin{bmatrix} 2+i & 3-i \\ 2i & 1 \end{bmatrix} = \begin{bmatrix} (2+i)(2+i)+(3-i)2i & (2+i)(3-i)+1(3-i) \\ 2i(2+i)+1(2i) & 2i(3-i)+1 \end{bmatrix} =$

$\begin{bmatrix} 5+10i & 10 \\ -2+6i & 3+6i \end{bmatrix}$ and $A^3 = A^2 \times A = \begin{bmatrix} 5+10i & 10 \\ -2+6i & 3+6i \end{bmatrix}\begin{bmatrix} 2+i & 3-i \\ 2i & 1 \end{bmatrix} =$

$\begin{bmatrix} (5+10i)(2+i)+10\times 2i & (5+10i)(3-i)+10 \\ (-2+6i)(2+i)+(3+6i)2i & (-2+6i)(3-i)+(3+6i) \end{bmatrix} = \begin{bmatrix} 45i & 35+25i \\ -22+16i & 3+26i \end{bmatrix}$. The computation is shown below:

```
>>A=[2+i 3-i;2i 1];  ↵
>>A^3 ↵
```

ans =
```
        0 +45.0000i    35.0000 +25.0000i
  -22.0000 +16.0000i     3.0000 +26.0000i
```

7.1.19 Product of complex polynomials

Recall that the subroutine 'conv' is conducted in the article 3.7 for the polynomial multiplication. That subroutine is still operational for the complex number coefficients (not applicable for symbolic variables). Multiplication of the complex polynomials $a+i+2iz+(3+i)z^2$ and iz^3-z-i is $-i[a+i+2iz+(3+i)z^2]-z$ $[a+i+2iz+(3+i)z^2]+iz^3[a+i+2iz+(3+i)z^2] = (3i-1)z^5-2z^4+(-4-i+ia)z^3+(1-5i)z^2+(2-i-a)z+1-ai$. Implement it as follows:

MATLAB Command

```
>>syms a z  ↵              ← Declare the variables as symbolic
>>p1=a+i+z*2*i+(3+i)*z^2;  ↵   ← Assign the first polynomial to p1
>>p2=i*z^3-z-i;  ↵          ← Assign the second polynomial to p2
>>O=expand(p1*p2);  ↵       ← The output is assigned to O
>>pretty(O) ↵               ← Display the contents of O in symbolic form
```

```
       3                3               4        2              5       5         3
i a z  - a z - i a - 4 z  - i z + 1 - 2 z  - 5 i z  + 2 z + 3 i z  - z  - i z

         2
       + z
```

The font style of the pretty output looks different. The reason behind is that the outcome is obtained from the camera capture of the output as displayed in the monitor screen.

7.1.20 Computations of functions with complex arguments

Argument of a trigonometric, hyperbolic, or exponential function can be a complex number. For instance, what is the value of $\sin(5+6i)$? On expansion, we have $\sin(5+6i) = \sin 5\cos(6i) + \cos 5\sin(6i)$. Utilize the trigonometric to hyperbolic conversion to write $\cos(6i) = \cosh 6$ and $\sin(6i) = i\sinh 6$. So, $\sin(5+6i)$ becomes $\sin 5\cosh 6 + i\cos 5\sinh 6 = -193.4300 + i\,57.2184$. The MATLAB response is as follows:

MATLAB Command
>>sin(5+6i) ↵

ans =

-1.9343e+002 +5.7218e+001i

For the matrix $A = \begin{bmatrix} 5+6i & 5i \\ 0 & 2+4i \end{bmatrix}$, 'sin(A)' performs similar computation for each element of A, that is, the

return is going to be $\begin{bmatrix} \sin(5+6i) & \sin(5i) \\ \sin(0) & \sin(2+4i) \end{bmatrix} = \begin{bmatrix} \sin 5\cosh 6 + i\cos 5\sinh 6 & i\sinh 5 \\ 0 & \sin 2\cosh 4 + i\cos 2\sinh 4 \end{bmatrix} =$

$\begin{bmatrix} -193.43 + i57.2184 & i74.2032 \\ 0 & 24.8313 - i11.3566 \end{bmatrix}$. MATLAB computation is as follows:

>>A=[5+6i 5i;0 2+4i]; ↵
>>sin(A) ↵

ans =

1.0e+002 *
-1.9343 + 0.5722i   0 + 0.7420i
0   0.2483 - 0.1136i

The multiplier '1.0e+002 *' (which is equivalent to 100) is to be multiplied with all elements of the matrix followed by the multiplier. Other trigonometric, hyperbolic, and exponential functions respond the computation in the same way. Let us see how we find the reciprocal of a complex number $A + iB$, that is, $\frac{1}{A+iB} \cdot \frac{1}{A+iB}$ can be written as $\frac{A-iB}{(A+iB)(A-iB)} = \frac{A-iB}{A^2+B^2} = \frac{A}{A^2+B^2} - i\frac{B}{A^2+B^2}$. If we have $8+i6$, reciprocal of $8+i6$ is $\frac{8}{8^2+6^2} - i\frac{6}{8^2+6^2} = 0.08 - i0.06$, whose implementation is shown below:

>>1/(8+6i) ↵

ans =
0.0800 - 0.0600i

Reciprocal of all elements of a complex matrix A can be calculated by the command '1./A'. Rational form output is seen by virtue of the command 'sym'.

7.1.21 Summation of a complex series

A series of complex numbers can be summed like real elements. To elucidate that, compute $\sum_{x=0}^{x=4} e^{\frac{i\pi x}{4}} \cos(xi+2)$. Expand the summation to write $\sum_{x=0}^{x=4} e^{\frac{i\pi x}{4}} \cos(xi+2) = \cos 2 + e^{\frac{i\pi}{4}}\cos(i+2) + e^{\frac{i2\pi}{4}}\cos(2i+2)$

$+e^{i\frac{3\pi}{4}}\cos(3i+2)+e^{i\frac{4\pi}{4}}\cos(4i+2)$. We know that $e^{i\theta}=\cos\theta+i\sin\theta$ and $\cos(ai+b)=\cos(ai)\cos b-\sin(ai)\sin b=\cosh a\cos b-i\sinh a\sin b$. Apply the last two identities to get $\sum_{x=0}^{x=4}e^{i\frac{\pi x}{4}}\cos(xi+2)=\cos 2+$ $(\cos\frac{\pi}{4}+i\sin\frac{\pi}{4})(\cosh 1\cos 2-i\sinh 1\sin 2)+(\cos\frac{\pi}{2}+i\sin\frac{\pi}{2})(\cosh 2\cos 2-i\sinh 2\sin 2)$ $+$ $(\cos\frac{3\pi}{4}+i\sin\frac{3\pi}{4})$ $(\cosh 3\cos 2-i\sinh 3\sin 2)+(\cos\pi+i\sin\pi)(\cosh 4\cos 2-i\sinh 4\sin 2)=-0.4161+0.3016-1.2097i+3.2979-1.5656i$ $+9.4037+3.4787i+11.3642+24.8147i=23.9513+25.5181i$. Compute that as follows:

MATLAB Command

```
>>x=0:4;  ↵
>>sum(exp(pi/4*x*i).*cos(x*i+2))  ↵

ans =
        23.9512 +25.5180i
```

Computations involving complex $\Sigma\Sigma$ can be handled also utilizing two 'sum's.

7.1.22 Solving equations of complex variables

Sometimes a set of algebraic equations may contain complex symbolic variables. Solution(s) of the set is(are) also some complex variables. As an example, solve the set $\begin{cases} ax+iy=a \\ xi+ya=1 \end{cases}$ for x and y. Divide the first and second equations by a and i respectively to write $\begin{cases} x+\dfrac{i}{a}y=1 \\ x+y\dfrac{a}{i}=\dfrac{1}{i} \end{cases}$ or $\begin{cases} x+\dfrac{i}{a}y=1 \\ x-iay=-i \end{cases}$. Subtraction of one

equation from the other and solving for y yield $y=\dfrac{1+i}{\dfrac{i}{a}+ia}=\dfrac{a(1+i)(-i)}{a^2+1}=\dfrac{a(1-i)}{a^2+1}$. Substitute $y=\dfrac{a(1-i)}{a^2+1}$ into the

second equation and solve for x to have $x=ia\dfrac{a(1-i)}{a^2+1}-i=i\left[\dfrac{a^2(1-i)}{a^2+1}-1\right]=i\left[\dfrac{a^2-ia^2-a^2-1}{a^2+1}\right]=\dfrac{a^2-i}{a^2+1}$. Bypass

the hassle of computation conducting the following steps:

MATLAB Command

```
>>syms a x y  ↵          ← Declare all variables of equations as symbolic
>>e1='a*x+i*y=a';  ↵     ← Define the first equation
>>e2='x*i+y*a=1';  ↵     ← Define the second equation
>>S=solve(e1,e2,x,y)  ↵  ← Apply the subroutine 'solve' and assign the output to S

S =                      ← Output indicates that S is a structured array,
    x: [1x1 sym]             which has two constituents x and y and
    y: [1x1 sym]             each of them is symbolic

>>pretty(S.x)  ↵         ← For x
              2
        -i + a
        -----------
              2
        1 + a

>>pretty(S.y)  ↵         ← For y
```

$$\frac{(1-i)a}{1+a^2}$$

As it is expected.

7.1.23 Limit of a complex function

We have pointed out the limits of real functions in article 5.1. The subroutine 'limit' is also applicable for the complex functions. Consider $\underset{x \to M}{Lt} \dfrac{1 - e^{i\frac{x-M}{M}}}{i(x-M) + \sin\frac{2\pi x}{M}}$ for specimen. When $x \to M$, the function takes

$\dfrac{0}{0}$ form. Differentiate the numerator and denominator w.r.to x to have $\underset{x \to M}{Lt} \dfrac{-\frac{i}{M}e^{i\frac{x-M}{M}}}{i + \frac{2\pi}{M}\cos\frac{2\pi x}{M}} = \dfrac{-\frac{i}{M}}{i + \frac{2\pi}{M}} =$

$\dfrac{-i}{iM + 2\pi}$. The implementation is conducted as follows:

MATLAB Command

```
>>syms x M ↵            ← Declare symbolic variables
>>N=1-exp(i*(x-M)/M); ↵  ← Assign numerator to N
>>D=i*(x-M)+sin(2*pi*x/M); ↵  ← Assign denominator to D
>>L=limit(N/D,x,M); ↵    ← Function is composed of N/D and the limiting value is
                              placed to L

>>pretty(L) ↵
                i
           - ---------------
              i M + 2 pi
```

7.1.24 Differentiation of a complex function/matrix

Computational aspect of the subroutine 'diff' is versatile (see chapter 5). Complex functions can also be differentiated by 'diff'. Find the derivative of the complex function $f(z) = 7z^3 + 2z + 3 + i(e^z + \sin z)$. The derivative of $f(z)$ is $\dfrac{d[f(z)]}{dz} = 21z^2 + 2 + i(e^z + \cos z)$. Ascertain it as follows:

MATLAB Command

```
>>syms z ↵
>>f=7*z^3+2*z+3+i*(exp(z)+sin(z)); ↵
>>pretty(diff(f,z)) ↵

                2
           21 z  + 2 + i (exp(z) + cos(z))
```

The subroutine responds element by element for a complex matrix function such as $A = \begin{bmatrix} e^{ix} & \tan(x+iy) \\ \cos xy & 7xi \end{bmatrix}$.

Compute the second derivative of each element of A with respect to x. For that, we have $\dfrac{dA}{dx} =$

$\begin{bmatrix} ie^{ix} & \sec^2(x+iy) \\ -y\sin xy & 7i \end{bmatrix}$, $\dfrac{d^2A}{dx^2} = \begin{bmatrix} -e^{ix} & 2\sec^2(x+iy)\tan(x+iy) \\ -y^2\cos xy & 0 \end{bmatrix}$. Have the differentiation as follows:

```
>>syms x y ↵
>>A=[exp(i*x) tan(x+i*y);cos(x*y) 7*x*i]; ↵
```

```
>>O=diff(A,2,x); ↵        ← The second derivative is assigned to O
>>pretty(O) ↵
```

```
[                                                    2 ]
[  -exp(i x)            2 tan(x + i y) (1 + tan(x + i y) )]
[                                                      ]
[            2                                         ]
[-cos(x y) y                       0                   ]
```

Following the differentiation, substitution can also be carried out. Calculate $\dfrac{d^2A}{dx^2}$ for $x=0$ and $y=\dfrac{i\pi}{4}$. For

$$x=0 \quad \text{and} \quad y=\frac{i\pi}{4}, \quad \begin{bmatrix} -e^{ix} & 2\sec^2(x+iy)\tan(x+iy) \\ -y^2\cos xy & 0 \end{bmatrix} \quad \text{becomes} \quad \begin{bmatrix} -1 & 2\sec^2\left(-\dfrac{\pi}{4}\right)\tan\left(-\dfrac{\pi}{4}\right) \\ -\left(\dfrac{i\pi}{4}\right)^2 & 0 \end{bmatrix} =$$

$\begin{bmatrix} -1 & -4 \\ \dfrac{\pi^2}{16} & 0 \end{bmatrix}$. In the last execution, assignee O is having $\dfrac{d^2A}{dx^2}$. Exercise the subroutine 'subs' on O for the substitution:

```
>>syms pi ↵              ← Declare pi as symbolic
>>subs(O,{x,y},{0,i*pi/4}) ↵

ans =

[      -1,      -4]
[ 1/16*pi^2,      0]
```

Conduct the command 'double' if you need the floating-point output.

7.1.25 Integration of a complex function/matrix

Like 'diff', the subroutine 'int' is also operational for the complex functions (see chapter 6 for the details of 'int'). Commence with one variable complex function $f(z)=\cosh z+2+i(2-z+z^3)$. Its integration w.r.to z (real and imaginary parts are treated separately) is $\int[\cosh z+2+i(2-z+z^3)]dz=\sinh z$ $+2z+i\left[2z-\dfrac{1}{2}z^2+\dfrac{1}{4}z^4\right]$. Execute it as follows:

MATLAB Command
```
>>syms z ↵
>>f=cosh(z)+2+i*(2-z+z^3); ↵
>>I=int(f); ↵
>>pretty(I) ↵
                          2      4
        sinh(z) + 2 z + i (2 z - 1/2 z + 1/4 z  )
```

What if we have the limits of integration. Compute the just mentioned integration with the limits $2i \sim 1+i$. So,

we have $\displaystyle\int_{z=2i}^{z=1+i}[\cosh z+2+i(2-z+z^3)]dz=\left[\sinh z+2z+i\left(2z-\dfrac{1}{2}z^2+\dfrac{1}{4}z^4\right)\right]_{z=2i}^{z=1+i}=\sinh(1+i)+2(1+i)+i\left[2(1+i)-\right.$

$\dfrac{1}{2}(1+i)^2+\dfrac{1}{4}(1+i)^4\Big]-\sinh(2i)-4i-i\left[4i-\dfrac{1}{2}(2i)^2+\dfrac{1}{4}(2i)^4\right] = \sinh(1+i)+2-2i+i(1+i)-\sinh(2i)-i(6+4i) =$

$\sinh(1+i) - \sinh(2i) + 5\ -7i = \dfrac{e^{1+i} - e^{-(1+i)}}{2} - i\sin 2 + 5 - 7i$ \qquad (applying \quad the \quad identities \quad $\sinh x = \dfrac{e^x - e^{-x}}{2}$ \quad and

$\sinh(ia) = i\sin a$). As long as we know the technique of computation, there is no need to go for such a long manipulation:

MATLAB Command
>>I=int(f,2i,1+i); ↵
>>pretty(I) ↵

$\qquad$ - 1/2 exp(-1) exp(-i) - 7 i + 1/2 exp(1) exp(i) + 5 - i sin(2)

Then, integrate a complex double integration such as $\displaystyle\int_{y=i}^{1}\int_{x=0}^{1}(x+iy)e^{ix}\,dxdy$. Step by step integration of

$\displaystyle\int_{y=i}^{1}\int_{x=0}^{1}(x+iy)e^{ix}\,dxdy$ is $\displaystyle\int_{y=i}^{1}[e^{ix}(y+1-xi)]_{x=0}^{x=1}\,dy$ $=$ $\displaystyle\int_{y=i}^{1}[e^{i}(y+1-i)-(y+1)]\,dy$ $=$ $\displaystyle\int_{y=i}^{1}[(e^{i}-1)y+e^{i}(1-i)-1]\,dy$ $=$

$\left[\dfrac{1}{2}(e^{i}-1)y^{2}\right]_{y=i}^{y=1} + [e^{i}(1-i)-1][y]_{y=i}^{y=1} = \dfrac{1}{2}(e^{i}-1)(1-i^{2}) + [e^{i}(1-i)-1](1-i) = (1-2i)e^{i}-2+i$. Following is the

convenient way of doing this:

MATLAB Command
>>syms x y ↵ $\qquad\qquad\qquad$ ← Declare variables of integrand as symbolic
>>f=(x+i*y)*exp(i*x); ↵ $\qquad\quad$ ← Assign the integrand to f
>>O=int(int(f,x,0,1),y,i,1); ↵ $\quad$ ← Output of double integration is assigned to O
>>pretty(O) ↵ $\qquad\qquad\qquad$ ← Display the contents of O

$\qquad\qquad$ -2 i exp(i) + exp(i) - 2 + i

Sharing the common trait, the triple integration of complex integrand can be performed too. A numerical integration (see chapter 6) involving the real constant limits and complex integrand is also possible. To enumerate by example, calculate $\displaystyle\int_{y=0}^{y=1}\int_{x=\pi}^{2\pi}(x-iy)\cos(x-iy)\,dxdy$. The analytical integration is as follows:

$\displaystyle\int_{y=0}^{y=1}\int_{x=\pi}^{2\pi}(x-iy)\cos(x-iy)\,dxdy = \int_{y=0}^{y=1}[(x-iy)\sin(x-iy)+\cos(x-iy)]_{x=\pi}^{x=2\pi}\,dy$ $=$ $\displaystyle\int_{y=0}^{y=1}[2\cosh y - 2y\sinh y - 3i\pi\sinh y]\,dy$ $=$

$[4\sinh y - 2y\cosh y - 3i\pi\cosh y]_{y=0}^{y=1} = -3e^{-1} + e - \dfrac{3i\pi e^{-1}}{2} - \dfrac{3i\pi e}{2} + 3i\pi = 1.6146 - 5.1184i$. Implement it as follows:

MATLAB Command
>>f=inline('(x-i*y).*cos(x-i*y)','x','y'); ↵ $\quad$ ← Assign the integrand to f
>>dblquad(f,pi,2*pi,0,1) ↵ $\qquad\qquad\qquad$ ← Numerical double integration is done by 'dblquad'

ans =
$\qquad$ 1.6146 - 5.1185i

Integration of complex matrix function is left as an exercise for the reader.

7.1.26 Partial fraction containing complex variables

$\qquad$ Numerical and symbolic forms of the partial fraction are addressed in article 3.16. We present here just one example of the symbolic partial fraction involving the complex number. Decompose the complex function $\dfrac{1}{x^{2}-5ix-6}$ into the partial fraction. The necessary manipulations are as follows: $\dfrac{1}{x^{2}-5ix-6} =$

$\dfrac{1}{x^{2}-5ix+6i^{2}} = \dfrac{1}{(x-2i)(x-3i)} = \dfrac{1}{(3i-2i)(x-3i)} + \dfrac{1}{(x-2i)(2i-3i)} = \dfrac{-i}{x-3i} + \dfrac{i}{x-2i}$. Obtain it as follows:

MATLAB Command

```
>>syms x ↵
>>f=1/(x^2-5*i*x-6); ↵          ← Given function is assigned to f
>>O=maple('convert',f,'parfrac',x); ↵   ← Maple output is assigned to O
>>pretty(O) ↵                   ← Print the contents of O
```

```
         i            i
     -----------  -  ------------
      -x + 3 i        -x + 2 i
```

7.1.27 Determinant of a complex square matrix

A square matrix may have complex elements, which can be numeric or symbolic. Determinant of $A = \begin{bmatrix} 2+i & i \\ -i & 3i \end{bmatrix}$ is $(2+i)3i + i^2 = -4 + 6i$. Find it as follows:

MATLAB Command

```
>>A=[2+i i;-i 3i]; ↵
>>det(A) ↵

ans =
        -4.0000 + 6.0000i
```

Carry out the implementation for $A = \begin{bmatrix} \cos x + iy & x + iy \\ x - iy & 4i \end{bmatrix}$. Its determinant is $(\cos x + iy)4i - (x + iy)(x - iy) = 4i\cos x - 4y - x^2 - y^2$. Apply the subroutine as follows:

MATLAB Command

```
>>syms x y ↵
>>A=[cos(x)+i*y x+i*y;x-i*y 4*i]; ↵
>>pretty(det(A)) ↵

                        2    2
          4 i cos(x) - 4 y - x  - y
```

7.1.28 Inverse of a complex number matrix

Elaborate discussion of the inverse of a square matrix is given in article 4.13. Without going to the details, it is given that the inverses of matrices $A = \begin{bmatrix} 1+4i & 4i \\ 7-5i & 3+2i \end{bmatrix}$ and $B = \begin{bmatrix} a+ic & b \\ c+i & d \end{bmatrix}$ are

$$\begin{bmatrix} -\dfrac{103}{821} - \dfrac{8i}{821} & \dfrac{56}{821} + \dfrac{100i}{821} \\ \dfrac{105}{821} - \dfrac{223i}{821} & -\dfrac{81}{821} - \dfrac{86i}{821} \end{bmatrix} \text{ and } \begin{bmatrix} \dfrac{d}{ad-bc+i(cd-b)} & \dfrac{-b}{ad-bc+i(cd-b)} \\ -\dfrac{c+i}{ad-bc+i(cd-b)} & \dfrac{a+ic}{ad-bc+i(cd-b)} \end{bmatrix}$$ respectively. If you know the

inverse, product of the inverse with the given matrix should return the identity matrix. Anyhow, responses of MATLAB on execution of 'inv' are as follows:

MATLAB Command for A,

```
>>A=sym([1+4i 4i;7-5i 3+2i]); ↵   ← Enter the matrix A
>>C=inv(A); ↵                     ← Inverse of A is put to C
>>pretty(C) ↵                     ← Display the contents of C
```

$$
\begin{bmatrix}
\dfrac{103}{821} - 8/821\,i & \dfrac{56}{821} + \dfrac{100}{821}\,i \\[2ex]
\dfrac{105}{821} - \dfrac{223}{821}\,i & -\dfrac{81}{821} - \dfrac{86}{821}\,i
\end{bmatrix}
$$

>>C*A ↵   ← Confirm that C×A is the identity matrix

ans =

[1,   0]
[0,   1]

for matrix B,
>>syms a b c d ↵
>>B=[a+i*c b;c+i d]; ↵
>>D=inv(B); ↵
>>pretty(D) ↵

$$
\begin{bmatrix}
\dfrac{d}{\%1} & -\dfrac{b}{\%1} \\[2ex]
-\dfrac{c+i}{\%1} & \dfrac{a+i\,c}{\%1}
\end{bmatrix}
$$

%1 := d a + i d c - b c - i b

The sub expression '%1' is equivalent to $ad + idc - bc - ib$ and that it is the denominator of each element of the inverse. Check the simplified product:
>>simplify(D*B) ↵

ans =

[1,   0]
[0,   1]

If you execute just D*B, a matrix having the long complex entries appears on the screen that means the output is not simplified. For that reason, we utilized the command 'simplify' following the multiplication.

With this we cease the discussion on the problems of complex variables. Not all functions that handle the complex variables are presented. We believe that the reader has had substantial guideline to proceed with. If the built-in M-files or the Maple functions do not appease the user's requirement, new M-files can be created to fulfil the demand utilizing the known functions of complex variables. In the next section, we concentrate on the problems of differential equations as we promised in the introduction.

7.2 Ordinary differential equations

To forecast the behavior of a physical system, we construct the mathematical models. We know that the derivatives measure the rates of change of a system. Frequently, the derivatives constitute the mathematical model of a system. The mathematical models or equations involving the derivatives are called differential equations. When we have one dependent and one independent variables, the differential equation is termed as the ordinary differential equation. For instance, $\frac{dy}{dx} + y = 9$ and $\frac{d^4y}{dx^4} + y = 9$ are the ordinary differential equations, where y is the dependent and x is the independent variables. Order of a differential equation is the highest order of derivatives present in the equation. Thus, a first order ordinary differential equation has the form $f\left(x, y, \frac{dy}{dx}\right) = 0$, a second order ordinary differential equation has the form $f\left(x, y, \frac{dy}{dx}, \frac{d^2y}{dx^2}\right) = 0 \ldots$ so forth.

The solution of a differential equation is a function that satisfies the differential equation over some domain of the independent variable. Method of finding the solution of a differential equation can be analytical or numerical. The analytical solution has two components – complementary function and particular integral. First, we concentrate on the analytical solution. Subroutine 'dsolve' finds almost all widely used differential equations' solutions but to acquire more experience and mastery of techniques, we touched the various differential equations in the following articles.

7.2.1 First order ordinary differential equations

A number of first order differential equations are seen as a mathematical model. We address few of them with and without the initial conditions. Commence with $\frac{dy}{dx} = 4x^4 y^2$. The equation is a separable one so write it as $\frac{dy}{y^2} = 4x^4 dx$. Integrate both sides w.r.to the respective variable to have $-\frac{1}{y} = \frac{4}{5}x^5 + C$, where C is a constant of integration. Rearrange the equation to have the general solution $y = \frac{-5}{4x^5 + 5C}$. For a first order equation, one should have one arbitrary constant. Operator D, which is used in argument of 'dsolve', is equivalent to $\frac{d}{dt}$ of calculus. Unless description of the independent variable is inserted, t is understood as the independent variable. From which, 'Dy' is equivalent to $\frac{dy}{dt}$ or $\frac{dy}{dx}$. A string represents the differential equation. As a string, the given equation is written as 'Dy=4*x^4*y^2'. See the solution below:

MATLAB Command

```
>>S=dsolve('Dy=4*x^4*y^2','x'); ↵      ← The output of 'dsolve' is assigned to S
>>pretty(S) ↵                          ← Display the contents of S
```

$$-\frac{5}{4x^5 - 5\,C1}$$

MATLAB assumed the arbitrary constant as $-C1$ instead of C. As par as the arbitrary constant is concerned, there is no mathematical detriment in assuming that. The equation we solved is without the initial condition. Solve $\frac{dy}{dx} = 4x^4 y^2$ with the initial condition: $y = 2$ when $x = 0$. Plug $x = 0$ and $y = 2$ into the general solution, from which, $C = -\frac{1}{2}$. Insert $C = -\frac{1}{2}$ in the general equation to get the particular solution $y = \frac{-10}{8x^5 - 5}$. The initial condition $y = 2$ when $x = 0$ is argumented as ' y(0)=2'. Have the solution with the initial condition as follows:

MATLAB Command

```
>>S=dsolve('Dy=4*x^4*y^2','y(0)=2','x'); ↵
>>pretty(S) ↵
```

$$-\frac{5}{4x^5 - 5/2}$$

Many first order differential equations are not separable. Some of them become a separable one subject to the change of variables. Homogeneous equation is such an example. A first order differential equation is said to be homogeneous if it can be written in the form $\frac{dy}{dx} = g\left(\frac{y}{x}\right)$. As an example, consider $2x\frac{dy}{dx} = \frac{y^2}{x} + 4y$. There is no way you can separate x and y on either side of the equation but the equation is of the form $\frac{dy}{dx} = g\left(\frac{y}{x}\right)$.

Substitute $y = zx$, it follows then, $\frac{dy}{dx} = z + x\frac{dz}{dx}$ and $2x(z + x\frac{dz}{dx}) = \frac{z^2 x^2}{x} + 4zx$. Until the general solution is achieved, do some manipulations as follows: $2x(z + x\frac{dz}{dx}) = \frac{z^2 x^2}{x} + 4zx \Rightarrow 2(z + x\frac{dz}{dx}) = z^2 + 4z \Rightarrow \frac{dz}{z^2 + 2z} = \frac{dx}{2x}$

$\Rightarrow \left[\frac{1}{2z} - \frac{1}{2(z+2)}\right]dz = \frac{dx}{2x} \Rightarrow \ln z - \ln(z+2) = \ln x + A$ (A is a constant of integration) $\Rightarrow \ln\frac{z}{z+2} = \ln x + A \Rightarrow$

$\frac{z}{z+2} = xe^A \Rightarrow \frac{z+2}{z} = \frac{1}{xe^A} \Rightarrow z = \frac{2xe^A}{1 - xe^A} \Rightarrow y = \frac{2x^2 e^A}{1 - xe^A} \Rightarrow y = \frac{2x^2}{e^{-A} - x}$. The string form of the equation is '2*x*Dy=y^2/x+4*y'. Following is the MATLAB's response:

MATLAB Command
>>S=dsolve('2*x*Dy=y^2/x+4*y','x'); ↵
>>pretty(S) ↵

```
              2
             x
    -2  ---------
         x - 2 C1
```

If you replace the term e^{-A} by $2C_1$, MATLAB's return is identical with the analytical one.

In differential form, a first order differential equation can be written as $M(x, y)dx + N(x, y)dy = 0$. A first order differential equation is said to be exact if the condition $\frac{\partial N(x, y)}{\partial x} = \frac{\partial M(x, y)}{\partial y}$ is satisfied. An important type of the first order differential equation, for which integrating factor can be found, is Bernoulli equation. Its general form is $P(x)\frac{dy}{dx} + Q(x)y = R(x)y^\alpha$, where $P(x)$, $Q(x)$, and $R(x)$ are the functions of x and α is a constant. The integrating factor is given by $I(x, y) = y^{-\alpha}\frac{e^{(1-\alpha)\int\frac{Q(x)}{P(x)}dx}}{P(x)}$. In general the Bernoulli equation is non-exact but multiplying the equation by the integrating factor makes the equation exact. As usual, verify the theory with the equation $x^3\frac{dy}{dx} + xy = y^2$. On comparison, we get $P(x) = x^3$, $Q(x) = x$, $R(x) = 1$, and $\alpha = 2$, on that, $I(x, y) = y^{-2}\frac{e^{-\int\frac{x}{x^3}dx}}{x^3} = x^{-3}y^{-2}e^{\frac{1}{x}}$. Multiply both sides of $x^3\frac{dy}{dx} + xy = y^2$ with $x^{-3}y^{-2}e^{\frac{1}{x}}$ to write

$\frac{dy}{dx}y^{-2}e^{\frac{1}{x}} + e^{\frac{1}{x}}x^{-2}y^{-1} = x^{-3}e^{\frac{1}{x}} \Rightarrow \frac{d}{dx}\left[y^{-1}e^{\frac{1}{x}}\right] = -x^{-3}e^{\frac{1}{x}} \Rightarrow y^{-1}e^{\frac{1}{x}} = \int -x^{-3}e^{\frac{1}{x}}dx \Rightarrow y^{-1}e^{\frac{1}{x}} = \frac{1}{x}e^{\frac{1}{x}} - e^{\frac{1}{x}} + A \Rightarrow$

$y^{-1} = \frac{1}{x} - 1 + Ae^{-\frac{1}{x}} \Rightarrow y = \frac{1}{\frac{1}{x} - 1 + Ae^{-\frac{1}{x}}} = \frac{x}{1 - x + xAe^{-\frac{1}{x}}}$. Argument the equation as 'x^3*Dy+x*y=y^2'. MATLAB solves the Bernoulli equation as follows:

MATLAB Command
>>S=dsolve('x^3*Dy+x*y=y^2','x'); ↵
>>pretty(S) ↵

```
              x
    -----------------------
    1 - x + exp(- 1/x) C1 x
```

Replacement of the arbitrary constant A by C1 gives the identical result. There are some Bernoulli equations for which we do not have the close form integration but at least we know the integration. Such an example is $\frac{dy}{dx} + xy = \sin x$. Utilize the concept of integrating factor to write $P(x) = 1$, $Q(x) = x$, $R(x) = \sin x$, $\alpha = 0$, and

$I(x, y) = y^{-\alpha} \dfrac{e^{(1-\alpha)\int \frac{Q(x)}{P(x)} dx}}{P(x)} = e^{\int x dx} = e^{\frac{x^2}{2}}$. One can find that the general solution is $y(x) = e^{-\frac{1}{2}x^2} \int e^{\frac{1}{2}x^2} \sin x \, dx + A e^{-\frac{1}{2}x^2}$.

MATLAB responds in the same way as shown below:

MATLAB Command
>>S=dsolve('Dy+x*y=sin(x)','x'); ↵
>>pretty(S) ↵

```
              2  /          2                      2
   exp(- 1/2 x ) |  exp(1/2 x ) sin(x) dx + exp(- 1/2 x ) C1
                 /
```

Our last example of the first order differential equation is Riccati equation. The general form of Riccati equation is given by $\frac{dy}{dx} = P(x)y^2 + Q(x)y + R(x)$. Right way we go for the example $\frac{dy}{dx} = \frac{y^2}{x} + \frac{y}{x} - \frac{2}{x}$ under the initial condition $y = 2$ when $x = 1$. A substitution is required to solve the equation, which is $y = S(x) + \frac{1}{z}$. Depending on the equation, $S(x)$ assumes different functions of x. Anyhow, substitute $y = 1 + \frac{1}{z}$ for the example at hand. Differentiate w.r.to x to get $\frac{dy}{dx} = -\frac{1}{z^2}\frac{dz}{dx}$. Apply some algebra to have $-\frac{1}{z^2}\frac{dz}{dx} = \frac{1}{x}\left(1 + \frac{1}{z}\right)^2 + \frac{1}{x}\left(1 + \frac{1}{z}\right) - \frac{2}{x} \Rightarrow -\frac{1}{z^2}\frac{dz}{dx} = \frac{3z+1}{z^2 x} \Rightarrow \frac{dz}{dx} + \frac{3z}{x} = -\frac{1}{x}$ (reduces to Bernoulli Equation, where $P(x) = 1$,

$Q(x) = \frac{3}{x}$, $R(x) = -\frac{1}{x}$, and $\alpha = 0$). The integrating factor is found as $I(x,y) = y^{-\alpha}\dfrac{e^{(1-\alpha)\int \frac{Q(x)}{P(x)} dx}}{P(x)} = e^{\int \frac{3}{x} dx} = e^{3\ln x} = x^3$, on that, the last equation becomes $\frac{d(x^3 z)}{dx} = -x^2 \Rightarrow x^3 z = -\frac{x^3}{3} + A \Rightarrow z = -\frac{1}{3} + \frac{A}{x^3} \Rightarrow y = 1 + \frac{1}{-\frac{1}{3} + \frac{A}{x^3}} = \frac{2x^3 + 3A}{3A - x^3}$.

Insert the initial condition to have $A = \frac{4}{3}$, from which, the particular solution is $y = \frac{2x^3 + 4}{4 - x^3}$. Let us see what MATLAB returns:

MATLAB Command
>>S=dsolve('Dy=y^2/x+y/x-2/x,y(1)=2','x'); ↵
>>pretty(S) ↵

```
         3
   1/2 x  + 1
 - -----------
             3
    -1 + 1/4 x
```

7.2.2 Second order ordinary differential equations

A second order differential equation has the form $f\left(x, y, \dfrac{dy}{dx}, \dfrac{d^2y}{dx^2}\right)=0$. We are able to solve only the equations having certain forms. Our focus will be on the linear equations. A second order linear differential equation has the form $\dfrac{d^2y}{dx^2}+P(x)\dfrac{dy}{dx}+Q(x)y=F(x)$, where $P(x)$ and $Q(x)$ are called the coefficient functions and $F(x)$ is called the forcing function. If $F(x)=0$, the equation is said to be homogeneous otherwise it is non-homogeneous. Start with the homogeneous equation with constant coefficients. An example can be $\dfrac{d^2y}{dt^2}-2\dfrac{dy}{dt}-15y=0$. To have the solution, we assume that $y=e^{mt}$ (this is termed as the trial solution) so $\dfrac{dy}{dt}=me^{mt}$, $\dfrac{d^2y}{dt^2}=m^2e^{mt}$. Insert all these in the given equation to write $m^2-2m-15=0$. This equation is called the auxiliary equation, roots of which are $m=-3$ and 5. An auxiliary equation may have real distinct (assume α, β), equal (assume each is α), or imaginary roots (assume $\alpha \pm j\beta$). Accordingly, the solutions are given by $Ae^{\alpha t}+Be^{\beta t}$, $(A+Bt)e^{\alpha t}$, or $e^{\alpha t}(A\cos\beta t+B\sin\beta t)$ respectively. The example falls into the first category so the general solution is $y(t)=Ae^{5t}+Be^{-3t}$, where A and B are two arbitrary constants. Since this is a second order equation, we should have two arbitrary constants. The second order derivative $\dfrac{d^2y}{dt^2}$ is represented by 'D2y'. The independent variable is t so there is no need to specify that. Anyhow, have the solution as follows:

MATLAB Command
```
>>S=dsolve('D2y-2*Dy-15*y=0'); ↵
>>pretty(S) ↵
```
```
C1 exp(-3 t) + C2 exp(5 t)
```

Instead of A and B, the arbitrary constants are returned as C1 and C2, which is not a problem at all.

As an example of the second order linear differential equation containing the initial condition, solve $u''+u'+u=0$, $u(0)=1$, and $u'(0)=\sqrt{3}$. Notation says that $u''=\dfrac{d^2u}{dt^2}$ and $u'=\dfrac{du}{dt}$ and the dependent variable is u. The trial solution is $u=e^{mt}$, therefore, the auxiliary equation and roots of the auxiliary equation are $m^2+m+1=0$ and $m=-\dfrac{1}{2}\pm j\dfrac{\sqrt{3}}{2}$ respectively. It suggests the general solution be $u(t)=e^{-\frac{t}{2}}\left(C_1\cos\dfrac{\sqrt{3}}{2}t+C_2\sin\dfrac{\sqrt{3}}{2}t\right)$. There are two arbitrary constants – C_1 and C_2 so two initial values are essential to remove them, which are $u(0)=1$ and $u'(0)=\sqrt{3}$. First initial value provides $C_1=1$, for this, evaluate $u(t)$ for $t=0$. The second one needs the first derivative of $u(t)$, $\therefore$ $u'(t)=-\dfrac{1}{2}e^{-\frac{t}{2}}C_1\cos\dfrac{\sqrt{3}}{2}t-\dfrac{\sqrt{3}}{2}e^{-\frac{t}{2}}C_1\sin\dfrac{\sqrt{3}}{2}t$ $-\dfrac{1}{2}e^{-\frac{t}{2}}C_2\sin\dfrac{\sqrt{3}}{2}t+\dfrac{\sqrt{3}}{2}e^{-\frac{t}{2}}C_2\cos\dfrac{\sqrt{3}}{2}t$ and $u'(0)=\sqrt{3}$ provides $\sqrt{3}=-\dfrac{1}{2}C_1+\dfrac{\sqrt{3}}{2}C_2$. Plugging $C_1=1$ into the last equation yields $C_2=2+\dfrac{1}{\sqrt{3}}$. Finally, the particular solution for the given initial value is $u(t)=e^{-\frac{t}{2}}\cos\dfrac{\sqrt{3}}{2}t$ $+e^{-\frac{t}{2}}\left(2+\dfrac{1}{\sqrt{3}}\right)\sin\dfrac{\sqrt{3}}{2}t$. MATLAB codes for $u(0)=1$ and $u'(0)=\sqrt{3}$ are 'u(0)=1' and 'Du(0)=sqrt(3)' respectively. To show the solution,

MATLAB Command

```
>>S=dsolve('D2u+Du+u=0','u(0)=1,Du(0)=sqrt(3)'); ↵
>>pretty(S) ↵
```

```
                          1/2
      exp(- 1/2 t) cos(1/2 3    t)

                     1/2  1/2                      1/2
      + 1/3 (1 + 2 3    ) 3    exp(- 1/2 t) sin(1/2 3    t)
```

To illustrate the non-homogeneous equation without the initial value, let us consider $y'' + 4y' + 4y = e^x + e^{-2x}$. The complementary function is obtained from $y'' + 4y' + 4y = 0$. The trial solution, auxiliary equation, and roots of the auxiliary equation are $y = e^{mx}$, $m^2 + 4m + 4 = 0$, and $m = -2, -2$ respectively. The complementary function is $y_{CF} = e^{-2x}(C_1 + C_2 x)$. The particular integral y_{PI} is found from $P_1 + P_2$, where $P_1 = \dfrac{e^x}{D^2 + 4D + 4}$ and $P_2 = \dfrac{e^{-2x}}{D^2 + 4D + 4}$. It is trivial to say that the operators D and D^2 are equivalent to $\dfrac{d}{dx}$ and $\dfrac{d^2}{dx^2}$ respectively. Various methods are there in the literature to find the particular integral. We follow the separation of exponential function, so, $P_1 = \dfrac{e^x}{D^2 + 4D + 4} = e^x \left[\dfrac{1}{1^2 + 4 \times 1 + 4} \right] = \dfrac{e^x}{9}$ (because the coefficient of the exponent x is 1). In P_2, $D^2 + 4D + 4$ becomes 0 for $D = -2$ (the coefficient of the exponent x is -2), on that, $P_2 = \dfrac{e^{-2x}}{D^2 + 4D + 4} = x \left[\dfrac{e^{-2x}}{\frac{\partial}{\partial D}(D^2 + 4D + 4)} \right] = x \left[\dfrac{e^{-2x}}{2D + 4} \right]$ (still $2D + 4$ is 0 for $D = -2$) $= x.x \left[\dfrac{e^{-2x}}{\frac{\partial}{\partial D}(2D + 4)} \right] = \dfrac{x^2 e^{-2x}}{2}$. We end up with the complete solution as $y = e^{-2x}(C_1 + C_2 x) + \dfrac{e^x}{9} + \dfrac{x^2 e^{-2x}}{2}$.

MATLAB representation of this equation is 'D2y+4*Dy+4*y=exp(x)+exp(-2*x)'. The execution is shown as follows:

MATLAB Command

```
>>S=dsolve('D2y+4*Dy+4*y=exp(x)+exp(-2*x)','x'); ↵
>>pretty(S) ↵
```

```
                                 2
      1/9 exp(x) + 1/2 exp(-2 x) x  + C1 exp(-2 x) + C2 exp(-2 x) x
```

7.2.3 Higher order ordinary differential equations

The methodology so described in the preceding articles can easily be extended to the higher order linear differential equations. The general form of the nth order non-homogeneous linear differential equation is $\dfrac{d^n y}{dx^n} + P_{n-1}(x) \dfrac{d^{n-1} y}{dx^{n-1}} + P_{n-2}(x) \dfrac{d^{n-2} y}{dx^{n-2}} + \ldots + P_1(x) \dfrac{dy}{dx} + P_0(x)y = F(x)$. The same subroutine is used to obtain the analytical solution but the arguments are different depending on the homogeneity/non-homogeneity, with/without the initial conditions, and the constant/variable coefficients. To furnish an example of higher order equation, solve $y''' + y' - 2y = 0$. The trial solution, auxiliary equation, and roots of auxiliary equation are $y = e^{mx}$, $m^3 + m - 2 = 0$, and $m = 1, \dfrac{-1 \pm j\sqrt{7}}{2}$ respectively. Hence, the general solution of the differential equation is $y(x) = C_1 e^x + e^{-\frac{x}{2}}\left[C_2 \cos \frac{\sqrt{7}x}{2} + C_3 \sin \frac{\sqrt{7}x}{2} \right]$, where C_1, C_2, and C_3 are the arbitrary constants. The

equation $y'' + y' - 2y = 0$ is represented as 'D3y+Dy-2*y=0'. Execute it as follows:

MATLAB Command
>>S=dsolve('D3y+Dy-2*y=0','x'); ↵
>>pretty(S) ↵

```
                              1/2                          1/2
C1 exp(x) + C2 exp(- 1/2 x) cos(1/2 7    x) + C3 exp(- 1/2 x) sin(1/2 7    x)
```

To learn about the insertion of the initial conditions for the higher order linear differential equations, solve $y''' - y'' = 0$, $y(1) = 1$, $y'(0) = 3$, and $y''(-1) = -1$ with the independent variable u. The trial solution, auxiliary equation, roots of the equation, and general solution are $y = e^{mu}$, $m^3 - m^2 = 0$, $m = 0, 0$, and 1, and $y(u) = C_1 + C_2 u + C_3 e^u$ respectively. The initial value $y(1) = 1$ provides $1 = C_1 + C_2 + C_3 e$, then, $y'(u) = C_2 + C_3 e^u$ and $y'(0) = 3$ let us write $C_2 + C_3 = 3$. The second derivative of $y(u)$ is needed to apply the third initial condition, $\therefore y''(u) = C_3 e^u$ and the third initial value $y''(-1) = -1$ implies $C_3 = -e$, therefore, $C_2 = 3 + e$ and $C_1 = 1 - (3 + e) - e(-e) = e^2 - e - 2$. Finally, the particular solution is $y(u) = e^2 - e - 2 + (3 + e)u - e^{u+1}$. Insertion of $y'(0) = 3$ and $y''(-1) = -1$ happens to be 'Dy(0)=3' and "D2y(-1)=-1' respectively. Have the solution as follows:

MATLAB Command
>>S=dsolve('D3y-D2y=0','Dy(0)=3,D2y(-1)=-1,y(1)=1','u'); ↵
>>pretty(S) ↵

```
      1 + 2 exp(-1) - exp(1)     (1 + 3 exp(-1)) u     exp(u)
    - ----------------------  +  -----------------  -  ------
            exp(-1)                   exp(-1)          exp(-1)
```

The return is not simplified yet. Simplify the output expression further:
>>pretty(simplify(S)) ↵

```
    -exp(1) - 2 + exp(2) + u exp(1) + 3 u - exp(1 + u)
```

Now the output is the same as we have had analytically. In a similar fashion, one can find the solution of the higher order non-homogeneous equation too.

7.2.4 Euler differential equations

The general form of the nth order Euler differential equation is given by $A_n t^n \dfrac{d^n y}{dt^n} + A_{n-1} t^{n-1} \dfrac{d^{n-1} y}{dt^{n-1}} + A_{n-2} t^{n-2} \dfrac{d^{n-2} y}{dt^{n-2}} + \ldots\ldots\ldots + A_1 t y^1 + A_0 y = 0$, where A_n, A_{n-1}, A_{n-2}, ..., A_1, and A_0 are constants. The Euler differential equation is the simplest form of the linear differential equation that has the variable coefficients. Since the leading coefficient (A_n) should not be zero, the equation is solved either for $t > 0$ or $t < 0$. The change of the independent variable by $t = \begin{cases} e^x & if \;\; t > 0 \\ e^{-x} & if \;\; t < 0 \end{cases}$ produces a differential equation with constant coefficients.

The fact is illustrated by the differential equation $2t^2 \dfrac{d^2 y}{dt^2} - 3t \dfrac{dy}{dt} - 3y = 0$. By making the use of the substitution $t = e^x$, we have $\dfrac{dt}{dx} = e^x = t$, $\dfrac{dx}{dt} = \dfrac{1}{e^x} = \dfrac{1}{t}$, $\dfrac{dy}{dt} = \dfrac{dy}{dx}\dfrac{dx}{dt} = \dfrac{dy}{dx}\dfrac{1}{t}$, $t\dfrac{dy}{dt} = \dfrac{dy}{dx}$, and $\dfrac{d^2 y}{dt^2} = \dfrac{d}{dt}\left(\dfrac{dy}{dt} \right) =$

$$\frac{d}{dt}\left(\frac{dy}{dx}\cdot\frac{1}{t}\right) = -\frac{1}{t^2}\frac{dy}{dx} + \frac{1}{t}\frac{d}{dt}\left(\frac{dy}{dx}\right) = -\frac{1}{t^2}\frac{dy}{dx} + \frac{1}{t}\frac{dx}{dt}\frac{d}{dx}\left(\frac{dy}{dx}\right) = -\frac{1}{t^2}\frac{dy}{dx} + \frac{1}{t^2}\frac{d^2y}{dx^2}$$, thereupon, $t^2\frac{d^2y}{dt^2} = \frac{d^2y}{dx^2} - \frac{dy}{dx}$. So,

the given equation $2t^2\frac{d^2y}{dt^2} - 3t\frac{dy}{dt} - 3y = 0$ transforms to $2\frac{d^2y}{dx^2} - 5\frac{dy}{dx} - 3y = 0$. By doing so, the independent

variable t is changed to x. The auxiliary equation of the last equation is $2m^2 - 5m - 3 = 0$. The roots of the

auxiliary equation are $-\frac{1}{2}$ and 3. The general solution is $y(x) = C_1 e^{-\frac{x}{2}} + C_2 e^{3x}$. But x must be changed to t.

Since $t = e^x$, $x = \ln t$ and the general solution becomes $y(t) = C_1 e^{-\frac{\ln t}{2}} + C_2 e^{3\ln t} = C_1 e^{\ln t^{-\frac{1}{2}}} + C_2 e^{\ln t^3} = C_1 t^{-\frac{1}{2}} + C_2 t^3 = $

$\dfrac{C_1 + C_2 t^{\frac{7}{2}}}{t^{\frac{1}{2}}}$. The string form of $2t^2\frac{d^2y}{dt^2} - 3t\frac{dy}{dt} - 3y = 0$ is '2*t^2*D2y-3*t*Dy-3*y=0'. The solution is as follows:

MATLAB Command
```
>>S=dsolve('2*t^2*D2y-3*t*Dy-3*y=0'); ↵
>>pretty(S) ↵
```

```
        7/2
C1 t         + C2
-----------------
       1/2
        t
```

Observe that the arbitrary constants are interchanged.

7.2.5 System of differential equations

A system of differential equations is given by $\begin{cases} \dfrac{dy_1}{dt} = f_1(t, y_1, y_2, \ldots, y_n) \\ \dfrac{dy_2}{dt} = f_2(t, y_1, y_2, \ldots, y_n) \\ \vdots \\ \dfrac{dy_n}{dt} = f_n(t, y_1, y_2, \ldots, y_n) \end{cases}$, where y_1, y_2, $\ldots$, and y_n are

the n dependent variables and t is the independent variable. Our intention is to find the functions y_1, y_2, $\ldots$, and y_n satisfying simultaneously the differential equations of the system. The system just defined is a general

one. For a linear system, the system of equations can be written as $\begin{cases} \dfrac{dy_1}{dt} = A_{11}y_1 + A_{12}y_2 + \cdots + A_{1n}y_n + b_1 \\ \dfrac{dy_2}{dt} = A_{21}y_1 + A_{22}y_2 + \cdots + A_{2n}y_n + b_2 \\ \vdots \\ \dfrac{dy_n}{dt} = A_{n1}y_1 + A_{n2}y_2 + \cdots + A_{nn}y_n + b_n \end{cases}$, where

A_{ij}'s are the functions of t and b's are the forcing functions. Write the system of equations in matrix form as

$Y' = AY + B$, where $Y = \begin{bmatrix} y_1 \\ y_2 \\ \vdots \\ y_n \end{bmatrix}$, $Y' = \begin{bmatrix} y_1' \\ y_2' \\ \vdots \\ y_n' \end{bmatrix}$, $A = \begin{bmatrix} A_{11} & A_{12} & \cdots & A_{1n} \\ A_{21} & A_{22} & \cdots & A_{2n} \\ & & \vdots & \\ A_{n1} & A_{n2} & \cdots & A_{nn} \end{bmatrix}$, and $B = \begin{bmatrix} b_1 \\ b_2 \\ \vdots \\ b_n \end{bmatrix}$. When $B = 0$, the system becomes

homogeneous, and its equation is given by $Y' = AY$.

Find the solution of the following system of equations that is homogeneous and that has constant

coefficients: $\left\{ \begin{array}{l} \dfrac{dy_1}{dt} = 5y_1 + 2y_2 \\ \dfrac{dy_2}{dt} = 2y_1 + 2y_2 \end{array} \right\}$ contingent to the initial conditions $\left\{ \begin{array}{l} y_1(0) = 1 \\ y_2(0) = 2 \end{array} \right\}$.

There are several methods to find the solution of a system of equations, namely, fundamental matrix method, Laplace transform method, and eigenvector method. We attempt to solve the system by the fundamental matrix method. In this method some algebra is performed until a differential equation of a single

dependent variable is found. Having expressed y_2 as $y_2 = \dfrac{1}{2}\left[\dfrac{dy_1}{dt} - 5y_1\right]$ from the first equation, the second

equation becomes $\dfrac{d}{dt}\left\{ \dfrac{1}{2}\left[\dfrac{dy_1}{dt} - 5y_1\right]\right\} = 2y_1 + 2 \times \dfrac{1}{2}\left[\dfrac{dy_1}{dt} - 5y_1\right]$. On simplification, one can write the last equation

as $\dfrac{d^2 y_1}{dt^2} - 7\dfrac{dy_1}{dt} + 6y_1 = 0$, whose trial solution, auxiliary equation, roots of the auxiliary equation, and the

general solution are $y_1 = e^{mt}$, $m^2 - 7m + 6 = 0$, $m = 1, 6$, and $y_1 = Ae^t + Be^{6t}$ respectively. Whence, the general

solution of y_2 is $y_2 = \dfrac{1}{2}\left[\dfrac{dy_1}{dt} - 5y_1\right] = -2Ae^t + \dfrac{B}{2}e^{6t}$. We are in a position to find the fundamental matrix Φ. As

you know that constants A and B are arbitrary, one can choose any values of A and B so that two vectors

formed by $\begin{bmatrix} y_1 \\ y_2 \end{bmatrix}$ are linearly independent. As regards, choose $\left\{ \begin{array}{l} A = 1 \\ B = 0 \end{array} \right\}$ and $\left\{ \begin{array}{l} A = 0 \\ B = 1 \end{array} \right\}$ to have the linearly

independent vectors $\begin{bmatrix} e^t \\ -2e^t \end{bmatrix}$ and $\begin{bmatrix} e^{6t} \\ \frac{1}{2}e^{6t} \end{bmatrix}$ respectively. These vector functions constitute the fundamental matrix

Φ, hence, for the example at hand, $\Phi = \begin{bmatrix} e^t & e^{6t} \\ -2e^t & \frac{1}{2}e^{6t} \end{bmatrix}$. Since the system of differential equations is a second

order one, there should be two arbitrary constant, C_1 and C_2. Matrix of constants is formed as $C = \begin{bmatrix} C_1 \\ C_2 \end{bmatrix}$. Our

so-called general solution of the system of differential equations is given by $Y = \Phi \times C$ or $\begin{bmatrix} y_1 \\ y_2 \end{bmatrix} =$

$\begin{bmatrix} e^t & e^{6t} \\ -2e^t & \frac{1}{2}e^{6t} \end{bmatrix}\begin{bmatrix} C_1 \\ C_2 \end{bmatrix} = \begin{bmatrix} C_1 e^t + C_2 e^{6t} \\ -2C_1 e^t + \frac{1}{2}C_2 e^{6t} \end{bmatrix}$. Apply the initial conditions $\begin{bmatrix} y_1(0) = 1 \\ y_2(0) = 2 \end{bmatrix}$ to write $\left\{ \begin{array}{l} C_1 + C_2 = 1 \\ -2C_1 + \frac{1}{2}C_2 = 2 \end{array} \right\}$, whose

solution is $\left\{ \begin{array}{l} C_1 = -\dfrac{3}{5} \\ C_2 = \dfrac{8}{5} \end{array} \right\}$. Therefore, the required solution is $\begin{bmatrix} y_1 \\ y_2 \end{bmatrix} = \begin{bmatrix} -\dfrac{3}{5}e^t + \dfrac{8}{5}e^{6t} \\ \dfrac{6}{5}e^t + \dfrac{4}{5}e^{6t} \end{bmatrix}$. In MATLAB terminology, the

dependent variables y_1 and y_2, the derivatives $\dfrac{dy_1}{dt}$ and $\dfrac{dy_2}{dt}$, and the initial conditions $y_1(0) = 1$ and $y_2(0) = 2$ are argumented as y1 and y2, Dy1 and Dy2, and y1(0)=1 and y2(0)=2 respectively. The whole methodology is as follows:

MATLAB Command

```
>>e1='Dy1=5*y1+2*y2'; ↵            ← Assign the first equation to e1
>>e2='Dy2=2*y1+2*y2'; ↵            ← Assign the second equation to e2
>>S=dsolve(e1,e2,'y1(0)=1,y2(0)=2') ↵   ← The output of 'dsolve' is assigned to S

S =
```

```
y1: [1x1 sym]
y2: [1x1 sym]
```

Above output says that the solution of the system of equations is returned as a structured array. The array has two components by name y1 and y2. Each one is a scalar symbolic. To see them:

>>pretty(S.y1) ↵   ← The output for y_1

```
- 3/5 exp(t) + 8/5 exp(6 t)
```

>>pretty(S.y2) ↵   ← The output for y_2

```
4/5 exp(6 t) + 6/5 exp(t)
```

Now we discuss a nonhomogeneous linear system $Y' = AY + B$. The solution of the homogeneous part $Y' = AY$ as well as convenience of the methods depends on the nature of eigenvalues of matrix A. The eigenvalues can be distinct, repetitive, or complex. We restrict our attention on the matrix A containing the real eigenvalues such that A can be diagonalized. The advantage of diagonalized system is that they can be transformed to the uncoupled system. Uncoupled means a nth order system is consisted of n first order differential equations and each first order differential equation can be solved independently. A square matrix A can be diagonalized by the similarity transform $P^{-1}AP$, where P can be obtained by the Jordan decomposition of a matrix (see article 4.32 for the Jordan form decomposition). To learn the theory, carry out some matrix algebra. $Y' = AY + B$ can be written as $\frac{dY}{dt} = AY + B$. Write $Y = PZ$, where P comes from the similarity transform and $Z = \begin{bmatrix} z_1 \\ z_2 \\ \vdots \\ z_n \end{bmatrix}$ are the variables following the transform. On assuming that, one can have

$$\frac{dY}{dt} = AY + B \Rightarrow \frac{d(PZ)}{dt} = APZ + B \Rightarrow P\frac{dZ}{dt} = APZ + B \text{ (since, } P \text{ is a constant matrix)} \Rightarrow \frac{dZ}{dt} = P^{-1}APZ + P^{-1}B$$

(left multiplication by P^{-1}). Since $P^{-1}AP$ is a diagonal matrix, the system represented by $\frac{dZ}{dt} = P^{-1}APZ + P^{-1}B$ is an uncoupled one and the general solution of the system is given by $Y = PZ$.

Do not be overwhelmed with the theory. Compute the solution of the nonhomogeneous system of equations given by $\begin{cases} \frac{dx}{dt} = 31x - 21y + 9z - e^{-3t} \\ \frac{dy}{dt} = 44x - 30y + 12z + 2t \\ \frac{dz}{dt} = -22x + 14y - 8z + \sin t \end{cases}$ satisfying the initial conditions $\begin{cases} x(0) = -2 \\ y(0) = 1 \\ z(0) = 0 \end{cases}$. The first step is to form the matrix A and B, which are $A = \begin{bmatrix} 31 & -21 & 9 \\ 44 & -30 & 12 \\ -22 & 14 & -8 \end{bmatrix}$ and $B = \begin{bmatrix} -e^{-3t} \\ 2t \\ \sin t \end{bmatrix}$ respectively. You can even use the subroutine 'jordan' to find the matrix P (see article 4.32). The response is as follows:

MATLAB Command

>>A=sym([31 -21 9;44 -30 12;-22 14 -8]); ↵   ← Assign the matrix A to A
>>[P J]=jordan(A) ↵   ← Apply the subroutine 'jordan'

P =

```
[ 371/11,    -33,   -3/11]
[     44,    -44,       0]
[    -21,     22,       1]
```

$J =$

```
[ -2,  0,  0]
[  0, -3,  0]
[  0,  0, -2]
```

From the return, one can write $P = \begin{bmatrix} \frac{371}{11} & -33 & -\frac{3}{11} \\ 44 & -44 & 0 \\ -21 & 22 & 1 \end{bmatrix}$ and $P^{-1}AP = \begin{bmatrix} -2 & 0 & 0 \\ 0 & -3 & 0 \\ 0 & 0 & -2 \end{bmatrix}$, hence,

$\frac{dZ}{dt} = P^{-1}APZ + P^{-1}B \quad \Rightarrow \quad \begin{bmatrix} \frac{dz_1}{dt} \\ \frac{dz_2}{dt} \\ \frac{dz_3}{dt} \end{bmatrix} = \begin{bmatrix} -2 & 0 & 0 \\ 0 & -3 & 0 \\ 0 & 0 & -2 \end{bmatrix} \begin{bmatrix} z_1 \\ z_2 \\ z_3 \end{bmatrix} + \begin{bmatrix} \frac{371}{11} & -33 & -\frac{3}{11} \\ 44 & -44 & 0 \\ -21 & 22 & 1 \end{bmatrix}^{-1} \begin{bmatrix} -e^{-3t} \\ 2t \\ \sin t \end{bmatrix}$ (assume that the

transformed variables are z_1, z_2, and z_3 corresponding to x, y, and z respectively) $\Rightarrow \begin{bmatrix} \frac{dz_1}{dt} \\ \frac{dz_2}{dt} \\ \frac{dz_3}{dt} \end{bmatrix} = \begin{bmatrix} -2z_1 \\ -3z_2 \\ -2z_3 \end{bmatrix} +$

$\begin{bmatrix} 1 & -\frac{27}{44} & \frac{3}{11} \\ 1 & -\frac{7}{11} & \frac{3}{11} \\ -1 & \frac{49}{44} & \frac{8}{11} \end{bmatrix} \begin{bmatrix} -e^{-3t} \\ 2t \\ \sin t \end{bmatrix} \Rightarrow \begin{bmatrix} \frac{dz_1}{dt} \\ \frac{dz_2}{dt} \\ \frac{dz_3}{dt} \end{bmatrix} = \begin{bmatrix} -2z_1 \\ -3z_2 \\ -2z_3 \end{bmatrix} + \begin{bmatrix} -e^{-3t} - \frac{27}{22}t + \frac{3}{11}\sin t \\ -e^{-3t} - \frac{14}{11}t + \frac{3}{11}\sin t \\ e^{-3t} + \frac{49}{22}t + \frac{8}{11}\sin t \end{bmatrix} \Rightarrow \begin{bmatrix} \frac{dz_1}{dt} + 2z_1 \\ \frac{dz_2}{dt} + 3z_2 \\ \frac{dz_3}{dt} + 2z_3 \end{bmatrix} = \begin{bmatrix} -e^{-3t} - \frac{27}{22}t + \frac{3}{11}\sin t \\ -e^{-3t} - \frac{14}{11}t + \frac{3}{11}\sin t \\ e^{-3t} + \frac{49}{22}t + \frac{8}{11}\sin t \end{bmatrix}$. As we

mentioned in the beginning, the equations can be solved independently. Pick up the first one from the last matrix equation, which is $\frac{dz_1}{dt} + 2z_1 = -e^{-3t} - \frac{27}{22}t + \frac{3}{11}\sin t$. The complementary part is $(z_1)_{CF} = C_1 e^{-2t}$. The particular

integrals are I_1, I_2, and I_3, where $I_1 = \frac{-e^{-3t}}{D+2} = e^{-3t}$, $I_2 = \frac{-27t}{22(D+2)} = -\frac{27}{44}\left(1 + \frac{D}{2}\right)^{-1}t = -\frac{27}{44}\left(1 - \frac{D}{2}\right)t =$

$-\frac{27}{44}\left(t - \frac{1}{2}\right)$, and $I_3 = \frac{3\sin t}{11(D+2)} = \text{imag}\left\{\frac{3e^{jt}}{11(D+2)}\right\} = \text{imag}\left\{\frac{3e^{jt}}{11(j+2)}\right\} = \text{imag}\left\{\frac{3e^{jt}(2-j)}{55}\right\} =$

$\text{imag}\left\{\frac{3(\cos t + j\sin t)(2-j)}{55}\right\} = \frac{6\sin t - 3\cos t}{55}$ respectively (operator $D \equiv \frac{d}{dt}$). Collecting all, we have $z_1 = C_1 e^{-2t} +$

$e^{-3t} - \frac{27}{44}\left(t - \frac{1}{2}\right) + \frac{6\sin t - 3\cos t}{55}$. In a similar fashion, the other two general solutions of z_2 and z_3 are $z_2 =$

$C_2 e^{-3t} - te^{-3t} - \frac{14t}{33} + \frac{14}{99} - \frac{3\cos t}{110} + \frac{9\sin t}{110}$ and $z_3 = C_3 e^{-2t} - e^{-3t} + \frac{49t}{44} - \frac{49}{88} + \frac{16\sin t}{55} - \frac{8\cos t}{55}$ respectively. In matrix

form, transformed equations are $\begin{bmatrix} z_1 \\ z_2 \\ z_3 \end{bmatrix} = \begin{bmatrix} C_1 e^{-2t} + e^{-3t} - \frac{27}{44}t + \frac{27}{88} + \frac{6}{55}\sin t - \frac{3}{55}\cos t \\ C_2 e^{-3t} - te^{-3t} - \frac{14}{33}t + \frac{14}{99} - \frac{3}{110}\cos t + \frac{9}{110}\sin t \\ C_3 e^{-2t} - e^{-3t} + \frac{49}{44}t - \frac{49}{88} + \frac{16}{55}\sin t - \frac{8}{55}\cos t \end{bmatrix}$, so, the required general

solution is $Y = PZ \Rightarrow \begin{bmatrix} x \\ y \\ z \end{bmatrix} = \begin{bmatrix} \frac{371}{11} & -33 & -\frac{3}{11} \\ 44 & -44 & 0 \\ -21 & 22 & 1 \end{bmatrix} \times \begin{bmatrix} C_1 e^{-2t} + e^{-3t} - \frac{27}{44}t + \frac{27}{88} + \frac{6}{55}\sin t - \frac{3}{55}\cos t \\ C_2 e^{-3t} - te^{-3t} - \frac{14}{33}t + \frac{14}{99} - \frac{3}{110}\cos t + \frac{9}{110}\sin t \\ C_3 e^{-2t} - e^{-3t} + \frac{49}{44}t - \frac{49}{88} + \frac{16}{55}\sin t - \frac{8}{55}\cos t \end{bmatrix}$. Following the multiplication, we get $\begin{bmatrix} \frac{371}{11}C_1e^{-2t} + 34e^{-3t} - 7t + \frac{35}{6} + \frac{9}{10}\sin t - \frac{9}{10}\cos t - 33C_2e^{-3t} + 33te^{-3t} - \frac{3}{11}C_3e^{-2t} \\ 44C_1e^{-2t} + 44e^{-3t} - \frac{25}{3}t + \frac{131}{18} + \frac{6}{5}\sin t - \frac{6}{5}\cos t - 44C_2e^{-3t} + 44te^{-3t} \\ -21C_1e^{-2t} - 22e^{-3t} + \frac{14t}{3} - \frac{35}{9} - \frac{1}{5}\sin t + \frac{2}{5}\cos t + 22C_2e^{-3t} - 22te^{-3t} + C_3e^{-2t} \end{bmatrix}$. That is not all.

Still we need to insert the initial conditions, which are $\begin{cases} x(0) = -2 \\ y(0) = 1 \\ z(0) = 0 \end{cases}$. Now we have

$\begin{cases} \frac{371}{11}C_1 - 33C_2 - \frac{3}{11}C_3 + \frac{584}{15} = -2 \\ 44C_1 - 44C_2 + \frac{4507}{90} = 1 \\ -21C_1 + 22C_2 + C_3 - \frac{1147}{45} = 0 \end{cases}$, solution of which is $C_1 = -\frac{1701}{440}$, $C_2 = -\frac{2723}{990}$, and $C_3 = \frac{2119}{440}$. In applying back

substitution, finally, the so expected particular solution of the given system is $\begin{bmatrix} x \\ y \\ z \end{bmatrix} =$

$\begin{bmatrix} (33t + \frac{3743}{30})e^{-3t} - \frac{1317}{10}e^{-2t} - 7t + \frac{35}{6} - \frac{9}{10}\cos t + \frac{9}{10}\sin t \\ (44t + \frac{7426}{45})e^{-3t} - \frac{1701}{10}e^{-2t} - \frac{25}{3}t + \frac{131}{18} - \frac{6}{5}\cos t + \frac{6}{5}\sin t \\ (-22t - \frac{3713}{45})e^{-3t} + 86e^{-2t} + \frac{2}{5}\cos t - \frac{1}{5}\sin t + \frac{14}{3}t - \frac{35}{9} \end{bmatrix}$. Following steps of MATLAB would not encourage someone

to carry out such laborious manipulation again:

MATLAB Command

```
>>e1='Dx=31*x-21*y+9*z-exp(-3*t)';           ← The first equation to e1
>>e2='Dy=44*x-30*y+12*z+2*t';                 ← The second equation to e2
>>e3='Dz=-22*x+14*y-8*z+sin(t)';              ← The third equation to e3
>>S=dsolve(e1,e2,e3,'x(0)=-2,y(0)=1,z(0)=0')

S =
        x: [1x1 sym]
        y: [1x1 sym]
        z: [1x1 sym]
>>pretty(S.x)                                 ← For x(t)
```

```
3743               1317
---- exp(-3 t) -  ---- exp(-2 t) + 35/6 + 33 t exp(-3 t) - 9/10 cos(t)
 30                 10

        + 9/10 sin(t) - 7 t
```

```
>>pretty(S.y)                                 ← For y(t)
```

```
1701             7426            131
- ---- exp(-2 t) + ---- exp(-3 t) + --- + 44 t exp(-3 t) - 6/5 cos(t)
  10                45              18

        + 6/5 sin(t) - 25/3 t
```

```
>>pretty(S.z)                                 ← For z(t)
```

```
                3713
86 exp(-2 t) -  ---- exp(-3 t) - 35/9 - 22 t exp(-3 t) + 2/5 cos(t)
                 45

        - 1/5 sin(t) + 14/3 t
```

7.2.6 Nonlinear differential equations

Mathematical formulations of many systems exhibit nonlinear differential equations. For nonlinear differential equations, the generalized theory and methods of solutions are not eminently developed. Study of such equations is usually confined to a variety of rather special problems. As a crude definition, the differential equations, which do not fall in the definition of the linear differential equations, are nonlinear differential equations. We present few examples of nonlinear differential equations.

Consider the nonlinear equation $x^2\left(\dfrac{dy}{dx}\right)^2 + xy\left(\dfrac{dy}{dx}\right) - 6y^2 = 0$. If it is possible, factorize the nonlinear

equation regarding $\dfrac{dy}{dx}$. On factorization, we have $\left(x\dfrac{dy}{dx} + 3y\right)\left(x\dfrac{dy}{dx} - 2y\right) = 0$. In fact there are two differential

equations $x\dfrac{dy}{dx} + 3y = 0$ and $x\dfrac{dy}{dx} - 2y = 0$, each of them is linear. Solution of the first one is found as follows:

$x\dfrac{dy}{dx} + 3y = 0 \Rightarrow \dfrac{dy}{y} + 3\dfrac{dx}{x} = 0 \Rightarrow \ln y + 3\ln x = A \Rightarrow \ln yx^3 = A \Rightarrow yx^3 = e^A = C_1 \Rightarrow y = \dfrac{C_1}{x^3}$. In a similar fashion,

the other solution is $y = \dfrac{C_2}{x^{-2}} = C_2 x^2$. Since the order of the nonlinear equation is 1 (because the order of the

highest order derivative is one), there can not be two arbitrary constants. C_1 and C_2 must be equal. With that,

the general solution is given by $\left(y - \dfrac{C_1}{x^3}\right)(y - C_1 x^2) = 0$. Next dexterity needs the writing style of the equation.

Powered derivative such as $\left(\dfrac{dy}{dx}\right)^2$ is written as 'Dy^2'. The string style is also observed to enter a nonlinear

differential equation. So, the example equation is stringed as 'x^2*Dy^2+x*y*Dy-6*y^2=0'. Carry out that in MATLAB as follows:

MATLAB Command

```
>>S=dsolve('x^2*Dy^2+x*y*Dy-6*y^2=0','x'); ↵
>>pretty(S) ↵

         [ C1  ]
         [---- ]
         [  3  ]
         [ x   ]
         [     ]
         [    2]
         [C1 x ]
```

As you see instead of the general solution, the individual solution is returned by MATLAB. Insertion of initial condition is not accepted. Because, several values of an arbitrary constant are possible for a particular x and y, thereby, giving a family of solutions.

Not all nonlinear equations can be linearly factored regarding $\dfrac{dy}{dx}$. To show by example, solve the

equation $y = x\dfrac{dy}{dx} + \left(\dfrac{dy}{dx}\right)^2$. Assume that $z = \dfrac{dy}{dx}$, on that, the given equation becomes $y = xz + z^2$. Differentiate

the last equation w.r.to x to write $\dfrac{dy}{dx} = x\dfrac{dz}{dx} + z + 2z\dfrac{dz}{dx} \Rightarrow z = x\dfrac{dz}{dx} + z + 2z\dfrac{dz}{dx} \Rightarrow (x + 2z)\dfrac{dz}{dx} = 0$. We have two

factors $x + 2z = 0$ and $\dfrac{dz}{dx} = 0$. The first factor $x + 2z = 0$ provides $z = -\dfrac{x}{2}$, from this cause, $y = xz + z^2$ turns to

$y = -\dfrac{x^2}{4}$, this is called the particular solution because no arbitrary constant is there. Integrate the second factor

$\dfrac{dz}{dx} = 0$ to have $z = C_1$. Insert $z = C_1$ into $y = xz + z^2$ to have the general solution $y = xC_1 + C_1^2$. Anyhow, the string form of the equation is 'y=x*Dy+ Dy^2' and execution of the subroutine is as follows:

MATLAB Command
>>S=dsolve('y=x*Dy+Dy^2','x'); ↵
>>pretty(S) ↵

```
[               2 ]
[ - 1/4 x       ]
[               ]
[      2        ]
[C1   + x  C1]
```

Not to be confined with the first order equations, consider the second order nonlinear equation $\dfrac{d^2 y}{dx^2} + 2\left(\dfrac{dy}{dx}\right)^3 = 0$. Taking $z = \dfrac{dy}{dx}$, the equation turns to $\dfrac{dz}{dx} + 2z^3 = 0$. Separation of variables makes us write

$\dfrac{dz}{2z^3} + dx = 0$, which becomes $-\dfrac{1}{4z^2} + x = A$ on integration. Solve for z to write $z = \pm\dfrac{1}{2\sqrt{x - A}}$, on back

substitution, $\dfrac{dy}{dx} = \pm\dfrac{1}{2\sqrt{x - A}}$. Finally, integrate to have $y = \pm\sqrt{x - A} + B$. The implementation is as follows:

MATLAB Command
>>S=dsolve('D2y+2*Dy^3=0','x'); ↵
>>pretty(S) ↵

```
[                    1/2          ]
[ 1/2 (4 x + C1)        + C2 ]
[                             ]
[                    1/2          ]
[- 1/2 (4 x + C1)       + C2]
```

If one writes C1=$-4A$ and C2=B, then the MATLAB return and the analytical solution are identical.

7.3 Numerical solution of differential equations

We have already employed the analytical method of various order differential equations. Solution of some differential equations may not fall into the known categories such as linear or nonlinear. Analytically unsolvable differential equations can be solved numerically by Taylor series approximation of the dependent variable up to certain order. In reality, exemplified simple linear differential equations may not occur but an understanding of it paves the way for an understanding of more complicated and practical systems that it would follow.

In the subsequent sections, we present the exact (analytical) and numerical solutions of various kinds of differential equations. Since the numerical solution as well as the exact solution has been found, one can gain some insight of the accuracy of the numerical methods. There are several methods to find the numerical solutions of differential equations such as Euler method, modified Euler method, Runge-Kutta methods, predictor-corrector methods…etc. All the while our approach is computationally subjectiveness for that reason we will not go through the theory of these methods. MATLAB is rich in having differential equation solvers. Following steps should be observed before executing a particular ODE (ordinary differential equation) subroutine:

☞ ☞ *Steps to be observed*

    Step A

        How an M-file is opened and executed (see chapter 1 regarding this),

    Step B

        An M-file, by the name 'f.m' (as we followed in the subsequent sections), must contain the differential equation (s),

    Step C

        Differential equation (s) must be rewritten in proper style $\left[\text{ for example, } \dfrac{dy}{dx} = f(x, y) \text{ or}\right.$

$$\left. \frac{d^2 y}{dx^2} = f\left(x, y, \frac{dy}{dx}\right) \right],$$

    Step D

        User's working path and the path containing the M-file 'f.m' must be identical,

    Step E

        An ODE solver is chosen based on the approximation order, stiffness of the dependent variable, and accuracy required,

    Step F

        Relative and absolute errors must be considered,

    Step G

        The initial values of the dependent variable (s) and the derivative (s) at a particular independent

variable must be known $\left[\text{ for instance, } y \text{ or } \dfrac{dy}{dx} \text{ at } x = 0 \right]$, and

    Step H

        In general, the arguments of the M-file 'f.m' are column vectors.

The acronyms and elaborations of some ODE subroutines are provided in table 7.A. Along with different options, changing solver's properties and having graphical plots of numerical outputs are also possible. Readers can have online help executing 'ode23', 'odeget', 'odeset', 'odeplot'... etc.

Table 7.A   Acronyms and descriptions of some ODE subroutines

Name of the ODE solver	Description of the subroutine
'ode23'	Can solve nonstiff differential equations using low order method [uses Runge-Kutta (2, 3) formula]
'ode45'	Can solve nonstiff differential equations using medium order method [uses Runge-Kutta (4, 5) formula]
'ode113'	Can solve nonstiff differential equations using variable order method
'ode23s'	Can solve stiff differential equations using low order method
'ode15s'	Can solve stiff differential equations using variable order method
'ode23t'	Can solve moderately stiff differential equations using trapezoidal rule
'ode23tb'	Can solve stiff differential equations using low order method

7.3.1 First order differential equations

Referring to article 7.2.1, we solved the differential equation $\frac{dy}{dx} = 4x^4 y^2$ with the initial condition $y = 2$ when $x = 0$. Its particular solution is found as $y = \frac{-10}{8x^5 - 5}$. Let us compute y from $x = 0$ to $x = 0.5$ with step size 0.1. Computations are $\left[\frac{-10}{8 \times 0 - 5} \quad \frac{-10}{8 \times 0.1^5 - 5} \quad \frac{-10}{8 \times 0.2^5 - 5} \quad \frac{-10}{8 \times 0.3^5 - 5} \quad \frac{-10}{8 \times 0.4^5 - 5} \right.$

$\left. \frac{-10}{8 \times 0.5^5 - 5} \right] = [2 \quad 2 \quad 2.001 \quad 2.0078 \quad 2.0333 \quad 2.1053]$ (these are the exact values of y up to 4 decimal places). We attempt to solve the differential equation utilizing the subroutine 'ode23', which uses the Runge-Katta 2 and 3 order method. To make the subroutine operational, the given differential equation must be rewritten as $\frac{dy}{dx} = f(x, y)$. Fortunately, the given equation is in that form. Implementation is shown as follows:

MATLAB Command
>>[x y]=ode23('f',[0:0.1:0.5],2); ↵
>>[x y] ↵

ans =

```
     0      2.0000
0.1000      2.0000
0.2000      2.0010
0.3000      2.0078
0.4000      2.0333
0.5000      2.1052
```

It is clear from the return that the first and the second columns of the output are the required x and y values. Except the last return [0.5000   2.1052], the exact and the numerical solutions are identical. As par as the fourth digit is concern, it can be taken as the exact as well. From this comparison, computational power of 'ode23' is easily understood. Attached right box contains the M-file description of the equation $\frac{dy}{dx} = 4x^4 y^2$ (where 'dy' refers to $\frac{dy}{dx}$). The M-file having the name 'f.m' must be present in your working path. The first argument of 'ode23' is 'f', which indicates that the subroutine invokes the M-file containing the differential equation. The second argument [0:0.1:0.5] specifies the required values of x. The third argument 2 represents the value of y at the first element of [0:0.1:0.5].

> **M-file for** $\frac{dy}{dx} = 4x^4 y^2$:
> function dy=f(x,y)
> dy=4*x^4*y^2;

Solve the following first order differential equations numerically:

A. $\frac{du}{dt} = \frac{-t^2 + e^t}{u}$ from $t = 1$ to $t = 1.5$ with step size 0.1 subject to $u(1) = -3$

B. $(x^2 + y^3)\frac{dy}{dx} = x + \sinh y$ from $x = 0$ to $x = 0.3$ with step size 0.05 subject to $y(0) = 5$

⌘ Problem A

The independent and dependent variables are t and u respectively. Apply the separation of variables to have $u\,du = (-t^2 + e^t)\,dt$, then, $\frac{1}{2}u^2 = -\frac{t^3}{3} + e^t + A$ on integration. Initial value $u(1) = -3$ results $A = \frac{29}{6} - e$ so the particular solution is $\frac{1}{2}u^2 = -\frac{t^3}{3} + e^t + \frac{29}{6} - e \Rightarrow u = \pm\sqrt{-\frac{2t^3}{3} + 2e^t + \frac{29}{3} - 2e}$. The given initial condition implies that $-$ve square root should be taken. For $t = 1$, 1.1, 1.2, 1.3, 1.4, and 1.5, we calculate $u =$

$$\left[-\sqrt{-\dfrac{2\times1^3}{3}+2e+\dfrac{29}{3}-2e} \qquad -\sqrt{-\dfrac{2\times1.1^3}{3}+2e^{1.1}+\dfrac{29}{3}-2e} \qquad -\sqrt{-\dfrac{2\times1.2^3}{3}+2e^{1.2}+\dfrac{29}{3}-2e} \right.$$

$$\left. -\sqrt{-\dfrac{2\times1.3^3}{3}+2e^{1.3}+\dfrac{29}{3}-2e} \qquad -\sqrt{-\dfrac{2\times1.4^3}{3}+2e^{1.4}+\dfrac{29}{3}-2e} \qquad -\sqrt{-\dfrac{2\times1.5^3}{3}+2e^{1.5}+\dfrac{29}{3}-2e} \right] = [-3$$

−3.058 −3.1174 −3.1787 −3.2421 −3.3081]. Let us have the MATLAB's return:

MATLAB Command

>>[t u]=ode23('f',[1:.1:1.5],-3); ↵
>>[t u] ↵

ans =

M-file for $\dfrac{du}{dt}=\dfrac{-t^2+e^t}{u}$:

function du=f(t,u)
du=(-t^2+exp(t))/u;

```
    1.0000    -3.0000
    1.1000    -3.0580
    1.2000    -3.1174
    1.3000    -3.1787
    1.4000    -3.2421
    1.5000    -3.3081
```

Even though we changed the dependent and independent variables, 'ode23' can perceive the changes.

⌑ Problem B

Not all first order differential equations have the analytical solution. $(x^2+y^3)\dfrac{dy}{dx}=x+\sinh y$ is such an example. Try to find the analytical solution of $(x^2+y^3)\dfrac{dy}{dx}=x+\sinh y$ in MATLAB:

MATLAB Command

>>dsolve('(x^2+y^3)*Dy=x+sinh(y)','x'); ↵

```
Warning: Explicit solution could not be found.
> In C:\MATLAB\toolbox\symbolic\dsolve.m at line 200
```

As you see, the look-up table of MATLAB does not return any analytical solution. Numerical method may not disappoint you but for that, the equation should be rewritten as $\dfrac{dy}{dx}=\dfrac{x+\sinh y}{x^2+y^3}$. Let us investigate that:

MATLAB Command

>>[x y]=ode23('f',[0:.05:0.3],5); ↵
>>[x y] ↵

ans =

M-file for $(x^2+y^3)\dfrac{dy}{dx}=x+\sinh y$:

function dy=f(x,y)
dy=(x+sinh(y))/(x^2+y^3);

```
         0     5.0000
    0.0500     5.0299
    0.1000     5.0601
    0.1500     5.0908
    0.2000     5.1218
    0.2500     5.1533
    0.3000     5.1852
```

It goes without saying that the numerical solution is possible. The only thing is we can not compare the output with the exact value. Having verified the other two equations, we have the reason to rely on the subroutine 'ode23'.

308

7.3.2 Second order differential equations

The subroutine 'ode23' can also solve numerically a second order equation of the form $\frac{d^2y}{dx^2} = f\left(x, y, \frac{dy}{dx}\right)$. Some mathematical juggling is associated with the M-file description of the differential

equation. To elucidate that, consider the second order nonhomogeneous equation $2\frac{d^2y}{dx^2} - 5\frac{dy}{dx} + 2y = x$ subject

to the initial conditions $\begin{Bmatrix} y(0) = 2 \\ y'(0) = -3 \end{Bmatrix}$. Having gone through the previous articles, we are able to find the

analytical solution using the subroutine 'dsolve'.

MATLAB Command
>>S=dsolve('2*D2y-5*Dy+2*y=x','y(0)=2,Dy(0)=-3','x'); ↵
>>pretty(S) ↵

```
                        31
    5/4 + 1/2 x - -- exp(2 x) + 10/3 exp(1/2 x)
                        12
```

From MATLAB output, the solution subject to the given initial

conditions is $y(x) = \frac{5}{4} + \frac{x}{2} - \frac{31}{12}e^{2x} + \frac{10}{3}e^{\frac{x}{2}}$. Rewrite the example

equation as $\frac{d^2y}{dx^2} = \frac{5}{2}\frac{dy}{dx} - y + \frac{x}{2}$. The so described mathematical

juggling is imperative to convert the given equation as a system of
two first order differential equations by making the substitution

$y_1 = y$ and $y_2 = \frac{dy}{dx}$. We have $\frac{dy_1}{dx} = \frac{dy}{dx} = y_2$ and $\frac{dy_2}{dx} = \frac{d^2y}{dx^2} = \frac{5}{2}\frac{dy}{dx} -$

> **M-file for** $2\frac{d^2y}{dx^2} - 5\frac{dy}{dx} + 2y = x$:
>
> function dy=f(x,y)
> dy(1)=y(2);
> dy(2)=5/2*y(2)-y(1)+x/2;
> dy=dy';

$y + \frac{x}{2} = \frac{5}{2}y_2 - y_1 + \frac{x}{2}$ thereby obtaining the system of differential equations as $\begin{Bmatrix} \frac{dy_1}{dx} = y_2 \\ \frac{dy_2}{dx} = \frac{5}{2}y_2 - y_1 + \frac{x}{2} \end{Bmatrix}$. Pay

attention to the first line of the M-file function of the preceding article, which is 'function dy=f(x,y)'. This time

the assignee 'dy' contains the system's derivatives $\begin{bmatrix} & \text{so, it is a two-element column vector, where dy(1)=}\frac{dy_1}{dx} \end{bmatrix}$

and dy(2)=$\frac{dy_2}{dx}$ $\Big]$. Accordingly, the argument 'y' of 'f(x,y)' is also a two-element vector, on MATLAB coding,

y(1)≡ y_1 and y(2)≡ y_2. Anyhow, the complete function file is shown in the above attached box. Referring to the
last line of the function file, MATLAB forms the matrix 'dy' as a row one by default but the subroutine needs a
column vector to be executable. For that reason dy is transposed to the column matrix by the command dy'.
Until now we did point out what the interval of x is. Let us assume that the numerical solution from $x = 0$ to
$x = 0.5$ with the step size 0.1 is necessary. The analytical expression of $y(x)$ has been found before, which is

$y(x) = \frac{5}{4} + \frac{x}{2} - \frac{31}{12}e^{2x} + \frac{10}{3}e^{\frac{x}{2}}$. The computed values of $y(x)$ as a row matrix for $x = 0, 0.1, 0.2, 0.3, 0.4$, and 0.5

are as follows: $\begin{bmatrix} \frac{5}{4} + 0 - \frac{31}{12} + \frac{10}{3} & \frac{5}{4} + \frac{0.1}{2} - \frac{31}{12}e^{0.2} + \frac{10}{3}e^{0.05} & \frac{5}{4} + \frac{0.2}{2} - \frac{31}{12}e^{0.4} + \frac{10}{3}e^{0.1} \end{bmatrix}$

$\frac{5}{4} + \frac{0.3}{2} - \frac{31}{12}e^{0.6} + \frac{10}{3}e^{0.15} \quad \frac{5}{4} + \frac{0.4}{2} - \frac{31}{12}e^{0.8} + \frac{10}{3}e^{0.2} \quad \frac{5}{4} + \frac{0.5}{2} - \frac{31}{12}e^{1} + \frac{10}{3}e^{0.25} \Big] = [2 \quad 1.6489 \quad 1.18 \quad 0.5656$

$-0.228 \quad -1.2421]$. The subroutine is also capable of returning the numerical $\frac{dy}{dx}$ for the same interval of x

when solving a second order equation. Now the derivative of $y(x)$ is $\dfrac{dy(x)}{dx} = \dfrac{1}{2} - \dfrac{31}{6}e^{2x} + \dfrac{5}{3}e^{\frac{x}{2}}$ which has the

value as a row matrix $\left[\begin{array}{ccccc} \dfrac{1}{2} - \dfrac{31}{6} + \dfrac{5}{3} & \dfrac{1}{2} - \dfrac{31}{6}e^{0.2} + \dfrac{5}{3}e^{0.05} & \dfrac{1}{2} - \dfrac{31}{6}e^{0.4} + \dfrac{5}{3}e^{0.1} & \dfrac{1}{2} - \dfrac{31}{6}e^{0.6} + \dfrac{5}{3}e^{0.15} \end{array} \right.$

$\dfrac{1}{2} - \dfrac{31}{6}e^{0.8} + \dfrac{5}{3}e^{0.2} \quad \left. \dfrac{1}{2} - \dfrac{31}{6}e^{1} + \dfrac{5}{3}e^{0.25} \right] = [-3 \quad -4.0585 \quad -5.3658 \quad -6.9779 \quad -8.963 \quad -11.4044]$ for $x = 0$,

0.1, 0.2, 0.3, 0.4, and 0.5 respectively. Before applying the numerical subroutine, the exact values of $y(x)$ and

$\dfrac{dy}{dx}$ are at the hand. Let us have the numerical output by virtue of 'ode23':

> **MATLAB Command**
> >>[x y]=ode23('f',[0:0.1:0.5],[2 -3]); ↵
> >>[x y] ↵

```
ans =

        0      2.0000    -3.0000
   0.1000      1.6490    -4.0584
   0.2000      1.1801    -5.3657
   0.3000      0.5657    -6.9777
   0.4000     -0.2278    -8.9626
   0.5000     -1.2419   -11.4039
```

The first, second, and third columns of MATLAB return correspond to the required x variation, numerical $y(x)$, and numerical $\dfrac{dy}{dx}$ respectively. It is conclusive that the output argument y of 'ode23' has two columns – the first and second of which are $y(x)$ and $\dfrac{dy}{dx}$ respectively. The third input argument of 'ode23' is [2 −3], which contains the initial values of $y(x)$ and $\dfrac{dy}{dx}$ respectively. Once again, the initial values of $y(x)$ and $\dfrac{dy}{dx}$ must correspond to the first element of the second input argument [0:0.1:0.5].

7.3.3 Higher order differential equations

The techniques employed in the previous articles can easily be extended to higher order equations. We demonstrate that on taking the third order Euler differential equation $6t^3 \dfrac{d^3u}{dt^3} - 7t^2 \dfrac{d^2u}{dt^2} + 9t \dfrac{du}{dt} - 4u = 3t - 3$

under the initial conditions $\begin{cases} u''(1) = 1 \\ u'(1) = 3 \\ u(1) = 0 \end{cases}$. Notice that the dependent variable is u. Following the method for the

Euler differential equation described in article 7.2.4, we have the particular solution of the differential equation

$u(t) = \left(\dfrac{1017}{484} - \dfrac{27}{22} \ln t \right) t^2 + \dfrac{3}{5}t - \dfrac{2088}{605}t^{\frac{1}{6}} + \dfrac{3}{4}$. It is the notability of the subroutine that the first and the second

numerical derivatives (since the order is three) of $u(t)$ will also be returned, hence, $\dfrac{du(t)}{dt} =$

$\left(\dfrac{360}{121} - \dfrac{27}{11} \ln t \right) t + \dfrac{3}{5} - \dfrac{348}{605}t^{-\frac{5}{6}}$ and $\dfrac{d^2u(t)}{dt^2} = \dfrac{63}{121} + \dfrac{58}{121}t^{-\frac{11}{6}} - \dfrac{27}{11} \ln t$. For implementation, compute $u(t)$, $\dfrac{du(t)}{dt}$,

and $\dfrac{d^2u(t)}{dt^2}$ from $t = 1$ to $t = 1.25$ with the step 0.05. The computation of the exact values is as follows: $u(t) \Rightarrow$

$\left[\left(\dfrac{1017}{484} - \dfrac{27}{22} \ln 1 \right) + \dfrac{3}{5} - \dfrac{2088}{605} + \dfrac{3}{4} \qquad \left(\dfrac{1017}{484} - \dfrac{27}{22} \ln 1.05 \right) 1.05^2 + \dfrac{3}{5} \times 1.05 - \dfrac{2088}{605} \times 1.05^{\frac{1}{6}} + \dfrac{3}{4} \right.$

$$\left(\frac{1017}{484}-\frac{27}{22}\ln 1.1\right)1.1^2+\frac{3}{5}\times 1.1-\frac{2088}{605}\times 1.1^{\frac{1}{6}}+\frac{3}{4}$$

$$\left(\frac{1017}{484}-\frac{27}{22}\ln 1.15\right)1.15^2+\frac{3}{5}\times 1.15-\frac{2088}{605}\times 1.15^{\frac{1}{6}}+\frac{3}{4}$$

$$\left(\frac{1017}{484}-\frac{27}{22}\ln 1.2\right)1.2^2+\frac{3}{5}\times 1.2-\frac{2088}{605}\times 1.2^{\frac{1}{6}}+\frac{3}{4}$$

$$\left(\frac{1017}{484}-\frac{27}{22}\ln 1.25\right)1.25^2+\frac{3}{5}\times 1.25-\frac{2088}{605}\times 1.25^{\frac{1}{6}}+\frac{3}{4}\Bigg]=[0$$

0.1512   0.3045   0.4595   0.6159   0.7733], $\dfrac{du(t)}{dt}\Rightarrow\Bigg[\left(\dfrac{360}{121}-\dfrac{27}{11}\ln 1\right)+\dfrac{3}{5}-\dfrac{348}{605}$

$$\left(\frac{360}{121}-\frac{27}{11}\ln 1.05\right)1.05+\frac{3}{5}-\frac{348}{605}\times 1.05^{-\frac{5}{6}}$$

$$\left(\frac{360}{121}-\frac{27}{11}\ln 1.1\right)1.1+\frac{3}{5}-\frac{348}{605}\times 1.1^{-\frac{5}{6}}$$

$$\left(\frac{360}{121}-\frac{27}{11}\ln 1.15\right)1.15+\frac{3}{5}-\frac{348}{605}\times 1.15^{-\frac{5}{6}}$$

$$\left(\frac{360}{121}-\frac{27}{11}\ln 1.2\right)1.2+\frac{3}{5}-\frac{348}{605}\times 1.2^{-\frac{5}{6}}$$

$$\left(\frac{360}{121}-\frac{27}{11}\ln 1.25\right)1.25+\frac{3}{5}-\frac{348}{605}\times 1.25^{-\frac{5}{6}}\Bigg]=[3 \quad 3.0459 \quad 3.0841 \quad 3.115 \quad 3.1391 \quad 3.1568],\ \text{and}\ \frac{d^2u(t)}{dt^2}\Rightarrow$$

$$\left[\frac{63}{121}+\frac{58}{121}-\frac{27}{11}\ln 1 \qquad \frac{63}{121}+\frac{58}{121}\times 1.05^{-\frac{11}{6}}-\frac{27}{11}\ln 1.05 \qquad \frac{63}{121}+\frac{58}{121}\times 1.1^{-\frac{11}{6}}-\frac{27}{11}\ln 1.1\right.$$

$$\frac{63}{121}+\frac{58}{121}\times 1.15^{-\frac{11}{6}}-\frac{27}{11}\ln 1.15 \qquad \frac{63}{121}+\frac{58}{121}\times 1.2^{-\frac{11}{6}}-\frac{27}{11}\ln 1.2 \qquad \frac{63}{121}+\frac{58}{121}\times 1.25^{-\frac{11}{6}}-\frac{27}{11}\ln 1.25\Bigg]=[1 \quad 0.8392$$

0.6892   0.5486   0.4163   0.2913]. Turn attention to the numerical solution. As indicated in the last article, perform some substitutions like $u_1=u$, $u_2=\dfrac{du}{dt}$, and $u_3=\dfrac{d^2u}{dt^2}$, on that, $\dfrac{du_1}{dt}=\dfrac{du}{dt}=u_2$ and $\dfrac{du_2}{dt}=\dfrac{d^2u}{dt^2}=u_3$.

Rewrite $6t^3\dfrac{d^3u}{dt^3}-7t^2\dfrac{d^2u}{dt^2}+9t\dfrac{du}{dt}-4u=3t-3$ to have $\dfrac{d^3u}{dt^3}=\dfrac{3t-3+7t^2\dfrac{d^2u}{dt^2}-9t\dfrac{du}{dt}+4u}{6\,t^3}=\dfrac{3t-3+7t^2u_3-9tu_2+4u_1}{6\,t^3}$, thereby, providing the system of differential equations pertaing to u_1, u_2, and u_3 as

$$\left.\begin{cases}\dfrac{du_1}{dt}=u_2 \\[2mm] \dfrac{du_2}{dt}=u_3 \\[2mm] \dfrac{du_3}{dt}=\dfrac{3t-3+7t^2u_3-9tu_2+4u_1}{6\,t^3}\end{cases}\right\}$$. Give the MATLAB codes to the system. The dependent variables u_1, u_2, and

u_3 and the derivatives $\dfrac{du_1}{dt}$, $\dfrac{du_2}{dt}$, and $\dfrac{du_3}{dt}$ are inserted by u(1), u(2), and u(3) and du(1), du(2), and du(3) respectively. The expression $\dfrac{3t-3+7t^2u_3-9tu_2+4u_1}{6\,t^3}$ is stringed as '(3*t-3+7*t^2*u(3)-9*t*u(2)+4*u(1))/t^3/6'. Following is the complete codes.

M-file for $6t^3\dfrac{d^3u}{dt^3}-7t^2\dfrac{d^2u}{dt^2}+9t\dfrac{du}{dt}-4u=3t-3$:

```
function du=f(t,u)
du(1)=u(2);
du(2)=u(3);
du(3)=(3*t-3+7*t^2*u(3)-9*t*u(2)+4*u(1))/t^3/6;
du=du';
```

Our immediate concern is to see the response of 'ode23' for the given third order Euler differential equation:

MATLAB Command

```
>>[t u]=ode23('f',[1:0.05:1.25],[0 3 1]); ↵
>>[t u] ↵
```

```
ans =

    1.0000         0    3.0000    1.0000
    1.0500    0.1512    3.0459    0.8392
    1.1000    0.3045    3.0841    0.6892
    1.1500    0.4595    3.1150    0.5486
    1.2000    0.6159    3.1391    0.4163
    1.2500    0.7733    3.1568    0.2913
```

$$\uparrow \qquad \uparrow \qquad \uparrow \qquad \uparrow$$

$$t \qquad u \qquad \frac{du}{dt} \qquad \frac{d^2u}{dt^2}$$

On comparison of the exact with the numerical solutions, it is obvious that the accuracy of numerical computation is greatly appreciated.

7.3.4 Some factors to be considered for the solution

Inserting the input and obtaining the output for all ODE subroutines presented in table 7.A are executed in the same way as we did for 'ode23' in the preceding articles. Of coarse, the higher order approximation of the dependent variable would give the result that is close to the exact. We know that the Runge-Kutta 4-5 order (whose MATLAB counterpart is 'ode45') is more accurate than the Runge-Kutta 2-3 ('ode23'). Computationally we proved that in this section. Pick up the example of article 7.3.2. On exercising 'ode23', we found the following:

| | $y(x)$ | | $\frac{dy}{dx}$ | |
x	Exact	Numerical	Exact	Numerical
0	2.0000	2.0000	-3.0000	-3.0000
0.1000	1.6489	1.6490	-4.0585	-4.0584
0.2000	1.1800	1.1801	-5.3658	-5.3657
0.3000	0.5656	0.5657	-6.9779	-6.9777
0.4000	-0.2280	-0.2278	-8.9630	-8.9626
0.5000	-1.2421	-1.2419	-11.4044	-11.4039

One can see that there is some deviation of the numerical computation from the exact one. Solve the equation with 'ode45'. The response is as follows:

MATLAB Command

```
>>[x y]=ode45('f',[0:0.1:0.5],[2 -3]); ↵
>>[x y] ↵

ans =

         0    2.0000   -3.0000
    0.1000    1.6489   -4.0585
    0.2000    1.1800   -5.3658
    0.3000    0.5656   -6.9779
    0.4000   -0.2280   -8.9630
    0.5000   -1.2421  -11.4044
```

As you see, the numerical output by making the use of 'ode45' becomes exact. This might not be true always because the choice of relative or absolute error is a factor of accuracy too. The default relative error of 'ode23' is 10^{-3}. What if one tries to find the solution with the relative error 10^{-6}. To argument the relative error, another subroutine 'odeset' is used. The relative error is a property of the differential equation solver, which is notified by 'RelTol' (abbreviation of Relative Tolerance). The execution is as follows:

MATLAB Command

      >>O=odeset('RelTol',1e-6); ↵   ← User's option is assigned to O

      >>[x y]=ode23('f',[0:0.1:0.5],[2 -3],O); ↵   ← O is the last argument of 'ode23'

      >>[x y] ↵

```
ans =

         0      2.0000     -3.0000
    0.1000      1.6489     -4.0585
    0.2000      1.1800     -5.3658
    0.3000      0.5656     -6.9779
    0.4000     -0.2280     -8.9630
    0.5000     -1.2421    -11.4044
```

Reducing the relative error from 10^{-3} to 10^{-6} makes the solution exact. Either increasing the approximation order or reducing the relative error can yield the better solution. There are instances when the numerical method would fail. Referring to article 7.3.1, we solved $\frac{dy}{dx} = 4x^4 y^2$ from interval $x=0$ to $x=0.5$. Try to solve the equation from $x=0$ to $x=1.5$ with the same step size:

      MATLAB Command

          >>[x y]=ode23('f',[0:0.1:1.5],2); ↵

      Warning: Failure at t=9.105613e-001. Unable to meet integration tolerances without reducing the step size below the smallest value allowed (3.234963e-015) at time t.

      > In C:\MATLAB\toolbox\matlab\funfun\ode23.m at line 334

We know the analytical solution, which is $y = \dfrac{-10}{8x^5 - 5}$. Setting the denominator $8x^5 - 5$ to 0 provides $x = \sqrt[5]{\dfrac{5}{8}} = 0.9103$. At $x = \sqrt[5]{\dfrac{5}{8}}$, y becomes undefined that is why the above warning message is appearing. Over the required interval of x, there must not be any x at which y is undefined.

      We desist the session of the ordinary differential equations with this. More than a dozen of options can be penned in to each solver. Execute 'odeset' to have the online help about the other options. The simplest problems are outlined as they happen to be in various academics. Thus the approach of having rudimentary concept is substantiated. In next article we briefly bring an introduction to implement the partial differential equations.

7.4 Solving the partial differential equations

      We have seen in the ordinary differential equations that the independent variable is only one. There are some physical process and laws, whose differential equations need to involve two or more independent variables. For instance, an electromagnetic wave is a function of distance and time as well. Likewise the dependent variable has more than one derivative, thereby, requiring partial derivatives. According to notation, if z is a function of two independent variables, that is $z = f(x, y)$, its different partial derivatives are denoted by $\frac{\partial z}{\partial x}$ (first order w.r.to x), $\frac{\partial z}{\partial y}$ (first order w.r.to y), $\frac{\partial^2 z}{\partial y^2}$ (second order w.r.to y), $\frac{\partial^2 z}{\partial y \partial x}$ (second order w.r.to first y, then x), ...etc. Any differential equation containing the partial derivative is termed as the partial differential equation. By resemblance with the theory of ordinary differential equation, the order of a partial differential equation is the highest order partial differential coefficient occurring in it. Thus,

$4y^2 \dfrac{\partial z}{\partial x} + \dfrac{\partial z}{\partial y} = x + y$ is a first order partial differential equation when z is a function of x and y,

$4v^2 \dfrac{\partial^2 f}{\partial u^2} + \dfrac{\partial f}{\partial v} + \dfrac{\partial f}{\partial w} = u \dfrac{\partial^2 f}{\partial v \partial u}$ is a second order partial differential equation when f is a function of u, v, and w, and

$\dfrac{\partial^3 g}{\partial p^3} = \left(\dfrac{\partial g}{\partial q}\right)^2 + p + q$ is a third order partial differential equation when g is a function of p and q.

Solutions of the partial differential equations encounter a much more difficult problem than the solutions of the ordinary differential equations do except certain linear or nonlinear partial differential equations. In Maple package, the provision for solving the partial differential equations has been accounted for. A great variety of partial differential equations occurring in physics, chemistry, or engineering can be solved analytically following the symbolism of MATLAB. Briefly, we outline the handling of several known types of partial differential equations in the following sections.

7.4.1 First order linear partial differential equations

If z is a function of two independent variables – x and y, a first order linear partial differential equation is given by $P(x,y,z) \dfrac{\partial z}{\partial x} + Q(x,y,z) \dfrac{\partial z}{\partial y} = R(x,y,z)$. The equation is called Lagrange's equation. The general solution of the Lagrange's equation is $F(u, v) = 0$, where F is an arbitrary function and $u(x,y,z) = c_1$ and $v(x,y,z) = c_2$ (c_1 and c_2 are arbitrary constants) are obtained from the solution of the equations $\dfrac{dx}{P(x,y,z)} = \dfrac{dy}{Q(x,y,z)} = \dfrac{dz}{R(x,y,z)}$ (called the auxiliary equation). To get more insight, solve the first order partial differential equation $x \dfrac{\partial z}{\partial x} + y \dfrac{\partial z}{\partial y} = 3(x+y)z$. On comparison, one can write $P(x,y,z) = x$, $Q(x,y,z) = y$, and $R(x,y,z) = 3(x+y)z$, from which, $\dfrac{dx}{x} = \dfrac{dy}{y} = \dfrac{dz}{3(x+y)z}$. The first two terms provide $\dfrac{dx}{x} = \dfrac{dy}{y}$ $\Rightarrow$ $\ln y = \ln x + A$ (integration) $\Rightarrow$ $\dfrac{y}{x} = e^A = C_1$. Exercise the ratio rule to write $\dfrac{dx+dy}{x+y} = \dfrac{dz}{3(x+y)z}$ $\Rightarrow$ $3 d(x+y) = \dfrac{dz}{z}$ $\Rightarrow$ $\ln z = 3(x+y) + A$ (integration) $\Rightarrow$ $z = e^{3(x+y)} e^B = C_2 e^{3(x+y)}$ $\Rightarrow$ $C_2 = \dfrac{z}{e^{3(x+y)}}$. The general solution of the partial differential equation is $F(C_1, C_2) = 0$ $\Rightarrow$ $F\left(\dfrac{y}{x}, \dfrac{z}{e^{3(x+y)}}\right) = 0$, where F is an arbitrary function and it can be written as $\dfrac{z}{e^{3(x+y)}} = h\left(\dfrac{y}{x}\right)$ (h is another arbitrary function). Hence the general solution becomes $z = e^{3(x+y)} h\left(\dfrac{y}{x}\right)$. The next agenda is how one can give the MATLAB codes to the partial differential equations. The dependent variable z and the first partial derivatives $\dfrac{\partial z}{\partial x}$ and $\dfrac{\partial z}{\partial y}$ are coded by 'f(x,y)' and 'diff(f(x,y),x)' and 'diff(f(x,y),y)' respectively. The entire equation is put as a string. For the partial differential equation, we are going to use the Maple function 'pdesolve' (abbreviation of partial differential equation solve). The implementation is as follows:

MATLAB Command
```
>>maple('e:=x*diff(f(x,y),x)+y*diff(f(x,y),y)=3*(x+y)*f(x,y)');  ↵
```

The equation is assigned to the variable e. The assignment operator is ':=' in Maple terminology.

```
>>maple('S:=pdesolve(e,f(x,y))');  ↵
```

The solution of the partial differential equation is assigned to the variable S. The second argument 'f(x,y)' of 'pdesolve' says that the dependent variable is $f(x,y)$.

>>maple('interface(prettyprint=1):simplify(S)'); ↵

```
F(x, y) = _F1(y/x) exp(3 x + 3 y)
```

The command 'interface(prettyprint=1)' of Maple executes a two dimensional formatter, which displays an expression or equation as close as mathematical form (similar to the command pretty of symbolic toolbox). Also notice that suppressing a Maple output is carried out by ':'. A Maple expression or equation is simplified by the command 'simplify'. Anyhow, '_F1(y/x)' means the arbitrary function $h\left(\dfrac{y}{x}\right)$. At this point, arbitrary means any function, which is a function of $\dfrac{y}{x}$, for example, $\cos\left(\dfrac{y}{x}\right)$, $\dfrac{y}{x}+\left(\dfrac{y}{x}\right)^2$ ... etc.

As another example, solve the partial differential equation $\dfrac{\partial u}{\partial \theta}=-5\sin\theta\dfrac{\partial u}{\partial r}+3\cos\theta$. Now the dependent and independent variables are u (function of r and θ) and r and θ respectively. Not having the provision of writing θ, t is used instead. Rearrange the equation to write $5\sin\theta\dfrac{\partial u}{\partial r}+\dfrac{\partial u}{\partial \theta}=3\cos\theta$. The auxiliary equation is $\dfrac{dr}{5\sin\theta}=\dfrac{d\theta}{1}=\dfrac{du}{3\cos\theta}$, from which, $\dfrac{dr}{5\sin\theta}=\dfrac{d\theta}{1}$ $\Rightarrow$ $r=-5\cos\theta+C_1$ $\Rightarrow$ $\theta=\cos^{-1}\dfrac{C_1-r}{5}$, $\dfrac{dr}{5\sin\theta}=$ $\dfrac{du}{3\cos\theta}$ $\Rightarrow$ $du=\dfrac{3}{5}\cot\theta\,dr$ $\Rightarrow$ $du=\dfrac{3}{5}\cot\cos^{-1}\dfrac{C_1-r}{5}\,dr=\dfrac{3}{5}\dfrac{C_1-r}{\sqrt{25-(C_1-r)^2}}\,dr$ $\Rightarrow$ $u=\dfrac{3}{5}\sqrt{25-(C_1-r)^2}+C_2$ $\Rightarrow$ $u-\dfrac{3}{5}\sqrt{25-r^2+2rC_1-C_1^2}=C_2$ $\Rightarrow$ $u-\dfrac{3}{5}\sqrt{25-r^2+2r(r+5\cos\theta)-(r+5\cos\theta)^2}=C_2$. So, the general solution of the equation is given by $F(C_1,C_2)=0$ (F is arbitrary function) $\Rightarrow$ $C_2=h(C_1)$ (h is another arbitrary function) $\Rightarrow$ $u-\dfrac{3}{5}\sqrt{25-r^2+2r(r+5\cos\theta)-(r+5\cos\theta)^2}=h(r+5\cos\theta)$ $\Rightarrow$ $u(r,\theta)=\dfrac{3}{5}\sqrt{25-r^2+2r(r+5\cos\theta)-(r+5\cos\theta)^2}$ $+h(r+5\cos\theta)$ that is what we are after. The string of the equation $\dfrac{\partial u}{\partial \theta}=-5\sin\theta\dfrac{\partial u}{\partial r}+3\cos\theta$ is 'diff(u(r,t),t)=-5*sin(t)*diff(u(r,t),r)+3*cos(t)'. Following is the implementation:

MATLAB Command
>>maple('S:=pdesolve(diff(u(r,t),t)=-5*sin(t)*diff(u(r,t),r)+3*cos(t), u(r,t))'); ↵
>>maple('interface(prettyprint=1):S'); ↵

```
                        2                              2 1/2
u(r, t) = 3/5 (25 - r  + 2 r (5 cos(t) + r) - (5 cos(t) + r) )

        + _F1(5 cos(t) + r)
```

7.4.2 Second order linear partial differential equations

We consider now the solutions of a second order linear partial differential equation with constant coefficients. Such an equation can be written in the form $U(D_x,D_y)z=g(x,y)$, where $D_x=\dfrac{\partial}{\partial x}$, $D_y=\dfrac{\partial}{\partial y}$, the dependent variable z is the function of independent variables x and y, and $g(x,y)$ is some other function of x and y. The corresponding homogeneous linear part is $U(D_x,D_y)z=0$.

Solve the equation $2\dfrac{\partial^2 z}{\partial x^2}+7\dfrac{\partial^2 z}{\partial x\partial y}+6\dfrac{\partial^2 z}{\partial y^2}+\dfrac{\partial z}{\partial x}+2\dfrac{\partial z}{\partial y}=0$. In factored form, one can write the equation as $\left(\dfrac{\partial}{\partial x}+2\dfrac{\partial}{\partial y}\right)\left(2\dfrac{\partial}{\partial x}+3\dfrac{\partial}{\partial y}+1\right)z=0$. Apply the just mentioned operator to write $(D_x+2D_y)(2D_x+3D_y+1)z=0$. A theorem pertaining to the partial differential equation of two independent variables x and y states that if aD_x+bD_y+c (a, b, and c are constants) is a factor of $U(D_x,D_y)z$, then, the complementary solution of $U(D_x,D_y)z=0$ is given by $e^{-\frac{cx}{a}}f(ay-bx)$. Utilize the theorem to have the general solution of the equation as $z=f_1(y-2x)+e^{-\frac{x}{2}}f_2(2y-3x)$. MATLAB solution is as follows:

MATLAB Command

```
>>maple('z:=f(x,y):'); ↵
>>maple('e:=2*diff(z,x$2)+7*diff(z,x,y)+6*diff(z,y$2)+diff(z,x)+2*diff(z,y)=0'); ↵
>>O=maple('pdesolve(e,f(x,y))'); ↵
>>pretty(sym(O)) ↵
```

```
f(x, y) = _F1(y - 2 x) + _F2(2 y - 3 x) exp(- 1/2 x)
```

Instead of typing 'f(x,y)' repeatedly, we assign that to z in the first line of the command. There is smarter way of writing $\dfrac{\partial^2 z}{\partial x^2}$ in Maple, which is 'diff(z,x$2)'. The subroutine is applied to the assigned equation to e and the output is put to the variable O. The command 'sym' is used for the symbolic conversion. Since the equation is a second order one, there should be two arbitrary functions, which are returned as _F1 and _F2.

Next consider the example of the nonhomogeneous equation $4\dfrac{\partial^2 z}{\partial x^2}-9\dfrac{\partial^2 z}{\partial y^2}=x^2-y$. Use the operator to write it as $(4D_x^2-9D_y^2)z=x^2-y$, then, factorize to get $(2D_x+3D_y)(2D_x-3D_y)z=x^2-y$. So, the complementary function is $f_1(2y+3x)+f_2(2y-3x)$. For particular integral, we can write $z_{PI}=\dfrac{x^2-y}{(4D_x^2-9D_y^2)}=\dfrac{1}{4D_x^2}\left(1-\dfrac{9D_y^2}{4D_x^2}\right)^{-1}(x^2-y)=\dfrac{1}{4D_x^2}(x^2-y)$ (the binomial expansion, keeping the necessary operators, and $\dfrac{1}{D_x}$ means the integration w.r.to x)$=\dfrac{1}{4}\left(\dfrac{1}{12}x^4-\dfrac{1}{2}x^2y\right)=\dfrac{1}{48}x^4-\dfrac{1}{8}x^2y$. Finally, the general solution is $z=f_1(2y+3x)+f_2(2y-3x)+\dfrac{1}{48}x^4-\dfrac{1}{8}x^2y$. The response of MATLAB is as follows:

MATLAB Command

```
>>maple('e:=4*diff(f(x,y),x$2)-9*diff(f(x,y),y$2)=x^2-y'); ↵
>>O=maple('pdesolve(e,f(x,y))'); ↵
>>pretty(sym(O)) ↵
```

```
              4         2
f(x, y) = 1/48 x  - 1/8 x  y + _F1(2 y + 3 x) + _F2(2 y - 3 x)
```

Anyhow we bring to an end the chapter with this example. The partial differential equation solver subroutine still can not handle all known partial differential equations. We hope that it will be more expanded like 'dsolve' in the future. Apart from the 'pdesolve' subroutine, there is a differential package. Have the online help executing 'mhelp DEtools'. Of coarse, this is Maple package and the package should be installed in your system.

Chapter 8

Problems of Fourier, Laplace, and Z Transforms

Integral transforms are important tools of applied mathematics. Popular transforms that are extensively used are Fourier, Laplace, and Z transforms. Fourier transform of a nonperiodic function $f(t)$ provides a continuous frequency resolution, which is called the frequency spectrum $F(\omega)$. $F(\omega)$ is complex in nature. A bounded condition is imposed on $f(t)$ in applying the Fourier transform, which necessitates the convergence of $\int\limits_{t=-\infty}^{t=\infty}|f(t)|\,dt$.

Different transform methods are employed for the analysis of linear systems. Fourier and Laplace transforms can be utilized to solve the problems of linear systems involving differential, integral, and/or integrodifferential equations. The nature of t-domain, when a function is a function of t, and the nature of excitation function determine the choice of transform methods. Laplace transform is especially suitable in solving constant coefficient linear ordinary differential equations where the excitation functions are discontinuous and have corners. A function f(t), which exists in t-domain, can be transformed into a function of s, F(s), by Laplace transform. Initial conditions are essential to find the s-domain equations of a linear system. A linear differential equation becomes an algebraic equation in s-domain. The analyzing function f(t) can also be a discrete one. Z transform is the discrete counterpart of Laplace transform.

8.1 Fourier transform

Fourier transform of a function $f(t)$ is defined as $F(\omega)=\int\limits_{t=-\infty}^{t=\infty}e^{-j\omega t}f(t)\,dt$. The transform is denoted by the operator $'F'$ so one can write $F(\omega)=F\ [f(t)]=\int\limits_{t=-\infty}^{t=\infty}e^{-j\omega t}f(t)\,dt$. The transform $F(\omega)$ is a complex function of ω (omega). IN MATLAB letter 'w' corresponds to the frequency variable ω. Independent variable of the function $f(t)$ is t. Other variable like x can also be the independent variable, then, $F(\omega)=\int\limits_{x=-\infty}^{x=\infty}e^{-j\omega x}f(x)\,dx$. All functions do not have the Fourier transforms (for example, t^t or e^{t^2}). Finding $F(\omega)$ is possible if $\int\limits_{t=-\infty}^{t=\infty}|f(t)|\,dt$ is finite. The MATLAB counterpart of Fourier transform is 'fourier'. The default return of the subroutine 'fourier' is in terms of w and in the string form. The output variable can also be others instead of w. The independent variable must be declared symbolically. To take the Fourier transform of a function,

MATLAB Step:
Use the command fourier(function in string form, independent variable, transform variable).

8.1.1 Fourier transforms of some functions

Let us see the Fourier transform of $f(t) = e^{-|t|}$. Before we apply the transform, check the convergence. $\int_{t=-\infty}^{t=\infty} |f(t)| dt = \int_{t=-\infty}^{t=\infty} e^{-|t|} dt = 2\int_{t=0}^{t=\infty} e^{-t} dt = 2$, which is a finite value so the transform is possible and it is obtained as

$$F[e^{-|t|}] = \int_{t=-\infty}^{t=\infty} e^{-j\omega t} f(t) dt = \int_{t=-\infty}^{t=\infty} e^{-j\omega t} e^{-|t|} dt = \int_{t=-\infty}^{t=0} e^{t} e^{-i\omega t} dt + \int_{t=0}^{t=\infty} e^{-t} e^{-i\omega t} dt = \left[\frac{e^{t(1-i\omega)}}{1-i\omega}\right]_{t=-\infty}^{t=0} + \left[\frac{-e^{-t(i\omega+1)}}{1+i\omega}\right]_{t=0}^{t=\infty} = \frac{1}{1-i\omega} + \frac{1}{1+i\omega} =$$

$\frac{2}{(1-i\omega)(1+i\omega)} = \frac{2}{1+\omega^2}$. The string form of the function $e^{-|t|}$ is 'exp(−abs(t))'. When implemented in MATLAB, we have the output of 'fourier' assigned to F as follows:

MATLAB Command

for string form,

```
>>syms t ↵
>>F=fourier(exp(-abs(t))) ↵

F =

2/(1+w^2)
```

for symbolic form,

```
>>pretty(F) ↵

      2
  ------------
        2
    1 + w
```

When the independent variable of the given function is x, i.e., $f(x) = e^{-|x|}$:

```
>>syms x ↵
>>F=fourier(exp(-abs(x))) ↵

F =

2/(1+w^2)          ← Still the transform is in terms of w
```

When the transform variable is required from w to z,

```
>>syms x z ↵
>>F=fourier(exp(-abs(x)),x,z) ↵

F =

2/(1+z^2)          ← That is what is expected
```

Ongoing tutorial gives you the idea of Fourier transform implementation in MATLAB. Table 8.A presents some functions' Fourier transforms and their MATLAB counterparts. Let us see the response of 'fourier' for a function which does not have the transform, for example, e^{t^2}. The string form of this function is 'exp(t^2)':

```
>>syms t ↵
>>F=fourier(exp(t^2)) ↵

F =

fourier(exp(t^2),t,w)     ← The ransform is not in the look-up table of MATLAB
                            and the return is just the definition.
```

Table 8.A Fourier transforms of some functions and their MATLAB counterparts

Mathematical form	MATLAB Command for symbolic form
$f(t) = e^{-\|at\|}$, where $a > 0$, $$F[f(t)] = \frac{2a}{a^2 + \omega^2}$$	```\n>>syms t a ↵\n>>pretty(fourier(exp(-abs(a*t)))) ↵\n |a|\n 2 ----------------\n 2 2\n |a| + w\n```
$f(t) = e^{-\|at+b\|}$, where $a > 0$, $b > 0$ $$F[f(t)] = \frac{2ae^{i\frac{b\omega}{a}}}{a^2 + \omega^2}$$	```\n>>syms t a b ↵\n>>pretty(fourier(exp(-abs(a*t+b)))) ↵\n i b w\n |a| exp(---------)\n a\n 2 ------------------------\n 2 2\n |a| + w\n```
$f(t) = e^{-t^2}$ $$F[f(t)] = \sqrt{\pi}\ e^{-\frac{1}{4}w^2}$$	```\n>>syms t ↵\n>>pretty(fourier(exp(-t^2))) ↵\n\n 1/2 2\n pi exp(- 1/4 w)\n```
$f(t) = \dfrac{1}{1+t^2}$ $$F[f(t)] =$$ $\pi[e^{\omega}u(-\omega) + e^{-\omega}u(\omega)]$	```\n>>syms t ↵\n>>f=1/(1+t^2); ↵\n>>F=fourier(f); ↵\n>>pretty(F) ↵\n\nexp(w) pi Heaviside(-w) + exp(-w) pi\nHeaviside(w)\n```

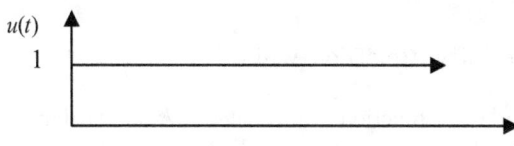

Figure 8.1 *Plot of unit step function, $u(t)$*

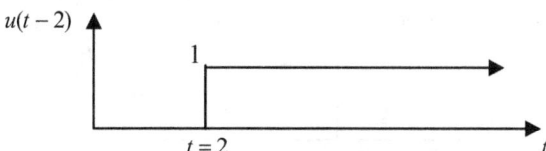

Figure 8.2 *Plot of the unit step function when shifted to the right at $t = 2$, $u(t-2)$*

8.1.2 Fourier transforms of unit step and Dirac delta functions

Figure 8.1 shows the plot of unit step function, mathematically, which is denoted by $u(t)$. It's MATLAB analogue is 'Heaviside(t)'. But the function has to be in symbolic form. So, we can say $u(t) \equiv$ sym('Heaviside(t)'). The Fourier transform of unit step function is $F[u(t)] = \pi\delta(\omega) - \dfrac{i}{\omega}$. It is obtained as follows:

MATLAB Command

>>fourier(sym('Heaviside(t)')) ↵

ans =

pi*Dirac(w)-i/w

320

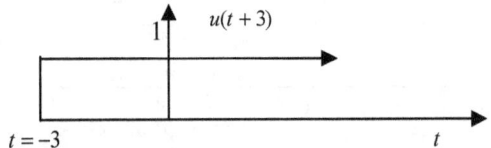

Figure 8.3   *Plot of unit step function when shifted to the left at $t=-3$, $u(t+3)$*

The unit step function can be shifted to the left or right. When $u(t)$ is shifted to the right as shown in figure 8.2, the transform is $F[u(t-2)] = e^{-2i\omega}\left(\pi\delta(\omega) - \dfrac{i}{\omega}\right)$, which is obtained as follows:

MATLAB Command
>>fourier(sym('Heaviside(t-2)')) ↵

ans =

exp(-2*i*w)*(pi*Dirac(w)-i/w)

Again, the transform of $u(t)$ shifted to the left at $t=-3$, $u(t+3)$, (shown in figure 8.3) is $e^{i3\omega}\left(\pi\delta(\omega) - \dfrac{i}{\omega}\right)$:

MATLAB Command
>>fourier(sym('Heaviside(t+3)')) ↵

ans =

exp(3*i*w)*(pi*Dirac(w)-i/w)

Now the terminology of Dirac delta function is presented. We know that the Dirac delta function located at $t=t_0$ on t axis is denoted by $\delta(t-t_0)$. This is also called the unit impulse function. It has infinite amplitude, zero width, but unity area. The plot of such function is as follows:

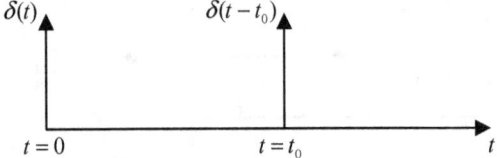

Figure 8.4   *Plot of Dirac delta function*

MATLAB representations of $\delta(t)$, $\delta(t-1)$, $\delta(t-2)$ ….. are sym('Dirac(t)'), sym('Dirac(t-1)'), sym('Dirac(t-2)'),…. respectively. Fourier transform of $\delta(t-t_0)$, $F[\delta(t-t_0)]$, is $e^{-i\omega t_0}$, for example, $F[\delta(t-2)] = e^{-i2\omega}$. Implementation is shown as follows:

MATLAB Command
>>fourier(sym('Dirac(t-2)')) ↵

ans =

exp(-2*i*w)

8.1.3 Inverse Fourier transform of $F(\omega)$

Given a complex function $F(\omega)$, its inverse Fourier transform is defined as $f(t) = F^{-1}[F(\omega)] = \frac{1}{2\pi}\int_{\omega=-\infty}^{\omega=\infty} F(\omega)e^{i\omega t}dt$, where F^{-1} is the inverse Fourier transform operator. MATLAB counterpart of F^{-1} is 'ifourier' (abbreviation of <u>in</u>verse <u>fourier</u>). The output of 'ifourier' by default is in terms of 'x'. To have the inverse Fourier transform of $F(\omega)$,

MATLAB Step:
   Use the command ifourier(frequency function $F(\omega)$ in string form, frequency variable, transform variable).

Computational examples are the best for concepts. Any function $f(t)$ that does not exist for $t < 0$ can be written in terms of the unit step function as $f(t)u(t)$. For example, function defined by

$$f(t) = \begin{cases} e^{-at} & t \geq 0, \quad for \ a > 0 \\ 0 & t < 0 \end{cases}$$

can be written as $f(t) = e^{-at}u(t)$. The Fourier transform of $e^{-at}u(t)$ is $\frac{1}{a+i\omega}$, which is utilized to find the inverse transform by inspection. Sometimes it is difficult to find directly from the integral defining the inverse Fourier transform. Utilization of the transform properties helps us find the inverse transform conveniently.

Find the inverse Fourier transforms of the following functions:

A. $F(\omega) = \dfrac{10}{6 + 5i\omega - \omega^2}$ 　　　 B. $X(\omega) = \dfrac{5e^{i(2\omega - 8)}}{6 - (4 - \omega)i}$ 　　　 C. $G(\omega) = \sin\omega$

D. $F(\omega) = \dfrac{\sin\omega}{\omega}$ 　　　 E. $Y(\omega) = \dfrac{15\sin(4\omega - 12)}{\omega - 3}$ 　　　 F. $H(\omega) = \begin{cases} 1 & if \ \omega > 0 \\ -1 & if \ \omega < 0 \end{cases}$

⊟ Example A

Start with factoring the denominator $6 + 5i\omega - \omega^2$, which provides $F(\omega) = \dfrac{10}{6 + 5i\omega - \omega^2} = \dfrac{10}{(2 + i\omega)(3 + i\omega)} = \dfrac{10}{2 + i\omega} - \dfrac{10}{3 + i\omega}$ (performing the partial fraction). From $F[e^{-at}u(t)]$, we can write $F^{-1}[F(\omega)] = 10[e^{-2t} - e^{-3t}]u(t)$. The string form of $F(\omega)$ is 10/(6+5*i*w-w^2). Implementation is shown below:

MATLAB Command

>>syms w ↵	← Defining ω as symbolic
>>F=10/(6+5*i*w-w^2); ↵	← Describing $F(\omega)$
>>f=ifourier(F) ↵	← Taking the inverse transform of $F(\omega)$

f =

10*Heaviside(x)*(exp(-2*x)-exp(-3*x))　　← Default return in terms of x

>>syms t ↵	← To see the output as the function of t
>>f=ifourier(F,w,t) ↵	

f =

10*Heaviside(t)*(exp(-2*t)-exp(-3*t))　　← Return in terms of t

322

⊞ Example B

Frequency shifting property of the transform states that $F[e^{i\omega_0 t} f(t)] = F(\omega - \omega_0)$ now $X(\omega) =$

$\dfrac{5 e^{i(2\omega - 8)}}{6 - (4 - \omega)i} = \dfrac{5 e^{2i(\omega - 4)}}{6 + (\omega - 4)i}$ and the property suggests that $F^{-1}[X(\omega)] = e^{i4t} F^{-1}\left[\dfrac{5 e^{2i\omega}}{6 + \omega i}\right]$. According to the time

shifting property, we have $F[f(t + t_0)] = e^{i\omega t_0} F(\omega)$, which implies $F^{-1}\left[\dfrac{5 e^{2i\omega}}{6 + \omega i}\right] = F^{-1}\left[\dfrac{5}{6 + \omega i}\right]\Bigg|_{t \to t+2} =$

$5 e^{-6t} u(t)\Bigg|_{t \to t+2} = 5 e^{-6(t+2)} u(t + 2)$. So, the complete inverse transform is $x(t) = 5 e^{i4t} e^{-6(t+2)} u(t + 2) = 5 e^{-6t - 12 + 4it} u(t + 2)$.

$X(\omega)$ string is given by 5*exp(i*(2*w-8))/(6-(4-w)*i):

MATLAB Command

>>syms w t ↵
>>X=5*exp(i*(2*w-8))/(6-(4-w)*i); ↵
>>x=ifourier(X,w,t); ↵
>>pretty(x) ↵

5 exp(-8 i) exp(-6 t - 12 + 4 i t + 8 i) Heaviside(t + 2)

We are able to simplify the output using the command 'simplify':

>>x=simplify(x); ↵   ← Simplification of contents of x and again assigning to x
>>pretty(x) ↵

5 Heaviside(t + 2) exp(-6 t - 12 + 4 i t)

⊞ Example C

We know from the transform of Dirac delta function (article 8.1.2) that $F[\delta(t - t_0)]$ is $e^{-i\omega t_0}$. Utilizing this, one obtains $F^{-1}[e^{i\omega}] = \delta(t + 1)$ and $F^{-1}[e^{-i\omega}] = \delta(t - 1)$. De Moivre's theorem says that $\sin \omega = \dfrac{e^{i\omega} - e^{-i\omega}}{2i}$.

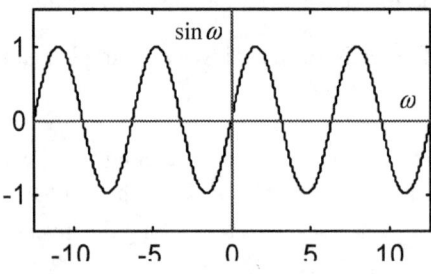

Figure 8.5   *Plot of* $\sin \omega$ *vs* ω

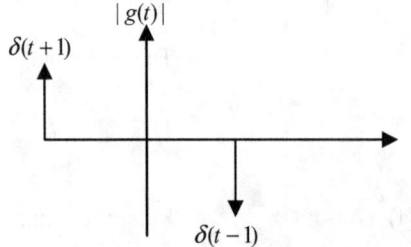

Figure 8.6 *Plot of* $|g(t)| = \dfrac{\delta(t + 1) - \delta(t - 1)}{2}$ *vs* t

Applying the linearity allows us to get $g(t) = F^{-1}[\sin\omega] = \dfrac{F^{-1}[e^{i\omega}] - F^{-1}[e^{-i\omega}]}{2i} = \dfrac{\delta(t+1) - \delta(t-1)}{2i} =$ $-i\dfrac{\delta(t+1) - \delta(t-1)}{2}$. Obtain it as follows:

MATLAB Command

```
>>syms w t ↵
>>g=ifourier(sin(w),w,t) ↵

g =

1/2*i*(-Dirac(t+1)+Dirac(t-1))
```

The conclusion from this example we can draw is the sinusoid in frequency (ω) domain translates to the Dirac delta function in time (t) domain. That is illustrated by the figures 8.5 and 8.6.

🖻 Example D

Having known the transform of $f(t) = e^{-at}u(t)$, we can write $F[e^{-at}u(t)] = \dfrac{1}{a+i\omega}$, $F[e^{at}u(-t)] = \dfrac{1}{a-i\omega}$,

and $F[e^{-at}u(t)] - F[e^{at}u(-t)] = \dfrac{1}{a+i\omega} - \dfrac{1}{a-i\omega} = \dfrac{-2i\omega}{(a+i\omega)(a-i\omega)} = \dfrac{-2i\omega}{a^2+\omega^2}$. Substituting $a = 0$ yields $F[u(t) -$

$u(-t)] = -\dfrac{2i}{\omega}$, therefore, $F^{-1}\left[\dfrac{1}{\omega}\right] = \dfrac{1}{2}i[u(t) - u(-t)]$. The inverse transform is applied as $F^{-1}\left[\dfrac{\sin\omega}{\omega}\right] =$

$F^{-1}\left[\dfrac{e^{i\omega} - e^{-i\omega}}{2i\omega}\right]$ (exponential form of sine function)$= -\dfrac{1}{2}i\left[F^{-1}\left(\dfrac{e^{i\omega}}{\omega}\right) - F^{-1}\left(\dfrac{e^{-i\omega}}{\omega}\right)\right]$ (applying the linearity)$=$

$-\dfrac{1}{2}i\left[F^{-1}\left(\dfrac{1}{\omega}\right)\Big|_{t\to t+1} - F^{-1}\left(\dfrac{1}{\omega}\right)\Big|_{t\to t-1}\right]$ (use of the time shifting property) $= -\dfrac{1}{2}i\left[\dfrac{1}{2}i\{u(t+1) - u(-t-1)\} - \right.$

$\dfrac{1}{2}i\{u(t-1) - u(-t+1)\}\Big] = \dfrac{1}{2}[u(t+1) - u(t-1)]$ (on simplification). The output is as follows:

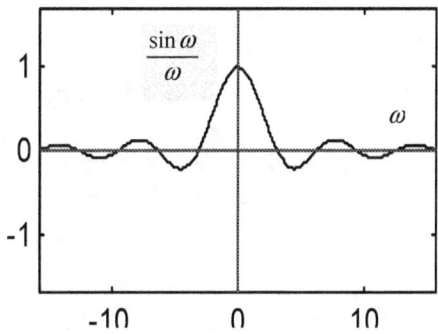

Figure 8.7   *Plot of* $\dfrac{\sin\omega}{\omega}$ *vs* ω

MATLAB Command

```
>>syms w t ↵
>>f=ifourier(sin(w)/w,w,t) ↵

f =
```

324

1/4*Heaviside(t+1)-1/4*Heaviside(-t-1)-1/4*Heaviside(t-1)+1/4*Heaviside(-t+1)

Simplify the contents of f further:
>>simplify(f) ↵

1/2*Heaviside(t+1)-1/2*Heaviside(t-1)

This conclusive example says that a sinc function $\left(\dfrac{\sin\omega}{\omega}\right)$ in the frequency domain becomes a rectangle function in the time domain (shown in figures 8.7 and 8.8).

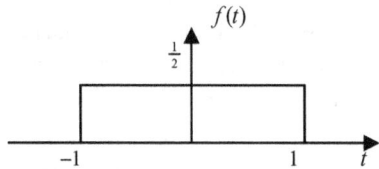

Figure 8.8 *Plot of* $f(t) = \dfrac{1}{2}[u(t+1) - u(t-1)]$ *vs* t

⊟ Example E
    Ideas and properties used in the previous examples are applied here in conjunction with the scaling property of Fourier transform {scaling property: $\dfrac{1}{a}F[f\left(\dfrac{t}{a}\right)] = F(a\omega)$ }. Inverse transform is obtained as $y(t) =$

$$F^{-1}\frac{15\sin(4\omega-12)}{\omega-3} = F^{-1}\frac{15\sin 4(\omega-3)}{\omega-3} = 15\,e^{i3t}\,F^{-1}\left[\frac{\sin 4\omega}{\omega}\right] = 60\,e^{i3t}\,F^{-1}\left[\frac{\sin 4\omega}{4\omega}\right] = 60\,e^{i3t}\,\frac{1}{4}\times\frac{1}{2}[u(t+$$

$1) - u(t-1)]\Big|_{t\to\frac{t}{4}} = \dfrac{15}{2}\,e^{i3t}\left[u\left(\dfrac{t}{4}+1\right)-u\left(\dfrac{t}{4}-1\right)\right]$, which is equivalent to $\dfrac{15}{2}\,e^{i3t}[\,u(t+4)-u(t-4)\,]$ because of the

nature of function $u(t)$:
MATLAB Command
    >>syms w t ↵
    >>Y=15*sin(4*w-12)/(w-3); ↵   ← The string form of $Y(\omega)$
    >>y=ifourier(Y,w,t); ↵
    >>y=simplify(y); ↵   ← Simplification is necessary otherwise long string will appear
    >>pretty(y) ↵   ← Displaying the simplified form of $y(t)$

    ans =

15/2 Heaviside(t + 4) exp(3 i t) - 15/2 Heaviside(t - 4) exp(3 i t)

⊟ Example F
    This is an example when the frequency function $F(\omega)$ is given in graphical form. $H(\omega)$ is graphed in figure 8.9. As a function, one can write $H(\omega) = u(\omega) - u(-\omega)$. Observe that the unit step function is applied to represent $H(\omega)$. Its inverse transform is

$\dfrac{i}{t\pi}$. MATLAB solution is shown as follows:

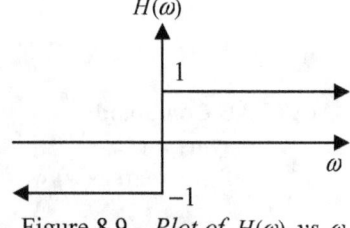

Figure 8.9 *Plot of* $H(\omega)$ *vs* ω

MATLAB Command
>>H=sym('Heaviside(w)-Heaviside(-w)'); ↵
>>syms w t ↵
>>h=ifourier(H,w,t); ↵
>>pretty(h) ↵

         i
    ----
      t pi

8.2 Laplace transform

Laplace transform of a function $f(t)$ is defined by $\int_{t=0}^{t=\infty} f(t)e^{-st}dt$, where $t>0$. Using the transform operator 'L', we can write $L\,[f(t)] = \int_{t=0}^{t=\infty} f(t)e^{-st}dt$. Independent variable t of $f(t)$ is usually used for time. This can be other variable x, for example, displacement. When independent variable is x, the transform is given by $L\,[f(x)] = \int_{x=0}^{x=\infty} f(x)e^{-sx}dx$. The transform is not applicable for the functions such as t^t or e^{t^2}. MATLAB function for the transform is 'laplace' that can conceive variable t or x. Default return of 'laplace' is a function of s. The general form of the transform is as follows:

MATLAB Step:
Use the command laplace(function in string form, independent variable, transform variable).

If you are willing to make the independent variable as x and the transform variable as z, that can be illustrated by $L\,[f(x)] = \int_{x=0}^{x=\infty} f(x)e^{-zx}dx \equiv$ laplace(f,x,z). Along with this the symbolic declaration of the variables is also necessary. If Laplace transform of a function exists, so does the inverse of the transform. Later in this chapter we present how the inverse Laplace transform can be found.

8.2.1 Laplace transforms of common functions

Begin with the Laplace transform of e^{at}, $L\,[e^{at}] = \int_{t=0}^{t=\infty} f(t)e^{-st}dt = \int_{t=0}^{t=\infty} e^{at}e^{-st}dt = \int_{t=0}^{t=\infty} e^{-(s-a)t}dt = \left[-\frac{e^{-(s-a)t}}{s-a}\right]_{t=0}^{t=\infty} = 0$

$-\left(-\frac{1}{s-a}\right) = \frac{1}{s-a}$. The transform by making the specific choice of independent variable (x) and the transform variable (z) is $L\,[e^{ax}] = \int_{x=0}^{x=\infty} e^{ax}e^{-zx}dx = \frac{1}{z-a}$. Implementation is shown as follows:

MATLAB Command

for string form,
>>syms a t ↵
>>laplace(exp(a*t)) ↵

ans =

1/(s-a)
for e^{ax} and the wanted transform variable z,
>>syms a x z ↵
>>laplace(exp(a*x),x,z) ↵

for nice form,
>>pretty(laplace(exp(a*t))) ↵
       1
    ----
       s - a
for e^{2t^2} (does not have the transform),
>>syms t ↵
>>laplace(exp(2*t^2)) ↵

ans =

laplace(exp(2*t^2),t,s)

ans =

1/(z-a)

Commonly used functions can be derived from e^{at}. For example, if $a=0$, e^{at} becomes the unit step function. Again, $\cos t$ is the real part of e^{at}, where $a=i$. Since the linearity exists in Laplace transform, the transform of just mentioned functions can be found easily. The output returned by MATLAB for the function like e^{2t^2} does not provide a close form function of s. Referring to the above implementation, laplace(exp(2*t^2),t,s) means just the definition of Laplace transform of e^{2t^2}. Our momentous concern is to expose the reader to some commonly used transforms implemented in MATLAB environment. A substantial number of transforms are given in table 8.B.

Table 8.B   Laplace transforms of some functions

Mathematical form	MATLLAB command for symbolic form

$L[\sin bt] = \dfrac{b}{s^2 + b^2}$	>>syms b t ↵
	>>pretty(laplace(sin(b*t))) ↵
	
	       b
	   --------------- 	
	     2 2
	    s + b

$L[\cos bt] = \dfrac{s}{s^2 + b^2}$	>>syms b t ↵
	>>pretty(laplace(cos(b*t))) ↵
	
	       s
	   --------------- 	
	     2 2
	    s + b

$L[\sinh bt] = \dfrac{b}{s^2 - b^2}$	>>syms b t ↵
	>>pretty(laplace(sinh(b*t))) ↵
	       b
	   --------------- 	
	     2 2
	    s - b

$L[\cosh bt] = \dfrac{s}{s^2 - b^2}$	>>syms b t ↵
	>>pretty(laplace(cosh(b*t))) ↵
	       s
	   --------------- 	
	     2 2
	    s - b

$L[e^{at} \cos bt] = \dfrac{s-a}{(s-a)^2 + b^2}$	>>syms a b t ↵
	>>pretty(laplace(exp(a*t)*cos(b*t))) ↵
	      s - a
	 --------------- 	
	      2 2
	  (s - a) + b

$L[e^{at} \sin bt] = \dfrac{b}{(s-a)^2 + b^2}$	>>syms a b t ↵
	>>pretty(laplace(exp(a*t)*sin(b*t))) ↵
	      b
	 --------------- 	
	      2 2
	  (s - a) + b

$L[u(t)] = \dfrac{1}{s}$	>>u=sym('Heaviside(t)'); ↵ >>pretty(laplace(u)) ↵ 1/s
$L[\delta(t)] = 1$	>>d=sym('Dirac(t)'); ↵ >>pretty(laplace(d)) ↵ 1

Continuation of the previous table:

Mathematical form	MATLAB command for symbolic form

$* L\left[\dfrac{1}{\sqrt{\pi(t+a)}}\right] = \dfrac{e^{as}erfc(\sqrt{as})}{\sqrt{s}}$	```>>syms a t ↵```
	```>>pretty(laplace(1/sqrt(pi*(t+a)))) ↵```
	
	``` 1/2```
	```exp(a s) erfc((a s))```
	```----------------------------```
	``` 1/2```
	``` s```

$* L\,[J_0(at)] = \dfrac{1}{\sqrt{s^2+a^2}}$	```>>syms a t ↵```
	```>>pretty(laplace(besselj(0,a*t))) ↵```
	
	``` 1```
	```--------------------```
	``` 2 2 1/2```
	``` (s +a)```

$* erfc(x)$ is complementary error function
$* J_0(x)$ is Bessel function of the first kind of order 0

8.2.2 Laplace transform of a polynomial of t

The transform also applies to a polynomial. To take the Laplace transform of a polynomial of t,

MATLAB Step:

Declare the variable t as symbolic and use the command laplace(polynomial as string).

Suppose the test polynomial is $-7 + 2t + t^4$. Term by term Laplace transform of the polynomial allows us to write $L\,[-7] = -\dfrac{7}{s}$, $L\,[2t] = \dfrac{2}{s^2}$, and $L\,[t^4] = \dfrac{24}{s^5}$. Gathering all terms, we have $L\,[-7 + 2t + t^4] = -\dfrac{7}{s} + \dfrac{2}{s^2} + \dfrac{24}{s^5}$. To show this:

MATLAB Command
```
>>syms t ↵
>>laplace(-7+2*t+t^4) ↵

ans =

-7/s+2/s^2+24/s^5
```

Simplification of the transform polynomial, which is $-\dfrac{7}{s} + \dfrac{2}{s^2} + \dfrac{24}{s^5} = \dfrac{-7s^4 + 2s^3 + 24}{s^5}$, can be carried out by the command 'simplify'. This is shown as follows:

```
>>simplify(laplace(-7+2*t+t^4)) ↵

ans =

-(7*s^4-2*s^3-24)/s^5
```

Display the simplified expression in symbolic form:
```
>>pretty(simplify(laplace(-7+2*t+t^4))) ↵
```

$$\frac{4}{7s} \quad \frac{3}{-2s} - 24$$

$$- \frac{7s^4 - 2s^3 - 24}{s^5}$$

Even if the polynomial coefficients are complex, the subroutine can handle that. Take the example of $L\left[-5+i+(2+3i)t+4it^2\right]=\dfrac{-5+i}{s}+\dfrac{2+3i}{s^2}+\dfrac{8i}{s^3}$ for implementation:

>>pretty(laplace(-5+i+(2+3i)*t+4i*t^2)) ↵

$$- 5/s + i/s + \frac{2+3i}{s^2} + 8\frac{i}{s^3}$$

8.2.3 Laplace transform of a function multiplied/divided by t^n

A particular type of function is the function $f(t)$ when it is multiplied or divided by t. For illustration, consider the functions $t\sin at$ and $\dfrac{\sinh mt}{t}$. It is often useful in finding the transforms of such functions by the use of $L\left[t^n f(t)\right]=(-1)^n\dfrac{d^n F(s)}{ds^n}$ and $L\left[\dfrac{f(t)}{t}\right]=\int\limits_{v=s}^{v=\infty}F(v)\,dv$, where $F(s)$ is the Laplace transform of $f(t)$. Using the multiplied form, the transform of $t\sin at$ is $L\left[t\sin at\right]=(-1)^1\dfrac{dF(s)}{ds}$, where $F(s)=L\left[\sin at\right]=\dfrac{a}{s^2+a^2}$ so $L\left[t\sin at\right]=-\dfrac{d}{ds}\left[\dfrac{a}{s^2+a^2}\right]=\dfrac{2sa}{(s^2+a^2)^2}$. The solution is shown as follows:

MATLAB Command
>>syms t a ↵
>>pretty(laplace(t*sin(a*t))) ↵

$$2\frac{a\,s}{(s^2+a^2)^2}$$

The next problem at hand is $L\left[\dfrac{\sinh mt}{t}\right]=?$, clearly, $L\left[\dfrac{\sinh mt}{t}\right]=\int\limits_{v=s}^{v=\infty}F(v)\,dv$, where $F(s)=L\left[\sinh mt\right]=\dfrac{m}{s^2-m^2}$.

We have $L\left[\dfrac{\sinh mt}{t}\right]=\int\limits_{v=s}^{v=\infty}\dfrac{m}{v^2-m^2}\,dv=\left[\dfrac{1}{2}\ln\dfrac{m-v}{m+v}\right]_{v=s}^{v=\infty}=\underset{v\to\infty}{Lt}\dfrac{1}{2}\ln\dfrac{m-v}{m+v}-\dfrac{1}{2}\ln\dfrac{m-s}{m+s}=\dfrac{1}{2}\ln(-1)-$

$\dfrac{1}{2}\ln(m-s)+\dfrac{1}{2}\ln(m+s)=\dfrac{1}{2}\ln(e^{i\pi})-\dfrac{1}{2}\ln(m-s)+\dfrac{1}{2}\ln(m+s)=i\dfrac{\pi}{2}-\dfrac{1}{2}\ln(m-s)+\dfrac{1}{2}\ln(m+s)$. The division example is shown as follows:

MATLAB Command
>>syms t m ↵
>>pretty(laplace(sinh(m*t)/t)) ↵

$$1/2\ i\ pi - 1/2\ \log(m-s) + 1/2\ \log(m+s)$$

8.2.4 Laplace transforms of unit step and Dirac delta functions

Table 8.B contains the transforms of unit step and Dirac delta functions. Some variants of these functions are presented in article 8.1.2. The illustrated functions are $u(t-2)$, $u(t+3)$, and $\delta(t-2)$ (see figures 8.2, 8.3, and 8.4 for their graphs). It is given that $L[u(t-2)] = \dfrac{e^{-2s}}{s}$ and $L[u(t+3)] = \dfrac{1}{s}$. Observe that the transform of $u(t+3)$ can be regarded as that of $u(t)$. The reason for this is that the transform is defined for $t > 0$. The last transform is $L[\delta(t-2)] = e^{-2s}$. These all are shown as follows:

MATLAB Command

for $L[u(t-2)]$,

    >>laplace(sym('Heaviside(t-2)')) ↵

    ans =

    exp(-2*s)/s

for $L[\delta(t-2)]$,

    >>laplace(sym('Dirac(t-2)')) ↵

    ans =

    exp(-2*s)

for $L[u(t+3)]$,

    >>laplace(sym('Heaviside(t+3)')) ↵

    ans =

    1/s

8.2.5 Laplace transforms of differential coefficients

Laplace transforms of different derivatives of a function $y(t)$ are given as follows:

$$L\left(\frac{dy}{dt}\right) = sY(s) - y(0)$$

$$L\left(\frac{d^2y}{dt^2}\right) = s^2Y(s) - sy(0) - y'(0) = s[sY(s) - y(0)] - y'(0)$$

$$L\left(\frac{d^3y}{dt^3}\right) = s^3Y(s) - s^2y(0) - sy'(0) - y''(0)$$

$$= s[s[sY(s) - y(0)] - y'(0)] - y''(0)$$

....so on.

In general, we have $L\left(\dfrac{d^n y}{dt^n}\right) = s^n Y(s) - s^{n-1}y(0) - s^{n-2}y'(0) - \ldots\ldots\ldots - sy^{n-2}(0) - y^{n-1}(0)$, where $Y(s)$ is the Laplace transform of $y(t)$. $Y(s)$ is equivalent to laplace(y(t),t,s) in MATLAB terminology and the return of the subroutine is a Horner polynomial (see article 3.10 for the Horner polynomial). The dependent variable y is a function of t that can be entered into MATLAB by writing the command y=sym('y(t)'). The nth derivative of y, that is, $\dfrac{d^n y(t)}{dt^n}$, is written as diff(y,n), where 'diff' is the differential operator $\left(\dfrac{d}{dt}\right)$ of MATLAB. Different order derivatives and their MATLAB counterparts are presented as follows:

Mathematical form: $L\left[\dfrac{dy(t)}{dt}\right] = sY(s) - y(0)$

MATLAB Counterpart:

    >>y=sym('y(t)'); ↵
    >>laplace(diff(y,1)) ↵

ans =

s*laplace(y(t),t,s)-y(0)

Mathematical form: $L\left[\dfrac{d^2y(t)}{dt^2}\right]=s[sY(s)-y(0)]-y'(0)$

MATLAB Counterpart:

```
>>y=sym('y(t)'); ↵
>>laplace(diff(y,2)) ↵
```

ans =

s*(s*laplace(y(t),t,s)-y(0))-D(y)(0)

Mathematical form: $L\left[\dfrac{d^3y(t)}{dt^3}\right]=s[s[sY(s)-y(0)]-y'(0)]-y''(0)$

MATLAB Counterpart:

```
>>y=sym('y(t)'); ↵
>>laplace(diff(y,3)) ↵
```

ans =

s*(s*(s*laplace(y(t),t,s)-y(0))-D(y)(0))-`@@`(D,2)(y)(0)

Mathematical form: $L\left[\dfrac{d^4y(t)}{dt^4}\right]=s[s[s[sY(s)-y(0)]-y'(0)]-y''(0)]-y'''(0)$

MATLAB Counterpart:

```
>>y=sym('y(t)'); ↵
>>laplace(diff(y,4)) ↵
```

ans =

s*(s*(s*(s*laplace(y(t),t,s)-y(0))-D(y)(0))-`@@`(D,2)(y)(0))-`@@`(D,3)(y)(0)

.... so as the other derivatives.

By virtue of the command 'simplify', simplification can be performed. See the following regarding the simplification of $L\left[\dfrac{d^4y(t)}{dt^4}\right]$:

```
>>y=sym('y(t)'); ↵
>>simplify(laplace(diff(y,4))) ↵
```

ans =

s^4*laplace(y(t),t,s)-s^3*y(0)-s^2*D(y)(0)-s*`@@`(D,2)(y)(0)-`@@`(D,3)(y)(0)

Command 'pretty' displays the form that is close to mathematical:

```
>>pretty(simplify(laplace(diff(y,4)))) ↵
```

```
      4                    3        2        (2)         (3)
     s  laplace(y(t), t, s) - s  y(0) - s  D(y)(0) - s (D   )(y)(0) - (D   )(y)(0)
```

We did not describe the entering syntax of the initial conditions so far. Their MATLAB analogue strings are

$y(t)\big|_{t=0}=y(0)\equiv y(0),\qquad \dfrac{dy(t)}{dt}\bigg|_{t=0}=y'(0)\equiv D(y)(0),\qquad \dfrac{d^2y(t)}{dt^2}\bigg|_{t=0}=y''(0)=\text{`@@`}(D,2)(y)(0),\qquad \dfrac{d^3y(t)}{dt^3}\bigg|_{t=0}=y'''(0)=$

`@@`(D,3)(y)(0), and so are the other order derivatives. *The characters in the string of any particular derivative are consecutive, there is no blank space between the characters.*

⊟ Example

Find the Laplace transform of $7\dfrac{d^3y}{dt^3} - 5\dfrac{dy}{dt} + 3y$ when $y(0) = 9$, $y'(0) = 3$, and $y''(0) = 2$.

Solution:

$$L\left(7\dfrac{d^3y}{dt^3} - 5\dfrac{dy}{dt} + 3y\right) = 7[\,s^3Y(s) - s^2y(0) - sy'(0) - y''(0)\,] - 5[\,sY(s) - y(0)\,] + 3\,Y(s) = [7s^3 - 5s + 3]Y(s) - 7s^2y(0)$$

$- 7sy'(0) - 7y''(0) + 5y(0) = [7s^3 - 5s + 3]Y(s) - 63s^2 - 21s + 31$. First, take the Laplace transform:

MATLAB Command

>>y=sym('y(t)'); ↵
>>L=laplace(7*diff(y,3)-5*diff(y,1)+3*y); ↵

Depending on the font size you choose for MATLAB Command Window, you may not see the whole string of the output. To see the rest of the string, slide the horizontal scroll bar to the right with the help of mouse. However, the Horner polynomial output is assigned to L. We simplify the contents of L and assign that to S as follows:

>>S=simplify(L) ↵

S =

7*s^3*laplace(y(t),t,s)-7*s^2*y(0)-7*s*D(y)(0)-7*`@@`(D,2)(y)(0)-5*s*laplace(y(t),t,s)+5*y(0)+3*laplace(y(t),t,s)

Observe that the capital 'S' and the small 's' are not the same. It is not nice that the long string 'laplace(y(t),t,s)' appears instead of $Y(s)$. What if we replace the whole string by the capital letter 'Y'. Subroutine 'subs' is capable of doing that. Before we do substitution, the symbolic declaration of Y is required. The whole output string is held in the variable S. Let us replace the string 'laplace(y(t),t,s)' by 'Y':

>>syms Y ↵
>>S=subs(S,{'laplace(y(t),t,s)'},{Y}) ↵

S =

7*s^3*Y-7*s^2*y(0)-7*s*D(y)(0)-7*`@@`(D,2)(y)(0)-5*s*Y+5*y(0)+3*Y

The replacement of 'laplace(y(t),t,s)' by 'Y' is rendered and the output is assigned to S. Substitute $y(0) = 9$ in the last S and assign the output to S again as follows:

>>S=subs(S,{'y(0)'},{9}) ↵

S =

7*s^3*Y-63*s^2-7*s*D(y)(0)-7*`@@`(D,2)(y)(0)-5*s*Y+45+3*Y

Then, replace $y'(0)$ by 3 and assign the output to S:

>>S=subs(S,{'D(y)(0)'},{3}) ↵

S =

7*s^3*Y-63*s^2-21*s-7*`@@`(D,2)(y)(0)-5*s*Y+45+3*Y

332

Next substitute 2 for $y''(0)$ as shown below:
>>S=subs(S,{'`@@`(D,2)(y)(0)'},{'2'}) ↵

S =

7*s^3*Y-63*s^2-21*s+31-5*s*Y+3*Y

Finally, take 'Y' common from the output string of S using the command 'collect':
>>S=collect(S,Y) ↵

S =

(7*s^3-5*s+3)*Y-63*s^2-21*s+31

The above string is what we are looking for. For understanding, we explained the different commands step by step. Following commands could have brought about all these:

MATLAB Command
>>y=sym('y(t)'); ↵
>>L=simplify(laplace(7*diff(y,3)-5*diff(y,1)+3*y)); ↵
>>S=subs(L,{'laplace(y(t),t,s)','y(0)','D(y)(0)','`@@`(D,2)(y)(0)'},{sym('Y'),9,3,2}); ↵
>>syms Y ↵
>>S=collect(S,Y) ↵

S =

(7*s^3-5*s+3)*Y-63*s^2-21*s+31

8.2.6 Laplace transforms of integrals

Laplace transform of integrals is also possible. Recall that the constant of integration is not returned by the subroutine 'int' and the independent variable of integrand can not be the limits of the subroutine. We know that $L\left[\int_{x=0}^{x=t} f(x)\,dx\right] = \frac{F(s)}{s}$, where $F(s)$ is the Laplace transform of $f(t)$. This is implemented as follows:

MATLAB Command
>>syms x t ↵
>>f=sym('f(x)'); ↵
>>laplace(int(f,x,0,t)) ↵

Warning: Explicit integral could not be found.
> In C:\MATLAB\toolbox\symbolic\@sym\int.m at line 58

ans =

laplace(f(t),t,s)/s

A warning message appears because no function is assigned to $f(t)$. We can ignore that. Anyhow the output is 'laplace(f(t),t,s)/s'. With the definition $F(s) \equiv \text{laplace(f(t),t,s)}$, the transform is $\frac{F(s)}{s}$. The output can be assigned to some variable for further manipulation. Let us see some other examples. Say, we want to find the Laplace transform of the sine integral, which is defined by $Si(t) = \int_{x=0}^{x=t} \frac{\sin x}{x}\,dx$, therefore, $L[Si(t)] = \frac{1}{s} L\left[\frac{\sin t}{t}\right] =$

$\dfrac{1}{s}\displaystyle\int_{r=s}^{r=\infty} F(r)\,dr$ [$F(r)$ is the Laplace transform of $\sin t$ with the dummy variable r]$=\dfrac{1}{s}\displaystyle\int_{r=s}^{r=\infty}\dfrac{1}{r^2+1}\,dr=\dfrac{1}{s}\cot^{-1}s$. The

transform is obtained as follows:

MATLAB Command
>>syms t ↵
>>laplace(sinint(t)) ↵

ans =

1/s*acot(s)

Finding the Laplace transform of cosine integral is another example. The transform is $L[Ci(t)]=-\dfrac{\ln(s^2+1)}{2s}$. See

the implementation below:
>>syms t ↵
>>laplace(cosint(t)) ↵

ans =
-1/s*log((1+s^2)^(1/2))
to see the symbolic form,
>>pretty(laplace(cosint(t))) ↵

```
                      2  1/2
          log((1 + s  )   )
        - ----------------------------
                    s
```

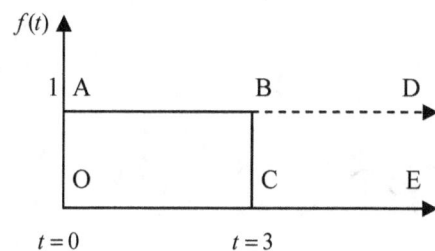

$f(t)$

1 | A   B   D

O   C   E

$t=0$ $t=3$

Figure 8.10   *A rectangular pulse*

8.2.7 Laplace transforms of graphical functions

Differential equations with nonsmooth forcing terms (the nonhomogeneous part) are often seen in applied mathematics and engineering problems. The nonsmooth graphical functions can be represented in terms of the unit step function. For instance the rectangular pulse $f(t)$ =OABC, as it is shown in figure 8.10, can be constructed from OADE [which is the unit step function $u(t)$] and ECBD [which is the unit step function but shifted to the right at $t=3$], that is, $f(t)$ =OABC=OADE−ECBD= $u(t)-u(t-3)$, and Laplace

transform of $f(t)$ is $F(s)=\dfrac{1}{s}-\dfrac{e^{-3s}}{s}$. Implementation is shown below:

MATLAB Command
>>f=sym('Heaviside(t)-Heaviside(t-3)'); ↵
>>laplace(f) ↵

ans =

334

1/s-exp(-3*s)/s

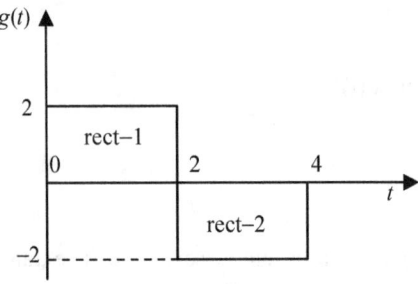

Figure 8.11   *A square pulse*

Next we have a square pulse $g(t)$ composed of rect–1 and rect–2 as labeled in the figure 8.11. By equation, we have $g(t) = \text{rect–1} + \text{rect–2} = 2[u(t) - u(t-2)] - 2[u(t-2) - u(t-4)] = 2u(t) - 4u(t-2) + 2u(t-4)$. The transform of $g(t)$ is $\dfrac{2}{s} - \dfrac{4e^{-2s}}{s} + \dfrac{2e^{-4s}}{s}$. Its MATLAB solution is as follows:

MATLAB Command

```
>>g=sym('2*Heaviside(t)-4*Heaviside(t-2)+2*Heaviside(t-4)');  ↵
>>pretty(laplace(g)) ↵
```

$$2/s - 4\ \frac{\exp(-2\ s)}{s} + 2\ \frac{\exp(-4\ s)}{s}$$

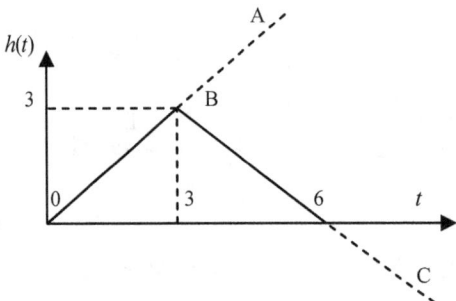

Figure 8.12   *A triangular pulse*

The last example is a triangular pulse $h(t)$ as depicted in figure 8.12. It can be constructed as $h(t) =$ lower triangle 0B3+triangle 3B6. Equations of the straight lines 0BA and B6C are $h(t) = t$ and $h(t) = -t + 6$ respectively. The triangle 0B3 is formed by $t[u(t) - u(t-3)]$ and 3B6 is formed by $(6-t)[u(t-3) - u(t-6)]$, so, $h(t) = t[u(t) - u(t-3)] + (6-t)[u(t-3) - u(t-6)] = tu(t) + (6-2t)u(t-3) - (6-t)u(t-6)$. The transform is $H(s) = \dfrac{1}{s^2} - \dfrac{2e^{-3s}}{s^2} + \dfrac{e^{-6s}}{s^2}$. We utilized here $L[f(t)u(t-t_0)] = e^{-st_0}L[f(t+t_0)]$. Have it as follows:

MATLAB Command

```
>>h=sym('t*Heaviside(t)+(6-2*t)*Heaviside(t-3)-(6-t)*Heaviside(t-6)');  ↵
>>pretty(laplace(h)) ↵
```

$$\frac{1}{s^2} \;-\; 2\,\frac{\exp(-3\,s)}{s^2} \;+\; \frac{\exp(-6\,s)}{s^2}$$

These techniques can be extended for the finite functions that follow the sinosoid or expoential variation for some interval.

8.2.8 Definition of inverse Laplace transform

Just like the symbol f^{-1}, which is used to denote the inverse of a function f, L^{-1} denotes the inverse Laplace transform. If $F(s)$ is a function of Laplace variable s, $L^{-1}[F(s)]$ is the inverse Laplace transform of $F(s)$. Inverse function is usually a function of t, that is, $f(t)$, but it can be a different one from t. By formal definition, we have $f(t) = L^{-1}[F(s)] = \dfrac{1}{2\pi i} \int_{s=c-i\infty}^{s=c+i\infty} e^{st} F(s)\,ds$, where c is a real number. The number c is selected in such a way that all singularities of $F(s)$ are to the left of the line $s=c$ in the s plane. As a working convenience, $F(s)$ is maneuvered to take the form of commonly used function or functions whose inverse Laplace transforms are known. MATLAB function that performs inverse Laplace transform is 'ilaplace'. To take inverse Laplace transform of F,

MATLAB Step:
    Use the command ilaplace(represent F in terms of string, variable of F, return variable).

Let us see the implementation for $\dfrac{1}{s^2+1}$. From the table 8.B, we read $L\,[\sin bt] = \dfrac{b}{s^2+b^2}$. Comparison tells us that $b=1$ and $L^{-1}\left[\dfrac{1}{s^2+1}\right] = \sin t$. The function 'ilaplace' is also operational for different input and output variables. Have it computed as follows:

MATLAB Command

when input variable is s,
    >>syms s ↵
    >>ilaplace(1/(s^2+1)) ↵

ans =

sin(t)

when input variable is w and output variable is x,
    >>syms w x ↵
    >>ilaplace(1/(w^2+1),w,x) ↵

ans =

sin(x)

8.2.9 Inverse Laplace transform of a polynomial of s

Recall that the Laplace transform of $\dfrac{d[f(t)]}{dt}$ is $sF(s) - f(0)$. Here $f(0)$ should be appropriately mentioned as $f(0_+)$ that means the right hand limit of $f(t)$ at $t=0$. For $f(t) = \delta(t)$ (Dirac delta function), $f(0_+)$ is 0 hence $L\left[\dfrac{d[\delta(t)]}{dt}\right] = sL[\delta(t)] - \delta(0_+) = s$. This result can be extended for the other derivatives of $\delta(t)$. If we perform that, we have $L\left[\dfrac{d^2[\delta(t)]}{dt^2}\right] = s^2$, $L\left[\dfrac{d^3[\delta(t)]}{dt^3}\right] = s^3$ ... so on. See the MATLAB implementation regarding this:

MATLAB Command
    >>syms s ↵
    >>ilaplace(s) ↵   ← Inverse Laplace transform of s

ans =

Dirac(1,t)   ← Which indicates the first derivative of $\delta(t)$
>>ilaplace(s^3) ↵   ← Inverse Laplace transform of s^3

ans =

Dirac(3,t)   ← Which indicates the third derivative of $\delta(t)$
>>ilaplace(3*s-7*s^2) ↵   ← Inverse Laplace transform of $3s - 7s^2$

ans =

3*Dirac(1,t)-7*Dirac(2,t)   ← Which indicates $3\dfrac{d[\delta(t)]}{dt} - 7\dfrac{d^2[\delta(t)]}{dt^2}$

8.2.10 Inverse Laplace transform of $\dfrac{P(s)}{Q(s)}$

We may have a rational function, which is given as the division of polynomials like $\dfrac{P(s)}{Q(s)}$. $Q(s)$ can be written as a product of some linear factors $(s - a)$ and/or irreducible quadratic factors $(s^2 + sa + b)$ employing the technique of partial fraction decomposition, where a and b are real constants. Decomposed in partial fraction, the inverse Laplace trasform can be applied term by term according to the table 8.B. The detail of partial fraction is presented in article 3.16.

Find the inverse Laplace transform of the following functions:

A. $F(s) = \dfrac{2s^3 + 3}{s^4 - 6s^3 + 32s}$ B. $Y(s) = \dfrac{s^5}{s^3 - 3s^2 + 4s - 12}$, the output is wanted in terms of x

C. $G(s) = \dfrac{4}{s(s^2 + 4s + 8)}$

⊟ Example A

Factorize the denominator to get $\dfrac{2s^3 + 3}{s^4 - 6s^3 + 32s} = \dfrac{2s^3 + 3}{s(s+2)(s-4)^2}$. Apply the partial fraction

decomposition to write $\dfrac{2s^3 + 3}{s(s+2)(s-4)^2} = \dfrac{3}{32s} + \dfrac{13}{72(s+2)} + \dfrac{497}{288(s-4)} + \dfrac{131}{24(s-4)^2}$. The transform of $t^n e^{at}$ is

$\dfrac{n!}{(s-a)^{n+1}}$. Utilizing the last formula, one can obtain $f(t) = L^{-1}[F(s)] = \dfrac{3}{32} + \dfrac{13}{72}e^{-2t} + \dfrac{497}{288}e^{4t} + \dfrac{131}{24}te^{4t}$. Have L^{-1} as follows:

MATLAB Command
```
>>syms s ↵
>>n=2*s^3+3; ↵                      ← The numerator string assigned to n
>>d=s^4-6*s^3+32*s; ↵               ← The denominator string assigned to d
>>f=ilaplace(n/d); ↵                ← 'f' contains string of f(t)
>>pretty(f) ↵
```

```
        13             131            497
3/32 + ------ exp(-2 t) + -------- t exp(4 t) + -------- exp(4 t)
        72              24             288
```

337

🔲 Example B

The degree of the numerator is greater than that of the denominator so a long division is required and that provides $\dfrac{s^5}{s^3-3s^2+4s-12}=s^2+3s+5+\dfrac{15s^2+16s+60}{s^3-3s^2+4s-12}$. Partial fraction of the rational part lets us write

$s^2+3s+5+\dfrac{15s^2+16s+60}{s^3-3s^2+4s-12}=s^2+3s+5+\dfrac{243}{13(s-3)}-\dfrac{16(3s-4)}{13(s^2+4)}\ =\ s^2+3s+5+\dfrac{243}{13(s-3)}-\dfrac{48s}{13(s^2+4)}+\dfrac{64}{13(s^2+4)}$.

From article 8.2.9 and table 8.B, the inverse transform is given by $y(x)=\dfrac{d^2[\delta(x)]}{dx^2}+3\dfrac{d[\delta(x)]}{dx}+5\,\delta(x)+\dfrac{243}{13}e^{3x}$

$-\dfrac{48}{13}\cos 2x+\dfrac{32}{13}\sin 2x$. See the solution below:

MATLAB Command
```
>>syms s x ↵
>>n=s^5; ↵
>>d=s^3-3*s^2+4*s-12; ↵
>>y=ilaplace(n/d,s,x); ↵
>>pretty(y) ↵
```

$$\text{Dirac(2, x) + 3 Dirac(1, x) + 5 Dirac(x) + }\frac{243}{13}\text{ exp(3 x) - }\frac{48}{13}\text{ cos(2 x)}$$
$$+\ \frac{32}{13}\text{ sin(2 x)}$$

🔲 Example C

Perform the partial fraction to get $\dfrac{4}{s(s^2+4s+8)}=\dfrac{1}{2s}-\dfrac{s+4}{2(s^2+4s+8)}$. The transform of the quadratic factor like s^2+4s+8 appears in table 8.B. We need to manipulate some algebra. Complete the square in the deniominator to have $\dfrac{1}{2s}-\dfrac{s+4}{2(s^2+4s+8)}=\dfrac{1}{2s}-\dfrac{s+2+2}{2[(s+2)^2+2^2]}=\dfrac{1}{2s}-\dfrac{s+2}{2[(s+2)^2+2^2]}-\dfrac{2}{2[(s+2)^2+2^2]}$. Read the

inverse transform from table 8.B to write $g(t)=\dfrac{1}{2}-\dfrac{1}{2}e^{-2t}\cos 2t-\dfrac{1}{2}e^{-2t}\sin 2t$. Get it implemented as follows:

MATLAB Command
```
>>syms s ↵
>>G=4/s/(s^2+4*s+8); ↵
>>g=ilaplace(G); ↵
>>pretty(g) ↵
```

$$1/2 - 1/2\ \exp(-2\ t)\ \cos(2\ t)\ -\ 1/2\ \exp(-2\ t)\ \sin(2\ t)$$

8.2.11 Inverse Laplace transform of $e^{-as}\,F(s)$

Shifting property of the Laplace transform says that if a function $f(t)$ is translated to $t=t_0$, its transform is $L[f(t-t_0)u(t-t_0)]=e^{-st_0}F(s)$. Inverse transform of $e^{-as}F(s)$ is $L^{-1}[F(s)]$, where t is replaced by $t-a$ and the whole inverse function is multiplied by $u(t-a)$. As usual, start with the example $e^{-3s}\dfrac{s+2}{s^2+2s+5}$, whence, $F(s)=\dfrac{s+2}{s^2+2s+5}=\dfrac{s+2}{(s+1)^2+2^2}=\dfrac{s+1}{(s+1)^2+2^2}+\dfrac{1}{(s+1)^2+2^2}$ and by inspection we have $L^{-1}[F(s)]=$

$e^{-t} \cos 2t + \dfrac{1}{2} e^{-t} \sin 2t$. One can write $L^{-1}\left[e^{-3s} \dfrac{s+2}{s^2 + 2s + 5}\right] = u(t-3) \left[e^{-t+3} \cos(2t-6) + \dfrac{1}{2} e^{-t+3} \sin(2t-6) \right]$. The string

form of $e^{-3s} \dfrac{s+2}{s^2 + 2s + 5}$ is exp(-3*s)*(s+2)/(s^2+2*s+5). The solution is shown as follows:

MATLAB Command
>>syms s ↵
>>F=exp(-3*s)*(s+2)/(s^2+2*s+5); ↵
>>pretty(ilaplace(F)) ↵

Heaviside(t - 3) exp(-t + 3) cos(2 t - 6)

+ 1/2 Heaviside(t - 3) exp(-t + 3) sin(2 t - 6)

8.2.12 Inverse Laplace transform using convolution

If $F(s)$ is the product of two functions, $H(s)$ and $G(s)$ [i.e., $F(s) = H(s)\,G(s)$], the inverse Laplace transform of $F(s)$ is obtained from the convolution of $h(t)$ and $g(t)$, where $h(t) = L^{-1}[H(s)]$ and $g(t) = L^{-1}[G(s)]$. The convolution of $h(t)$ and $g(t)$ is defined as $h(t)*g(t) = g(t)*h(t) = \int_{\tau=0}^{\tau=t} h(t-\tau)g(\tau)d\tau = \int_{\tau=0}^{\tau=t} g(t-\tau)h(\tau)d\tau$, where $*$ is the convolution operator (order of the function is immaterial). To proceed with an example, say, $F(s) = \dfrac{1}{(s^3+1)^3}$. Using the partial fraction of $F(s)$ to find L^{-1} might be clumsy instead convolution makes the finding

easier. By operator one can write $L^{-1}[F(s)] = \left(L^{-1}\left[\dfrac{1}{s^2+1}\right]\right)*\left(L^{-1}\left[\dfrac{1}{s^2+1}\right]\right)*\left(L^{-1}\left[\dfrac{1}{s^2+1}\right]\right) = \sin t * \sin t * \sin t =$

$\sin t * \int_{\tau=0}^{\tau=t} \sin(t-\tau)\sin\tau \, d\tau = \sin t *[\dfrac{1}{2}\sin t - \dfrac{1}{2}t\cos t] = \int_{\tau=0}^{\tau=t} \sin(t-\tau)[\dfrac{1}{2}\sin\tau - \dfrac{1}{2}\tau\cos\tau]d\tau = \dfrac{3}{8}\sin t - \dfrac{3}{8}t\cos t - \dfrac{1}{8}t^2\sin t$.

The string form of $\dfrac{1}{(s^3+1)^3}$ is 1/(s^2+1)^3. See the direct solution of MATLAB:

MATLAB Command
>>syms s ↵
>>F=1/(s^2+1)^3; ↵
>>pretty(ilaplace(F)) ↵

$\qquad$ 2
- 1/8 t sin(t) - 3/8 cos(t) t + 3/8 sin(t)

8.2.13 Solving differential equations using Laplace transforms

Subroutine 'dsolve' can directly find the analytical solutions of the differential equations. But if one needs the Laplace transform method of solving the differential equations, the following model examples can provide a guideline about the procedure to be adopted. Finding the Laplace transform of the differential coefficients is discussed in article 8.2.5. Subroutine 'ilaplace' is utilized to find the inverse Laplace transform as implemented before. It is better if an M–file executes all statements. Since we are going to assign the different returns to one variable, going back by one statement might show some error. This problem can be overcome by using different assignment variables. How an M–file is opened and executed is presented in chapters 1 and 11.

Solve the following differential equations using the Laplace transform:
  A. $y'' + 9y = 2\cos t$, $y(0)=1$, and $y'(0)=1$
  B. $y'' + 2y' - y = -x - e^{-x}$, $y(0)=0$, and $y'(0)=3$

C. $\begin{cases} x' = -2x + y, & x(0) = 1 \\ y' = 3x + 2y, & y(0) = 2 \end{cases}$

D. $\begin{cases} x' = x + 3y + t, & x(0) = 2 \\ y' = -3x - 5y + \sin t\, e^{-t}, & y(0) = 1 \end{cases}$

⊟ Example A

Taking the Laplace transform on both sides of the equation $y'' + 9y = 2\cos t$ provides $[s^2 Y(s) - sy(0) - y'(0)] + 9\,Y(s) = \dfrac{2s}{s^2 + 1}$. Substituting the initial conditions, the equation becomes $s^2 Y(s) - s - 1 + 9\,Y(s) = \dfrac{2s}{s^2 + 1}$. Solving for $Y(s)$ provides $Y(s) = \dfrac{s^3 + s^2 + 3s + 1}{(s^2 + 9)(s^2 + 1)} = \dfrac{3s}{4(s^2 + 9)} + \dfrac{s}{4(s^2 + 1)} + \dfrac{1}{s^2 + 9}$ (after partial fraction). By inspection the inverse transform is $y(t) = \dfrac{3}{4}\cos 3t + \dfrac{1}{4}\cos t + \dfrac{1}{3}\sin 3t$. The given equation can be organized as $y'' + 9y - 2\cos t = 0$. See the implementation as follows:

MATLAB Command

```
>>syms t ↵          ← Defining independent variable t as symbolic
>>y=sym('y(t)'); ↵  ← Defining dependent variable y(t) as symbolic
```

Take the Laplace transform of left side of the organized equation and assign that to T:
```
>>T=laplace(diff(y,2)+9*y-2*cos(t)); ↵
```

Substitute the initial conditions and put the output to T again:
```
>>T=subs(T,{'y(0)','D(y)(0)'},{'1','1'}); ↵
```

Substitute the transform string 'laplace(y(t),t,s)', which is equivalent to $Y(s)$, by 'Y' and assign the output to T:
```
>>T=subs(T,'laplace(y(t),t,s)','Y'); ↵
```

Find $Y(s)$ by forming an equation T=0 and assign the output of the subroutine 'solve' to Y:
```
>>Y=solve(T,'Y') ↵

Y =

(s^3+3*s+s^2+1)/(s^2+1)/(s^2+9)
```

Display $Y(s)$ in symbolic form:
```
>>pretty(Y) ↵
            3        2
          s + 3 s + s + 1
         ----------------------
              2       2
          (s + 1) (s + 9)
```

Take the inverse Laplace transform of Y and assign the output to y:
```
>>y=ilaplace(Y) ↵

y =

1/4*cos(t)+3/4*cos(3*t)+1/3*sin(3*t)
```

Finally, the symbolic form of $y(t)$ is seen as:
```
>>pretty(y) ↵
```

$$1/4 \cos(t) + 3/4 \cos(3 \ t) + 1/3 \sin(3 \ t)$$

⊟ **Example B**

Apply the Laplace transform on both sides of $y'' + 2y' - y = -x - e^{-x}$ to get $[s^2 Y(s) - sy(0) - y'(0)] +$ $2[sY(s) - y(0)] - Y(s) = -\dfrac{1}{s^2} - \dfrac{1}{s+1}$. Notice that the independent variable is x instead of t. Plugging the initial

conditions and solving for $Y(s)$ give us $Y(s) = \dfrac{3s^3 + 2s^2 - s - 1}{s^2(s+1)(s^2 + 2s - 1)}$. Partial fraction of $Y(s)$ yields $Y(s) =$

$\dfrac{2}{s} + \dfrac{1}{s^2} + \dfrac{1}{2(s+1)} - \dfrac{5(s+1)}{2(s^2 + 2s - 1)} = \dfrac{2}{s} + \dfrac{1}{s^2} + \dfrac{1}{2(s+1)} - \dfrac{5}{2}\dfrac{s+1}{(s+1)^2 - (\sqrt{2})^2}$. Comparing with the standard transforms,

we have $y(x) = 2 + x + \dfrac{1}{2}e^{-x} - \dfrac{5}{2}e^{-x} \cosh \sqrt{2} \ x$. Arrange the given equation to write $y'' + 2y' - y + x + e^{-x} = 0$. The

solution is as follows:

MATLAB Command

```
>>syms x ↵            ← Defining independent variable x as symbolic
>>y=sym('y(x)'); ↵    ← Defining dependent variable y(x) as function of x
>>T=laplace(diff(y,x,2)+2*diff(y,x)-y+x+exp(-x)); ↵  ← L on left side of organized equation
>>T=subs(T,{'y(0)','D(y)(0)'},{'0','3'}); ↵          ← Insert the initial conditions
>>T=subs(T,'laplace(y(x),x,s)','Y'); ↵    ← Replace transform string 'laplace(y(x),x,s)' by 'Y'
>>Y=solve(T,'Y'); ↵   ← Obtain Y(s) by forming equation T=0
>>pretty(Y) ↵         ← Display Y(s) in symbolic form
```

```
             3     2
    -s - 1 + 3 s  + 2 s
  ---------------------------------
     2                2
    s  (s + 1) (2 s - 1 + s  )
```

```
>>y=ilaplace(Y,x); ↵    ← L⁻¹ of Y(s) is assigned to y
>>pretty(y) ↵           ← Display symbolic form of y(x)
```

```
                         1/2
    x + 2 + 1/2 exp(-x) - 5/2 exp(-x) cosh(2    x)
```

⊟ **Example C**

This example has a constant coefficient system of two linear differential equations. We must apply the transform to each equation. Letting $X(s) = L[x(t)]$, $Y(s) = L[y(t)]$, and applying the transform, we have

$\begin{cases} sX(s) - x(0) = -2X(s) + Y(s) \\ sY(s) - y(0) = 3X(s) + 2Y(s) \end{cases}$. Inserting the initial data results $\begin{cases} sX(s) - 1 = -2X(s) + Y(s) \\ sY(s) - 2 = 3X(s) + 2Y(s) \end{cases}$. The last set is a system

of linear algebraic equations for $X(s)$ and $Y(s)$, solution of which is $X(s) = \dfrac{s}{s^2 - 7}$ and $Y(s) = \dfrac{2s + 7}{s^2 - 7} =$

$\dfrac{2s}{s^2 - 7} + \dfrac{7}{s^2 - 7}$, hence, L^{-1} is $\begin{cases} x(t) = \cosh \sqrt{7} \ t \\ y(t) = 2\cosh \sqrt{7} \ t + \sqrt{7} \sinh \sqrt{7} \ t \end{cases}$. The given equations can be organized as

$\begin{cases} x' + 2x - y = 0 \\ y' - 3x - 2y = 0 \end{cases}$. The transform is applied as follows:

MATLAB Command

```
>>syms t ↵           ← Declaration of the independent symbolic variable t
>>x=sym('x(t)'); ↵   ← Declaration of x(t)
```

>>y=sym('y(t)'); ↵   ← Declaration of $y(t)$

Laplace transforms on left sides of the organized equations are assigned to eqn1 and eqn2 respectively:
>>eqn1=laplace(diff(x)+2*x-y); ↵
>>eqn2=laplace(diff(y)-3*x-2*y); ↵

Insert the initial data and put the outputs to eqn1 and eqn2 again:
>>eqn1=subs(eqn1,{'x(0)'},{'1'}); ↵
>>eqn2=subs(eqn2,{'y(0)'},{'2'}); ↵

Replace the strings 'laplace(x(t),t,s)' by 'X' and 'laplace(y(t),t,s)' by 'Y' and assign the outputs to eqn1 and eqn2 repetively:
>>eqn1=subs(eqn1,{'laplace(x(t),t,s)','laplace(y(t),t,s)'},{'X','Y'}); ↵
>>eqn2=subs(eqn2,{'laplace(x(t),t,s)','laplace(y(t),t,s)'},{'X','Y'}); ↵

Have $X(s)$ and $Y(s)$ forming the equations eqn1=0 and eqn2=0 respectively and assign the output of the subroutine 'solve' to R:
>>R=solve(eqn1,eqn2,'X','Y') ↵

R =
          X: [1x1 sym]
          Y: [1x1 sym]   ← It indicates R is a structure array having two members X and Y
>>X=(R.X); ↵   ← Assigning the first member of R to X
>>Y=(R.Y); ↵   ← Assigning the second member of R to Y
>>pretty(X) ↵   ← Displaying contents of X, which is $X(s)$

```
            s
        ------------
             2
          s  - 7
```
>>pretty(Y) ↵   ← Displaying contents of Y, which is $Y(s)$
```
          7 + 2 s
        --------------
             2
          s  - 7
```

Inverse transforms of X and Y are assigned to x and y respectively:
>>x=ilaplace(X); ↵
>>y=ilaplace(Y); ↵

The mathematical form is seen as:
>>pretty(x) ↵
```
               1/2
        cosh(7    t )
```
>>pretty(y) ↵
```
              1/2        1/2        1/2
        2 cosh(7    t) + 7    sinh(7    t)
```

⊞ Example D

Utilize the transform on both sides of each equation {with $X(s) = L[x(t)]$ and $Y(s) = L[y(t)]$} to get

$$\begin{cases} sX(s) - x(0) = X(s) + 3Y(s) + \dfrac{1}{s^2} \\ sY(s) - y(0) = -3X(s) - 5Y(s) + \dfrac{1}{(s+1)^2 + 1} \end{cases}$$. We therefore obtain $$\begin{cases} sX(s) - 2 = X(s) + 3Y(s) + \dfrac{1}{s^2} \\ sY(s) - 1 = -3X(s) - 5Y(s) + \dfrac{1}{(s+1)^2 + 1} \end{cases}$$ (insersion of

the initial conditions). The last set of linear algebraic equations has the solution $X(s) = \dfrac{2s^5 + 17s^4 + 31s^3 + 36s^2 + 12s + 10}{s^6 + 6s^5 + 14s^4 + 16s^3 + 8s^2}$ and $Y(s) = \dfrac{s^5 - 5s^4 - 11s^3 - 18s^2 - 6s - 6}{s^6 + 6s^5 + 14s^4 + 16s^3 + 8s^2}$. The mathematical juggling of partial

fraction brings forth $X(s) = -\dfrac{1}{s} + \dfrac{5}{4s^2} + \dfrac{9}{2(s+2)} + \dfrac{45}{4(s+2)^2} - \dfrac{3(s+1)}{2(s^2 + 2s + 2)}$ and $Y(s) =$

$\dfrac{3}{4s} - \dfrac{3}{4s^2} - \dfrac{3}{4(s+2)} - \dfrac{45}{4(s+2)^2} + \dfrac{s+1}{s^2 + 2s + 2} + \dfrac{1}{2(s^2 + 2s + 2)}$. Comparison with the standard transforms allows us

to write $x(t) = -1 + \dfrac{5}{4}t + \dfrac{45}{4}te^{-2t} + \dfrac{9}{2}e^{-2t} - \dfrac{3}{2}e^{-t}\cos t$ and $y(t) = \dfrac{3}{4} - \dfrac{3}{4}t - \dfrac{45}{4}te^{-2t} - \dfrac{3}{4}e^{-2t} + e^{-t}\cos t + \dfrac{1}{2}e^{-t}\sin t$. The

given differential equations should be arranged before you apply the transform:

MATLAB Command

```
>>syms t ↵              ← Declaration of t
>>x=sym('x(t)'); ↵      ← Declaration of x(t)
>>y=sym('y(t)'); ↵      ← Declaration of y(t)
>>eqn1=laplace(diff(x)-x-3*y-t); ↵                     ← Transform on equation 1
>>eqn2=laplace(diff(y)+3*x+5*y-sin(t)*exp(-t)); ↵      ← Transform on equation 2
>>eqn1=subs(eqn1,{'x(0)'},{'2'}); ↵    ← Inserting initial data on equation 1
>>eqn2=subs(eqn2,{'y(0)'},{'1'}); ↵    ← Inserting initial data on equation 2
```

Change the strings 'laplace(x(t),t,s)' to X and 'laplace(y(t),t,s)' to Y:

```
>>eqn1=subs(eqn1,{'laplace(x(t),t,s)','laplace(y(t),t,s)'},{'X','Y'}); ↵
>>eqn2=subs(eqn2,{'laplace(x(t),t,s)','laplace(y(t),t,s)'},{'X','Y'}); ↵
```

Obtain $X(s)$ and $Y(s)$ forming the equations eqn1=0 and eqn2=0 respectively:

```
>>R=solve(eqn1,eqn2,'X','Y') ↵

R =
    X: [1x1 sym]
    Y: [1x1 sym]        ← R is a structure array having two members X and Y
>>X=(R.X); ↵            ← Assign the first member of R to X
>>Y=(R.Y); ↵            ← Assign the second member of R to Y
>>pretty(X) ↵           ← Display the contents of X


        4     5      2      3
    17 s  + 2 s  + 36 s  + 31 s  + 12 s + 10
    ---------------------------------------------
        2  4     3      2
    s  (s  + 6 s  + 14 s  + 16 s + 8)

>>pretty(Y) ↵           ← Display the contents of Y


      5        3      4      2
    s  - 11 s  - 5 s  -18 s  - 6 s - 6
    ---------------------------------------------
        2  4     3      2
    s  (s  + 6 s  + 14 s  + 16 s + 8)

>>x=ilaplace(X); ↵          ← L⁻¹ on X(s)
>>y=ilaplace(Y); ↵          ← L⁻¹ on Y(s)
```

The final solution is displayed as:

```
>>pretty(x) ↵
```

5/4 t - 1 + 45/4 t exp(-2 t) + 9/2 exp(-2 t) - 3/2 exp(-t) cos(t)

>>pretty(y) ↵

- 3/4 t + 3/4 - 45/4 t exp(-2 t) - 3/4 exp(-2 t) + exp(-t) cos(t)

+ 1/2 sin(t) exp(-t)

8.2.14 Solving intgrodifferential equations using Laplace transforms

An equation can have both the differential $\frac{d}{dx}$ and integral $\int .... dx$ operators. Such type of equation is called an integrodifferential equation. For example, $\frac{dy}{dx} + \int_{t=0}^{t=x} y\,dt + y = 0$ is an integrodifferential equation. Articles 8.2.6 and 8.2.13 illustrate the implementation of Laplace transform on the differential and integral equations. We combine these two techniques to solve an integrodifferential equation.

Solve the following integrodifferential equations using the Laplace transform:

A. $y' + \int_{x=0}^{x=t} y(x)dx = 2u(t)$, $y(0) = 4$

B. $8y'' + 12y' + 6y + \int_{x=0}^{x=t} y(x)dx = 2\delta(t-2)$, $y(0) = 0$, and $y'(0) = 3$

⊟ Example A

The equation becomes $[sY(s) - y(0)] + \frac{Y(s)}{s} = \frac{2}{s}$ after taking the transform. Use the initial data to get $Y(s) = \frac{2(2s+1)}{s^2+1} = \frac{4s}{s^2+1} + \frac{2}{s^2+1}$, from which $y(t) = 4\cos t + 2\sin t$. To apply the subroutine, an integrodifferential equation is rearranged in two parts – the first is the nonintegral part and the other is the rest of the equation. So, $y' + \int_{x=0}^{x=t} y(x)dx = 2u(t)$ is written as $[y' - 2u(t)] + \left[\int_{x=0}^{x=t} y(x)dx\right] = 0$. The transform of the first part is assigned to part1.

The dummy dependent variable $y(x)$ needs declaration before one takes the transform of the second part. The step by step procedure is shown below:

MATLAB Command

```
>>syms t x ↵          ← Declaration of independent variable t and dummy variable x
>>y=sym('y(t)'); ↵     ← Declaration of y(t) for the first part of the equation
```

Transform on the first part of the equation is assigned to part1:

```
>>part1=laplace(diff(y)-2*sym('Heaviside(t)')); ↵
>>y=sym('y(x)'); ↵     ← Defining y(x) as function of x
```

Transform on the second part of the equation is assigned to part2:

```
>>part2=laplace(int(y,x,0,t)); ↵
```

```
Warning: Explicit integral could not be found.
> In C:\MATLAB\toolbox\symbolic\@sym\int.m at line 58          ← Ignore the warning
```

Form the complete equation adding part1 and part2 and assign that to eqn:

```
>>eqn=part1+part2; ↵
```

Initial condition is inserted to eqn:

```
>>eqn=subs(eqn,{'y(0)'},{'4'}); ↵
```

```
>>eqn=subs(eqn,{'laplace(y(t),t,s)'},{'Y'});⏎      ← Change the transform string
>>Y=solve(eqn,'Y');⏎                               ← Y(s) is obtained from Y=0
>>pretty(Y)⏎                                        ← Displaying Y(s)
```

$$2 \; \frac{2s+1}{s^2+1}$$

```
>>y=ilaplace(Y);⏎
>>pretty(y)⏎                                        ← Displaying y(t)
```

$$4\cos(t)+2\sin(t)$$

⊟ Example B

Proceeding like previous examples, we have $Y(s)=\dfrac{2s(12+e^{-2s})}{8s^3+12s^2+6s+1}=\dfrac{3}{\left(s+\dfrac{1}{2}\right)^2}-\dfrac{3}{2\left(s+\dfrac{1}{2}\right)^3}+$

$e^{-2s}\left[\dfrac{1}{4\left(s+\dfrac{1}{2}\right)^2}-\dfrac{1}{8\left(s+\dfrac{1}{2}\right)^3}\right]$ (after partial fraction). Utilize $L^{-1}[e^{-t_0 s}F(s)]=f(t-t_0)\,u(t-t_0)$ to have $y(t)=3te^{-\frac{t}{2}}-$

$\dfrac{3t^2 e^{-\frac{t}{2}}}{4}+\dfrac{1}{4}(t-2)e^{-\frac{t-2}{2}}u(t-2)-\dfrac{1}{16}(t-2)^2 e^{-\frac{t-2}{2}}u(t-2)=3te^{-\frac{t}{2}}-\dfrac{3t^2 e^{-\frac{t}{2}}}{4}-\dfrac{1}{16}t^2 e^{1-\frac{t}{2}}u(t-2)+\dfrac{1}{2}te^{1-\frac{t}{2}}u(t-2)-\dfrac{3}{4}e^{1-\frac{t}{2}}u(t-2)$

(after simplification). The rearranged equation is $[8y''+12y'+6y-2\delta(t-2)]+[\int_{x=0}^{x=t}y(x)dx]=0$. Implement it as follows:

MATLAB Command

```
>>syms t x ⏎
>>y=sym('y(t)');⏎
>>part1=laplace(8*diff(y,2)+12*diff(y)+6*y-2*sym('Dirac(t-2)'));⏎    ← L on the part1
>>y=sym('y(x)');⏎
>>part2=laplace(int(y,x,0,t));⏎

Warning: Explicit integral could not be found.
> In C:\MATLAB\toolbox\symbolic\@sym\int.m at line 58
>>eqn=part1+part2;⏎                                   ← L of the complete equation
>>eqn=subs(eqn,{'y(0)','D(y)(0)'},{'0','3'});⏎
>>eqn=subs(eqn,{'laplace(y(t),t,s)'},{'Y'});⏎
>>Y=solve(eqn,'Y');⏎
>>pretty(Y)⏎                                          ← Showing Y(s)
```

$$2 \; \frac{s\,(12+\exp(-2s))}{8s^3+12s^2+6s+1}$$

```
>>y=ilaplace(Y);⏎
>>pretty(y)⏎                                          ← See the final output y(t)
```

$$-3/4\ t^2\ exp(-\ 1/2\ t) + 3\ t\ exp(-\ 1/2\ t)$$

$$- 1/16\ Heaviside(t - 2)\ exp(-\ 1/2\ t + 1)\ t^2$$

$$+ 1/2\ Heaviside(t - 2)\ exp(-\ 1/2\ t + 1)\ t$$

$$- 3/4\ Heaviside(t - 2)\ exp(-\ 1/2\ t + 1)$$

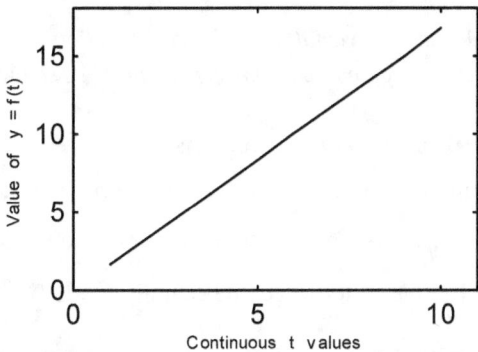

Figure 8.13 *Plot of continuous y = f(t)*

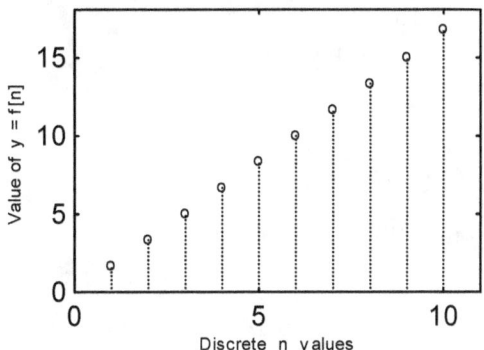

Figure 8.14 *Plot of discrete y = f[n]*

8.3 Z transform

Recall that the Laplace transform of a function $f(t)$ is given by $L\ [f(t)] = \int_{t=0}^{t=\infty} f(t)e^{-st}\ dt$. To have the transform, function $f(t)$ should be continuous, that means, the Laplace transform is applicable for the continuous functions. Function $f(t)$ can be in discrete form too. We call the discrete functions as sequences. A sequence is denoted by $f[n]$, where n is the discrete independent variable. Following figures explain the difference between a continuous and a discrete functions. In figure 8.13, $f(t) = \frac{5}{3}t$ is plotted for $1 \le t \le 10$, where t can assume any integer or fractional value between 1 and 10. On the contrary, $f[n] = \frac{5}{3}n$ is plotted for $1 \le n \le 10$ in figure 8.14, but here, n assumes only the integer values from 1 to 10. The independent variable n can not be a fractional number that is the meaning of discrete. For discrete n , $f[n]$ is generally discrete but the envelope of $f[n]$ may follow some continuous function. Z transform is the discrete counterpart of the Laplace transform. Z transform is denoted by the operator 'Z', that is, $Z\{f[n]\} = \sum\limits_{n=-\infty}^{n=\infty} f[n]z^{-n}$. The transform is called the bilateral Z transform or Z transform of discrete function $f[n]$. *But in MATLAB, the Z transform is unilateral that is given by* $Z\{f[n]\} = \sum\limits_{n=0}^{n=\infty} f[n]z^{-n}$ *and unilateral Z transfotrm incorporates only the values of* $f[n]$ *for* $n \ge 0$. The Z transform of any discrete function $f[n]$ in general is a function of variable z, so, we can say $Z\{f[n]\} = F(z)$. MATLAB counterpart of Z transform is 'ztrans' (abbreviation of z transform). To have the Z transform of a discrete function f,

MATLAB Steps:
1. *Declare the variables of discrete function f as symbolic and*
2. *Use the command ztrans(represent f as string, variable of f, transform variable).*

8.3.1 Z transforms of some discrete functions

Verify the transform with $f[n] = e^{na}$. Its Z transform, $F(z)$, is $Z\{[e^{an}]\} = \sum\limits_{n=0}^{n=\infty} e^{an}z^{-n} = \sum\limits_{n=0}^{n=\infty}(e^{a}z^{-1})^{n} = 1 + (e^{a}z^{-1}) + (e^{a}z^{-1})^{2} + (e^{a}z^{-1})^{3} + \ldots + \infty$. The last series is similar to the series of $(1-x)^{-1}$ for $|x| \le 1$, which is $1 + x + x^{2} + x^{3} + \ldots + \infty$. Utilize the series to get the sum as $(1 - e^{a}z^{-1})^{-1} = \dfrac{1}{1 - e^{a}z^{-1}} = \dfrac{z}{z - e^{a}}$. We therefore have $Z\{[e^{an}]\} = \dfrac{z}{z - e^{a}}$ over the region $|e^{a}z^{-1}| \le 1$ or $z \ge |e^{a}|$. Implementation of Z transform is very similar to those of Fourier and Laplace transforms. Obtain the Z transform for the example at hand:

MATLAB Command
```
>>syms n a ↵
>>F=ztrans(exp(n*a)) ↵

F =

z/(z-exp(a))
```
Nice form is seen by:
```
>>pretty(F) ↵
        z
    ---------------
      z - exp(a)
```

When the independent variable is p and the return is wanted in terms of w, that is, $f[p] = e^{ap}$ and $Z\{f[p]\} = \sum\limits_{p=0}^{p=\infty} f[p]w^{-p} = \dfrac{w}{w - e^{a}}$. To do so:

```
>>syms p a w ↵
>>F=ztrans(exp(p*a),p,w); ↵
>>pretty(F) ↵
          w
    ----------------
       w - exp(a)
```

Close form summation of all functions may not be obtained by 'ztrans'. Example of $f[n] = e^{-n^{2}}$ can be mentioned in this regard. Try with this:

```
>>syms n ↵
>>ztrans(exp(-n^2)) ↵
```

ans =

ztrans(exp(-n^2),n,z) ← Output is just the definition of $Z \{[e^{-n^2}]\}$

Shown table 8.C presents the mathematical and MATLAB correspondence of some unilateral Z transforms.

Table 8.C   Unilateral Z transforms of some discrete functions

Mathematical form	MATLAB command for symbolic form

$Z\{\sin[na]\}=\dfrac{z\sin a}{z^2-2z\cos a+1}$	`>>syms a n ↵`
	`>>pretty(ztrans(sin(n*a))) ↵`
	
	`            sin(a) z`
	`    ---------------------------`
	`                  2`
	`        - 2 z cos(a) + z + 1`

$Z\{\cos[na]\}=\dfrac{z(z-\cos a)}{z^2-2z\cos a+1}$	`>>syms a n ↵`
	`>>pretty(ztrans(cos(n*a))) ↵`
	
	`            z (-cos(a) + z)`
	`    ---------------------------`
	`                  2`
	`        - 2 z cos(a) + z + 1`

$Z\{[a^n]\}=\dfrac{z}{z-a}$	`>>syms a n ↵`
	`>>pretty(ztrans(a^n)) ↵`
	
	`              z`
	`         - -----------`
	`             -z + a`

$Z\{u[n]\}=\dfrac{z}{z-1}$	`>>u=sym('Heaviside(n)');  ↵`
	`>>pretty(ztrans(u)) ↵`
	
	`              z`
	`           ----------`
	`             z - 1`

$Z\{\delta[n]\}=1$	`>>d=sym('charfcn[0](n)');  ↵`
	`>>pretty(ztrans(d)) ↵`
	`              1`

$Z\{[r^n\sin na]\}=\dfrac{rz\sin a}{z^2-2zr\cos a+r^2}$	`>>syms a n r ↵`
	`>>F=ztrans(r^n*sin(n*a));  ↵`
	`>>pretty(F) ↵`
	
	`              r sin(a) z`
	`    -----------------------------`
	`                  2     2`
	`        -2 z r cos(a) + z + r`

Notice that $\delta[n]$ is represented by 'charfcn[0](n)' not by 'Dirac(n)'	

8.3.2 Z transforms of $nf[n]$, $f[n-n_0]\,u[n-n_0]$, and $z_0^{\,n}f[n]$

In this article, we verify some theorems and properties of Z transforms, many of which are often useful in manipulating the Z transforms. The Z transforms of $nf[n]$, $f[n-n_0]\,u[n-n_0]$, and $z_0^{\,n}f[n]$ are given

by $-z\dfrac{dF(z)}{dz}$, $z^{-n_0}F(z)$, and $F\left(\dfrac{z}{z_0}\right)$ respectively, where $F(z)=Z\{f[n]\}$. Take an example of $f[n]=a^n$. We

wish to find $Z\{[n^2a^n]\}$ therefore $Z\{[n^2a^n]\}=-z\dfrac{dZ\{[na^n]\}}{dz}=-z\dfrac{d}{dz}\left[-z\dfrac{dZ\{[a^n]\}}{dz}\right]=-z\dfrac{d}{dz}\left[-z\dfrac{d}{dz}\left(\dfrac{z}{z-a}\right)\right]=$

$-z\dfrac{d}{dz}\left[\dfrac{za}{(z-a)^2}\right]=\dfrac{za(z+a)}{(z-a)^3}$. The verification is as follows:

MATLAB Command

```
>>syms n a ↵
>>F=ztrans(n^2*a^n); ↵
```

>>pretty(F) ↵

$$- \frac{z\,a\,(z+a)}{(-z+a)^3}$$

Then, we are interested in finding $Z\{a^{n-3}\,u[n-3]\}$, which is $z^{-3}\,Z\{[a^n]\}=z^{-3}\,\dfrac{z}{z-a}=\dfrac{1}{z^2(z-a)}$ and is implemented as follows:

MATLAB Command

>>syms n a ↵
>>f=a^(n-3)*sym('Heaviside(n-3)'); ↵
>>F=ztrans(f); ↵
>>pretty(simplify(F)) ↵

$$- \frac{1}{(-z+a)^2\,z}$$

Remember that $Z\{[a^{n-3}]\,u[n-3]\}$ and $Z\{[a^{n-3}]\}$ are different, $Z\{[a^{n-3}]\}=\dfrac{z}{a^3(z-a)}$. The next verification is

for $Z\{[3^n a^n]\}$, which is calculated as $Z\{[a^n]\}\Big|_{z\to\frac{z}{3}}=\dfrac{z}{(z-a)}\Big|_{z\to\frac{z}{3}}=\dfrac{\frac{z}{3}}{\frac{z}{3}-a}=\dfrac{z}{z-3a}$:

MATLAB Command

>>syms n a ↵
>>F=ztrans(3^n*a^n); ↵
>>pretty(F) ↵

z (-3 | a | cos(1/2 (-1 + signum(a)) pi) + z

   - 3 i | a | sin(1/2 (-1 + signum(a)) pi))/(

-6 z | a | cos(1/2 (-1 + signum(a)) pi) + z^2 + 9 | a |^2)

The return of the transform might make you doubtful. Actually, we did not specify anything about the symbolic constant a. Function 'ztrans' assumes a as a complex number along with this both the real and imaginary parts of a belong to the set of all real numbers. $Z\{[3^n a^n]\}=\dfrac{z}{z-3a}$ is valid if $a>0$. We take the help from Maple package to assume $a>0$. Continue with the following:

>>clear all ↵   ← Removing previous variables
>>syms a n ↵
>>maple('assume','a>0'); ↵   ← Applying $a>0$ using 'assume' facility of Maple
>>F=ztrans(3^n*a^n); ↵
>>pretty(F) ↵

$$\frac{z}{z-3\,a\!\sim}$$

                                 ← 'a~' indicates the real part of symbolic constant a

8.3.3 Z transforms of finite sequences

Discrete functions discussed so far can conceive any n where $n\geq 0$. Functions may be seen that exist only for few values of n. These types of functions are called the finite sequence. We present two examples of finite sequence. Within the given interval, a sequence can follow some function. Shown figure 8.15 is the plot

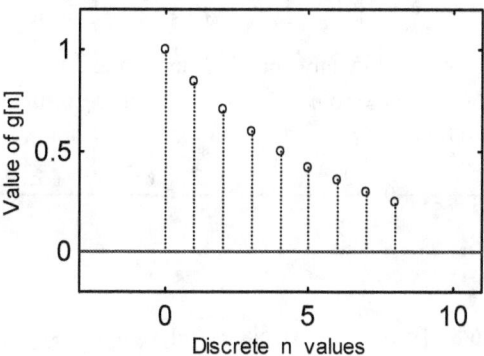

Figure 8.15 *Plot of g[n] vs n*

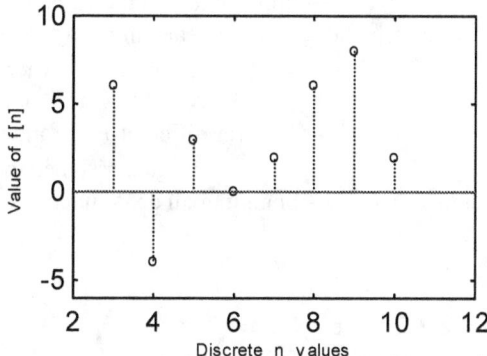

Figure 8.16 *Plot of f[n] vs n*

of such a function, which is given by $g[n]=\begin{cases} 2^{-\frac{n}{4}} & \text{for } 0 \le n \le 8 \\ \\ 0 & \text{elsewhere} \end{cases}$. By dint of the unit step sequence, $g[n]$ is

expressed as $g[n]=[2^{-\frac{n}{4}}]\{u[n]-u[n-8]\}$, on that, $Z\{g[n]\}=Z\{[2^{-\frac{n}{4}}]u[n]\}-Z\{[2^{-\frac{n}{4}}]u[n-8]\}=\dfrac{2^{\frac{1}{4}}z}{2^{\frac{1}{4}}z-1}-$

$\dfrac{1}{4}\dfrac{2^{\frac{1}{4}}}{(2^{\frac{1}{4}}z-1)z^7}=\dfrac{2^{\frac{1}{4}}(4z^8-1)}{4(2^{\frac{1}{4}}z-1)z^7}=\dfrac{1}{4z^7}\left[4z^7+2\times2^{\frac{3}{4}}z^6+2\times2^{\frac{1}{2}}z^5+2\times2^{\frac{1}{4}}z^4+2z^3+2^{\frac{3}{4}}z^2+2^{\frac{1}{2}}z+2^{\frac{1}{4}}\right]$ (performing long

division). The solution is as follows:

MATLAB Command

```
>>w=sym('Heaviside(n)-Heaviside(n-8)');  ↵          ← Defining  u[n]−u[n−8]
>>syms n  ↵

>>g=2^(-n/4)*w;  ↵                        ← g[n] is formed from the product of 2^(-n/4) and u[n]−u[n−8]
>>G=ztrans(g);  ↵
>>pretty(G)  ↵
              7     3/4  6      1/2  5     1/4  4     3    3/4  2    1/2        1/4
      1/4 (4 z  + 2 2    z  + 2 2    z  + 2 2    z  + 2 z + 2    z  + 2    z + 2

                                            /   7
                                          ) /  z
                                          /
```

The next illustrative finite sequence is $\begin{Bmatrix} f[n] & \rightarrow & 6 & -4 & 3 & 0 & 2 & 6 & 8 & 2 \\ n & \rightarrow & 3 & 4 & 5 & 6 & 7 & 8 & 9 & 10 \end{Bmatrix}$ (shown in figure 8.16). The

sequence tells us that $f[n]$ does not follow any specific function and that $f[n]$ exists for $3 \leq n \leq 10$. At $n = 3$, $f[n] = 6$ can be represented as $f[3] = 6\delta[n-3]$. As a closed form function, we can write $f[n] = 6\delta[n-3] - 4\delta[n-4] + 3\delta[n-5] + 0\delta[n-6] + 2\delta[n-7] + 6\delta[n-8] + 8\delta[n-9] + 2\delta[n-10]$. Applying $Z\{\delta[n-n_0]\} =$

$\dfrac{1}{z^{n_0}}$ makes us get $\dfrac{6}{z^3} - \dfrac{4}{z^4} + \dfrac{3}{z^5} + 0 + \dfrac{2}{z^7} + \dfrac{6}{z^8} + \dfrac{8}{z^9} + \dfrac{2}{z^{10}} = \dfrac{6z^7 - 4z^6 + 3z^5 + 2z^3 + 6z^2 + 8z + 2}{z^{10}}$. To have the

transform,

MATLAB Command

>>y=sym([6 -4 3 0 2 6 8 2]); ↵   ← Declare $f[n]$ values as symbolic and put to a row matrix y

>>d=sym('charfcn[m](n)'); ↵   ← Define the general delta function $\delta[n-m]$ as symbolic

>>syms m ↵   ← Define m as symbolic

>>f=[]; ↵   ← Initialization of the variable f

>>for k=3:10 f=[f subs(d,m,k)]; end ↵   ← Create all delta functions from $\delta[n-3]$ to $\delta[n-10]$ and assign those to f, where f is a row matrix. For loop index k gives the control on the location of the delta functions

>>f=y.*f; ↵   ← $f[n]$ for different n is formed from the scalar product of y and f and again assigned to f

>>Z=ztrans(sum(f)); ↵   ← Summing all delta functions of f is done by 'sum'

>>pretty(simplify(Z)) ↵

$$\frac{6z^7 - 4z^6 + 3z^5 + 2z^3 + 6z^2 + 8z + 2}{z^{10}}$$

That is what is expected.

8.3.4 Inverse Z transform

Once we know the Z transform $F(z)$ of a discrete function $f[n]$, its inverse Z transform can be obtained in terms of the contour integral. The formal definition of the inverse Z transform (by operator, Z^{-1}) is given by $f[n] = \dfrac{1}{2\pi i}\oint_C F(z)z^{n-1}dz$, where C is a counterclockwise closed contour in the region of the convergence of $F(z)$ and the contour encircles the origin of the z-plane. If $F(z)z^{n-1}$ is a rational function of z, it can be expressed as $F(z)z^{n-1} = \dfrac{\phi(z)}{(z-p)^m}$, where $F(z)z^{n-1}$ has m poles at $z = p$ and $\phi(z)$ does not have any pole at $z = p$. By formula the residue of $F(z)z^{n-1}$ at $z = p$ is given by $\dfrac{1}{(m-1)!}\dfrac{d^{m-1}\phi(z)}{dz^{m-1}}\bigg|_{z=p}$. But finding the residues of nonrational function is somewhat difficult. Having found the residues, the sequence can be obtained from $f[n] = \dfrac{1}{2\pi i}\oint_C F(z)z^{n-1}dz = \Sigma[\text{Residues of } F(z)z^{n-1} \text{ at the poles inside } C]$. The MATLAB counterpart of the inverse Z transform is 'iztrans' (abbreviation of inverse z transform). To perform inverse Z transform,

    MATLAB Step:

         Use the command iztrans(function in string form, transform variable, return variable).

From table 8.C, we have $Z\{[a^n]\} = \dfrac{z}{z-a}$ so the inverse Z transform of $\dfrac{z}{z-a}$ should return us $[a^n]$. The

string form of $\dfrac{z}{z-a}$ is z/(z-a). Implementation is shown as follows:

MATLAB Command

        >>syms z a ↵   ← Defining variables of $\dfrac{z}{z-a}$ as symbolic

        >>Z=z/(z-a); ↵   ← Assigning $\dfrac{z}{z-a}$ to Z

        >>iztrans(Z) ↵   ← Applying 'iztrans' on Z

        ans =

        a^n   ← Default return in terms of n

When the transform is given in terms of the variable w (i.e., $Z\{[a^n]\}=\dfrac{w}{w-a}$) and the return is wanted in terms of x (i.e., $Z^{-1}[\dfrac{w}{w-a}]=a^x$):

        >>syms w a x ↵   ← Defining variables of $\dfrac{w}{w-a}$ as well as x as symbolic

        >>Z=w/(w-a); ↵   ← Assigning $\dfrac{w}{w-a}$ to Z

        >>iztrans(Z,w,x) ↵

        ans =

        a^x

Find the inverse Z transforms of the following functions:

A. $X(z)=\dfrac{z^2}{(z-a)(z-c)}$ B. $F(z)=\dfrac{z^{-1}+z^{-2}}{\left(1-\dfrac{1}{2}z^{-1}\right)\left(1+\dfrac{1}{3}z^{-1}\right)}$

C. $F(z)=\ln(1-4z)$ D. $F(z)=\dfrac{3}{(2-\frac{2}{3}z^{-1})^2(2-3z^{-1})(1-4z^{-1})}$

▣ Example A

Poles of $X(z)z^{n-1}=\dfrac{z^{n+1}}{(z-a)(z-c)}$ are $z=a$ and $z=c$. Residues of $X(z)z^{n-1}$ at $z=a$ and $z=c$ are $\varphi(a)$ and $\psi(c)$ respectively, where $\varphi(a)=\dfrac{z^{n+1}}{z-c}\Big|_{z=a}=\dfrac{a^{n+1}}{a-c}$ and $\psi(c)=\dfrac{z^{n+1}}{z-a}\Big|_{z=c}=\dfrac{c^{n+1}}{c-a}$, hence, $Z^{-1}[X(z)]=x[n]=$

$\sum[\text{residues}]=\dfrac{a^{n+1}}{a-c}+\dfrac{c^{n+1}}{c-a}=\dfrac{a^{n+1}-c^{n+1}}{a-c}$. To have it done:

MATLAB Command

        >>syms z a c ↵
        >>X=z^2/(z-a)/(z-c); ↵   ← Assigning string of $X(z)$ to X
        >>x=iztrans(X) ↵

        x =

        (a*a^n-c*c^n)/(-c+a)
        >>pretty(x) ↵

```
          n     n
      a a  - c c
      -------------
        -c + a
```

352

⊟ Example B

This type of representation of $F(z)$ arises in digital filter synthesis. By means of rationalization,

$F(z)\ z^{n-1}$ turns out to be $\dfrac{z^n(1+z)}{z\left(z-\dfrac{1}{2}\right)\left(z+\dfrac{1}{3}\right)}$ from which the poles are $z=0$, $z=\dfrac{1}{2}$, and $z=-\dfrac{1}{3}$. Residue of

$F(z)z^{n-1}$ at $z=0$ is $R_1=\dfrac{z^n(1+z)}{\left(z-\dfrac{1}{2}\right)\left(z+\dfrac{1}{3}\right)}\Bigg|_{z=0}=-6\times 0^n=\begin{cases}-6 & for\ \ n=0\\ 0 & for\ \ n\neq 0\end{cases}=-6\delta[n]$. The second and third residues

are $R_2=\dfrac{z^n(1+z)}{z\left(z+\dfrac{1}{3}\right)}\Bigg|_{z=\frac{1}{2}}=\dfrac{18}{5}\left(\dfrac{1}{2}\right)^n$ and $R_3=\dfrac{z^n(1+z)}{z\left(z-\dfrac{1}{2}\right)}\Bigg|_{z=-\frac{1}{3}}=\dfrac{12}{5}\left(-\dfrac{1}{3}\right)^n$, whence, $Z^{-1}[F(z)]=f[n]=\sum R=$

$-6\delta[n]+\dfrac{18}{5}\left(\dfrac{1}{2}\right)^n+\dfrac{12}{5}\left(-\dfrac{1}{3}\right)^n$. For this specific example, we assign the numerator and denominator separately.

Notice that $z^{n-1}=\dfrac{z^n}{z}$ means one pole is located at $z=0$. Following is the implementation:

MATLAB Command
>>syms z ↵
>>n=1/z+1/z^2; ↵   ← Assigning the numerator $z^{-1}+z^{-2}$ to n

>>d=(1-1/2/z)*(1+1/3/z); ↵   ← Assigning the denominator $\left(1-\dfrac{1}{2}z^{-1}\right)\left(1+\dfrac{1}{3}z^{-1}\right)$ to d

>>f=iztrans(n/d); ↵   ← $F(z)$ is formed by n/d
>>pretty(f) ↵

-6 charfcn[0](n) + 18/5 (1/2)^n + 12/5 (-1/3)^n

⊟ Example C

Inverse Z transform look-up table of MATLAB does not contain the inverse of function like $\ln(1-4z)$. So, the response for the example is as follows:

MATLAB Command
>>syms z ↵
>>Z=log(1-4*z); ↵
>>iztrans(Z) ↵

ans =

iztrans(log(1-4*z),z,n)   ← Which is the definition of Z^{-1}

⊟ Example D

Arrange $F(z)z^{n-1}$ as $\dfrac{3z^{n-1}}{(2-\frac{2}{3}z^{-1})^2(2-3z^{-1})(1-4z^{-1})}=\dfrac{3z^{n+3}}{(2z-\frac{2}{3})^2(2z-3)(z-4)}$ (rationalization by z^4)=

$\dfrac{3z^{n+3}}{8(z-\frac{1}{3})^2(z-\frac{3}{2})(z-4)}$. The poles of $F(z)z^{n-1}$ are located at $z=\dfrac{1}{3}$ (of multiplicity 2), $z=\dfrac{3}{2}$, and $z=4$. The

residue R_1 of $F(z)z^{n-1}$ at $z=\dfrac{1}{3}$ is $\dfrac{1}{(2-1)!}\dfrac{d[\phi(z)]}{dz}\Bigg|_{z=\frac{1}{3}}$, where $\phi(z)=\dfrac{3z^{n+3}}{8(z-\frac{3}{2})(z-4)}$ and from which we have

$$\frac{d[\phi(z)]}{dz} = \frac{-\frac{3}{4}[-2z^{n+4} + 22z^{n+3} - 36z^{n+2} - 2nz^{n+4} + 11nz^{n+3} - 12nz^{n+2}]}{(2z-3)^2(z-4)^2} \quad \text{and} \quad R_1 = \frac{1}{(2-1)!} \frac{d[\phi(z)]}{dz}\bigg|_{z=\frac{1}{3}} = \frac{195}{5929}3^{-n} + \frac{3}{308}n3^{-n}.$$

The other residues (R_2 and R_3) are calculated as $R_2 = \phi(z)\big|_{z=\frac{3}{2}} = -\frac{729}{1960}\left(\frac{3}{2}\right)^n$ and $R_3 = \psi(z)\big|_{z=4} = \frac{432 \times 4^n}{605}$,

where $\phi(z) = \frac{3z^{n+3}}{8(z-\frac{1}{3})^2(z-4)}$ and $\psi(z) = \frac{3z^{n+3}}{8(z-\frac{1}{3})^2(z-\frac{3}{2})}$ respectively. Summing all, one obtains $Z^{-1}[F(z)] =$

$$f[n] = \frac{195}{5929}3^{-n} + \frac{3}{308}n3^{-n} - \frac{729}{1960}\left(\frac{3}{2}\right)^n + \frac{432 \times 4^n}{605}.$$ It is better if we can put the numerator and denominator

strings separately to MATLAB. To show all these:

MATLAB Command

```
>>syms z ↵
>>n=3; ↵                        ← Assigning numerator string to n
>>d=(2-2/3/z)^2*(2-3/z)*(1-4/z); ↵   ← Assigning denominator string to d
>>f=iztrans(n/d) ↵              ← Z⁻¹ of F(z) =n/d
```

f =

195/5929*(1/3)^n+3/308*(1/3)^n*n-729/1960*(3/2)^n+ 432/605*4^n

Symbolic form is better for verification:
```
>>pretty(f) ↵
```

```
 195              n              n     729          n     432    n
--------- (1/3)  + 3/308 (1/3)  n - --------- (3/2)  + --------- 4
 5929                               1960                605
```

8.3.5 Representations of difference equations

Differential equations are applicable if the rates of change of the dependent variables with respect to the independent variables are known (in other words, continuous variation of the dependent variable w.r.to the independent one). Sometimes we may come across a relationship between the changes or differences rather than the rates of changes. The equation $y_k = y_{k-1} + y_{k-2}$ is an example of the difference equation. The equation says that for any k, the output y is the sum of the preceding two terms. That is,

when	$k = 2$,	$y_2 = y_1 + y_0$,
	$k = 3$,	$y_3 = y_2 + y_1$,
	$k = 4$,	$y_4 = y_3 + y_2$,
	$k = 5$,	$y_5 = y_4 + y_3$, and
		.... so on.

Another style of writing the difference equation $y_k = y_{k-1} + y_{k-2}$ is $y[k] = y[k-1] + y[k-2]$. In addition to the mathematical notation, MATLAB style of representing the discrete dependent variable y_k or $y[k]$ is also provided. y_k is a symbolic variable and it is entered in MATLAB as follows:

MATLAB Command

```
>>yk=sym('y(k)');  ↵    ← Describing yk as symbolic and assign that to yk
>>yk1=sym('y(k+1)');  ↵  ← Describing yk+1 as symbolic and assign that to yk1
>>yk2=sym('y(k+2)');  ↵  ← Describing yk+2 as symbolic and assign that to yk2
                        ........... so on.
```

Order of a difference equation is the difference between the largest and smallest indices appearing in the equation. Thus the order of difference equation $y_k = y_{k-1} + y_{k-2}$ is $k - (k-2) = 2$. The notions of the derivatives of differential and difference equations are little different. The discrete first derivative of any sequence y_k is defined as $\dot{y}_k$ or $Dy_k = y_{k+1} - y_k$. The second derivative is given by $\ddot{y}_k$ or $D^2 y_k = D(Dy_k) = D(y_{k+1} - y_k) = y_{k+2} - y_{k+1} - (y_{k+1} - y_k) = y_{k+2} - 2y_{k+1} - y_k$. In a similar fashion, higher order derivatives can be obtained, which are as follows:

$$D^3 y_k = y_{k+3} - 3y_{k+2} + 3y_{k+1} - y_k ,$$
$$D^4 y_k = y_{k+4} - 4y_{k+3} + 6y_{k+2} - 4y_{k+1} + y_k ,$$
$$D^5 y_k = y_{k+5} - 5y_{k+4} + 10y_{k+3} - 10y_{k+2} + 5y_{k+1} - y_k , \text{ and}$$

.... so on.

8.3.6 Solving difference equations using unilateral Z transforms

The procedure is very similar to the problems we encountered in solving the differential equations using the Laplace transforms as stated in article 8.2.13. Z transforms to be determined for the difference equations do not apply on the derivatives instead they do on the dependent variables of the equation (which are termed as sequences, described in article 8.3.3). This is a lucid difference between the differential and difference equations. How Z transform applies to sequences is derived taking the example of $Z\{y_{k+2}\}$:

Thus, $Z\{y_{k+2}\} = \sum\limits_{k=0}^{k=\infty} y[k+2]z^{-k}$ ← From definition

$\qquad = \sum\limits_{p=2}^{p=\infty} y[p]z^{-(p-2)}$ ← Substitute $k+2$ by p

$\qquad = z^2 \sum\limits_{p=2}^{p=\infty} y[p]z^{-p}$

$\qquad = z^2\{y[2]z^{-2} + y[3]z^{-3} + y[4]z^{-4} + y[5]z^{-5} + ...\}$ ← Expansion

$\qquad = z^2\{y[0] + y[1]z^{-1} + y[2]z^{-2} + y[3]z^{-3} + y[4]z^{-4} + y[5]z^{-5} + ...-(y[0] + y[1]z^{-1})\}$

         (algebraic manipulation)

$\qquad = z^2\{\sum\limits_{n=0}^{n=\infty} y[n]z^{-n} - (y[0] + y[1]z^{-1})\}$

$\qquad = z^2(Z\{y[n]\} - y[0] - y[1]z^{-1})$ ← Apply the definition

$\qquad = z^2 Y(z) - z^2 y[0] - y[1]z$

Performing similar manipulations can find the unilateral Z transforms of other sequences. Different seqences and their MATLAB counterparts are presented as follows:

MATLAB Command

for $Z\{y_k\} = Y(z)$,

    `>>yk=sym('y(k)');` ↵
    `>>ztrans(yk)` ↵

    ans =

    ztrans(y(k),k,z)

for $Z\{y_{k+1}\} = zY(z) - y[0]z$,

    `>>yk1=sym('y(k+1)');` ↵
    `>>ztrans(yk1)` ↵

    ans =

    z*ztrans(y(k),k,z)-y(0)*z

for $Z\{y_{k+2}\} = z^2 Y(z) - y[0]z^2 - y[1]z$,

    `>>yk2=sym('y(k+2)');` ↵
    `>>ztrans(yk2)` ↵

    ans =

z^2*ztrans(y(k),k,z)-y(0)*z^2-y(1)*z

for $Z\{y_{k+3}\}=z^3Y(z)-y[0]z^3-y[1]z^2-y[2]z$,

>>yk3=sym('y(k+3)'); ↵
>>ztrans(yk3) ↵

ans =

z^3*ztrans(y(k),k,z)-y(0)*z^3-y(1)*z^2-y(2)*z

……….. so on.

So far our representations have focused on the positive indices. Situation may come when we will have negative indices. We carry out representations for the negative indices as follows:

for $Z\{y_{k-1}\}=\dfrac{1}{z}Y(z)$,

>>yk_1=sym('y(k-1)'); ↵
>>ztrans(yk_1) ↵

ans =

1/z*ztrans(y(k),k,z)

for $Z\{y_{k-2}\}=\dfrac{1}{z^2}Y(z)$,

>>yk_2=sym('y(k-2)'); ↵
>>ztrans(yk_2) ↵

ans =

1/z^2*ztrans(y(k),k,z)

for $Z\{y_{k-3}\}=\dfrac{1}{z^3}Y(z)$,

>>yk_3=sym('y(k-3)'); ↵
>>ztrans(yk_3) ↵

ans =

1/z^3*ztrans(y(k),k,z)

for $Z\{y_{k-4}\}=\dfrac{1}{z^4}Y(z)$,

>>yk_4=sym('y(k-4)'); ↵
>>ztrans(yk_4) ↵

ans =

1/z^4*ztrans(y(k),k,z)

……….. so on.

In the various transforms, we see the transform string appears as 'ztrans(y(k),k,z)', which is equivalent to $Y(z)=Z\{y[k]\}$. That string can be substituted for the working convenience. Now we pay attention to the initial conditions. The equivalence between the symbolic notation and MATLAB counterparts are $y[0]\equiv y(0)$, $y[1]\equiv y(1)$, $y[2]\equiv y(3)$, … etc. Replacement of the transform string and the insertion of initial conditions by any suitable symbolic constant or variables are accomplished by the use of command 'subs' (for step by step procedure go through article 8.2.13).

Solve the following difference equations using unilateral Z transform:

A. $y_{k+2}+4y_{k+1}+4y_k=0$, $y_0=1$, and $y_1=2$

B. $y[n+2]-5y[n+1]+6y[n]=3-n^2$, $y[0]=0$, and $y[1]=2$

C. $\begin{cases} y[n+1]-x[n-1]=2^n \\ x[n]-9y[n]=3\delta[n] \end{cases}$ $y[0]=3$

⊟ Example A

This is a homogeneous difference equation because the right side does not have any forcing function. Apply Z transforms on both sides of the equation to write $z^2Y(z)-y[0]z^2-y[1]z+4\{zY(z)-y[0]z\}+4Y(z)=0$.

Substitute the initial conditons, from which, we have $z^2Y(z)-z^2-2z+4\{zY(z)-z\}+4\,Y(z)=0$. Solving for $Y(z)$

yields $Y(z)=\dfrac{z^2+6z}{z^2+4z+4}=\dfrac{z^2+6z}{(z+2)^2}$. The poles of $z^{k-1}\,Y(z)=\dfrac{z^k(z+6)}{(z+2)^2}$ are $z=-2$ (multiplicity 2). The residue of

$z^{k-1}\,Y(z)$ is $(-2)^k-2(-2)^k k$ (see article 8.3.4 for residue). So, the solution of the given equation is

$y[k]=(-2)^k-2(-2)^k k$ as follows:

MATLAB Command

>>eqn=sym('y(k+2)+4*y(k+1)+4*y(k)'); ↵   ← Forming the left side of the equation
           $y_{k+2}+4y_{k+1}+4y_k=0$ and assign that to eqn

>>T=ztrans(eqn); ↵   ← Z transform on eqn and assign that to T

>>syms Y ↵   ← Defining Y as symbolic, where $Y \equiv Y(z)$

>>T=subs(T,{'ztrans(y(k),k,z)'},Y); ↵   ← Replacing 'ztrans(y(k),k,z)' by Y and
          assigning output to T again

>>T=subs(T,{'y(0)','y(1)'},{'1','2'}); ↵   ← Substituting $y_0=1$ and $y_1=2$

>>Y=solve(T,Y); ↵   ← Finding $Y(z)$ by forming an equation T=0

>>pretty(Y) ↵   ← Displaying $Y(z)$

```
      z (z + 6)
   ------------------
          2
      z + 4 z + 4
```

>>y=iztrans(Y,sym('k')); ↵   ← Taking inverse transform of $Y(z)$.
         Adding extra argument sym('k'), we ask
         MATLAB to return the output in terms of k

>>pretty(y) ↵

```
          k        k
   -2 (-2) k + (-2)
```

⊟ Example B

The equation contains a forcing function on the right side so it represents a nonhomogeneous system. Still the method is useful. As usual, the transform provides $z^2Y(z)-y[0]z^2-y[1]z-5\{zY(z)-y[0]z\}+$

$6\,Y(z)=\dfrac{3z}{z-1}-\dfrac{z(z+1)}{(z-1)^3}$, hence, $Y(z)=\dfrac{z^2(2z^2-3z-1)}{(z-1)^3(z-2)(z-3)}$ (applying the initial conditions and simplification).

The poles of $z^{n-1}\,Y(z)$ are located at $z=1$ (multiplicity 3), $z=2$, and $z=3$. The residues of $z^{n-1}\,Y(z)$ are

$-\dfrac{n^2}{2}-\dfrac{3n}{2}-1$, -2^{n+1}, and 3^{n+1} for $z=1$, $z=2$, and $z=3$ respectively. The final output should be $y[n]=$

$-\dfrac{n^2}{2}-\dfrac{3n}{2}-1-2^{n+1}+3^{n+1}$ that is what is presented below:

MATLAB Command

>>eqn=sym('y(n+2)-5*y(n+1)+6*y(n)-3+n^2'); ↵   ← Left side of the equation
         $y[n+2]-5y[n+1]+6y[n]-3+n^2=0$ is assigned to eqn

>>T=ztrans(eqn); ↵   ← Z transform on eqn

>>eqn=subs(T,{'ztrans(y(n),n,z)','y(0)','y(1)'},{'Y','0','2'}); ↵   ← Change the transform string and
         apply the initial data

>>Y=solve(eqn,'Y'); ↵   ← Find $Y(z)$ from eqn=0

>>pretty(Y) ⏎   ← Readable form of $Y(z)$

$$\frac{z^2 (-1 + 2 z - 3 z^2)}{(z - 1)(z^2 - 5 z + 6)}$$

>>y=iztrans(Y); ⏎   ← Z^{-1} on $Y(z)$
>>pretty(y) ⏎   ← Showing the output $y[n]$

$$- 1/2\, n^2 - 3/2\, n - 1 + 3 \cdot 3^n - 2 \cdot 2^n$$

⊟ Example C

This system has two sequences, which are representing the nonhomogeneous system. Both equations need Z transforms, we therefore have $\begin{cases} z\,Y(z) - 3z - \dfrac{1}{z} X(z) = \dfrac{z}{z-2} \\ X(z) - 9 Y(z) = 3 \end{cases}$, solution of which is

$X(z) = \dfrac{3 z^2 (10 z - 17)}{(z-2)(z+3)(z-3)}$ and $Y(z) = \dfrac{3 z^3 - 5 z^2 + 3 z - 6}{(z-2)(z+3)(z-3)}$. It goes without saying that the poles of $z^{n-1} X(z)$ and $z^{n-1} Y(z)$ are $z = -3, 0, 2,$ and 3. The computed residues are $\dfrac{141}{10}(-3)^n$, 0, $-\dfrac{18}{5} 2^n$, and $\dfrac{39}{2} 3^n$ and $\dfrac{47}{30}(-3)^n$, $-\dfrac{1}{3}\delta[n]$, $-\dfrac{2}{5} 2^n$, and $\dfrac{13}{6} 3^n$ for the just mentioned poles respectively. At the end, we have $x[n] = \dfrac{141}{10}(-3)^n - \dfrac{18}{5} 2^n + \dfrac{39}{2} 3^n$ and $y[n] = \dfrac{47}{30}(-3)^n - \dfrac{1}{3}\delta[n] - \dfrac{2}{5} 2^n + \dfrac{13}{6} 3^n$. Solve the set as follows:

MATLAB Command

```
>>eqn1=sym('y(n+1)-x(n-1)-2^n');        ← Enter equation 1
>>eqn2=sym('x(n)-9*y(n)-3*charfcn[0](n)');  ← Enter equation 2
>>eqn1=ztrans(eqn1);                     ← Z transform on the equation 1
>>eqn2=ztrans(eqn2);                     ← Z transform on the equation 2
>>syms X Y
>>eqn1=subs(eqn1,{'ztrans(y(n),n,z)','ztrans(x(n),n,z)'},{Y,X});
>>eqn2=subs(eqn2,{'ztrans(y(n),n,z)','ztrans(x(n),n,z)'},{Y,X});
>>eqn1=subs(eqn1,{'y(0)'},'3');          ← Initial condition only on the equation 1
>>R=solve(eqn1,eqn2,X,Y)
```

```
R =
    X: [1x1 sym]
    Y: [1x1 sym]                         ← X(z) and Y(z) are returned as a structured array
```

```
>>X=R.X;                                 ← Separate X(z)
>>Y=R.Y;                                 ← Separate Y(z)
>>x=iztrans(X);
>>y=iztrans(Y);
>>pretty(x)
```

$$39/2 \cdot 3^n - 18/5\, 2^n + \frac{141}{10}(-3)^n$$

358

>>pretty(y) ↵

$$- \frac{1}{3} \, charfcn[0](n) + \frac{13}{6} \, 3^n - \frac{2}{5} \, 2^n + \frac{47}{30} \, (-3)^n$$

As it is expected.

Apparently, the method seems to be cumbersome but there are some operational skills. If you perform the execution in the command window, take the help of 'dos shell key' (see chapter 1 regarding this). It prevents you from typing the repetitive words or statements. Or the other way around, if you use the M–file programming, take the copy, cut, and paste facility of the editor. Certainly you can have it accomplished faster in doing so.

Chapter 9

Problems of Statistics

Before we commence our exploration on the subroutines of statistical problems, it might be helpful to give a detail of the importance of such study. Statistics and probability are becoming increasingly important in natural science and in social science as well. To be illustrative, the fields of statistics compass communications engineering, computer communications, networking theory, artificial intelligence, robotics, growth-yield treatment in agriculture and plant science, demographic prediction, meteorological prediction...etc. The problems selected for this chapter cover the level of a first coarse in engineering and scientific statistics. Out of numerous features of statistical toolbox, only a few such as random number generations, descriptive statistics, probability distributions, and regression analysis are chosen just to confer the computational capability of the toolbox. Among other appreciative features, non-linear regression models, hypothesis test, statistical process control (SPC), design of experiments, and statistical plots can be cited. Make sure that the toolbox is installed in your system.

9.1 Random number generators

Generating random numbers is the first step of any statistical analysis. Programmed random number generators in MATLAB are rand, randn, sprand, randperm,...etc for which the following discussions are presented. Each random number generator produces scalar or matrix with random entry/entries decided on the type of input argument/arguments. The distribution required for the random number typifies the selection of random number generator. There are several distributions that are used, for instance, continuous uniform, discrete uniform, normal etc. For better understanding, plot of the probability density functions for different generators are also furnished.

⊟ Subroutine rand

It generates uniformly distributed random floating-point numbers between 0 and 1 (0 and 1 inclusive). It can have no/one/two argument(s). The random numbers in row, column, or rectangular matrix form can be generated by two arguments — the first one is for the number of rows and the second one is for the number of columns. If no argument is specified, a single number is returned by the subroutine. In all cases, the output is between 0 and 1. Implementation of 'rand' is shown in the following page:

360

MATLAB Command

generation of a single number S,

>>S=rand ↵

S =
0.9501

generation of a column matrix C of length 3,

>>C= rand(3,1) ↵

C =
0.7919
0.9218
0.7382

generation of a row matrix R of length 4,

>>R= rand(1,4) ↵

R =
0.0185 0.8214 0.4447 0.6154

generation of a rectangular matrix A of order 2×3,

>>A= rand(2,3) ↵

A =
0.2311 0.4860 0.7621
0.6068 0.8913 0.4565

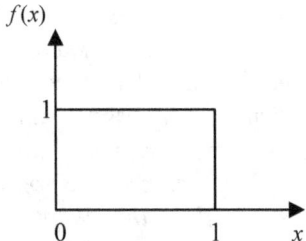

Figure 9.1 *Probability density function of the subroutine 'rand'*

When you execute the commands, these output numbers may not appear in the command window due to the randomness of the generation but the numbers will be between 0 and 1 that is for sure. Figure 9.1 depicts the probability density function of 'rand'.

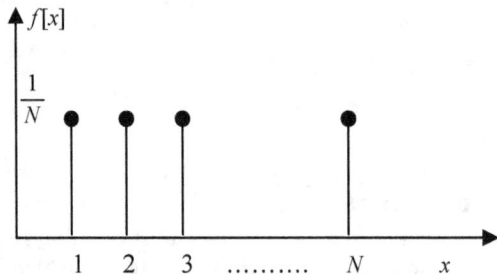

Figure 9.2 *Probability density function of the subroutine 'unidrnd'*

⌗ Subroutine unidrnd

The elaboration of 'unidrnd' is <u>uni</u>form <u>d</u>iscrete <u>rand</u>om. It returns the uniformly distributed random discrete numbers between 1 and N (inclusive), where N is any positive integer. It must have some argument, at least one, to run the subroutine. If no argument is provided with the subroutine, an error message is printed by MATLAB. The main difference between the 'rand' and 'unidrnd' is that the former is continuous and the later is discrete. Another difference is that the output variation of 'rand' is from 0 to 1 but the variation is from 1 to N for 'unidrnd'. The return can also be in a matrix form. To obtain a matrix of integers between 1 and N ,

MATLAB Step:

Use the command unidrnd(N, row number, column number).

Figure 9.2 shows the probability density function of the subroutine. By uniform distribution, we mean that any number from 1 to N is equally likely. One may look for the matrix elements to be other than 1 to N. Illustrative cases are shown for better clarification. Say, any integer from −20 to −3 is necessary. The difference of the two limits is 17. The output of 'unidrnd' is not less than 1. For this example, the subroutine should be written as −2−unidrnd(18). Again, any integers from −20 to 0, from −30 to 5, from −10 to 10, and from 0 to 10 are obtained from 1−unidrnd(21), 6−unidrnd(36), 11−unidrnd(21), and 11−unidrnd(11) respectively. So far we mentioned how a single integer is obtained. For matrices of integers, the numbers of rows and columns follow the second and third arguments of the subroutine respectively. See all implementations as follows:

MATLAB Command
generation of a single number S between 1 and 5,
>>S=unidrnd(5) ⏎

S =
4

generation of a row matrix R of length 4 in which each element is between 1 and 10,
>>R=unidrnd(10,1,4) ⏎

R =
7 3 5 1

generation of a column matrix C of length 3 in which each element is between 1 and 5,
>>C=unidrnd(5,3,1) ⏎

C =
5
2
4

generation of a rectangular matrix A of order 2×4 in which each element is between 1 and 4,
>>A= unidrnd(4,2,4) ⏎

A =
3 4 4 4
4 3 1 2

generation of a single number S from −20 to −3,
>>S=-2-unidrnd(18) ⏎

S =
-14

generation of a row matrix R of length 4 in which each element is from −20 to −3,
>>R=-2-unidrnd(18,1,4) ⏎

R =
-17 -20 -12 -18

generation of a column matrix C of length 3 in which each element is from −20 to −3,
>>C=-2-unidrnd(18,3,1) ⏎

C =
-6
-20
-7

generation of a rectangular matrix A of order 2×4 in which each element is from −20 to −3,
>>A=-2-unidrnd(18,2,4) ⏎

A =

$$\begin{array}{rrrr} -7 & -16 & -3 & -6 \\ -18 & -5 & -19 & -8 \end{array}$$

Next, the random integers according to a matrix of integers are found. Implement that on $X = \begin{bmatrix} 12 & 5 & 9 \\ 23 & 2 & 50 \end{bmatrix}$. In

the new matrix, the element (1, 1) will be any integer from 1 to 12 [since the element (1, 1) of X is 12], the element (1, 2) will be any integer from 1 to 5, the element (1, 3) will be any integer from 1 to 9, and so will be the others. Order of the newly formed matrix A is the same as that of X. See the formation as follows:

>>X=[12 5 9;23 2 50]; ↵
>>A=unidrnd(X) ↵

A =
$$\begin{array}{rrr} 12 & 4 & 9 \\ 6 & 1 & 39 \end{array}$$

⌑ Subroutine unifrnd
This is the continuous counterpart of 'unidrnd'. It generates uniformly distributed continuous random numbers between some interval specified by the input arguments of the subroutine. The elaboration of 'unifrnd' is uniformly distributed floating point random numbers. Its probability density function is graphed in figure 9.3. To generate a matrix of the floating-point numbers from A to B (where $B > A$),

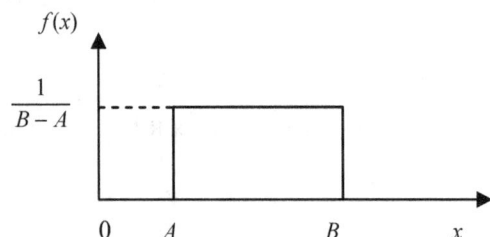

Figure 9.3 *Probability density function of the subroutine 'unifrnd'*

MATLAB Step:
Use the command unifrnd(A, B, row number, column number).

Say, we want to generate matrix R of order 3×4 in which the elements are floating-point numbers ranging from −3 to 5. Generation of a single number from A to B needs two input arguments. Implementation is presented below:

MATLAB Command
generation of 3×4 matrix R
for interval −3 to 5,
>>R=unifrnd(-3,5,3,4) ↵

R =
$$\begin{array}{rrrr} 4.6010 & 0.8879 & 0.6517 & 0.5576 \\ -1.1509 & 4.1304 & -2.8520 & 1.9235 \\ 1.8547 & 3.0968 & 3.5713 & 3.3355 \end{array}$$

for a single number S from -3 to 5,
>>S=unifrnd(-3,5) ↵

S =
4.3745

⌑ Subroutine exprnd
Shown figure 9.4 is the plot of probability density function of the generator 'exprnd'. The elaboration of 'exprnd' is exponentially distributed floating-point random numbers. The probability density function of the distribution with parameter θ is given by $f(x) = \begin{cases} \dfrac{1}{\theta} e^{-\frac{x}{\theta}} & for \ x > 0 \\ \\ 0 & elsewhere \end{cases}$.

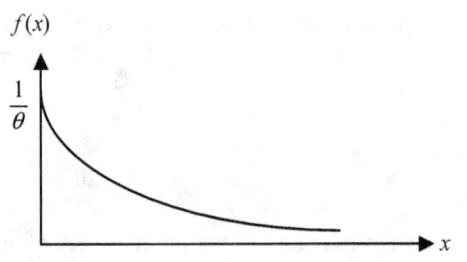

Figure 9.4 *Probability density function of the subroutine 'exprnd'*

To generate a matrix of floating-point numbers whose elements are exponentially distributed with parameter θ,

MATLAB Step:
Use the command exprnd(θ , row number, column number).

Generate a 5×4 matrix of exponentially distributed floating-point random numbers with parameter θ =3, where the necessary command is 'exprnd(3,5,4)'. To have a single number, only one argument (which is θ) is required. Both executions are shown below:

MATLAB Command

generation of 5×4 matrix with parameter θ =3,
>>A=exprnd(3,5,4) ↵

A =

4.4345	4.0991	0.1835	3.6270
6.2750	0.3254	5.8117	10.1409
1.7182	0.0337	0.5714	0.8725
2.0635	0.7988	1.6934	6.8906
2.5264	0.6243	3.0555	1.0177

generation of a single number with parameter $\theta = 3$,
>>S=exprnd(3) ↵

S =

1.9150

⯐ Question of fairness

A question of fairness comes whether these generators follow the specified statistics (that is, mean and variance). Let us verify this. Take the example of subroutine 'rand'. From figure 9.1, mean of the distribution is 0.5. We generate 20 random numbers in a column matrix and find the mean of these generated numbers as follows:

MATLAB Command
>>C=rand(20,1); ↵
>>mean(C) ↵

ans =
0.4790

Let us perform the trial for 2000 numbers:

>>C=rand(2000,1); ↵
>>mean(C) ↵

ans =
0.5073

From the above output, it is clear that 2000 numbers are giving mean 0.5073 that is close to 0.5. What conclusion can be drawn? The random number generators provide better statistics for the higher number of trials. But the generation of higher numbers may not be feasible due to not having sufficient memory. You can overcome the memory problem using the for loop. As another alternative of fair random numbers, estimate the mean of the generated random numbers, find the difference between estimated and desired means, and add or subtract the difference with the estimated mean to fulfill the proper statistics.

Preceding discussion lets you know how to implement the random number generators to generate discrete or continuous random numbers. We mentioned only four in details still there are several generators found in the *Statistical Toolbox*. Their brief descriptions are presented in table 9.A. Prior to generating them, be introduced with the statistics of the probability density function in particular. We presented statistics of some known probability functions in the subsequent sections.

Table 9.A Random number generators and their MATLAB counterparts

Name of the random number	MATLAB generator	Executable form
Beta random numbers with parameters α and β	betarnd	A=betarnd(α,β,M,N) for matrix S=betarnd(α,β) for single number
Binomial random numbers with parameters N and p	binornd	A=binornd(N,p,M,N) for matrix S=binornd(N,p) for single number
Chi square random numbers with parameter n	chi2rnd	A=chi2rnd(n,M,N) for matrix S=chi2rnd(n) for single number
F random numbers with parameters n_1 and n_2	frnd	A=frnd(n_1,n_2,M,N) for matrix S=frnd(n_1,n_2) for single number
Gamma random numbers with parameters α and β	gamrnd	A=gamrnd(α,β,M,N) for matrix S=gamrnd(α,β) for single number
Normal (Gaussian) random numbers with parameters μ and σ	normrnd	A=normrnd(μ,σ,M,N) for matrix S=normrnd(μ,σ) for single number
Poisson random numbers with parameter λ	poissrnd	A=poissrnd(λ,M,N) for matrix S=poissrnd(λ) for single number
T random numbers with parameter n	trnd	A=trnd(n,M,N) for matrix S=trnd(n) for single number

Some other generators:

Name	Counterpart
Geometric random numbers	geornd
Hypergeometric random numbers	hygernd
Lognormal random numbers	lognrnd
Multivariate normal random numbers	mvnrnd
Negative binomial random numbers	nbinrnd
Noncentral F random numbers	ncfrnd
Noncentral t random numbers	nctrnd
Noncentral Chi-square random numbers	ncx2rnd
Rayleigh random numbers	raylrnd
Weibull random numbers	weibrnd

* Argument M means the number of rows required to form matrix A
* Argument N means the number of columns required to form matrix A
* A is a matrix of random numbers and S is a single random number

Another style of finding the specified random number is calling the function 'random'. Let us generate matrix A of order 3×5 whose elements are the Poisson random numbers with parameter 4. Its implementation is as follows:

MATLAB Command
>>A=random('poiss',4,3,5) ↵

A =

$$\begin{array}{ccccc} 3 & 10 & 4 & 4 & 5 \\ 5 & 5 & 4 & 5 & 5 \\ 3 & 3 & 4 & 5 & 3 \end{array}$$

As you see, the name of the generator is 'poiss' instead of 'poissrnd' located in table 9.A. Other generators also need modification in the names if you want to use this style. Table 9.B gives the necessary modification.

Table 9.B   Generators using function random and their MATLAB counterparts

Distribution	First argument of random	Distribution	First argument of random
Beta	beta	Noncentral F	ncf
Binomial	bino	Exponential	exp
Chisquare	chi2	Geometric	geo
Gamma	gam	Hypergeometric	hyge
F	f	Normal	norm
T	t	Log normal	logn
Uniform discrete	unid	Negative Binomial	nbin
Uniform continuous	unif	Weibull	weib

9.2 Preparing a frequency table of positive integers

If we have a vector V of positive integers (0 is not a positive integer), the frequency table can be found by using the subroutine 'tabulate'. Frequency means the number of occurrences of each element in V. Return of the subroutine can also be in percent form. Say, a vector is given as $V =$[2   4   2   3   3   5   6   2   4   8]. In V, number 2 is appearing three times, so, its frequency is 3. There are 10 elements in V, as a percent, 2 occurs $\frac{3}{10} \times 100 = 30\%$ of V. The number of occurrences of different elements is as follows:

MATLAB Command
>>V=[2 4 2 3 3 5 6 2 4 8]; ↵
>>T=tabulate(V) ↵

T =

$$\begin{array}{lll} 1 & 0 & 0 \\ 2 & 3 & 30 \quad \leftarrow \text{Appearance of 2 three times and occupying 30\%} \\ 3 & 2 & 20 \\ 4 & 2 & 20 \\ 5 & 1 & 10 \\ 6 & 1 & 10 \\ 7 & 0 & 0 \quad \leftarrow \text{Even though 7 is not in } V \text{, but it is occupying the table} \\ 8 & 1 & 10 \end{array}$$

The first column of T is just the serial number of integers, the second column is the frequency, and the third one is the percentage of occurrence. V can not be a rectangular matrix. If it is, convert it to a row or column matrix using the command 'V(:)' and apply the function.

9.3 Mean of a sample

Mean is the average of all observations on a random variable. If X is a random variable, mean of X is defined as $\dfrac{\sum_{n=1}^{N} X_n}{N}$, where N is the number of observations on X. In MATLAB, a random variable is entered as a matrix. To compute the mean of a random variable,

MATLAB Step:
> *Use the command mean (matrix name) for row/column or mean(mean(matrix name)) for rectangular matrices.*

Suppose X is given as row matrix R, column matrix C, and rectangular matrix A, where $R = [21 \quad 12 \quad 13 \quad 7$

$6 \quad 3 \quad 17]$, $C = \begin{bmatrix} 5 \\ 6 \\ -1 \\ 0 \\ 2 \\ 8 \end{bmatrix}$, and $A = \begin{bmatrix} 4 & 0 & 8 \\ 6 & 5 & 2 \\ 6 & 0 & 1 \\ 2 & 1 & 6 \end{bmatrix}$. The means are MR$= \dfrac{\sum_{n=1}^{N} R_n}{N} = \dfrac{21+12+13+7+6+3+17}{7} = \dfrac{79}{7} =$

11.28571429, MC$= \dfrac{\sum_{n=1}^{N} C_n}{N} = \dfrac{5+6-1+0+2+8}{6} = \dfrac{20}{6} = 3.333333$, and MA$= \dfrac{\frac{4+6+6+2}{4} + \frac{0+5+0+1}{4} + \frac{8+2+1+6}{4}}{3} = \dfrac{41}{3 \times 4} =$

3.416666667 for R, C, and A respectively. Computations are displayed in the following:

MATLAB Command

for the row matrix,
>>R=[21 12 13 7 6 3 17]; ↵
>>MR=mean(R) ↵

MR =
      11.2857

for the column matrix,
>>C=[5 6 -1 0 2 8]'; ↵
>>MC=mean(C) ↵

MC =
      3.3333

for the rectangular matrix,
>>A=[4 0 8;6 5 2;6 0 1; 2 1 6]; ↵
>>MA=mean(mean(A)) ↵

MA =
      3.4167

9.4 Geometric mean of a sample

Geometric mean of a sample is defined as $\sqrt[N]{\prod_{i=1}^{N} X_i}$, where X is the random variable and N is the number of elements in the sample. Symbol $\prod_{i=1}^{N} X_i$ means the product of all elements in the sample. MATLAB subroutine for the computation of geometric mean is 'geomean' (abbreviation of geometric mean). To compute the geometric mean of a sample,

MATLAB Step:
> *Use the command geomean (matrix name) for row or column and geomean (geomean (matrix name)) for rectangular matrix.*

Calculate the geometric means of $R = [21 \quad 12 \quad 13 \quad 7 \quad 6 \quad 3 \quad 17]$, $C = \begin{bmatrix} 6 \\ 70 \\ 41 \\ 9 \end{bmatrix}$, and $A = \begin{bmatrix} 3 & 5 & 7 \\ 8 & 9 & 2 \\ 1 & 2 & 4 \\ 5 & 1 & 7 \end{bmatrix}$. The

geometric means of R and C are $\sqrt[7]{21 \times 12 \times 13 \times 7 \times 6 \times 3 \times 17} = 9.506558764$ and $\sqrt[4]{6 \times 70 \times 41 \times 9} = 19.84124474$

respectively. For A, the geometric means over columns are $\sqrt[4]{3\times8\times1\times5}=\sqrt[4]{120}$, $\sqrt[4]{5\times9\times2\times1}=\sqrt[4]{90}$, and $\sqrt[4]{7\times2\times4\times7}=\sqrt[4]{392}$ respectively. The outer 'geomean' finds the geometric means of $\sqrt[4]{120}$, $\sqrt[4]{90}$, and $\sqrt[4]{392}$, which is $\sqrt[3]{\sqrt[4]{120}\sqrt[4]{90}\sqrt[4]{392}}$ =3.566365242. See all calculations as follows:

MATLAB Command

for the row matrix,
>>R=[21 12 13 7 6 3 17]; ↵
>>GR=geomean(R) ↵

GR =
    9.5066

for the column matrix,
>>C=[6 70 41 9]'; ↵
>>GC=geomean(C) ↵

GC =
    19.8412

for the rectangular matrix,
>>A=[3 5 7;8 9 2;1 2 4;5 1 7]; ↵
>>GA=geomean(geomean(A)) ↵

GA =
    3.5664

The matrix elements must be positive numbers and can be floating-point numbers too.

9.5 Harmonic mean of a sample

Definition of the harmonic mean is given by $H=\dfrac{N}{\sum\limits_{i=1}^{N}\dfrac{1}{X_i}}$, where X and N have their usual meanings.

Its corresponding MATLAB subroutine is 'harmmean' (abbreviation of <u>harm</u>onic <u>mean</u>). To calculate the harmonic mean of a sample,

    MATLAB Step:
        Use the command harmmean (matrix name) for row/column and harmmean (harmmean (matrix name)) for rectangular matrix.

Check the subroutine with matrices $R=[13\quad 7\quad 6\quad 3]$, $C=\begin{bmatrix}5\\6\\-1\end{bmatrix}$, and $A=\begin{bmatrix}4&5&8\\6&5&2\\6&-4&1\end{bmatrix}$. The harmonic means

of R and C are $HR=\dfrac{4}{\frac{1}{13}+\frac{1}{7}+\frac{1}{6}+\frac{1}{3}}=\frac{728}{131}=5.557251908$ and $HC=\dfrac{3}{\frac{1}{5}+\frac{1}{6}+\frac{1}{-1}}=-\frac{90}{19}=-4.736842105$ respectively.

The harmonic means of the first, second, and third columns of A are $\dfrac{3}{\frac{1}{4}+\frac{1}{6}+\frac{1}{6}}=\frac{36}{7}$, $\dfrac{3}{\frac{1}{5}+\frac{1}{5}-\frac{1}{4}}=20$, and

$\dfrac{3}{\frac{1}{8}+\frac{1}{2}+\frac{1}{1}}=\frac{24}{13}$ respectively and for the whole matrix is $HA=\dfrac{3}{\frac{7}{36}+\frac{1}{20}+\frac{13}{24}}=\frac{1080}{283}=3.816254417$. Rational form

output is also available by means of the subroutine 'sym'. Verification is shown below:

MATLAB Command

for the row matrix,
>>R=[13 7 6 3]; ↵
>>HM=harmmean(R) ↵

HR =
    5.5573

for the column matrix,
>>C=[5 6 -1]'; ↵

for the rectangular matrix,
>>A=[4 5 8;6 5 2;6 -4 1]; ↵
>>HA=harmmean(harmmean(A)) ↵

HA =
    3.8163

in rational form,
>>HA=harmmean(harmmean(sym(A))) ↵

>>HC=harmmean(C) ↵

HA =

HC =

    -4.7368   1080/283

9.6 Range of a sample

The range of a sample is the difference between the maximum and minimum values of a random variable. Subroutine 'range' can find the range of a sample. For row or column matrices, the output of the subroutine is just a single number and for rectangular matrices, the output is a row vector that contains the range for each column. Elements of the matrix can be integer or floating-point numbers. To find the range of a matrix,

MATLAB Step:
> *Use the command range (matrix name) for row/column or range (matrix name (:)) for rectangular matrix.*

For illustration, say, $R = \begin{bmatrix} 5 & 6 & 0 & -3 & 5 & 7 \end{bmatrix}$, $C = \begin{bmatrix} 45 \\ -78 \\ 32 \\ 98 \end{bmatrix}$, and $A = \begin{bmatrix} 4 & -6 & 0 \\ 3 & -3 & -1 \\ 5 & 0 & 1 \\ -7 & 3 & 2 \end{bmatrix}$. The maximum

values are 7, 98, and 5 in R, C, and A respectively. The respective minimum values are −3, −78, and −7. Therefore, the ranges of R, C, and A are 7−(−3)=10, 98−(−78)=176, and 5−(−7)=12 respectively. Notice that 'range(range(A))' is not applicable here for A. Ranges of the first, second, and third columns are 12, 9, and 3 respectively and the same for 12, 9, and 3 is 9, which is not equal to the actual range 12. Finding the ranges of different matrices is shown below:

MATLAB Command

for the row matrix,
>>R=[5 6 0 -3 5 7]; ↵
>>range(R) ↵

   ans =
           10
for the column matrix,
>>C=[45 -78 32 98]'; ↵
>>range(C) ↵

   ans =
           176

for the rectangular matrix,
>>A=[4 -6 0;3 -3 -1;5 0 1;-7 3 2]; ↵
>>range(A(:)) ↵

   ans =
           12

9.7 Variance of a sample

Assume that Y is a random variable with mean m. Variance of Y is defined as $V(Y) = \dfrac{\sum_{i=1}^{N}(Y_i - m)^2}{N-1}$,

where $m = \dfrac{\sum_{i=1}^{N} Y_i}{N}$ and N is the number of observations on Y. For row or column matrix, variance is computed

by $V(Y) = \dfrac{\sum_{i=1}^{N}(Y_i - m)^2}{N-1}$. For rectangular matrix, the variance is computed over each column using the same

formula. Elements of the matrix can be floating-point or integer numbers. To compute the variance of a random variable,

MATLAB Step:
Use the command var(observations of the variable in a matrix).

Test matrices are $R = [15 \quad 0 \quad -1 \quad 20 \quad 6 \quad 8]$, $C = \begin{bmatrix} -2 \\ -8 \\ 0 \\ -14 \end{bmatrix}$, and $A = \begin{bmatrix} -2 & 0 & 8 \\ 3 & 3 & -7 \\ 9 & 12 & 13 \\ 2 & 1 & -6 \end{bmatrix}$. The means of row and

column matrices are $\frac{15+0-1+20+6+8}{6} = 8$ and $\frac{-2-8+0-14}{4} = -6$ respectively. The numbers of observations are 6 and 4

for R and C respectively. Variances of R and C are computed as $V(R) = \frac{(15-8)^2+(0-8)^2+(-1-8)^2+(20-8)^2+(6-8)^2+(8-8)^2}{6-1} =$

68.4 and $V(C) = \frac{(-2+6)^2+(-8+6)^2+(0+6)^2+(-14+6)^2}{4-1} = 40$. For A, the means of the first, second, and third columns are

$\frac{-2+3+9+2}{4} = 3$, $\frac{0+3+12+1}{4} = 4$, and $\frac{8-7+13-6}{4} = 2$ respectively. The respective variances for the first, second, and

third columns are $\frac{(-2-3)^2+(3-3)^2+(9-3)^2+(2-3)^2}{4-1} = \frac{62}{3} = 20.6667$, $\frac{(0-4)^2+(3-4)^2+(12-4)^2+(1-4)^2}{4-1} = 30$, and $\frac{(8-2)^2+(-7-2)^2+(13-2)^2+(-6-2)^2}{4-1} = \frac{302}{3}$

=100.6667. See all implementations in the following:

MATLAB Command

for the row matrix,   for the column matrix,

    >>R=[15 0 -1 20 6 8]; ↵   >>C=[-2 -8 0 -14]'; ↵

    >>var(R) ↵   >>var(C) ↵

    ans =   ans =

        68.4000   40

for the rectangular matrix,

    >>A=[-2 0 8; 3 3 -7;9 12 13;2 1 -6]; ↵

    >>var(A) ↵

    ans =

        20.6667   30.0000   100.6667

in rational form,

    >>var(sym(A)) ↵

    ans =

    [62/3,   30,   302/3]

9.8 Standard deviation of a sample

Standard deviation of a random variable is defined as the positive square root of the variance.

Referring to article 9.7, the standard deviation is given by $\sigma = \sqrt{\dfrac{\sum\limits_{i=1}^{N}(Y_i - m)^2}{N-1}}$, where the symbols have their usual

meanings and σ (sigma) is the standard deviation of Y. Its MATLAB counterpart is 'std' (abbreviation of standard deviation). To compute the standard deviation,

    MATLAB Step:
    Use the command std (matrix name).

Choose the same examples of R, C, and A as we did for the variance. Their standard deviations are

$\sqrt{68.4} = 8.270429251$, $\sqrt{40} = 6.32455532$, and $\begin{bmatrix} \sqrt{\frac{62}{3}} & \sqrt{30} & \sqrt{\frac{302}{3}} \end{bmatrix} = [4.546060566 \qquad 5.477225575$

10.03327796] for R, C, and A respectively. That is exercised as follows:

MATLAB Command

for the row matrix,   for the column matrix,

    >>R=[15 0 -1 20 6 8]; ↵   >>C=[-2 -8 0 -14]'; ↵
    >>std(R) ↵   >>std(C) ↵

  ans =   ans =

      8.2704   6.3246

for the rectangular matrix,

    >>A=[-2 0 8; 3 3 -7;9 12 13;2 1 -6]; ↵
    >>std(A) ↵

    ans =

      4.5461   5.4772   10.0333

9.9 Mean absolute deviation of a sample

Computation of mean absolute deviation of a random variable X is obtained from $\dfrac{\sum_{i=1}^{N}|X-\overline{X}|}{N}$. Its MATLAB counterpart is 'mad' (abbreviation of <u>m</u>ean <u>a</u>bsolute <u>d</u>eviation). To have the mean absolute deviation,

MATLAB Step:
    Use the command mad (matrix name).

Test the subroutine with $R=[2\quad -5\quad -1\quad 0]$, $C=\begin{bmatrix}8\\-3\\-4\end{bmatrix}$, and $A=\begin{bmatrix}4 & -4 & -2\\5 & 0 & 6\end{bmatrix}$. The means of R, C, and the first, second, and third columns of A are -1, $\frac{1}{3}$, and $\frac{9}{2}$, -2, and 2 respectively. Mean absolute deviations of R, C, and the columns of A (rational form for A) are $\dfrac{|2+1|+|-5+1|+|-1+1|+|0+1|}{4}=2$, $\dfrac{|8-\frac{1}{3}|+|-3-\frac{1}{3}|+|-4-\frac{1}{3}|}{3}=\frac{46}{9}=$ 5.1111, and $\dfrac{|4-\frac{9}{2}|+|5-\frac{9}{2}|}{2}=\frac{1}{2}$, $\dfrac{|-4+2|+|0+2|}{2}=2$, and $\dfrac{|-2-2|+|6-2|}{2}=4$ respectively. See all these in the following:

MATLAB Command

for the row matrix,   for the column matrix,

    >>R=[2 -5 -1 0]; ↵   >>C=[8 -3 -4]'; ↵
    >>MR=mad(R) ↵   >>MC=mad(C) ↵

  MR =   MC =
      2   5.1111

for the rectangular matrix,

    >>A=sym([4 -4 -2;5 0 6]); ↵
    >>MA=mad(A) ↵

    MA =

    [1/2,   2,   4]

9.10 Median value of a sample

In ascending order, the middle element of some observations of a random variable is called the median. Assume that there are N sorted elements in a sample, the median is the $\left(\dfrac{N+1}{2}\right)^{th}$ element if the number of elements is odd or the average of the $\left(\dfrac{N}{2}\right)^{th}$ and $\left(\dfrac{N}{2}+1\right)^{th}$ elements if the number of elements is even. To obtain the median,

MATLAB Step:
Use the command median (matrix name).

For illustration, take $R = [4 \quad 7 \quad 5 \quad -5 \quad 0 \quad 1 \quad 9]$, $C = \begin{bmatrix} 7 \\ -3 \\ -4 \\ -2 \\ -1 \\ 0 \end{bmatrix}$, and $A = \begin{bmatrix} -3 & 3 & 9 \\ 8 & 7 & 5 \\ 6 & -2 & -1 \\ 0 & 54 & 10 \end{bmatrix}$. The numbers of

elements in R and C are odd and even respectively. Sort them to have $R = [-5 \quad 0 \quad 1 \quad 4 \quad 5 \quad 7 \quad 9]$

and $C = \begin{bmatrix} -4 \\ -3 \\ -2 \\ -1 \\ 0 \\ 7 \end{bmatrix}$. So, their medians are the 4th and the average of the 3rd and 4th elements of R and C respectively

that are 4 and $-\dfrac{3}{2}$. In a similar fashion, one can find that the medians of the columns of A are 3, 5, and 7 respectively. These all are shown as follows:

MATLAB Command

for the row matrix,

```
>>R=[4 7 5 -5 0 1 9]; ↵
>>MR=median(R) ↵
```

MR =
 4

for the column matrix,

```
>>C=[7 -3 -4 -2 -1 0]'; ↵
>>MC=median(C) ↵
```

MC =
 -1.5000

for the rectangular matrix,

```
>>A=[-3 3 9;8 7 5;6 -2 -1;0 54 10]; ↵
>>MA=median(A) ↵
```

MA =
 3   5   7

9.11 Moment of a sample

The Pth moment of a random variable X is defined as $M_X^P = \dfrac{\sum\limits_{i=1}^{N}(X - \overline{X})^P}{N}$, where $\overline{X}$ is the mean of

X, N is the number of elements in X, and P is the order of the moment. To compute the Pth moment,
 MATLAB Step:
 Use the command moment (matrix name, P).

Sample values are entered as matrices. Take as examples $R = [-1 \quad 2 \quad 90 \quad 34]$, $C = \begin{bmatrix} 50 \\ 8 \\ -3 \end{bmatrix}$, and

$A = \begin{bmatrix} 1 & 2 & 4 \\ 0 & -1 & 7 \\ 3 & 1 & -12 \end{bmatrix}$. Means of R and C are $\overline{R} = \dfrac{-1+2+90+34}{4} = 31.25$ and $\overline{C} = \dfrac{50+8-3}{3} = \dfrac{55}{3} = 18.3333$ respectively.

Compute the 4th and 5th moments of R and C respectively. We have $M_R^4 = \dfrac{\sum_{i=1}^{4}(R-\overline{R})^4}{4} =$

$\dfrac{(-1-31.25)^4 + (2-31.25)^4 + (90-31.25)^4 + (34-31.25)^4}{4} = 3431764.52$ and $M_C^5 = \dfrac{\sum_{i=1}^{3}(C-\overline{C})^5}{3} = \dfrac{(50-18.3333)^5 + (8-18.3333)^5 + (-3-18.3333)^5}{3} = 9102110.29$. For

A, the computation is carried out on columns. Find the third moment of A in the rational form. The means of the first, second, and third columns of A are $\frac{4}{3}$, $\frac{2}{3}$, and $-\frac{1}{3}$ respectively. The third moments of the first,

second, and third columns of A are $\dfrac{\left(1-\frac{4}{3}\right)^3 + \left(0-\frac{4}{3}\right)^3 + \left(3-\frac{4}{3}\right)^3}{3} = \frac{20}{27}$, $\dfrac{\left(2-\frac{2}{3}\right)^3 + \left(-1-\frac{2}{3}\right)^3 + \left(1-\frac{2}{3}\right)^3}{3} = -\frac{20}{27}$, and $\dfrac{\left(4+\frac{1}{3}\right)^3 + \left(7+\frac{1}{3}\right)^3 + \left(-12+\frac{1}{3}\right)^3}{3} = -\frac{10010}{27}$

respectively. Because of the higher number digits, MATLAB displays the outputs in exponential forms. All examples are shown in the following:

MATLAB Command

for the 4th moment of R,
>>R=[-1 2 90 34]; ↵
>>MR=moment(R,4) ↵

for the 5th moment of C,
>>C=[50 8 -3]'; ↵
>>MC=moment(C,5) ↵

MR =
   3.4318e+006

MC =
   9.1021e+006

for the 3rd moment of A,
>>A=sym([1 2 4;0 -1 7;3 1 -12]); ↵
>>MA=moment(A,3) ↵

MA =

[20/27, -20/27, -10010/27]

9.12 Skewness of a sample

The skewness of a random variable X is given by $S = \dfrac{\dfrac{\sum_{i=1}^{N}(X-\overline{X})^3}{N}}{\left(\dfrac{\sum_{i=1}^{N}(X-\overline{X})^2}{N}\right)^{\frac{3}{2}}}$. In words, the skewness is the

3th central moment divided by the cube of standard deviation of the variable X, where $\overline{X}$ is the mean of X, N is the number of observations, and S is the skewness of X. There is some difference between the definitions of the divider number of standard deviation and that of skewness. In the standard deviation, the divider number is $N-1$ but in skewness, it is N. To compute the skewness,

MATLAB Step:
    Use the command skewness (matrix name).

To proceed with, consider $R = [2 \quad -1 \quad -7]$, $C = \begin{bmatrix} 7 \\ -3 \\ -4 \\ 0 \end{bmatrix}$, and $A = \begin{bmatrix} 4 & -4 \\ 5 & 0 \\ -2 & 7 \end{bmatrix}$. The means of R, C, and

columns of A are -2, 0, and $\dfrac{7}{3}$ and 1 respectively. The skewness of R and C are computed as SR=

$$\frac{\dfrac{(2+2)^3+(-1+2)^3+(-7+2)^3}{3}}{\left(\dfrac{(2+2)^2+(-1+2)^2+(-7+2)^2}{3}\right)^{\frac{3}{2}}}=-\frac{20}{14^{\frac{3}{2}}}=-0.3818 \quad \text{and} \quad SC=\frac{\dfrac{(7-0)^3+(-3-0)^3+(-4-0)^3+(0-0)^3}{4}}{\left(\dfrac{(7-0)^2+(-3-0)^2+(-4-0)^2+(0-0)^2}{4}\right)^{\frac{3}{2}}}=\frac{63}{\left(\dfrac{37}{2}\right)^{\frac{3}{2}}}=$$

0.7917. In a similar fashion, the columns of A have the skewness $-\dfrac{520\sqrt{27}}{258^{\frac{3}{2}}}=-0.6520$ and $\dfrac{30}{\left(\dfrac{62}{3}\right)^{\frac{3}{2}}}=0.3193$

respectively. Implementation is shown as follows:

MATLAB Command

for the row matrix,   for the column matrix,

    >>R=[2 -1 -7]; ↵   >>C=[7 -3 -4 0]'; ↵

    >>SR=skewness(R) ↵   >>SC=skewness(C) ↵

    SR =   SC =

      - 0.3818   0.7917

for the rectangular matrix,

         >>A=[4 -4;5 0;-2 7]; ↵

         >>SA=skewness(A) ↵

      SA =

         -0.6520   0.3193

9.13 Kurtosis of a sample

Kurtosis of a random variable X is defined as the 4th moment divided by the 4th power of the

standard deviation, that is, kurtosis $K=\dfrac{\dfrac{\sum\limits_{i=1}^{N}(X-\overline{X})^4}{N}}{\left(\dfrac{\sum\limits_{i=1}^{N}(X-\overline{X})^2}{N}\right)^2}$, where $\overline{X}$, X, and N have their usual meanings. To

calculate the kurtosis of X,

    MATLAB Step:

        Use the command kurtosis (matrix name).

Test matrices are $R=[-3 \quad 4 \quad -13]$, $C=\begin{bmatrix}7\\-3\\-4\end{bmatrix}$, and $A=\begin{bmatrix}4 & -4\\5 & 8\end{bmatrix}$. R, C, and columns of A have the means

-4, 0, and $\frac{9}{2}$ and 2 respectively. The kurtosis of R and C are KR$=\dfrac{\dfrac{(-3+4)^4+(4+4)^4+(-13+4)^4}{3}}{\left(\dfrac{(-3+4)^2+(4+4)^2+(-13+4)^2}{3}\right)^2}=\dfrac{\dfrac{10658}{3}}{\left(\dfrac{146}{3}\right)^2}$

$=\dfrac{3}{2}$ and $\dfrac{\dfrac{(7-0)^4+(-3-0)^4+(-4-0)^4}{3}}{\left(\dfrac{(7-0)^2+(-3-0)^2+(-4-0)^2}{3}\right)^2}=\dfrac{\dfrac{2738}{3}}{\left(\dfrac{74}{3}\right)^2}=\dfrac{3}{2}$. Similarly, the kurtosis of the columns of A are 1 and 1

respectively. Computations are shown below:

MATLAB Command

for the row matrix,

  >>R=[-3 4 -13]; ↵
  >>KR=kurtosis(R) ↵

   KR =
    1.5000

for the column matrix,

  >>C=[7 -3 -4]'; ↵
  >>KC=kurtosis(C) ↵

   KC =
    1.5000

for the rectangular matrix,

  >>A=[4 -4;5 8]; ↵
  >>KA=kurtosis(A) ↵

   KA =
    1   1

9.14 Covariance of random variables

  As indicated earlier, the random variables are entered as row or column matrices. In a rectangular matrix usually the columns represent the observations on the random variables. The number of columns can be taken as the number of random variables. Mean, median, or variance of a random variable describes the information about the variable itself. If we have two or more random variables placed in a rectangular matrix, covariance provides a relationship between the random variables or about their tendency to vary together rather than independently. The covariance of a matrix is a matrix. Elements of the covariance matrix

$$V = \begin{bmatrix} V_{11} & V_{12} & \cdots & V_{1N} \\ V_{21} & V_{22} & \cdots & V_{2N} \\ \vdots & \vdots & \ddots & \vdots \\ V_{N1} & V_{N2} & \cdots & V_{NN} \end{bmatrix} \quad \text{of a matrix} \quad A = \begin{bmatrix} A_{11} & A_{12} & \cdots & A_{1N} \\ A_{21} & A_{22} & \cdots & A_{2N} \\ \vdots & \vdots & \ddots & \vdots \\ A_{M1} & A_{M2} & \cdots & A_{MN} \end{bmatrix} \quad \text{is defined as}$$

$$V_{ij} = \frac{\sum_{k=1}^{M}(A_{ki} - \overline{A_i})(A_{kj} - \overline{A_j})}{M-1}, \text{ where } A_1 = \begin{bmatrix} A_{11} \\ A_{21} \\ \vdots \\ A_{M1} \end{bmatrix}, A_2 = \begin{bmatrix} A_{12} \\ A_{22} \\ \vdots \\ A_{M2} \end{bmatrix} \dots, \text{ and } A_N = \begin{bmatrix} A_{1N} \\ A_{2N} \\ \vdots \\ A_{MN} \end{bmatrix} \text{ and } \overline{A_j} \text{ is the mean of the } j^{th}$$

column of A. The diagonal elements of V are variances and the off-diagonal elements are covariances that is why V is called the variance-covariance matrix. Another name of V is dispersion matrix and it is a symmetric matrix. Orders of A and V are $M \times N$ and $N \times N$ respectively. Elements of A can be floating-point or integer numbers. To evaluate the covariance of a matrix,

  MATLAB Step:

   Use the command cov(matrix name).

Consider $A = \begin{bmatrix} -2 & 6 & 6 \\ 4 & 30 & 1 \\ 1 & -4 & -4 \\ 5 & 0 & 5 \end{bmatrix}$ for elucidation, where M =4 and N =3. The order of V must be 3×3 and V

prescribes the matrix form $\begin{bmatrix} V_{11} & V_{12} & V_{13} \\ V_{21} & V_{22} & V_{23} \\ V_{31} & V_{32} & V_{33} \end{bmatrix}$. Since V is a symmetric matrix, we have $V_{12} = V_{21}$, $V_{13} = V_{31}$,

and $V_{23} = V_{32}$. Different means are $\overline{A_1} = \frac{1}{4}(-2+4+1+5)=2$, $\overline{A_2} = \frac{1}{4}(6+30-4+0)=8$, and $\overline{A_3} = \frac{1}{4}(6+1-4+5)=2$, on

that, $V_{11} = \frac{\sum_{k=1}^{4}(A_{k1} - \overline{A_1})^2}{4-1} = \frac{(-2-2)^2 + (4-2)^2 + (1-2)^2 + (5-2)^2}{4-1} = \frac{30}{3} = 10$, $V_{22} = \frac{\sum_{k=1}^{4}(A_{k2} - \overline{A_2})^2}{4-1} = \frac{(6-8)^2 + (30-8)^2 + (-4-8)^2 + (0-8)^2}{4-1} =$

$\frac{696}{3} = 232$, $V_{33} = \frac{\sum_{k=1}^{4}(A_{k3} - \overline{A_3})^2}{4-1} = \frac{(6-2)^2 + (1-2)^2 + (-4-2)^2 + (5-2)^2}{4-1} = \frac{62}{3} = 20.6667$, $V_{21} = V_{12} = \frac{\sum_{k=1}^{4}(A_{k1} - \overline{A_1})(A_{k2} - \overline{A_2})}{4-1} =$

$$\frac{(-2-2)(6-8)+(4-2)(30-8)+(1-2)(-4-8)+(5-2)(0-8)}{4-1}=\frac{40}{3}=13.3333,$$

$$V_{31}=V_{13}=\frac{\sum\limits_{k=1}^{4}(A_{k1}-\overline{A_1})(A_{k3}-\overline{A_3})}{4-1}=$$

$$\frac{(-2-2)(6-2)+(4-2)(1-2)+(1-2)(-4-2)+(5-2)(5-2)}{4-1}=-1,$$ and

$$V_{32}=V_{23}=\frac{\sum\limits_{k=1}^{4}(A_{k2}-\overline{A_2})(A_{k3}-\overline{A_3})}{4-1}=$$

$$\frac{(6-8)(6-2)+(30-8)(1-2)+(-4-8)(-4-2)+(0-8)(5-2)}{4-1}=6.$$ Arranging the elements in a matrix form yields $V=$

$$\begin{bmatrix} 10 & 40/3 & -1 \\ 40/3 & 232 & 6 \\ -1 & 6 & 62/3 \end{bmatrix}=\begin{bmatrix} 10 & 13.3333 & -1 \\ 13.3333 & 232 & 6 \\ -1 & 6 & 20.6667 \end{bmatrix}.$$ If A were a row or column matrix, the

command 'cov(A)' would return just the variance of A. Implementation accompanying the rational form is shown as follows:

MATLAB Command

for decimal form,
```
>>A=[-2 6 6;4 30 1;1 -4 -4;5 0 5]; ↵
>>cov(A) ↵
```

ans =

10.0000	13.3333	-1.0000
13.3333	232.0000	6.0000
-1.0000	6.0000	20.6667

for rational form,
```
>>cov(sym(A)) ↵
```

ans =

[10,	40/3,	-1]
[40/3,	232,	6]
[-1,	6,	62/3]

9.15 Correlation of random variables

Correlation coefficient gives a measure of the linear association between two random variables. Correlation of the matrix formed by placing the random variables in columns is a matrix, which is called the correlation matrix. As a general rule, it is applicable for more than two random variables. It can be expressed in terms of the variance-covariance matrix V, which is defined in the last article. The correlation matrix R of the rectangular matrix A is defined as $R=D\times V\times D$, where '$\times$' indicates multiplication of matrices. From A, the diagonal matrix D is obtained by taking the reciprocals of standard deviations of columns. With regard to the

variance-covariance matrix of article 9.14, diagonal matrix is given by $D=\begin{bmatrix} \frac{1}{\sqrt{V_{11}}} & 0 & 0 & 0 \\ 0 & \frac{1}{\sqrt{V_{22}}} & 0 & 0 \\ \vdots & \vdots & \vdots & \ddots & \vdots \\ 0 & 0 & 0 & \frac{1}{\sqrt{V_{NN}}} \end{bmatrix}$. The

diagonal elements of R are unity because a variable is perfectly correlated with itself. Each element of R satisfies $-1\le R_{ij}\le1$, where R_{ij} is any element of R. The order of R is $N\times N$, where A is of order $M\times N$ and V is of order $N\times N$. R is a symmetric matrix too. To obtain the correlation matrix of a matrix,

MATLAB Step:
Use the command corrcoef(matrix name).

Consider the previous article's A, from that, $V=\begin{bmatrix} 10 & 40/3 & -1 \\ 40/3 & 232 & 6 \\ -1 & 6 & 62/3 \end{bmatrix}$ and the reciprocals of standard

deviations of the first, second, and third columns of A are $\frac{1}{\sqrt{10}}$, $\frac{1}{\sqrt{232}}$, and $\frac{1}{\sqrt{62/3}}$ respectively. The

376

diagonal matrix is formed as $D = \begin{bmatrix} \dfrac{1}{\sqrt{10}} & 0 & 0 \\ 0 & \dfrac{1}{\sqrt{232}} & 0 \\ 0 & 0 & \dfrac{1}{\sqrt{62/3}} \end{bmatrix}$. The correlation matrix R is computed as

$$\begin{bmatrix} \dfrac{1}{\sqrt{10}} & 0 & 0 \\ 0 & \dfrac{1}{\sqrt{232}} & 0 \\ 0 & 0 & \dfrac{1}{\sqrt{62/3}} \end{bmatrix} \times \begin{bmatrix} 10 & 40/3 & -1 \\ 40/3 & 232 & 6 \\ -1 & 6 & 62/3 \end{bmatrix} \times \begin{bmatrix} \dfrac{1}{\sqrt{10}} & 0 & 0 \\ 0 & \dfrac{1}{\sqrt{232}} & 0 \\ 0 & 0 & \dfrac{1}{\sqrt{62/3}} \end{bmatrix} = \begin{bmatrix} \dfrac{1}{\sqrt{10}} & 0 & 0 \\ 0 & \dfrac{1}{\sqrt{232}} & 0 \\ 0 & 0 & \dfrac{1}{\sqrt{62/3}} \end{bmatrix} \times$$

$$\begin{bmatrix} \sqrt{10} & \dfrac{40}{3\sqrt{232}} & \dfrac{-1}{\sqrt{62/3}} \\ \dfrac{40}{3\sqrt{10}} & \sqrt{232} & \dfrac{6}{\sqrt{62/3}} \\ \dfrac{-1}{\sqrt{10}} & \dfrac{6}{\sqrt{232}} & \sqrt{62/3} \end{bmatrix} = \begin{bmatrix} 1 & \dfrac{2\sqrt{145}}{87} & -\dfrac{\sqrt{465}}{310} \\ \dfrac{2\sqrt{145}}{87} & 1 & \dfrac{3\sqrt{2697}}{1798} \\ -\dfrac{\sqrt{465}}{310} & \dfrac{3\sqrt{2697}}{1798} & 1 \end{bmatrix} = \begin{bmatrix} 1 & 0.2768 & -0.0696 \\ 0.2768 & 1 & 0.0867 \\ -0.0696 & 0.0867 & 1 \end{bmatrix}.$$

Following is the implementation:

MATLAB Command
```
>>A=[-2 6 6;4 30 1;1 -4 -4;5 0 5]; ↵
>>R=corrcoef(A) ↵

R =
        1.0000    0.2768   -0.0696
        0.2768    1.0000    0.0867
       -0.0696    0.0867    1.0000
```

You can have the symbolic form as follows:
```
>>R=corrcoef(sym(A)); ↵
>>pretty(R) ↵
```

```
[                        1/2                    1/2  ]
[        1          2/87 145        - 1/310   465    ]
[                                                    ]
[        1/2                                   1/2   ]
[ 2/87 145              1           3/1798   2697    ]
[                                                    ]
[        1/2                   1/2                   ]
[- 1/310 465        3/1798  2697             1       ]
```

The correlation coefficient of row or column matrix will be just unity. Verify that with $B = \begin{bmatrix} 7 & -3 & -2 & 6 \end{bmatrix}$:
```
>>B=[7 -3 -2 6]; ↵
>>R=corrcoef(B) ↵

R =
        1
```

9.16 Probability density functions

Probability density function (very often abbreviated as pdf) $f(x)$ is related with the frequency of occurrence of a random variable. It can be obtained from the cumulative distribution function $F(x)$ by taking a derivative, that is, $f(x) = \dfrac{d\,F(x)}{dx}$, where $F(x)$ is a non decreasing function with the characteristics $\begin{cases} 0 \le F(x) \le 1 \\ F(-\infty) = 0 \\ F(\infty) = 1 \end{cases}$.

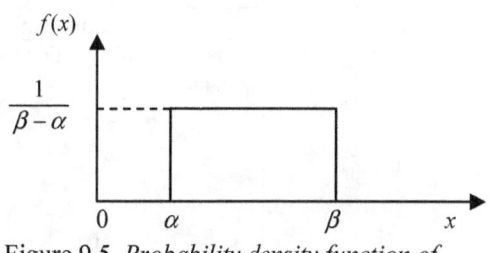

Figure 9.5 *Probability density function of the continuous uniform random variable*

The definition just mentioned is applicable when the random variable X is continuous. For discrete variable, probability density function is denoted by $f[x]$ and is obtained by taking the difference of the cumulative distribution functions at any two consecutive points. Begin with the probability density function of the continuous uniform random variable. A random variable X is said to have a continuous uniform distribution with parameters α and β if its probability density function is given by $f(x) = \begin{cases} \dfrac{1}{\beta - \alpha} & for \quad \alpha < x < \beta \\ \\ 0 & elsewhere \end{cases}$,

where $\alpha > 0$. The pdf is depicted in figure 9.5. Occurrence of any value of X from α to β is equally likely. MATLAB has the pdf function by name 'unifpdf' (<u>unif</u>orm <u>f</u>loating-point <u>p</u>robability <u>d</u>ensity <u>f</u>unction). What 'unifpdf' does is it returns the functional value of $f(x)$ at any x. Think about the pdf with parameters $\alpha = 3$ and $\beta = 5$, on that, $f(2)$, $f(4.1)$, and $f(5.5)$ should be 0, $\dfrac{1}{5-3} = 0.5$, and 0 respectively. Conduct that as follows:

MATLAB Command

for $f(2)$,

>>unifpdf(2,3,5) ↵

ans =
0

for $f(4.1)$,

>>unifpdf(4.1,3,5) ↵

ans =
0.5000

for $f(5.5)$,

>>unifpdf(5.5,3,5) ↵

ans =
0

The function is well suited to a matrix too. To show that, take $A = \begin{bmatrix} 3 & 5 \\ 4 & -1 \end{bmatrix}$. Our output should be $\begin{bmatrix} f(3) & f(5) \\ f(4) & f(-1) \end{bmatrix} = \begin{bmatrix} 0.5 & 0.5 \\ 0.5 & 0 \end{bmatrix}$:

>>A=[3 5;4 -1]; ↵
>>unifpdf(A,3,5) ↵

ans =
0.5000   0.5000
0.5000   0

Anyhow, more than a dozen of pdfs are available in the *Statistical Toolbox*. We present a brief list of them accompanying the definition in table 9.C.

Table 9.C Different probability density functions and their MATLAB counterparts

Definition	Example

Exponential distribution with parameter θ : $$f(x) = \begin{cases} \dfrac{1}{\theta} e^{-\frac{x}{\theta}} & \text{for } x > 0 \\ \\ 0 & \text{elsewhere} \end{cases}$$ Type: Continuous Executable form: exppdf(x , θ), where θ or x can be a scalar or matrix	⊟ Compute $f(3.5)$ for parameter θ =3
	 Solution: $f(3.5) = \dfrac{1}{3} e^{-\frac{3.5}{3}} = 0.1038$
	 MATLAB Command
	       `>>exppdf(3.5,3)` ↵
	
	 ans =
	      0.1038

Normal distribution with parameter μ an σ : $$f(x) = \dfrac{1}{\sigma\sqrt{2\pi}} e^{-\frac{(x-\mu)^2}{2\sigma^2}} \ ,$$ $$\text{for } -\infty < x < \infty$$ Type: Continuous Executable form: normpdf(x , μ , σ), where x , μ , or σ can be a scalar or matrix	⊟ Compute $f(3.5)$ and $f(-1)$ for parameters μ =2 and σ =4
	
	 Solution: $[\ f(3.5) \quad f(-1)\ ] =$
	 $\left[\dfrac{1}{4\sqrt{2\pi}} e^{-\frac{(3.5-2)^2}{2\times 4^2}} \quad \dfrac{1}{4\sqrt{2\pi}} e^{-\frac{(-1-2)^2}{2\times 4^2}} \right] =$
	 [0.0930 0.0753]
	
	 MATLAB Command
	       `>>normpdf([3.5 -1],2,4)` ↵
	
	 ans =
	      0.0930 0.0753

Discrete uniform distribution with parameter N (N is any positive integer): $$f[x] = \begin{cases} \dfrac{1}{N} & \text{for } 1 \le x \le N \\ \\ 0 & \text{elsewhere} \end{cases}$$ Type: Discrete Executable form: unidpdf(x , N), where x (integer) and N can be scalar or matrix	⊟ Compute $\begin{bmatrix} f[1] & f[7] \\ f[-3] & f[10] \end{bmatrix}$ for parameter N =7
	
	 Solution: $\begin{bmatrix} f[1] & f[7] \\ f[-3] & f[10] \end{bmatrix} = \begin{bmatrix} \frac{1}{7} & \frac{1}{7} \\ 0 & 0 \end{bmatrix} =$
	 $\begin{bmatrix} 0.1429 & 0.1429 \\ 0 & 0 \end{bmatrix}$
	
	 MATLAB Command
	       `>>unidpdf([1 7;-3 10],7)` ↵
	
	 ans =
	      0.1429 0.1429
	      0 0

Binomial distribution with parameter N and p : $$f[x] = {}^{N}C_{x} \, p^{x} q^{N-x}$$ where x =0, 1, 2, 3, 4, …, and N , $0 < p < 1$, and $q = 1 - p$ Type: Discrete Executable form: binopdf(x , N , p), where x , N , or p can be scalar or matrix	⊟ Compute $f[4]$ for parameters N =10 and p =0.6
	 Solution: $f[4] = {}^{10}C_{4} \, 0.6^4 \, 0.4^6 = 0.1115$
	
	 MATLAB Command
	       `>>binopdf(4,10,.6)` ↵
	
	 ans =
	      0.1115

Continuation of table 9.C:

Definition	Example
Chi-square distribution with n degrees of freedom: $$f(x) = \begin{cases} \dfrac{x^{\frac{n-2}{2}} e^{-\frac{x}{2}}}{2^{\frac{n}{2}}\Gamma\left(\dfrac{n}{2}\right)} & \text{for } x>0 \\ 0 & \text{elsewhere} \end{cases}$$ Type: Continuous Executable form: chi2pdf(x, n), where x and n can be scalar or matrix	⊟ Compute $f(3)$ for $n=5$ degrees of freedom Solution: $f(3) = \dfrac{3^{\frac{3}{2}} e^{-\frac{3}{2}}}{2^{\frac{5}{2}}\Gamma\left(\dfrac{5}{2}\right)} = 0.1542$ MATLAB Command $\quad$>>chi2pdf(3,5) ↵ ans = $\qquad$ 0.1542
t distribution with n degrees of freedom: $$f(x) = \begin{cases} \dfrac{\Gamma\left(\dfrac{n+1}{2}\right)\left(1+\dfrac{x^2}{n}\right)^{-\frac{n+1}{2}}}{\sqrt{\pi n}\ \Gamma\left(\dfrac{n}{2}\right)}, \\ \quad \text{for } -\infty < t < \infty \end{cases}$$ Type: Continuous Executable form: tpdf(x, n), where x and n can be scalar or matrix and n is an integer	⊟ Find $f(-4.5)$ for $n=4$ degrees of freedom Solution: $$f(-4.5) = \dfrac{\Gamma\left(\dfrac{5}{2}\right)\left(1+\dfrac{(-4.5)^2}{4}\right)^{-\frac{5}{2}}}{\sqrt{4\pi}\ \Gamma 2} = 0.0041$$ MATLAB Command $\quad$>>tpdf(-4.5,4) ↵ ans = $\qquad$ 0.0041
F distribution with parameters n_1 and n_2 (n_1 and n_2 are called the degrees of freedom): $$f(x) = \begin{cases} \dfrac{\Gamma\left(\dfrac{n_1+n_2}{2}\right)}{\Gamma\left(\dfrac{n_1}{2}\right)\Gamma\left(\dfrac{n_2}{2}\right)}\left(\dfrac{n_1}{n_2}\right)^{\frac{n_1}{2}} \times \\ x^{\frac{n_1}{2}-1}\left(1+\dfrac{n_1}{n_2}x\right)^{-\frac{n_1+n_2}{2}} & \text{for } x>0 \\ 0 & \text{elsewhere} \end{cases}$$ Type: Continuous Executable form: fpdf(x, n_1, n_2), where arguments can be scalar or matrix	⊟ Calculate $f(2)$ for degrees of freedom $n_1=4$ and $n_2=5$ Solution: $$f(2) = \dfrac{\Gamma\left(\dfrac{9}{2}\right)}{\Gamma 2\ \Gamma\left(\dfrac{5}{2}\right)}\left(\dfrac{4}{5}\right)^2 2\left(1+\dfrac{8}{5}\right)^{-\frac{9}{2}} = 0.152$$ MATLAB Command $\quad$>>fpdf(2,4,5) ↵ ans = $\qquad$ 0.1520
Beta distribution with parameters α and β: $$f(x) = \begin{cases} \dfrac{\Gamma(\alpha+\beta)}{\Gamma(\alpha)\Gamma(\beta)}x^{\alpha-1}(1-x)^{\beta-1} \\ \quad \text{for} \quad 0<x<1 \\ 0 \qquad \text{elsewhere} \end{cases}$$ where $\alpha>0$, $\beta>0$ Type: Continuous Executable form: betapdf(x, α, β)	⊟ Calculate $f(0.4)$ for $\alpha=4$ and $\beta=5$ Solution: $f(0.4) = \dfrac{\Gamma 9}{\Gamma 4 \Gamma 5}0.4^3 0.6^4 = 2.3224$ MATLAB Command $\quad$>>betapdf(0.4,4,5) ↵ ans = $\qquad$ 2.3224

Continuation of table 9.C:

Definition	Example

Poisson distribution with parameter λ :
 	
 $$f[x] = \frac{\lambda^x e^{-\lambda}}{x!} \quad \text{for} \quad \lambda > 0$$
 $x = 0, 1, 2, 3, 4, \dots.$
 	
 Type: Discrete   Executable form: poisspdf(x , λ)
	⊟ Calculate $f[5]$ for $\lambda = 3$
	
	 Solution: $f[5] = \dfrac{3^5 e^{-3}}{5!} = 0.1008$
	
	 MATLAB Command
	          >>poisspdf(5,3) ↵
	
	 ans =
	       0.1008

Log normal distribution with parameters μ and σ :
 	
 $$f(x) = \frac{e^{-\frac{(\ln x - \mu)^2}{2\sigma^2}}}{x \sigma \sqrt{2\pi}} \quad x > 0$$
 	
 Type: Continuous   Executable form: lognpdf(x , μ , σ)
	⊟ Compute $f(2.9)$ for parameter $\mu = 3$ and $\sigma = 1$
	
	 $f(2.9) = \dfrac{e^{-\frac{(\ln 2.9 - 3)^2}{2}}}{2.9\sqrt{2\pi}} = 0.0211$
	
	 MATLAB Command
	          >>lognpdf(2.9,3,1) ↵
	
	 ans =
	       0.0211

That is not all. Seeking for the rest pdfs is left as an exercise for the reader. Table 9.D provides the acronyms of the other subroutines. To know about any pdf distribution, execute 'help acronym'.

Table 9.D   Other probability density functions that are available in the statistical toolbox

Name of the probability density function	MATLAB counterpart	Name of the probability density function	MATLAB counterpart
Gamma	gampdf	Noncentral t	nctpdf
Geometric	geopdf	Non central chi	ncx2pdf
Hypergeometric	hygepdf	square	
Negative binomial	nbinpdf	Rayleigh	raylpdf
Noncentral F	ncfpdf	Weibull	weibpdf

9.17 Statistics from a given distribution

Understandably, finding the statistics of a random variable means finding the mean and variance of the variable from the given probability density function. The mean (m) and variance (V) of a continuous random variable, whose probability density function is $f(x)$, are given by $m = \int_{x=-\infty}^{x=\infty} x f(x)\, dx$ and $V = \int_{x=-\infty}^{x=\infty} (x-m)^2 f(x)\, dx$ respectively. On the contrary, the mean and variance of a discrete counterpart (discrete probability density function is $f[x]$) are obtained from $m = \sum_{x=-\infty}^{x=\infty} x f[x]$ and $V = \sum_{x=-\infty}^{x=\infty} (x-m)^2 f[x]$ respectively. We present one example for each. For the continuous case, take the beta distribution's probability density function,

$$f(x) = \begin{cases} \dfrac{\Gamma(\alpha+\beta)}{\Gamma(\alpha)\Gamma(\beta)} x^{\alpha-1}(1-x)^{\beta-1} & \text{for} \quad 0 < x < 1 \\ \\ 0 & \text{elsewhere} \end{cases}$$

, where $\alpha > 0$, $\beta > 0$, and α and β are called the parameters of the beta distribution.

Calculate the beta statistics, i.e., mean and variance of the distribution, for $\alpha=3$ and $\beta=4$. The mean is calculated by $m=\int_{x=-\infty}^{x=\infty}x\,f(x)\,dx=\int_{x=0}^{x=1}x\frac{\Gamma 7}{\Gamma 3\Gamma 4}x^2(1-x)^3\,dx=60\int_{x=0}^{x=1}x^3(1-x)^3\,dx=\frac{3}{7}=0.4286$ and the variance

computation is $V=\int_{x=-\infty}^{x=\infty}(x-m)^2\,f(x)\,dx=\int_{x=0}^{x=1}\left(x-\frac{3}{7}\right)^2\frac{\Gamma 7}{\Gamma 3\Gamma 4}x^2(1-x)^3\,dx=60\int_{x=0}^{x=1}\left(x-\frac{3}{7}\right)^2x^2(1-x)^3\,dx=\frac{3}{98}=0.0306$. As a

matter of fact, one can find $m=\dfrac{\alpha}{\alpha+\beta}$ and $V=\dfrac{\alpha\beta}{(\alpha+\beta)^2(\alpha+\beta+1)}$ for any x, α, and β. We could have also

used these formulas to compute m and V. Anyhow, MATLAB subroutine that finds the beta statistics is 'betastat' (abbreviation of <u>beta</u> <u>stat</u>istics). It has two input arguments, namely, α and β and two output arguments – m and V. We should get $m=0.4286$ and $V=0.0306$ for $\alpha=3$ and $\beta=4$. See the implementation below:

MATLAB Command
>>[m V]=betastat(3,4) ↵

m =
    0.4286
V =
    0.0306

Table 9.E   Other statistics functions that are available in the statistical toolbox

Name of the statistics whose mean and variance are required	MATLAB counterpart
Uniform discrete	unidstat
Uniform floating	unifstat
Binomial	binostat
Normal	normstat
Exponential	expstat
Gamma	gamstat
Chi square	chi2stat
t	tstat
F	fstat
Geometric	geostat
Hypergeometric	hygestat
Lognormal	lognstat
Negative binomial	nbinstat
Noncentral F	ncfstat
Noncentral t	nctstat
Non central chi square	ncx2stat
Rayleigh	raylstat
Weibull	weibstat

The next example is a discrete one. We chose the Poisson distribution for that. Its probability density function with parameter λ is given by $f[x]=\dfrac{\lambda^x e^{-\lambda}}{x!}$, where $\lambda>0$ and $x=0, 1, 2, 3, 4, \ldots$etc. The mean and variance of

the distribution are calculated by $m=\sum_{x=-\infty}^{x=\infty}xf[x]=\sum_{x=0}^{x=\infty}x\frac{\lambda^x e^{-\lambda}}{x!}=\lambda e^{-\lambda}\sum_{x=0}^{x=\infty}\frac{\lambda^{x-1}}{(x-1)!}=\lambda e^{-\lambda}\sum_{x=1}^{x=\infty}\frac{\lambda^{x-1}}{(x-1)!}=\lambda e^{-\lambda}e^{\lambda}=\lambda$ and $V=$

$\sum_{x=-\infty}^{x=\infty}(x-m)^2\,f[x]=\sum_{x=0}^{x=\infty}(x-\lambda)^2\frac{\lambda^x e^{-\lambda}}{x!}=\lambda$ (after simplification). The corresponding MATLAB function is 'poisstat' (abbreviation of <u>poi</u>sson <u>stat</u>istics). The output should be $m = 3$ and $V = 3$ for the parameter $\lambda = 3$. That is what is shown below:

MATLAB Command
>>[m v]=poisstat(3) ↵

m =
    3
V =
    3

We believe that the last two examples of computing the statistics for continuous and random variables make the reader familiar with the style how one can find the statistics of some distribution in MATLAB. Table 9.E gives the other known distributions including the MATLAB counterparts.

9.18 Cumulative distribution functions

The formal definition of cumulative distribution function (cdf) of a random variable X is given by $F(x) = P(X \leq x)$ for $-\infty < x < \infty$. $F(x)$ is obtained from $\int_{v=-\infty}^{x} f(v)\,dv$ (if the variable is continuous) or $\sum_{v=-\infty}^{v=x} f[v]$ (if the variable is discrete), where $f(x)$ and $f[x]$ are the probability density functions for continuous and discrete cases respectively. Two examples, one for each, are presented. As indicated earlier, probability density functions have some parameters. To compute a cumulative distribution function, these parameters must be provided. As an example of continuous distribution, we chose the exponential one, which is parameterized by

θ. Its probability density function is given by $f(x) = \begin{cases} \dfrac{1}{\theta} e^{-\frac{x}{\theta}} & for \ x > 0 \\ \\ 0 & elsewhere \end{cases}$. The cumulative distribution function

of this distribution is $F(x) = \int_{v=-\infty}^{x} f(v)\,dv = \int_{v=0}^{x} \dfrac{1}{\theta} e^{-\frac{v}{\theta}}\,dv = 1 - e^{-\frac{x}{\theta}}$. Compute $F(3)$ for $\theta = 2$. We have $F(3) = 1 - e^{-\frac{3}{2}} = 0.7769$. MATLAB counterpart of the cumulative function is 'expcdf' (abbreviation of <u>ex</u>ponential <u>c</u>umulative <u>distribution f</u>unction). It has two input arguments – θ and x and the executable form is 'expcdf(x, θ)'. See the implementation as follows:

MATLAB Command
>>expcdf(3,2) ↵

ans =
        0.7769

For discrete example, we chose the binomial distribution with parameter N and p. The distribution is formulated by $f[x] = {}^{N}C_{x}\,p^{x}\,q^{N-x}$, where $x = 0, 1, 2, 3, 4, ..., N$, $0 < p < 1$, and $q = 1 - p$. Its cumulative distribution function is given by $F(x) = \sum_{v=-\infty}^{v=x} f[v] = \sum_{v=0}^{v=x} {}^{N}C_{v}\,p^{v}\,q^{N-v}$. Calculate $F(3)$ in relation to parameters $N = 5$ and $p = 0.4$. The computations are $F(3) = \sum_{v=0}^{v=3} {}^{5}C_{v}\,(0.4)^{v}\,(0.6)^{5-v} = {}^{5}C_{0}\,(0.4)^{0}\,(0.6)^{5-0} + {}^{5}C_{1}\,(0.4)^{1}\,(0.6)^{5-1} + {}^{5}C_{2}\,(0.4)^{2}\,(0.6)^{5-2}$ $+ {}^{5}C_{3}\,(0.4)^{3}\,(0.6)^{5-3} = 0.9130$. The MATLAB counterpart for the cdf is 'binocdf' (abbreviation of <u>bino</u>mial <u>c</u>umulative <u>distribution f</u>unction). Its executable form is 'binocdf(x, N, p)'. The output of MATLAB is presented below:

MATLAB Command
>>binocdf(3,5,0.4) ↵

ans =
          0.9130

Similar problems for the other distributions can be solved by the function such as the one that follows in table 9.F.

Table 9.F   Other cumulative distribution functions
that are available in the statistical toolbox

Name of the cumulative distribution function	MATLAB counterpart
Beta	betacdf
Chi square	chi2cdf
F	fcdf
Gamma	gamcdf
Geometric	geocdf
Hypergeometric	hygecdf
Lognormal	logncdf
Negative binomial	nbincdf
Noncentral F	ncfcdf
Noncentral t	nctcdf
Noncentral Chi-square	ncx2cdf
Normal (Gaussian)	normcdf
Poisson	poisscdf
Rayleigh	raylcdf
T	tcdf
Discrete uniform	unidcdf
Uniform	unifcdf
Weibull	weibcdf

9.19 Inverse cumulative distribution functions or critical values

In the last article we defined the cumulative distribution function of a random variable X as $F(x) = P(X \leq x)$, where $F(x) = \int_{v=-\infty}^{x} f(v)\,dv$ or $\sum_{v=-\infty}^{v=x} f[v]$ (if the variable is continuous or discrete). In addition to the parameters of a particular distribution, the variable x was given and we computed $F(x)$. Now $F(x)$ is given, the variable x is to be computed for a particular distribution. This is called the inverse cumulative distribution function or finding the critical value. Consider the same examples as we chose for article 9.18. For exponential distribution with parameter $\theta = 2$, we had $F(3) = 0.7769$. Now provide $F(x) = 0.7769$ to get $x = 3$ as follows:

MATLAB Command
          >>expinv(0.7769,2) ↵   ← 'expinv' means e<u>x</u>ponential distribution's <u>inv</u>erse

      ans =
              3.0003

The output is 3.0003 instead of 3 because of the round off error. For the binomial distribution with parameters $N = 5$ and $p = 0.4$, we found $F(3) = 0.9130$. We are supposed to get 3 if we insert $F(x) = 0.9130$. The response is as follows:

      >>binoinv(0.9130,5,0.4) ↵   ← 'binoinv' means <u>bino</u>mial distribution's <u>inv</u>erse

      ans =
          4

Maybe you are displeased with the output. The discrepancy is coming from the numeric format. For computation, we have used the short numeric format, which rounds the output up to four decimal places. With long format, one would get $F(3) = 0.91296$. Try with this:

>>binoinv(0.9129,5,0.4) ↵

ans =

3

Now the return is the same as the expected output. Attention should be given to the numeric format and output as well when one has to deal with the discrete inverse cdfs. Anyhow, the other inverse cdfs are provided in table 9.G.

Table 9.G   Other inverse cumulative distribution functions that are available in the statistical toolbox

Name of the cumulative distribution	MATLAB counterpart
Beta	betainv
Chi square	chi2inv
F	finv
Gamma	gaminv
Geometric	geoinv
Hypergeometric	hygeinv
Lognormal	logninv
Negative binomial	nbininv
Noncentral F	ncfinv
Noncentral t	nctinv
Noncentral Chi-square	ncx2inv
Normal (Gaussian)	norminv
Poisson	poissinv
Rayleigh	raylinv
T	tinv
Discrete uniform	unidinv
Uniform	unifinv
Weibull	weibinv

9.20 Best fit straight line/curve of higher degree

Given is a set of x and y data in x-y plane, the best fit straight line or curve of higher degree from these points can be obtained by using the subroutine 'polyfit' (which is the abbreviation of <u>poly</u>nomial <u>fit</u>ting). The output of 'polyfit' is a polynomial. To obtain the best fit polynomial coefficients,

MATLAB Step:
Use the command polyfit (x data, y data, assumed degree of polynomial).

Two examples are mentioned — one for the straight line and the other for the curve of degree 2, which is a parabola. Suppose we have five points given as (3, 8), (4, 7), (5, 7), (6, 10), and (7, 8). These data can be organized as $x = \begin{bmatrix} 3 \\ 4 \\ 5 \\ 6 \\ 7 \end{bmatrix}$ and $y = \begin{bmatrix} 8 \\ 7 \\ 7 \\ 10 \\ 8 \end{bmatrix}$. Shown 'o' marks in figure 9.6 are the plots of the given points. The line AB of figure 9.6 is assumed to be the best fit straight line passing through these points. We would like to find its equation. The best fit line has the equation $\hat{y} = ax + b$. Due to this assumption, the error is $\varepsilon = y - \hat{y} = y - ax - b$.

Since the error can be positive or negative, take the square of the error, which is $\varepsilon^2 = (y - ax - b)^2$. The mean square error is given by $\bar{\varepsilon}^2 = \frac{1}{N}\sum_{i=1}^{N}(y_i - ax_i - b)^2 = \frac{1}{N}\sum(y - ax - b)^2$ (for simplicity, summation indexes are avoided), where N is the number of points in the given set. For the best fit line, $\bar{\varepsilon}^2$ is minimum and accordingly, $\frac{\partial(\bar{\varepsilon}^2)}{\partial a} = 0$ and $\frac{\partial(\bar{\varepsilon}^2)}{\partial b} = 0$, on that cause, $\frac{\partial(\bar{\varepsilon}^2)}{\partial a} = \frac{\partial[\frac{1}{N}\sum(y - ax - b)^2]}{\partial a} = \frac{1}{N}\sum 2(y - ax - b)(-x) = 0$ and

$\frac{\partial(\bar{\varepsilon}^2)}{\partial b} = \frac{\partial[\frac{1}{N}\sum(y - ax - b)^2]}{\partial b} = \frac{1}{N}\sum 2(y - ax - b)(-1) = 0$. Simplification yields $\begin{Bmatrix} a\sum x + bN = \sum y \\ a\sum x^2 + b\sum x = \sum xy \end{Bmatrix}$, which are just like the simultaneous linear equations relating to a and b (when b is a constant, $\sum b = bN$ is used in last set of equations). Solve the last set of equations to have $a = \frac{\sum x \sum y - N\sum xy}{(\sum x)^2 - N\sum x^2}$ and $b = \frac{\sum x \sum xy - \sum y \sum x^2}{(\sum x)^2 - N\sum x^2}$. The computations of a and b are our objective. We rendered the theoretical discussion. For the given set of data,

we have $x = \begin{bmatrix} 3 \\ 4 \\ 5 \\ 6 \\ 7 \end{bmatrix}$, $\sum x = 3+4+5+6+7=25$, $y = \begin{bmatrix} 8 \\ 7 \\ 7 \\ 10 \\ 8 \end{bmatrix}$, $\sum y = 40$, $\sum xy = 3\times8+4\times7+5\times7+6\times10+7\times8=203$, $\sum x^2 = 3^2+4^2+$

$5^2+6^2+7^2=135$, $a = \frac{25\times40 - 5\times203}{25^2 - 5\times135} = 0.3$ and $b = \frac{25\times203 - 40\times135}{25^2 - 5\times135} = 6.5$. Thereupon, the equation of the best fit line is $\hat{y} = 0.3x + 6.5$. The degree of polynomial is 1 if it is a straight line. The subroutine 'polyfit' should return [0.3   6.5] as the output for the given set of data. See the implementation in the following:

MATLAB Command

for the coefficient form,
```
>>x=[3 4 5 6 7]; ↵
>>y=[8 7 7 10 8]; ↵
>>polyfit(x,y,1) ↵

ans =
      0.3000    6.5000
```

to display as a polynomial,
```
>>p=polyfit(x,y,1); ↵
>>poly2str(p,'x') ↵

ans =
      0.3 x + 6.5
```

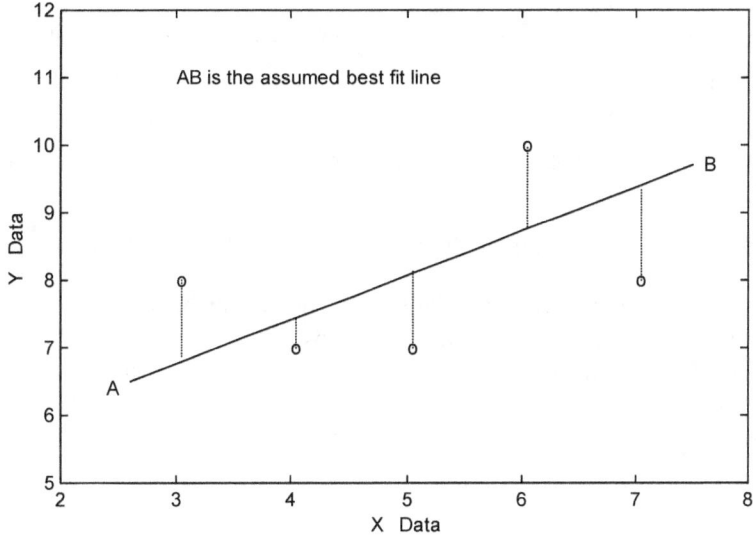

Figure 9.6   *Plot of X and Y data and the best fit line*

386

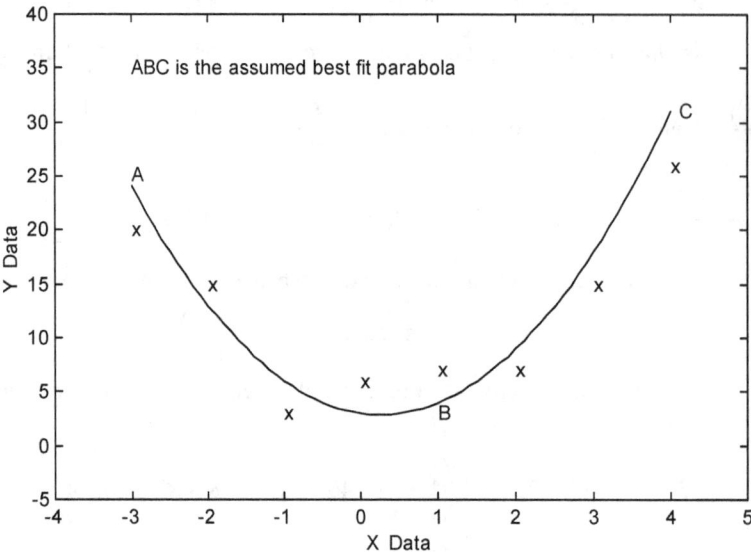

Figure 9.7 *Plot of X and Y data and the best fit parabola*

The next example is to find the best fit curve of degree 2 or parabola. Assumed parabolic equation is $\hat{y} = ax^2 + bx + c$, the error is $\varepsilon = y - \hat{y} = y - ax^2 - bx - c$, and the mean square error is $\bar{\varepsilon}^2 = \frac{1}{N}\sum_{i=1}^{N}(y_i - ax_i^2 - bx_i - c)^2$

$= \frac{1}{N}\sum(y - ax^2 - bx - c)^2$. The minimum error allows us to write $\frac{\partial(\bar{\varepsilon}^2)}{\partial a} = 0$, $\frac{\partial(\bar{\varepsilon}^2)}{\partial b} = 0$, and $\frac{\partial(\bar{\varepsilon}^2)}{\partial c} = 0$, from which,

$\dfrac{\partial\left[\frac{1}{N}\sum(y - ax^2 - bx - c)^2\right]}{\partial a} = 0$, $\dfrac{\partial\left[\frac{1}{N}\sum(y - ax^2 - bx - c)^2\right]}{\partial b} = 0$, and $\dfrac{\partial\left[\frac{1}{N}\sum(y - ax^2 - bx - c)^2\right]}{\partial c} = 0$. Differentiation and

simplification provide $\begin{cases} a\sum x^4 + b\sum x^3 + c\sum x^2 = \sum x^2 y \\ a\sum x^3 + b\sum x^2 + c\sum x = \sum xy \\ a\sum x^2 + b\sum x + cN = \sum y \end{cases}$. Write the equations in the matrix form as

$\begin{bmatrix} \sum x^2 & \sum x & N \\ \sum x^3 & \sum x^2 & \sum x \\ \sum x^4 & \sum x^3 & \sum x^2 \end{bmatrix}\begin{bmatrix} a \\ b \\ c \end{bmatrix} = \begin{bmatrix} \sum y \\ \sum xy \\ \sum x^2 y \end{bmatrix}$. Our unknowns are a, b, and c, in matrix form, $\begin{bmatrix} a \\ b \\ c \end{bmatrix} =$

$\begin{bmatrix} \sum x^2 & \sum x & N \\ \sum x^3 & \sum x^2 & \sum x \\ \sum x^4 & \sum x^3 & \sum x^2 \end{bmatrix}^{-1}\begin{bmatrix} \sum y \\ \sum xy \\ \sum x^2 y \end{bmatrix}$. Notice that for the degree 2 curve, the number of unknowns is 3 (which are

the polynomial coefficients), for the degree 3 curve, the number of unknowns is 4, and so on.

Consider the following points to find the best fit parabola numerically. The points are $(-3, 20)$, $(-2, 15)$, $(-1, 3)$, $(0, 6)$, $(1, 7)$, $(2, 7)$, $(3, 15)$, and $(4, 26)$. The given points (indicated by 'x' marks) and the best fit parabola (named by ABC) are depicted in figure 9.7. Put the coordinates together to have $x = [-3 \quad -2 \quad -1 \quad 0 \quad 1 \quad 2 \quad 3 \quad 4]$ and $y = [20 \quad 15 \quad 3 \quad 6 \quad 7 \quad 7 \quad 15 \quad 26]$, so, $\sum x = 4$, $\sum y = 99$, $\sum x^2 = (-3)^2 + (-2)^2 + (-1)^2 + 0^2 + 1^2 + 2^2 + 3^2 + 4^2 = 44$, $\sum xy = -3\times20 - 2\times15 - 1\times3 + 0\times6 + 1\times7 + 2\times7 + 3\times15 + 4\times26 = 77$, $\sum x^2 y = (-3)^2 20 + (-2)^2 15 + (-1)^2 3 + 0^2 6 + 1^2 7 + 2^2 7 + 3^2 15 + 4^2 26 = 829$, $\sum x^3 = (-3)^3 + (-2)^3 + (-1)^3 + 0^3 + 1^3 + 2^3 + 3^3 + 4^3 = 64$, and $\sum x^4 = (-3)^4 + (-2)^4 +$

$(-1)^4+0^4+1^4+2^4+3^4+4^4=452$. Hence, the matrices $\begin{bmatrix} \sum x^2 & \sum x & N \\ \sum x^3 & \sum x^2 & \sum x \\ \sum x^4 & \sum x^3 & \sum x^2 \end{bmatrix}$ and $\begin{bmatrix} \sum y \\ \sum xy \\ \sum x^2 y \end{bmatrix}$ turn to

$\begin{bmatrix} 44 & 4 & 8 \\ 64 & 44 & 4 \\ 452 & 64 & 44 \end{bmatrix}$ and $\begin{bmatrix} 99 \\ 77 \\ 829 \end{bmatrix}$ respectively. The inverse of $\begin{bmatrix} 44 & 4 & 8 \\ 64 & 44 & 4 \\ 452 & 64 & 44 \end{bmatrix}$ is

$\begin{bmatrix} -\dfrac{5}{168} & -\dfrac{1}{168} & \dfrac{1}{168} \\ \dfrac{1}{56} & \dfrac{5}{168} & -\dfrac{1}{168} \\ \dfrac{47}{168} & \dfrac{1}{56} & -\dfrac{5}{168} \end{bmatrix}$. On having that, the unknown polynomial coefficients are $\begin{bmatrix} a \\ b \\ c \end{bmatrix} =$

$\begin{bmatrix} -\dfrac{5}{168} & -\dfrac{1}{168} & \dfrac{1}{168} \\ \dfrac{1}{56} & \dfrac{5}{168} & -\dfrac{1}{168} \\ \dfrac{47}{168} & \dfrac{1}{56} & -\dfrac{5}{168} \end{bmatrix} \begin{bmatrix} 99 \\ 77 \\ 829 \end{bmatrix} = \begin{bmatrix} \dfrac{257}{168} \\ -\dfrac{7}{8} \\ \dfrac{739}{168} \end{bmatrix} = \begin{bmatrix} 1.5298 \\ -0.875 \\ 4.3988 \end{bmatrix}$ and the equation of the best fit parabola is

$\hat{y} = 1.5298 x^2 - 0.875 x + 4.3988$. That is what is exhibited by MATLAB:

MATLAB Command

for coefficient form,

```
>>x=[-3 -2 -1 0 1 2 3 4]; ↵
>>y=[20 15 3 6 7 7 15 26]; ↵
>>polyfit(x,y,2) ↵

ans =
      1.5298  -0.8750   4.3988
```

to display as a polynomial,

```
>>p=polyfit(x,y,2); ↵
>>poly2str(p,'x') ↵

ans =
     1.5298 x^2 - 0.875 x + 4.3988
```

In a similar fashion, one can find the best fit curve of degree 3 or higher in least mean square error sense using the 'polyfit'.

9.21 Regression analysis

Suppose y is a dependent variable and it is a function of n independent variables $x_1, x_2, x_3, \ldots\ldots\ldots, x_n$. These independent variables are called predictors. Widely used linear relationship between the dependent variable y and the predictors is as follows:

$$y = \beta_1 x_1 + \beta_2 x_2 + \beta_3 x_3 + \beta_4 x_4 + \ldots\ldots\ldots\ldots\ldots + \beta_n x_n + \varepsilon \qquad \ldots\text{eqn(9.1)}$$

The coefficients $\beta_1, \beta_2, \beta_3, \beta_4, \ldots\ldots\ldots\ldots, \beta_n$ are termed as the regression coefficients and ε is called an error due to the regression analysis. Equation 9.1 is referred to as the linear regression model. The model represents one dependent variable. If we have many dependent variables, say, m, then the equation for the ith dependent variable can be written as follows:

$$y_i = \beta_1 x_{i1} + \beta_2 x_{i2} + \beta_3 x_{i3} + \beta_4 x_{i4} + \ldots\ldots\ldots\ldots\ldots + \beta_n x_{in} + \varepsilon_i \qquad \ldots\text{eqn(9.2)}$$

where $i = 1, 2, 3, \ldots, m$

In matrix form, organize the variables as $\begin{bmatrix} y_1 \\ y_2 \\ y_3 \\ \vdots \\ y_m \end{bmatrix} = \begin{bmatrix} x_{11} & x_{12} & x_{13} & & x_{1n} \\ x_{21} & x_{22} & x_{23} & & x_{2n} \\ x_{31} & x_{32} & x_{33} & \cdots & x_{3n} \\ & & \vdots & & \\ x_{m1} & x_{m2} & x_{m3} & & x_{mn} \end{bmatrix} \begin{bmatrix} \beta_1 \\ \beta_2 \\ \beta_3 \\ \vdots \\ \beta_n \end{bmatrix} + \begin{bmatrix} \varepsilon_1 \\ \varepsilon_2 \\ \varepsilon_3 \\ \vdots \\ \varepsilon_m \end{bmatrix}$ or $Y = X\beta + \varepsilon$, where Y

$= \begin{bmatrix} y_1 \\ y_2 \\ y_3 \\ \vdots \\ y_m \end{bmatrix}$, $X = \begin{bmatrix} x_{11} & x_{12} & x_{13} & & x_{1n} \\ x_{21} & x_{22} & x_{23} & & x_{2n} \\ x_{31} & x_{32} & x_{33} & \cdots & x_{3n} \\ & & \vdots & & \\ x_{m1} & x_{m2} & x_{m3} & & x_{mn} \end{bmatrix}$, $\beta = \begin{bmatrix} \beta_1 \\ \beta_2 \\ \beta_3 \\ \vdots \\ \beta_n \end{bmatrix}$, and $\varepsilon = \begin{bmatrix} \varepsilon_1 \\ \varepsilon_2 \\ \varepsilon_3 \\ \vdots \\ \varepsilon_m \end{bmatrix}$. Objective of the regression analysis is to

estimate the regression coefficients $\beta = \begin{bmatrix} \beta_1 \\ \beta_2 \\ \beta_3 \\ \vdots \\ \beta_n \end{bmatrix}$ from the given $Y = \begin{bmatrix} y_1 \\ y_2 \\ y_3 \\ \vdots \\ y_m \end{bmatrix}$ and $X = \begin{bmatrix} x_{11} & x_{12} & x_{13} & & x_{1n} \\ x_{21} & x_{22} & x_{23} & & x_{2n} \\ x_{31} & x_{32} & x_{33} & \cdots & x_{3n} \\ & & \vdots & & \\ x_{m1} & x_{m2} & x_{m3} & & x_{mn} \end{bmatrix}$. It is

not to be mentioned that the orders of Y, X, β, and ε are $m \times 1$, $m \times n$, $n \times 1$, and $m \times 1$ respectively. Error

due to the regression is $\varepsilon = Y - X\beta$. The square error is given by $S = \sum_{i=1}^{m} \varepsilon_i^2 = \varepsilon^T \varepsilon = (Y - X\beta)^T (Y - X\beta) =$

$Y^T Y - (X\beta)^T Y - [Y^T - (X\beta)^T]X\beta = Y^T Y - \beta^T X^T Y - Y^T X\beta + (X\beta)^T X\beta = Y^T Y - \beta^T X^T Y - Y^T X\beta + \beta^T X^T X\beta$. According

to our assumption, the orders of β^T, X^T, and Y are $1 \times n$, $n \times m$, and $m \times 1$ respectively, on account of that,

the order of $\beta^T X^T Y$ is 1×1 that means a scalar. The transpose of $\beta^T X^T Y$ is $(\beta^T X^T Y)^T = Y^T (\beta^T X^T)^T = Y^T X\beta$.

$Y^T X\beta$ is a scalar too and $\beta^T X^T Y = Y^T X\beta$. Hence, S can be written as $S = Y^T Y - 2\beta^T X^T Y + \beta^T X^T X\beta$. To

achieve the least square error, the minimum S is taken observing $\dfrac{\partial S}{\partial \beta} = 0$, therefore, $\dfrac{\partial S}{\partial \beta} =$

$\dfrac{\partial[Y^T Y - 2\beta^T X^T Y + \beta^T X^T X\beta]}{\partial \beta} = \dfrac{\partial[-2\beta^T X^T Y + \beta^T X^T X\beta]}{\partial \beta}$. At this point, if P is a column vector and A is a

rectangular matrix, the partial derivative of quadratic form $P^T A P$ is $\dfrac{\partial(P^T A P)}{\partial P} = AP + A^T P$ and $\dfrac{\partial(P^T A)}{\partial P} = A$.

Using the last two identities helps us write $\dfrac{\partial S}{\partial \beta} = -2X^T Y + \dfrac{\partial[\beta^T (X^T X)\beta]}{\partial \beta} = -2X^T Y + (X^T X)\beta +$

$(X^T X)^T \beta = -2X^T Y + X^T X\beta + X^T X\beta = -2X^T Y + 2X^T X\beta$. Equate $\dfrac{\partial S}{\partial \beta}$ to 0 to write $X^T X\beta = X^T Y$, from which, $\beta =$

$(X^T X)^{-1} X^T Y$ that is what we are looking for. Clearly, $(X^T X)^{-1}$ is the inverse of $X^T X$. The subroutine that

performs the regression analysis is 'regress'. To perform the regression analysis,

MATLAB Step:

Use the command regress (y data as column vector, x data in rectangular matrix form).

Another amenable consideration is the consistency property. If A is a rectangular matrix and B is a column vector, the system of equations $A X = B$ is said to be consistent provided that the rank of augmented matrix $[A \quad B]$ is equal to the rank of A. Regression analysis is pertinent to the inconsistent equations. For inconsistent equations, the rank of $[A \quad B]$ is greater than that of A. We may seek for an approximate solution of $A X = B$ that minimizes the distance between $A X$ and B in spite of their inconsistency.

To exemplify this, take $X = \begin{bmatrix} 5 & 1 & 2 \\ 4 & -2 & 1 \\ -1 & 3 & -1 \\ 0 & 2 & -3 \end{bmatrix}$ and $Y = \begin{bmatrix} 7 \\ -2 \\ 8 \\ 0 \end{bmatrix}$. Related computations are $X^T =$

$\begin{bmatrix} 5 & 4 & -1 & 0 \\ 1 & -2 & 3 & 2 \\ 2 & 1 & -1 & -3 \end{bmatrix}$, $X^T X = \begin{bmatrix} 5 & 4 & -1 & 0 \\ 1 & -2 & 3 & 2 \\ 2 & 1 & -1 & -3 \end{bmatrix} \times \begin{bmatrix} 5 & 1 & 2 \\ 4 & -2 & 1 \\ -1 & 3 & -1 \\ 0 & 2 & -3 \end{bmatrix} = \begin{bmatrix} 42 & -6 & 15 \\ -6 & 18 & -9 \\ 15 & -9 & 15 \end{bmatrix}$, and

$$\begin{bmatrix} 42 & -6 & 15 \\ -6 & 18 & -9 \\ 15 & -9 & 15 \end{bmatrix}^{-1} = \begin{bmatrix} \dfrac{7}{184} & -\dfrac{5}{552} & -\dfrac{1}{23} \\ -\dfrac{5}{552} & \dfrac{15}{184} & \dfrac{4}{69} \\ -\dfrac{1}{23} & \dfrac{4}{69} & \dfrac{10}{69} \end{bmatrix}.$$ With that, the regression coefficients are $\beta = \begin{bmatrix} \beta_1 \\ \beta_2 \\ \beta_3 \end{bmatrix} =$

$$(X^T X)^{-1} X^T Y = \begin{bmatrix} \dfrac{7}{184} & -\dfrac{5}{552} & -\dfrac{1}{23} \\ -\dfrac{5}{552} & \dfrac{15}{184} & \dfrac{4}{69} \\ -\dfrac{1}{23} & \dfrac{4}{69} & \dfrac{10}{69} \end{bmatrix} \begin{bmatrix} 5 & 4 & -1 & 0 \\ 1 & -2 & 3 & 2 \\ 2 & 1 & -1 & -3 \end{bmatrix} \begin{bmatrix} 7 \\ -2 \\ 8 \\ 0 \end{bmatrix} = \begin{bmatrix} \dfrac{16}{69} \\ \dfrac{67}{23} \\ \dfrac{41}{23} \end{bmatrix} = \begin{bmatrix} 0.2319 \\ 2.9130 \\ 1.7826 \end{bmatrix}.$$ That is what is displayed

as the output of regression analysis in the following:

MATLAB Command
```
>>y=[7 -2 8 0]';
>>x=[5 1 2;4 -2 1;-1 3 -1;0 2 -3];
>>regress(y,x)

ans =
      0.2319
      2.9130
      1.7826
```

Consistent equations show some errors during the execution. $X = \begin{bmatrix} 5 & 1 & 2 & 0 \\ 4 & -2 & 1 & -4 \\ -1 & 3 & -1 & -2 \end{bmatrix}$ and $Y = \begin{bmatrix} 7 \\ -2 \\ 8 \end{bmatrix}$ are the

examples of consistent equations. Following error will be encountered:

MATLAB Command
```
>>y=[7 -2 8]';
>>x=[5 1 2 0;4 -2 1 -4;-1 3 -1 -2];
>>regress(y,x)
??? Error using ==> \
Matrix dimensions must agree.

Error in ==> C:\MATLAB\toolbox\stats\regress.m
On line 49  ==> RI = R\eye(p);
```

⊡ **Regression analysis when the variance matrix of error is given**

In the preceding discussion, the error due to the regression analysis is given by $\varepsilon = Y - X\beta$. The error assumes a predetermined variation in many applications. For instance, in antenna array processing it takes the Gaussian distribution. The square error is given by $S = (Y - X\beta)^T V^{-1} (Y - X\beta)$ when the variance-covariance or dispersion matrix V is provided. Recall that V is a symmetric matrix and its order is $n \times n$ when the order of X is $m \times n$. In a similar fashion, the error minimization purveys $X^T V^{-1} X\beta = X^T V^{-1} Y$. This equation is called *the generalized least square equation* from which the regression coefficients can be obtained as $\beta = (X^T V^{-1} X)^{-1} X^T V^{-1} Y$. MATLAB counterpart of the regression with the variance is 'lscov' (abbreviation of least square with covariance). To perform this computation,

MATLAB Step:

Use the command regress (x data in rectangular matrix, y data as column vector, covariance matrix).

Insertion of the arguments for the subroutines 'regress' and 'lscov' does not happen in the same way. In 'regress', the first argument is y vector and the second one is the x matrix but the order is altered in 'lscov'.

Consider the same X and Y as we chose for the 'regress' and take the covariance matrix V as

$$\begin{bmatrix} 2 & -1 & 0 & 4 \\ -1 & 6 & 2 & -2 \\ 0 & 2 & 1 & 6 \\ 4 & -2 & 6 & 3 \end{bmatrix}.$$ Then, the regression coefficients with the variance are $\beta = (X^T V^{-1} X)^{-1} X^T V^{-1} Y$. We

have $V^{-1} = \begin{bmatrix} \frac{262}{411} & \frac{89}{411} & -\frac{142}{411} & -\frac{2}{137} \\ \frac{89}{411} & \frac{82}{411} & -\frac{20}{411} & -\frac{8}{137} \\ -\frac{142}{411} & -\frac{20}{411} & \frac{55}{411} & \frac{22}{137} \\ -\frac{2}{137} & -\frac{8}{137} & \frac{22}{137} & -\frac{1}{137} \end{bmatrix}$, $X^T V^{-1} X = \begin{bmatrix} 5 & 4 & -1 & 0 \\ 1 & -2 & 3 & 2 \\ 2 & 1 & -1 & -3 \end{bmatrix} \times \begin{bmatrix} \frac{262}{411} & \frac{89}{411} & -\frac{142}{411} & -\frac{2}{137} \\ \frac{89}{411} & \frac{82}{411} & -\frac{20}{411} & -\frac{8}{137} \\ -\frac{142}{411} & -\frac{20}{411} & \frac{55}{411} & \frac{22}{137} \\ -\frac{2}{137} & -\frac{8}{137} & \frac{22}{137} & -\frac{1}{137} \end{bmatrix}$

$\begin{bmatrix} 5 & 1 & 2 \\ 4 & -2 & 1 \\ -1 & 3 & -1 \\ 0 & 2 & -3 \end{bmatrix} = \begin{bmatrix} \frac{13057}{411} & -\frac{899}{137} & \frac{5830}{411} \\ -\frac{899}{137} & \frac{355}{137} & -\frac{596}{137} \\ \frac{5830}{411} & -\frac{596}{137} & \frac{2734}{411} \end{bmatrix}$, $(X^T V^{-1} X)^{-1} = \begin{bmatrix} \frac{13057}{411} & -\frac{899}{137} & \frac{5830}{411} \\ -\frac{899}{137} & \frac{355}{137} & -\frac{596}{137} \\ \frac{5830}{411} & -\frac{596}{137} & \frac{2734}{411} \end{bmatrix}^{-1} = \begin{bmatrix} \frac{347}{10602} & \frac{1237}{3534} & \frac{1687}{10602} \\ \frac{1237}{3534} & -\frac{231}{1178} & -\frac{3091}{3534} \\ \frac{1687}{10602} & -\frac{3091}{3524} & -\frac{4034}{5301} \end{bmatrix}$,

$X^T V^{-1} Y = \begin{bmatrix} \frac{4310}{411} \\ -\frac{184}{137} \\ \frac{251}{411} \end{bmatrix}$, and $\beta = (X^T V^{-1} X)^{-1} X^T V^{-1} Y = \begin{bmatrix} \frac{347}{10602} & \frac{1237}{3534} & \frac{1687}{10602} \\ \frac{1237}{3534} & -\frac{231}{1178} & -\frac{3091}{3534} \\ \frac{1687}{10602} & -\frac{3091}{3524} & -\frac{4034}{5301} \end{bmatrix} \begin{bmatrix} \frac{4310}{411} \\ -\frac{184}{137} \\ \frac{251}{411} \end{bmatrix} = \begin{bmatrix} -\frac{35}{1178} \\ \frac{4005}{1178} \\ \frac{1401}{589} \end{bmatrix} = \begin{bmatrix} -0.0297 \\ 3.3998 \\ 2.3786 \end{bmatrix}$. See the

computation with the covariance matrix as follows:

MATLAB Command
```
>>y=[7 -2 8 0]'; ↵
>>x=[5 1 2;4 -2 1;-1 3 -1;0 2 -3]; ↵
>>v=[2 -1 0 4;-1 6 2 -2;0 2 1 6;4 -2 6 3]; ↵
>>lscov(x,y,v) ↵

ans =
        -0.0297
         3.3998
         2.3786
```

Indeed, it is amazing that using the word 'lscov' does so much calculation.

9.22 Principal component analysis

Principal component analysis is a kind of multivariate analysis. If N random variables form a data matrix A of order $M \times N$ (whose rows represent M observations of each variable), the matrix can be transformed to N orthogonal random variables. The first few transformed variables (say, P) will carry nearly all the information possessed by the given N variables. The P variables are termed as the principal components of A. Since $P < N$, the analysis is a dimension reduction approach. In most cases the given N random variables are correlated but the principal components hold orthogonality.

The first step of the analysis is to have the variance-covariance matrix of A. In article 9.14, we have

mentioned how the variance-covariance matrix can be found. Consider $A = \begin{bmatrix} 21 & 5 \\ 18 & 7 \\ 25 & 9 \\ 20 & 5 \\ 22 & 7 \end{bmatrix}$ for illustrative example.

Another method of finding the variance-covariance matrix is introduced, which is the matrix method. It can be formulated as $V = \dfrac{A^T C A}{M-1}$, where V is the variance-covariance matrix and C is the centering matrix defined by

$C = I - \dfrac{O}{M}$. In C, the matrix I is the identity matrix of order $M \times M$ and O is the matrix of ones of order

$M \times M$ ($M = 5$ for A), $\therefore$ $C = \begin{bmatrix} 1 & 0 & 0 & 0 & 0 \\ 0 & 1 & 0 & 0 & 0 \\ 0 & 0 & 1 & 0 & 0 \\ 0 & 0 & 0 & 1 & 0 \\ 0 & 0 & 0 & 0 & 1 \end{bmatrix} - \dfrac{1}{5}\begin{bmatrix} 1 & 1 & 1 & 1 & 1 \\ 1 & 1 & 1 & 1 & 1 \\ 1 & 1 & 1 & 1 & 1 \\ 1 & 1 & 1 & 1 & 1 \\ 1 & 1 & 1 & 1 & 1 \end{bmatrix} = \begin{bmatrix} \frac{4}{5} & -\frac{1}{5} & -\frac{1}{5} & -\frac{1}{5} & -\frac{1}{5} \\ -\frac{1}{5} & \frac{4}{5} & -\frac{1}{5} & -\frac{1}{5} & -\frac{1}{5} \\ -\frac{1}{5} & -\frac{1}{5} & \frac{4}{5} & -\frac{1}{5} & -\frac{1}{5} \\ -\frac{1}{5} & -\frac{1}{5} & -\frac{1}{5} & \frac{4}{5} & -\frac{1}{5} \\ -\frac{1}{5} & -\frac{1}{5} & -\frac{1}{5} & -\frac{1}{5} & \frac{4}{5} \end{bmatrix}$, on

that, we have $V = \dfrac{1}{5-1} \times \begin{bmatrix} 21 & 18 & 25 & 20 & 22 \\ 5 & 7 & 9 & 5 & 7 \end{bmatrix} \begin{bmatrix} \frac{4}{5} & -\frac{1}{5} & -\frac{1}{5} & -\frac{1}{5} & -\frac{1}{5} \\ -\frac{1}{5} & \frac{4}{5} & -\frac{1}{5} & -\frac{1}{5} & -\frac{1}{5} \\ -\frac{1}{5} & -\frac{1}{5} & \frac{4}{5} & -\frac{1}{5} & -\frac{1}{5} \\ -\frac{1}{5} & -\frac{1}{5} & -\frac{1}{5} & \frac{4}{5} & -\frac{1}{5} \\ -\frac{1}{5} & -\frac{1}{5} & -\frac{1}{5} & -\frac{1}{5} & \frac{4}{5} \end{bmatrix} \begin{bmatrix} 21 & 5 \\ 18 & 7 \\ 25 & 9 \\ 20 & 5 \\ 22 & 7 \end{bmatrix} = \begin{bmatrix} \frac{67}{10} & \frac{13}{5} \\ \frac{13}{5} & \frac{14}{5} \end{bmatrix}$.

The second step of the principal component analysis is to decompose V as $E \times D \times E^T$, where D is a diagonal matrix containing the ordered eigenvalues of V and $E = [E_1 \ E_2 \ E_3 \ ... \ E_N]$ ($E_1 \ E_2 \ E_3 \ ... \ E_N$ are the normalized eigenvectors of V). Detail discussion of the eigenvalues and eigenvectors is presented in article

4.16. Go through the article to have the eigenvalues and normalized eigenvectors of V as $\dfrac{3}{2}$, 8 and $\begin{bmatrix} \frac{1}{\sqrt{5}} \\ -\frac{2}{\sqrt{5}} \end{bmatrix}$,

$\begin{bmatrix} \frac{2}{\sqrt{5}} \\ \frac{1}{\sqrt{5}} \end{bmatrix}$ respectively. Having known the eigenvalues and eigenvectors, one can write $D = \begin{bmatrix} 8 & 0 \\ 0 & \frac{3}{2} \end{bmatrix} = \begin{bmatrix} 8 & 0 \\ 0 & 1.5 \end{bmatrix}$

and $E = \begin{bmatrix} \frac{2}{\sqrt{5}} & \frac{1}{\sqrt{5}} \\ \frac{1}{\sqrt{5}} & -\frac{2}{\sqrt{5}} \end{bmatrix} = \begin{bmatrix} 0.8944 & 0.4472 \\ 0.4472 & -0.8944 \end{bmatrix}$, hence, the required decomposition is brought about. The

necessary MATLAB subroutine is 'princomp' (abbreviation of <u>prin</u>cipal <u>comp</u>onent analysis). Implementation is shown below:

MATLAB Command

```
>>A=[21 5;18 7;25 9;20 5;22 7]; ↵
>>[E Q D S]=princomp(A); ↵
>>E ↵

E =

    0.8944  -0.4472
    0.4472   0.8944
>>D ↵

D =

    8.0000
    1.5000
```

The subroutine has four output arguments as indicated by [E Q D S]. The second (Q) and fourth (S) outputs are called the component scores and Hotelling's T^2 respectively. We excluded their descriptions. Recall that the eigenvectors are not unique. Multiplying an eigenvector with a negative sign does not make any difference. Compare the second column of the computed E with that of MATLAB return. They are identical following a multiplication of negative sign. Anyhow, the transformed data can be computed by $A \times E$. Compute the variance-covariance matrix of $A \times E$. Use MATLAB to avoid the computational hassle:

>>cov(A*sym(E)) ↵

ans =

[8,   0]
[0, 3/2]

Thereupon, the variance-covariance matrix of $A \times E$ is the same as the matrix D. The variance is a measure of the information content of a random variable. Since the diagonal elements of variance-covariance matrix are the variances of the columns, it can be said that $\dfrac{8}{8+\dfrac{3}{2}}$ =84.21% of the information contained in A is retained

due to the transformation even if the second component $\left(\dfrac{3}{2}\right)$ is ignored.

9.23 Convolution of two discrete random processes

An indexed family of random variables defines a random process, which is $\{X_n\}$. Usually the subscript n is associated with the time index. So, the random process $\{X_n\}$ is dependent on the family variables X and the time index n as well. If $x[n]$ and $y[n]$ denote two discrete random processes, their convolution is given by $c[n] = \sum_{k=-\infty}^{k=\infty} x[k]y[n-k]$, where $c[n]$ is the resulting convolution process. The formula can be viewed as the multiplication of two polynomials $x[n]$ and $y[n]$. MATLAB counterpart for the convolution is 'conv'. Find the convolution of the random processes $x[n]$ and $y[n]$ as presented in the following table:

n	−2	−1	0	1
$x[n]$	−1	4	3	−6

n	−1	0	1
$y[n]$	7	2	−5

Apply the polynomial multiplication as follows:

```
   −1    4    3    −6      ← x[n]
         7    2    −5      ← y[n]
   ─────────────────────
   −7   28   21   −42
        −2    8     6    −12
              5   −20   −15    30
   ─────────────────────────────────
   −7   26   34   −56   −27    30    ←c[n]
```

If the numbers of observations for the processes $x[n]$ and $y[n]$ are M and N respectively, the length of the convolution process $c[n]$ is $M + N - 1$(4+3−1=6 for the given processes). The length of $x[n]$ (which is 4) is greater than that of $y[n]$ (which is 3). The indexes of n are from −3 to 2 with the increment 1. That is, n and $c[n]$ are as follows:

n	−3	−2	−1	0	1	2
$c[n]$	−7	26	34	−56	−27	30

Have it computed as follows:

MATLAB Command

>>x=[-1 4 3 -6]; ↵
>>y=[7 2 -5]; ↵
>>c=conv(x,y) ↵

c =

-7   26   34   -56   -27   30

9.24 Crosscorrelation of two random processes

Mean and variance do not provide the complete information about a random process. The reason is that a random process is not only a function of random variable but also a function of time. A more informative approach is crosscorrelation and crosscovariance. Only the discrete case of crosscorrelation is highlighted.

Crosscorrelation $C_{xy}[n]$ of two discrete random processes $x[n]$ and $y[n]$ is defined as $C_{xy}[m] = E\{x[n]y[n-m]\}$, where E is the expectation operator. Observe the expression for $C_{xy}[m]$:

for $m=0$, $C_{xy}[0]$ is the sum of the product of $x[n]$ and $y[n]$ divided by the number of observations,

for $m=1$, $C_{xy}[1]$ is the sum of the product of $x[n]$ and $y[n-1]$ divided by the number of observations, that is, slide the sequence $y[n]$ to the right by 1 and take the product,

for $m=-1$, $C_{xy}[-1]$ is the sum of the product of $x[n]$ and $y[n+1]$ divided by the number of observations, that is, slide the sequence $y[n]$ to the left by 1 and take the product,

................, and so on for other m's.

The conception will be clear from the following example. We compute the crosscorrelation of random processes $x[n]$ and $y[n]$ as presented in the following table:

n	-2	-1	0	1
$x[n]$	-1	6	-4	-5
$y[n]$	9	3	-10	8

For each $C_{xy}[m]$, compute only the product of the dotted elements and consider only the number of observations inside the dotted mark as presented below.

Central:

$$
\begin{array}{cccc}
\vdots-1 & 6 & -4 & -5\vdots \\
\vdots\ 9 & 3 & -10 & 8\vdots
\end{array}
\quad\begin{array}{c}\leftarrow x[n]\\ \leftarrow y[n]\end{array}
$$

$x[n]\times y[n] = -9\quad 18\quad 40\quad -40$ $\Sigma=9$ $C_{xy}[0] = \dfrac{9}{4} = 2.25$

Left shifts:

$$
\begin{array}{cccc}
\vdots-1 & 6 & -4\vdots & -5 \\
9 & \vdots\ 3 & -10 & 8\vdots
\end{array}
\quad\begin{array}{c}\leftarrow x[n]\\ \leftarrow y[n+1]\end{array}
$$

$x[n]\times y[n+1] = -3\quad -60\quad -32$ $\Sigma=-95$ $C_{xy}[-1] = -\dfrac{95}{3} = -31.6667$

$$
\begin{array}{cccc}
\vdots-1 & 6\vdots & -4 & -5 \\
9 & 3 & \vdots-10 & 8\vdots
\end{array}
\quad\begin{array}{c}\leftarrow x[n]\\ \leftarrow y[n+2]\end{array}
$$

$x[n]\times y[n+2] = 10\quad 48$ $\Sigma=58$ $C_{xy}[-2] = \dfrac{58}{2} = 29$

$$
\begin{array}{cccccc}
 & \vdots -1 \vdots & 6 & -4 & -5 & \leftarrow x[n] \\
9 \quad 3 \quad -10 & \vdots\ 8\ \vdots & & & & \leftarrow y[n+3]
\end{array}
$$

$$x[n] \times y[n+3] = -8 \qquad\qquad \Sigma = -8 \qquad C_{xy}[-3] = -\frac{8}{1} = -8$$

Right shifts:

$$
\begin{array}{cccccc}
-1 & \vdots\ 6 & -4 & -5\ \vdots & & \leftarrow x[n] \\
 & \vdots\ 9 & 3 & -10\ \vdots & 8 & \leftarrow y[n-1]
\end{array}
$$

$$x[n] \times y[n-1] = 54 \quad -12 \quad 50 \qquad \Sigma = 92 \qquad C_{xy}[1] = \frac{92}{3} = 30.6667$$

$$
\begin{array}{cccccc}
-1 & 6 & \vdots -4 & -5\ \vdots & & \leftarrow x[n] \\
 & & \vdots\ 9 & 3\ \vdots & -10 \quad 8 & \leftarrow y[n-2]
\end{array}
$$

$$x[n] \times y[n-2] = -36 \quad -15 \qquad \Sigma = -51 \qquad C_{xy}[2] = -\frac{51}{2} = -25.5$$

$$
\begin{array}{cccccc}
-1 & 6 & -4 & \vdots -5 \vdots & & \leftarrow x[n] \\
 & & & \vdots\ 9\ \vdots & 3 \quad -10 \quad 8 & \leftarrow y[n-3]
\end{array}
$$

$$x[n] \times y[n-3] = -45 \qquad \Sigma = 45 \qquad C_{xy}[3] = -\frac{45}{1} = -45$$

Collecting all $C_{xy}[m]$'s, we have the computed crosscorrelation coefficients in tabular form as follows:

m	-3	-2	-1	0	1	2	3
$C_{xy}[m]$	-8	29	-31.6667	2.25	30.6667	-25.5	-45

The function that is dedicated for the purpose is 'xcorr' (abbreviation of cross (x) correlation). The computation is shown below:

MATLAB Command

```
>>x=[-1 6 -4 -5]; ↵
>>y=[9 3 -10 8]; ↵
>>C=xcorr(x,y,'unbiased') ↵

C =
        -8.0000  29.0000 -31.6667   2.2500  30.6667 -25.5000 -45.0000
```

The above crosscorrelation is called the unbiased crosscorrelation. There is one more, which does not divide the sum of the product of $x[n]y[n-m]$ by the number of observations. Conformably, $C_{xy}[m]$ is the discrete convolution of $x[n]$ and $y[-n]$ computed from the polynomial multiplication of the given $x[n]$ and $y[n]$ as follows:

$$
\begin{array}{cccccccc}
-1 & 6 & -4 & -5 & & & & \leftarrow x[n] \\
8 & -10 & 3 & 9 & & & & \leftarrow y[-n] \\
\hline
-8 & 48 & -32 & -40 & & & & \\
 & 10 & -60 & 40 & 50 & & & \\
 & & -3 & 18 & -12 & -15 & & \\
 & & & -9 & 54 & -36 & -45 & \\
\hline
-8 & 58 & -95 & 9 & 92 & -51 & -45 & \leftarrow C_{xy}[m]
\end{array}
$$

For the problem at hand, exclude the argument 'unbiased' of 'xcorr'. The analysis is termed as the biased crosscorrelation of the processes $x[n]$ and $y[n]$ and seen in signal processing applications. Anyhow the output is as follows:

>>C=xcorr(x,y) ↵

C =

    -8.0000 58.0000 -95.0000   9.0000 92.0000 -51.0000 -45.0000

The crosscorrelation becomes autocorrelation when one uses $x[n]=y[n]$. The autocorrelation of a random process is a measure of the dependence between the values of the random process at different times. It also accounts the time variation of a random process. However, the autocorrelation of $x[n]$ is computed by using the commands 'xcorr(x,'unbiased')' and 'xcorr(x)' for the unbiased and biased cases respectively. Execute 'help xcorr' to learn more about the other variants of the subroutine.

9.25 Mahalanobis distance

Mahalanobis distance is given by $d^2(Y_i,\overline{X})=(Y_i-\overline{X})^TV^{-1}(Y_i-\overline{X})$, where Y_i is any k dimensional vector (that is, the order of matrix Y_i is $m\times k$), V is the variance-covariance matrix as defined in article 9.14 (obtained from another k dimensional vector X, which is the reference of computation), and $\overline{X}$ is the mean vector of the reference vector X. It gives a measure of the distance between Y_i and $\overline{X}$ with respect to the variance-covariance matrix formed by X. The equation $(X-\overline{X})^TV^{-1}(X-\overline{X})=d^2$ represents a k dimensional ellipsoid in a k dimensional space, where $\overline{X}$ is the mean of X and $\overline{X}$ is called the center of the ellipsoid. All points lying on the surface of the ellipsoid $(X-\overline{X})^TV^{-1}(X-\overline{X})=d^2$ are having the equal distances from $\overline{X}$.

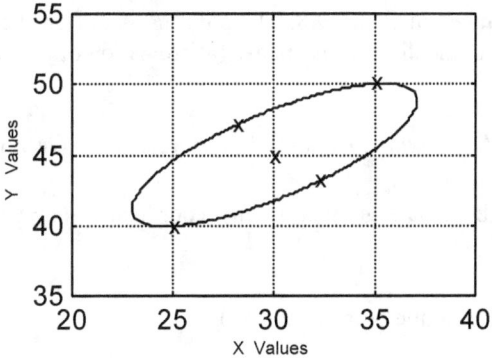

Figure 9.8 *Ellipse formed by variance-covariance matrix V*

V is a positive definite matrix that means V has positive eigenvalues and it is symmetric and that d is a constant. An ellipse is nothing but a two dimensional ellipsoid. One important application of the distance is to analyze the multivariate data locally.

We describe the concept taking 5 points in the $x-y$ plane, which are given by $(x,y)=A(28,47)$, $B(32,43)$, $C(25,40)$, $D(35,50)$, and $E(30,45)$. Shown 'x' marks of figure 9.8 indicate the location of these points. We have 2 dimensions x and y so $k=2$, that means X must have two columns. The given data can be arranged as $x=\begin{bmatrix}28\\32\\25\\35\\30\end{bmatrix}$ and $y=\begin{bmatrix}47\\43\\40\\50\\45\end{bmatrix}$. Just mentioned vector X is formed as $X=[x\quad y]=\begin{bmatrix}28&47\\32&43\\25&40\\35&50\\30&45\end{bmatrix}$. Following the method of article 9.14, the variance-covariance matrix of X is computed as $V = \begin{bmatrix} \dfrac{29}{2} & \dfrac{21}{2} \\ \dfrac{21}{2} & \dfrac{29}{2} \end{bmatrix}$. V^{-1}, inverse

of V, is found as $\begin{bmatrix} \dfrac{29}{200} & -\dfrac{21}{200} \\ -\dfrac{21}{200} & \dfrac{29}{200} \end{bmatrix}$ (see article 4.14 for matrix inverse). The means of x and y are 30 and 45

respectively, from which, $\bar{X} = \begin{bmatrix} 30 \\ 45 \end{bmatrix}$. Y_i can be any point including the given A, B, C, D, or E. For point A,

we have $Y_1 = \begin{bmatrix} 28 \\ 47 \end{bmatrix}$, $Y_1 - \bar{X} = \begin{bmatrix} -2 \\ 2 \end{bmatrix}$, and the Mahalanobis distance is given by $d^2(Y_1, \bar{X}) = (Y_1 - \bar{X})^T V^{-1}(Y_1 - \bar{X}) =$

$\begin{bmatrix} -2 \\ 2 \end{bmatrix}^T \begin{bmatrix} \dfrac{29}{200} & -\dfrac{21}{200} \\ -\dfrac{21}{200} & \dfrac{29}{200} \end{bmatrix} \begin{bmatrix} -2 \\ 2 \end{bmatrix} = \begin{bmatrix} -2 & 2 \end{bmatrix} \begin{bmatrix} -\dfrac{1}{2} \\ \dfrac{1}{2} \end{bmatrix} = 2$. Point B gives the distance as $\begin{bmatrix} 2 \\ -2 \end{bmatrix}^T \begin{bmatrix} \dfrac{29}{200} & -\dfrac{21}{200} \\ -\dfrac{21}{200} & \dfrac{29}{200} \end{bmatrix} \times$

$\begin{bmatrix} 2 \\ -2 \end{bmatrix} = 2$. In a similar fashion, we get 2, 2, and 0 for the points C, D, and E respectively. The points A, B,

C, and D are providing the same distance 2 from $\bar{X}$ relative to the matrix V therefore they are located on an ellipse, and the ellipse is graphed in figure 9.8. The point E is not on the ellipse instead coincides with $\bar{X}$. MATLAB function adopted for the distance is 'mahal' (abbreviation of mahalanobis). To have the distance,

MATLAB Step:
Use the command mahal(arbitrary vector Y_i, given vector X for computation of V).

For the example at hand, the first argument is any coordinates (x, y) and the second argument is $X = \begin{bmatrix} 28 & 47 \\ 32 & 43 \\ 25 & 40 \\ 35 & 50 \\ 30 & 45 \end{bmatrix}$. To find the distance for the point A (28, 47),

MATLAB Command
>>X=[28 47;32 43;25 40;35 50;30 45]; ↵
>>A=[28 47]; ↵
>>mahal(A,X) ↵

ans =
         2.0000

Form the matrix Y from $\begin{bmatrix} A \\ B \\ C \\ D \\ E \end{bmatrix}$ to calculate the distances for all points – A, B, C, D, and E. The

computation is as follows:
>>Y=[28 47;32 43;25 40;35 50;30 45]; ↵
>>mahal(Y,X) ↵

ans =
                2.0000
                2.0000
                2.0000
                2.0000
                     0

We computed the distance considering A as a column vector but in MATLAB, the insersion style is a row one.

9.26 Help about other subroutines

Our objective is not to explain the whole *Statistical Toolbox*. Numerous functions are included in the toolbox from where we presented just the flavor of the toolbox. Some clues are divulged to continue with. Synopsis of the other functionality and comprehensive help inventory of the toolbox can be found by executing the following:

MATLAB Command
>>help stats ⏎

Statistics Toolbox.
Version 2.1.1 21-Nov-1997

New Features
  Readme   - Version 2.1.0 synopsis of new functionality.

Distributions.
Parameter estimation.
  betafit   - Beta parameter estimation.
  binofit   - Binomial parameter estimation.
  expfit   - Exponential parameter estimation.
  gamfit   - Gamma parameter estimation.
  mle   - Maximum likelihood estimation (MLE).
  normfit   - Normal parameter estimation.
  poissfit   - Poisson parameter estimation.
  unifit   - Uniform parameter estimation.
  weibfit   - Weibull parameter estimation.

Probability density functions (pdf).
  betapdf   - Beta density.
  binopdf   - Binomial density.
  chi2pdf   - Chi square density.
  exppdf   - Exponential density.
  fpdf   - F density.
  gampdf   - Gamma density.
  geopdf   - Geometric density.
  hygepdf   - Hypergeometric density.
  lognpdf   - Lognormal density.
  nbinpdf   - Negative binomial density.
  ncfpdf   - Noncentral F density.
  nctpdf   - Noncentral t density.
  ncx2pdf   - Noncentral Chi-square density.
  normpdf   - Normal (Gaussian) density.
  pdf   - Density function for a specified distribution.
  poisspdf   - Poisson density.

```
raylpdf   - Rayleigh density.
tpdf      - T density.
unidpdf   - Discrete uniform density.
unifpdf   - Uniform density.
weibpdf   - Weibull density.
```

Cumulative Distribution functions (cdf).
```
betacdf   - Beta cdf.
binocdf   - Binomial cdf.
cdf       - Specified cumulative distribution function.
chi2cdf   - Chi square cdf.
expcdf    - Exponential cdf.
fcdf      - F cdf.
gamcdf    - Gamma cdf.
geocdf    - Geometric cdf.
hygecdf   - Hypergeometric cdf.
logncdf   - Lognormal cdf.
nbincdf   - Negative binomial cdf.
ncfcdf    - Noncentral F cdf.
nctcdf    - Noncentral t cdf.
ncx2cdf   - Noncentral Chi-square cdf.
normcdf   - Normal (Gaussian) cdf.
poisscdf  - Poisson cdf.
raylcdf   - Rayleigh cdf.
tcdf      - T cdf.
unidcdf   - Discrete uniform cdf.
unifcdf   - Uniform cdf.
weibcdf   - Weibull cdf.
```

Critical Values of Distribution functions.
```
betainv   - Beta inverse cumulative distribution function.
binoinv   - Binomial inverse cumulative distribution function.
chi2inv   - Chi square inverse cumulative distribution function.
expinv    - Exponential inverse cumulative distribution function.
finv      - F inverse cumulative distribution function.
gaminv    - Gamma inverse cumulative distribution function.
geoinv    - Geometric inverse cumulative distribution function.
hygeinv   - Hypergeometric inverse cumulative distribution function.
icdf      - Specified inverse cdf.
logninv   - Lognormal inverse cumulative distribution function.
nbininv   - Negative binomial inverse distribution function.
ncfinv    - Noncentral F inverse cumulative distribution function.
nctinv    - Noncentral t inverse cumulative distribution function.
ncx2inv   - Noncentral Chi-square inverse distribution function.
norminv   - Normal (Gaussian) inverse cumulative distribution
                                        function.
poissinv  - Poisson inverse cumulative distribution function.
raylinv   - Rayleigh inverse cumulative distribution function.
tinv      - T inverse cumulative distribution function.
unidinv   - Discrete uniform inverse cumulative distribution function.
unifinv   - Uniform inverse cumulative distribution function.
weibinv   - Weibull inverse cumulative distribution function.
```

Random Number Generators.
```
betarnd   - Beta random numbers.
binornd   - Binomial random numbers.
```

chi2rnd   - Chi square random numbers.
exprnd   - Exponential random numbers.
frnd   - F random numbers.
gamrnd   - Gamma random numbers.
geornd   - Geometric random numbers.
hygernd   - Hypergeometric random numbers.
lognrnd   - Lognormal random numbers.
mvnrnd   - Multivariate normal random numbers.
nbinrnd   - Negative binomial random numbers.
ncfrnd   - Noncentral F random numbers.
nctrnd   - Noncentral t random numbers.
ncx2rnd   - Noncentral Chi-square random numbers.
normrnd   - Normal (Gaussian) random numbers.
poissrnd   - Poisson random numbers.
random   - Random numbers from specified distribution.
raylrnd   - Rayleigh random numbers.
trnd   - T random numbers.
unidrnd   - Discrete uniform random numbers.
unifrnd   - Uniform random numbers.
weibrnd   - Weibull random numbers.

Statistics.
betastat   - Beta mean and variance.
binostat   - Binomial mean and variance.
chi2stat   - Chi square mean and variance.
expstat   - Exponential mean and variance.
fstat   - F mean and variance.
gamstat   - Gamma mean and variance.
geostat   - Geometric mean and variance.
hygestat   - Hypergeometric mean and variance.
lognstat   - Lognormal mean and variance.
nbinstat   - Negative binomial mean and variance.
ncfstat   - Noncentral F mean and variance.
nctstat   - Noncentral t mean and variance.
ncx2stat   - Noncentral Chi-square mean and variance.
normstat   - Normal (Gaussian) mean and variance.
poisstat   - Poisson mean and variance.
raylstat   - Rayleigh mean and variance.
tstat   - T mean and variance.
unidstat   - Discrete uniform mean and variance.
unifstat   - Uniform mean and variance.
weibstat   - Weibull mean and variance.

Descriptive Statistics.
bootstrp   - Bootstrap statistics for any function.
corrcoef   - Correlation coefficient.
cov   - Covariance
crosstab   - Cross tabulation.
geomean   - Geometric mean.
grpstats   - Summary statistics by group.
harmmean   - Harmonic mean.
iqr   - Interquartile range.
kurtosis   - Kurtosis.
mad   - Median Absolute Deviation.
mean   - Sample average
median   - 50th percentile of a sample.

```
moment    - Moments of a sample.
nanmax    - Maximum ignoring NaNs.
nanmean   - Mean ignoring NaNs.
nanmedian - Median ignoring NaNs.
nanmin    - Minimum ignoring NaNs.
nanstd    - Standard deviation ignoring NaNs.
nansum    - Sum ignoring NaNs.
prctile   - Percentiles.
range     - Range.
skewness  - Skewness.
tabulate  - Frequency table.
trimmean  - Trimmed mean.
var       - Variance.
```

Linear Models.
```
anova1     - One-Way Analysis of Variance.
anova2     - Two-Way Analysis of Variance.
dummyvar   - Dummy-variable coding.
leverage   - Regression diagnostic.
lscov      - Least-squares estimates with known covariance matrix.
polyfit    - Least-squares polynomial fitting.
polyval    - Predicted values for polynomial functions.
regress    - Multivariate linear regression.
regstats   - Regression diagnostics.
ridge      - Ridge regression.
rstool     - Multidimensional response surface visualization (RSM).
stepwise   - Interactive tool for stepwise regression.
x2fx       - Factor settings matrix (x) to design matrix (fx).
```

Nonlinear Models
```
nlinfit   - Nonlinear least-squares data fitting (Newton's method).
nlintool  - Interactive graphical tool for prediction in nonlinear
                                              models.
nlpredci  - Confidence intervals for prediction.
nlparci   - Confidence intervals for parameters.
nnls      - Non-negative least-squares.
```

Design of Experiments (DOE)
```
cordexch   - D-optimal design (coordinate exchange algorithm).
daugment   - Augment D-optimal design.
dcovary    - D-optimal design with fixed covariates.
ff2n       - Two-level full-factorial design.
fullfact   - Mixed-level full-factorial design.
hadamard   - Hadamard matrices (orthogonal arrays).
rowexch    - D-optimal design (row exchange algorithm).
```

Statistical Process Control (SPC)
```
capable    - Capability indices.
capaplot   - Capability plot.
ewmaplot   - Exponentially weighted moving average plot.
histfit    - Histogram with superimposed normal density.
normspec   - Plot normal density between specification limits.
schart     - S chart for monitoring variability.
xbarplot   - Xbar chart for monitoring the mean.
```

Principal Components Analysis

barttest   - Bartlett's test for dimensionality.
pcacov   - Principal components from covariance matrix.
pcares   - Residuals from principal components.
princomp   - Principal components analysis from raw data.

Multivariate Statistics.
classify   - Linear Discriminant Analysis.
mahal   - Mahalanobis distance.

Hypothesis Tests.
ranksum   - Wilcoxon rank sum test (independent samples).
signrank   - Wilcoxon sign rank test (paired samples).
signtest   - Sign test (paired samples).
ztest   - Z test.
ttest   - One sample t test.
ttest2   - Two sample t test.

Statistical Plotting.
boxplot   - Boxplots of a data matrix (one per column).
fsurfht   - Interactive contour plot of a function.
gline   - Point, drag and click line drawing on figures.
gname   - Interactive point labelling in x-y plots.
lsline   - Add least-square fit line to scatter plot.
normplot   - Normal probability plot.
qqplot   - Quantile-Quantile plot.
refcurve   - Reference polynomial curve.
refline   - Reference line.
surfht   - Interactive contour plot of a data grid.
weibplot   - Weibull probability plot.

Statistics Demos.
disttool   - GUI tool for exploring probability distribution functions.
polytool   - Interactive graph for prediction of fitted polynomials.
randtool   - GUI tool for generating random numbers.
rsmdemo   - Reaction simulation (DOE, RSM, nonlinear curve fitting).

File Based I/O
tblread   - Read in data in tabular format.
tblwrite   - Write out data in tabular format to file.
caseread   - Read in case names.
casewrite - Write out case names to file.

The above output is the comprehensive inventory of the statistical toolbox. From the displayed list, select any function and execute 'help function' to learn the details of the function.

Chapter 10

Miscellaneous Functions

Although the chapters already discussed are focused for widely known problems, it is sometimes useful to perform computations in some other classified disciplines. To furnish by examples, dot and cross products of vectors and vector transformations are used in antenna pattern synthesis, sinc function is used in signal processing problems, orthogonal polynomials such as hermites are used in image processing problems, number system conversions are used in digital electronics, conversion of measurement units is a real life problem, inverse error function is seen in communication engineering ... etc. Out of numerous built in subroutines, only a few are chosen to show the computational capability of MATLAB in various disciplines. Even though we are not going to discuss the complete problem of a particular field in details, at least the reader will have an outline of the specialized functions and matrices used in that field.

10.1 Special functions

There are some functions, which are not used frequently but these functions are useful in specialized disciplines. For example, Fresnel sine and cosine integrals are used in antenna engineering. These special functions are located in 'mfun' list of Maple toolbox. Each of these functions is applicable for the numerical evaluation. Maple's evaluation is up to 16 digit accurate. The input arguments of these functions are numerical quantity. NaN (not a number) notifies any singularity of the execution. There are over fifty functions that can be found in the toolbox. Most of them are used in classical applied mathematics. Perform the following command in the command window of MATLAB.

MATLAB Command
>>help mfunlist ⏎

> MFUNLIST Special functions for MFUN.
>> The following special functions are listed in alphabetical order
>> according to the third column. n denotes an integer argument,
>> x denotes a real argument, and z denotes a complex argument. For
>> more detailed descriptions of the functions, including any
>> argument restrictions, see the Reference Manual, or use MHELP.

bernoulli	n	Bernoulli Numbers
bernoulli	n,z	Bernoulli Polynomials
BesselI	x1,x	Bessel Function of the First Kind
BesselJ	x1,x	Bessel Function of the First Kind
BesselK	x1,x	Bessel Function of the Second Kind
BesselY	x1,x	Bessel Function of the Second Kind
Beta	z1,z2	Beta Function
binomial	x1,x2	Binomial Coefficients
LegendreKc	x	Complete Elliptic Integral of First Kind
LegendreEc	x	Complete Elliptic Integral of Second Kind
LegendrePic	x1,x	Complete Elliptic Integral of Third Kind
LegendreKc1	x	LegendreKc using Complementary Modulus
LegendreEc1	x	LegendreEc using Complementary Modulus
LegendrePic1	x1,x	LegendrePic using Complementary Modulus
erfc	z	Complementary Error Function
erfc	n,z	Complementary Error Function's Iterated Integrals
Ci	z	Cosine Integral
dawson	x	Dawson's Integral
Psi	z	Digamma Function
dilog	x	Dilogarithm Integral
erf	z	Error Function
euler	n	Euler Numbers
euler	n,z	Euler Polynomials
Ei	x	Exponential Integral
Ei	n,z	Exponential Integral
FresnelC	x	Fresnel Cosine Integral
FresnelS	x	Fresnel Sine Integral
GAMMA	z	Gamma Function
harmonic	n	Harmonic Function
Chi	z	Hyperbolic Cosine Integral
Shi	z	Hyperbolic Sine Integral
hypergeom	X1,X2	(Generalized) Hypergeometric Function
LegendreF	x,x1	Incomplete Elliptic Integral of First Kind
LegendreE	x,x1	Incomplete Elliptic Integral of Second Kind
LegendrePi	x,x2,x1	Incomplete Elliptic Integral of Third Kind
GAMMA	z1,z2	Incomplete Gamma Function
W	z	Lambert's W Function
W	n,z	Lambert's W Function
lnGAMMA	z	Logarithm of the Gamma function
Li	x	Logarithmic Integral
Psi	n,z	Polygamma Function
Ssi	z	Shifted Sine Integral
Si	z	Sine Integral
Zeta	z	(Riemann) Zeta Function
Zeta	n,z	(Riemann) Zeta Function
Zeta	n,z,x	(Riemann) Zeta Function

Orthogonal Polynomials (Extended Symbolic Math Toolbox only)

T	n,x	Chebyshev of the First Kind
U	n,x	Chebyshev of the Second Kind
G	n,x1,x	Gegenbauer
H	n,x	Hermite
P	n,x1,x2,x	Jacobi
L	n,x	Laguerre

L   n,x1,x   Generalized Laguerre
P   n,x   Legendre

See also MFUN, MHELP.

Actually what we executed is we asked the help from MATLAB about the 'mfunlist'. MATLAB displayed a number of functions available in the 'mfun' toolbox. The number and nature of arguments of a function that can dwell beside each function are also displayed. We are not going to place all of them. Along with the examples some special functions and their MATLAB counterparts are shown as follows:

⊡ Binomial coefficients, ^mC_n

$$^mC_n = \frac{m!}{n!(m-n)!}$$, where m and n are integers and $m > n$, if $m < n$, $^mC_n = 0$.   Mfun subroutine name:

binomial   Executable form: mfun ('binomial',integer m, integer n)   Examples: $^5C_3 = \frac{5!}{3!(5-3)!} = 10$ and $^3C_5 = 0$.

MATLAB Command

for $^5C_3$,   command for $^3C_5$,

>>mfun('binomial',5,3) ↵   >>mfun('binomial',3,5) ↵

ans =   ans =
         10   0

A set of binomial coefficients can also be found. Say, we wish to find $\begin{bmatrix} ^8C_2 \\ ^8C_0 \\ ^8C_1 \\ ^8C_4 \end{bmatrix}$. The result should look like $\begin{bmatrix} \frac{8!}{2!6!} \\ \frac{8!}{0!8!} \\ \frac{8!}{1!7!} \\ \frac{8!}{4!4!} \end{bmatrix}$

$= \begin{bmatrix} 28 \\ 1 \\ 8 \\ 70 \end{bmatrix}$. In this example, we have $m = 8$ and $n = \begin{bmatrix} 2 \\ 0 \\ 1 \\ 4 \end{bmatrix}$. The command is as follows:

MATLAB Command
                >>m=8; ↵
                >>n=[2 0 1 4]'; ↵
                >>mfun('binomial',m,n) ↵

                ans =
                        28
                         1
                         8
                        70

⊡ Fresnel sine integral, $S(x)$

The definition of $S(x)$ is $\int_{t=0}^{t=x} \sin\left(\frac{\pi}{2}t^2\right) dt$, where x can be any real/complex numbers. This is an odd

function (shown in figure 10.1) and oscillatory in nature about $x = \pm 0.5$.   Mfun subroutine name: FresnelS Executable form: mfun('FresnelS',x), where x can be a scalar or matrix   Examples: We want to find the Fresnel sine integral for each element of the row matrix R=[0 1.5 −2 4], the output is going to be O, where

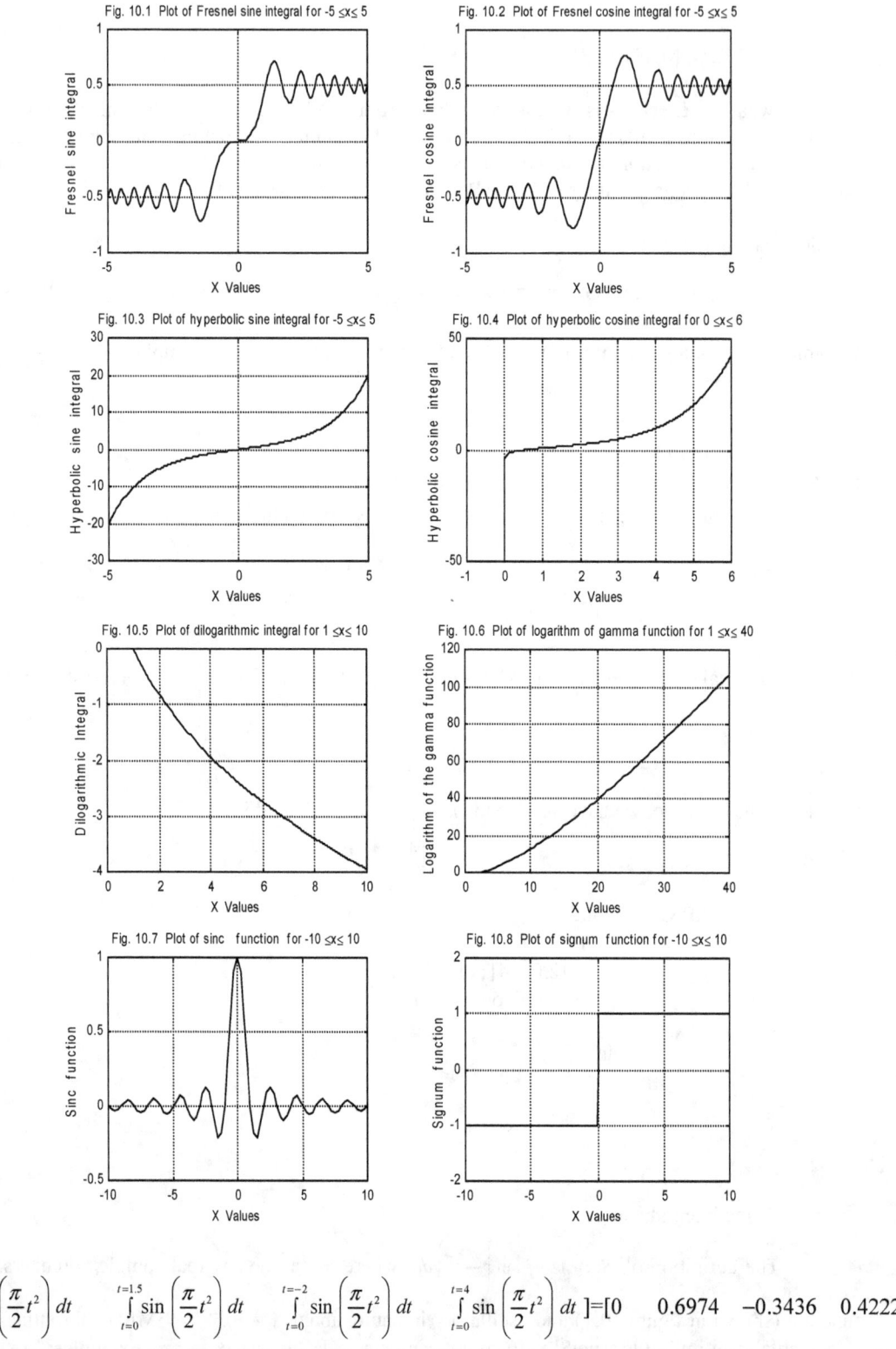

Fig. 10.1 Plot of Fresnel sine integral for -5 ≤x≤ 5

Fig. 10.2 Plot of Fresnel cosine integral for -5 ≤x≤ 5

Fig. 10.3 Plot of hyperbolic sine integral for -5 ≤x≤ 5

Fig. 10.4 Plot of hyperbolic cosine integral for 0 ≤x≤ 6

Fig. 10.5 Plot of dilogarithmic integral for 1 ≤x≤ 10

Fig. 10.6 Plot of logarithm of gamma function for 1 ≤x≤ 40

Fig. 10.7 Plot of sinc function for -10 ≤x≤ 10

Fig. 10.8 Plot of signum function for -10 ≤x≤ 10

$$O=[\ \int_{t=0}^{t=0} \sin\left(\frac{\pi}{2}t^2\right)dt \quad \int_{t=0}^{t=1.5} \sin\left(\frac{\pi}{2}t^2\right)dt \quad \int_{t=0}^{t=-2} \sin\left(\frac{\pi}{2}t^2\right)dt \quad \int_{t=0}^{t=4} \sin\left(\frac{\pi}{2}t^2\right)dt\ ]=[0 \quad 0.6974 \quad -0.3436 \quad 0.4222]$$

(numerical computation using the trapezoidal rule taking 100 steps for each, see chapter 6 for the trapezoidal numerical computation). The solution is shown as follows:

MATLAB Command

>>R=[0 1.5 -2 4]; ↵

>>O=mfun('FresnelS',R) ↵

O =

   0   0.6975   -0.3434   0.4205

⊟ Fresnel cosine integral, $C(x)$

$C(x)$ is given by $\int_{t=0}^{t=x} \cos\left(\frac{\pi}{2}t^2\right) dt$. The expression for the Fresnel cosine integral is very similar to the sine one, just sine will be replaced by the cosine. This integral also shows the oscillatory behavior about $x = \pm 0.5$. But the function's rise from $x = 0$ is faster than that of the sine counterpart (shown in figure 10.2). Like sine integral, the argument x can be any real or complex number. Mfun subroutine name: FresnelC Executable form: mfun ('FresnelC',x), where x can be a scalar or matrix   Examples: We wish to find the Fresnel cosine integral for each element of the matrix $A = \begin{bmatrix} 3 & -2 \\ 0 & 1 \end{bmatrix}$. Our output should be

$$\begin{bmatrix} \int_{t=0}^{t=3} \cos\left(\frac{\pi}{2}t^2\right) dt & \int_{t=0}^{t=-2} \cos\left(\frac{\pi}{2}t^2\right) dt \\ \int_{t=0}^{t=0} \cos\left(\frac{\pi}{2}t^2\right) dt & \int_{t=0}^{t=1} \cos\left(\frac{\pi}{2}t^2\right) dt \end{bmatrix} = \begin{bmatrix} 0.6055 & -0.4883 \\ 0 & 0.7799 \end{bmatrix}$$ (using 200 steps trapezoidal integration for each).

MATLAB computation is as follows:

MATLAB Command

>>A=[3 -2;0 1]; ↵

>>mfun('FresnelC',A) ↵

ans =

   0.6057   -0.4883

   0   0.7799

⊟ Hyperbolic sine integral, $shi(x)$

The integral is defined as $shi(x) = \int_{t=0}^{t=x} \frac{\sinh t}{t} dt$, where x can be any real or complex number. This is an odd function (shown in figure 10.3).   Mfun subroutine name: Shi(x)   Executable form: mfun('Shi',x), where x can be a scalar or matrix   Examples: Find the hyperbolic sine integral for each element of the row matrix R=[0 3 −4 4], so, the output is going to be $[\int_{t=0}^{t=0} \frac{\sinh t}{t} dt \quad \int_{t=0}^{t=3} \frac{\sinh t}{t} dt \quad \int_{t=0}^{t=-4} \frac{\sinh t}{t} dt \quad \int_{t=0}^{t=4} \frac{\sinh t}{t} dt]$ =[0   4.9735   −9.8175   9.8175] (using the trapezoidal rule with 200 steps). See the implementation in MATLAB:

MATLAB Command

>>R=[0 3 -4 4]; ↵

>>mfun('Shi',R) ↵

ans =

   0   4.9734   -9.8173   9.8173

⊟ Hyperbolic cosine integral, $chi(x)$

The definition of the integral is $chi(x) = \gamma + \ln x + \int_{t=0}^{t=x} \frac{\cosh t - 1}{t} dt$, where γ is the Euler's constant (given by 0.57721566490) and the argument x can be any real or complex number except $x = 0$. At $x = 0$, $chi(x)$ is

408

not defined. The plot of hyperbolic cosine integral is shown in figure 10.4.   Mfun subroutine name: Chi(x)
Executable form: mfun('Chi',x), where x can be a scalar or matrix   Example: We want to find $chi(4.5)$.

$chi(4.5) = 0.5772156649 + \ln 4.5 + \int_{t=0}^{t=4.5} \frac{\cosh t - 1}{t} dt = 13.9661$ (trapezoidal integration with 200 steps). MATLAB's

mfun solution as well as the 200 steps trapezoidal computation is as follows:

MATLAB Command

for mfun solution,   for the trapezoidal integration,
>>mfun('Chi',4.5) ↵   >>ga=0.57721566490; ↵
                                                >>t=[4.5/200:4.5/200:4.5]; ↵
ans =   >>t1=[0 t]; ↵
        13.9658   >>y=(cosh(t)-1)./t; ↵
                                                >>y1=[0 y]; ↵
                                                >>ga+log(4.5)+trapz(t1,y1) ↵

                                                ans =
                                                        13.9661

From this illustration, you can compare how advantageous the mfun function is. So many computational steps
can be bypassed conveniently. Belonging to the trapezoidal computation, we treated $t = 0$ separately. Why is
that? Because at $t = 0$, the integrand $\frac{\cosh t - 1}{t}$ takes $\frac{0}{0}$ form that can not be handled by MATLAB but

$\underset{t \to 0}{Lt} \frac{\cosh t - 1}{t} = 0$. However, at $x = 0$, $chi(x)$ is not defined and MATLAB returns a not a number as follows:

        >>mfun('Chi',0) ↵

    ans =
        NaN

⊟ Logarithm of the Gamma function, $F(x)$

The function is defined as $F(x) = \ln(\Gamma x) = \ln\left[\int_{t=0}^{t=\infty} t^{x-1} e^{-t} dt\right]$, where x can not be zero or negative

integers. See the plot of the function in figure 10.6.   Mfun subroutine name: lnGAMMA(x)   Executable
form: mfun ('lnGAMMA',x), where x can be a scalar or matrix   Example: Compute [$F(4)$ $F(3)$]
Solution: [$F(4)$ $F(3)$]=[$\ln \Gamma 4$ $\ln(\Gamma 3)$]=[$\ln 6$ $\ln 2$]=[1.7918   0.6931]. The solution is shown below:

        MATLAB Command
        >>x=[4 3]; ↵
        >>mfun('lnGAMMA',x) ↵

        ans =
                1.7918   0.6931

⊟ Dilogarithmic integral, $F(x)$

The definition of the dilogarithmic integral is $F(x) = \int_{t=1}^{t=x} \frac{\ln t}{1-t} dt$, where $x \geq 1$. The function is depicted

in figure 10.5.   Mfun subroutine name: dilog(x)   Executable form: mfun('dilog',x), where x can be a
scalar or matrix   Example: We wish to compute the dilogarithmic integral for each element of the column

matrix $C = \begin{bmatrix} 2 \\ 3 \end{bmatrix}$. The output should be $\begin{bmatrix} F(2) \\ F(3) \end{bmatrix} = \begin{bmatrix} \int_{t=1}^{t=2} \frac{\ln t}{1-t} dt \\ \int_{t=1}^{t=3} \frac{\ln t}{1-t} dt \end{bmatrix} = \begin{bmatrix} -0.8225 \\ -1.4368 \end{bmatrix}$ (trapezoidal rule with 100 steps). The

solution is shown below:

MATLAB Command
>>C=[2 3]'; ↵
>>mfun('dilog',C) ↵

ans =
-0.8225
-1.4367

The preceding discussions show how one can implement the 'mfun' functions in MATLAB. Table 10.A is provided with some other 'mfun' functions along with the definitions, executable forms, and examples.

Table 10.A   Some mfun functions' definitions and their MATLAB analogues

Definition	Example
Incomplete gamma function, $\Gamma(a,x) = \int_{t=x}^{t=\infty} e^{-t} t^{a-1} dt$ Executable form: mfun('GAMMA',a,x), where x can be a scalar or matrix	⊟ Compute $\Gamma(2,3)$ Solution: $\Gamma(2,3) = \int_{t=3}^{t=\infty} e^{-t} t \, dt = 4e^{-3} = 0.1991$ MATLAB Command >>mfun('GAMMA',2,3) ↵ ans = 0.1991
Shifted sine integral, $Ssi(x) = Si(x) - \frac{\pi}{2}$, where $Si(x) = \int_{t=0}^{t=x} \frac{\sin t}{t} dt$ is the sine integral of x Executable form: mfun('Ssi',x), where x can be a scalar or matrix	⊟ Compute $Ssi(3)$ Solution: $Ssi(3) = Si(3) - \frac{\pi}{2} = \int_{t=0}^{t=3} \frac{\sin t}{t} dt - \frac{\pi}{2} = 0.2779$ MATLAB Command >>mfun('Ssi',3) ↵ ans = 0.2779
Harmonic function, $h(x) = \sum_{k=1}^{k=x} \frac{1}{k}$, where $x > 0$ Executable form: mfun('harmonic', x), where x can be a scalar or matrix	⊟ Compute $h(4)$ Solution: $h(4) = \sum_{k=1}^{k=4} \frac{1}{k} = 2.0833$ MATLAB Command >>mfun('harmonic',4) ↵ ans = 2.0833
Factorial of an integer n, which is given by $n! = n(n-1)(n-2) \dots 4.3.2.1$ Executable form: mfun('factorial', x), where x can be a scalar or matrix	⊟ Compute [4! 6!] Solution: [4! 6!]=[4.3.2.1 6.5.4.3.2.1]= [24 720] MATLAB Command >>mfun('factorial',[4 6]) ↵ ans = 24 720

410

Continuation of the last table:

Definition	Example

Exponential integral is defined as $Ei(n, x) = \int_{t=1}^{t=\infty} \frac{e^{-xt}}{t^n} dt$, where $n \geq 0$ and $real(x) > 0$. It is related with the incomplete gamma function by $Ei(n, x) = x^{n-1} \Gamma(1 - n, x)$
 	
Executable form: mfun('Ei',n,x)
	⊟ Compute $Ei(2, 4)$
	Solution:
	$Ei(2, 4) = 4^{2-1} \Gamma(1 - 2, 4) = 4\Gamma(-1, 4) = 0.0032$
	MATLAB Command
	  >>mfun('Ei',2,4) ↵
	
	ans =
	   0.0032

Surd of a number x is denoted by $x^{\frac{1}{n}}$ or $\sqrt[n]{x}$, where n is an integer and x can also be a complex number
 	
Executable form: mfun('surd',x,n),
	⊟ Compute $\sqrt[3]{32}$
	Solution: $\sqrt[3]{32} = 32^{\frac{1}{3}} = 3.1748$
	MATLAB Command
	  >>mfun('surd',32,3) ↵
	
	ans =
	   3.1748
	
	⊟ Compute $\sqrt[3]{3 - 4i}$
	Solution: $\sqrt[3]{3 - 4i} = \left(5e^{-i\tan^{-1}\frac{4}{3}}\right)^{\frac{1}{3}} =$
	$5^{\frac{1}{3}} e^{-i\frac{1}{3}\tan^{-1}\frac{4}{3}} = 1.6289 - i\,0.5202$
	MATLAB Command
	  >>mfun('surd',3-4i,3) ↵
	
	ans =
	   1.6289 - 0.5202i

Integer part of logarithm of a number N with respect to base 10, which is defined as the integer portion of the number $\log_{10} N$
 	
Executable form: mfun('ilog10',N), where N is a single element or matrix
	⊟ Compute the integer part of $\log_{10} 516$
	Solution: $\log_{10} 516 = 2.7126$, so, the integer part is 2
	MATLAB Command
	  >>mfun('ilog10',516) ↵
	
	ans =
	   2

Fractional part of a floating point number f, which is obtained by excluding the integer part from the number
 	
Executable form: mfun('frac',f), where f is a single element or matrix
	⊟ Compute the fractional part of each element of the row matrix R=[7.4 −4.9 0.2 6]
	Solution: Operation of 'frac' mfunction on R should return the output matrix as [0.4 −0.9 0.2 0]
	MATLAB Command
	  >>R=[7.4 -4.9 0.2 6]; ↵
	  >>mfun('frac',R) ↵
	
	ans =
	  0.4000 -0.9000 0.2000 0

Continuation of the last table:

Definition	Example

Digamma function is defined as $\psi(x) = \frac{d}{dx}[\ln(\Gamma x)] = \frac{\frac{d(\Gamma x)}{dx}}{\Gamma x}$, where Γx is the gamma function
 	
Executable form: mfun('Psi',x), where x can be a scalar or matrix
	⊟ Compute $\psi(4)$
	
	Solution: $\psi(4) = \frac{d}{dx}[\ln(\Gamma x)]\Big|_{x=4} = \frac{\frac{d(\Gamma x)}{dx}\Big|_{x=4}}{\Gamma 4} =$
	$\frac{7.5367}{6} = 1.2561$
	MATLAB Command
	    >>mfun('Psi',4) ↵
	
	ans =
	    1.2561

The nth polygamma function is defined as the nth derivative of the digamma function, which is given by $P_n(x) = \frac{d^n}{dx^n}[\psi(x)]$, where $\psi(x)$ is the digamma function as mentioned before and $n \geq 0$
 	
Executable form: mfun('Psi',n,x), where n can be nonnegative integer
	⊟ Compute $P_2(3)$
	
	Solution: $P_2(3) = \frac{d^2}{dx^2}[\psi(x)]\Big|_{x=3}$, $\Gamma x = \int_{t=0}^{t=\infty} e^{-t} t^{x-1} dt$,
	$(\Gamma x)' = \int_{t=0}^{t=\infty} e^{-t} t^{x-1} \ln t \, dt$, $(\Gamma x)'' = \int_{t=0}^{t=\infty} e^{-t} t^{x-1} (\ln t)^2 \, dt$,
	$(\Gamma x)''' = \int_{t=0}^{t=\infty} e^{-t} t^{x-1} (\ln t)^3 \, dt$, $\psi(x) = \frac{(\Gamma x)'}{\Gamma x}$, $\frac{d}{dx}[\psi(x)] =$
	$\frac{(\Gamma x)''}{\Gamma x} - \psi^2(x)$, and $\frac{d^2}{dx^2}[\psi(x)] = \frac{(\Gamma x)'''}{\Gamma x} - \frac{(\Gamma x)''(\Gamma x)'}{(\Gamma x)^2} -$
	$2\psi(x)\psi'(x)$. Plugging $x=3$ provides $\Gamma 3 = 2$,
	$(\Gamma 3)' = 1.8456$, $(\Gamma 3)'' = 2.4929$, $(\Gamma 3)''' = 3.45$,
	$\psi(3) = 0.9228$, $\psi'(3) = 0.395$, and $\psi''(3) = -0.1542$.
	MATLAB Command
	    >>mfun('Psi',2,3) ↵
	
	ans =
	    -0.1541

Euler numbers and polynomial: The generating function of the polynomial is $\sec ht$ or $\frac{1}{\cosh t}$. If we perform the Taylor series expansion of $\sec ht$ up to the 10th power, we have $\sec ht = 1 - \frac{t^2}{2} + \frac{5t^4}{24} - \frac{61t^6}{720}$
$+ \frac{277 t^8}{8064} - \frac{50521 t^{10}}{3628800} + \ldots = (1)\frac{t^0}{0!} + (0)\frac{t^1}{1!} +$
$(-1)\frac{t^2}{2!} + (0)\frac{t^3}{3!} + (5)\frac{t^4}{4!} + (0)\frac{t^5}{5!} + (-61)\frac{t^6}{6!} +$
$(0)\frac{t^7}{7!} + (1385)\frac{t^8}{8!} + (0)\frac{t^9}{9!} + (-50521)\frac{t^{10}}{10!} +$
$\ldots = \sum_{n=0}^{n=\infty} E_n \frac{t^n}{n!}$. E_n is called the Euler number of order n. Comparison provides that $E_0 = 1$, $E_1 = 0$, $E_2 = -1$, $E_3 = 0$, $E_4 = 5$, $E_5 = 0$, $E_6 = -61$, $E_7 = 0$, $E_8 = 1385$, $E_9 = 0$, $E_{10} = -50521$ etc.
Executable form: mfun('euler',n) for the nth euler number where $n \geq 0$. The argument n can be a scalar or matrix
	⊟ Compute E_0, E_1, E_3, and E_8
	Solution: First we form a row matrix of the required orders, then, apply the subroutine 'euler'
	MATLAB Command
	    >>R=[0 1 3 8]; ↵
	    >>mfun('euler',R) ↵
	
	ans =
	    1 0 0 1385
	To see the Euler numbers from 0 to 10th order,
	    >>N=[0:10]'; ↵
	    >>E=mfun('euler',N); ↵
	    >>[N E] ↵
	
	ans =
	    0 1
	    1 0
	    2 -1
	    3 0
	    4 5
	    5 0
	    6 -61
	    7 0
	    8 1385
	    9 0
	    10 -50521

Continuation of the last table:

Definition	Example
Incomplete elliptic integral of the first kind, $$EllipticF(x,\,k) = \int\limits_{t=0}^{t=x} \frac{dt}{\sqrt{1-t^2}\sqrt{1-k^2t^2}}$$ Executable form: mfun('EllipticF',x,k)	⊟ Compute $EllipticF\left(\frac{1}{2},1\right)$ Solution: $EllipticF\left(\frac{1}{2},1\right) = \int\limits_{t=0}^{t=\frac{1}{2}} \frac{dt}{1-t^2} = \tanh^{-1}\left(\frac{1}{2}\right) =$ 0.5493 MATLAB Command $\qquad$>>mfun('EllipticF',1/2,1) ↵ $\qquad$ ans = $\qquad\qquad$ 0.5493

⊟ Help for 'mfun' functions

We made you acquainted with some 'mfun' functions in detailed and tabular form still there are so many. Regarding the first page of this chapter, a brief list of other mfun functions is presented. If you need any of them, just use the command *mhelp('function name')*. The called function will appear before you in details. Just to give an example, the first function of the page is 'bernoulli', we need to know about it. Perform the following to get the specific help:

MATLAB Command
>>mhelp('bernoulli') ↵ $\qquad$ you will see,

Function: bernoulli or numtheory[B] - Bernoulli numbers and polynomials

Calling Sequence:
$\qquad$bernoulli(n)
$\qquad$bernoulli(n, x)
$\qquad$numtheory[B](n)
$\qquad$numtheory[B](n, x)

Parameters:
$\qquad$n - a non-negative integer
$\qquad$x - an expression

Description:

- The bernoulli function computes the nth Bernoulli number, or the nth Bernoulli polynomial, in the expression x. The nth Bernoulli polynomial B(n,x) is defined by the exponential generating function:
$\qquad$t*exp(x*t)/(exp(t)-1) = sum(B(n,x)/n!*t^n, n=0..infinity) .

- The nth Bernoulli number B(n) is defined as B(n,0).

Examples:
> bernoulli(4);
$\qquad\qquad$-1
$\qquad\qquad$-----
$\qquad\qquad$30
> bernoulli(4,x);
$\qquad\qquad\qquad$2$\quad$4$\quad$3
$\qquad\qquad$- 1/30 + x + x - 2 x
> bernoulli(4,1/2);
$\qquad\qquad\qquad$7/240
> with(numtheory):
Warning, new definition for order
> B(6);
$\qquad\qquad$1/42

```
> B(6,x);
                  2        4    6     5
      1/42 - 1/2 x  + 5/2 x  + x  - 3 x
> B(6,1/2);
                    -31
                  --------
                   1344
```
                See Also: euler, inifcns

Following functions are not in the 'mfun' list. You can call them from the command window:

⊡ **Sinc function**

The sinc function is defined as $\sin c(x) = \dfrac{\sin \pi x}{\pi x}$. The plot of $\sin c\ (x)$ is depicted in figure 10.7 and it is an even function. With increasing x, $\sin c\ (x)$ decreases to 0 in an oscillatory manner. Its MATLAB counterpart is 'sinc(x)', where x can be a single element or matrix. Say, we wish to find $[\sin c\ (0)\quad\quad \sin c\ (2)$

$\sin c\ (0.5)\ ]$. The output should be $\left[\ \underset{x \to 0}{Lt}\ \dfrac{\sin \pi x}{\pi x}\quad \dfrac{\sin 2\pi}{2\pi}\quad \dfrac{\sin 0.5\ \pi}{0.5\ \pi}\ \right] = [1\quad 0\quad 0.6366]$.

MATLAB Command
>>sinc([0 2 0.5]) ↵

ans =

    1.0000   -0.0000   0.6366

⊡ **Signum function**

The plot of signum function is shown in figure 10.8. The function is given by $\text{sgn}(x) =$ $\begin{cases} 1 & when\ x > 0 \\ 0 & when\ x = 0 \\ -1 & when\ x < 0 \end{cases}$ and it is an odd function. The MATLAB correspondent is 'sign(x)', where x can be a

single element or matrix. If A=$\begin{bmatrix} 4 & -3.3 & 0 \\ 8 & -2 & -5 \end{bmatrix}$, sign(A) should return $\begin{bmatrix} 1 & -1 & 0 \\ 1 & -1 & -1 \end{bmatrix}$.

MATLAB Command
>>A=[4 -3.3 0;8 -2 -5]; ↵
>>sign(A) ↵

ans =

    1 -1   0
    1 -1 -1

⊡ **Inverse error function**

The error function is defined as $erf(x) = \dfrac{2}{\sqrt{\pi}} \int\limits_{t=0}^{t=x} e^{-t^2} dt$. If x is given, $erf(x)$ can be computed. What if $erf(x)$ is given, but we need x. That means we are looking for the inverse error function. Its corresponding MATLAB function is 'erfinv(x)' (abbreviation of <u>er</u>ror <u>f</u>unction's <u>inv</u>erse), where x can be a scalar or matrix. We know that $erf(0) = 0$, $erf(\infty) = 1$, and $erf(x)$ is not greater that 1. If we provide [0 1 3] as the argument of function 'erfinv', we should get [0 ∞ undefined]. That is what is displayed below:

MATLAB Command
>>R=[0 1 3]; ↵
>>erfinv(R) ↵

ans =

0 Inf NaN

⊟ Scaled complementary error function

The scaled complementary error function is defined as $erfcx(x) = e^{x^2} erfc(x)$, where $erfc(x)$ is the complementary error function. The resembling function of MATLAB is 'erfcx(x)', where x can be a scalar or matrix. Compute $erfcx(2)$. We have $erfcx(2) = e^{2^2} erfc(2) = 54.5981 \times 0.0047 = 0.2566$.

MATLAB Command
>>erfcx(2) ↵

ans =
0.2554

10.2 Conversion of numbers from one base to other

The most common and widely used number system is decimal. The base of the decimal number is 10. In this system, any number is consisted of the integers 0 1 2 3 4 5 6 7 8 9. Say, a number in decimal is 789 now $789 = 7 \times 10^2 + 8 \times 10^1 + 9 \times 10^0$. The number upon which the power is appearing is 10 that is why the base of the decimal number is 10. In binary number, the base is 2. Any number in the binary system is consisted of 0 and 1. To show by example, number 42 in the binary system is written as 101010. We can write $42 = 1 \times 2^5 + 0 \times 2^4 + 1 \times 2^3 + 0 \times 2^2 + 1 \times 2^1 + 0 \times 2^0$. It can be said that if the base of a number system is N, the number of that system is formed from the integers 0 to $N-1$. There are several subroutines in MATLAB that convert a number from one base to the other, namely,

dec2bin — for conversion from decimal to(2) binary,
bin2dec — for conversion from binary to(2) decimal,
hex2dec — for conversion from hexa decimal to(2) decimal,
dec2hex — for conversion from decimal to(2) hexadecimal,
dec2base — for conversion from decimal to(2) any other base, and
base2dec — for conversion from any other base to(2) decimal.

In hexadecimal system, the constituent numbers are 1 2 3 4 5 6 7 8 9 A B C D E F, where A, B, C, D, E, and F correspond to 10, 11, 12, 13, 14, and 15 of the decimal respectively. Let us see how one can convert a number from one base to other. Say, a number in decimal system is 42. We wish to convert it to binary (base 2). Perform the following long division in this regard:

2)42
2)21 − 0
2)10 − 1
2)5 − 0
2)2 − 1
2)1 − 0
0 − 1

Recall that the integer division of 1 by 2 is 0 and the remainder is $1 - 2 \times 0 = 1$. Gathering the remainders from down to up of the above division yields the number in binary system as 101010. The conversion of the binary number 101010 to the decimal is mentioned earlier in this article. *Numbers other than decimal must be in string form when they are put as the arguments of the conversion subroutines.* As different example, say, the decimal number 85043 is to be converted in hexadecimal system (base 16). The required long division is as follows:

$$
\begin{array}{r}
16\overline{)85043} \\
16\overline{)5315} - 3 \\
16\overline{)332} - 3 \\
16\overline{)20} - C(12) \\
16\overline{)1} - 4 \\
0 - 1
\end{array}
$$

In the last division, integer division of 1 by 16 is zero so the decimal number 85043 is equivalent to the hexadecimal number 14C33. The reverse conversion of the hexadecimal 14C33 to the decimal is $1\times16^4+4\times16^3+(12)\times16^2+3\times16^1+3\times16^0=85043$. All conversions are shown as follows:

MATLAB Command

from decimal number 42 to binary,

>>dec2bin(42) ↵

ans =

101010

the conversion of decimal number 85043 to hexadecimal,

>>dec2hex(85043) ↵

ans =

14C33

the reverse conversion of binary 101010 to decimal,

>>bin2dec('101010') ↵

ans =

42

the reverse conversion of hexadecimal 14C33 to decimal,

>>hex2dec('14C33') ↵

ans =

85043

In general, to convert a number from one base to the other,

MATLAB Step:

Use the command given base 2 required base (number in the given base).

To deal with many numbers, we provide the table 10.B where some numbers are given in different bases.

Table 10.B   Numbers in different bases

Decimal Base=10	Binary Base=2	Other base, say, Base=6	Octal Base=8	Hexadecimal Base=16
7	111	11	7	7
13	1101	21	15	D
197	11000101	525	305	C5
333	101001101	1313	515	14D

We assign all decimal numbers to row matrix N. See the different conversions as follows:

MATLAB Command

for all elements of N to binary,

>>N=[7 13 197 333]; ↵
>>dec2bin(N) ↵

ans =

000000111
000001101
011000101
101001101

for all elements of N to base 6,

>>dec2base(N,6) ↵

ans =

0011
0021
0525
1313

for all elements of N to octal,	for all elements of N to hexadecimal,
>>dec2base(N,8) ⏎	>>dec2hex(N) ⏎
ans =	ans =
007	007
015	00D
305	0C5
515	14D

Notice that the converted numbers are preceded by the zero/zeroes up to the maximum number of digits. The subroutine 'dec2base' can be applied to any base from 2 to 36.

10.3 Coordinate transformation

In this article, we mention the coordinate transformation from Cartesian to cylindrical (and vice versa) and from Cartesian to spherical (and vice versa). In Cartesian, cylindrical, and spherical systems, any point in 3-D space is specified by (x, y, z), (ρ, φ, z), and (r, θ, φ) respectively.

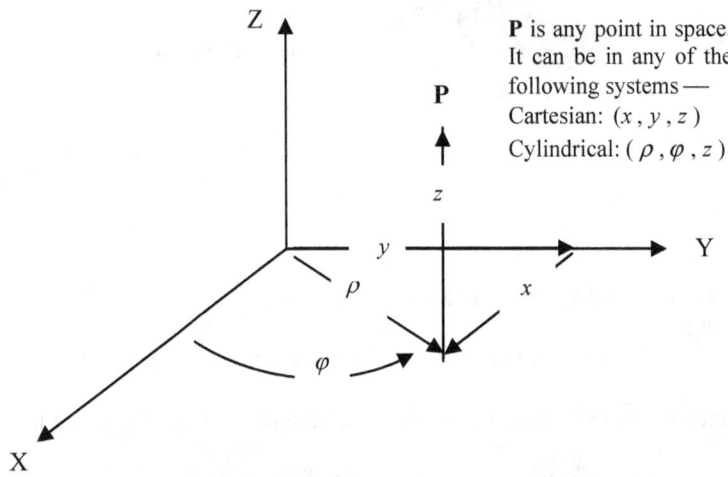

Figure 10.9 *Geometry of the coordinate systems defined for the Cartesian and cylindrical*

▱ Cylindrical to Cartesian

The coordinate systems are depicted in figure 10.9. Cylindrical to Cartesian conversion is given by the equations $\begin{cases} x = \rho\cos\varphi \\ y = \rho\sin\varphi \\ z = z \end{cases}$ .. (10–1) . Let us take a point in the cylindrical system whose coordinates are (ρ, φ,

z)=(2, 120⁰, –3). Using the equation (10–1) can evaluate the Cartesian coordinates for this point, which are (x, y, z)=($2\cos120^0$, $2\sin120^0$, –3)=(–1, 1.7321, –3). Since the z coordinate is common to both the Cartesian and cylindrical systems, it can be avoided. The subroutine 'pol2cart' (abbreviation of polar to (2) Cartesian) can convert the coordinates of a point from cylindrical to Cartesian. It has two output arguments if the z coordinate is not considered. If you want to keep the z coordinate, the number of output arguments is three (x, y, and z). Here in the given point the azimuth angle φ is in degrees (120^0) but in MATLAB φ has to be in radians. Instead of 120^0, write 120*pi/180 to have φ in radians. To convert the coordinates from cylindrical to Cartesian,

MATLAB Step:
  Use the command [x y z]=pol2cart (φ in radians, ρ, z).

The subroutine can also be applied to convert a set of points from cylindrical to Cartesian. By the way, the symbols ρ and φ are not found in the workspace. To represent them, 'rho' and 'phi' are written instead of ρ and φ respectively. Table 10.C shows the coordinates of several points in cylindrical system and in Cartesian as well.

Table 10.C   Cylindrical and Cartesian coordinates of several points

Points in cylindrical coordinates (ρ, φ, z)	Coordinates in Cartesian system using the equation (10–1), (x, y, z)
(15, 90°, −5)	(0, 15, −5)
(12, 30°, −6)	(10.3923, 6, −6)
(8, 180°, 10)	(−8, 0, 10)
(9, 0°, 35)	(9, 0, 35)
(17, 120°, −3).	(−8.5, 14.7224, −3)

Implementations of both examples are shown in the following:
  MATLAB Command
  for a single point,
  >>[x y z]=pol2cart(120*pi/180,2,-3) ↵

  x =
        -1.0000
  y =
        1.7321
  z =
        -3
  for a set of points,
  >>rho=[15 12 8 9 17]; ↵
  >>phi=[90 30 180 0 120]*pi/180; ↵
  >>z=[-5 -6 10 35 -3]; ↵
  >>[x y z]=pol2cart(phi,rho,z) ↵

  x =
        0.0000   10.3923   -8.0000   9.0000   -8.5000
  y =
        15.0000   6.0000   0.0000   0   14.7224
  z =
        -5   -6   10   35   -3

⌕ Cartesian to cylindrical

The reverse conversion that is Cartesian to cylindrical is given by $\left\{\begin{array}{l} \rho = \sqrt{x^2 + y^2} \\ \varphi = \tan^{-1}\dfrac{y}{x} \\ z = z \end{array}\right\}$ .. (10−2). To perform the conversion,

  MATLAB Step:
    Use the command [phi rho z]=cart2pol (x,y, z).

Cart2pol' is the abbreviation of <u>Cart</u>esian to (2) <u>pol</u>ar. The coordinates of example point in the Cartesian system are (x, y, z)=(−1, 2, −3). By virtue of equation (10−2), the cylindrical coordinates for this point are

418

$(\rho,\varphi,z)=(\sqrt{(-1)^2+2^2}$, $\tan^{-1}\frac{2}{-1}$, $-3)=(2.2361,\ 2.0344^c,\ -3)$. Notice that $\tan^{-1}\frac{2}{-1}$ is the principal value. The computation is shown below:

MATLAB Command
>>[phi,rho,z]=cart2pol(-1,2,-3) ↵

phi =
        2.0344
rho =
        2.2361
z =
        -3

Execute 'phi*180/pi' to have φ in degrees. The subroutine can convert several points too and the variation of φ can be from $-\pi$ to π with $\rho \geq 0$.

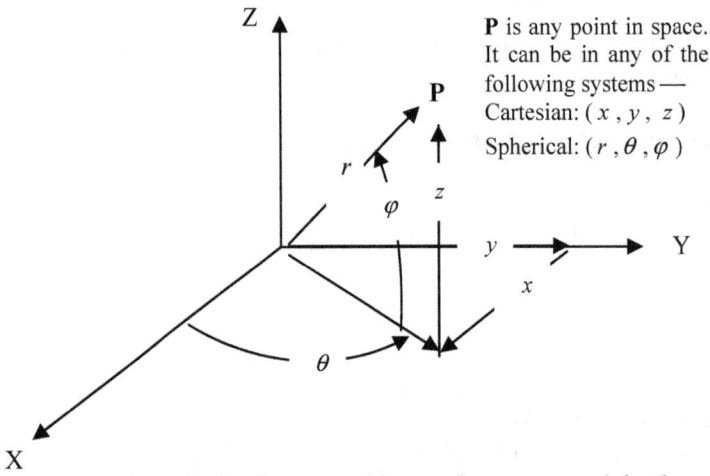

P is any point in space. It can be in any of the following systems —
Cartesian: (x,y,z)
Spherical: (r,θ,φ)

Figure 10.10 *Geometry of the coordinate systems defined for the Cartesian and spherical*

⊟ Spherical to Cartesian
Geometry of this conversion is shown in figure 10.10. In MATLAB spherical to Cartesian conversion is given by $\begin{cases} x = r\cos\theta\cos\varphi \\ y = r\cos\varphi\sin\theta \\ z = r\sin\varphi \end{cases}$ ....(10−3). Take a point in spherical system whose coordinates are $(r,\theta,\varphi)=(4,$ $120^0,\ 75^0)$. From equation (10−3), we have $(x,y,z)=(4\cos120^0\cos75^0,\ 4\cos75^0\sin120^0,\ 4\sin75^0)=(-0.5176,\ 0.8966, 3.8637)$. To perform the conversion,

MATLAB Step:
Use the command [x y z]=sph2cart (theta in radians, phi in radians, r), where sph2cart is the abbreviation of spherical to(2) Cartesian.

See the conversion as follows:
MATLAB Command
>>[x y z]=sph2cart(120*pi/180,75*pi/180,4) ↵

x =
        -0.5176
y =

$$z = \begin{matrix} 0.8966 \\ 3.8637 \end{matrix}$$

The subroutine can also deal with several points.

⊡ Cartesian to spherical

The last conversion is Cartesian to spherical, which is given by $\left\{\begin{matrix} r = \sqrt{x^2 + y^2 + z^2} \\ \varphi = \sin^{-1} \dfrac{z}{\sqrt{x^2 + y^2 + z^2}} \\ \theta = \tan^{-1} \dfrac{y}{x} \end{matrix}\right\}$ ....$(10-4)$.

To have the conversion,

MATLAB Step:
Use the command [theta phi r]=cart2sph (x, y, z), where theta and phi are in radians and cart2sph is the abbreviation of Cartesian to(2) spherical.

For illustration, take the Cartesian coordinates as (x,y,z)=(–2, 4, –3). Use $(10-4)$ to write

$(r,\theta,\varphi)=\left(\sqrt{(-2)^2 + 4^2 + (-3)^2} , \tan^{-1}\frac{4}{-2} , \sin^{-1}\frac{-3}{5.3852} \right)$=(5.3852, 2.0344ᶜ, –0.5909ᶜ). Its implementation is

shown below:

MATLAB Command
>>[theta phi r]=cart2sph(-2,4,-3) ↵

    theta =
        2.0344
    phi =
        -0.5909
    r =
        5.3852

Angle θ can be seen in degrees by executing 'theta*180/pi'. Like previous counterparts, 'cart2sph' can also handle the conversion of several points.

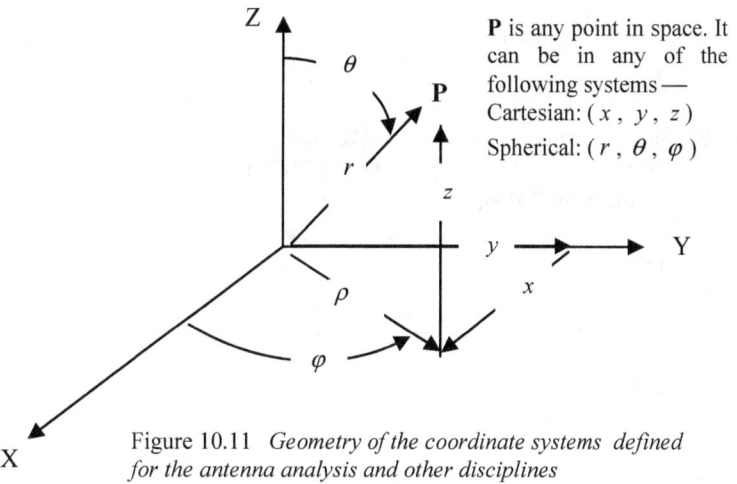

Figure 10.11 *Geometry of the coordinate systems defined for the antenna analysis and other disciplines*

To cover the whole space about the origin of the coordinate systems, the different domains are $0 \leq \theta \leq 360^0$ or $-180^0 \leq \theta \leq 180^0$, $-90^0 \leq \varphi \leq 90^0$, $r \geq 0$, $-\infty \leq x \leq \infty$, $-\infty \leq y \leq \infty$, and $-\infty \leq z \leq \infty$. In some disciplines, different geometry is used in the coordinate system. Most of the antenna engineering analysis follows the geometry of the coordinate system as depicted in figure 10.11, which do not match with that of MATLAB. What is the difference between the MATLAB co-ordinate system and the one of figure 10.11? In later, φ becomes θ and vice versa. Angle θ is being measured from the positive z axis instead of the $x-y$ plane.

The conversion equations are as follows: spherical to Cartesian —
$$\begin{cases} x = r\sin\theta\cos\varphi \\ y = r\sin\theta\sin\varphi \\ z = r\cos\theta \end{cases} \quad ...(10-5)$$
and Cartesian to

spherical —
$$\begin{cases} r = \sqrt{x^2 + y^2 + z^2} \\ \theta = \cos^{-1}\dfrac{z}{\sqrt{x^2 + y^2 + z^2}} \\ \varphi = \tan^{-1}\dfrac{y}{x} \end{cases} \quad ...(10-6).$$
In the last system, the variation of θ should be from 0^0 to 180^0

and that of φ should be from 0^0 to 360^0 or from -180^0 to 180^0. With the help of some substitutions we can have the coordinates corresponding to figure 10.11 from the MATLAB one. Convert the spherical coordinates $(r, \theta, \varphi) = (2, 80^0, 120^0)$ to Cartesian according to the coordinate system of figure 10.11. From the equation $(10-6)$, we have $(x, y, z) = (2\sin80^0\cos120^0, 2\sin80^0\sin120^0, 2\cos80^0) = (-0.9848, 1.7057, 0.3473)$. *Use MATLAB Command [x y z]=sph2cart (phi in radians, $\dfrac{\pi}{2}$ – theta in radians, r).* The reverse conversion of the

same point is $\left(\sqrt{(-0.9848)^2 + 1.7057^2 + 0.3473^2} \ , \ \cos^{-1}\dfrac{0.3473}{\sqrt{(-0.9848)^2 + 1.7057^2 + 0.3473^2}} \ , \ \tan^{-1}\dfrac{1.7057}{-0.9848} \right) = (2, 80^0,$

$120^0) = (2, 1.3963^c, 2.0944^c)$. *Use MATLAB Command [phi theta r]=cart2sph (x, y, z) and theta=$\dfrac{\pi}{2}$ - theta for the reverse conversion.* See the implementations of the last two conversions as follows:

MATLAB Command
for spherical to Cartesian conversion according to figure 10.11,
```
>>[x y z]=sph2cart(120*pi/180,10*pi/180,2) ↵
```

    x =
          -0.9848
    y =
          1.7057
    z =
          0.3473

for Cartesian to spherical conversion according to figure 10.11,
```
>>[phi theta r]=cart2sph(-.9848,1.7057,0.3473); ↵
>>theta=pi/2-theta ↵
```

    theta =
          1.3963
```
>>phi ↵
```

    phi =
          2.0944
```
>>r ↵
```

    r =
          2.0000

10.4 Dot and cross products of two vectors

Two types of products are seen for two vectors – dot or scalar and cross or vector. The dot product of two vectors $\overline{A} = A_1 i + A_2 j + A_3 k$ and $\overline{B} = B_1 i + B_2 j + B_3 k$ is defined as $D = A_1 B_1 + A_2 B_2 + A_3 B_3$ and is a scalar quantity. To evaluate the dot product of two vectors,

MATLAB Steps:
1. *Assign each vector to matrices A and B respectively and*
2. *Use the command dot(A,B).*

To illustrate, the dot product of two vectors $\overline{A} = -3i + 2j + 9k$ and $\overline{B} = i - 7j + 2k$ is given by $D = (-3i + 2j + 9k) \circ (i - 7j + 2k) = (-3).1 + 2.(-7) + 9.2 = 1$. Since $\overline{A}.\overline{B} = \overline{B}.\overline{A}$, it does not matter whichever command 'dot(A,B)' or 'dot(B,A)' is used. The subroutine is also operational for the dot products of sets of two vectors. Table 10.D shows the dot products of sets of two vectors. A matrix is formed for the first vector $\overline{A}$ as $A = \begin{bmatrix} 1 & 1 & 0 & 1 \\ 0 & 0 & 2 & 1 \\ -3 & 0 & 3 & -1 \end{bmatrix}$ (vectors are arranged as columnwise), and so is for $\overline{B}$, where $B = \begin{bmatrix} 0 & 0 & -1 & 2 \\ 1 & 1 & 7 & 5 \\ 2 & 9 & 2 & 2 \end{bmatrix}$. The dot outputs will be in a row matrix.

Table 10.D   Dot products of several vectors

Vector $\overline{A}$	Vector $\overline{B}$	Dot product $D = \overline{A}.\overline{B}$
i–3k	j+2k	1×0+0×1–3×2=–6
i	j+9k	1×0+0×1+0×9=0
2j+3k	–i+7j+2k	0×(–1)+2×7+3×2=20
i+j–k	2i+5j+2k	1×2+1×5–1×2=5

The cross or vector product of two vectors can be performed by using the subroutine 'cross'. As an example, the cross product $\overline{C}$ of just mentioned two vectors $\overline{A}$ and $\overline{B}$ is given by $\overline{C} = \overline{A} \times \overline{B} = \begin{bmatrix} i & j & k \\ -3 & 2 & 9 \\ 1 & -7 & 2 \end{bmatrix} = i \begin{bmatrix} 2 & 9 \\ -7 & 2 \end{bmatrix} - j \begin{bmatrix} -3 & 9 \\ 1 & 2 \end{bmatrix} + k \begin{bmatrix} -3 & 2 \\ 1 & -7 \end{bmatrix} = 67i + 15j + 19k$. The product is a vector quantity and the commands 'cross(A,B)' and 'cross(B,A)' do not produce the same result since $\overline{A} \times \overline{B} \neq \overline{B} \times \overline{A}$. That is, if $\overline{A} \times \overline{B}$ is to be found, vector $\overline{A}$ must come first. To take the cross product $\overline{A} \times \overline{B}$,

MATLAB Steps:
1. *Assign the first vector to A,*
2. *Assign the second vector to B, and*
3. *Use the command cross(A,B).*

Like 'dot' the subroutine can be used to find the cross products of sets of two vectors. Let us consider the same sets of vectors as we used for the dot products. Table 10.E shows the computation.

Table 10.E   Cross product of several vectors

Vector $\overline{A}$	Vector $\overline{B}$	Cross product $\overline{C} = \overline{A} \times \overline{B}$	Cross product $\overline{C} = \overline{A} \times \overline{B}$ in MATLAB Notation
i–3k	j+2k	3i–2j+k	[3 –2 1]
i	j+9k	–9j+k	[0 –9 1]
2j+3k	–i+7j+2k	–17i–3j+2k	[–17 –3 2]
i+j–k	2i+5j+2k	7i–4j+3k	[7 –4 3]

See the implementation of dot and cross products as follows:

MATLAB Command

for dot product of two vectors,
>>A=[-3 2 9]; ↵
>>B=[1 -7 2]; ↵
>>dot(A,B) ↵

ans =
1

for the sets of two vectors of table 10.D,
>>A=[1 1 0 1;0 0 2 1;-3 0 3 -1]; ↵
>>B=[0 0 -1 2;1 1 7 5;2 9 2 2]; ↵
>>dot(A,B) ↵

ans =
-6   0   20   5

cross product of two vectors,
>>A=[-3 2 9]; ↵
>>B=[1 -7 2]; ↵
>>cross(A,B) ↵

ans =
67   15   19

cross products of the sets of two vectors,
>>A=[1 1 0 1;0 0 2 1;-3 0 3 -1]; ↵
>>B=[0 0 -1 2;1 1 7 5;2 9 2 2]; ↵
>>cross(A,B) ↵

ans =
3   0   -17   7
-2   -9   -3   -4
1   1   2   3

Following implementation the columns display the cross products of the respective sets of two vectors. That is,

the first column $\begin{bmatrix} 3 \\ -2 \\ 1 \end{bmatrix}$ corresponds to vector 3i–2j+k, the second column $\begin{bmatrix} 0 \\ -9 \\ 1 \end{bmatrix}$ corresponds to –9j+k, and so

do the others. Notice that the numbers of rows of matrices A and B are three.

⊟ Composite products

Three vector fields are given by $\overline{P} = 5a_x + 9a_y + 2a_z$, $\overline{Q} = -3a_x - 7a_y + a_z$, and $\overline{R} = a_x - 6a_y - 8a_z$.
Determine the following:

A. $(\overline{P} \times \overline{Q}) \times \overline{R}$

B. $\overline{P} \circ (\overline{Q} \times \overline{R})$

C. $\dfrac{\overline{P} \circ [(\overline{P} - \overline{Q}) \times (\overline{P} - \overline{R})]}{\left| (\overline{P} - \overline{Q}) \times (\overline{P} - \overline{R}) \right|}$ (actually, this is the shortest distance to the origin from the plane formed

by the three points $\overline{P}$, $\overline{Q}$, and $\overline{R}$)

Problem A

First, we find $\overline{P} \times \overline{Q} = \begin{bmatrix} a_x & a_y & a_z \\ 5 & 9 & 2 \\ -3 & -7 & 1 \end{bmatrix} = 23a_x - 11a_y - 8a_z$, $\therefore$ $(\overline{P} \times \overline{Q}) \times \overline{R} = \begin{bmatrix} a_x & a_y & a_z \\ 23 & -11 & -8 \\ 1 & -6 & -8 \end{bmatrix} =$

$40a_x + 176a_y - 127a_z$.

Problem B

We have $\overline{Q} \times \overline{R} = \begin{bmatrix} a_x & a_y & a_z \\ -3 & -7 & 1 \\ 1 & -6 & -8 \end{bmatrix} = 62a_x - 23a_y + 25a_z$ and $\overline{P} \circ (\overline{Q} \times \overline{R}) = (5a_x + 9a_y + 2a_z) \circ$

$(62a_x - 23a_y + 25a_z) = 153$.

Problem C

Computations are $\overline{P} - \overline{Q} = 8a_x + 16a_y + a_z$, $\overline{P} - \overline{R} = 4a_x + 15a_y + 10a_z$, $(\overline{P} - \overline{Q}) \times (\overline{P} - \overline{R}) =$

$145a_x - 76a_y + 56a_z$, $|(\overline{P} - \overline{Q}) \times (\overline{P} - \overline{R})| = \sqrt{29937}$, $\overline{P} \circ (\overline{P} - \overline{Q}) \times (\overline{P} - \overline{R}) = (5a_x + 9a_y + 2a_z) \circ (145a_x - 76a_y + 56a_z)$

$=153$, and $|(\overline{P}-\overline{Q})\times(\overline{P}-\overline{R})|=\sqrt{145^2+76^2+56^2}=\sqrt{29937}$, therefore, $\dfrac{\overline{P}\circ[(\overline{P}-\overline{Q})\times(\overline{P}-\overline{R})]}{\left|(\overline{P}-\overline{Q})\times(\overline{P}-\overline{R})\right|}=\dfrac{153}{\sqrt{29937}}=0.8843.$

These all are shown in the following:

MATLAB Command

for problem A,

```
>>P=[5 9 2]; ↵
>>Q=[-3 -7 1]; ↵
>>R=[1 -6 -8]; ↵
>>cross(cross(P,Q),R) ↵

ans =
      40  176  -127
```

for problem B,

```
>>dot(P,cross(Q,R)) ↵

ans =
      153
```

for problem C,

```
>>C=cross(P-Q,P-R); ↵
>>D=sqrt(sum(C.^2)); ↵
>>N=dot(P,C); ↵
>>N/D ↵

ans =
      0.8843
```

So far we congested only the numerical dot and cross products. Situation may occur when one has to deal with the dot and cross products of symbolic functions. You can enjoy the facility of symbolic dot and cross products as well. To exemplify, the cross product of $\overline{A}=(x+y)\overline{a}_x-zx\overline{a}_y$ and $\overline{B}=x\overline{a}_x-\cos x\overline{a}_y+e^x\overline{a}_z$ is $\overline{A}\times\overline{B}=$

$$\begin{bmatrix} \overline{a}_x & \overline{a}_y & \overline{a}_z \\ x+y & -zx & 0 \\ x & -\cos x & e^x \end{bmatrix}=-zxe^x\overline{a}_x-(x+y)e^x\overline{a}_y+[-(x+y)\cos x+zx^2]\overline{a}_z.$$ The solution is as follows:

MATLAB Command

```
>>syms x y z ↵              ← Declare concern variables symbolically
>>A=[x+y -z*x 0]; ↵         ← Assign the first vector to A
>>B=[x -cos(x) exp(x)]; ↵   ← Assign the second vector to B
>>C=cross(A,B); ↵           ← Cross product is assigned to C
>>pretty(C) ↵               ← Display the nice form

  [                                                    2]
  [-z x exp(x)    -(x + y) exp(x)    -(x + y) cos(x) + z x  ]
```

The subroutine 'dot' is also functional for symbolic expressions. For instance, we wish to see $\overline{A}\circ(\overline{A}\times\overline{B})$ for the last vector functions. Just now we have found that $\overline{A}\times\overline{B}=-zxe^x\overline{a}_x-(x+y)e^x\overline{a}_y+[-(x+y)\cos x+zx^2]\overline{a}_z$ so $\overline{A}\circ(\overline{A}\times\overline{B})=0$. It is executed as follows:

MATLAB Command

```
>>syms x y z ↵
>>A=[x+y -z*x 0]; ↵
>>B=[x -cos(x) exp(x)]; ↵
>>dot(A,cross(A,B)) ↵

ans =

-conj(x+y)*z*x*exp(x)+conj(z*x)*(x+y)*exp(x)
```

The output seems to be unsatisfactory because it is not equal to 0. Since the symbolic variables are taken as complex, the conjugates are appearing as the output. This can be overcome if we declare the variables as real.

```
>>syms x y z real ↵
>>A=[x+y -z*x 0]; ↵
>>B=[x -cos(x) exp(x)]; ↵
>>dot(A,cross(A,B)) ↵


ans =


0
```

10.5 Vector transformation

A vector can be represented in any coordinate system, which is orthogonal or orthonormal. An orthogonal system is one in which the coordinates are mutually perpendicular. Nonorthogonal systems are hard to deal with. Typical examples of orthogonal systems include rectangular (Cartesian), cylindrical (circular), spherical, elliptical cylindrical, conical, … etc. Choosing a particular coordinate system that best states a given problem may save a lot of clumsy work and valuable time. Sometimes it may be necessary to transform a vector in one system to the other. Belonging to the coordinate system of figure 10.11, transformation of a vector from Cartesian to cylindrical is given by $\begin{bmatrix} A_\rho \\ A_\phi \\ A_z \end{bmatrix} = \begin{bmatrix} \cos\phi & \sin\phi & 0 \\ -\sin\phi & \cos\phi & 0 \\ 0 & 0 & 1 \end{bmatrix} \begin{bmatrix} A_x \\ A_y \\ A_z \end{bmatrix}$. Say, the example vector

is $\overline{A} = xy\overline{a}_x - y^2\overline{a}_y - 9(z+x)\overline{a}_z$. We wish to have $\overline{A}$ in the cylindrical system. It is obvious that $A_x = xy$, $A_y = -y^2$, and $A_z = -9(z+x)$ from which $\begin{bmatrix} A_\rho \\ A_\phi \\ A_z \end{bmatrix} = \begin{bmatrix} \cos\phi & \sin\phi & 0 \\ -\sin\phi & \cos\phi & 0 \\ 0 & 0 & 1 \end{bmatrix} \times \begin{bmatrix} xy \\ -y^2 \\ -9(z+x) \end{bmatrix}$. Along with this one has to use

$\begin{cases} x = \rho\cos\varphi \\ y = \rho\sin\varphi \end{cases}$. After matrix multiplication and substitution, we have $\begin{bmatrix} A_\rho \\ A_\phi \\ A_z \end{bmatrix} = \begin{bmatrix} \rho^2\cos^2\varphi\sin\varphi - \rho^2\sin^3\varphi \\ -2\rho^2\sin^2\varphi\cos\varphi \\ -9z - 9\rho\cos\varphi \end{bmatrix}$. The

variables in the given expressions are x, y, z, ρ, and φ. Unavailability of ρ and φ makes us use r and p instead of ρ and φ respectively. The transformation is as follows:

MATLAB Command

```
>>syms r p x y z real ↵          ← Declare concern variables as real
>>T=[cos(p) sin(p) 0;-sin(p) cos(p) 0;0 0 1]; ↵ ← Assign the transformation matrix to T
>>A=[x*y -y^2 -9*(z+x)]'; ↵      ← Assign Cartesian column vector to A
>>B=T*A; ↵                        ← Multiplication of T and A is assigned to B
>>x=r*cos(p); ↵         ← Insert x = ρcosφ
>>y=r*sin(p); ↵         ← Insert y = ρsinφ
>>O=subs(B); ↵          ← Substitute x and y and assign the output to O
>>pretty(O) ↵           ← Display the contents of O
```

```
[          2    2                        3    2]
[cos(p~)  r~  sin(p~) -  sin(p~)  r~  ]
[                                              ]
[                  2    2                      ]
[         -2 sin(p~)  r~   cos(p~)             ]
[                                              ]
[         -9 z~ - 9 r~  cos(p~)                ]
```

Don't be perplexed with the return. Referring to the output, 'p~', 'z~', and 'r~' mean φ, z, and r respectively. For reader's convenience, the other transformation matrices corresponding to the coordinate system of figure 10.11 are given as follows:

from rectangular to cylindrical $\Rightarrow \begin{bmatrix} A_\rho \\ A_\varphi \\ A_z \end{bmatrix} = \begin{bmatrix} \cos\varphi & \sin\varphi & 0 \\ -\sin\varphi & \cos\varphi & 0 \\ 0 & 0 & 1 \end{bmatrix} \begin{bmatrix} A_x \\ A_y \\ A_z \end{bmatrix}$

from cylindrical to rectangular $\Rightarrow \begin{bmatrix} A_x \\ A_y \\ A_z \end{bmatrix} = \begin{bmatrix} \cos\varphi & -\sin\varphi & 0 \\ \sin\varphi & \cos\varphi & 0 \\ 0 & 0 & 1 \end{bmatrix} \begin{bmatrix} A_\rho \\ A_\varphi \\ A_z \end{bmatrix}$

from rectangular to spherical $\Rightarrow \begin{bmatrix} A_r \\ A_\theta \\ A_\varphi \end{bmatrix} = \begin{bmatrix} \sin\theta\cos\varphi & \sin\theta\sin\varphi & \cos\theta \\ \cos\theta\cos\varphi & \cos\theta\sin\varphi & -\sin\theta \\ -\sin\varphi & \cos\varphi & 0 \end{bmatrix} \begin{bmatrix} A_x \\ A_y \\ A_z \end{bmatrix}$

from spherical to rectangular $\Rightarrow \begin{bmatrix} A_x \\ A_y \\ A_z \end{bmatrix} = \begin{bmatrix} \sin\theta\cos\varphi & \cos\theta\cos\varphi & -\sin\varphi \\ \sin\theta\sin\varphi & \cos\theta\sin\varphi & \cos\varphi \\ \cos\theta & -\sin\theta & 0 \end{bmatrix} \begin{bmatrix} A_r \\ A_\theta \\ A_\varphi \end{bmatrix}$

from cylindrical to spherical $\Rightarrow \begin{bmatrix} A_r \\ A_\theta \\ A_\varphi \end{bmatrix} = \begin{bmatrix} \sin\theta & 0 & \cos\theta \\ \cos\theta & 0 & -\sin\theta \\ 0 & 1 & 0 \end{bmatrix} \begin{bmatrix} A_\rho \\ A_\varphi \\ A_z \end{bmatrix}$

from spherical to cylindrical $\Rightarrow \begin{bmatrix} A_\rho \\ A_\varphi \\ A_z \end{bmatrix} = \begin{bmatrix} \sin\theta & \cos\theta & 0 \\ 0 & 0 & 1 \\ \cos\theta & -\sin\theta & 0 \end{bmatrix} \begin{bmatrix} A_r \\ A_\theta \\ A_\varphi \end{bmatrix}.$

10.6 Inverse of a function

If a function $y = f(x)$ is given, its inverse function can be obtained by exercising the subroutine 'finverse'. Inverse function of $f(x)$ is denoted by $f^{-1}(x)$. The question is how one can obtain the inverse function of a function $f(x)$. Manipulate $y = f(x)$ so long as you express that as $x = f(y)$ then replace y by x

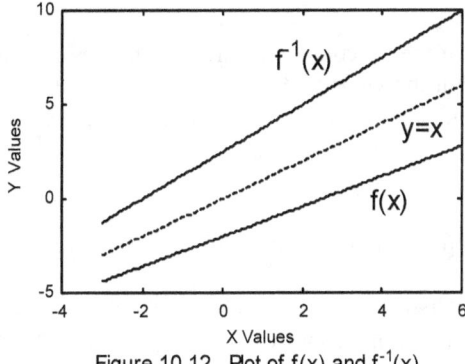

Figure 10.12   Plot of f(x) and f⁻¹(x)

to obtain the inverse. As usual, proceed with $f(x) = \dfrac{4}{5}x - 2 \Rightarrow y = \dfrac{4}{5}x - 2 \Rightarrow x = \dfrac{5(y+2)}{4}$, hence, $f^{-1}(x) = \dfrac{5(x+2)}{4}$. As a matter of fact, $f(x)$ and $f^{-1}(x)$ are the image of each other about the line $y = x$. It becomes clear viewing the figure 10.12. Anyhow the implementation is as follows:

MATLAB Command

>>syms x ↵   ← Declare variable of $f(x)$ as symbolic

>>y=4/5*x-2; ↵   ← Define $y = f(x)$

>>finverse(y) ↵   ← Conduct the subroutine 'finverse'

ans =

5/2+5/4*x

There are some functions, which do not have the unique inverse. Find the inverse of $y = x^2 - 3x - 7$. Rearrange the equation to write $x^2 - 3x - 7 - y = 0$. Calculate the solution of the quadratic equation relating to x so $x = \dfrac{3 \pm \sqrt{9 - 4(-7 - y)}}{2} = \dfrac{3 \pm \sqrt{37 + 4y}}{2}$ from which $f^{-1}(x) = \dfrac{3 \pm \sqrt{37 + 4x}}{2}$. Following is the MATLAB's response:

MATLAB Command

>>syms x ↵

>>y=x^2-3*x-7; ↵

>>O=finverse(y); ↵   ← The output of 'finverse' is put to O

Warning: finverse(x^2-3*x-7) is not unique.
> In C:\MATLAB\toolbox\symbolic\@sym\finverse.m at line 43

>>pretty(O) ↵   ← Print the contents of O

```
                           1/2
     3/2 + 1/2 (37 + 4 x)
```

It is evident that only one inverse is returned and that the warning is appearing due to the nonuniqueness of the inverse function.

10.7 Piecewise continuous function

It is possible to define a piecewise continuous function using the Maple package. The name of the function is 'piecewise'. In Maple terminology, a function $f(x)$, which is a function of x, is assigned as 'f:= x–>'. An example of the piecewise continuous function can be $f(x) = \begin{cases} x & x \le -2 \\ 3x & -2 < x < 6 \\ x^2 & \text{otherwise} \end{cases}$. To argument the function, the interval description of the function comes first then does the function over the interval (s). The condition '≤' is notified by '<='. For the second interval $-2 < x < 6$, the insertion of $x < 6$ is enough because it is understood that x is increasing. The function does not need to have the condition for 'otherwise'. See the implementation below:

MATLAB Command

>>maple('f:=x->piecewise(x<=-2,x,x<6,3*x,x^2)'); ↵

Once $f(x)$ is in the workspace, the assignee can be invoked to know the functional value of $f(x)$, for instance, $f(-2) = -2$, $f(6) = 36$ (since, $x = 6$ is not in the interval $-2 < x < 6$), and $f(7) = 49$. Obtain these as follows:

>>maple('f(-2),f(6),f(7)') ↵

ans =

-2, 36, 49

That is not the whole that the subroutine can perform. A variety of applications in respect to the piecewise continuous functions are supported through the medium of Maple package. Piecewise integration of $f(x)$ is found as $\int f(x)\,dx$ = $\begin{cases} \int x\,dx & x \le -2 \\ \int_{x=-2}^{x=x} 3x\,dx + \text{functional value of } \int f(x)\,dx \ \text{at} \ x=-2 & -2 < x < 6 \\ \int_{x=6}^{x=x} x^2\,dx + \text{functional value of } \int f(x)\,dx \ \text{at} \ x=6 & \text{otherwise} \end{cases}$ =

$\begin{cases} \dfrac{x^2}{2} & x \le -2 \\ \dfrac{3x^2}{2} - \dfrac{3\times4}{2} + \dfrac{4}{2} & -2 < x < 6 \\ \dfrac{x^3}{3} - \dfrac{6^3}{3} + 50 & \text{otherwise} \end{cases} = \begin{cases} \dfrac{x^2}{2} & x \le -2 \\ \dfrac{3x^2}{2} - 4 & -2 < x < 6 \\ \dfrac{x^3}{3} - 22 & \text{otherwise} \end{cases}$. Still the assignee 'f(x)' is having the function.

Carry out the integration of $f(x)$ as follows:

MATLAB Command
```
>>maple('interface(prettyprint=1):');  ↵          ← To see the output in symbolic form
>>maple('int(f(x),x)');  ↵
```

```
{            2
{     1/2 x                   x <= -2
{
{            2
{    3/2 x   - 4              x <= 6
{
{                  3
{   -22 + 1/3 x               6 < x
```

The differentiation of $f(x)$ can also be carried out. We have two discontinuities at $x=-2$ and $x=6$ respectively. At the discontinuous points, the derivatives become ∞ or $-\infty$. The derivatives of the piecewise functions are 1, 3, and $2x$ respectively. The MATLAB return is as follows:

```
>>maple('diff(f(x),x)');  ↵
```

```
{       1            x < -2
{
{  undefined         x = -2
{
{       3            x < 6
{
{  undefined         x = 6
{
{      2 x           6 < x
```

The piecewise continuous function can be the nonhomogeneous part of a differential equation and the subroutine 'dsolve' is still effective with the piecewise nonhomogeneous input.

10.8 Solution of a difference or recurrence equation

In article 8.3, we introduced the notions of the difference equation and its derivatives. To find the solution of a difference equation, we employed Z transforms. One may not appreciate the lengthy process of carrying out Z transform. We have a master subroutine 'rsolve' (a cognate of 'dsolve') which can find the solution of a difference equation directly. Located in the Maple package, 'rsolve' is the abbreviation of recurrence equation solve. The domain of the solution is $k \ge 0$ when the independent variable of the difference equation is k (integer). Start with the difference equation $y[k+1]=4^k y[k]$. The method of induction solves the problem on substitution of $k=0, 1, 2, 3, \ldots$ etc:

$$y[1] = 4^0 y[0],$$

$$y[2] = 4^1 y[1] = 4^1 4^0 y[0],$$

$$y[3] = 4^2 y[2] = 4^2 4^1 4^0 y[0],$$

$$y[4] = 4^3 y[3] = 4^3 4^2 4^1 4^0 y[0],$$

$$......,$$

hence, $y[k] = 4^{0+1+2+3+....+(k-1)} y[0] = 4^{\frac{k(k-1)}{2}} y[0] = 2^{k(k-1)} y[0]$.

So, the general solution of the given equation is $y[k] = 2^{k(k-1)} y[0]$. Now pay attention to the MATLAB terminology. The equation is stringed as 'y(k+1)= 4^k*y(k)'. We invoke the 'rsolve' as follows:

MATLAB Command
>>O=maple('rsolve(y(k+1)=4^k*y(k),y(k))'); ↵   ← The output is assigned to O
>>pretty(sym(O)) ↵

```
     (k (k - 1))
    2              y(0)
```

To pursue the case of difference equation with the initial condition, solve the foregoing equation with $y[0] = -3$, then, the particular solution should be $y[k] = -3 \times 2^{k(k-1)}$. The MATLAB code of $y[0] = -3$ is 'y(0)=-3'. Implement it as follows:

>>O=maple('rsolve({y(k+1)=4^k*y(k),y(0)=-3},y(k))'); ↵
>>pretty(sym(O)) ↵

```
        (k (k - 1))
   -3 2
```

Under the second brace, the whole equation including the initial condition is inserted. The argument 'y(k)' indicates that the dependent variable is $y[k]$.

Next, find the solution of the second order constant coefficient linear difference equation $y[n-2] - 2y[n-1] - 3y[n] = -10 + 2n$ subject to the initial conditions $\begin{cases} y[0] = -2 \\ y[1] = 1 \end{cases}$. The difference equation is much like differential equations. Assume that the trial solution is of the form $y[n] = a^n$ therefore $y[n-2] = a^{n-2}$ and $y[n-1] = a^{n-1}$. The homogeneous part of the difference equation becomes $a^{n-2} - 2a^{n-1} - 3a^n = 0$ $\Rightarrow$ $a^{n-2}(1 - 2a - 3a^2) = 0$ $\Rightarrow$ $(1-3a)(1+a) = 0$, from which, the roots of the characteristic equation are -1 and $\frac{1}{3}$ and the complementary solution is $y[n]_C = C_1(-1)^n + C_2\left(\frac{1}{3}\right)^n$, where C_1 and C_2 are the arbitrary constants. Then, the nonhomogeneous part is a first degree polynomial in n whose solution takes the form $A + Bn$. With that, we obtain $y[n-2] = A + B(n-2)$, $y[n-1] = A + B(n-1)$, and $y[n] = A + Bn$. Substitute the last three equations in the given equation and simplify to have $-2A + 5 = n + 2Bn$. Equate the coefficients of the constant term and n to provide $A = \frac{5}{2}$ and $B = -\frac{1}{2}$. On having the constants, the associated nonhomogeneous solution is $y[n]_P = \frac{5}{2} - \frac{n}{2}$ and the complete solution becomes $y[n] = y[n]_C + y[n]_P = C_1(-1)^n + C_2\left(\frac{1}{3}\right)^n + \frac{5}{2} - \frac{n}{2}$. Interject the initial conditions to have $\begin{cases} -2 = C_1 + C_2 + \frac{5}{2} \\ 1 = -C_1 + \frac{C_2}{3} + 2 \end{cases}$, on solving, $C_1 = -\frac{3}{8}$ and $C_2 = -\frac{33}{8}$. Finally, the particular solution is $y[n] = -\dfrac{3}{8}(-1)^n - \dfrac{33}{8}\left(\dfrac{1}{3}\right)^n + \dfrac{5}{2} - \dfrac{n}{2}$. The independent and dependent variables are n and $y[n]$ respectively.

Following is the MATLAB execution:

MATLAB Command
>>O=maple('rsolve({y(n-2)-2*y(n-1)-3*y(n)=-10+2*n,y(0)=-2,y(1)=1},y(n))'); ↵
>>pretty(sym(O)) ↵

$$- 3/8\ (-1)^n\ -\ 33/8\ (1/3)^n\ -\ 1/2\ n\ +\ 5/2$$

In a similar fashion the higher order difference equations can also be solved. Not only that, a system of difference equations can be the arguments of 'rsolve' too. Just to show one example, solve the system of difference equations $\left\{\begin{array}{l} x[n] - 2y[n-1] = 2n \\ x[n+1] + 3y[n] = 2^n \end{array}\right\}$. Replace n of the first equation by $n+1$ to have $x[n+1] - 2y[n] = 2(n+1)$, then, subtract the last equation from the given second equation which provides $y[n] = \dfrac{2^n}{5} - \dfrac{2n}{5} - \dfrac{2}{5}$. As the equation of $y[n]$ is at hand, $y[n-1]$ can be written as $\dfrac{2^{n-1}}{5} - \dfrac{2(n-1)}{5} - \dfrac{2}{5}$. Back substitute $y[n-1]$ in the given first equation and simplify to obtain $x[n] = \dfrac{2^n}{5} + \dfrac{6n}{5}$. Finally, the solution of the system is $\left\{\begin{array}{l} x[n] = \dfrac{2^n}{5} + \dfrac{6n}{5} \\ y[n] = \dfrac{2^n}{5} - \dfrac{2n}{5} - \dfrac{2}{5} \end{array}\right\}$. Have the solution of the system by 'rsolve' as follows:

MATLAB Command

>>maple('e1:=x(n)-2*y(n-1)=2*n'); ↵   ← Assign the first equation to e1
>>maple('e2:=x(n+1)+3*y(n)=2^n'); ↵   ← Assign the second equation to e2
>>maple('S:=rsolve({e1,e2},{x(n),y(n)})'); ↵   ← Conduct the subroutine
>>maple('interface(prettyprint=1):S'); ↵   ← Set the format for the symbolic output

$$\{x(n) = 6/5\ n + 1/5\ 2^n\ ,\ y(n) = -\ 2/5\ n\ -\ 2/5\ +\ 1/5\ 2^n\ \}$$

10.9 Orthogonal polynomials

Orthogonal polynomials are found in the Maple package 'orthopoly'. Activate the package before you invoke any orthogonal polynomial. Available orthogonal polynomials are Chebyshev first and second kinds, Hermite, Laguerre, Legendre, and Jacobi. Out of many applications of the orthogonal polynomials, designing the filters using the Chebyshev polynomials and image processing by the Hermite polynomials can be mentioned. Inherent perspicuity of the orthogonal polynomials is that they are the solution of the linear differential equations.

⊟ Chebyshev polynomial of the first kind

The polynomial is defined by the recursive formula $T_n(x) = 2xT_{n-1}(x) - T_{n-2}(x)$ with $T_0(x) = 1$ and $T_1(x) = x$, where $n > 1$, n is the order of the polynomial, and n should be nonnegative integer. Polynomials of different orders are as follows:

$$T_0(x) = 1$$
$$T_1(x) = x$$
$$T_2(x) = 2x^2 - 1$$

$$T_3(x) = 4x^3 - 3x$$
$$T_4(x) = 8x^4 - 8x^2 + 1$$
$$T_5(x) = 16x^5 - 20x^3 + 5x$$
.......so on.

The polynomials are orthogonal on the interval $[-1, 1]$ with respect to the weight function $w(x) = \dfrac{1}{\sqrt{1-x^2}}$. That is, the integration $\int_{x=-1}^{x=1} w(x)T_n(x)T_m(x)dx$ is equal to 0 for $m \neq n$. Prove the property considering $m = 3$ and $n = 5$.

The integration becomes $\displaystyle\int_{x=-1}^{x=1} \frac{1}{\sqrt{1-x^2}}(4x^3 - 3x)(16x^5 - 20x^3 + 5x)dx = \int_{x=-1}^{x=1} \frac{64x^8 - 128x^6 + 80x^4 - 15x^2}{\sqrt{1-x^2}}dx =$

$\displaystyle\int_{\theta=-\frac{\pi}{2}}^{\theta=\frac{\pi}{2}} (64\sin^8\theta - 128\sin^6\theta + 80\sin^4\theta - 15\sin^2\theta)d\theta$ (for $x = \sin\theta$) $= 2\displaystyle\int_{\theta=0}^{\theta=\frac{\pi}{2}} (64\sin^8\theta - 128\sin^6\theta + 80\sin^4\theta - 15\sin^2\theta)d\theta$

$= 2\left[64 \times \dfrac{35\pi}{256} - 128 \times \dfrac{5\pi}{32} + 80 \times \dfrac{3\pi}{16} - 15 \times \dfrac{\pi}{4}\right] = 0$. Anyhow, the name of the function is 'T' (comes from the earlier style of writing Chebyshev, which is Tchebyshev). The function has two arguments; the first is the order and the second one is x. To see $T_0(x)$ and $T_5(x)$,

MATLAB Command

```
>>maple('with(orthopoly)');  ⏎        ← Activate the package
>>maple('T(0,x)')  ⏎                  ← Invoke the function

ans =

1
>>pretty(sym(maple('T(5,x)')))  ⏎     ← To display the symbolic form

       5        3
16 x  - 20 x  + 5 x
```

⊟ Chebyshev polynomial of the second kind

The recursive formula of this polynomial is $U_n(x) = 2xU_{n-1}(x) - U_{n-2}(x)$, where n is the order of the polynomial, $n > 1$, and it should be nonnegative integer. The starting polynomials are $U_0(x) = 1$ and $U_1(x) = 2x$. Its weight function is $\sqrt{1-x^2}$. The orthogonality happens to be on the interval $[-1, 1]$. The MATLAB analogue is 'U'. Various polynomials are as follows:

$$U_0(x) = 1,$$
$$U_1(x) = 2x,$$
$$U_2(x) = 4x^2 - 1,$$
$$U_3(x) = 8x^3 - 4x,$$
$$U_4(x) = 16x^4 - 12x^2 + 1,$$
$$U_5(x) = 32x^5 - 32x^3 + 6x,$$
.......so on.

Find $U_5(x) = 32x^5 - 32x^3 + 6x$ as follows:

MATLAB Command

```
>>O=maple('with(orthopoly):U(5,x)');  ⏎
>>pretty(sym(O))  ⏎
```

$$32\ x^5\ -\ 32\ x^3\ +\ 6\ x$$

⊟ Legendre polynomial

The third member of the 'orthopoly' family is Legendre polynomial. Its recursive formula is given by

$$\begin{cases} P_0(x) = 1 & for\ n = 0 \\ P_1(x) = x & for\ n = 1 \\ P_n(x) = \dfrac{(2n-1)xP_{n-1}(x) - (n-1)P_{n-2}(x)}{n} & for\ n > 1 \end{cases}$$. The interval for the orthogonality is $[-1, 1]$. The MATLAB

counterpart is 'P'. Exercise the recursive relationship to have

$$P_0(x) = 1,$$
$$P_1(x) = x,$$
$$P_2(x) = \frac{3}{2}x^2 - \frac{1}{2},$$
$$P_3(x) = \frac{5}{2}x^3 - \frac{3}{2}x,$$
$$P_4(x) = \frac{35}{8}x^4 - \frac{15}{4}x^2 + \frac{3}{8},$$
$$P_5(x) = \frac{63}{8}x^5 - \frac{35}{4}x^3 + \frac{15}{8}x,$$
.......so on.

Compute $P_3(x) = \frac{5}{2}x^3 - \frac{3}{2}x$ at $x = -\frac{1}{2}$. The result should be $P_3(x) = -\frac{5}{2} \times \frac{1}{8} + \frac{3}{2} \times \frac{1}{2} = \frac{7}{16}$ as follows:

MATLAB Command

>>maple('with(orthopoly):L:=P(3,x)'); ↵   ← The polynomial is put to Maple assignee L

>>maple('subs(x=-1/2,L)') ↵   ← Substitute $x = -\frac{1}{2}$ in L

ans =

7/16

⊟ Hermite polynomial

The next member of the family is Hermite polynomial that has the recursive relationship

$$\begin{cases} H_0(x) = 1 & for\ n = 0 \\ H_1(x) = 2x & for\ n = 1 \\ H_n(x) = 2xH_{n-1}(x) - 2(n-1)H_{n-2}(x) & for\ n > 1 \end{cases}$$ and that has the weight function $w(x) = e^{-x^2}$ on the interval

$[-\infty, \infty]$ $\left[\text{the orthogonality necessitates } \int_{x=-\infty}^{x=\infty} w(x)H_m(x)H_n(x)dx = 0 \text{ for } m \neq n \right]$. Different polynomials are as follows:

$$H_0(x) = 1$$
$$H_1(x) = 2x$$
$$H_2(x) = 4x^2 - 2$$
$$H_3(x) = 8x^3 - 12x$$
$$H_4(x) = 16x^4 - 48x^2 + 12$$

$$H_5(x) = 32x^5 - 160x^3 + 120x$$

.......so on.

Compute $\frac{d}{dx}[H_5(x)]$ at $x = -1$. The derivative of $H_5(x)$ is $160x^4 - 480x^2 + 120$ and at $x = -1$, $\frac{d}{dx}[H_5(x)] = -200$. The computation is as follows:

MATLAB Command

```
>>maple('with(orthopoly):L:=H(5,x)');  ↵        ← H_5(x) is put to Maple assignee L
>>maple('subs(x=-1,diff(L,x))')  ↵             ← Substitution following differentiation
```

ans =

-200

⊟ Laguerre and Gegenbauer polynomials

The last two members of the 'orthopoly' family are Laguerre and Gegenbauer polynomials with parameters a. Their recursive consanguinities are

$$\begin{cases} L_0(a,x) = 1 & \text{for } n = 0 \\ L_1(a,x) = a + 1 - x & \text{for } n = 1 \\ L_n(a,x) = \dfrac{2n + a - 1 - x}{n} L_{n-1}(a,x) - \dfrac{n + a - 1}{n} L_{n-2}(a,x) & \text{for } n > 1 \end{cases}$$

and

$$\begin{cases} G_0(a,x) = 1 & \text{for } n = 0 \\ G_1(a,x) = 2ax & \text{for } n = 1 \\ G_n(a,x) = \dfrac{2(n + a - 1)x}{n} G_{n-1}(a,x) - \dfrac{n + 2a - 2}{n} G_{n-2}(a,x) & \text{for } n > 1 \end{cases}$$

respectively. The weight functions are $e^{-x}x^a$

and $(1-x^2)^{a-\frac{1}{2}}$ respectively. The intervals of the orthogonality are $[0, \infty]$ and $[-1, 1]$ respectively. They follow the similar style of implementation as the other orthogonal counterparts do associating with one more argument for the parameter a. Various order polynomials are as follows:

Order	Laguerre Polynomial
$n = 0$	$L_0(a,x) = 1$
$n = 1$	$L_1(a,x) = 1 + a - x$
$n = 2$	$L_2(a,x) = 1 + \dfrac{3a}{2} + \dfrac{a^2}{2} - 2x - ax + \dfrac{x^2}{2}$
$n = 3$	$L_3(a,x) = 1 + \dfrac{11a}{6} + a^2 + \dfrac{a^3}{6} - 3x - \dfrac{5ax}{2} - \dfrac{a^2 x}{2} + \dfrac{3x^2}{2} + \dfrac{ax^2}{2} - \dfrac{x^3}{6}$

.... so on.

Order	Gegenbauer polynomial
$n = 0$	$G_0(a,x) = 1$
$n = 1$	$G_1(a,x) = 2ax$
$n = 2$	$G_2(a,x) = -a + 2x^2 a + 2x^2 a^2$
$n = 3$	$G_3(a,x) = -2ax - 2xa^2 + \dfrac{8x^3 a}{3} + 4x^3 a^2 + \dfrac{4}{3}x^3 a^3$

.... so on.

Just to show one example with the parameter a, find $G_3(a, x)$.

MATLAB Command

```
>>O=maple('with(orthopoly):G(3,a,x)');  ↵
>>pretty(sym(O))  ↵
```

```
         3          3  2                 3  3          2
8/3 x    a  + 4 x    a  - 2 a x + 4/3 x    a  - 2 x a
```

10.10 Polynomial interpolation

We know that a polynomial of degree n can be written as $y = a_0 + a_1 x + a_2 x^2 + \ldots + a_{n-1} x^{n-1} + a_n x^n$. There are $n+1$ coefficients in the polynomial. Polynomial interpolation means finding the polynomial of degree n passing through $n+1$ distinct points (x, y).

Find the polynomial passing through the points (0, 1), (1, 2), (2, –3), (3, –4), and (4, 5). We have 5 points so the degree of interpolation polynomial should be 5–1=4. Assumed polynomial takes the form $y = a_0 + a_1 x + a_2 x^2 + a_3 x^3 + a_4 x^4$. Upon satisfying by each point, the set of equations relating a_0, a_1, a_2, a_3, and

a_4 is $\begin{cases} 1 = a_0 \\ 2 = a_0 + a_1 + a_2 + a_3 + a_4 \\ -3 = a_0 + 2a_1 + 4a_2 + 8a_3 + 16a_4 \\ -4 = a_0 + 3a_1 + 9a_2 + 27a_3 + 81a_4 \\ 5 = a_0 + 4a_1 + 16a_2 + 64a_3 + 256a_4 \end{cases}$. Solve the equations to have $a_0 = 1$, $a_1 = \dfrac{25}{3}$, $a_2 = -\dfrac{59}{6}$, $a_3 = \dfrac{8}{3}$, and $a_4 =$

$-\dfrac{1}{6}$. Having known the coefficients, the interpolation polynomial is $y = 1 + \dfrac{25}{3}x - \dfrac{59}{6}x^2 + \dfrac{8}{3}x^3 - \dfrac{1}{6}x^4$. For this purpose, we are going to use the Maple's function 'interp' (abbreviation of <u>interp</u>olation). Arrange the x and y

values to write $x = \begin{bmatrix} 0 \\ 1 \\ 2 \\ 3 \\ 4 \end{bmatrix}$ and $y = \begin{bmatrix} 1 \\ 2 \\ -3 \\ -4 \\ 5 \end{bmatrix}$ respectively. The procedure is as follows:

MATLAB Command
```
>>maple('x:=[0,1,2,3,4]'); ↵
>>maple('y:=[1,2,-3,-4,5]'); ↵
>>P=maple('interp(x,y,z)'); ↵
>>pretty(sym(P)) ↵
```

```
        4         3          2
- 1/6 z   + 8/3 z   - 59/6 z   + 25/3 z + 1
```

Since the Maple assignee x is containing the x coordinates, the polynomial is expressed in terms of z (the third argument of the function).

10.11 Conversions of measurement units

Metric conversion is carried out by the function 'metric', which is located in the Maple Toolbox. Convert 5 feet and 9 inches to meters. We know that 1 inch=2.54 cm and 12 inches=1 feet, hence, 5 feet and 9 inches=5×12+9= 69 inches. In meters, the measurement is given by $\dfrac{69 \times 2.54}{100} = 1.7526$. Have it as follows:

MATLAB Command
```
>>maple('convert(5*ft+9*inches,metric)') ↵

ans =

1.7526000000000000000000000000000*m
```

It is obvious that the unit 'm' stands for meter. Following metrics are known to the convert functions:

acre	acres	bu	bushel	bushels	chain
chains	cm	cord	cords	feet	foot
ft	furlong	furlongs	gal	gallon	gallons
Gals	gill	gills	gr	hr	ins
inch	inches	kg	km	Lb	lbs
light_year	light_years	mi	Mile	miles	MPG
MPH	ounce	Ounces	oz	Ozs	pint
pints	pole	poles	pound	pounds	quart
quarts	yard	yards	yd	yds	

There are two kinds of metric conversions – imperial and U. S. Any one of them is notified by a third argument. For example, 1 gallon is equal to 4.5460 liters in imperial system but it is equal to 3.7854 liters in US system. Both of them are shown as follows:

MATLAB Command

>>maple('convert(gallon,metric,imp)') ↵   ← For the imperial system

ans =

4.5459631*lt   ← Unit 'lt' stands for liter
>>maple('convert(gallon,metric,US)') ↵   ← For the U. S. system

ans =

3.785411784*lt

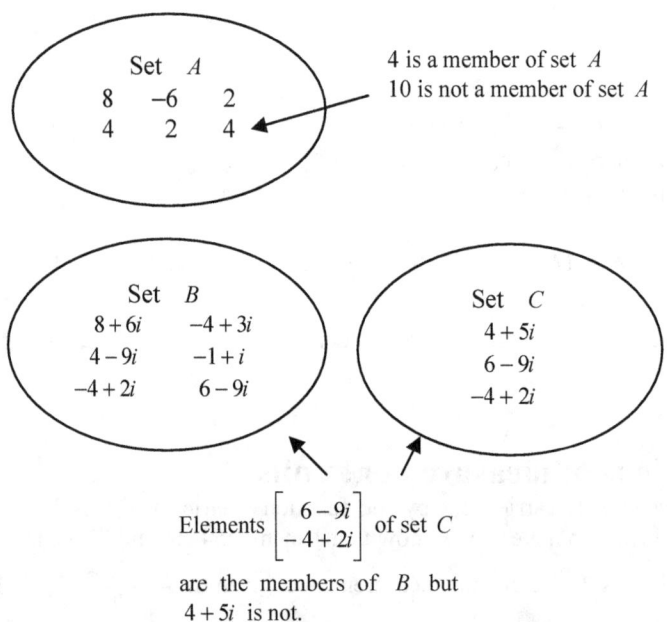

4 is a member of set A
10 is not a member of set A

Elements $\begin{bmatrix} 6-9i \\ -4+2i \end{bmatrix}$ of set C

are the members of B but $4+5i$ is not.

10.12 Set theory functions

A group of similar things is termed as a set. The members of a set can be numbers, characters, objects…etc. But in MATLAB, we deal only with the members that are numbers and characters. Any set is represented by a matrix — row, column, or rectangular. The numbers can be integers, floating-points, or complex. Set operations such as union or intersection can not be carried out on the symbolic variables.

⊟ Is an element member of a set?

Suppose a set is given, an element may be or may not be in that set. To determine whether an element is in a set,

MATLAB Step:

Use the command ismember (element(s), given set as row, column, or rectangular matrix).

In the previous page, 4 is a member of set A but 10 is not. Again, comparison of one set's member with the other can occur as shown for the sets B and C. The return of the subroutine 'ismember' is 1 or 0 depending on the presence or absence of an element. The subroutine does not operate on columns instead on the whole matrix. See all implementations below:

MATLAB Command for checking 4 as a member of set A,
>>A=[8 -6 2 4 2 4]; ↵
>>ismember(4,A) ↵

ans =
1
10 as a member of set A,
>>ismember(10,A) ↵

ans =
0
checking the elements of set C as members of B,
>>B=[8+6i -4+3i;4-9i -1+i;-4+2i 6-9i]; ↵
>>C=[4+5i 6-9i -4+2i]; ↵
>>ismember(C,B) ↵

ans =
0   1   1

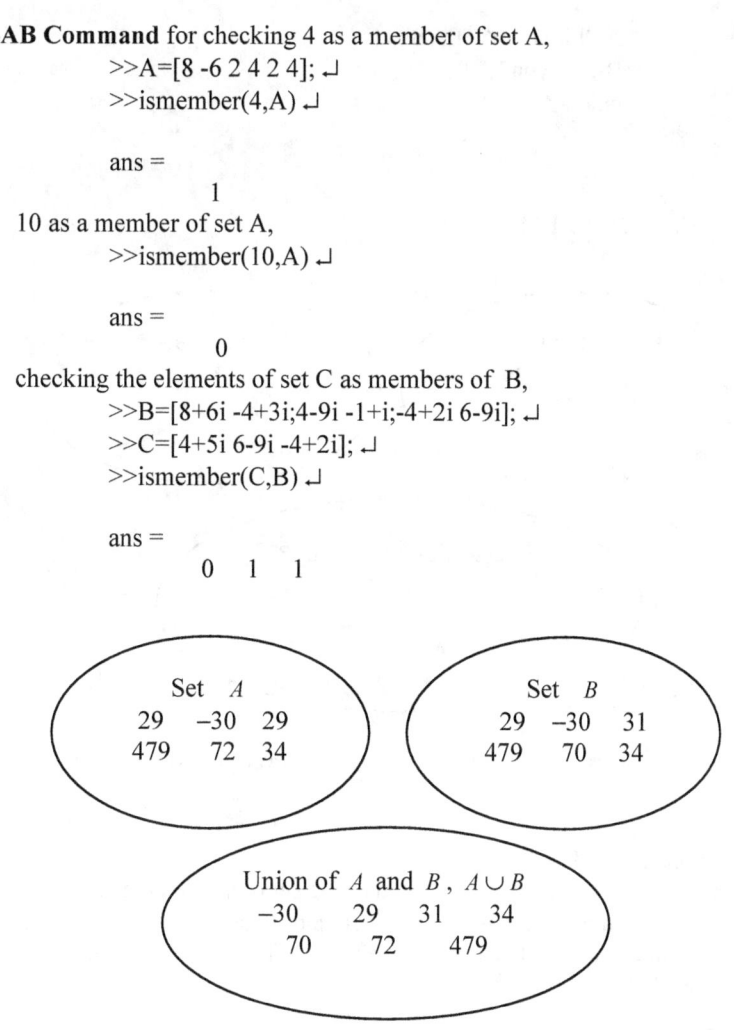

Figure 10.13   *Union of two sets A and B*

⊟ Union of two sets A and B

Union of two sets A and B is defined as the set of all elements consisting to A 'or' to B. The word 'or' means inclusive so that the members consisting to both A and B are also in the union. It is written as $A \cup B$. See the schematic representation of the union of two sets A and B in figure 10.13. We enter the members of A and B as row or column matrix. Union of sets A and B as indicated in figure 10.13 is found by the subroutine 'union' as follows:

MATLAB Command

      >>A=[29 -30 29 479 72 34]; ↵
      >>B=[29 -30 31 479 70 34]; ↵
      >>union(A,B) ↵

    ans =
          -30 29 31 34 70 72 479

One of the repeatedly occurring elements comes in the union. The output matrix following the union is a sorted one and the subroutine is also operational on character sets. Say, there are two character strings: 'I love MATLAB' and 'Do you?'. Union of these two character strings is 'I l o v e M A T L B D y u ?'. In ascending order, the characters in the set become '? A B D I L M T e l o u v y'. Following is the implementation:

    >>A='I love MATLAB'; ↵
    >>B='Do you?'; ↵
    >>union(A,B) ↵

    ans =

    ?ABDILMTelouvy

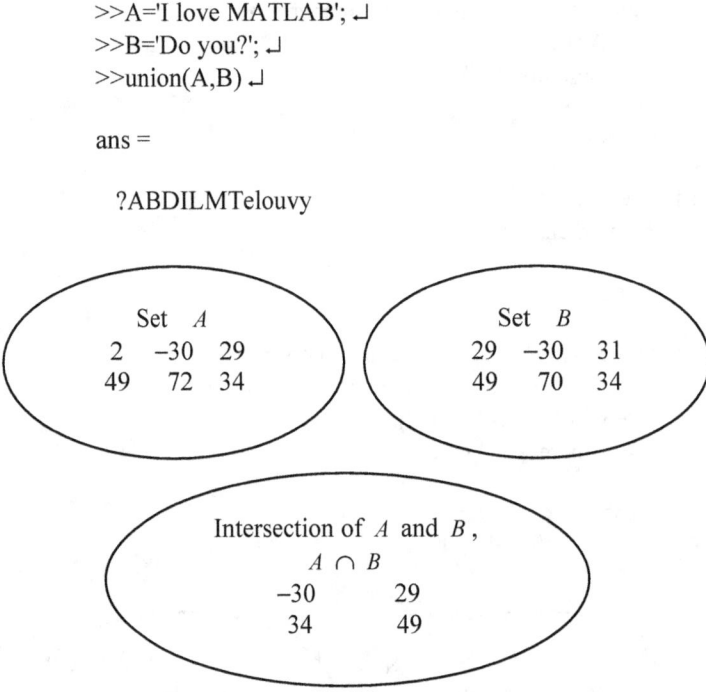

Figure 10.14   *Intersection of two sets A and B*

⊟ Intersection of two sets A and B

    Intersection of two sets A and B is defined as the set of all elements consisting to A 'and' to B. The word 'and' means only the members common to both A and B are in the intersection. It is written as $A \cap B$. Figure 10.14 shows the intersection of two sets A and B. The resembling MATLAB function is 'intersect'. Implementation is shown below:

MATLAB Command

      >>A=[2 -30 29 49 72 34]; ↵
      >>B=[29 -30 31 49 70 34]; ↵
      >>intersect(A,B) ↵

    ans =
          -30 29 34 49

The character strings 'I love MATLAB' and 'Do you?' have only one letter common that is 'o'.

>>A='I love MATLAB'; ↵
>>B='Do you?'; ↵
>>intersect(A,B) ↵

ans =

o

⬚ Determining the unique elements in a set

In a given set, there may be repetition of some members. Conduct the command 'unique' to find the uniquely occurred elements in a set. Our test matrix is $A = \begin{bmatrix} 2 & 4 & 5 & 6 \\ 4 & 4 & 6 & 9 \\ 0 & 9 & 2 & 4 \end{bmatrix}$. In A, the element 4 is occurring 4 times and 9 and 6 are occurring twice. The unique elements in A are [0 2 4 5 6 9]. Character string 'Working in MATLAB is easy' has the unique characters 'W o r k i n g n M A T L B s e a y'. In ascending order, the characters are 'A B L M T W a e g i k n o r s y'. These all are shown as follows:

MATLAB Command for finding the unique members of A,

>>A=[2 4 5 6;4 4 6 9;0 9 2 4]; ↵
>>unique(A) ↵

ans =

0
2
4
5
6
9

finding the unique characters in the character string,

>>A='Working in MATLAB is easy'; ↵
>>unique(A) ↵

ans =

ABLMTWaegiknorsy

⬚ Finding the difference between two sets

If two sets A and B have identical elements, $A - B$ means the elements of A which are not in B except from the common elements of A and B. The necessary MATLAB function is 'setdiff'. Test sets are the rectangular matrices $A = \begin{bmatrix} 2 & 4 \\ 5 & 6 \end{bmatrix}$ and $B = \begin{bmatrix} 4 & 5 \\ 0 & -1 \end{bmatrix}$. The common elements of A and B are 4 and 5. Then, $A - B$ should be $\begin{bmatrix} 2 \\ 6 \end{bmatrix}$. The procedure is as follows:

MATLAB Command

>>A=[2 4;5 6]; ↵
>>B=[4 5;0 −1]; ↵
>>setdiff(A(:),B(:)) ↵   ← Convert A and B to column matrices

ans =

2
6

438

⊟ Finding the complement set of intersection of two sets

The complement set of two sets means all elements except the common elements of the two sets. Referring to figure 10.14, we have 2 and 72 from A and 31 and 70 from B. In ascending order, the complement set is [2   31   70   72]. MATLAB function 'setxor' can perform this:

MATLAB Command
>>A=[2 -30 29 49 72 34]; ↵
>>B=[29 -30 31 49 70 34]; ↵
>>setxor(A,B) ↵

ans =
                2   31   70   72

10.13 Number theory functions

The objective of number theory is to investigate the properties of the natural numbers. There exist many simple rules regarding the numbers. Earlier culture, civilization, and society required the concept of numbers. Even today, the applications of numerals can not be bypassed. To see some uses of the number theory, coding of the electronic communication and the data encryption can be cited. Numerous number theory functions are overlaid in the Maple package of MATLAB, which are addressed by 'numtheory'. We implement some members of the package.

⊟ Fermat numbers

Fermat number is defined as $f(n) = 2^{2^n} + 1$, for example, $f(3) = 2^{2^3} + 1 = 257$.

MATLAB Command
>>maple('with(numtheory):fermat(3)') ↵

Warning: Warning, new definition for order

ans =

257

⊟ Sum to square of an integer

Sum to square of an integer is found by the subroutine 'sum2sqr'. One can easily verify that $1000 = 18^2 + 26^2$ and $1000 = 10^2 + 30^2$.

MATLAB Command
>>maple('with(numtheory):sum2sqr(1000)') ↵   ← Ignore the warning

ans =

[[18, 26], [10, 30]]

⊟ Divisors and sum of divisors

All possible positive divisors of a positive integer and sum of all divisors are given by the functions 'divisors' and 'sigma' respectively. For instance, all possible positive divisors of 75 are 1, 3, 5, 15, 25, and 75 and the sum is 1+3+5+15+25+75=124.

MATLAB Command
>>maple('with(numtheory):divisors(75)') ↵   ← To know all possible divisors

ans =

{1, 3, 5, 15, 25, 75}
>>maple('with(numtheory):sigma(75)') ↵   ← Sum of all possible divisors

ans =

124

There is another function $\tau(n)$ which counts the number of all possible divisors of a positive integer n. For instance, just mentioned number 75 has 6 such divisors. Count it by exercising function 'tau' as follows:

>>maple('with(numtheory):tau(75)') ↵   ← The number of divisors

ans =

6

⊟ Mersenne prime

Mersenne prime has the form $M_p = 2^p - 1$, where M_p is the Mersenne prime and exponent p is itself a prime. One example can be $M_5 = 2^5 - 1 = 31$.

MATLAB Command
>>maple('with(numtheory):mersenne(5)') ↵

ans =

31

⊟ Computations regarding the finite field arithmetic

In finite field arithmetic problems, computations are performed on modulus m (m is a positive integer, called modulo m or mod m). All rational coefficients are reduced to integers in the range from 0 to $m-1$. The field does not exist for any arbitrary numbers, in general, it exists only when m is a prime number. Let us consider the operation with modulo 5. Convert the numbers 26 and −43 in modulo 5. There can not be any number which is greater than 5. The numbers greater than 5 are replaced by the remainder of division by 5. When the numbers 26 and −43 are divided by 5, the remainders are 1 and −3 respectively. The negative numbers are turned to positive ones by adding modulus. The remainder −3 is written as −3+5=2. That is, 26 and −43 become 1 and 2 in modulo 5. The necessary function is 'modp'.

MATLAB Command
>>maple('with(numtheory):modp([26,-43],5)') ↵

ans =

[1, 2]

Next, transform the polynomial $f(x) = (2x+4)(3x-4)(7x-6)$ in modulus 3. Multiply the factors to write $f(x) = (2x+4)(3x-4)(7x-6) = 42x^3 - 8x^2 - 136x + 96$. All coefficients are 0, 1, or 2 in modulo 3 operation. The coefficients 42, −8, −136, and 96 have the remainders 0, −2, −1, and 0 when divided by 3. Therefore, they become 0, −2+3=1, −1+3=2, and 0 respectively. The converted polynomial in modulo 3 is $0 \times x^3 + 1 \times x^2 + 2 \times x + 0$ or $x^2 + 2x$.

MATLAB Command
>>maple('f:=expand((2*x+4)*(3*x-4)*(7*x-6))'); ↵   ← Expanded $f(x)$ is put to f

```
>>maple('with(numtheory):modp(f,3)')  ↵
```

ans =

x^2+2*x

After that, see the division of modular numbers. Let us compute $\frac{3}{4}$ and $\frac{2}{3}$ in modulo 5. The rationals $\frac{3}{4}$ and $\frac{2}{3}$ can be written as 3×4^{-1} and 2×3^{-1}, where 4^{-1} and 3^{-1} mean the modular inverses of 4 and 3 respectively. Check that which multiplication of the field elements gives 1 in modulo 5 operation.

for 4^{-1},   for 3^{-1},
  4×0=0 $\Rightarrow$ 0 (mod 5)   3×0=0 $\Rightarrow$ 0 (mod 5)
  4×1=4 $\Rightarrow$ 4 (mod 5)   3×1=3 $\Rightarrow$ 3 (mod 5)
  4×2=8 $\Rightarrow$ 3 (mod 5)   3×2=6 $\Rightarrow$ 1 (mod 5)
  4×3=12 $\Rightarrow$ 2 (mod 5)   3×3=9 $\Rightarrow$ 4 (mod 5)
  4×4=16 $\Rightarrow$ 1 (mod 5)   3×4=12 $\Rightarrow$ 2 (mod 5)

From the last row of the first and the third row of the second computations, one can say that $4^{-1} \equiv 4$ (mod 5) and $3^{-1} \equiv 2$ (mod 5) respectively. So, 3×4^{-1} and 2×3^{-1} become $3 \times 4 = 12 \equiv 2 (\text{mod } 5)$ and $2 \times 2 = 4$ respectively. See this as follows:

MATLAB Command
```
>>maple('with(numtheory):modp([3/4,2/3],5)')  ↵
```

ans =

[2, 4]

⊡ Congruence and modulus

Two integers a and b are said to be congruent for modulus m if their difference $a - b$ is divisible by the integer m. This is expressed in symbolic statement as $a \equiv b(\text{mod } m)$. The definition of congruence is exemplified by $36 \equiv 26 (\text{mod } 5)$. The difference $36 - 26 = 10$ is divisible by 5. Again, the congruence $2 \equiv 29$ (mod 9) has the difference -27 which is divisible by 9. What can we say about $2 \equiv 29$ (mod 10)? The difference -27 is not divisible by 10, so, the equation is not congruent. If the equation $a \equiv b(\text{mod } m)$ is given, one can find its modular solution by 'msolve'. Suppose we have the congruence $3x^3 - 4x + 1 \equiv 0 (\text{mod } 5)$. Let us compute the polynomial for different integers:

x	$3x^3 - 4x + 1$	$3x^3 - 4x + 1 (\text{mod } 5)$	Does x satisfy?
0	1	$1 \equiv 1$	no
1	0	$0 \equiv 0$	yes
2	17	$17 \equiv 2$	no
3	70	$70 \equiv 0$	yes
4	177	$177 \equiv 2$	no

We do not need to compute after $x = 4$ because the output should be in modulo 5. From the above tabular computation, it is plain that the solution of the congruence $3x^3 - 4x + 1 \equiv 0 (\text{mod } 5)$ is $x = 1$ and $x = 3$. Conduct it as follows:

MATLAB Command
```
>>maple('with(numtheory):msolve(3*x^3-4*x+1=0,5)')  ↵
```

ans =

$$\{x = 3\},\ \{x = 1\}$$

Euler's φ function

To illustrate the idea, find the relative prime numbers less than 20. The relative prime means 20 must not be divisible by the prime number and it can not be a number constituted from the factors of 20 except 1. The relative prime numbers to 20 are 1, 3, 7, 9, 11, 13, 17, and 19. Numbers 2, 4, 5, and 10 can not be the relative primes because they are the factors of 20. Also the number 6 can not be a relative prime because it has a factor 2 that is factor of 20 as well. Anyhow, there are 8 relative prime numbers to 20. This is called Euler's φ function. For the example at hand, we say $\varphi(20) = 8$. Obtain this by the intermediacy of function 'phi' as follows:

MATLAB Command

>>maple('with(numtheory):phi(20)') ↵

ans =

8

We found $\varphi(20) = 8$, then, the inverse function is $20 = \varphi^{-1}(8)$. It questions that for which number, there are 8 relative primes to. This can be obtained by the function 'invphi':

>>maple('with(numtheory):invphi(8)') ↵

ans =

[15, 16, 20, 24, 30]

From the return, the number 20 is not the only one that gives 8 relative primes. Numbers 15, 16, 24, and 30 provide 8 relative primes as well.

Mobius function, $\mu(n)$

The possible values of Mobius function are −1, 0, and +1. One important property of the function is that the Kronecker delta function can be represented by using the function. Mobius function $\mu(n)$ is a function of the positive integer n and it is formulated by $\mu(n) =$

$$\begin{cases} 1 & \text{for } n = 1, \\ (-1)^m & \text{for } n = p_1 \times p_2 \times p_3 ..... \times p_m, \\ & \text{where } p_1,\ p_2,\ p_3, ..., \ p_m \text{ are distinct primes of } n,\ \text{and}\ . \\ 0 & \text{for } n \text{ when the division } \dfrac{n}{p^2} \text{ does not leave any} \\ & \text{remainder for any prime } p \end{cases}$$

Consider the integers 3, 15, and 20. They can be factored as 3, 3×5, and 2×2×5 respectively, hence, $\mu(3) = (-1)^1 = -1$, $\mu(15) = (-1)^2 = 1$, and $\mu(20) = 0$ (because $\dfrac{20}{2^2} = 5$ does not leave any remainder). The MATLAB function is having the same name as the theory does:

MATLAB Command

>>maple('with(numtheory):mobius(3),mobius(15),mobius(20)') ↵

ans =

-1, 1, 0

Mobius function is used in the signal processing problems.

⊟ Help about the other number theory functions

One dedicated chapter is essential to explicate the whole 'numtheory' package. Other functions affiliated with the package are as follows:

```
B            F            GIgcd        J            L
M            bernoulli    bigomega     cfrac        cfracpol
cyclotomic   divisors     euler        factorEQ     factorset
fermat       ifactor      ifactors     imagunit     index
invcfrac     invphi       isolve       isprime      issqrfree
ithprime     jacobi       kronecker    lambda       legendre
mcombine     mersenne     minkowski    mipolys      mlog
mobius       mroot        msqrt        nearestp     nextprime
nthconver    nthdenom     nthnumer     nthpow       order
pdexpand     phi          pprimroot    prevprime    primroot
quadres      rootsunity   safeprime    sigma        sq2factor
sum2sqr      tau          thue
```

To learn about any function (say the third function of the first column), execute 'mhelp cyclotomic' in command window.

10.14 Logic gate functions

Only a small number of operations are performed in a digital system regardless of the complexities of the system. Logic gates perform these operations. There are many logic gates, such as NOT, OR, NOR, AND, NAND, XOR, XNOR, etc. The advent of integrated circuits (ICs) made it possible to fabricate complex digital circuits, namely, microprocessors, silicon chips, DSP chips, etc. Thousands of similar or composite gates function in the heart of these digital systems. In this section, an attempt has been made for the realization to various logic gates and the complicated circuits derived from them. To be applicable for the logic gates, a digital signal has two levels – either 1 or 0. The next problem is how one can generate these signals. The subroutine 'randint' (abbreviation of <u>rand</u>om <u>int</u>egers) of Communication Toolbox produces random integers. Let us generate 8 random 0 and 1's.

MATLAB Command

>>A=randint(8,1) ↵

A =

       0
       0
       1
       0
       1
       1
       1
       1

As you see, the output is in column matrix form. To obtain 10 random 0 and 1's in a row matrix,

>>B=randint(10,1)' ↵

B =

    1 0 1 0 1 1 0 0 1 0

Matrices A and B become the inputs of the logic gates.

⊟ Inverter

An inverter (NOT) changes the state of the input from 1 to 0 or vice versa. The symbol and truth table of an inverter are given as follows:

| Truth table ||
Input	Output
1	0
0	1

Input ———▷o——— Output

In MATLAB, NOT operation is carried out by '~'. If input is [1 0 1 0], the output should be [0 1 0 1].

MATLAB Command

        >>A=[1 0 1 0]; ↵   ← Input signal is assigned to A
        >>B=~A ↵

    B =   ← The output is in B
        0   1   0   1

⊟ AND gate

In an AND gate, the output is 1 if all inputs are 1 otherwise the output is 0. The symbol and truth table of a two input AND gate are shown below:

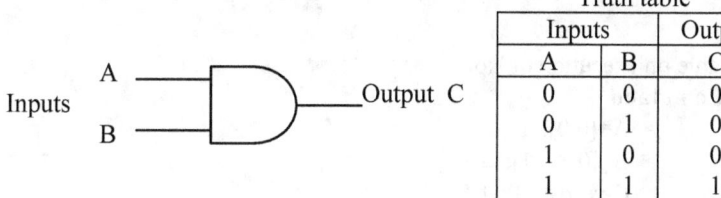

| Truth table |||
| Inputs || Output |
A	B	C
0	0	0
0	1	0
1	0	0
1	1	1

We need at least two inputs and the MATLAB analogue of the gate is 'and'. Take the inputs from the truth table.

MATLAB Command

        >>A=[0 0 1 1]; ↵
        >>B=[0 1 0 1]; ↵
        >>C=and(A,B) ↵

    C =
        0 0 0 1

⊟ OR gate

The logic behind the OR gate is the output is 1 if any 1 is there as the input otherwise the output is 0. The symbol and truth table for two input OR gate are as follows:

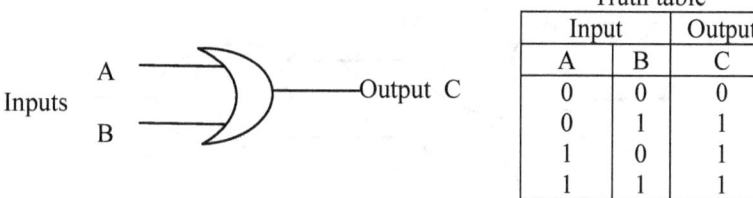

| Truth table |||
| Input || Output |
A	B	C
0	0	0
0	1	1
1	0	1
1	1	1

Implement the truth table for OR gate.

MATLAB Command

>>A=[0 0 1 1]; ↵
>>B=[0 1 0 1]; ↵
>>C=or(A,B) ↵

C =

    0   1   1   1

⊟ XOR gate

Exclusive OR (XOR) gate returns 1 if odd number of inputs is there otherwise the output is 0. Its schematic representation and the truth table are shown in the following:

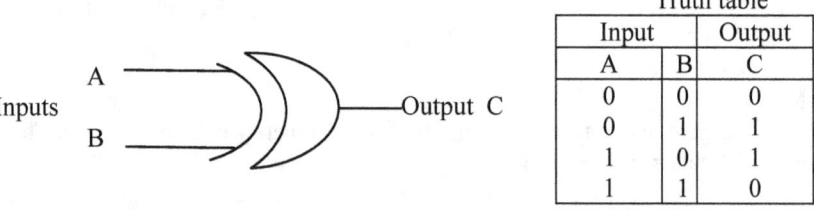

Truth table

Input		Output
A	B	C
0	0	0
0	1	1
1	0	1
1	1	0

Implement the truth table on execution of 'xor':

MATLAB Command

>>A=[0 0 1 1]; ↵
>>B=[0 1 0 1]; ↵
>>C=xor(A,B) ↵

C =

    0   1   1   0

⊟ Other gates

The aforementioned three gates are the basic gates of the digital logic circuits. Other members of the gate family are NAND, NOR, and XNOR. NAND, NOR, and XNOR are the NOT ('~') outputs of AND, OR, and XOR respectively. To show by example, the last XOR operation returned [0 1 1 0]. If XNOR operation of A and B were required, the output would be [1 0   0   1]. Have it as follows:

>>C=~xor(A,B) ↵

C =

    1   0   0   1

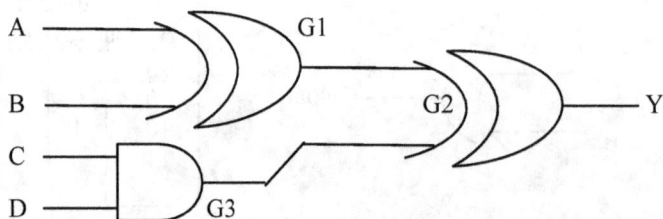

Figure 10.15   *A composite logic gate with 4 inputs and 1 output.*

⊟ Composite gates

Multi-input and composite gates simulate a practical digital logic problem. We present one simplest example of composite gates in figure 10.15. As you see, there are four inputs – A, B, C, and D and one output Y. The gates are labeled by G1(XOR), G2(XOR), and G3(AND). The inputs can be any set of ones and zeroes. The outputs of gates G1 and G3 are the inputs of gate G2. Apply the concept of different gates to have the following table. We assumed some random inputs for instance.

inputs				output of G1	output of G3	output of G2
A	B	C	D			Y
1	0	1	0	1	0	1
0	1	0	1	1	0	1
1	0	1	1	1	1	0
1	1	1	1	0	1	1

See the MATLAB implementation as follows:

MATLAB Command
```
>>A=[1 0 1 1]; ↵
>>B=[0 1 0 1]; ↵
>>C=[1 0 1 1]; ↵
>>D=[0 1 1 1]; ↵
>>G1=xor(A,B); ↵
>>G3=and(C,D); ↵
>>Y=xor(G1,G3) ↵

Y =
    1    1    0    1
```

This is just a model example. To simulate a complicated digital logic circuit, we will have hundreds of ones and zeroes which are generated by the subroutine 'randint' as mentioned earlier.

10.15 Different kinds of substitutions, conversions, and simplifications

Different kinds of substitutions are required in calculus, algebraic problems, and differential equations. Symbolic Toolbox and Maple package are very convenient for symbolic substitution, conversion, or simplification. We delve few computations on that.

⊟ Change of variable in integration

Change of variable can be accomplished by the Maple function 'changevar' (found in student package). Let us see how the function works for the definite integration $\int_{x=1}^{x=2} \frac{dx}{(x+3)\sqrt{3x+2}}$. The Maple code of $\int_{x=1}^{x=2} \frac{dx}{(x+3)\sqrt{3x+2}}$ is 'Int(1/(x+3)/sqrt(3*x+2),x=1..2)'. We want to substitute $3x+2=z^2$, hence, $3dx=2zdz$ and $x=\frac{z^2-2}{3}$. The limits are changed from $x=1$ and $x=2$ to $z=\sqrt{5}$ and $z=\sqrt{8}$ respectively. After simplification, the integration with respect to variable z becomes $\int_{z=\sqrt{5}}^{z=\sqrt{8}} \frac{2dz}{7+z^2}$. Have it as follows:

MATLAB Command
```
>>maple('with(student):'); ↵          ← Activate the package
>>maple('y:=Int(1/(x+3)/sqrt(3*x+2),x=1..2)'); ↵   ← Integration is assigned to y
>>maple('interface(prettyprint=1):'); ↵    ← Set the formatter for the symbolic output
>>maple('changevar(3*x+2=z^2,y,z)'); ↵
```

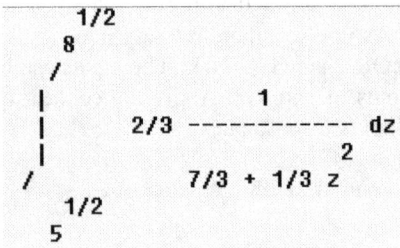

```
   1/2
   8
   /
   |              1
   |    2/3 ----------- dz
   |              2
   |         7/3 + 1/3 z
   /
  1/2
   5
```

This type of substitution is also possible in double integration. Consider the double integration $\iint \frac{x}{x^2+y^2} dxdy$ subject to the substitution $\begin{Bmatrix} x = r\cos\theta \\ y = r\sin\theta \end{Bmatrix}$. Our intention is to see the integration in terms of r and θ. Since x and y are functions of two variables, we need jacobian to have the substitution. A detail of jacobian is presented in article 5.9. In terms of r and θ, the integration turns to $\iint \frac{r\cos\theta}{r^2\cos^2\theta + r^2\sin^2\theta}|J|drd\theta$, where $|J|$

is the jacobian due to the change of coordinates and $|J| = \begin{bmatrix} \frac{\partial x}{\partial r} & \frac{\partial x}{\partial \theta} \\ \frac{\partial y}{\partial r} & \frac{\partial y}{\partial \theta} \end{bmatrix} = \begin{bmatrix} \cos\theta & -r\sin\theta \\ \sin\theta & r\cos\theta \end{bmatrix} = |r|$.

Simplification provides $\iint \frac{r\cos\theta}{r^2\cos^2\theta + r^2\sin^2\theta}|J|drd\theta = \iint \frac{|r|\cos\theta}{r}drd\theta$. Give the Maple code of the double integration, which is 'Doubleint(x/(x^2+y^2),x,y)'. Use 't' instead of θ and following is the implementation:

MATLAB Command
```
        >>maple('with(student):'); ↵
        >>maple('R:=Doubleint(x/(x^2+y^2),x,y)'); ↵ ← Integration is assigned to R
        >>maple('interface(prettyprint=1):'); ↵
        >>maple('changevar({x=r*cos(t),y=r*sin(t)},R,[r,t])'); ↵
```

```
  /  /
  |  |   cos(t) | r |
  |  |   ------------- dr dt
  |  |        r
  /  /
```

The concept can easily be extended for a triple integration. We suggest you remove all variables and objects from the workspace executing command 'clear all; close all' before going to the next implementation.

⊡ Change of variable in a differential equation

Sometimes it is necessary to convert a differential equation by means of the substitution. Conversion turns the differential equation to a linear or familiar one. The Maple function 'Dchangevar' (located in DEtools package) helps us carry out that. As usual, proceed with the first order differential equation $x^2 \frac{dy}{dx} = \frac{y^3}{x} + y^2$.

The equation is a nonseparable one. Substitute $y = zx$. The equation turns to $x\frac{dz}{dx} = z^3 + z^2 - z$ (which is a separable one) on substitution and simplification. Anyhow the former differential equation is stringed as 'x^2*diff(y(x),x)=y(x)^3/x+y(x)^2'. It is implemented as follows:

MATLAB Command
```
        >>maple('with(DEtools):'); ↵
        >>maple('e:=x^2*diff(y(x),x)=y(x)^3/x+y(x)^2'); ↵          ← Equation is assigned to e
```

>>O=maple('Dchangevar({y(x)=x*z(x),y(x)=z(x)},e,x)'); ↵   ← Maple output is put to O
>>pretty(sym(O)) ↵   ← To see the symbolic form

```
 2 /          /d     \\    2      3     2     2
x  |z(x) + x |-- z(x)|| = x  z(x)  + x  z(x)
   \          \dx     //
```

The return is not simplified because simplification is not the function of the subroutine. Divide both sides of the above equation by x^2 to have the simplified form. Notice that the dependent variables y and z are represented as $y(x)$ and $z(x)$ respectively.

⊟ Rationalization of a surd expression

A function containing surds can be rationalized by the Maple function 'rationalize'. Convert the surd numeric $\dfrac{1}{2-\sqrt{3}-\sqrt{5}}$ to a rational form. The necessary manipulation is as follows: $\dfrac{1}{2-\sqrt{3}-\sqrt{5}}$ ⇒

$\dfrac{2-\sqrt{3}+\sqrt{5}}{(2-\sqrt{3}-\sqrt{5})\,(2-\sqrt{3}+\sqrt{5})}$ ⇒ $\dfrac{2-\sqrt{3}+\sqrt{5}}{2-4\sqrt{3}}$ ⇒ $\dfrac{(2-\sqrt{3}+\sqrt{5})(2+4\sqrt{3})}{(2-4\sqrt{3})\,(2+4\sqrt{3})}$ ⇒ $\dfrac{-8+6\sqrt{3}+2\sqrt{5}+4\sqrt{15}}{4-48}$ ⇒ $\dfrac{2}{11}-\dfrac{3\sqrt{3}}{22}-$

$\dfrac{\sqrt{5}}{22}-\dfrac{\sqrt{15}}{11}$. Let us see the response of MATLAB:

MATLAB Command

>>maple('e:=1/(2-sqrt(3)-sqrt(5))'); ↵   ← Assign the numeric to e
>>O=maple('rationalize(e)'); ↵   ← The output is kept in O
>>pretty(sym(O)) ↵   ← Display the symbolic form

```
                1/2    1/2          1/2
1/22 (-2 + 3      - 5    ) (1 + 2 3    )
```

It seems that the output is not in the expanded form. Use the command 'expand' to have the equivalence:
>>pretty(expand(sym(O))) ↵

```
             1/2          1/2         1/2  1/2
2/11 - 3/22 3    - 1/22 5    - 1/11 5    3
```

The subroutine is so smart that it can handle an expression too. As an illustration, rationalize the expression $\dfrac{x}{x-\sqrt[3]{x}}$. Obviously, one needs to apply $a^3-b^3=(a-b)(a^2+ab+b^2)$, hence, $\dfrac{x}{x-\sqrt[3]{x}}$ ⇒

$\dfrac{x[x^2+x\sqrt[3]{x}+(\sqrt[3]{x})^2]}{(x-\sqrt[3]{x})\,[x^2+x\sqrt[3]{x}+(\sqrt[3]{x})^2]}$ ⇒ $\dfrac{x[x^2+x^{\frac{4}{3}}+x^{\frac{2}{3}}]}{x^3-x}$ ⇒ $\dfrac{x^2+x^{\frac{4}{3}}+x^{\frac{2}{3}}}{x^2-1}$. Following is the execution:

MATLAB Command

>>maple('e:=x/(x-x^(1/3))'); ↵   ← Assign the expression to e
>>O=maple('rationalize(e)'); ↵   ← Maple output is kept in O
>>pretty(sym(O)) ↵

```
 2    4/3    2/3
x  + x    + x
----------------
        2
       x   - 1
```

⊟ Partial fraction when the denominator factors have the surd roots

In section 3.16, we discussed how the symbolic form of partial fraction can be obtained. If the denominator has real root and surd factor, the subroutine 'parfrac' does not work. For example, the denominator of $\dfrac{2}{2x^2 - 2x - 3}$ can be factored as $2\left(x - \dfrac{1+\sqrt{7}}{2}\right)\left(x - \dfrac{1-\sqrt{7}}{2}\right)$, which is having the real roots. Once the factors of the denominator are known, the function can be converted to a partial one whence $\dfrac{2}{2x^2 - 2x - 3} =$

$\dfrac{2}{\sqrt{7}}\left[\dfrac{1}{2x - 1 - \sqrt{7}} - \dfrac{1}{2x - 1 + \sqrt{7}}\right]$. This conversion is accomplished by the agency of the Maple function 'rgf_pfrac' (located in 'genfunc' package):

MATLAB Command
```
>>syms x  ↵
>>y=2/(2*x^2-2*x-3);  ↵
>>maple('convert',y,'parfrac',x)  ↵
```

```
ans =

2/(2*x^2-2*x-3)
```
← As we mentioned, the 'parfrac' does not work

```
>>maple('with(genfunc):');  ↵
>>pretty(maple('rgf_pfrac',y,x))  ↵
```

```
             1/2                    1/2
            7                      7
    2/7  --------------  -  2/7  -------------
                   1/2                    1/2
        2 x  -  1  - 7          2 x  -  1  + 7
```

⊟ Converting a higher order differential equation to a system of the first order linear equations

In section 7.3.3, we converted the third order differential equation $6t^3\dfrac{d^3u}{dt^3} - 7t^2\dfrac{d^2u}{dt^2} + 9t\dfrac{du}{dt} - 4u = 3t - 3$

to a system of first order linear differential equations as $\begin{cases}\dfrac{du_1}{dt} = u_2 \\ \dfrac{du_2}{dt} = u_3 \\ \dfrac{du_3}{dt} = \dfrac{3t - 3 + 7t^2u_3 - 9tu_2 + 4u_1}{6t^3}\end{cases}$. The conversion of

this type is accomplished through the medium of 'convertsys' associated with the package 'DEtools'. Implement the conversion as follows:

MATLAB Command
```
>>maple('a:=u(t)');  ↵          ← The dependent variable u(t) is assigned to a

>>maple('e:=6*t^3*diff(a,t$3)-7*t^2*diff(a,t$2)+9*t*diff(a,t)-4*a=3*t-3');  ↵
                                        (The equation is in e)

>>S=maple('with(DEtools):convertsys(e,[],u(t),t,U,Up)');  ↵ ← Maple output is kept in S

>>pretty(sym(S))  ↵
```

```
[[Up[1] = U[2], Up[2] = U[3],

                        2
              3 t + 7 t  U[3] - 9 t U[2] + 4 U[1] - 3
Up[3] = 1/6  -------------------------------------------],
                              3
                             t


                      d                    d
[U[1] = u(t), U[2] = -- u(t), U[3] = --- u(t)], undefined, []]
                     dt                   2
                                        dt
```

The assignee S contains the output of the 'convertsys' and the subroutine has many input arguments. The first (e), second ([ ]), third (u(t)), fourth (t), fifth (U), and sixth (Up) arguments represent the given higher order differential equation, the initial conditions (which is empty for this case), the dependent variable, the independent variable, variable of the required first order system, and the first order derivative of the system variable respectively. The output may seem to be weird and has three lines. The first and second lines represent the equation of the system and the third line describes the assignees. Be familiarized with the symbolism: U[1] $\Rightarrow u_1$, U[2] $\Rightarrow u_2$, U[3] $\Rightarrow u_3$, Up[1] $\Rightarrow \dfrac{du_1}{dt}$, Up[2] $\Rightarrow \dfrac{du_2}{dt}$, and Up[3] $\Rightarrow \dfrac{du_3}{dt}$. Since the first order derivative of u is denoted by u' (u prime), the derivatives of the system are distinguished by Up's.

⌧ Translate a polynomial at some point

Suppose a polynomial $f(x) = 7x^3 - 4x + 1$ is given and is to be translated at $x = 5$, so, the converted polynomial is going to be $f(x+5) = 7(x+5)^3 - 4(x+5) + 1 = 7x^3 + 105x^2 + 521x + 856$. Notice that the MATLAB convention is opposite to that of the geometry (i.e., at $x = -5$). Obtain it by dint of the function 'translate' as follows:

MATLAB Command

```
>>maple('p:=7*x^3-4*x+1'); ↵          ← Polynomial is kept in p
>>O=maple('readlib(translate):translate(p,x,5)'); ↵   ← O contains the output
>>pretty(sym(O)) ↵
```

```
                  2        3
856 + 521 x + 105 x  + 7 x
```

⌧ Conversion of a powered sine and cosine expression to multiple angles

Consider the powered sine and cosine expression $\sin^3\theta\cos^3\theta$. It can be converted to the multiple angle function as follows: $\sin^3\theta\cos^3\theta \Rightarrow \dfrac{1}{8}(2\sin\theta\cos\theta)^3 \Rightarrow \dfrac{1}{8}(\sin 2\theta)^3 \Rightarrow \dfrac{1}{8}\left(\dfrac{3}{4}\sin 2\theta - \dfrac{1}{4}\sin 6\theta\right)$. We used two identities — $\sin 2A = 2\sin A\cos A$ and $\sin 3A = 3\sin A - 4\sin^3 A$. Maple function 'combine' can perform the conversion:

MATLAB Command

```
>>maple('f:=sin(t)^3*cos(t)^3'); ↵     ← sin³θ cos³θ is assigned to f
>>O=maple('combine(f,trig)'); ↵        ← 'trig' stands for trigonometry
>>pretty(sym(O)) ↵
```

```
- 1/32 sin(6 t) + 3/32 sin(2 t)
```

Very often simplification of an expression or equation is obligatory. Simplification can be of different types, for example, trigonometric, radical, exponential, power...etc. In Symbolic Toolbox, there are two variants for the simplification – 'simple' and 'simplify'. On the other hand, Maple expressions are simplified by 'simplify'. Insertion style of a particular simplification can be found from the online help. To learn about the subroutines, execute 'help simple', 'help simplify', or, 'mhelp simplify' in Command Window.

Chapter 11

M-file Programming and Some Utilities

Mostly in the previous chapters, attentions have been given to the built-in functions or programs in MATLAB. Many computations and simulations beside the built-in subroutines require that an M–file be written for a specific purpose. Contents of the M–files are the language codes of MATLAB. Like other high–level structured programming languages such as FORTRAN / C / PASCAL, MATLAB also has its own code of writing program statements. This chapter highlights the commonly used syntaxes of the M–file programming. The M–files have the ability to write very concise source programs. A syntactically correct M–file does not need to build an execution file like FORTRAN or C. The M–file is accessible to everyone as long as it is executed in the command window. To be an efficient M–file programmer, the knowledge of string's handling is necessary. Strings are nothing but a set of characters and can take many forms. These forms can be useful for mathematical computation, for writing input and output, and for representation of some functions. The strings can be static or generated when an M–file is being run. The string functions are also appended in this chapter.

11.1 What is an M–file?

An M–file is a script file and nothing but a sequence of executable MATLAB statements. If the MATLAB commands are just two or three lines, they can be executed in the command window. But if we have many executable commands, a file is required where we put all commands together so that any modification or editing can be performed according to user's convenience. Thus M–files are written externally by users. Next question is how we write and execute an M–file. See chapter 1 regarding this.

11.2 Control statements

Most of the MATLAB Command statements encountered so far invoked some M-file or Maple functions. These built-in functions are inherently composed of program statements, logical verification, looping operations, conditional execution of the group statements... etc. There are a number of control statements in MATLAB. The control statements followed by the specific syntax observe the sequence of operations, the limit of repetitive computation, and the selection of multiple objective tasks. Understanding of these statements would divulge the ingenuity of computation.

11.2.1 Relational operators

The relational operators are mainly used for comparisons. In MATLAB, we have six relational operators as presented follows:

	Mathematical Notation	MATLAB Notation
equal to	$=$	$==$
not equal to	$\neq$	$\sim=$
greater than	$>$	$>$
greater than or equal to	$\geq$	$>=$
less than	$<$	$<$
less than or equal to	$\leq$	$<=$

The output of relational operators is either true or false. The output '1' represents true and '0' does false. For example, when A=3 and B=4, the comparisons A=B, A$\neq$B, A>B, A$\geq$B, A<B, and A$\leq$B are false, true, false, false, true, and true respectively. Responses of some relational operators are as follows:

MATLAB Command

```
>>A=3; ↵              >>A>B ↵
>>B=4; ↵
>>A==B ↵              ans =
                             0
ans =                 >>A>=B ↵
     0
>>A~=B ↵              ans =
                             0
ans =
     1
```

One operand of the relational operators can be a matrix while the other is a scalar. In that case, the scalar operand is compared to all elements of the matrix one. For instance, if A=3 and B=$\begin{bmatrix} 2 & 1 & 5 \\ -2 & 0 & 7 \end{bmatrix}$, A>B should

be $\begin{bmatrix} 3>2 & 3>1 & 3>5 \\ 3>-2 & 3>0 & 3>7 \end{bmatrix}$ or $\begin{bmatrix} 1 & 1 & 0 \\ 1 & 1 & 0 \end{bmatrix}$. That is what is shown below:

MATLAB Command

```
>>A=3; ↵
>>B=[2 1 5;-2 0 7]; ↵
>>A>B ↵

ans =
     1    1    0
     1    1    0
```

11.2.2 Logical operators

The basic logical operations performed are NOT, OR, and AND. The truth tables for the different logical operators are given as follows:

NOT	OR
NOT True → False	True OR True → True
NOT False → True	False OR True → True

True OR False → True

False OR False → False

AND

True AND True → True

True AND False → False

False AND True → False

False AND False → False

The characters '~', '|', and '&' are used for the logical NOT, OR, and AND respectively. '1' stands for true and '0' does for false. Different logical operations are as follows:

Logical operator NOT (~):
When A = 0, ~A = 1
When A = 1, ~A = 0
Logical operator OR (|):
When A = 0 and B = 0, A|B = 0
When A = 0 and B = 1, A|B = 1
When A = 1 and B = 0, A|B = 1
When A = 1 and B = 1, A|B = 1
Logical operator AND (&):
When A = 0 and B = 0, A&B = 0
When A = 0 and B = 1, A&B = 0
When A = 1 and B = 0, A&B = 0
When A = 1 and B = 1, A&B = 1

The logical operators apply to matrices too. Suppose A=[1   0   1   0] and B=[1   0   0   1], logically, NOT A, A OR B, and A AND B should return [0   1   0   1], [1   0   1   1], and [1   0   0   0] respectively. Implementation is shown below:

MATLAB Command

for NOT A,

>>A=[1 0 1 0]; ↵

>>~A ↵

ans =

    0   1   0   1

for A OR B,

>>B=[1 0 0 1]; ↵

>>A|B ↵

ans =

    1   0   1   1

for A AND B,

>>A&B ↵

ans =

    1   0   0   0

Some other logical operations such as XOR can also be performed. Any odd number of trues returns true as presented below:

Logical function XOR (xor):
When A = 0 and B = 0, xor(A,B) = 0
When A = 0 and B = 1, xor(A,B) = 1
When A = 1 and B = 0, xor(A,B) = 1
When A = 1 and B = 1, xor(A,B) = 0.

11.2.3 Suppressing output

Go to MATLAB Command Window and execute the following:

MATLAB Command

>>A=[2 3 4 8 3 8] ↵ ← Assign the row matrix [2   3   4   8   3   8] to A

A =

$\quad\quad$ 2 $\quad$ 3 $\quad$ 4 $\quad$ 8 $\quad$ 3 $\quad$ 8 $\quad\quad$ ← Assignment is displayed

>>A=[2 3 4 8 3 8]; ↵

>> $\quad\quad\quad\quad\quad\quad\quad\quad\quad\quad\quad$ ← Assignment is not displayed

As you see, the line ending with a semicolon (;) stops displaying the assignment. If user is sure about the command, displaying steps in screen during execution can be suppressed by appending one semicolon at the end of MATLAB statement for longer matrix or other execution. It is applicable for an M-file too.

11.2.4 For loop structure

For loop performs the similar operations for a specific number of times.

Program syntax: for *counter* = starting value : increment or decrement
$\quad\quad\quad\quad\quad\quad\quad\quad\quad\quad\quad$ of the counter value: final value
$\quad\quad\quad\quad\quad\quad\quad\quad$ *Executable MATLAB command(s)*
$\quad\quad\quad$ end

Example: We wish to compute $y = \cos x$ for $x = 10^0$ to 70^0 with the increment 10^0. Assign the output value to matrix y, where y should be $[\cos 10^0 \quad \cos 20^0 \quad \cos 30^0 \quad \cos 40^0 \quad \cos 50^0 \quad \cos 60^0 \quad \cos 70^0] = [0.9848 \quad 0.9397 \quad 0.866 \quad 0.766 \quad 0.6428 \quad 0.5 \quad 0.342]$.

Executable M-file:
```
for k=1:1:7
        y(k)=cos(k*10*pi/180);
end
```

Steps: Open the MATLAB editor (see articles 1.2 and 1.3 and figure 1.2.1 and choose File ⇒ New ⇒ M-file), type the executable statements, and save the editor contents as a file (which is an M-file) by the name 'test' in your working path.

Interactive sessions with the command window:
>>test ↵
>>y ↵

y =

$\quad\quad$ 0.9848 $\quad$ 0.9397 $\quad$ 0.8660 $\quad$ 0.7660 $\quad$ 0.6428 $\quad$ 0.5000 $\quad$ 0.3420

Example: Many computational problems need writing multiple for loops. As an example, one has to write a program containing two for loops in image processing problems. To illustrate the implementation, compute $f(x, y) = x + y$ for $-1 \le x \le 0$ and $0 \le y \le 1$ with the step 0.5. Of coarse, this is a two dimensional problem and the output is returned in a rectangular matrix, which should have

$$\begin{bmatrix} f(-1, 0) & f(-1, 0.5) & f(-1, 1) \\ f(-0.5, 0) & f(-0.5, 0.5) & f(-0.5, 1) \\ f(0, 0) & f(0, 0.5) & f(0, 1) \end{bmatrix} = \begin{bmatrix} -1 & -0.5 & 0 \\ -0.5 & 0 & 0.5 \\ 0 & 0.5 & 1 \end{bmatrix}.$$

Executable M-file:
```
r=[];
c=[];
for x=-1:.5:0
        for y=0:.5:1
                r=[r x+y];
        end
                c=[c;r];
                r=[];
end
```

Sessions with the command window:
>>test ↵
>>c ↵

```
c =

    -1.0000   -0.5000         0
    -0.5000         0    0.5000
         0    0.5000    1.0000
```

Pertaining to the M-file program, 'r' and 'c' refer to the row and column of the rectangular matrix respectively. The initialization of 'r=[];' and 'c=[];' means that r and c are empty matrices. To understand the terminology of 'r=[r x+y];' and 'c=[c;r];', go through the article 2.23. Before the end of the second loop's end, another command 'r=[];' is inserted. This is necessary otherwise 'r' would contain all computations in a row matrix. In doing so, a mark of distinction for the consecutive rows of the rectangular matrix is accomplished.

11.2.5 Simple if / if-else / nested if structure

The conditional commands are performed by if-else statements. We can have different if-else structures, namely, simple-if, if-else, or nested-if.

⊟ Simple if

Program syntax: if *logical expression*
                         Executable MATLAB command(s)
                 end

Example: If $x \geq 1$, $y = \sin\dfrac{\pi}{2} = 1$

Executable M–file:
    x=1;
    if x>=1
            y=sin(pi/2);
    end

Steps: Follow the steps of the for loop structure to open, save, and execute the M-file 'test'.

Check from the command window after running the M–file:
    >>y ↵

    y =
        1

⊟ If-else

Program syntax: if *logical expression*
                         Executable MATLAB command(s)
                 else
                         Executable MATLAB command(s)
                 end

Example: When $x = 1$, $y = \sin\dfrac{x\pi}{2} = 1$, otherwise, $y = \cos\dfrac{x\pi}{2}$

Executable M–file when $x = 1$:
            x=1;
        if x==1
            y=sin(x*pi/2);

456

```
        else
            y=cos(x*pi/2);
        end
```

Steps: Same as before.

Check from the command window after running the M–file:
```
        >>y ↵

        y =
            1
```

Executable M–file when $x = 2$ *:*
```
            x=2;
        if x==1
            y=sin(x*pi/2);
        else
            y=cos(x*pi/2);
        end
```

Check from the command window after running the M–file:
```
        >>y ↵

        y =
            −1
```

🗗 Nested-if

Program syntax:
```
        if logical expression
                Executable MATLAB command(s)
        elseif logical expression
                Executable MATLAB command(s)
        elseif logical expression
                Executable MATLAB command(s)
                    ⋮
        elseif logical expression
                Executable MATLAB command(s)
        else
                Executable MATLAB command(s)
        end
```

Example: The best example can be taking the decision of grades based on the achieved number of a student out of 100. The grading policy is stated as if the achieved number of a student is greater than or equal 90, greater than or equal to 80 but less than 90, greater than or equal to 70 but less than 80, greater than or equal to 60 but less than 70, greater than or equal to 50 but less than 60, and less than 50, then the grade falls into A, B, C, D, E, and F respectively.

In the following program, N and g refer to the number achieved and the grade respectively. If the number N is 77, the grade g should be C.

Executable M–file:
```
            N=77;
        if N>=90
            g='A';
        elseif (N<90)&(N>=80)
```

```
        g='B';
elseif (N<80)&(N>=70)
        g='C';
elseif (N<70)&(N>=60)
        g='D';
elseif (N<60)&(N>=50)
        g='E';
else
        g='F';
end
```

Steps: Same as before.

Check from the command window after running the M–file:

```
>>g ↵
```

g =

C

11.2.6 User input during run time

Sometimes it is necessary to have some input from the user when a program is being run. It can be accomplished by the command 'input'. Assume that we need any integer from 1 to 10 from the user. Its implementation can be conducted without opening an M-file as follows:

MATLAB Command

```
>>A=input('Enter any integer from 1 to 10: '); ↵
>>Enter any integer from 1 to 10: 5 ↵          ← Suppose, 5 is typed
>>A ↵                                          ← To make sure, what there is in A

A =
        5
```

Integers are not the only input that the subroutine accept. As another example, a column matrix of three floating-point elements is to be asked from the user:

```
>>A=input('Enter a floating-point column matrix of 3 elements: '); ↵
>>Enter a floating-point column matrix of 3 elements: [2.1 3.2 1.5]' ↵
                                                    (Type the above)
>>A ↵                        ← To see the contents of A

A =
        2.1000
        3.2000
        1.5000
```

Next example shows a string input, say, the name of a student is needed:

```
>>A=input('What is the name of the student:  '); ↵
>>What is the name of the student:  'Rebeca' ↵          ← Type 'Rebeca'
>>A ↵                                                   ← Display the contents of A

A =

Rebeca
```

The string can even be a file name. One can use the 'input' statement in an M-file if some data is required during execution.

11.2.7 Switch-case-otherwise structure

The switch-case-otherwise structure provides the programming technique to choose a particular set of executable commands from several sets. The switch requires a key to make the structure operational and the key is compared to each available case. The structure executes the set of commands only with a case that matches with the key. It has the basic form as follows:

Program syntax: switch   *key for opening the switch*

case I

Executable MATLAB command(s)

case II

Executable MATLAB command(s)

⋮

otherwise

Executable MATLAB command(s)

end

Example: Suppose, a university library has the policy that teachers, researchers, and students can borrow 10, 8, and 5 books respectively. Other people of the university can not borrow a book. The teachers, researchers, and students have codes 'T', 'R', and 'S' respectively. To check how many books one can borrow, run the following M-file.

Executable M-file:
```
I=input('Enter your code : ');
switch I                          % I is the key to switch
    case 'T', disp('You can borrow 10 books');
    case 'R', disp('You can borrow 8 books');
    case 'S', disp('You can borrow 5 books');
    otherwise, disp('You are not supposed to borrow any books')
end
```

Steps: Follow the same steps of opening and saving the M-file as we did before.

Interactive sessions with the command window:
```
>>test ↵
Enter your code : 'R'  ↵
You can borrow 8 books
>>test ↵
Enter your code : 'D' ↵
You are not supposed to borrow any books
```

The previous example key is a character and it can be a numeric as well. See the next example where a numeric key is used.

Example: The binary numbers can represent a digital signal. The possible signal values are −1, 0, and 1. If the signal has the value other than the specified ones, it is termed as a noise. A practical digital signal is composed of thousands of zeroes and ones. For simplicity, we enter the signal values from the command window to see the implementation of the structure.

Executable M-file:
```
I=input('Enter the signal value : ');
```

```
switch I
        case −1, disp('The is a negative digital signal');
        case 0, disp('The signal value is 0');
        case 1, disp('This is a positive digital signal');
        otherwise, disp('This is a noise')
end
```

Steps: Same as before.

Interactive sessions with the command window:
>>test ↵
Enter the signal value : 7 ↵
This is a noise
>>test ↵
Enter the signal value : 1 ↵
This is a positive digital signal

11.2.8 While-end structure

While-end structure also performs the looping operations. Inside the while-end a set of similar commands is carried out until the logical expression beside the while is satisfied. Its general form is given as follows:

Program syntax: while *logical expression*
                          Executable MATLAB command(s)
                    end

Example: A positive integer greater than 1 will be asked from the user. Sum of the squares from 1 to that integer is required to compute.

Executable M-file:
```
I=input('Enter any integer greater than 1 : ');
k=0;
s=0;
while ~(k>I)
    s=s+k^2;
    k=k+1;
end
```

Explanations: I is not known beforehand and that is the user's choice. The counter index k inside the while-end is increased by 1 for each looping operation using the command 'k=k+1;'. Variable 's' adds consecutively the sum of squares for all integers less than I, which is achieved by 's=s+k^2;'. Just to have a check, we input 7 at that the output should be $1^2 + 2^2 + 3^2 + 4^2 + 5^2 + 6^2 + 7^2 = 140$.

Steps: Same as before.

Interactive sessions with the command window:
>>test ↵
Enter any integer greater than 1 : 7 ↵
>>s ↵

s =
        140

Notice that the logical expression 'k is not greater than I' is written as '~(k>I)'.

11.2.9 Comment on executable statements

There is no hard and fast rule for naming the assignees. You can choose any variable to assign the output of an executable statement. This creates a problem of understanding for an M-file accessible to multiple users. If an M-file programmer writes some comment beside each executable statement, that can help the others to go through the program. This is accomplished by '%'. We have already used '%' in section 11.2.7 to explain what assignee I is.

11.2.10 Break statement

Break statement is used to terminate the loop operation such as 'for loop' or 'while loop' subject to certain condition.

Example: Compute the sum of all squares from 1 to 20. But as soon as the sum is greater than 400, terminate the computation and display the sum. The sums of squares up to 10 and up to 11 are 385 and 506 respectively.

Executable M-file:
```
                s=0;
        for k=1:20
                s=s+k^2;
                if s>400
                        break;
                end
        end
        disp(s)
```
Steps: Same as before.

Interactive sessions with the command window:
```
        >>test ↵
        506
```
Notice that 'simple if' is used to check the sums. For multiple or nested loop, the 'break' statement terminates the innermost loop.

11.3 What is a string?

The string is a set of characters that are placed consecutively. Each character has the unique numeric code stored in MATLAB depending on the character set encoding of a given font. Usually, we do not access these values. The characters can appear from the first 127 codes of ASCII. We work on the characters as they are displayed on the screen. Typical examples of strings include naming a variable or an array, arguments of an M–file, MATLAB codes of a symbolic expression etc. The strings can be evaluated, compared, or split as if they are numbers and they can be static or dynamic. The static strings are preassigned set of characters on the other hand the dynamic strings may be created when an M–file or some executable command is run.

11.3.1 Strings for computations

The strings used for computation can be divided into two classes. The first one is used for the scalar or point to point computation and the second is for the vector computation. Point to point computation results the order of the output matrix same as that of the variable matrix. On the contrary, the order for the vector computation is followed in accordance with matrix algebra rules. Before writing strings, one should know the MATLAB codes of the symbolic functions. A list of symbolic functions and their MATLAB counterparts is presented in table 3.B. The operators used for arithmetic computations are as follows:

addition by	+
subtraction by	−
multiplication by	*
division by	/
power by	^

⬛ **Strings for the point to point computation**

Underlying concept of the point to point computation is that one has to assume the variable of a string being a vector (either a column or row matrix) or rectangular matrix having the same dimensions. The operators '*', '/', and '^' are sensitive for the computation. The syntax of the computation urges to use '.*', './', and '.^' in lieu of '*', '/', and '^' respectively.

Write the point to point form of the following functions:

$A.$ $\sin^3 x \cos^5 x$　　　$B.$ $2 + \ln x$　　　$C.$ $x^4 + 3x - 5$　　　$D.$ $\dfrac{x^3 - 5}{x^2 - 7x - 7}$

$E.$ $\sqrt{|x^3| + \sec^{-1} x}$　　　$F.$ $(1 + e^{\sin x})^{x^2 + 3}$　　　$G.$ $\dfrac{\cosh x + 3}{\sqrt{\dfrac{x + 4}{\log_{10}(x^3 - 6)}}}$

Example A

$\sin x$ and $\cos x$ are represented by sin(x) and cos(x) respectively. As we said, x is a vector, so is sin(x) or cos(x). $\sin^3 x$ is written as sin(x).^3 for the point to point computation. Similarly, $\cos^5 x$ is written as cos(x).^5. Each of sin(x).^3 and cos(x).^5 is a vector. Finally, $\sin^3 x \cos^5 x$ is written as sin(x).^3.*cos(x).^5. On either side of '.*' operator we have a vector.

Example B

$\ln x$ is written as log(x). Signs '+' or '−' is not sensitive for the point to point computation. If x is a vector, so is log(x). Hence, $2 + \ln x$ is written as 2+log(x), which is also a vector.

Example C

This is a polynomial of x and x is a vector, so, the power term x^4 is written as x.^4. $3x$ is written as 3*x not as 3.*x because 3 is a scalar. The complete representation is x.^4+3*x-5.

Example D

The numerator and denominator of this example are the polynomials of x. Proceeding like example C, the numerator and denominator polynomials are written as x.^3-5 and x.^2-7*x-7 respectively. The point to point division is performed by the operator './' so the whole string is written as (x.^3-5)./(x.^2-7*x-7).

Example E

x^3 is written as x.^3 and the MATLAB counterpart of modulus sign (| |) is 'abs'. If x^3 is a vector, $|x^3|$ is also a vector and written as abs(x.^3). $\sec^{-1} x$ is a vector as per as x is a vector and written as asec(x). $|x^3| + \sec^{-1} x$ is just the sum of abs(x.^3) and asec(x). The argument of 'sqrt' $\left(\sqrt{}\right)$ is a vector and the point to point string is sqrt(abs(x.^3)+asec(x)).

Example F

The power expression $x^2 + 3$ is written as x.^2+3. Sin(x) and exp(x) are vectors when x is a vector. So, $1 + e^{\sin x}$ is written as 1+exp(sin(x)). The point to point power is indicated by the operator '.^'. Therefore, the complete string is (1+exp(sin(x))).^(x.^2+3).

Example G

The MATLAB code of $\cosh x$ is cosh(x). The numerator is written as cosh(x)+3. The denominator is again having numerator and denominator. $\log_{10} x$ is written as log10(x). The argument of logarithm is a vector, which is x.^3-6. There is point to point division under the square root so the whole string is coded as (cosh(x)+3)./sqrt((x+4)./log10(x.^3-6)). Notice that the 'sqrt' does not need another brace ( ). The function 'sqrt' must be followed by the opening and closing braces that is understood by MATLAB.

462

⊟ Strings for vector computation

Contents of the variable matrices in the string must have the proper order to apply addition, subtraction, multiplication, division, or powering according to the rules of matrix algebra. Unlike point to point computation the operators '*', '/', or '^' are never preceded by '.' for vector computation. This string is the MATLAB code of any symbolic expression or function often found in mathematics. Starting from the simplest one, we present some examples for writing the complicated or long expressions. Our style of writing is not the only representation, there may be other form. The operation sequence of different operators in a vector string is the following:

enclosing braces	( )	first
power operator	^	then
division operator	/	next
multiplication operator	*	after that
addition operator	+	then
subtraction operator	−	finally

Following examples illustrate some symbolic expressions followed by the MATLAB codes:

$\dfrac{1}{x}$ can be written as 1/x,

$\dfrac{1}{xy}$ can be written as 1/x/y,

$\dfrac{1}{xyz}$ can be written as 1/x/y/z ,

$\dfrac{1}{(x-3)(x+4)(x-2)}$ can be written as 1/(x-3)/(x+4)/(x-2) (No multiplication (*) is necessary),

x^3 can be written as x^3,

$x^3 y^4$ can be written as x^3*y^4,

$x^3 y^4 z^7$ can be written as x^3*y^4*z^7 (No brace for y^4 is necessary),

$\dfrac{1}{x^4}$ can be written as 1/x^4,

$\dfrac{1}{x^4 y^7}$ can be written as 1/x^4/y^7,

$\dfrac{1}{x^4 y^7 z^6}$ can be written as 1/x^4/y^7/z^6,

$\dfrac{w^9}{x^4 y^7 z^6}$ can be written as w^9/x^4/y^7/z^6,

$\dfrac{u^2 w^9}{x^4 y^7 z^6}$ can be written as u^2*w^9/x^4/y^7/z^6 (No brace for u^2 is necessary),

$\dfrac{u^2 v^3 w^9}{x^4 y^7 z^6}$ can be written as u^2*v^3*w^9/x^4/y^7/z^6,

$\dfrac{a}{x+a} + \dfrac{b}{y+b}$ can be written as a/(x+a)+b/(y+b),

$\dfrac{a}{x+a} + \dfrac{b}{y+b} + \dfrac{c}{z+c}$ can be written as a/(x+a)+b/(y+b)+c/(z+c),

$\dfrac{1}{1+\dfrac{1}{x}}$ can be written as 1/(1+1/x), and

$\dfrac{1}{1+\dfrac{1}{1+\dfrac{1}{x}}}$ can be written as 1/(1+1/(1+1/x)).

Write the symbolic forms of the following string representations:

A. x^(n+1)/(n+1) $\qquad$ B. 1/log(a)*a^x

C. 1/4*exp(x^2+y^2) $\qquad$ D. u*(9-u^2-v^2)^(1/3)

String A

The order is $(\ldots) \to \,^\wedge \to /$, so, x^(n+1)/(n+1) becomes $x^{n+1}/(n+1) \to \dfrac{x^{n+1}}{n+1}$.

String B

It has the operation order $(\,) \to \,^\wedge \to / \to *$. Observing the order, we have $1/\ln x *a\,^\wedge x \to 1/\ln x * a^x \to$ $\dfrac{1}{\ln x} * a^x \to \dfrac{a^x}{\ln x}$.

String C

Carry out the first brace operation first to write 1/4*exp(x^2+y^2) $\to$ $1/4*\exp(x^2+y^2)$, which becomes $\dfrac{1}{4} * e^{x^2+y^2} \to \dfrac{1}{4}\,e^{x^2+y^2}$.

String D

Writing code on the first brace says that u*(9-u^2-v^2)^(1/3) $\to$ $u*(9-u^2-v^2)\,^\wedge\dfrac{1}{3}$, then, we have $u*(9-u^2-v^2)^{\frac{1}{3}} \to u(9-u^2-v^2)^{\frac{1}{3}}$.

⊟ Computation of an arithmetic string

A string can only be evaluated if the variables of the string possess the proper order and is syntactically correct. The subroutine 'eval' can compute a mathematical string coded as point to point or vector form. Computations of different functions are presented in section 3.19. We have a variant of 'eval', which is 'subs' and takes care of the symbolic function strings. We have also implemented 'subs' in the previous chapters.

11.3.2 Changing case from lower to upper or vice versa

Suppose a string is given as 'MATLAB is a computational software'. We want to see all letters of this string set in upper case. This is accomplished with the subroutine 'upper' as follows:

MATLAB Command

>>s='MATLAB is a computational software'; ↵ $\qquad$ ← Assign the string to s

>>upper(s) ↵

ans =

MATLAB IS A COMPUTATIONAL SOFTWARE

The reverse problem is if a string is given as 'USA IS FINANCIALLY STRONG', all characters in the string can be converted to lower case by using the subroutine 'lower' as shown below:

MATLAB Command

>>s='USA IS FINANCIALLY STRONG'; ↵

>>lower(s) ↵

ans =

usa is financially strong

You can even assign the output string to some variable p for further manipulation:

>>p=lower(s) ↵

p =

usa is financially strong

11.3.3 Placing strings horizontally/vertically one after another

Placing the strings one after another is called catenation. There can be two types of catenation — one is horizontal and the other is vertical. For horizontal catenation, the command is 'strcat' (abbreviation of <u>str</u>ing <u>cat</u>enation and the same for vertical is 'strvcat' (abbreviation of <u>str</u>ing <u>v</u>ertically <u>cat</u>enated). Say, we have two strings 'I love MATLAB' and 'Do you?'. We wish to see them as 'I love MATLAB, Do you?'. Application of 'strcat' is as follows:

MATLAB Command

>>s1='I love MATLAB, '; ↵   ← Assign the first string to s1
>>s2='Do you?'; ↵   ← Assign the second string to s2
>>strcat(s1,s2) ↵

ans =

I love MATLAB,Do you?

It does not look good that the output word 'Do' sits right after 'MATLAB'. Referring to the first command (assignment to s1), one blank space is there following ',' but that is not appearing in the concatenated string. What if we put one blank space before the string 'Do you?'.

>>s2=' Do you?'; ↵
>>strcat(s1,s2) ↵

ans =

I love MATLAB, Do you?

The blank space located at the end of the string is not taken but before the string is taken into consideration by the 'strcat'. Sometimes the numeric needs to be converted as strings. Say, z is an integer variable. We want to form a string in which 'value' will be followed by the integers assigned to z. The subroutine 'int2str' (abbreviation of conversion from <u>int</u>egers to(<u>2</u>) <u>str</u>ings) turns out the integer to strings. It is performed as follows:

>>z=1000; ↵   ← Assume that z=1000
>>s1='value'; ↵   ← Assign the value to s1
>>s2=int2str(z); ↵   ← Convert the contents of z to string
>>strcat(s1,s2) ↵

ans =

value1000

Next we present the vertical catenation of strings. Suppose we have three strings 'I love MATLAB', 'Researchers do', and 'Do you?'. We would like to see them one after another vertically. Following operation lets us attain that:

MATLAB Command

>>s1='I love MATLAB'; ↵
>>s2='Researchers do'; ↵
>>s3='Do you?'; ↵
>>strvcat(s1,s2,s3) ↵

ans =

I love MATLAB
Researchers do
Do you?

11.3.4 Replacing some part of a string by other

Say, we have a string 'Computations in MATLAB is done in matrix form'. Replace the word 'done' by 'performed' from the string. The purpose is served by the subroutine 'strrep' (abbreviation of string's replacement) as follows:

MATLAB Command
>>s='Computations in MATLAB is done in matrix form'; ↵   ← Assign the string to s
>>strrep(s,'done','performed') ↵

ans =

Computations in MATLAB is performed in matrix form

The function finds another application when one types a string for the numeric computation and does some mistakes in typing. The correction can be made very easily by using 'strrep'. To illustrate, compute $\sin^3 x \cos^3 x$ for $x = [\frac{\pi}{4} \quad \pi \quad 2\pi]$. The point to point string for $\sin^3 x \cos^3 x$ is 'sin(x).^3.*cos(x).^3'. The computation is accomplished as follows:

MATLAB Command
>>x=[pi/4 pi 2*pi]; ↵
>>s='sin(x)^3.*cos(x)^3'; ↵

No error message is seen as long as the assignee is a string. When one tries to evaluate the function for different x, the response is as follows:
>>eval(s) ↵
??? Error using ==> ^
Matrix must be square.

You discovered that the mistake is in the string. Mistakenly, we typed 'sin(x)^3.*cos(x)^3' instead of 'sin(x).^3.*cos(x).^3'. The assignee s is having the string and correct it as follows:

>>s=strrep(s,'^','.^'); ↵
>>eval(s) ↵

ans =
    0.1250 -0.0000 -0.0000

Now the execution is all right.

11.3.5 Comparison of strings

If two strings are given, we can compare whether they are identical by the subroutine 'strcmp' (abbreviation of string comparison). The subroutine behaves as a logical statement and the return of 'strcmp' is either 0 or 1. Compare the strings 'I love MATLAB' and 'I love MATLAb'. The comparison is shown below:

MATLAB Command
>>s1='I love MATLAB'; ↵
>>s2='I love MATLAb'; ↵
>>strcmp(s1,s2) ↵

465

ans =

0

The return is 0 since the last letters of the given strings are not having the same case and the subroutine is case sensitive. There is another subroutine called 'strcmpi' (abbreviation of <u>str</u>ing <u>c</u>omparison <u>i</u>gnoring case), which is not case sensitive. Use of the later should return 1. Verify that as follows:

>>strcmpi(s1,s2) ↵

ans =

1

Arguments s1 and s2 can subsist the character arrays too. To check that, consider $s1=\begin{bmatrix} \text{saturday} \\ \text{sunday} \end{bmatrix}$ and $s2=$

$\begin{bmatrix} \text{saturday} \\ \text{sunday} \end{bmatrix}$. The comparison is shown as follows:

>>s1=strvcat('saturday','sunday'); ↵
>>s2=strvcat('saturday','sunday'); ↵
>>strcmp(s1,s2) ↵

ans =

1

There are two more variants of the comparison subroutines. Comparison up to the certain number of characters is obtained by 'strncmp' (not ignoring case) and 'strncmpi' (ignoring case). Each of the first two strings has 13 characters including the blank space. If they are compared up to 12 characters, return should be 1:

>>s1='I love MATLAB'; ↵
>>s2='I love MATLAb'; ↵
>>strncmp(s1,s2,12) ↵

ans =

1

11.3.6 Finding one string in other

Given is the string 'You know about MATLAB' from where check whether the string 'Do you?' is within. The case sensitive subroutine 'findstr' (abbreviation of <u>find</u>ing <u>str</u>ing) can be applied in this regard. For the strings at hand, there is no match therefore the return should be empty. Implement it as follows:

MATLAB Command

>>s1='You know about MATLAB'; ↵
>>s2='Do you?'; ↵
>>findstr(s1,s2) ↵

ans =

[ ]

When a match is found, the position index of the search string is returned. To see this, the position indexes of all characters for the first string is given in the following table:

Y	o	u		k	n	o	w		a	b	o	u	t	
1	2	3	4	5	6	7	8	9	10	11	12	13	14	15

M	A	T	L	A	B
16	17	18	19	20	21

The string 'about' is a part of the first string and appears from the 10th character. In assignee s1, we have the string. Find the match of 'about' as follows:

>>s2='about'; ↵
>>findstr(s1,s2) ↵

ans =
  10

The character 'o' is appearing in the 2nd, 7th, and 12th characters:

>>findstr(s1,'o') ↵

ans =
  2   7   12

The blank space, which occurs in the 4th, 9th, and 15th characters, is also dealt with the subroutine:

>>findstr(s1,' ') ↵

ans =
  4   9   15

11.3.7 Conversions of integers and numbers to characters and vice versa

Given a matrix of integers, it can be converted to a matrix of characters by the subroutine 'int2str' (abbreviation of <u>int</u>eger to(<u>2</u>) <u>str</u>ing). As usual, consider the rectangular matrix $A = \begin{bmatrix} 1 & 2 & 3 \\ 6 & 0 & 4 \end{bmatrix}$. Enter it as follows:

MATLAB Command

```
>>A=[1 2 3;6 0 4]; ↵
>>class(A) ↵                    ← To know what the contents of A are

ans =

double                          ← Elements of A are double precision values
>>B=int2str(A); ↵               ← Converted matrix is assigned to B
>>class(B) ↵                    ← Check the element type of B

ans =

char                            ← Elements of B are characters
```

What if we have the matrix of floating-point numbers. Conduct the subroutine 'num2str' (abbreviation of <u>num</u>ber to(<u>2</u>) <u>str</u>ing) for the conversion on $A = \begin{bmatrix} -0.433 & -0.123 \\ 3.56 & 3.23 \end{bmatrix}$:

```
>>A=[-0.433 -0.123;3.56 3.23]; ↵
>>class(num2str(A)) ↵

ans =

char
```

There is one more variant of the conversion subroutines by name 'mat2str' (abbreviation of <u>mat</u>rix to(<u>2</u>) <u>str</u>ing). The reverse conversion, that is string to numbers, is also possible by the intermediacy of 'str2num' (abbreviation of <u>str</u>ing to(<u>2</u>) <u>num</u>ber). Convert the string '1 +2i 3.3' to numbers:

$$>>str2num('1 +2i 3.3')\ \lrcorner$$

ans =

    1.0000   0 + 2.0000i   3.3000

The string characters must be consecutive otherwise they are treated as individual ones. In the last string, there is a space gap between 1 and +2i that is why we have two different number elements 1.0000 and 0+2.0000i.

If you know the code of an ASCII character, it can be converted to the character by the subroutine 'char'. For example, the characters #, $, %, &, ', (,), *, and + have the numeric codes 35, 36, 37, 38, 39, 40, 41, 42, and 43 respectively. Have them as follows:

$$>>char([35:43])\ \lrcorner$$

ans =

#$%&'()*+

11.3.8 Examining letters and blank spaces in a string

A string can have letters, characters, or numerics. The decimal numbers 1, 2, 3, 8, 0…etc are numerics. A, a, x, g, t, c…etc are letters. The symbols /, $, *, ., @…etc are characters. We can check whether the characters of a string are letters by 'isletter' (abbreviation of is the character a letter?). Return of the function is 1 or 0 depending on being or not being a letter. Verify the subroutine taking the string s='MATLAB 5.2'. There are 10 characters in the string including the blank space where the first six are letters. Implementation is as follows:

MATLAB Command

$$>>s='MATLAB 5.2';\ \lrcorner$$
$$>>isletter(s)\ \lrcorner$$

ans =

    1   1   1   1   1   1   0   0   0   0

The blank space is checked by 'isspace' (abbreviation of is the character a space?). A blank space appears in the 7th character of the given string. Verify that as follows:

$$>>isspace(s)\ \lrcorner$$

ans =

    0   0   0   0   0   0   1   0   0   0

11.4 Three dimensional, structure, and cell arrays

In the previous chapters, mostly we manipulated the data which are given in a matrix form like row, column, or rectangular. This matrix-oriented arrangement of data is not convenient for multidimensional and group-related problems. There are three more data types called three dimensional, structure, and cell arrays for which the following discourse is dedicated.

11.4.1 Three dimensional arrays

A rectangular matrix has two dimensions. The reason we say two dimensions is any element position of the rectangular matrix needs two indexes to describe it – row and column. For example, a 4×4 rectangular matrix has the following position indexes:

(1, 1)	(1, 2)	(1, 3)	(1, 4)
(2, 1)	(2, 2)	(2, 3)	(2, 4)
(3, 1)	(3, 2)	(3, 3)	(3, 4)
(4, 1)	(4, 2)	(4, 3)	(4, 4)

Column

Row

Suppose, the above 4×4 rectangular matrix fits in one page. If we have two more pages each containing a 4×4 rectangular matrix, how can one accommodate the three pages in one variable? This necessitates the use of a three dimensional array. The position indexes of the rectangular matrices contained in the three pages can be labeled as follows:

Column

(1, 1, 1)	(1, 2, 1)	(1, 3, 1)	(1, 4, 1)
(2, 1, 1)	(2, 2, 1)	(2, 3, 1)	(2, 4, 1)
(3, 1, 1)	(3, 2, 1)	(3, 3, 1)	(3, 4, 1)
(4, 1, 1)	(4, 2, 1)	(4, 3, 1)	(4, 4, 1)

Row   Page 1

Column

(1, 1, 2)	(1, 2, 2)	(1, 3, 2)	(1, 4, 2)
(2, 1, 2)	(2, 2, 2)	(2, 3, 2)	(2, 4, 2)
(3, 1, 2)	(3, 2, 2)	(3, 3, 2)	(3, 4, 2)
(4, 1, 2)	(4, 2, 2)	(4, 3, 2)	(4, 4, 2)

Row   Page 2

Column

(1, 1, 3)	(1, 2, 3)	(1, 3, 3)	(1, 4, 3)
(2, 1, 3)	(2, 2, 3)	(2, 3, 3)	(2, 4, 3)
(3, 1, 3)	(3, 2, 3)	(3, 3, 3)	(3, 4, 3)
(4, 1, 3)	(4, 2, 3)	(4, 3, 3)	(4, 4, 3)

Row   Page 3

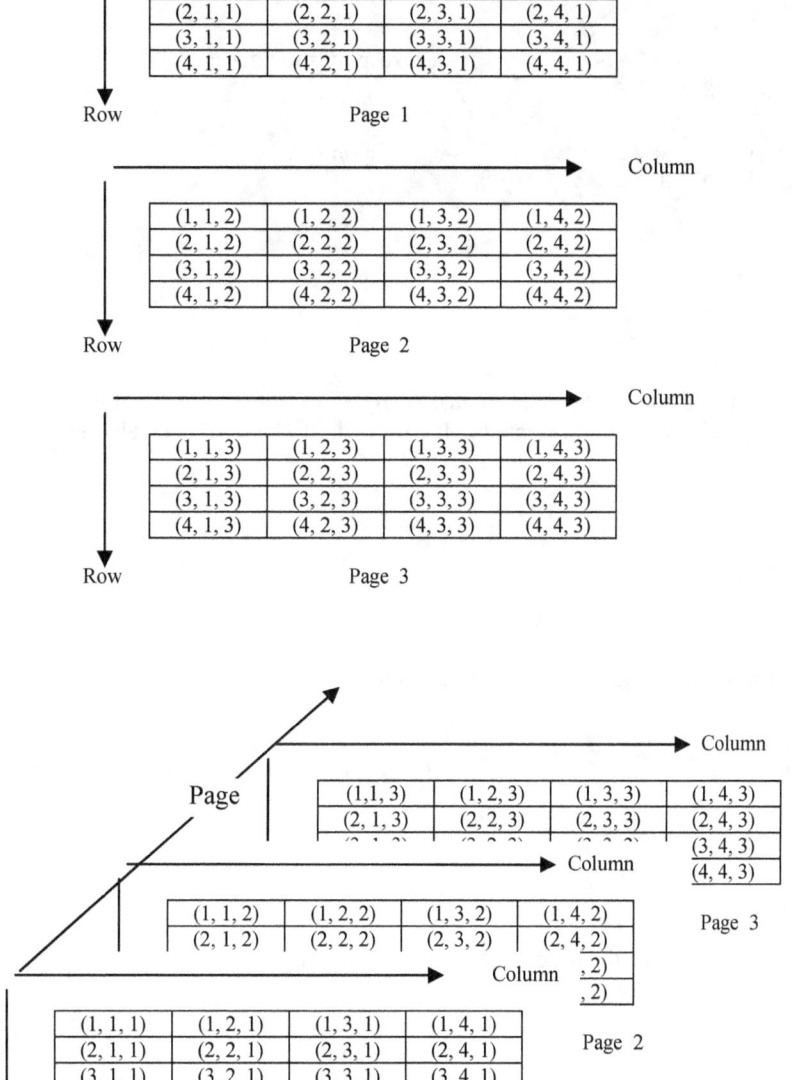

Figure 11.1   *A three dimensional array*

If one gathers the three pages one after another, the three dimensional block of figure 11.1 is formed. That is how a three dimensional array is created. One can assign any integer or floating-point values to these position indexes. There are three position indexes of an element in the three dimensional array – dimension 1 (row), dimension 2 (column) and dimension 3 (page). Now we enter the data of a three dimensional array. Assume that the three page data is given as follows; page 1: $\begin{bmatrix} 8 & 3 & 6 \\ 2 & 2 & 1 \end{bmatrix}$, page 2: $\begin{bmatrix} 0 & 4 & 4 \\ 5 & 3 & 8 \end{bmatrix}$, and page 3:

$\begin{bmatrix} -1 & 2 & 7 \\ -5 & 5 & 6 \end{bmatrix}$. Name the three dimensional array as A. Following is the implementation:

MATLAB Command

```
>>A(:,:,1)=[8 3 6;2 2 1]; ↵        ← Enter the elements of page 1
>>A(:,:,2)=[0 4 4;5 3 8]; ↵        ← Enter the elements of page 2
>>A(:,:,3)=[-1 2 7;-5 5 6]; ↵      ← Enter the elements of page 3
>>A ↵                              ← To see what in A is
```

ans(:,:,1) =   ← Corresponds to page 1

          8   3   6
          2   2   1

ans(:,:,2) =   ← Corresponds to page 2

          0   4   4
          5   3   8

ans(:,:,3) =   ← Corresponds to page 3

          -1   2   7
          -5   5   6

Most manipulations of the rectangular matrix can be extended for the three dimensional array. Some manipulations pertaining to A are presented in the following. The third page element 7 has the index (1, 3). It is called by

```
>>A(1,3,3) ↵
```

ans =
      7

You can change the value, say, by 10. Have it as follows:

```
>>A(1,3,3)=10; ↵
>>A(:,:,3) ↵                       ← To see only the third page
```

ans =
      -1   2   10
      -5   5   6

Suppose, you want to remove the third page from A. Carry out the following:

```
>>A(:,:,3)=[]; ↵
>>                                 ← It is done
```

A long row or column matrix can be converted to a three dimensional array by using the command 'reshape'. For example, choose R=[1 8 61 11 40 68 34 12 45 32 89 43]. There are twelve elements in R. The product of row, column, and page numbers must be 12. Formation is shown below:

```
>>R=[1 8 61 11 40 68 34 12 45 32 89 43]; ↵
```

>>reshape(R,2,3,2) ↵

ans(:,:,1) =   ← Displays the first page

$$\begin{array}{ccc} 1 & 61 & 40 \\ 8 & 11 & 68 \end{array}$$

ans(:,:,2) =   ← Displays the second page

$$\begin{array}{ccc} 34 & 45 & 89 \\ 12 & 32 & 43 \end{array}$$

Notice that the conversion is columnwise. Again, a three dimensional array A can be converted to a column matrix by using the command 'A(:)'. Now A is containing the first two pages. Squares of all elements of A are $\begin{bmatrix} 64 & 9 & 36 \\ 4 & 4 & 1 \end{bmatrix}$ (page 1) and $\begin{bmatrix} 0 & 16 & 16 \\ 25 & 9 & 64 \end{bmatrix}$ (page 2) respectively. You can have that on execution of 'A.^2' as follows:

>>A.^2 ↵

ans(:,:,1) =

$$\begin{array}{ccc} 64 & 9 & 36 \\ 4 & 4 & 1 \end{array}$$

ans(:,:,2) =

$$\begin{array}{ccc} 0 & 16 & 16 \\ 25 & 9 & 64 \end{array}$$

One can add one more page by assigning a 2×3 matrix to A(:,:,3). Adding some scalar, say 5, to each element of A is accomplished by A+5. Multiplication of a three dimensional array is not defined. Pages of the array can be multiplied according to the rules of matrix algebra, for example, page 2 with page 3 by 'A(:,:,2)*A(:,:,3)'. For simplicity, we have shown all manipulations taking the integer elements but the elements can be floating-points, characters, or even symbolic variables.

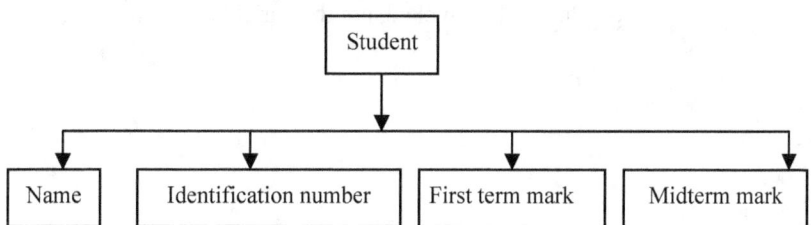

Figure 11.2   *Hierarchy of a structure array*

11.4.2 Structure arrays

We encountered these arrays in chapter 3 to find the solution of a set of algebraic equations containing several variables. If the solution of the set is one, there is no need to use a structure array. But if the set consists of multiple outcomes, the reason for applying the structure arrays is obvious. All elements of an ordinary array are identical data. Data can be integers, floating-points, complex numbers, or symbolic variables. A structure array may have mixed type of arrays as individuals and the individuals are termed as members. The ideology becomes transparent from the following example. Suppose, a physics teacher teaches two classes. For each class, he wants to keep the examination records of all students applying the hierarchy of figure 11.2.

Consider that the names, identification numbers, first term marks out of 30, and midterm marks out of 30 are

$$\begin{bmatrix} Reza \\ Shameem \\ John \\ Rebeca \\ Richard \end{bmatrix}, \begin{bmatrix} 91 \\ 92 \\ 89 \\ 96 \\ 95 \end{bmatrix}, \begin{bmatrix} 23.5 \\ 29.7 \\ 23 \\ 9 \\ 12 \end{bmatrix}, \text{ and } \begin{bmatrix} 25.5 \\ 27.7 \\ 21 \\ 20 \\ 19 \end{bmatrix}$$ respectively. The data we have is of mixed type, for instance, the names

and the identification numbers are characters and integers respectively. Enter the various data as follows:

MATLAB Command

>>N=strvcat('Reza','Shameem','John','Rebeca','Richard'); ↵   ← Assign the names to N

>>I=[91 92 89 96 95]'; ↵   ← Assign the identification numbers to I

>>F=[23.5 29.7 23 9 12]'; ↵   ← Assign the first term grades to F

>>M=[25.5 27.7 21 20 19]'; ↵   ← Assign the midterm grades to M

We entered all values and characters as column matrices. As par as simplicity is concerned, the names, identification numbers, first term marks, and midterm marks are assigned to N, I, F, and M respectively. The command 'struct' can build a structure array from the various members. In MATLAB terminology, the members are called fields. We name the various fields as Name, ID, Fgrade, and Mgrade. Following is the structure formation:

>>S=struct('Name',N,'ID',I,'Fgrade',F,'Mgrade',M) ↵ ← Name the structure as S

S =

       Name: [5x7 char   ]   ←'Name' is a 2 dimensional array

           ID: [5x1 double]

     Fgrade: [5x1 double]

    Mgrade: [5x1 double]

To invoke a particular member from the structure S, execute the command 'S.member'. For instance, only the identification numbers are seen by:

>>S.ID ↵   ← The field is invoked, not the variable

ans =

    91

    92

    89

    96

    95

See the names by,   Sort the identification numbers by,

>>S.Name ↵   >>sort(S.ID) ↵

ans =   ans =

                                       89

Reza   91

Shameem   92

John   95

Rebeca   96

Richard

A specific element of the field 'ID' (say, the second one, which is 92), can be accessed by S.ID(2):

>>S.ID(2) ↵

ans =

    92

Since the field 'Name' is a two dimensional (5×7) character array, the maximum number of characters in a row can be 7. The second name 'Shameem' is addressed by:

>>S.Name(2,:) ↵

ans =

Shameem

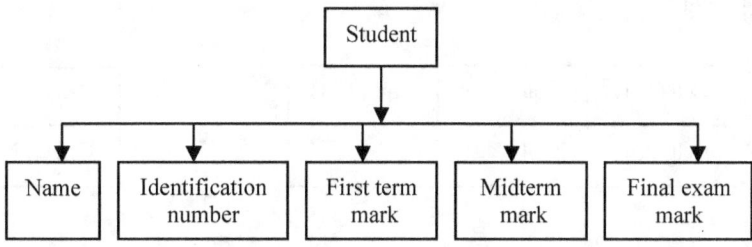

Figure 11.3   *Adding a field final exam mark to the existing hierarchy*

Suppose the final examination marks out of 40, which are $\begin{bmatrix} 35.5 \\ 37.7 \\ 24 \\ 30 \\ 23 \end{bmatrix}$ for the respective students, are available at the end of the semester. The hierarchy is pictured in figure 11.3. He wants to add the final marks with the existing structure. Name the field as 'Final'. The subroutine 'setfield' can add a field to the existing structure S:

>>FN=[35.5 37.7 24 30 23]'; ↵   ← Assign the final examination marks to FN
>>Physics1=setfield(S,'Final',FN) ↵   ← The new structure is named as 'Physics1'

Physics1 =
        Name: [5x7 char   ]
          ID: [5x1 double]
      Fgrade: [5x1 double]
      Mgrade: [5x1 double]
       Final: [5x1 double]   ← The field 'Final' is added here

Sometimes removal of a field may be required. Implement it by the subroutine 'rmfield'. To show that, remove the last field 'Final' as follows:

>>rmfield(Physics1,'Final') ↵

ans =
        Name: [5x7 char   ]
          ID: [5x1 double]
      Fgrade: [5x1 double]
      Mgrade: [5x1 double]

Referring to the output, the variable name is not necessary for removal only the field name is enough. The assignee Physics1 is still having the first term, midterm, and final grades. Finally, aggregate all marks to have the total grades out of 100 and put them to another field 'Total' as follows:

>>T=Physics1.Fgrade+Physics1.Mgrade+Physics1.Final; ↵   ← Sums of all marks are assigned to T
>>Physics1=setfield(Physics1,'Total',T); ↵   ← The total grade is put to field 'Total'
>>Physics1.Total ↵   ← To see the total grades out of 100

ans =
        84.5000
        95.1000
        68.0000
        59.0000
        54.0000

Cell (1, 1)	Cell (1, 2)	Cell (1, 3)		Cell (1, N)
Cell (2, 1)	Cell (2, 2)		------	Cell (2, N)

⋮

Cell (N, 1)	Cell (N, 2)	Cell (N, 3)		Cell (N, N)

Figure 11.4   *A two dimensional cell array of order N×N*

11.4.3 Cell arrays

A cell array is composed of cells, where the cells can contain ordinary arrays (of real, integer, or complex numbers), structure arrays, multidimensional arrays, character arrays...etc. Figure 11.4 shows a two dimensional cell array of order N×N. To be specific, we build a cell array of order 2×3. The cells of a cell array are indexed like a rectangular matrix but using the second brace {..}. The position indexes of different cells are A{1, 1}, A{1, 2}, A{1, 3}, A{2, 1}, A{2, 2}, and A{2, 3} respectively (consider that the name of the cell array

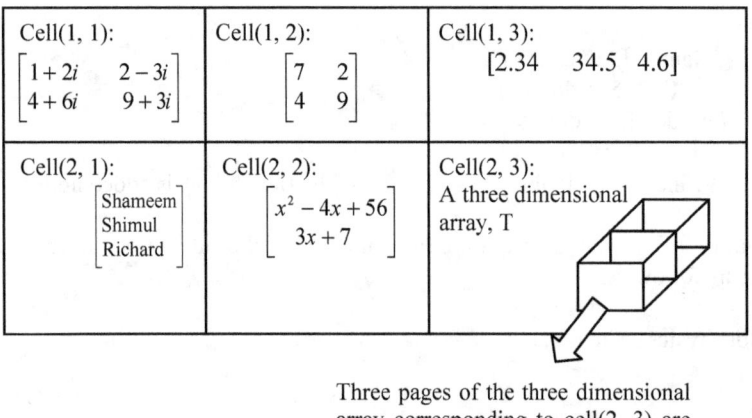

Cell(1, 1): $\begin{bmatrix} 1+2i & 2-3i \\ 4+6i & 9+3i \end{bmatrix}$	Cell(1, 2): $\begin{bmatrix} 7 & 2 \\ 4 & 9 \end{bmatrix}$	Cell(1, 3): [2.34 34.5 4.6]
Cell(2, 1): $\begin{bmatrix} Shameem \\ Shimul \\ Richard \end{bmatrix}$	Cell(2, 2): $\begin{bmatrix} x^2 - 4x + 56 \\ 3x + 7 \end{bmatrix}$	Cell(2, 3): A three dimensional array, T

Three pages of the three dimensional array corresponding to cell(2, 3) are

$$\begin{bmatrix} 2 & 3 \\ 4 & 5 \end{bmatrix}, \quad \begin{bmatrix} 11.1 & 30.3 \\ 12.2 & 51.9 \end{bmatrix}, \quad \text{and}$$

$$\begin{bmatrix} 19 & 36 \\ 2.2 & 55 \end{bmatrix} \text{ respectively.}$$

Figure 11.5   *Two dimensional cell array A of order 2×3*

is A). For example, assign the matrices $\begin{bmatrix} 1+2i & 2-3i \\ 4+6i & 9+3i \end{bmatrix}$, $\begin{bmatrix} 7 & 2 \\ 4 & 9 \end{bmatrix}$, [2.34   34.5   4.6], $\begin{bmatrix} \text{Shameem} \\ \text{Shimul} \\ \text{Richard} \end{bmatrix}$, and

$\begin{bmatrix} x^2 - 4x + 56 \\ 3x + 7 \end{bmatrix}$ and a three dimensional array with three pages $\begin{bmatrix} 2 & 3 \\ 4 & 5 \end{bmatrix}$, $\begin{bmatrix} 11.1 & 30.3 \\ 12.2 & 51.9 \end{bmatrix}$, and $\begin{bmatrix} 19 & 36 \\ 2.2 & 55 \end{bmatrix}$

respectively to the different cells. Schematic representation of the different cells' content is shown in figure 11.5. Implementation is presented as follows:

MATLAB Command

>>A{1,1}=[1+2i 2-3i;4+6i 9+3i]; ↵ ← Assign the complex matrix $\begin{bmatrix} 1+2i & 2-3i \\ 4+6i & 9+3i \end{bmatrix}$ to cell(1,1)

>>A{1,2}=[7 2;4 9]; ↵   ← Assign the integer matrix $\begin{bmatrix} 7 & 2 \\ 4 & 9 \end{bmatrix}$ to cell(1,2)

>>A{1,3}=[2.34 34.5 4.6]; ↵   ← Assign the floating-point matrix [2.34   34.5   4.6] to cell(1,3)

>>A{2,1}=strvcat('Shameem','Shimul','Richard'); ↵   ← Assign the two dimensional character array

$\begin{bmatrix} \text{Shameem} \\ \text{Shimul} \\ \text{Richard} \end{bmatrix}$ to cell (2,1)

>>syms x ↵

>>A{2,2}=[x^2-4*x+56;3*x+7]; ↵   ← Assign the symbolic matrix $\begin{bmatrix} x^2 - 4x + 56 \\ 3x + 7 \end{bmatrix}$ to cell(2,2)

>>T(:,:,1)=[2 3;4 5]; ↵   ← Assign the first page of T
>>T(:,:,2)=[11.1 30.3;12.2 51.9]; ↵   ← Assign the second page of T
>>T(:,:,3)=[19 36;2.2 55]; ↵   ← Assign the third page of T
>>A{2,3}=T ↵   ← Assign the three dimensional array T to cell(2,3)

A =

[2x2 double]	[2x2 double]	[1x3 double]
[3x7 char]	[2x1 sym]	[2x2x3 double]

Instead of displaying the contents, A is showing the type of the component cells. Anyhow some maneuverings of the cell arrays are presented in the following. The cell(1, 2) has the 2×2 integer matrix. Element having the position index (2, 1) of this matrix is 4. Access the element as follows:

>>A{1,2}(2,1) ↵

ans =
     4

Cell(1, 1):	Cell(1, 2):
A	[47 31]

Figure 11.6   *1×2 cell array B showing cell inside cell*

Placing cell inside cell is also possible. Suppose, another cell array B of order 1×2 is to be built, where the cell(1, 1) and the cell(1, 2) of B contain the previous mentioned 2×3 cell array A and a row matrix [47   31] respectively (see figure 11.6). It is just the matter of assignment as follows:

>>B{1,1}=A; ↵
>>B{1,2}=[47 31]; ↵

>>B ↵

B =
        {2x3 cell}   [1x2 double]

Cell indexing similar to an ordinary array can access to the subset of a cell. For instance, the cells of the cell array A taken from the intersection of the first and second rows and the second column, which are shown in figure 11.7, are invoked as follows:

>>A(1:2,2) ↵

ans =
        [2x2   double]
        [2x1   sym   ]

Delete the cell(1,3) and the cell(2,3) from A as follows:

Cell(1, 2): $\begin{bmatrix} 7 & 2 \\ 4 & 9 \end{bmatrix}$

Cell(2, 2):
$\begin{bmatrix} x^2 - 4x + 56 \\ 3x + 7 \end{bmatrix}$

Figure 11.7   *Subset of cell array A*

>>A{1,3}=[]; ↵
>>A{2,3}=[] ↵

A =   2 deleted cells
        [2x2 double]   [2x2 double]   [ ]
        [3x7 char   ]   [2x1 sym   ]   [ ]

Reshaping, catenating, and forming three dimensional arrays of the cell arrays can be accomplished too. Some functions, which handle different types of arrays, are supplied in table 11.A.

Table 11.A Some functions pertaining to multidimensional, structure, and cell arrays

Purpose	Function	Purpose	Function
To concatenate arrays	cat	To see structure field names	fieldnames
To know the number of array dimensions	ndims	To check whether a field is in a structure array	isfield
To permute array dimensions	permute	To display contents of a cell array	celldisp
To shift array dimensions	shiftdim	To convert a numeric array to cell array	num2cell
To remove singleton dimensions	squeeze	To convert a cell array to structure array	cell2struct
To check whether an array is a structure array	isstruct	To convert a structure array to cell array	struct2cell
To create a cell array	cell	To check whether an array is a cell array	iscell

11.5 Saving and loading data

User can save workspace variables or data in a binary file having the extension .mat and the command 'save' allows us to carry out that. The reverse operation, that is retrieval of the data, is performed by the command 'load'. A variety of options regarding the 'load' and 'save' are available. Few of them are presented.

⊟ Saving some matrices in a MAT file

Suppose we have two matrices $A=\begin{bmatrix} 3 & 4 & 8 \\ 0 & 2 & 1 \end{bmatrix}$ and $B=\begin{bmatrix} -2 & 5 \\ 9 & 3 \end{bmatrix}$. We want to save them in a MAT file named first.mat. Enter the matrices into the workspace of MATLAB as follows:

MATLAB Command
>>A=[3 4 8;0 2 1]; ↲
>>B=[-2 5;9 3]; ↲

Save the matrices A and B in first.mat as follows:
>>save first A B ↲

To see whether the file exists,
>>dir *.mat ↲

first.mat

Depending on available MAT files in the working path, you may or may not see other MAT files.

⊟ Saving the whole workspace
If the workspace contains many variables, it is not feasible that you save all of them typing the names. Save the whole workspace by the name 'allmat.mat' in the current working path as follows:

>>save allmat ↲

You can quit MATLAB and work later loading the file 'allmat.mat'.

⊟ Loading an existing MAT file
Just now you saved the matrices A and B in the file first.mat. The matrices A and B are existing in the MATLAB workspace. Remove them from the workspace using the command 'clear'.

>>clear ↲
>>

To check whether A and B exist,
>>who ↲
>>   ← It means no variables are present in the workspace

But you are sure that the MAT file first.mat contains the matrices A and B and it is in the current path. To load first.mat file in the workspace,

MATLAB Command
>>load first ↲   ← Extension '.mat' is not necessary
>>

To see the workspace variables,
>>who ↲

Your variables are:

A   B

To make sure the contents of A,
>>A ↲

$$A =$$

$$\begin{matrix} 3 & 4 & 8 \\ 0 & 2 & 1 \end{matrix}$$

In the command window, you will find the 'Import Data' and 'Save Workspace As' options by clicking the File menu. One can import (load) data clicking this option. This is advantageous because you will see the workspace browser along with the data import wizard. The save option also prompts with a dialog box where you can enter the workspace file name.

⊟ Saving MAT files with run time names

When an M–file is being executed, some data or matrix has been created. These data or matrix can take any value according to the program's parameter. Likewise the file name of a MAT file can be dependent on the program's parameter. Following example explains this. Say, M and N are two random integers. M and N can assume any value from 0 to 4 and from 4 to 8 respectively. We can generate M and N using the commands 'M=randint(1,1,5)' and 'N=4+randint(1,1,5)' respectively. The function 'randint' generates uniformly distributed discrete random integers. Before the execution of 'randint', we are not sure about the values of M and N. For instance, save the matrix $A=\begin{bmatrix} M & M \\ N & N \end{bmatrix}$ in MAT file fileMN.mat, where 'file' is the first four letters of the MAT file and M and N are integers from 0 to 4 and from 4 to 8 respectively once 'randint' is run. Different steps are as follows:

MATLAB Command

>>M=randint(1,1,5); ↵	← Generating a random integer M from 0 to 4
>>N=4+randint(1,1,5); ↵	← Generating a random integer N from 4 to 8
>>A=[M M;N N]; ↵	← Forming matrix A
>>assignin('base','A',A); ↵	← Forming matrix A's name as string
>>f=strcat('file',int2str(M),int2str(N)); ↵	← Forming fileMN as string
>>save(f,'A') ↵	← Saving A in f
>>dir *.mat ↵	← To see all MAT files

file25.mat

Above execution indicates that M and N were 2 and 5 respectively during the execution.

⊟ Appending some variables to an existing MAT file

Suppose you saved the matrix $A=\begin{bmatrix} 2 & 3 & 4 \\ 8 & 3 & 8 \end{bmatrix}$ in a MAT file 'data.mat' by the following:

MATLAB Command

>>A=[2 3 4;8 3 8]; ↵	
>>save data A ↵	

After saving, you remembered that another matrix $B=\begin{bmatrix} x & 7x \\ 4x & 56 \end{bmatrix}$ would have been saved in 'data.mat'. Have B appended with 'data.mat' as follows:

>>syms x ↵	← Since x is symbolic
>>B=[x 7*x;4*x 56]; ↵	← Enter B
>>save data B -append ↵	← Append B with 'data.mat'
>>clear ↵	← Remove all variables
>>load data ↵	← Loading 'data.mat'
>>who ↵	← To make sure, A and B are there

Your variables are:

A   B

11.6 Importing and exporting data

One may work in other software, for example, you have data in Microsoft Excel worksheets. You want the data to be imported for manipulation in MATLAB. Or, some matrices are generated in MATLAB but you want to export them in Excel for plotting a graph. MATLAB also has the data exchanging capabilities. As

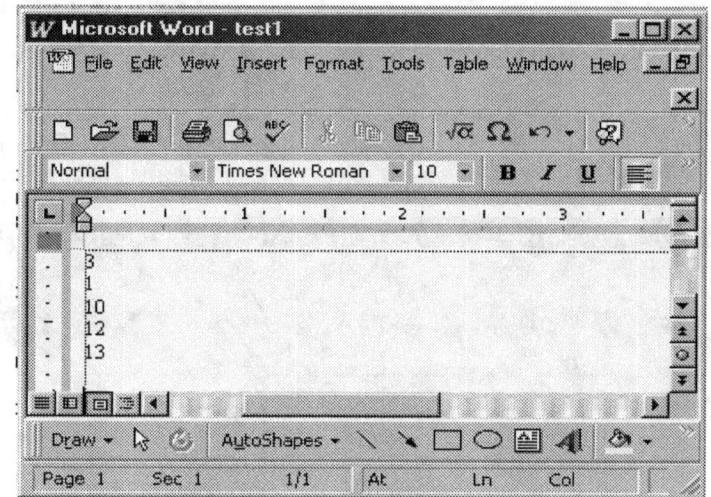

Figure 11.8 *Matrix A is typed in Microsoft Word*

par as data format is known, it can be imported or exported very conveniently from one platform to other. We present some data exchanging in the following.

⊟ Exchanging data between Microsoft Word and MATLAB

Suppose the column matrix $A = \begin{bmatrix} 3 \\ 1 \\ 10 \\ 12 \\ 13 \end{bmatrix}$ is typed in Microsoft Word 7 as shown in figure 11.8. When one

attempts to save the Word file from the File menu or from the Save icon, a window prompts the save operation. In that window, select your working path (If you do not have a path, follow the procedure mentioned in article 1.3. Let us say that the path is c:\mfile) clicking 'Save in', type the file name as 'test1' in File name box, choose 'Text Only' format in 'Save as Type' box, and remember the path and file name. So, you have the word file 'test1.txt' in the folder 'c:\mfile'. Now move on to MATLAB command prompt and execute the following:

MATLAB Command

```
>>cd c:\mfile ↵          ← Same action can be performed with path browser icon
>>dir test1.txt ↵        ← To make sure, file 'test1.txt' is in the path 'c:\mfile'

test1.txt
>>load test1.txt ↵       ← Load the file 'test1.txt'
>>test1 ↵                ← MATLAB matrix name is the same as file name

test1 =
```
3

480

$$\begin{array}{c} 1 \\ 10 \\ 12 \\ 13 \end{array}$$

Enter the matrix B=$\begin{bmatrix} 63.4 & 72.6 \\ 72.4 & 73 \\ 32.5 & 62 \end{bmatrix}$ into MATLAB to proceed with the reverse exchange (i.e., from MATLAB

to Microsoft Word):

>>B=[63.4 72.6;72.4 73;32.5 62]; ↵

>>save test2 B -ascii -double -tabs ↵ ← Save B as ASCII and tab separated data

The matrix B is saved as ASCII and tab separated double-precision data in the file 'test2' and lives in the current directory 'c:\mfile'. Now, go to Microsoft Word and click File Menu down Open File or Open File icon to prompt Open File window. Click 'Look in' box to select the path 'c:\mfile', click 'Files of Types' box to put

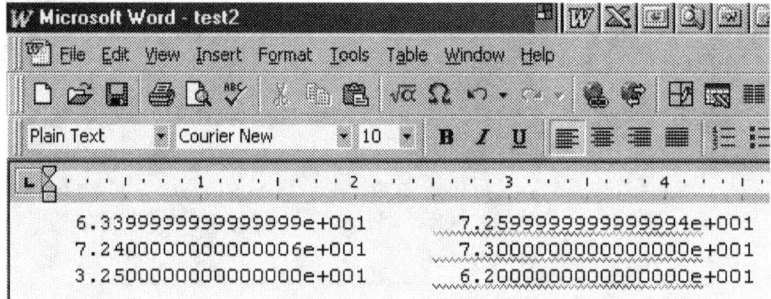

Figure 11.9   *Matrix B is opened in Microsoft Word*

'All Files', type 'test2' in 'File Name' box, and click 'OK' in File Open window. Microsoft Word shows the data of figure 11.9. As you see, the data is put in exponential form.

⊟   Exchanging data between Microsoft Excel and MATLAB

Type matrix C=$\begin{bmatrix} 23.4 & 12.5 \\ 2.4 & 23 \\ 2.5 & 12 \end{bmatrix}$ in Microsoft Excel (shown in figure 11.10). Prompt the save window

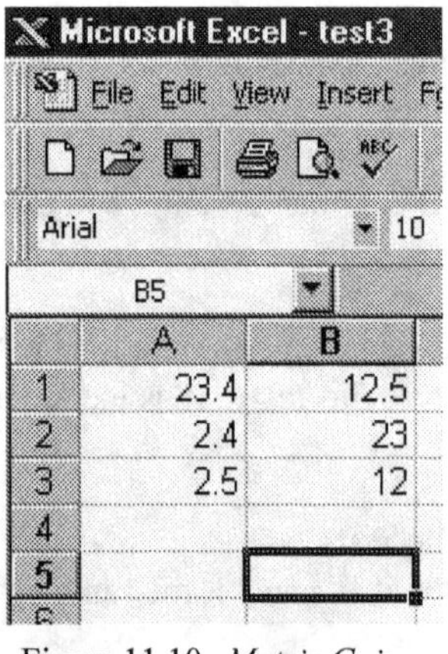

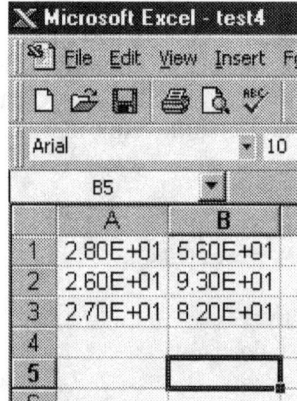

Figure 11.10 *Matrix C is typed in Microsoft Excel*

Figure 11.11 *Matrix D is opened in Excel*

clicking save icon, select the path 'c:\mfile' in 'Save in' box, type 'test3' in 'File name' box, click 'Text (Tab Delimited)' in 'Save as type' box, and click 'Save'. Perform the following in command window:

MATLAB Command

>>cd c:\mfile ↵
>>load test3.txt ↵
>>test3 ↵

test3 =

23.4000	12.5000	← Same as matrix C
2.4000	23.0000	
2.5000	12.0000	

To open the data in Excel from MATLAB, consider $D=\begin{bmatrix} 28 & 56 \\ 26 & 93 \\ 27 & 82 \end{bmatrix}$. Enter and save D as follows:

>>D=[28 56;26 93;27 82]; ↵
>>save test4 D -ascii -double -tabs ↵

From Excel, open the file with $\left\{\begin{array}{l} \text{Select path} \Rightarrow c:\backslash mfile \\ \text{Files of types} \Rightarrow \text{All Files} \\ \text{Filename} \Rightarrow test4 \end{array}\right\}$ to have the exponential form data of figure 11.11. In like manner, the data exchange is practicable between MATLAB and other packages. You can also import the saved data in other software by clicking the file menu of the command window.

11.7 Formatted writing and reading

In the event that an M–file is being run, the generated data by the M–file may be necessary to print or store in a specific format. Not only the format is required for writing, but also it is required for reading. The command 'disp' prints the text or contents of a matrix enclosed in single quotation marks such as 'Text displayed by disp'.

MATLAB Command

>>disp('Text displayed by disp') ↵
Text displyed by disp

The characters are displayed from the first column of the monitor screen. Some texts may have one single quotation, for example, 'It's nice to work in MATLAB'. See this exception as follows:

>>disp('It''s nice to work in MATLAB') ↵
It's nice to work in MATLAB

One numeric example is displaying the contents of A, where $A=\begin{bmatrix} 3 & 9 & -7 \\ 1 & -4 & 0 \end{bmatrix}$.

>>A=[3 9 -7;1 -4 0]; ↵
>>disp(A) ↵
    3   9   -7
    1   -4   0

By default the command window execution is assigned to 'ans' but 'disp' just shows the contents without assigning. It does not provide enough control over the output. Formatted output is printed by taking the help of the commands 'sprintf' or 'fprintf'. The command 'sprintf' gives control over the MATLAB command window whereas 'fprintf' writes the formatted data in a file. These two functions are for writing. The reading or scanning counterparts are 'sscanf' (from window screen) and 'fscanf' (from a file). Suppose, a string 'MATLAB IS A NICE SOFTWARE' is assigned to the variable A in command window. Exercise the following:

>>A='MATLAB IS A NICE SOFTWARE'; ↵
>>O=sprintf('%s',A) ↵

O =

MATLAB IS A NICE SOFTWARE

Whatever is in A is taken as a string (notified by the specifier '%s') and is assigned to O. Insert one blank character between the two successive characters of the string held by A. Have it through the specifier '%c':

>>O=sprintf('%2c',A) ↵

O =

M A T L A B   I S   A   N I C E   S O F T W A R E

Three names 'Shameem', 'Julia', and 'Rafa' are assigned to the variables A, B, and C respectively. Put them as one in a row (left justified) by the specifier '\n' (called the line feed) so that we have $\begin{Bmatrix} \text{Shameem} \\ \text{Julia} \\ \text{Rafa} \end{Bmatrix}$:

>>A='Shameem'; ↵
>>B='Julia'; ↵
>>C='Rafa'; ↵
>>O=sprintf('%s\n%s\n%s',A,B,C) ↵

O =

Shameem
Julia
Rafa

The assignee O is not a two dimensional character array instead a one dimensional one. Including the line feed, there are 18 characters. Ascertain that as follows:

>>size(O) ↵

ans =
        1   18

Like the line feed, we have other feeds as well. The carriage return, tab, backspace, and formfeed are denoted by '\r', '\t', '\b', and '\f' respectively. The composite outputs (that is, numeric and characters together) are also printed. Write $\begin{Bmatrix} \text{Order of matrix A is} \\ 5 \times 5 \end{Bmatrix}$ as follows:

>>A='Order of matrix A is'; ↵   ← The first line is assigned to A
>>O=sprintf('%s\n\t%dx%d',A,5,5); ↵

The formatted output is stored in O. Now conduct the command 'disp' to see the contents of O. This technique might be beneficial in the M-file programming:

```
>>disp(O) ↵
Order of matrix A is
        5x5
```

One character gap is inserted by pressing the Space Bar from keyboard once but more than one character gaps are set by the tabs. We used the tab specifier '\t' to bring 5x5 in the middle of the second line. The characters '%' and '\' are employed as specifiers and different feeds respectively. They are understood to MATLAB by writing twice. To show this, implement the second brace lines $\begin{Bmatrix} \text{Left division is performed by A} \setminus \text{B} \\ \text{20\% of 40 is 8} \end{Bmatrix}$ as follows:

```
>>F='Left division is performed by A'; ↵          ← Part of the 1ˢᵗ string is assigned to F
>>O=sprintf('%s\\%c\n%d%% %s',F,'B',20,'of 40 is 8'); ↵
>>disp(O) ↵
Left division is performed by A\B
20% of 40 is 8
```

Remember that whatever data is printed by the 'sprintf' on the monitor screen can be written in a file by 'fprintf'. Assume that we will write the data $\begin{Bmatrix} \text{Left division is performed by A} \setminus \text{B} \\ \text{20\% of 40 is 8} \end{Bmatrix}$ in a file by the name 'test5.txt'.

Before writing the formatted data in the file, one needs to know about two more functions by the names 'fopen' and 'fclose'. The former function has the general form 'I=fopen(desired file name, type of access)', where I is called the identifier of the file to be created and the type of access can be the kind of access user requires. The possible accesses are given by $\begin{Bmatrix} \text{only for reading by r} \\ \text{only for writing by w} \\ \text{only for appending by a} \\ \text{both for reading and writing by r+} \\ \text{binary file for reading by rb} \end{Bmatrix}$. The later function just closes the file identifier on completion of writing. Anyhow, write the last formatted data in file test5.txt as follows:

MATLAB Command

```
>>cd c:\mfile ↵                                    ← Work in your path
>>I=fopen('test5.txt','w'); ↵                      ← Hold file 'test5.txt' by I
>>F='Left division is performed by A'; ↵
>>fprintf(I,'%s\\%c\n%d%% %s',F,'B',20,'of 40 is 8'); ↵   ← Write data to I
>>fclose(I); ↵                                     ← Close identifier I
>>dir *.txt ↵                                      ← To see whether the file exists

test5.txt
>>type('test5.txt') ↵                              ← To make sure the contents of the file
Left division is performed by A\B
20% of 40 is 8
```

That is what is obtained by exercising the 'sprintf'.

In a similar fashion, one can store the numeric data also. As an illustration, compute $y = x(x+4)$ considering $x = \begin{bmatrix} 2 \\ 3 \\ 4 \\ 5 \end{bmatrix}$. Following the computation, we have $y = \begin{bmatrix} 12 \\ 21 \\ 32 \\ 45 \end{bmatrix}$. Put them in a file by the name 'test6.txt' so that they are arranged as $\begin{bmatrix} 2 & 12 \\ 3 & 21 \\ 4 & 32 \\ 5 & 45 \end{bmatrix}$. The whole procedure is as follows:

MATLAB Command

```
>>cd c:\mfile ↵
>>I=fopen('test6.txt','w'); ↵
>>x=[2 3 4 5]; ↵
>>y=x.*(x+4); ↵
>>for k=1:length(x)  fprintf(I,'%d    %d\n',x(k),y(k));  end ↵
>>fclose(I); ↵
>>dir  *.txt ↵

test5.txt test6.txt
>>load test6.txt  ↵              ← Loading the file
>>test6  ↵                       ← Name of the matrix is 'test6'

test6 =
          2    12
          3    21
          4    32
          5    45
```

Notice that a 'for loop' is required to have the formatted data. Why is that? For each row of the expected matrix $\begin{bmatrix} 2 & 12 \\ 3 & 21 \\ 4 & 32 \\ 5 & 45 \end{bmatrix}$, we have two formats (each one is decimal integer). The for loop counter provides the control on the line numbers.

Following example elucidates how one can handle with a mixed type of data. Presume that we have identification numbers, names, and grades of three students as A= $\begin{bmatrix} 9801 \\ 9802 \\ 9803 \end{bmatrix}$, B= $\begin{bmatrix} \text{Rebeca} \\ \text{Rafa} \\ \text{Julia} \end{bmatrix}$, and C= $\begin{bmatrix} C \\ A \\ B \end{bmatrix}$ respectively. We want them to be printed as follows:

```
                 Grades of Semester 991
==============================================
ID#                NAME                Grades
==============================================
9801               Rebeca                C
9802               Rafa                  A
9803               Julia                 B
```

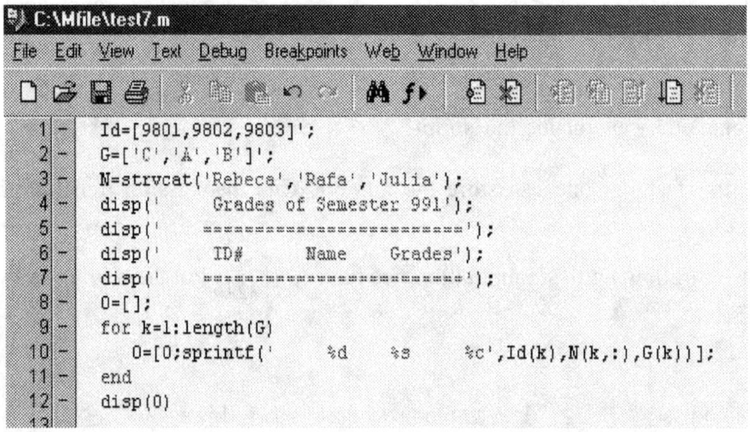


```
Id=[9801,9802,9803]';
G=['C','A','B']';
N=strvcat('Rebeca','Rafa','Julia');
disp('     Grades of Semester 991');
disp('    ==========================');
disp('     ID#      Name    Grades');
disp('    ==========================');
O=[];
for k=1:length(G)
   O=[O;sprintf('     %d     %s      %c',Id(k),N(k,:),G(k))];
end
disp(O)
```

Figure 11.12 *M-file for displaying mixed type of formatted data*

Open a new M-file editor and type all MATLAB statements in the new M-file as depicted in figure 11.12. Referring to the figure, the first three lines are the assignment of the identification numbers, grades, and names to variables Id, G, and N respectively. The lines 4 through 7 are just the presentation of the static texts. We add outcome of 'sprintf' to O one by one that is why initialization of O requires an empty matrix. The matrix N is a two dimensional character array. To pick up the kth row of the array, the command 'N(k,:)' is exercised. Save the file as 'test7' and execute that in your path to exhibit the following:

>>test7 ↵

```
      Grades of Semester 991
     ===========================
       ID#     Name     Grades
     ===========================
       9801    Rebeca      C
       9802    Rafa        A
       9803    Julia       B
```

If just mentioned formatted data is needed to write in a file, exercise the commands 'fopen', 'fprintf', and 'fclose' as we did previously. We stop the discourse of writing the formatted data here.

Now move on to the reading of the formatted data and this is just the reverse transaction of writing. The numeric string '2332432384120' is read as the four decimal digits at a time by the specifier '%4d':

>>A='2332432384129'; ↵
>>O=sscanf(A,'%4d') ↵

O =   ← The output is a column matrix
        2332
        4323
        8412
           9

A tab or blank space separated numeric string '34.23   2.44   233.44' is read as the floating-point numbers by '%f':

>>A='34.23 2.44   233.44'; ↵
>>O=sscanf(A,'%f') ↵

O =   ← The numbers are in a column
       34.2300
        2.4400
      233.4400

Other specifiers are i (decimal, hexadecimal, or octal integer), o (octal integer), u (unsigned decimal integer), x (hexadecimal integer), e (floating-point or exponential form), and g (floating-point). More than one floating-point specifiers are there and the conversion is carried out based on which one is shorter. Be aquatinted with more functions on execution of 'help iofun'.

11.8 Creating a function file

A function file is a special type of M-file which has some input and output arguments. Both arguments can be single or multiple. A function file starts with the name 'function'. For convenience, the long and clumsy programs can be split into smaller modules and these modules are written in a function file. The basic structure of a function file is as follows:

MATLAB Command Prompt   Function file

$\gg g$ =call f $\Longrightarrow$ $g(y_1, y_2, ...y_m) = f(x_1, x_2, x_3....x_n)$

We present two examples of devising the function files – the first one is multi-input and single output and the other is multi-input and multi-output. Remember that the arguments' order and types of the caller and the function files must be identical.

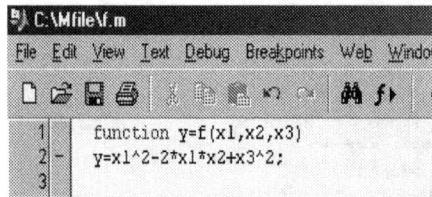

Figure 11.13   *Function written in a M-file for the single output and multi-input*

⧉ Multi-input and single output

Assume that a three variable function $f(x_1, x_2, x_3) = x_1^2 - 2x_1 x_2 + x_3^2$ is to be written in a function file. The inputs are just x_1, x_2, and x_3. Input variables are single values not a vector or a matrix.

MATLAB Command

$\gg$cd c:\mfile ⏎   ← Work in your path

Then, open a new M-file, type the statements in the file as shown in figure 11.13, and save the file by the name 'f'. Test the function with x_1=3, x_2=4, and x_3=5. We should get the output as $f(3,4,5) = 3^2 - 2 \times 3 \times 4 + 5^2 = 10$. Verify this as follows:

MATLAB Command

$\gg$f(3,4,5) ⏎

ans =

    10

After that, we have a set of values, which are $x_1 = \begin{bmatrix} 2 \\ 3 \\ 4 \end{bmatrix}$, $x_2 = \begin{bmatrix} -2 \\ 2 \\ 5 \end{bmatrix}$, and $x_3 = \begin{bmatrix} 1 \\ 0 \\ 3 \end{bmatrix}$. The outputs should look like

$\begin{bmatrix} 2^2 - 2 \times 2 \times (-2) + 1^2 \\ 3^2 - 2 \times 3 \times 2 + 0 \\ 4^2 - 2 \times 4 \times 5 + 3^2 \end{bmatrix} = \begin{bmatrix} 13 \\ -3 \\ -15 \end{bmatrix}$. Call the function from the prompt after changing the operators * and ^ by .* and

.^ respectively to see the computation as follows:

MATLAB Command

$\gg$x1=[2 3 4]'; ⏎
$\gg$x2=[-2 2 5]'; ⏎
$\gg$x3=[1 0 3]'; ⏎
$\gg$f(x1,x2,x3) ⏎

ans =

    13
    -3
    -15

⊟ Multi-input and multi-output

To illustrate the multi-input and multi-output function file, consider that the partial derivatives $\dfrac{\partial f}{\partial x_1}$, $\dfrac{\partial f}{\partial x_2}$, and $\dfrac{\partial f}{\partial x_3}$ are to be found from any function $f(x_1, x_2, x_3)$. The input arguments are the multivariable function $f(x_1, x_2, x_3)$ and the three symbolic variables. Open another M-file and write the statements as shown in figure 11.14. Save the file in the same path by the name 'f1'. We want to verify the function taking

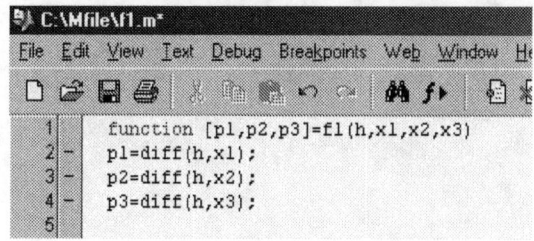

Figure 11.14 *Function written in a M-file for the multi-input and multi-output*

$f(x, y, z) = \dfrac{x+y}{z}$. The partial derivatives of $f(x, y, z) = \dfrac{x+y}{z}$ w.r.to x, y, and z are $\dfrac{1}{z}$, $\dfrac{1}{z}$, and $-\dfrac{x+y}{z^2}$ respectively. The execution is as follows:

MATLAB Command

```
>>syms x y z ↵           ← Declare the variables as symbolic
>>f=(x+y)/z; ↵           ← Define the function
>>[y1 y2 y3]=f1(f,x,y,z) ↵  ← Call the function
```

y1 =

1/z

y2 =

1/z

y3 =

-(x+y)/z^2

There are three output arguments $-\dfrac{\partial f}{\partial x}$, $\dfrac{\partial f}{\partial y}$, and $\dfrac{\partial f}{\partial z}$. We assigned them to y1, y2, and y3 respectively. Even though the variables are different from the function ones, the function file is working.

11.9 Time functions

Many time-related functions are available in the command window. One can invoke them just by typing few words. We present some of them. Suppose you want to see the calendar of the month you are working in. Following is the MATLAB solution:

MATLAB Command

```
>>calendar ↵
```

```
                   Oct 2002
    S    M    Tu    W    Th    F    S
    0    0     1    2     3    4    5
    6    7     8    9    10   11   12
   13   14    15   16    17   18   19
   20   21    22   23    24   25   26
   27   28    29   30    31    0    0
    0    0     0    0     0    0    0
```

To see the specific calendar, for instance, March (which is the third month of the year) 2003:

>>calendar(2003,3) ↵

```
              Mar 2003
    S     M    Tu    W    Th    F     S
    0     0     0    0     0    0     1
    2     3     4    5     6    7     8
    9    10    11   12    13   14    15
   16    17    18   19    20   21    22
   23    24    25   26    27   28    29
   30    31     0    0     0    0     0
```

To know the today's date:
>>date ↵

ans =

23-Oct-2002

What is the weekday on February 15, 2001?
>>[D W]=weekday('15-Feb-20001') ↵

D =   ← D means the numeric day
       5
W =   ← W is the name of the day
     Thu

The day codes of MATLAB are shown in table 11.B. There are more functions affiliated with the package. Online help is seen on execution of 'help timefun'. The exploration is left on the reader.

MATLAB resources for the technical programming are opulent. Applications dependent programming such as graphical user interface, simulink modeling...etc are not covered in the various chapters. That is beyond the scope of the book. We are confined to the scheme of delineating the introductory programming for computations. Once the preliminary steps of computations and programming are known, the advanced programming techniques can facilely be developed including file handling and communicating with other packages. The toolboxes and easy accessible graphics added one more dimension in versatility of the package. Frequent examples in MATLAB's easy to use environment would inspire the reader to grasp the technique in a typical implementation or programming.

Table 11.B Day codes of MATLAB

1	Sun
2	Mon
3	Tue
4	Wed
5	Thu
6	Fri
7	Sat

References

[1] Peter V. O'Neil, *"Advanced Engineering Mathematics"*, Third Edition, 1991, Wadsworth Publishing Company, Belmont, California.

[2] Ali S. Hadi, " *Matrix Algebra — As A Tool"*, First edition, 1996, Wadsworth Publishing Company.

[3] Serge Lang, " *Calculus of Several Variables"*, Second Edition, 1979, Addison–Wesley Publishing Company.

[4] Rogers, Gerald Stanley, *"Matrix Derivatives"*, 1980, New York: M. Dekker.

[5] Marcus, Marvin, *"Matrices and MATLAB — A Tutorial"*, 1993, Englewood Cliffs, N. J. Prentice Hall.

[6] Ogata, Katsuhiko, *"Solving Control Engineering Problems with MATLAB"*, 1994, Englewood Cliffs, N. J. Prentice Hall.

[7] Part-Enander, Eva, *"The MATLAB Handbook"*, 1998, Harlow: Addisson Wesley.

[8] Prentice Hall, Inc., *"The Student Edition of MATLAB for MS-DOS Personal Computers"*, 1992, Englewood Cliffs, N. J. Prentice Hall.

[9] Saadat, Hadi., *"Computational Aids in Control Systems Using MATLAB"*, 1993, New York: McGraw–Hill.

[10] Gander, Walter. and Hrebicek, Jiri. (coauthor), *"Solving Problems in Scientific Computing Using MAPLE and MATLAB"*, 1997, Third, Expanded, and Revised Edition, New York Springer–Verlag.

[11] Biran, Adrian B and Breiner, Moshe (coauthor), *"MATLAB for Engineers"*, Harlow, Eng.: Addison–Wesley, Reprinted 1997, 1995.

[12] D. M. Etter, *"Engineering Problem Solving with MATLAB"*, 1993, Englewood Cliffs, N. J. Prentice Hall.

[13] Shahian, Bahram. and Hassul, Michael., *"Control System Design Using MATLAB"*, 1993, Englewood Cliffs, N. J. Prentice Hall.

[14] Prentice Hall, Inc., *"The Student Edition of MATLAB for Macintosh Computers"*, 1992, Englewood Cliffs, N. J. Prentice Hall.

[15] Ogata, Katshuiko, *"Designing Linear Control Systems with MATLAB"*, 1994, Englewood Cliffs, N. J., Prentice Hall.

[16] Bishop, Robert H., *"Modern Control Systems Analysis and Design Using MATLAB"*, 1993, Reading, MA:Addsison – Wesley.

[17] Moscinski, Jerzy (Ed.) and Ogonowski, Zbigniew. (Coed.), *"Advanced Control with MATLAB and Simulink"*, 1995, Chichester, Eng. :E. Horwood.

[18] Gene Howard Golub and Charles F. Van Loan, *"Matrix Computations"*, 1983, Johns Hopkins University Press, Baltimore.

[19] Alberto Cavallo, Roberto Setola, and Francesco Vasca., *"Using MATLAB Simulink and Control Systems Toolbox —A Practical Approach"*, 1996, Prentice Hall, London.

[20] I. Gohberg, P. Lancaster, and L. Rodman, *"Matrix Polynomials"*, 1982, New York: Academic Press.

[21] Jackson, Leland B., *"Digital Filters and Signal Processing with MATLAB Exercises"*, Third Edition, 1996, Kluwer Academic Publishers, Boston.

[22] Kuo, Benjamin C. and Hanselman, Duanec., *"MATLAB Tools for Control System Analysis and Design"*, 1994, Englewood Cliffs, N. J., Prentice Hall.

[23] Chipperfield, A. J. and Fleming, P. J., *"MATLAB Toolboxes and Applications for Control"*, 1993, London, New York: Peter Peregrinus on Behalf of the Institute of Electrical Engineers.

[24] Math Works Inc., *"MATLAB Reference Guide"*, MathWorks Inc., 1993, Natick, Massachusets.

[25] Cleve Moler and Peter J. Costa, *"MATLAB Symbolic Math Toolbox"*, User's Guide, Version 2.0, May 1997, Natick, Massachusets.

[26] R. Braae, *"Matrix Algebra for Electrical Engineers"*, 1963, London: I. Pitman.